LONDON MATHEMATICAL SOCIETY
MONOGRAPHS NEW SERIES

Series Editors

H. G. Dales   Peter M. Neumann

LONDON MATHEMATICAL SOCIETY MONOGRAPHS NEW SERIES

Previous volumes of the LMS Monographs were published by Academic Press, to whom all enquiries should be addressed. Volumes in the New Series will be published by Oxford University Press throughout the world.

NEW SERIES

1. *Diophantine inequalities* R. C. Baker
2. *The Schur multiplier* Gregory Karpilovsky
3. *Existentially closed groups* Graham Higman and Elizabeth Scott
4. *The asymptotic solution of linear differential systems* M. S. P. Eastham
5. *The restricted Burnside problem* Michael Vaughan-Lee
6. *Pluripotential theory* Maciej Klimek
7. *Free Lie algebras* Christophe Reutenaur
8. *The restricted Burnside problem* (2nd edition) Michael Vaughan-Lee
9. *The geometry of topological stability* Andrew du Plessis and Terry Wall
10. *Spectral decompositions and analytic sheaves* J. Eschmeier and M. Putinar
11. *An atlas of Brauer characters* C. Jansen, K. Lux, R. Parker, and R. Wilson
12. *Fundamentals of semigroup theory* John M. Howie
13. *Area, lattice points, and exponential sums* M. H. Huxley
14. *Super-real fields* H. G. Dales and W. H. Woodin
15. *Integrability, self-duality, and twistor theory* L. Mason and N. M. J. Woodhouse
16. *Categories of symmetries and infinite-dimensional groups* Yu. A. Neretin
17. *Interpolation, identification, and sampling* Jonathan R. Partington
18. *Metric number theory* Glyn Harman
19. *Profinite groups* John S. Wilson
20. *An introduction to local spectral theory* K. B. Laursen and M. M. Neumann
21. *Characters of finite Coxeter groups and Iwahori-Hecke algebras*
   M. Geck and G. Pfeiffer
22. *Classical harmonic analysis and locally compact groups*
   Hans Reiter and Jan D. Stegeman

Preface

Finite Coxeter groups are finite groups of orthogonal transformations which are generated by reflections. Classical examples are the dihedral groups, which are the symmetry groups of regular polygons in dimension 2, and the symmetry groups of regular polyhedra in dimension 3; another standard example is the symmetric group $\mathfrak{S}_n$. The classification of finite Coxeter groups shows that there are several infinite series, which include all finite dihedral groups and the so-called classical types A_n, B_n, D_n (for any integer $n \geq 1$), and six exceptional types denoted by H_3, H_4, F_4, E_6, E_7 and E_8.

One reason why finite Coxeter groups are important is that they may be regarded as the "skeleton" of other algebraic structures, for example, as Weyl groups of Lie algebras and algebraic groups. In the representation theory of groups of Lie type, finite Coxeter groups also arise as natural indexing sets for the bases of certain endomorphism algebras, which are known as Iwahori–Hecke algebras. These algebras can be regarded as deformations of the group algebras of finite Coxeter groups.

The standard reference for the general theory of finite Coxeter groups is [Bourbaki 1968]; the connections with the representation theory of groups of Lie type are treated in the relevant chapters of [Curtis and Reiner 1987] and [Carter 1985]. The aim of *this* book is to develop (a substantial part of) the theory of conjugacy classes and irreducible characters, both for finite Coxeter groups and the associated Iwahori–Hecke algebras, covering all types of finite Coxeter groups according to the classification, in a systematic way. More specifically, our aim is to provide a frame in which the following topics can be treated.

- The classification of the conjugacy classes of finite Coxeter groups; the first uniform approach was given by [Carter 1972a].
- The existence of elements with special properties in the classes of finite Coxeter groups, following [Geck and Pfeiffer 1993] and [Geck and Michel 1997].
- The theory of characters of "parabolic type" and rationality properties of the irreducible characters, as developed in [Benson and Curtis 1972].
- Lusztig's theory of "special" characters and "families"; see [Lusztig 1979a], [Lusztig 1982] and [Lusztig 1984].
- Character formulas for Iwahori–Hecke algebras, following [Pfeiffer 1995], [Halverson and Ram 1996] and [Geck and Michel 1997].
- The algorithmic theory of finite Coxeter groups and related structures, and its implementation in the CHEVIE system; see [Geck et al. 1996].

Each of the above topics involves statements about the classes and the characters of finite Coxeter groups whose proofs require a case-by-case argument according to the classification. Even if there is a general proof, it is still useful to have the concrete data available for each type of group, for example: tables describing the conjugacy classes, formulas for character values and other invariants attached to the irreducible characters and so on. In general, the idea would be to obtain combinatorial or inductive descriptions for the classical types, and to present explicit tables or efficient algorithms for the exceptional types. And this is what we will do in this book.

We mostly adopt the "abstract" point of view, where the main objects of the theory are given by presentations with generators and relations. This has the advantage of being completely elementary and accessible with a minimum of prerequisites. The book includes an introduction to the theory of finite Coxeter groups, the associated braid groups and Iwahori–Hecke algebras. The representation theory is developed in the general framework of symmetric algebras. A detailed study of parabolic subgroups and the parabolic Burnside ring is also included. Furthermore, we discuss the computation of generic degrees and Kazhdan–Lusztig "W-graphs". We describe applications to the theory of knots and links, the representation theory of groups of Lie type and to modular representations.

We assume a certain familiarity with finite groups, associative algebras and character theory. (Our standard references are the books by Isaacs and Curtis–Reiner, which are largely sufficient.) Thus, the interested reader will find an elementary introduction to a field of active research, together with some exercise in the translation of mathematical theory into efficient (computer) algorithms, and the application of such algorithms in the study of the mathematical objects.

Let us now describe in more detail the content of the various chapters. Chapters 1–6 deal with finite Coxeter groups, parabolic subgroups, conjugacy classes and characters, while Chapters 7–11 deal with Iwahori–Hecke algebras and their characters.

Chapter 1 contains an introduction to Coxeter groups. The material in this chapter is more or less standard but we adopt a new approach which is similar to that in the theory of Kac–Moody algebras, where the basic objects are defined in terms of Cartan matrices. Such a matrix encodes in a direct and compact way the information needed to define a Coxeter group in terms of generators and relations on the one hand and via its reflection representation and the action on a corresponding root system on the other hand—which is important for algorithmic aspects. We explain the classification of finite Coxeter groups and discuss in detail properties of the longest elements, which are substantial for later applications to normal forms in braid groups and to character formulas for Iwahori–Hecke algebras.

Chapter 2 is about parabolic subgroups, those subgroups of a Coxeter group which are generated by subsets of the original generating set. We provide a detailed analysis of certain sets of distinguished coset representatives, illustrated in the case of finite Coxeter groups by means of the coset graph. These representatives give rise to a subalgebra of the group algebra of the Coxeter group which was first discovered by Solomon and is now known as the descent algebra. A first step towards the classification of the conjugacy classes is the result (originally due to Deodhar, Howlett, Richardson, and Lusztig–Spaltenstein) which describes how the subsets of the generating set fall into classes under the conjugation action of the group. We also study the permutation action of the group on the cosets of all its parabolic subgroups and collect, as a first step towards the character table, numerical invariants of these actions in the "parabolic table of marks".

Chapter 3 develops a new approach to the conjugacy classes of a finite Coxeter group, extending and putting into a systematic conceptual framework the earlier results of the authors and of Michel. Concentrating on "cuspidal classes", i.e., those conjugacy classes which do not contain elements of proper parabolic subgroups, we establish strong properties of the conjugation action of a finite Coxeter group on itself. For example, it is shown that all elements of minimal length of a cuspidal class can be obtained from one another by cyclically permuting reduced expressions. The proofs involve an algorithmic construction of a list of representatives of elements of minimal length of the cuspidal classes. Our methods also yield a new way of classifying the conjugacy classes of a finite Coxeter group, in terms of parabolic subgroups and cuspidal classes.

Chapter 4 introduces the braid monoid and the braid group associated with a finite Coxeter group. Following recent work of Michel, we present the main results of Brieskorn–Saito and Deligne on a normal form of the elements in the braid group. This is used in the definition of the notion of "good" elements in the classes of finite Coxeter groups, whose existence for all types we prove following Geck–Michel. Finally, we introduce Iwahori–Hecke algebras as quotients of group algebras of braid groups. As a first application and as a motivation for the results on trace functions in Chapter 8, we describe Jones's construction of an invariant of knots and links (with some simplifications suggested by Michel).

Chapter 5 provides a self-contained introduction to the character theory of finite Coxeter groups, where we work with "ordinary" complex representations throughout. In particular, we prove Frobenius's classical results on the parametrization of the irreducible characters of the symmetric group $\mathfrak{S}_n$. The main tools are the induced actions of a Coxeter group on the exterior and on the symmetric powers of the natural reflection representation. This also includes a detailed study of the Macdonald–Lusztig–Spaltenstein induction of characters. Our approach yields a quite straightforward proof of the fact (due originally to Young, Specht, Kondo, Benard and Benson–Grove) that every irreducible character of a finite Coxeter group is rational over the field of definition of the reflection representation.

Chapter 6 presents the main results of Benson–Curtis about characters of "parabolic type" and develops a purely axiomatic approach to Lusztig's theory of "special" characters and "families". These results are particularly important for applications to the representation theory of finite groups of Lie type. As far as the classical types are concerned, we rely in an essential way on the Littlewood–Richardson rule about the induction of characters in symmetric groups, which is stated without proof in Section 6.1. For a proof of that rule, as well as for the connections of the representation theory of the symmetric group with the theory of symmetric functions, we refer to the books by Macdonald and Fulton.

Chapter 7 lays the foundations for the subsequent treatment of Iwahori–Hecke algebras and their representations. It is a common feature of group algebras of finite groups and Iwahori–Hecke algebras that they are both examples of symmetric algebras. Our aim here is to present a self-contained introduction to the representation theory of symmetric algebras, with a particular emphasis on the Schur relations for irreducible representations. The main results of this chapter are Tits's deformation theorem, a general semisimplicity criterion for symmetric algebras and the existence theorem for decomposition maps.

Chapter 8 develops the general structure theory of Iwahori–Hecke algebras. We introduce generic algebras and specializations. Using the results on elements of minimal length in the classes of finite Coxeter groups (see Chapter 3), we introduce the notion of the "character table" for an Iwahori–Hecke algebra. This is a table analogous to that for finite groups, but where the entries are polynomial expressions in the parameters of the algebra. Following Kilmoyer–Solomon, the dihedral types are discussed in full detail. Finally, we give a brief account of Iwahori's theorem which shows that abstract Iwahori–Hecke algebras (as introduced in Chapter 4) occur as endomorphism algebras in groups of Lie type.

Chapter 9 studies the irreducible characters of generic Iwahori–Hecke algebras. The main tools for doing this are Tits's deformation theorem, a result of Springer concerning the action of the longest element, and the compatibility of induction of representations with specializations. One of the main results of this chapter is Theorem 9.3.5 (due originally to Benson–Curtis, Lusztig, Alvis–Lusztig and Digne–Michel), which describes a splitting field for the irreducible characters of a generic Iwahori–Hecke algebra.

Chapter 10 is concerned with explicit formulas for the values in the character tables of the Iwahori–Hecke algebras of classical types. The prototype of such a formula is the classical Murnaghan–Nakayama rule for the character values of the symmetric group. We use a variant of the representing matrices originally set up by Hoefsmit to derive these formulas, following recent work by Halverson–Ram and Pfeiffer. We also discuss the determination of the generic degrees for classical types, which is due to Steinberg, Hoefsmit and Benson–Gay. In the final section, we present without proofs the main ingredients of the Lascoux–Leclerc–Thibon conjecture about decomposition numbers of Iwahori–Hecke algebras of type A. This conjecture and its proof by Ariki constitute a major achievement in the modular representation theory of Iwahori–Hecke algebras.

Chapter 11 describes methods for computing generic degrees and character tables of generic Iwahori–Hecke algebras of any type. These methods are partly presented in the form of algorithms. Applying these algorithms to the exceptional types H_3, H_4, F_4, E_6, E_7, and E_8, we obtain complete results for these cases in a straightforward manner. (The results on generic degrees are due originally to Surowski, Benson, Lusztig and Alvis–Lusztig; those on character tables are due to Alvis–Lusztig, Lusztig, Geck and Geck–Michel.) As an application, we determine the blocks of modular characters in exceptional types. We also discuss the Kazhdan–Lusztig "W-graphs", which provide explicit matrix realizations for irreducible characters. One of our main results describes a complete list of W-graphs for the generic two-parameter algebra of type F_4.

Finally, the Appendix contains tables with basic information about the exceptional types, their parabolic subgroups, conjugacy classes, irreducible characters, generic degrees, and blocks of modular characters.

We have tried as much as possible to write the chapters in such a way that they can be read independently of each other. Each chapter has an introduction which explains in more detail the content and the main results; at the end, the reader will find bibliographic remarks (with precise references to the literature) and exercises. Usually, we give hints to solutions, at least for those exercises which are used in the text.

There are (at least) three areas where we do not go into too much detail: (a) the theory of Kazhdan–Lusztig cells, (b) non-semisimple Iwahori–Hecke algebras and (c) complex reflection groups and cyclotomic Hecke algebras. As far as (a) is concerned, we only discuss the related concept of W-graphs and the construction of the Kazhdan–Lusztig basis of an Iwahori–Hecke algebra; the only aspect of (b) which we treat is the determination of blocks of characters for exceptional types in Chapter 11; while (c) only makes a short appearance in Chapter 6 (in connection with the Macdonald–Lusztig–Spaltenstein induction).

There are various reasons why we do not develop these aspects further: for example, any deeper understanding of Kazhdan–Lusztig cells seems to require the use of geometric methods (intersection cohomology) in order to establish certain positivity properties (see the survey article [Lusztig 1991]). Furthermore, the situation for non-semisimple algebras is far from being understood. The most significant recent progress in this area appears to be the Lascoux–Leclerc–Thibon conjecture and its proof by Ariki. (A brief introduction will be given in Section 10.6.) Any more substantial treatment of these matters would certainly have gone far beyond the scope of the present volume. Finally, the theory of cyclotomic algebras is an active area of current research where a general theory is beginning to take shape (see, for example, [Broué et al. 1998]). Nevertheless, we try to indicate connections to these new areas and to point out suitable references for further reading in the bibliographic remarks at the end of each chapter.

We have included about a dozen algorithms in the text, each of which translates a piece of general theory into a practical solution to a general question like:

how do I find reduced expressions for a set of distinguished coset representatives, or an element of minimal length in a conjugacy class? How can I tell whether two elements in a finite Coxeter group are conjugate or not? How do I compute the Deligne normal form of an element in the braid group? Each of these questions looks "innocent" from a theoretical point of view but in our applications, e.g., in the proof of the fundamental Theorem 3.2.7 on properties of cuspidal classes, it is essential to have efficient practical solutions to these questions. (Note that the properties in question have to be checked, for example, in the exceptional group of type E_8, with 696 729 600 elements.)

For the formulation of the algorithms, we follow the style in [Knuth 1968]. An algorithm is presented as a sequence of steps, each described in plain English together with the usual mathematical notation. The only piece of special notation is an expression like "$x \leftarrow 3$" to indicate the assignment of the value 3 to the variable x. All algorithms in this book have been implemented in the computer algebra system GAP (see [Schönert 1995]). The programs also require the functions and library files contained in the CHEVIE share package of GAP. For more information on CHEVIE, see [Geck et al. 1996] and the home page at

`http://www.math.rwth-aachen.de/~CHEVIE`

The contributions directory on the above home page contains files with the actual implementation of the algorithms concerning conjugacy classes (see Chapter 3), generic degrees and character tables (see Chapter 11). For the computation of the irreducible representations of the generic algebra of type F_4 in Section 11.3, we have also made use of an implementation of Parker's MEATAXE package (see [Parker 1984]) in the computer algebra system MAPLE (see [Char et al. 1991]). To some extent we can say that the work on actual implementations of the algorithmic theory of Coxeter groups also helped to clarify some ideas concerning the general theory.

Work on this book was started while the first author enjoyed the hospitality of the Newton Institute (Cambridge, UK) during the special semester on algebraic groups (January to June 1997). Part of the material in this book grew out of a *Habilitationsschrift* [Geck 1995], a Ph.D. thesis [Pfeiffer 1995], and a series of lectures given by the second author at the University of St Andrews (Scotland) in 1995. The whole work was done while the first author held a CNRS research position in the "Équipe des Groupes Finis" at the Université Paris 7 (France). Part of the material in this book was used by the second author for a lecture course at the National University of Galway (Ireland).

We are indebted to a number of colleagues and friends for reading parts of the manuscript and suggesting improvements. Above all we wish to thank Gunter Malle who carefully read several preliminary versions of all chapters and made numerous and detailed comments. We believe that the manuscript greatly benefited from taking into account his remarks and suggestions. We are also grateful for the critical comments by Thomas Breuer, Gerhard Hiss, Frank

Lübeck, Jürgen Müller, and Max Neunhöffer, who worked through parts of this book during a seminar held at Lehrstuhl D für Mathematik, RWTH Aachen, in 1998. Erzsebet Horvath read several chapters and sent a long list of misprints. Further errors were discovered by Rex Dark and Dane Flannery. Thanks are due to Lacrimioara Iancu for help with the graphical presentation of the W-graphs in Figure 11.3 and many helpful and critical comments on various parts of the text, especially Chapter 1. Of course, all remaining errors are ours.

We thank Jean Michel for allowing us to use a preprint version of his note on words in braid monoids and for helpful discussions on that subject, on which the content of Section 4.2 depends. We also thank Bob Howlett for allowing us to use parts of his unpublished notes on elements of minimal length in the classes of finite Coxeter groups (which appear here as Corollary 3.1.11 and Proposition 3.2.10).

Our understanding of the general theory of symmetric algebras, braid groups and Hecke algebras also benefited from numerous informal discussions with the members of the "Équipe des Groupes Finis" at the Université Paris 7, especially, Michel Broué, Jean Michel and Raphael Rouquier. In the same spirit, we are also grateful to the further members of the CHEVIE team, Frank Lübeck, Gerhard Hiss, Gunter Malle and Jean Michel, for a number of years of fruitful discussions and collaboration.

We are obliged to the creators of various pieces of free software, which we used for the preparation of the book, the Emacs editor (`http://www.gnu.org`), the CVS version control system (`http://www.cyclic.com`) which allowed us to work simultaneously on the same document from different countries, and the TeX typesetting system together with many of its extensions (`http://www.ctan.org`).

We wish to thank Sönke Adlung and the people at OUP for supporting the project of writing this book.

Galway and Lyon, January 2000

Contents

1	**Cartan matrices and finite Coxeter groups**	**1**
1.1	Cartan matrices and reflection groups	4
1.2	Matsumoto's theorem	10
1.3	Cartan matrices of finite type	15
1.4	Coxeter groups of classical type	20
1.5	The longest element	26
1.6	Bibliographical remarks and exercises	33
2	**Parabolic subgroups**	**38**
2.1	Distinguished coset representatives	39
2.2	Longest representatives and the coset graph	45
2.3	Coxeter classes	53
2.4	The parabolic table of marks	60
2.5	Bibliographical remarks and exercises	67
3	**Conjugacy classes and special elements**	**73**
3.1	Cuspidal classes	74
3.2	The conjugacy graph	79
3.3	Algorithms and reduction theorems	87
3.4	Minimal elements in classical types	91
3.5	Bibliographical remarks and exercises	100
4	**The braid monoid and good elements**	**105**
4.1	The braid monoid and the braid group	106
4.2	Divisibility in braid monoids	111
4.3	Good elements in braid monoids	116
4.4	Iwahori–Hecke algebras	122
4.5	Invariants of knots and links	127
4.6	Bibliographical remarks and exercises	133
5	**Irreducible characters of finite Coxeter groups**	**137**
5.1	Exterior powers of the reflection representation	138
5.2	Macdonald–Lusztig–Spaltenstein induction	141
5.3	Applications to finite Coxeter groups	148
5.4	The characters of the symmetric group	154
5.5	The characters of Coxeter groups of type B	163
5.6	The characters of Coxeter groups of type D	169
5.7	Bibliographical remarks and exercises	172

6	**Parabolic subgroups and induced characters**	177
6.1	Induction of characters in classical types	178
6.2	Central characters in classical types	183
6.3	Characters of parabolic type	190
6.4	Symbols for the classical types	193
6.5	Families and a-functions	202
6.6	Bibliographical remarks and exercises	212
7	**Representation theory of symmetric algebras**	217
7.1	Trace functions	218
7.2	Schur relations and Schur elements	223
7.3	Grothendieck groups and integrality	228
7.4	Specializations and Tits's deformation theorem	234
7.5	Properties of the decomposition map	239
7.6	Bibliographical remarks and exercises	246
8	**Iwahori–Hecke algebras**	250
8.1	Generic Iwahori–Hecke algebras	251
8.2	Class polynomials	261
8.3	Iwahori–Hecke algebras of dihedral type	267
8.4	Iwahori's theorem	274
8.5	Bibliographical remarks and exercises	281
9	**Characters of Iwahori–Hecke algebras**	286
9.1	Induction from parabolic subalgebras	287
9.2	Character values and the longest element	294
9.3	Splitting fields for Iwahori–Hecke algebras	302
9.4	Character values and Schur elements	309
9.5	Bibliographical remarks and exercises	317
10	**Character values in classical types**	321
10.1	Hoefsmit's matrices	322
10.2	Character values of type A	327
10.3	Character values of type B	332
10.4	Character values of type D	336
10.5	On the generic degrees of the classical types	345
10.6	The Lascoux–Leclerc–Thibon conjecture	353
10.7	Bibliographical remarks and exercises	357
11	**Computing character values and generic degrees**	361
11.1	W-graphs and representing matrices	362
11.2	W-graphs for the irreducible characters of type H_4	370
11.3	W-graphs for the irreducible characters of type F_4	375
11.4	On the computation of the generic degrees	380

11.5 On the computation of character values 383
11.6 Bibliographical remarks and exercises 393

Appendix: Tables for the exceptional types 398
A Coxeter classes and parabolic tables of marks 398
B Good representatives of cuspidal classes 406
C Irreducible characters 411
D Truncated induction 417
E Generic degrees 422
F Blocks at roots of unity 428

References 435

Index 445

1
Cartan matrices and finite Coxeter groups

A *Coxeter group* is a group W which has a presentation of a very special form: we have a set of generators $S \subseteq W$ such that the defining relations are just relations of the form $s^2 = 1$ and $(st)^{m(s,t)} = 1$, where $s \neq t$ in S and $m(s,t) = m(t,s) > 1$ are positive integers. (It can be shown that $m(s,t)$ is in fact the order of st.) Thus, the group is generated by involutions, and the only additional relations are those prescribing the order of the product of any two generators. The standard example is the symmetric group $\mathfrak{S}_n$ on n letters, with generators given by the basic transpositions $\{(i, i+1) \mid 1 \leq i \leq n-1\}$.

It is a remarkable fact that every Coxeter group W with generating set S can be embedded into $\mathsf{GL}(V)$ for some finite-dimensional $\mathbb{R}$-vector space V, in such a way that the elements of S act as reflections on V. This result provides a link between the theory of Coxeter groups, reflection groups, and root systems. There are now a number of textbooks on these subjects available: for example, [Bourbaki 1968], [Benson and Grove 1985], [Curtis and Reiner 1987, §64], [Hiller 1982] and [Humphreys 1990]. So we certainly do not want to add another one to this list. Therefore, we only discuss in this chapter those topics which will be particularly important for the subsequent discussion of characters, braid groups and Hecke algebras. Each of these topics forms the core of one section.

In Section 1.1, we introduce the objects of our study as certain subgroups $W \subset \mathsf{GL}(V)$ generated by *reflections*, where the action of the generators is encoded in a *Cartan matrix*. For finite W, the action of W on V will provide the main tool for constructing the irreducible characters of W in Chapter 5.

The main result of Section 1.2 is *Matsumoto's theorem*, which yields a presentation for W and shows that W is a Coxeter group as defined above. That presentation will occur again in a modified form in the study of braid groups and Iwahori–Hecke algebras; see Chapter 4.

In Section 1.3, we present the main results about the classification of finite Coxeter groups, in terms of *Coxeter graphs* and *Dynkin diagrams*. (This will be done partly without proof.)

Section 1.4 describes explicitly the *Coxeter groups of classical type* in terms of their natural matrix representations. This will provide the frame in which we can work with these groups in the subsequent chapters.

Finally, Section 1.5 is devoted to the study of properties of the *longest element* of W, when W is finite. The main result describes that element as a power of a *Coxeter element* of W, and this will be one of the key ingredients in the computation of character values of Iwahori–Hecke algebras; see Chapter 9.

Some essential features of the general theory already appear in the simplest non-trivial example: that of finite reflection groups in dimension 2. It will be quite useful to always keep in mind this example. It can be treated in elementary geometric terms as follows.

Let $m \geq 2$ and consider a regular m-gon in the real plane $\mathbb{R}^2$ with centre at the origin and arranged such that one corner has coordinates $(0,1)$. Let W be the group of all rotations and reflections in $\mathbb{R}^2$ which leave our m-gon invariant. Thus, W consists of the m clockwise rotations through the angles $2k\pi/m$ (for $0 \leq k < m$) around the origin, and the m reflections in the hyperplanes orthogonal to the vectors $e_0, e_1, \ldots, e_{m-1}$, where we define

$$e_k := (\cos(k\pi/m), \sin(k\pi/m)), \qquad 0 \leq k \leq 2m-1,$$

as in the following picture:

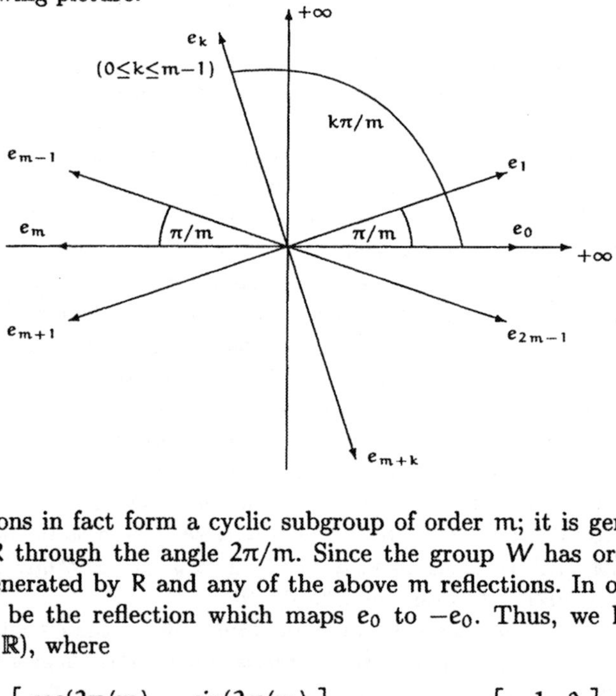

The rotations in fact form a cyclic subgroup of order m; it is generated by the rotation R through the angle $2\pi/m$. Since the group W has order $2m$, it is therefore generated by R and any of the above m reflections. In order to fix things, let S_0 be the reflection which maps e_0 to $-e_0$. Thus, we have $W = \langle R, S_0\rangle \subseteq GL_2(\mathbb{R})$, where

$$R = \begin{bmatrix} \cos(2\pi/m) & -\sin(2\pi/m) \\ \sin(2\pi/m) & \cos(2\pi/m) \end{bmatrix} \quad \text{and} \quad S_0 = \begin{bmatrix} -1 & 0 \\ 0 & 1 \end{bmatrix},$$

where we use the convention that matrices act from the right on $\mathbb{R}^2$.

On the other hand, W can also be generated by two suitably chosen reflections. We take S_0 as the first reflection. Then there is an optimal choice for a second reflection: we take the reflection S_1 which maps e_{m-1} to $-e_{m-1}$. Note that $R = S_0S_1$, so we certainly have $W = \langle S_0, S_1 \rangle$. The point about choosing S_1 is that then the following properties are satisfied:

(1) *We have* $\{e_k \mid 0 \leq k \leq 2m-1\} = \{e_0.w, e_{m-1}.w \mid w \in W\}$.
(2) *Every* e_k *can be written as* $e_k = xe_0 + ye_{m-1}$ *with either* $x, y \geq 0$ *or* $x, y \leq 0$.
(3) *Let* $e_k = xe_0 + ye_{m-1}$ *be such that* $x, y > 0$, *i.e.,* e_k *strictly lies in the cone spanned by* e_0, e_{m-1}. *Then both* $e_k.S_0$ *and* $e_k.S_1$ *are non-negative linear combinations of* e_0 *and* e_{m-1}.

(These properties are geometrically clear from the above picture.) If we express the transformations S_0, S_1 with respect to the basis $\{e_0, e_{m-1}\}$, the corresponding matrices are

$$\begin{bmatrix} -1 & 0 \\ 2\cos(\pi/m) & 1 \end{bmatrix} \quad \text{and} \quad \begin{bmatrix} 1 & 2\cos(\pi/m) \\ 0 & -1 \end{bmatrix}, \quad \text{respectively.}$$

Up to now we have only considered vectors which have norm 1. It will be convenient to relax also this condition. Thus, let α_0 and α_{m-1} be positive multiples of e_0 and e_{m-1}, respectively. Then the matrices of S_0, S_1 with respect to the basis $\{\alpha_0, \alpha_{m-1}\}$ are

$$\begin{bmatrix} -1 & 0 \\ -c & 1 \end{bmatrix} \quad \text{and} \quad \begin{bmatrix} 1 & -c' \\ 0 & -1 \end{bmatrix}, \quad \text{respectively,}$$

where $c, c' \in \mathbb{R}$ are such that $c, c' \leq 0$ and $cc' = 4\cos^2(\pi/m)$. In this context, the above remarks about the particular choice of the generators S_0, S_1 are formulated as follows.

(1′) *Let* $\Phi := \{\alpha_0.w, \alpha_{m-1}.w \mid w \in W\}$. *Then for every* $\alpha \in \Phi$ *there is a reflection in* W *which maps* α *to* $-\alpha$.
(2′) *If* $\alpha \in \Phi$, *we have* $\alpha = x\alpha_0 + y\alpha_{m-1}$ *with either* $x, y \geq 0$ *or* $x, y \leq 0$.
(3′) *Let* $\alpha \in \Phi$ *be such that* $\alpha = x\alpha_0 + y\alpha_{m-1}$ *with* $x, y > 0$. *Then both* $\alpha.S_0$ *and* $\alpha.S_1$ *lie in the cone spanned by* α_0 *and* α_{m-1}.

The general theory will be built up from these two-dimensional examples: we consider a finite-dimensional $\mathbb{R}$-vector space V with a given basis $\{\alpha_s \mid s \in S\}$, and a subgroup $W \subset \mathrm{GL}(V)$ which is generated by "reflections" corresponding to the elements in S. This is done in such a way that if we choose any pair $\{s, t\} \subseteq S$, then the corresponding reflections leave the subspace generated by $\{\alpha_s, \alpha_t\}$ invariant and the action is given by matrices as above. The whole situation is most economically encoded in a corresponding "Cartan matrix" which basically records the real numbers c, c' determining the above matrices.

1.1 CARTAN MATRICES AND REFLECTION GROUPS

The basic object from which we build up the theory is that of a Cartan matrix. This contains exactly the information that is needed to determine a root system and the action of a corresponding reflection group on it.

1.1.1. Definition. Let S be a finite non-empty index set and $C = (c_{st})_{s,t \in S}$ be a matrix with entries in $\mathbb{R}$. We say that C is a *Cartan matrix* if the following conditions are satisfied:

(C1) For $s \neq t$ we have $c_{st} \leq 0$; furthermore, $c_{st} \neq 0$ if and only if $c_{ts} \neq 0$.
(C2) We have $c_{ss} = 2$ and, for $s \neq t$, we have $c_{st}c_{ts} = 4\cos^2(\pi/m_{st})$, where $m_{st} \geq 2$ is an integer or $m_{st} = \infty$.

We always have $0 \leq c_{st}c_{ts} \leq 4$. Here are some values for the product $c_{st}c_{ts}$:

m_{st}	2	3	4	5	6	8	∞
$c_{st}c_{ts}$	0	1	2	$(3+\sqrt{5})/2$	3	$2+\sqrt{2}$	4

Using the relation $\cos(2\theta) = 2\cos^2(\theta) - 1$ ($\theta \in \mathbb{R}$) and Exercise 1.9, we see that $2, 3, 4, 6, \infty$ are the only values of m_{st} for which the product $c_{st}c_{ts}$ is rational.

Let $C = (c_{st})_{s,t \in S}$ be a Cartan matrix and V be an $\mathbb{R}$-vector space of dimension $|S|$, with a fixed basis $\{\alpha_s \mid s \in S\}$. An element $g \in GL(V)$ acts on V from the right, and we write this action in the form $(v, g) \mapsto v.g$ ($v \in V$). We wish to associate with C a subgroup of $GL(V)$. For this purpose, we define a linear action of each $s \in S$ on V as follows:

$$s\colon V \to V, \quad \alpha_t \mapsto \alpha_t - c_{st}\alpha_s \quad (t \in S).$$

This linear map will be called the *reflection* with root α_s. This notation is justified by the following result:

1.1.2. Lemma. *The linear map $s\colon V \to V$ has the following properties:*

$$\alpha_s.s = -\alpha_s, \quad \mathrm{Tr}(s) = |S| - 2, \quad \text{and} \quad s^2 = \mathrm{id}_V.$$

Consequently, each s is a diagonalizable automorphism of V, with precisely $|S|-1$ eigenvalues equal to 1 and precisely one eigenvalue equal to -1.

Proof. The formula $\alpha_s.s = -\alpha_s$ is clear by the defining equation, since $c_{ss} = 2$. To compute $\mathrm{Tr}(s)$, we consider the coefficient of α_t ($t \in S$) in the basis decomposition of $\alpha_t.s$. By the defining formula for s, this coefficient is 1 for $s \neq t$ and -1 for $s = t$. Hence $\mathrm{Tr}(s) = (|S| - 1) + (-1) = |S| - 2$. Finally, we have $\alpha_t.s^2 = (\alpha_t - c_{st}\alpha_s).s = \alpha_t - c_{st}\alpha_s + c_{st}\alpha_s = \alpha_t$, as desired.

It follows from the relation $s^2 = \mathrm{id}_V$ that s is a diagonalizable automorphism of V, with all eigenvalues equal to ± 1. The assertion about the number of eigenvalues $+1$ and -1 now follows from the fact that the sum of these eigenvalues equals $\mathrm{Tr}(s) = |S| - 2$. □

1.1.3. Definition. Let $C = (c_{st})_{s,t \in S}$ be a Cartan matrix. We regard S as a subset of $\mathsf{GL}(V)$ (via the linear action defined above) and call the group

$$W = W(C) := \langle S \rangle \subset \mathsf{GL}(V)$$

the *reflection group* associated with C. The subset

$$\Phi = \Phi(C) := \{\alpha_s.w \mid w \in W, s \in S\} \subseteq V$$

is called the *root system* associated with C; its elements are called *roots*. Thus, Φ is a subset of $V - \{0\}$ which contains a basis of V and which is invariant under the action of W. The roots $\{\alpha_s \mid s \in S\} \subseteq \Phi$ are called *simple roots* or *fundamental roots*. The integer $|S|$ is called the *rank* of W or of Φ.

Note that $\Phi = -\Phi$ (since $\alpha_s.s = -\alpha_s$ for all $s \in S$). Furthermore, since Φ contains a basis of V, the permutation representation of W on Φ is faithful. Hence Φ is finite if and only if W is finite.

1.1.4. Example. The matrices

$$\begin{bmatrix} 2 & -1/2 \\ -2 & 2 \end{bmatrix}, \quad \begin{bmatrix} 2 & -1 & -1 \\ -1 & 2 & -1 \\ -1 & -1 & 2 \end{bmatrix}, \quad \begin{bmatrix} 2 & -\lambda & 0 & 0 \\ -\lambda & 2 & -1 & 0 \\ 0 & -1 & 2 & -1 \\ 0 & 0 & -1 & 2 \end{bmatrix}$$

where $\lambda := (1+\sqrt{5})/2$, are Cartan matrices. Consider the first matrix. We let $S = \{s, t\}$ be the row and column indices, so that $\alpha_s.t = \alpha_s + 2\alpha_t$ and $\alpha_t.s = \alpha_t + (1/2)\alpha_s$. We represent α_s by $(1,0)$ and α_t by $(0,1)$. Now we compute Φ by an orbit algorithm as follows: first take α_s and apply t to it, next apply s to the resulting vector, then apply again t and so on. It turns out that, after three applications of this process, we do not obtain any new roots. We do the same with α_t, and find that:

$$\begin{aligned} \Phi &= \{\pm(1,0), \pm(1,2), \pm(0,2), \pm(0,1), \pm(1/2,1), \pm(1/2,0)\}, \\ W &= \{1, s_1, s_2, s_1s_2, s_2s_1, s_1s_2s_1 = s_2s_1s_2\} \quad \cong \text{symmetric group } \mathfrak{S}_3. \end{aligned}$$

In principle, one could also compute Φ and W for the third matrix; using a computer, one can check that the results are $|W| = 14\,400$ and $|\Phi| = 120$ in this case. Finally, consider the second matrix. Then it is not difficult to check that Φ is infinite; see Exercise 1.1.

We fix a reflection group $W \subset \mathsf{GL}(V)$ associated with a Cartan matrix C as above. A basic tool for working with W is the *length function* $l\colon W \to \mathbb{N}_0$, which is defined as follows. Let $w \in W$. If $w = 1$ we set $l(w) = 0$. Otherwise, there exists some $k \geq 1$ and elements $s_1, \ldots, s_k \in S$ such that $w = s_1 \cdots s_k$. If k is chosen minimal with this property we say that we have a *reduced expression* for w and set $l(w) = k$.

1.1.5. Remark. Recall that W is defined as a subgroup of $\mathrm{GL}(V)$. Hence taking the determinant defines a group homomorphism $\varepsilon\colon W \to \mathbb{R}^\times$ which we call the *sign representation* (or character) of W. Using Lemma 1.1.2, we have $\varepsilon(s) = -1$ for all $s \in S$ and $\varepsilon(w) = (-1)^k$ if $w = s_1 \cdots s_k$ with $s_1, \ldots, s_k \in S$. In particular,

$$\varepsilon(w) = (-1)^{l(w)} \qquad \text{for all } w \in W.$$

Now let $w \in W$ and $s \in S$. Then we certainly have $l(sw) \leq l(w) + 1$. Applying this to the element sw also yields that $l(w) = l(s(sw)) \leq l(sw) + 1$. Thus, we always have $l(w) - 1 \leq l(sw) \leq l(w) + 1$. Since ε is a homomorphism, $\varepsilon(sw) = -\varepsilon(w)$ and, hence, $l(sw) \equiv l(w) + 1 \bmod 2$. It follows that

$$l(sw) = l(w) \pm 1 \qquad \text{for all } s \in S \text{ and } w \in W.$$

Note also that if $l(sw) < l(w)$ then there exists a reduced expression for w which starts with s. This remark will be used frequently in what follows.

Our next aim is to characterize the condition that $l(sw) > l(w)$ in terms of the action of W on Φ. Such a characterization will turn out to be an effective tool both for computational and theoretical purposes. The proof will be by reduction to the rank 2 case. Therefore, we take a closer look at this case first.

1.1.6. Lemma. *Let $s, t \in S$, $s \neq t$, and $V' := \mathbb{R}\alpha_s \oplus \mathbb{R}\alpha_t \subseteq V$. Then V' is invariant under s and t. The action of s and t on V' with respect to the basis $\{\alpha_s, \alpha_t\}$ is given by the matrices*

$$M_s := \begin{bmatrix} -1 & 0 \\ -c_{st} & 1 \end{bmatrix} \quad \textit{and} \quad M_t := \begin{bmatrix} 1 & -c_{ts} \\ 0 & -1 \end{bmatrix}, \quad \textit{respectively.}$$

Furthermore, if $m_{st} < \infty$, there exists a subspace $U \subseteq V$ such that $V = V' \oplus U$ and each of s, t acts as the identity on U.

Proof. The defining equations show that $V'.s \subseteq V'$ and $V'.t \subseteq V'$, and that the action is given by the above matrices. Now assume that $m_{st} < \infty$. The desired subspace U is constructed as follows. For each $u \in S - \{s, t\}$ consider the equations $2x_s^u + c_{st}x_t^u = -c_{su}$ and $c_{ts}x_s^u + 2x_t^u = -c_{tu}$, with unknowns x_s^u, x_t^u. Since $m_{st} < \infty$ we have $c_{st}c_{ts} < 4$. It follows that the above system of linear equations has a unique solution for x_s^u, x_t^u. Now set

$$U := \langle \alpha_u' := \alpha_u + x_s^u\alpha_s + x_t^u\alpha_t \mid u \in S - \{s, t\}\rangle \subseteq V.$$

It is clear that $V = V' \oplus U$, so it remains to show that s and t act as the identity on U. Let $u \in S - \{s, t\}$. Then

$$\begin{aligned} \alpha_u'.s &= (\alpha_u - c_{su}\alpha_s) - x_s^u\alpha_s + x_t^u(\alpha_t - c_{st}\alpha_s) \\ &= \alpha_u - (c_{su} + x_s^u + x_t^u c_{st})\alpha_s + x_t^u\alpha_t. \end{aligned}$$

Inserting the defining equations for x_s^u and x_t^u, we see that the coefficient of α_s is just x_s^u. Hence we have $\alpha_u'.s = \alpha_u'$. A similar computation also shows that $\alpha_u'.t = \alpha_u'$. Thus, both s and t act on U as the identity, as desired. $\square$

We introduce the following notation. If $(\mathcal{M}, \cdot)$ is a monoid (i.e., a set together with an associative multiplication and an identity element), we denote

$$\mathsf{Prod}(a, b; m) := \underbrace{aba\cdots}_{m \text{ factors}} \qquad \text{for any } a, b \in \mathcal{M} \text{ and } m \geq 0.$$

1.1.7. Subsystems of rank 2. Let $s, t \in S$, $s \neq t$. Let $W' := \langle s, t\rangle \subseteq W$ and $\Psi := \{\alpha_s.w', \alpha_t.w' \mid w' \in W'\}$. Since s, t have order 2, the group W' is a dihedral group (cf. Exercise 1.2). We claim that

(a) the element st has order m_{st};
(b) for each $\alpha \in \Psi$, we have either $\alpha \geq 0$ or $\alpha \leq 0$;
(c) if $\alpha \in \Psi$ and $u \in \{s, t\}$ are such that $\alpha \geq 0$ but $\alpha.u \leq 0$, then α must be a multiple of α_u.

Here, we write $v \geq 0$ (for any $v \in V$) if $v = v_s\alpha_s + v_t\alpha_t$ with $v_s, v_t \in \mathbb{R}$ such that $v_s \geq 0$ and $v_t \geq 0$; similarly, we write $v \leq 0$ if $v_s \leq 0$ and $v_t \leq 0$.

Let M_s, M_t be the matrices of s, t as in Lemma 1.1.6.

Case 1. Assume that $m_{st} = \infty$. The above statements are not affected if we replace M_s, M_t by PM_sP^{-1}, PM_tP^{-1}, where P is a diagonal matrix with positive entries. Hence, choosing P suitably, we may assume without loss of generality that $c_{st} = c_{ts} = -2$. Then we can compute:

$$\alpha_s.\mathsf{Prod}(M_t, M_s; k) = \begin{cases} k\alpha_s + (k+1)\alpha_t & \text{if } k \text{ is odd,} \\ (k+1)\alpha_s + k\alpha_t & \text{if } k \text{ is even.} \end{cases}$$

A symmetrical statement also holds for $\alpha_t.\mathsf{Prod}(M_s, M_t; k)$. In particular, we see that the orbit of α_s under the action of W' is infinite. Hence so is Ψ. Thus, W' is the infinite dihedral group, and (a) holds. The above formula for the action on α_s and α_t also proves (b) and (c).

Case 2. Assume that $m_{st} < \infty$. By Exercise 1.4, there exists an invertible matrix $P \in GL_2(\mathbb{R})$ such that

$$M_s' := PM_sP^{-1} = \begin{bmatrix} -1 & 0 \\ 0 & 1 \end{bmatrix}, \quad M_t' := PM_tP^{-1} = \begin{bmatrix} -\cos(\theta) & \sin(\theta) \\ \sin(\theta) & \cos(\theta) \end{bmatrix},$$

where $\theta = 2\pi/m_{st}$. The same exercise also gives a relation for $(M_s'M_t')^k$ $(k \geq 0)$, which shows that $M_s'M_t'$, and hence also M_sM_t, has order m_{st}. By Lemma 1.1.6, the action of st on V can be represented by a block diagonal matrix, where one block is the matrix M_sM_t and the other block is the identity matrix. Hence st has order m_{st}, which proves (a). Thus, W' is the dihedral group of order $2m_{st}$, and we are exactly in the same situation as discussed in the introduction to this chapter. Statement (b) is just a reformulation of property (2′) in that introduction. Finally, consider (c) and assume, for example, that $u = s$. Since $\alpha_t.M_s = \alpha_t - c_{st}\alpha_s \geq 0$, the given root α cannot be a multiple of α_t. If it were not a multiple of α_s either, then α would strictly lie in the cone spanned by α_s and α_t, and hence $\alpha.M_s \geq 0$ by property (3′), a contradiction.

The reduction to the rank 2 case is based on the following technical result.

1.1.8. Lemma. *Let $w \in W$ and $J \subseteq S$. Then $w = w'x$ with $w' \in \langle J \rangle$ and $x \in W$ such that $l(w) = l(w') + l(x)$ and $l(sx) > l(x)$ for all $s \in J$.*

Proof. We proceed by induction on $l(w)$. If $l(w) = 0$, then $w = 1$ and we can take $w' = x = 1$. Now let $w \neq 1$. If $l(sw) > l(w)$ for all $s \in J$, we can take $w' = 1$, $x = w$. Otherwise, there exists some $t \in J$ and $v \in W$ such that $w = tv$ with $l(tv) > l(v)$. By induction, we can write $v = v'x$ where $v' \in \langle J \rangle$ such that $l(v) = l(v') + l(x)$ and $l(sx) > l(x)$ for all $s \in J$. So we have $w = tv'x$ with $l(w) = l(v) + 1 = l(v') + l(x) + 1$. This also shows that $l(tv') = l(v') + 1$. Hence we are done if we set $w' := tv'$. □

We can now obtain a basic result relating the length function on W to the action of W on the root system. For any $v \in V$, we write $v \geq 0$ if $v = \sum_{s \in S} v_s \alpha_s$ with $v_s \in \mathbb{R}$ and $v_s \geq 0$ for all $s \in S$; similarly, we write $v \leq 0$ if $-v \geq 0$.

1.1.9. Theorem. *Let $s \in S$ and $w \in W$. Then $\alpha_s.w \geq 0$ if and only if $l(sw) = l(w) + 1$. (Equivalently, we have $\alpha_s.w \leq 0$ if and only if $l(sw) = l(w) - 1$.)*

Proof. We first prove by induction on $l(w)$ that, if $l(sw) = l(w) + 1$ for some $s \in S$, then $\alpha_s.w \geq 0$. If $l(w) = 0$, then $w = 1$ and there is nothing to prove. Now let $l(w) \geq 1$. Taking any reduced expression for w we can write $w = tw_1$ with $t \in S$, $w_1 \in W$ and $l(tw_1) > l(w_1)$. If we had $s = t$ then $sw = w_1$ and hence $l(sw) = l(w_1) < l(w)$, contrary to our assumption. So $s \neq t$. Applying Lemma 1.1.8 with $J = \{s, t\}$, we can write $w = w'x$ with $w' \in \langle J \rangle$ and $x \in W$ such that $l(w) = l(w') + l(x)$, $l(sx) > l(x)$, $l(tx) > l(x)$. Note that $w' \neq 1$.

Since $l(sw) = l(w) + 1$ and $w = w'x$, we must have $l(sw') = l(w') + 1$. Since $w' \in \langle J \rangle$, the defining formula for the reflections s, t shows that $\alpha_s.w'$ is a linear combination of α_s and α_t. Assume now that $l(w') < l(w)$. Then, by induction, $\alpha_s.w' \geq 0$. So we can write $\alpha_s.w' = c_s\alpha_s + c_t\alpha_t$ with $c_s, c_t \geq 0$. Since $w' \neq 1$, we have $l(x) < l(w)$. The conditions that $l(sx) = l(x) + 1$ and $l(tx) = l(x) + 1$ therefore imply, by induction, that $\alpha_s.x \geq 0$ and $\alpha_t.x \geq 0$. It follows that

$$\alpha_s.w = (\alpha_s.w').x = (c_s\alpha_s + c_t\alpha_t).x = c_s\alpha_s.x + c_t\alpha_t.x \geq 0,$$

as desired. It remains to consider the case where $l(w') = l(w)$, i.e., $w = w'$. We are thus reduced to a computation in rank 2.

Let $W' = \langle J \rangle$ and $V' := \mathbb{R}\alpha_s \oplus \mathbb{R}\alpha_t$. We are then in the situation of (1.1.7). If $m_{st} = \infty$ then every element in W' has a unique reduced expression, and hence we must have $w = \mathsf{Prod}(t, s; k)$ for some $k \geq 1$. The desired result in this case is obvious from the formula in (1.1.7). Now assume that $m_{st} < \infty$. Since $w \neq 1$, we can write $w = w'u$ with $w' \in W'$ and $u \in \{s, t\}$ such that $l(w'u) > l(w')$. Then we still have $l(sw') > l(w')$. So, by induction, $\alpha := \alpha_s.w' \geq 0$. Assume, if possible, that $\alpha.u = \alpha_s.w'u = \alpha_s.w \leq 0$. By (1.1.7)(c), this implies that α is a scalar multiple of α_u. Now set $y := (w')^{-1}sw'u \in W'$. We claim that $y = 1$. Indeed, y has finite order and, hence, is diagonalizable (over $\mathbb{C}$). So it is enough to show that y has all eigenvalues equal to 1. Now, by Lemma 1.1.6, we have $V = V' \oplus U$ where every element in W' acts as the identity on U. So we only need to show that y has two eigenvalues equal to 1 on V'. Now, a straightforward computation shows that $\alpha_u.y = \alpha_u$. It remains to note that $\det(y) = 1$, and our claim is proved. Thus, $y = 1$ and so $sw' = w'u = w$. This implies $l(sw) < l(w)$, a contradiction. So our assumption was wrong, and we have completed the proof of the assertion that $l(sw) > l(w)$ implies $\alpha_s.w \geq 0$.

Conversely, assume that $l(sw) < l(w)$. Then we can write $w = sw'$ with $l(sw') > l(w')$. We have just proved that then $\alpha_s.w' \geq 0$. Since $\alpha_s.s = -\alpha_s$, this yields $-\alpha_s.w = -\alpha_s.(sw') = \alpha_s.w' \geq 0$, as desired.

Since either $l(sw) < l(w)$ or $l(sw) > l(w)$, the proof is complete. □

1.1.10. Remark. Let $\alpha \in \Phi$ and write $\alpha = \alpha_s.w$ where $s \in S$ and $w \in W$. The above result shows that if $l(sw) > l(w)$ then $\alpha \geq 0$; on the other hand, if $l(sw) < l(w)$, then we can set $w' = sw$ and conclude that $-\alpha = (\alpha_s.s).w = \alpha_s.w' \geq 0$, since $l(sw') > l(w')$. Thus, we see that every root in Φ either lies in $\Phi^+ := \{\alpha \in \Phi \mid \alpha \geq 0\}$ or in $\Phi^- := \{-\alpha \mid \alpha \in \Phi^+\}$. Accordingly, we will say that α is a *positive root* or a *negative root*.

Now we can also formulate our first fundamental algorithm.

Algorithm A (Reduced word). Given a reflection group $W = W(C) \subset GL(V)$ associated with a Cartan matrix $C = (c_{st})_{s,t \in S}$ as in Definition 1.1.3 and an element $w \in W$, a reduced expression for w in terms of the generators $S \subseteq W$ is returned. This is done in such a way that if we fix a labelling $S = \{s_1, \ldots, s_n\}$, the lexicographically smallest reduced expression is computed.

A1. [Initialize.] Set $y \leftarrow w$ and $w_{red} \leftarrow []$.

A2. [Next y.] Given $1 \neq y \in W$, find a generator $s \in S$ with $l(sy) < l(y)$. Consider the matrix (a_{ij}) of y with respect to the basis $\{\alpha_{s_i}\}_{1 \leq i \leq n}$ of V and let $k := \min\{i \mid a_{ij} < 0$ for some $j\}$. (By Theorem 1.1.9, we have $\alpha_{s_i}.y \leq 0$ if $l(s_i y) < l(y)$.) Writing $w_{red} = [s_{i_1}, \ldots, s_{i_l}]$, set

$$y \leftarrow s_k y \quad \text{and} \quad w_{red} \leftarrow [s_k, s_{i_1}, \ldots, s_{i_l}].$$

A3. [Loop.] Repeat step A2 until $y = \mathrm{id}_V$. Then w_{red} is the desired reduced expression for w.

Note that one can even apply this algorithm to any $g \in GL(V)$. If, in each loop, an index k as in step A2 is found and if the algorithm terminates after a finite number of loops, one can compare the element given by w_{red} with g and decide if g lies in W or not. For an example, see the matrices in step 1 of Exercise 6.10.

1.2 MATSUMOTO'S THEOREM

Let $C = (c_{st})_{s,t \in S}$ be a Cartan matrix and $W = \langle S \rangle \subset \mathrm{GL}(V)$ be the corresponding reflection group, as in Section 1.1. In this section, we prove the basic structure theorems about W.

We begin with an extension of Lemma 1.1.8 in the case where $|J| = 2$. Let $s, t \in S$, $s \neq t$, and consider the subgroup $W_{st} := \langle s, t \rangle$. By (1.1.7)(a), the element st has order m_{st}, and hence the group W_{st} is either infinite dihedral or has order $2m_{st}$. In the latter case, by Exercise 1.2(b), W_{st} contains a unique element of maximal length:

$$\Delta_{st} := \mathrm{Prod}(s, t; m_{st}) = \mathrm{Prod}(t, s; m_{st}).$$

1.2.1. Lemma. *Let $J = \{s, t\} \subseteq S$, $s \neq t$. Let $w \in W$ and $w = w'x$ be as in Lemma 1.1.8. If $l(sw) < l(w)$ and $l(tw) < l(w)$, then $m_{st} < \infty$ and $w' = \Delta_{st}$.*

Proof. We claim that $l(sw') < l(w')$ and $l(tw') < l(w')$. Assume, if possible, that $l(sw') > l(w')$. By Theorem 1.1.9, this implies that $\alpha_s.w' \geq 0$. Now $\alpha_s.w'$ is a linear combination of α_s and α_t, with non-negative coefficients. Since $l(sx) > l(x)$ and $l(tx) > l(x)$ (see Lemma 1.1.8), we also have $\alpha_s.x \geq 0$ and $\alpha_t.x \geq 0$ by Theorem 1.1.9. Hence $\alpha_s.w = (\alpha_s.w').x \geq 0$ and so $l(sw) > l(w)$, using Theorem 1.1.9 once more. Hence our assumption was wrong and we must have $l(sw') < l(w')$. Similarly, we also have that $l(tw') < l(w')$.

Assume, if possible, that $m_{st} = \infty$. Then W_{st} is an infinite dihedral group. By Exercise 1.2(a), every element of W_{st} has a unique reduced expression, which either starts with s or with t. Hence there is no element $w' \in W_{st}$ with $l(sw') < l(w')$ and $l(tw') < l(w')$, a contradiction.

So we must have $m_{st} < \infty$. By (1.1.7), W_{st} is the dihedral group of order $2m_{st}$. Exercise 1.2(b) now shows that every element of W_{st} of length strictly smaller than m_{st} has a unique reduced expression, which either starts with s or with t. Hence $l(w') = m_{st}$ and we have in fact $w' = \Delta_{st}$. □

The importance of the following result can hardly be overestimated. Almost all the results to come on Coxeter groups and Iwahori–Hecke algebras rely in their proofs in some way on this fact.

1.2.2. Theorem (Matsumoto). *Let $(\mathcal{M}, \cdot)$ be a monoid and $f\colon S \to \mathcal{M}$ be a map such that*

$$\mathrm{Prod}(f(s), f(t); m_{st}) = \mathrm{Prod}(f(t), f(s); m_{st}) \qquad \text{for } s, t \in S,\ s \neq t,\ m_{st} < \infty.$$

Then there exists a unique map $F\colon W \to \mathcal{M}$ such that $F(w) = f(s_1) \cdots f(s_k)$ whenever $w = s_1 \cdots s_k$ ($s_i \in S$) is a reduced expression.

Proof. First note that if F exists, then it is uniquely determined. So the problem is to show existence. For each $k \geq 1$ let $W_k := \{w \in W \mid l(w) \leq k\}$, so that $W = \bigcup_k W_k$. We define F inductively on each W_k. If $k = 1$ then W_k consists of the elements of S and the identity element. We set $F(s) = f(s)$ for $s \in S$, and map the identity element of W to that of $\mathcal{M}$. Now assume that $k \geq 2$ and that F is already defined on W_{k-1}, having the desired property. We must show the following: let $w \in W$ with $l(w) = k$ and assume that $w = s_1 \cdots s_k = t_1 \cdots t_k$ ($s_i, t_j \in S$) are two reduced expressions; then $f(s_1) \cdots f(s_k) = f(t_1) \cdots f(t_k)$.

If $s_1 = t_1$, we also have $s_2 \cdots s_k = t_2 \cdots t_k$, and we are done by induction. Now assume that $s_1 \neq t_1$. Then $l(s_1 w) < l(w)$ and $l(t_1 w) < l(w)$. So, by Lemma 1.2.1, we have $m_{s_1 t_1} < \infty$ and $w = \Delta_{s_1 t_1} x$ for some $x \in W$ with $l(w) = m_{s_1 t_1} + l(x)$. Writing $\Delta_{s_1 t_1} = \mathsf{Prod}(s_1, t_1; m_{s_1 t_1})$ and taking some reduced expression for x, we obtain a reduced expression for $s_1 w = \mathsf{Prod}(t_1, s_1; m_{s_1 t_1} - 1)x$. Since $s_1 w \in W_{k-1}$, we can apply induction and conclude that

$$f(s_2) \cdots f(s_k) = F(s_1 w) = \mathsf{Prod}(f(t_1), f(s_1); m_{s_1 t_1} - 1) \cdot F(x). \tag{1}$$

Similarly, we also obtain a reduced expression for $t_1 w = \mathsf{Prod}(s_1, t_1; m_{s_1 t_1} - 1)x$. As above, we conclude that

$$f(t_2) \cdots f(t_k) = F(t_1 w) = \mathsf{Prod}(f(s_1), f(t_1); m_{s_1 t_1} - 1) \cdot F(x). \tag{2}$$

Using the assumption on f and combining the above relations eventually yields:

$$\begin{aligned} f(s_1) \cdot f(s_2) \cdots f(s_k) &= f(s_1) \cdot \mathsf{Prod}(f(t_1), f(s_1); m_{s_1 t_1} - 1) \cdot F(x) \qquad \text{by (1)} \\ &= \mathsf{Prod}(f(s_1), f(t_1); m_{s_1 t_1}) \cdot F(x) = \mathsf{Prod}(f(t_1), f(s_1); m_{s_1 t_1}) \cdot F(x) \\ &= f(t_1) \cdot \mathsf{Prod}(f(s_1), f(t_1); m_{s_1 t_1} - 1) \cdot F(x) = f(t_1) \cdot f(t_2) \cdots f(t_k) \quad \text{by (2)}. \end{aligned}$$

Thus, we can extend F to W_k as desired. □

There exists in fact a "universal" monoid to which Matsumoto's theorem applies: this is called the *braid monoid* and will be studied in more detail in Chapter 4. We will now give a number of applications of Matsumoto's theorem. A first and very typical application is the following:

1.2.3. Corollary. *Let $w \in W$. Then there is a well-defined subset $J(w) \subseteq S$ such that if $w = s_1 \cdots s_k$ with $s_i \in S$ is a reduced expression then $J(w) = \{s_1, \ldots, s_k\}$.*

Proof. Let $\mathcal{M}$ be the monoid whose elements are the subsets of S and where the product is given by the union of two subsets. Consider the map $f\colon S \to \mathcal{M}$, $s \mapsto \{s\}$. Then the assumptions of Matsumoto's theorem are satisfied: if $s, t \in S$ are such that $s \neq t$ and $m_{st} < \infty$, then

$$\mathsf{Prod}(f(s), f(t); m_{st}) = \{s\} \cdot \{t\} \cdots = \{s, t\} = \mathsf{Prod}(f(t), f(s); m_{st}).$$

Let $F\colon W \to \mathcal{M}$ be the extension of f as in Theorem 1.2.2. Then, if $w \in W$ and $w = s_1 \cdots s_k$ ($s_i \in S$) is any reduced expression, we have $F(w) = \{s_1, \ldots, s_k\}$, i.e., this set does not depend on the choice of a reduced expression. □

1.2.4. Subexpressions. Let $w \in W$ and take a reduced expression $w = s_1 \cdots s_k$ ($s_i \in S$). We say that $v \in W$ is a *subexpression* of w, and write

$$v \leq w, \quad \text{if } v = s_{i_1} \cdots s_{i_l} \text{ where } l \geq 0 \text{ and } 1 \leq i_1 < \cdots < i_l \leq k.$$

Despite its appearance, this definition does not depend on the choice of a reduced expression for w. Indeed, let $\mathcal{M}$ be the monoid whose elements are the subsets of W and where the product is given by $A \cdot B := \{ab \mid a \in A, b \in B\}$ (for $A, B \subseteq W$). We check that the assumptions of Matsumoto's theorem are satisfied for the map $f\colon S \to \mathcal{M}$, $s \mapsto \{1, s\}$.

Let $s, t \in S$ be such that $s \neq t$ and $m_{st} < \infty$. Then $\mathsf{Prod}(f(s), f(t); m_{st})$ is the subset of W consisting of all elements which can be formed by taking products $st\cdots$ or $ts\cdots$, where the number of factors is at most m_{st}. Hence, since s, t generate the dihedral group of order $2m_{st}$, we have $\mathsf{Prod}(f(s), f(t); m_{st}) = \langle s, t\rangle$. We certainly obtain the same result if we compute $\mathsf{Prod}(f(t), f(s); m_{st})$. So the assumptions of Theorem 1.2.2 are indeed satisfied.

It follows that there exists a unique map $F\colon W \to \mathcal{M}$ such that $F(w) = f(s_1) \cdots f(s_k)$ whenever $w \in W$ and $w = s_1 \cdots s_k$ ($s_i \in S$) is a reduced expression. Writing out the product $f(s_1) \cdots f(s_k)$, we see that

$$F(w) = \{s_{i_1} \cdots s_{i_l} \mid 1 \leq i_1 < \cdots < i_l \leq k \text{ for some } l \geq 0\}.$$

Thus, the set of subexpressions of $w \in W$ is independent of the choice of a reduced expression for w.

Note that a subexpression of an element $w \in W$ is not necessarily reduced. But the following result shows that we can always extract a reduced expression from a given expression of an element as a product of generators.

1.2.5. Theorem (Cancellation law). *Let $s_1, \ldots, s_k \in S$ (for some $k \geq 2$) and assume that $l(s_1 \cdots s_k) < k$. Then there exist indices $i < j$ such that $s_1 \cdots s_k = s_1 \cdots \hat{s_i} \cdots \hat{s_j} \cdots s_k$, where the notation indicates that the ith and jth factors have been "cancelled".*

Proof. Since s_1 is reduced and $s_1 \cdots s_k$ is not reduced, there exists some $j \geq 2$ such that $w := s_1 \cdots s_{j-1}$ is reduced and $l(ws_j) = l(w) - 1 = j - 2$. We write $ws_j = t_1 \cdots t_{j-2}$ with $t_i \in S$, so that $w = t_1 \cdots t_{j-2}s_j$ is a reduced expression for w which ends with s_j. Then $ws_j \leq w$, i.e., ws_j is a subexpression of w. But the relation $ws_j \leq w$ is independent of a choice of reduced expression for w. We now take the reduced expression $w = s_1 \cdots s_{j-1}$ that we started with. So ws_j is given by taking a subexpression of $s_1 \cdots s_{j-1}$. Thus, since $l(ws_j) = j - 2$, there exists some $i < j$ such that $ws_j = s_1 \cdots \hat{s_i} \cdots s_{j-1}$. Hence we have

$$s_1 \cdots s_k = (ws_j)s_{j+1} \cdots s_k = (s_1 \cdots \hat{s_i} \cdots s_{j-1})s_{j+1} \cdots s_k,$$

as desired. □

We remark that the *exchange condition* and the *Bruhat–Chevalley order* are immediate consequences of the above results; see Exercises 1.6 and 1.7. Another application of the above discussion is the following result, which may seem quite technical but which will turn up at several places in the subsequent development, e.g., in the proof of Deodhar's Lemma 2.1.2 or Bourbaki's construction of Iwahori–Hecke algebras in Theorem 4.4.6.

1.2.6. Lemma. *Let $w \in W$ and $s, t \in S$ be such that $l(sw) = l(wt)$ and $l(swt) = l(w)$. Then $sw = wt$.*

Proof. Assume first that $l(wt) > l(w)$. Since $l(swt) = l(w) < l(wt)$, we can write $wt = sv$ with $v \in W$ and $l(sv) > l(v)$. Let $v = s_1 \cdots s_k$ ($s_i \in S$) be a reduced expression. Now we have $w \leq wt = ss_1 \cdots s_k$ and $l(w) = k$. Thus, we obtain a reduced expression for w by taking a suitable subexpression of $ss_1 \cdots s_k$, where exactly one factor has to be dropped. Since $l(sw) > l(w)$, that subexpression cannot start with s. Hence it is the first factor s which has to be dropped. So we have $w = s_1 \cdots s_k = v$, whence $sw = sv = wt$.

If $l(wt) < l(w)$, let $w' = wt$. Then we have $l(sw't) = l(w')$ and $l(sw') = l(w't) > l(w')$. Hence we can apply the previous argument to conclude that $sw' = w't$. Inserting the relation $w' = wt$ again yields the desired result. □

1.2.7. Theorem. *Recall that we have $s^2 = 1$ for all $s \in S$ and that m_{st} is the order of st for $s \neq t$. Then we have a presentation of the form*

$$W = \langle s \in S \mid s^2 = 1 \text{ for } s \in S \text{ and } (st)^{m_{st}} = 1 \text{ for } s \neq t,\ m_{st} < \infty \rangle.$$

Proof. Let G be the group defined by the above presentation. By (1.1.7)(a), the elements of $S \subseteq W$ satisfy the relations $(st)^{m_{st}} = 1$. So we have a group homomorphism $\pi \colon G \to W$, sending the generators of G to the generators of W. This homomorphism is clearly surjective. Conversely, if $s \neq t$ and $m_{st} < \infty$, the relation $(st)^{m_{st}} = 1$ is equivalent to the relation $\text{Prod}(s, t; m_{st}) = \text{Prod}(t, s; m_{st})$ (since all $s \in S$ are involutions). Hence Matsumoto's theorem shows that we also have a map $F \colon W \to G$ sending the generators of W to the generators of G. Now F is in fact a group homomorphism. Indeed, it is enough to check that $F(sw) = F(s)F(w)$ for all $s \in S$ and $w \in W$. If $l(sw) > l(w)$ then this holds by construction of F. If $l(sw) < l(w)$, we set $w' = sw$ and conclude that $F(w) = F(sw') = F(s)F(w') = F(s)F(sw)$, since $l(sw') > l(w')$. So we again have $F(sw) = F(s)F(w)$ (note that $F(s)$ is an involution). The map F being a group homomorphism which sends the generators of W to those of G, we now see that $F \circ \pi = \text{id}_G$. It follows that π is injective, and hence an isomorphism. □

1.2.8. Coxeter groups. Let X be any finite index set and $M = (m_{xy})_{x,y \in X}$ be a symmetric matrix whose entries are positive integers (or ∞) such that $m_{xx} = 1$ and $m_{xy} > 1$ for all $x, y \in X$, $x \neq y$. Define a group $G(M)$ by the presentation:

$$G(M) = \langle x \in X \mid x^2 = 1 \text{ for } x \in X \text{ and } (xy)^{m_{xy}} = 1 \text{ for } x \neq y,\ m_{xy} < \infty \rangle.$$

Then the pair $(G(M), X)$ is called a *Coxeter system* and the group $G(M)$ is called a *Coxeter group*. We have just seen in Theorem 1.2.7 that reflection groups as in Section 1.1 are Coxeter groups.

Conversely, note that the matrix $C = (-2\cos(\pi/m_{xy}))_{x,y \in X}$ satisfies the requirements of Definition 1.1.1, i.e., it is a Cartan matrix. So Theorem 1.2.7 shows that $G(M)$ arises as the reflection group associated with the Cartan matrix C. This also shows that m_{xy} is precisely the order of $xy \in G(M)$; see (1.1.7). The fact that the (abstract) Coxeter group $G(M)$ also is a reflection group as in Section 1.1 means that we have an embedding $G(M) \subset GL(V)$, where V is a real vector space of dimension $|X|$ on which the elements of X act as reflections. The embedding $G(M) \subset GL(V)$ will be called the *reflection representation* of $G(M)$.

1.2.9. Parabolic subgroups. Let $J \subseteq S$ be any subset and consider the subgroup $W_J := \langle J \rangle \subseteq W$. Then W_J is called a *parabolic subgroup* of W.

Let $V_J := \bigoplus_{s \in J} \mathbb{R}\alpha_s$. The defining equations show that $V_J.s \subseteq V_J$ for all $s \in J$. Hence, by restricting the maps $s\colon V \to V$ (for $s \in J$) to V_J, we get a canonical surjective group homomorphism

$$\pi\colon W_J \to W(C_J), \qquad \text{where } C_J := (c_{st})_{s,t \in J}.$$

On the other hand, since $W(C_J)$ has a presentation as in Theorem 1.2.7 and the generators of W_J certainly satisfy the defining relations, we also get a map from $W(C_J)$ to W_J by identifying corresponding generators. We conclude that π is also injective, and hence an isomorphism. It follows, in particular, that (W_J, J) is a Coxeter system.

Now we have two length functions on W_J. First, we can restrict the length function $l\colon W \to \mathbb{N}_0$ to W_J. Thus, given $w \in W_J$, the integer $l(w)$ is the smallest $k \geq 0$ such that there exist $s_1, \ldots, s_k \in S$ with $w = s_1 \cdots s_k$. On the other hand, via the canonical isomorphism $\pi\colon W_J \to W(C_J)$, we have a second length function $l_J\colon W_J \to \mathbb{N}_0$: given $w \in W_J$, the integer $l_J(w)$ is the smallest $n \geq 0$ such that there exist $s_1, \ldots, s_n \in J$ with $w = s_1 \cdots s_n$. The following result shows that these two length functions coincide.

1.2.10. Proposition. *Let $J \subseteq S$ and $W_J \subseteq W$ be the corresponding parabolic subgroup. Then $l(w) = l_J(w)$ for all $w \in W_J$. Moreover, for $w \in W_J$, all reduced expressions for w (in terms of the generators S) have all their factors in J.*

Proof. Let $w \in W_J$ and write $w = s_1 \cdots s_k$ with $s_i \in J$. A repeated application of the cancellation law shows that, by possibly cancelling some factors, we obtain a reduced expression for w which has all its factors in J. Corollary 1.2.3 shows that every reduced expression for w has its factors in the same set as the above reduced expression. So these factors always lie in J and, hence, $l(w) = l_J(w)$. □

Parabolic subgroups and, more generally, cosets of parabolic subgroups will be considered in more detail in Chapter 2.

1.3 CARTAN MATRICES OF FINITE TYPE

In the previous two sections, we have developed the basic theory of reflection groups and Coxeter groups in completely general terms. We have made no finiteness assumptions; even the order of the product of two generators could have been infinite. In this section, we discuss the classification of Cartan matrices of so-called finite type; see the exact definition in (1.3.2) below.

1.3.1. Irreducible Coxeter systems. Let $C = (c_{st})_{s,t \in S}$ be a Cartan matrix and assume that C is decomposable, i.e., we have a partition $S = S_1 \amalg S_2$ with $S_1, S_2 \neq \varnothing$ and such that $c_{st} = 0$ whenever $s \in S_1$, $t \in S_2$. Note that (C1) implies that then also $c_{ts} = 0$ whenever $s \in S_1$, $t \in S_2$. Then we can write C as a block matrix with diagonal blocks $C_1 = (c_{st})_{s,t \in S_1}$ and $C_2 = (c_{st})_{s,t \in S_2}$. Correspondingly, we also have a direct sum decomposition $V = V_1 \oplus V_2$ where V_1 has basis $\{\alpha_s \mid s \in S_1\}$ and V_2 has basis $\{\alpha_s \mid s \in S_2\}$. The defining equations show that V_1, V_2 are invariant under all $s \in S$. Moreover, if $s \in S_1$ then $s|_{V_2} = \mathrm{id}_{V_2}$, and if $s \in S_2$ then $s|_{V_1} = \mathrm{id}_{V_1}$. This implies, in particular, that s and t commute with each other whenever $s \in S_1$ and $t \in S_2$. It follows that we have an isomorphism

$$W(C) \xrightarrow{\sim} W(C_1) \times W(C_2), \quad w \mapsto \left(w|_{V_1}, w|_{V_2}\right),$$

and a partition $\Phi(C) = \Phi(C_1) \amalg \Phi(C_2)$ with $\Phi(C_i) = \Phi(C) \cap V_i$ for $i = 1, 2$. Moreover, we have $l(w_1 w_2) = l(w_1) + l(w_2)$ for all $w_1 \in W(C_1)$, $w_2 \in W(C_2)$.

In this way, the study of reflection groups $W(C)$ as in Definition 1.1.3 is reduced to the case where C is an *indecomposable Cartan matrix*, i.e., a Cartan matrix for which no partition as above exists. In this case, we call the corresponding Coxeter system (W, S) an *irreducible Coxeter system.*

1.3.2. W-invariant forms. Let $C = (c_{st})_{s,t \in S}$ be a Cartan matrix and $W = \langle S \rangle \subset \mathrm{GL}(V)$ be the corresponding reflection group.

(a) Assume that we have a W-invariant symmetric bilinear form $(\ ,\ )$ on V such that $(\alpha_s, \alpha_s) > 0$ for all $s \in S$. Let $s, t \in S$. Then $(\alpha_s, \alpha_t) = (\alpha_s.s, \alpha_t.s) = -(\alpha_s, \alpha_t.s)$. Inserting the equation $\alpha_t.s = \alpha_t - c_{st}\alpha_s$ yields that

$$c_{st} = 2\frac{(\alpha_s, \alpha_t)}{(\alpha_s, \alpha_s)} \qquad \text{for all } s, t \in S.$$

Setting $d_s = \sqrt{(\alpha_s, \alpha_s)} > 0$, we have $d_s^2 c_{st} = 2(\alpha_s, \alpha_t) = d_t^2 c_{ts}$. Using also the relation $c_{st}c_{ts} = 4\cos^2(\pi/m_{st})$ and the fact that $c_{st} \leq 0$ for $s \neq t$, we conclude that $d_s d_t^{-1} c_{st}/2 = -\cos(\pi/m_{st})$ for all $s, t \in S$.

(b) Conversely, if there exist positive real numbers d_s $(s \in S)$ such that $d_s d_t^{-1} c_{st}/2 = -\cos(\pi/m_{st})$ for all $s, t \in S$, then we have a W-invariant symmetric bilinear form on V given by $(\alpha_s, \alpha_t) = d_s^2 c_{st}/2$. Indeed, to check the W-invariance, we compute for any $s, t, u \in S$:

$$(\alpha_s.u, \alpha_t.u) = (\alpha_s - c_{us}\alpha_u, \alpha_t - c_{ut}\alpha_u)$$
$$= (\alpha_s, \alpha_t) - c_{us}d_u^2 c_{ut}/2 - c_{ut}d_s^2 c_{su}/2 + d_u^2 c_{us}c_{ut} = (\alpha_s, \alpha_t),$$

where the last equality holds since $d_s d_u^{-1} c_{su} = -2\cos(\pi/m_{su}) = d_u d_s^{-1} c_{us}$. For example, if C is a symmetric matrix, we may take $d_s = 1$ for all $s \in S$ and then the form defined by $(\alpha_s, \alpha_t) = -\cos(\pi/m_{st})$ is W-invariant (cf. also (1.2.8)).

(c) We say that C is a *Cartan matrix of finite type* if there exists a bilinear form (,) as in (a) which is also positive-definite. This is equivalent to the condition that the matrix $(-\cos(\pi/m_{st}))_{s,t\in S}$ is positive-definite, that is, we have $\det(-\cos(\pi/m_{st}))_{s,t\in J} > 0$ for every subset $J \subseteq S$. It is a standard fact about finite groups that such a form exists, for example, if W is finite; see Exercise 1.3.

We can now describe the classification of the indecomposable Cartan matrices of finite type. With any Cartan matrix $C = (c_{st})_{s,t\in S}$, we associate a graph, called the *Coxeter graph*, as follows. It has vertices labelled by the elements of S, and two vertices labelled by $s \neq t$ are joined by an edge if $m_{st} \geq 3$ (with m_{st} as in Definition 1.1.1(C2)). Moreover, if $m_{st} \geq 4$, we label the edge by m_{st}. The Coxeter graph encodes precisely the information needed in the presentation of the corresponding reflection group $W = W(C)$ as in Theorem 1.2.7.

1.3.3. Theorem. *The Coxeter graphs of the indecomposable Cartan matrices of finite type are given in Table 1.1.*

Table 1.1 Coxeter graphs of Cartan matrices of finite type

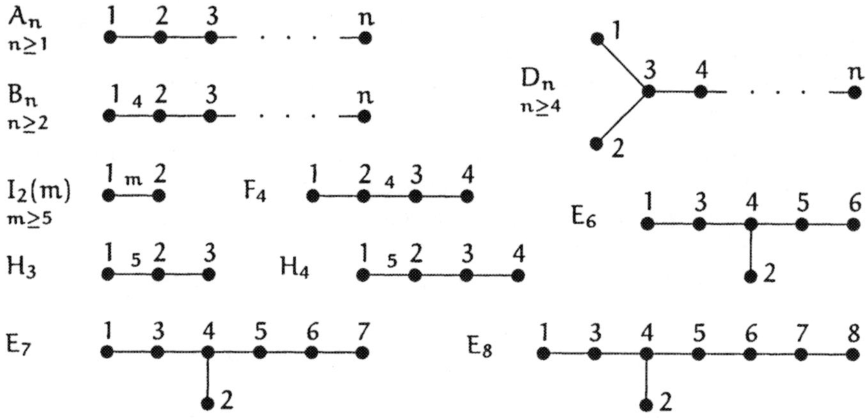

The numbers on the vertices correspond to a chosen labelling of the elements of S.

Proof. Let $C = (c_{st})_{s,t\in S}$ be a Cartan matrix of finite type. Then the discussion in (1.3.2) shows that the matrix $(-\cos(\pi/m_{st}))_{s,t\in S}$ must be positive-definite. Thus, we first classify all matrices $(m_{st})_{s,t\in S}$ (for some finite index set S) having the following properties:

(a) we have $m_{ts} = m_{st} \in \mathbb{Z} \cup \{\infty\}$ for all $s, t \in S$;
(b) we have $m_{ss} = 1$ and $m_{st} \geq 2$ for $s \neq t$;
(c) the matrix $(-\cos(\pi/m_{st}))$ is positive-definite.

This classification is a standard result in the theory of Coxeter groups; see [Bourbaki 1968, Chap. VI, §4, No. 1] or [Humphreys 1990, §2.7]. For a somewhat different approach to such a classification, see [Kac 1985, Chap. 4].

Conversely, assume that the Coxeter graph of C occurs in the above list. Then we note that this graph does not contain circles. It is easily checked that this implies that there exists a diagonal matrix D, whose diagonal entries d_s are positive real numbers, such that DC is symmetric. Since also $c_{st}c_{ts} = 4\cos^2(\pi/m_{st})$, an argument similar to that in (1.3.2)(a) shows that we must have $d_s d_t^{-1} c_{st}/2 = -\cos(\pi/m_{st})$ for all $s, t \in S$. So we have an invariant symmetric bilinear form as in (1.3.2), which is positive-definite since (m_{st}) satisfies (c). □

We have seen in (1.2.8) that any Coxeter system (W, S) is defined in terms of a suitable Cartan matrix C, such that $W = \langle S \rangle \subset \mathsf{GL}(V)$ is a reflection group as in Section 1.1. Now, in general, C is not uniquely determined by (W, S) and the question arises whether there are choices that are "better" than others. For example, the root system associated with the first Cartan matrix in Example 1.1.4 had the property that it contained various scalar multiples of the same root. This leads us to the following notion.

1.3.4. Definition. We say that Φ is a *reduced root system* if, for any $\alpha \in \Phi$ and $x \in \mathbb{R}$, we have $x\alpha \in \Phi$ only for $x = \pm 1$.

Note that every Coxeter system (W, S) arises from a Cartan matrix C whose root system is reduced. For example, we may take a symmetric Cartan matrix as in (1.2.8) and then (1.3.2)(b) shows that we have a W-invariant symmetric bilinear form $(\,,\,)$ on V such that $(\alpha_s, \alpha_s) = 1$ for all $s \in S$. Consequently, we have $(\alpha, \alpha) = 1$ for all $\alpha \in \Phi$, which certainly implies that Φ is reduced.

1.3.5. Proposition. *Assume that Φ is reduced. For each $w \in W$, let $N(w) := \{\alpha \in \Phi^+ \mid \alpha.w \leq 0\}$. Let $w = s_1 \cdots s_k$ with $s_i \in S$ be a reduced expression. Then $|N(w)| = l(w)$ and $N(w)$ consists precisely of the roots*

$$\alpha_1, \quad \alpha_2.s_1, \quad \alpha_3.s_2s_1, \quad \ldots, \quad \alpha_k.s_{k-1}s_{k-2}\cdots s_1,$$

where α_i denotes the simple root corresponding to the generator s_i $(1 \leq i \leq k)$.

Proof. First we check that the above set of roots is contained in $N(w)$. Let $\alpha := \alpha_i.s_{i-1}\cdots s_1$ with $1 \leq i \leq k$. Then, since $s_1 \cdots s_k$ is reduced, so is $s_k \cdots s_1$ and hence also $s_i s_{i-1} \cdots s_1$. It follows that $\alpha = \alpha_i.s_{i-1}\cdots s_1 \geq 0$, by Theorem 1.1.9. Moreover, $\alpha.w = \alpha_i.(s_{i-1}\cdots s_1)(s_1 \cdots s_k) = \alpha_i.s_i \cdots s_k = -\alpha_i.s_{i+1}\cdots s_k$. Again, since $s_i s_{i+1} \cdots s_k$ is reduced, we have $\alpha_i.s_{i+1}\cdots s_k \geq 0$, and so $\alpha.w \leq 0$. Thus $\alpha \in N(w)$.

Conversely, let $\alpha \in N(w)$. Since $\alpha \geq 0$ and $\alpha.s_1 \cdots s_k \leq 0$, there exists some $1 \leq i \leq k$ such that $\alpha' := \alpha.s_1 \cdots s_{i-1} \geq 0$ and $\alpha'.s_i \leq 0$. Write $\alpha' = \sum_{s \in S} v_s \alpha_s$

with $\nu_s \geq 0$ for all $s \in S$. The defining equation for s_i shows that $\alpha'.s_i$ is a linear combination of the simple roots where all coefficients are the same as those for α', except possibly that of α_i. Hence, since $\alpha'.s_i \leq 0$, all coefficients at roots α_s with $s \neq s_i$ must be 0. So α' is a scalar multiple of α_i. Since Φ is reduced and $\alpha' \geq 0$, we conclude that $\alpha' = \alpha_i$. It follows that $\alpha.s_1 \cdots s_{i-1} = \alpha_i$ and hence $\alpha_i.s_{i-1} \cdots s_1 = \alpha$, as desired.

Finally, we must check that all roots in the above set are distinct. So assume, if possible, that $\alpha_i.s_{i-1} \cdots s_1 = \alpha_j.s_{j-1} \cdots s_1$ for some $1 \leq i < j \leq k$. Then $\alpha_j.s_{j-1} \cdots s_1 = \alpha_j.(s_{j-1} \cdots s_i)(s_{i-1} \cdots s_1)$ and hence $\alpha_i = \alpha_j.s_{j-1} \cdots s_i$. It follows that $\alpha_j.s_{j-1} \cdots s_{i+1} \leq 0$ which is a contradiction by Theorem 1.1.9 since, again, $s_j s_{j-1} \cdots s_{i+1}$ is reduced.

Thus, $N(w)$ consists precisely of the roots in the above set. Since there are $n = l(w)$ such roots, we have $|N(w)| = l(w)$. □

The following result shows that the property of having a reduced root system can be easily checked from the Cartan matrix.

1.3.6. Proposition. *Assume that* C *is of finite type. Recall that, for* $s \neq t$, *we have* $c_{st}c_{ts} = 4\cos^2(\pi/m_{st})$ *for some* $m_{st} \geq 2$. *Then the root system* Φ *is reduced if and only if* $c_{st} = c_{ts}$ *for all* $s \neq t$ *with* m_{st} *odd.*

Proof. Let $(\ ,\ )$ be a W-invariant scalar product as in (1.3.2). The formula $c_{st} = 2(\alpha_s, \alpha_t)/(\alpha_s, \alpha_s)$ shows that $c_{st} = c_{ts}$ if and only if $(\alpha_s, \alpha_s) = (\alpha_t, \alpha_t)$. Moreover, since $(\ ,\ )$ is W-invariant, two roots have the same norm (with respect to $(\ ,\ )$) if they lie in the same W-orbit.

Assume first that Φ is reduced, and let $s, t \in S$ with $s \neq t$ and m_{st} odd. Then s, t are conjugate by Exercise 1.2(b). Let $w \in W$ be such that $s = wtw^{-1}$. Then $(\alpha_s.w).t = \alpha_s.(sw) = -\alpha_s.w$, i.e., $\alpha_s.w$ is an eigenvector of t with eigenvalue -1. By Lemma 1.1.2, this means that $\alpha_s.w = x\alpha_t$ for some $x \in \mathbb{R}$, i.e., α_s and $x\alpha_t$ lie in the same W-orbit. Since Φ is reduced, we have $x = \pm 1$, and so α_s and α_t have the same norm. We have remarked above that this implies $c_{st} = c_{ts}$, as desired.

Conversely, assume that the above condition on C is satisfied and let $\alpha \in \Phi$, $x \in \mathbb{R}$ be such that $x\alpha \in \Phi$. We must show that $x = \pm 1$. By definition, we can write $\alpha = \alpha_s.w_1$ and $x\alpha = \alpha_t.w_2$ for some $s, t \in S$, $w_1, w_2 \in W$. Hence $x\alpha_s$ and α_t lie in the same W-orbit. If $s = t$, this means that $x\alpha_s$ and α_s have the same norm, and hence that $x = \pm 1$, as desired.

Now assume that $s \neq t$. By Exercise 1.10, the condition that $x\alpha_s$ and α_t lie in the same W-orbit implies that s, t are conjugate in W, and this can only happen if the vertices corresponding to s, t are joined by a path all of whose edges are labelled by odd numbers. But if two simple roots are joined by an edge with an odd label, the corresponding Cartan numbers are equal, and hence the two simple roots have the same norm. It follows that α_s and α_t have the same norm, and hence $x = \pm 1$. □

1.3.7. Dynkin diagrams. Let us fix a graph Γ as in Table 1.1 and consider the set $\mathcal{C}(\Gamma)$ of all Cartan matrices having Γ as their Coxeter graph. We wish to single out certain "standard" Cartan matrices in $\mathcal{C}(\Gamma)$.

Let $C = (c_{st})_{s,t \in S} \in \mathcal{C}(\Gamma)$ and m_{st} be as in Definition 1.1.1. First, we require that C satisfies the conditions in Proposition 1.3.6, i.e., we have $c_{st} = c_{ts}$ when m_{st} is odd. This already determines a large part of C. Next we require that

$$c_{st} \in \mathbb{Z}[2\cos(2\pi/m_{s't'}) \mid s', t' \in S] \qquad \text{for all } s, t \in S. \tag{$*$}$$

(This property will play a role, for example, in the proof of Theorem 5.3.8.) Note that we always have $c_{st}c_{ts} = 4\cos^2(\pi/m_{st}) = 2+2\cos(2\pi/m_{st})$. It is easily seen that $(*)$ certainly holds if m_{st} is odd and $c_{st} = c_{ts}$. On the other hand, if m_{st} is even, the easiest way to make sure that $(*)$ holds is by requiring that $c_{st} = -1$ or $c_{ts} = -1$.

Taking this as a general definition, we say that $C \in \mathcal{C}(\Gamma)$ is "standard" if $c_{st} = c_{ts}$ whenever m_{st} is odd and $c_{st} = -1$ or $c_{ts} = -1$ whenever $m_{st} \geq 4$ is even. A "standard" Cartan matrix will be encoded in a graph, called the *Dynkin diagram*, as follows. We take the Coxeter graph and put an arrow on the edge between the nodes labelled by s, t according to the following scheme:

	Edge between $s \neq t$	Values for c_{st}, c_{ts}
m_{st} odd	no arrow	$c_{st} = c_{ts} = -2\cos(\pi/m_{st})$
$m_{st} = 4$		$c_{st} = -1 \quad c_{ts} = -2$
$m_{st} = 6$		$c_{st} = -1 \quad c_{ts} = -3$
$m_{st} = 2k \geq 8$	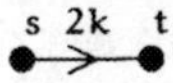	$c_{st} = -1 \quad c_{ts} = -4\cos^2(\pi/(2k)) \notin \mathbb{Z}$

Thus, since $c_{st} = 2(\alpha_s, \alpha_t)/(\alpha_s, \alpha_s)$ where $(\ ,\ )$ is an invariant scalar product as in (1.3.2), an arrow on the edge between the roots α_s, α_t and pointing towards α_t indicates that the norm of α_s is bigger than that of α_t.

The list of Dynkin diagrams encoding indecomposable Cartan matrices of finite type all of whose entries are rational integers is given in Table 1.2. In this case, we say that C is of *crystallographic type*, and that the corresponding reflection group $W = W(C)$ is a *Weyl group*. The significance of this notion is that there exists a corresponding simple Lie algebra over $\mathbb{C}$ and, for every field k, a corresponding *Chevalley group* over k; see [Carter 1972b] and also Section 8.4.

Otherwise, the Dynkin diagram of C is one of the following, and we say that C is of *non-crystallographic type*.

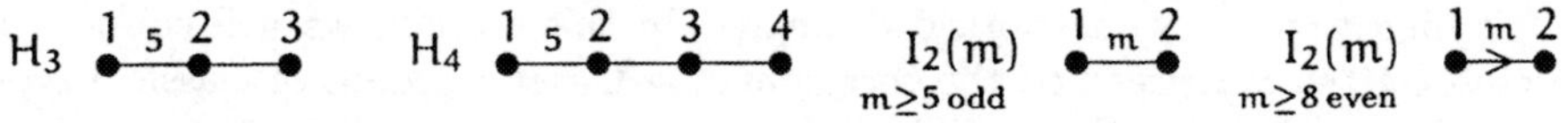

Finally, we can settle the question of which Coxeter groups are finite:

1.3.8. Proposition. *Let C be a Cartan matrix and (W, S) be the corresponding Coxeter system. Then W is finite if and only if C is of finite type.*

Table 1.2 Dynkin diagrams of Cartan matrices of crystallographic type

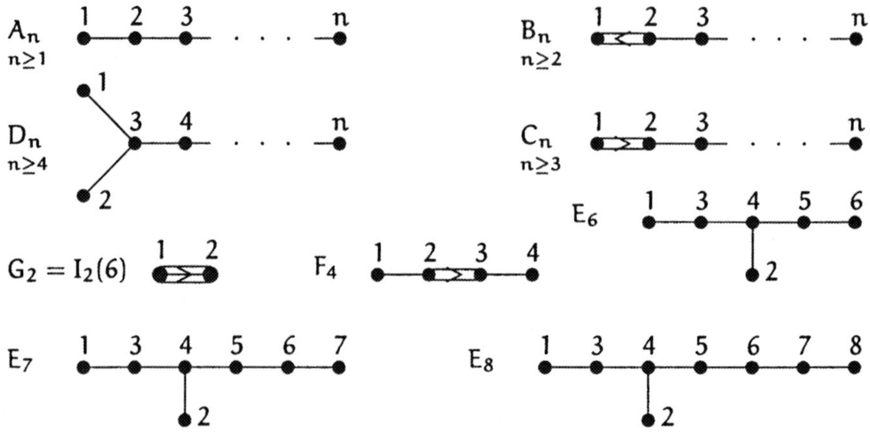

Proof. We have already remarked in (1.3.2)(c) that if W is finite then C is of finite type. To prove the converse, we may certainly assume that C is indecomposable (see (1.3.1)). Now, if one is willing to use a case-by-case argument and some explicit computations based on Algorithm A (p. 9), we can proceed as follows. Let C be a Cartan matrix whose Coxeter graph occurs in Table 1.1. We may assume that C is chosen such that the corresponding root system is reduced. If that graph is of type A_n, B_n or D_n, we will construct the corresponding group W explicitly in the following section and see that W is finite. For type $I_2(m)$, see Exercise 1.2. Next, consider the graphs of type H_4, F_4 and E_8. Using Algorithm A, one can check that in each of these cases we have $-\mathrm{id}_V \in W$. Thus, W contains an element which maps all positive roots to negative roots. Then Proposition 1.3.5 shows that there can only be finitely many roots. Since W acts faithfully on the root system, W itself must be finite. Finally, each of the graphs of type H_3, E_6 and E_7 occurs as a subgraph of the graphs that we have already considered. Hence the corresponding groups are parabolic subgroups of Coxeter groups which are already known to be finite.

For a general argument, based on the fact that $W \subset \mathrm{GL}(V)$ is a bounded and discrete set, see [Humphreys 1990, §6.2]. □

1.4 COXETER GROUPS OF CLASSICAL TYPE

It is high time for some worked examples! In this section, we will consider in detail Cartan matrices and the corresponding Coxeter groups of classical type, i.e., of type A_n, B_n, C_n and D_n. This will also be a good illustration of the general theory developed in the previous sections. For each of the above types, we will determine the root system and the isomorphism type of the corresponding Coxeter group. This will set the frame in which we can work with these groups in the subsequent chapters. All this can be done in a rather systematic way by analysing certain monomial subgroups of $GL_n(\mathbb{R})$, which we will now introduce.

1.4.1. The groups W_n. Let $n \geq 1$ and $W_n \subset GL_n(\mathbb{R})$ be the subgroup consisting of all matrices which have precisely one non-zero entry in each row and each column, and where this non-zero entry is ± 1. It is clear that W_n is a finite group, of order $2^n n!$.

Let $H_n \subseteq W_n$ be the subgroup of all permutation matrices in W_n, i.e., matrices in W_n whose entries are just 0 and 1. Then each $w \in H_n$ determines a permutation of the standard basis of $\mathbb{R}^n$. Since all permutations can be realized in this way, we can identify H_n with $\mathfrak{S}_n$, the symmetric group on $\{1, \ldots, n\}$.

Let $N_n \subseteq W_n$ be the subgroup of all diagonal matrices in W_n. Clearly, this is an abelian group. Since each diagonal entry can be ± 1, we see that $|N_n| = 2^n$. Moreover, we obviously have $N_n \cap H_n = \{1\}$ and hence $|N_n \cdot H_n| = |N_n||H_n|/|H_n \cap N_n| = 2^n n! = |W_n|$. It follows that $N_n \cdot H_n = W_n$.

If we conjugate any element of N_n (i.e., a diagonal matrix) by an element of H_n (i.e., a permutation matrix), the resulting matrix will again lie in N_n. This shows that N_n is a normal subgroup of W_n and hence we have a semidirect product decomposition $W_n = N_n \rtimes H_n$.

Now we can find a convenient set of generators of W_n. First recall that each element of $\mathfrak{S}_n$ is a product of disjoint cycles, each cycle is a product of transpositions, and each transposition is a product of basic transpositions, i.e., transpositions which just interchange two consecutive numbers. Thus, if we let $s_i \in W_n$ (for $1 \leq i \leq n-1$) be the matrix obtained from the identity matrix by interchanging the ith and $(i+1)$th row, then $s_1, \ldots, s_{n-1}$ will generate the subgroup H_n.

Moreover, let $t_i \in W_n$ $(0 \leq i \leq n-1)$ be the diagonal matrix whose $(i+1)$th diagonal entry is -1 while all other diagonal entries are 1. Then $t_0, \ldots, t_{n-1}$ generate N_n. (In fact, they form a basis for the elementary abelian 2-group N_n.) Now note that the elements t_i are conjugate in W_n: we have

$$t_i = s_i t_{i-1} s_i = s_i s_{i-1} \cdots s_1 t s_1 \cdots s_{i-1} s_i \qquad \text{for } 1 \leq i \leq n-1. \qquad (1)$$

Thus, if we set $t := t_0$, then the elements $S := \{t, s_1, \ldots, s_{n-1}\}$ generate W_n. It is readily checked that the following relations hold:

$$\begin{aligned} ts_1ts_1 &= s_1ts_1t, & ts_i &= s_it & &\text{for } i > 1, \\ s_is_{i+1}s_i &= s_{i+1}s_is_{i+1}, & s_is_j &= s_js_i & &\text{for } |i-j| > 1. \end{aligned}$$

This already shows that W_n is a quotient of a Coxeter group of type B_n.

1.4.2. Example. Let $n = 3$. Then W_3 has 48 elements, and it is generated by the three matrices

$$t = \begin{bmatrix} -1 & 0 & 0 \\ 0 & 1 & 0 \\ 0 & 0 & 1 \end{bmatrix}, \quad s_1 = \begin{bmatrix} 0 & 1 & 0 \\ 1 & 0 & 0 \\ 0 & 0 & 1 \end{bmatrix}, \quad s_2 = \begin{bmatrix} 1 & 0 & 0 \\ 0 & 0 & 1 \\ 0 & 1 & 0 \end{bmatrix}.$$

The subgroup N_3 consists of the eight matrices with diagonal entries ± 1, and H_3 consists of the six permutation matrices corresponding to the elements of $\mathfrak{S}_3$.

We have just seen that W_n is generated by n elements which satisfy a set of relations as required by a Coxeter system of type B_n. Our next aim is to show that these are in fact defining relations. The method for establishing this is by using Theorem 1.2.7: we show that W_n is a reflection group associated with a Cartan matrix of type B_n. Let $\{e_1, \ldots, e_n\}$ be the standard basis of $\mathbb{R}^n$ and $(\ ,\ )$ be the standard scalar product on $\mathbb{R}^n$, i.e., we have $(e_i, e_j) = \delta_{ij}$ for all i, j.

1.4.3. Proposition. *Let $n \geq 2$ and $W_n = \langle S \rangle \subset GL_n(\mathbb{R})$ as above. We define a matrix $C = (c_{ij})_{0 \leq i,j \leq n-1}$ by $c_{ij} = 2(\alpha_i, \alpha_j)/(\alpha_i, \alpha_i)$, where*

$$\alpha_0 := e_1, \quad \alpha_1 := e_2 - e_1, \quad \ldots, \quad \alpha_{n-1} := e_n - e_{n-1}.$$

Then C is the Cartan matrix given by the Dynkin diagram

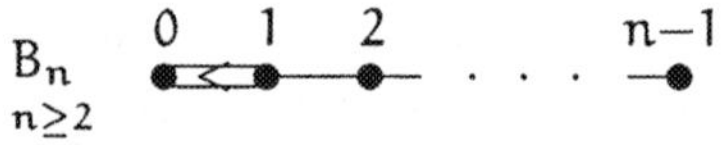

and the generators $t, s_1, \ldots, s_{n-1}$ are reflections with respect to C and with corresponding simple roots $\alpha_0, \alpha_1, \ldots, \alpha_{n-1}$, respectively. Thus, up to a change of basis of $\mathbb{R}^n$, the group $W_n \subset GL_n(\mathbb{R})$ is the reflection group associated with a Cartan matrix of type B_n.

Proof. First, one checks that $c_{i,i} = 2$, $c_{i,j} = 0$ for $|i-j| > 1$, $c_{i,i+1} = c_{i+1,i} = -1$ for $i > 0$ and $c_{1,0} = -1$, $c_{0,1} = -2$. Thus, the Dynkin diagram of C is as required. Then one checks that the elements $t, s_1, \ldots, s_{n-1}$ (as defined in (1.4.1)) act precisely as the reflections defined from the Cartan matrix in Lemma 1.1.2. $\square$

1.4.4. Corollary. *For each $n \geq 2$, the pair $(W_n, \{t, s_1, \ldots, s_{n-1}\})$ is a Coxeter system of type B_n.*

Proof. By Proposition 1.4.3, W_n is the reflection group associated with a Cartan matrix of type B_n, and $t, s_1, \ldots, s_{n-1}$ are the reflections corresponding to the simple roots. The result now follows from Theorem 1.2.7. $\square$

Having established that W_n is the reflection group associated with a Cartan matrix of type B_n, we will also want to know the corresponding root system. Thus, using the notation of Proposition 1.4.3, consider

$$\Phi_n = \{\alpha_i.w \mid 0 \leq i \leq n-1,\ w \in W_n\} \subseteq \mathbb{R}^n.$$

Note that the Cartan matrix that we constructed satisfies the requirements of Proposition 1.3.6, and so Φ_n is a reduced root system.

1.4.5. Proposition. *For each $n \geq 2$, we have*

$$\Phi_n^+ = \{e_j \pm e_i \mid 1 \leq i < j \leq n\} \cup \{e_i \mid 1 \leq i \leq n\}.$$

Thus, there are precisely n^2 positive roots in a reduced root system of type B_n.

Proof. We have to compute the orbits of the vectors $\alpha_0, \ldots, \alpha_{n-1}$ under the action of W_n. Let us start with the orbit of $\alpha_0 = e_1$. The effect of an arbitrary element $w \in W_n$ is to permute the basis vectors $e_1, \ldots, e_n$ and to attach signs ± 1 to them. Thus, the orbit of e_1 is precisely the set $\{\pm e_i \mid 1 \leq i \leq n\}$. This set has cardinality $2n$.

Now consider any $\alpha_k = e_{k+1} - e_k$, for $1 \leq k \leq n-1$. Again, the two basis vectors e_k, e_{k+1} can be mapped to any other pairs of vectors e_i, e_j, possibly with a change of signs. Thus, the orbit of α_k is precisely the set $\{\pm e_i \pm e_j \mid 1 \leq i < j \leq n\}$. This set has cardinality $2n(n-1)$.

It follows that Φ_n is just the union of the above two sets. Its cardinality is $2n + 2n(n-1) = 2n^2$. Since $|\Phi_n| = |\Phi_n^+| + |\Phi_n^-|$ and $|\Phi_n^+| = |\Phi_n^-|$, there are precisely n^2 positive roots. Writing each root as a linear combination of simple roots, we can determine which are positive and which are negative roots. □

1.4.6. Example. Consider the element $w_0 := t_0 t_1 \cdots t_{n-1} \in W_n$. Recall that t_i is a diagonal matrix with entry -1 at the $(i+1)$th position. It follows that $w_0 = -\mathrm{id}_{\mathbb{R}^n}$. Now, by the definition in (1.4.1)(1), $l(t_i) \leq 2i+1$ and so $l(w_0) = l(t_0 t_1 \cdots t_{n-1}) \leq 1+3+\cdots+(2n-1) = n^2$. But w_0 maps any positive root to its negative, so Proposition 1.3.5 and the above description of the root system show that $l(w_0) = |\Phi^+| = n^2$. Hence the expression $w_0 = t_0 t_1 \cdots t_{n-1}$ is reduced. We will see in Section 1.5 that w_0 is the unique element of maximal length in W_n.

By slightly modifying the above constructions, we can also realize W_n as a reflection group corresponding to the Cartan matrix of type C_n; see Exercise 1.11.

Since the diagram of type A_{n-1} is a subdiagram of that of type B_n, we automatically have a description of Coxeter groups of type A_{n-1}:

1.4.7. Proposition. *Let $n \geq 2$ and $H_n = \langle s_1, \ldots, s_{n-1} \rangle \subseteq W_n$ be the subgroup isomorphic to $\mathfrak{S}_n$, as in (1.4.1). Then the pair $(H_n, \{s_1, \ldots, s_{n-1}\})$ is a Coxeter system associated with the Cartan matrix given by the Dynkin diagram*

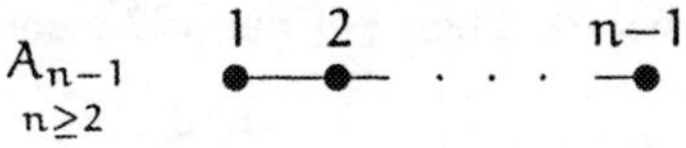

Denoting as above the simple roots corresponding to s_i by α_i (for $1 \leq i \leq n-1$), the positive roots are given by $\alpha_i + \alpha_{i+1} + \cdots + \alpha_j$ ($1 \leq i \leq j \leq n-1$). Thus, there are $n(n-1)/2$ positive roots in a reduced root system of type A_{n-1}.

Proof. The fact that H_n is generated by a subset of the generators of W_n means that H_n is a parabolic subgroup of W_n. Hence the pair $(H_n, \{s_1, \ldots, s_{n-1}\})$ is a Coxeter system by (1.2.9). It remains to prove the assertion about the root system. We have to compute the orbits of $\alpha_1, \ldots, \alpha_{n-1}$ under the action of H_n. Since the effect of an arbitrary element $w \in H_n$ is to permute the basis vectors $e_1, \ldots, e_n$, we see that this orbit consists precisely of the vectors $e_i - e_j$ where $1 \leq i, j \leq n$, $i \neq j$. Rewriting these vectors as linear combinations of $\alpha_1, \ldots, \alpha_{n-1}$ yields the above description. □

The group W_n can also be used to construct Coxeter groups of type D_n. This is done as follows.

1.4.8. The groups W'_n. Let $n \geq 2$ and $W_n \subset GL_n(\mathbb{R})$ be as in (1.4.1). By Corollary 1.4.4, the pair $(W_n, \{t, s_1, \ldots, s_{n-1}\})$ is a Coxeter system. So, by checking the defining relations, we see that there exists a unique group homomorphism $\varepsilon' \colon W_n \to \{\pm 1\}$ such that $\varepsilon'(t) = -1$ and $\varepsilon'(s_i) = 1$ for all i. We let W'_n denote the kernel of ε'. Thus, W'_n is a normal subgroup of W_n of index 2, and an element $w \in W_n$ belongs to W'_n if and only if it can be written as a product of generators $t, s_1, \ldots, s_{n-1}$ with an even number of factors equal to t.

We are now looking for generators of W'_n and a similar semidirect product decomposition as for W_n. First note that all elements s_i lie in W'_n, and hence $H_n = \langle s_1, \ldots, s_{n-1} \rangle \subseteq W'_n$. Next, let

$$N'_n := N_n \cap W'_n \quad \text{and} \quad u := ts_1t \in W'_n.$$

Note that $u^2 = 1$. We define the diagonal matrices (with -1 at positions $1, i+1$):

$$u_1 = us_1 \quad \text{and} \quad u_i = s_i u_{i-1} s_i \quad \text{for } 2 \leq i \leq n-1. \tag{1}$$

Inserting the equation $u = ts_1t$, we find that $u_i = tt_i \in N'_n$ for all $i \geq 1$. Since all elements t_i commute (they are diagonal matrices), we conclude that every product in the t_i with an even number of factors can be rewritten as a product in the u_j, and hence lies in N'_n. It follows that N'_n is a subgroup of index 2 in N_n, and it is generated by $u_1, \ldots, u_{n-1}$. Thus, we have a semidirect product decomposition $W'_n = N'_n \rtimes H_n$.

This implies that W'_n is generated by $S' := \{u, s_1, s_2, \ldots, s_{n-1}\}$. We have the following relations:

$$us_1 = ts_1ts_1 = s_1ts_1t = s_1u, \quad us_2u = s_2us_2, \quad us_i = s_iu \quad \text{for } i \geq 3.$$

This already shows that W'_n is a quotient of a Coxeter group of type D_n.

1.4.9. Example. Let $n = 4$. Then W'_4 has 192 elements, and is generated by

$$u = \begin{bmatrix} 0 & -1 & 0 & 0 \\ -1 & 0 & 0 & 0 \\ 0 & 0 & 1 & 0 \\ 0 & 0 & 0 & 1 \end{bmatrix},$$

together with the three permutation matrices s_1, s_2, s_3 corresponding to the basic transpositions $(1,2)$, $(2,3)$, $(3,4)$ in $\mathfrak{S}_4$. The subgroup N'_4 consists of the eight diagonal matrices with an even number of -1's on the diagonal.

1.4.10. Proposition. *Let $n \geq 4$ and $W'_n = \langle S' \rangle \subset GL_n(\mathbb{R})$ as above. We define a matrix $C' = (c'_{ij})_{0 \leq i,j \leq n-1}$ by $c'_{ij} = 2(\alpha'_i, \alpha'_j)/(\alpha'_i, \alpha'_i)$, where*

$$\alpha'_0 := e_1 + e_2, \quad \alpha'_1 := e_2 - e_1, \quad \ldots, \quad \alpha'_{n-1} := e_n - e_{n-1}.$$

Then C' is the Cartan matrix given by the Dynkin diagram

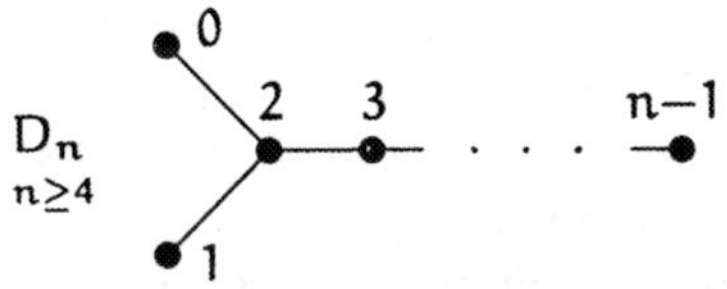

and the generators $u, s_1, \ldots, s_{n-1}$ *are reflections with respect to* C' *and with corresponding simple roots* $\alpha'_0, \alpha'_1, \ldots, \alpha'_{n-1}$, *respectively. Thus,* (W'_n, S') *is a Coxeter system of type* D_n. *Moreover, the map* $u \mapsto ts_1t$, $s_i \mapsto s_i$ $(i \geq 1)$ *defines an embedding of* W'_n *into* W_n, *where* (W_n, S) *is of type* B_n; *see Corollary 1.4.4.*

Proof. The proof is completely analogous to that of Proposition 1.4.3 and its corollary. The last statement follows from (1.4.8), where W'_n was constructed as a subgroup of index 2 in W_n. □

Finally, let us compute the corresponding root system. Let

$$\Phi'_n = \{\alpha'_i.w \mid 0 \leq i \leq n-1,\ w \in W'_n\} \subseteq \mathbb{R}^n.$$

Again, note that the Cartan matrix that we constructed satisfies the requirements of Proposition 1.3.6, and so Φ'_n is a reduced root system.

1.4.11. Proposition. *For each* $n \geq 4$, *we have*

$$(\Phi'_n)^+ = \{e_j \pm e_i \mid 1 \leq i < j \leq n\}.$$

Thus, there are $n(n-1)$ *positive roots in a reduced root system of type* D_n.

Proof. The proof is similar to that of Proposition 1.4.5. We have to compute the orbits of the vectors $\alpha'_0, \ldots, \alpha'_{n-1}$, where we have to keep in mind that the effect of an arbitrary element $w \in W'_n$ is to act by signed permutations on the basis vectors $e_1, \ldots, e_n$ where the number of negative signs must be even.

The orbit of $\alpha'_0 = e_1 + e_2$ consists precisely of all vectors $e_i + e_j$ and $-e_i - e_j$, for $1 \leq i < j \leq n$. Furthermore, the orbit of $\alpha'_k = e_{k+1} - e_k$ (for $1 \leq k \leq n-1$) consists precisely of all vectors $e_i - e_j$ and $-e_i + e_j$ for $1 \leq i < j \leq n$. □

Since W'_n is not a parabolic subgroup of W_n, the length of an element $w \in W'_n$ may depend on whether we write w as a reduced word in the generators $t, s_1, \ldots, s_{n-1}$ or in the generators $u, s_1, \ldots, s_{n-1}$. The following result shows exactly the relation between the two length functions.

1.4.12. Lemma. *There is a well-defined map* $l_t \colon W_n \to \mathbb{N}_0$ *such that* $l_t(w)$ *is the number of occurrences of the generator* t *in a reduced expression of* w. *Denoting by* l *and* l' *the length functions on* W_n *and* W'_n, *respectively, we have:*

(a) $l_t(w) = l_t(s_i w) = l_t(w s_i)$ *for all* $w \in W_n$ *and* $1 \leq i \leq n-1$;
(b) $l(w) = l'(w) + l_t(w)$ *for all* $w \in W'_n$.

Proof. Regard $\mathbb{N}_0$ as a monoid under addition and consider the map $S \to \mathbb{N}_0$, $t \mapsto 1$, $s_i \mapsto 0$ $(1 \leq i \leq n-1)$. Using the defining relations for type B_n we see that the assumptions of Matsumoto's Theorem 1.2.2 are satisfied for this map. Hence we have a well-defined map l_t as desired.

Now (a) is proved as follows. Assume first that $l(s_i w) > l(w)$. Then we clearly have $l_t(s_i w) = l_t(w)$. On the other hand, if $l(s_i w) < l(w)$ then there exists a reduced expression for w which starts with s_i. So we see again that $l_t(w) = l_t(s_i(s_i w)) = l_t(s_i w)$. The proof for right multiplication with s_i is similar.

To prove (b), we use the interpretation of the length function in Proposition 1.3.5. Thus, for any $w \in W'_n$, $l(w)$ is the number of positive roots $\alpha \in \Phi_n$ such that $\alpha.w$ is negative and $l'(w)$ is the number of positive roots $\alpha \in \Phi'_n$ such that $\alpha.w$ is negative. Now the descriptions of the root systems in Propositions 1.4.5 and 1.4.11 show that the positive roots in Φ_n are precisely the positive roots in Φ'_n plus the n roots $e_1, \ldots, e_n$. So it suffices to show that

$$l_t(w) = |\{1 \le i \le n \mid e_i.w < 0\}| \qquad \text{for all } w \in W_n. \tag{1}$$

This can be seen as follows. Using the notation of (1.4.1), we have $W_n = N_n \rtimes H_n$; hence we can write any $w \in W_n$ uniquely in the form

$$w = t_{i_1} \cdots t_{i_r} h, \quad \text{where } 0 \le i_1 < \cdots < i_r \le n-1 \text{ and } h \in H_n. \tag{2}$$

Since each element of H_n permutes the e_i and each t_j is a diagonal matrix with one entry -1, we conclude that (for w as in (2)) the right-hand side of (1) equals r. So it remains to show that $l_t(w) = r$ if w is written as in (2). To see this, first note that since h is a product of generators s_i, we have $l_t(w) = l_t(t_{i_1} \cdots t_{i_r})$ by (a). Further, we have seen in Example 1.4.6 that $w_0 = t_0 t_1 \cdots t_{n-1}$ is a reduced expression. Since N_n is commutative, we can reorder the factors and obtain $w_0 = (t_{i_1} \cdots t_{i_r})x$, where x is the product of all t_j with $j \notin \{i_1, \ldots, i_r\}$. Consequently, $t_{i_1} \cdots t_{i_r}$ is also reduced and so $l_t(t_{i_1} \cdots t_{i_r}) = r$. □

1.5 THE LONGEST ELEMENT

Let (W, S) be a Coxeter system. We also fix a corresponding Cartan matrix such that $W = \langle S \rangle \subset \mathsf{GL}(V)$ is a reflection group as in Section 1.1. We can and will assume that C is chosen such that the corresponding root system Φ is reduced; see Definition 1.3.4. One basic property pertaining to finite Coxeter groups is the existence of a unique element of maximal length. This longest element will prove useful in various situations in the subsequent chapters.

1.5.1. Proposition. *The following three conditions are equivalent:*

(a) *The group W is finite.*
(b) *The set $\{l(w) \mid w \in W\}$ is finite.*
(c) *There exists an element $w_0 \in W$ such that $l(sw_0) < l(w_0)$ for all $s \in S$.*

Moreover, if these conditions hold, w_0 is uniquely determined, we have $w_0^2 = 1$ and $l(w) < l(w_0) = |\Phi^+|$ for all $w \neq w_0$. We call w_0 the longest element of W.

Proof. The implication from (a) to (b) is trivial. Now let $w_0' \in W$ be such that $l(w) \leq l(w_0')$ for all $w \in W$. Then $l(sw_0') \leq l(w_0')$ and, hence, $l(sw_0') < l(w_0')$ for all $s \in S$ (see Remark 1.1.5). So w_0' also satisfies condition (c). Now assume that $w_0 \in W$ is as in (c). Then, by Theorem 1.1.9, we have $\alpha_s.w_0 \leq 0$ for all $s \in S$. Consequently, every positive root is made negative by w_0. By Proposition 1.3.5, the number of positive roots made negative by any element of W is finite, and so $|\Phi^+| < \infty$. Thus, since W acts faithfully on Φ, condition (a) holds. This completes the proof of the equivalence of (a), (b) and (c).

Now assume that these conditions are satisfied. We first show that $w_0 \in W$ as in (c) is unique. As above, we have $\alpha_s.w_0 \leq 0$ for all $s \in S$. Now consider another element $w \in W$ which also satisfies $l(sw) < l(w)$ for all $s \in S$. Then consider the product ww_0. Since $\alpha_s.w \leq 0$ and since w_0 maps each simple root to a negative root, we have $\alpha_s.ww_0 \geq 0$ for all $s \in S$. By Theorem 1.1.9, this implies $l(sww_0) > l(ww_0)$ for all $s \in S$. But this can only happen if $ww_0 = 1$, for otherwise we can take any reduced expression of ww_0, and if this reduced expression starts with s then $l(sww_0) < l(ww_0)$.

Thus, we have $w = w_0^{-1}$. We can apply the whole argument with $w = w_0$ and conclude that $w_0 = w_0^{-1}$, i.e., w_0 has order 2. The above argument then also shows that w_0 is unique.

We have seen above that an element satisfying (b) automatically satisfies (c). Hence the uniqueness in (c) implies that $l(w) < l(w_0)$ for all $w \neq w_0$. The fact that $l(w_0) = |\Phi^+|$ follows from Proposition 1.3.5. □

For inductive arguments, the following result concerning parabolic subgroups will be useful. Recall that for each $J \subseteq S$ we have a parabolic subgroup $W_J \subseteq W$. By (1.2.9), the pair (W_J, J) is again a Coxeter system. We shall denote by w_J the longest element in W_J, if W_J is finite.

1.5.2. Lemma. *Let $w \in W$ and $J \subseteq \{s \in S \mid l(sw) < l(w)\}$. Then W_J is finite and we have $w = w_J x$, with $x \in W$ such that $l(w) = l(w_J) + l(x)$ and $l(sx) > l(x)$ for all $s \in J$.*

Proof. By Lemma 1.1.8, we can write $w = w'x$ with $w' \in W_J$, $l(w) = l(w') + l(x)$ and $l(sx) > l(x)$ for all $s \in J$. We must show that W_J is finite and that $w' = w_J$. Now, since $l(sw) < l(w)$ for all $s \in J$, an argument similar to that in the proof of Lemma 1.2.1 shows that we must have $l(sw') < l(w')$ for all $s \in J$. Applying Proposition 1.5.1 to (W_J, J) yields that W_J is finite and that $w' = w_J$, as desired. □

We assume for the remainder of this section that W is finite. Denote by $w_0 \in W$ the longest element.

1.5.3. Lemma. *We have $l(ww_0) = l(w_0) - l(w)$ for all $w \in W$. Moreover, conjugation by w_0 defines an automorphism of W which preserves the set S.*

Proof. By Proposition 1.5.1, we have $l(w) \leq l(w_0)$, with equality only for $w = w_0$. Now assume that $w \in W$ is such that $l(w) < l(w_0)$. We claim that there exists a sequence $s_1, \ldots, s_k \in S$ such that $w_0 = (s_k \cdots s_1)w$ and $l(w_0) = k + l(w)$. This sequence is constructed as follows. Again by Proposition 1.5.1, there exists some $s_1 \in S$ such that $l(s_1 w) > l(w)$. If $l(s_1 w) = l(w_0)$, then $k = 1$ and we are done. Otherwise, there exists some $s_2 \in S$ such that $l(s_2 s_1 w) > l(s_1 w)$, and so on.

Since $w_0 w^{-1} = s_k \cdots s_1$ has length k and each element of W has the same length as its inverse, the first assertion is proved. Now let $s \in S$. Then we have $l(w_0 s w_0) = l(w_0) - l(w_0 s) = l(w_0) - (l(w_0) - l(s)) = l(s) = 1$. This shows that $w_0 s w_0 \in S$, and we are done. □

1.5.4. Example. Let (W, S) be of type A_n $(n \geq 1)$, with a labelling of the Coxeter graph as in Theorem 1.3.3. We know from Propositions 1.5.1 and 1.4.7 that $l(w_0) = n(n+1)/2$. We claim that

$$w_0 = s_1(s_2 s_1)(s_3 s_2 s_1) \cdots (s_n s_{n-1} \cdots s_1).$$

If $n = 1$, this is clear. Now let $n \geq 2$. By Lemma 1.5.2, we can write $w_0 = w_J x$ where $J = \{1, \ldots, n-1\}$. Since W_J is of type A_{n-1} we know by induction that $w_J = s_1(s_2 s_1)(s_3 s_2 s_1) \cdots (s_{n-1} s_{n-2} \cdots s_1)$. Now consider the element x. Since $l(w_J) = n(n-1)/2$, we know already that $l(x) = n$. Let $x = s_{i_1} \cdots s_{i_n}$ be a reduced expression. Since $l(s_i x) > l(x)$ for $i < n$, the only possibility is that $i_1 = n$. All generators s_i with $i < n-1$ commute with s_n. Again, since $l(s_i x) > l(x)$ for $i < n-1$, we must have $i_2 = n-1$. Continuing this way we find that $i_3 = n-2, \ldots, i_n = 1$. Thus, w_0 has a reduced expression as desired.

One could in principle apply similar inductive arguments to other types as well, but there are better ways of finding w_0: for example, it may be possible to see directly that $-\mathrm{id}_V \in W \subset \mathrm{GL}(V)$; then the element $-\mathrm{id}_V$ maps all positive roots to negative roots, and so $w_0 = -\mathrm{id}_V$ by Proposition 1.5.1.

1.5.5. Example. Let $n \geq 2$ and $(W_n, \{t, s_1, \ldots, s_{n-1}\})$ be a Coxeter system of type B_n, as in Corollary 1.4.4. The matrix description in (1.4.1) shows that $w_0 = -\mathrm{id}_{\mathbb{R}^n} \in W_n$. A reduced expression for w_0 is given in Example 1.4.6.

A similar discussion also applies to a Coxeter system of type D_n with $n \geq 4$. We already know that $l(w_0) = n(n-1)$, using the formula for the number of positive roots in Proposition 1.4.11. Now consider the elements $u_1, \ldots, u_{n-1}$ defined in (1.4.8). We claim that $w_0 = u_1 \cdots u_{n-1}$. Indeed, let $w := u_1 \cdots u_{n-1}$ and recall that u_i is a diagonal matrix with entries -1 at the first and the $(i+1)$th positions. Hence, if n is even, then $w = -\mathrm{id}_{\mathbb{R}^n}$ and it follows as above that $w = w_0$. On the other hand, if n is odd, then w is the diagonal matrix with diagonal entries $1, -1, \ldots, -1$. Using the description of simple roots in Proposition 1.4.10, we see that each simple root is mapped to a negative root by w. Hence it follows again that $w = w_0$.

There is in fact a general approach to finding reduced expressions for the longest element. The clue to this approach is given by the following observation:

1.5.6. Lemma. *Assume that* (W, S) *is irreducible and* W *is finite. Then, up to interchanging the roles of* S_1, S_2*, there exists a unique partition* $S = S_1 \amalg S_2$ *such that* $st = ts$ *for all* $s, t \in S_1$ *and for all* $s, t \in S_2$.

Proof. This follows by inspection of the diagrams in Theorem 1.3.3; we have:

Type	S_1	S_2	Type	S_1	S_2
A_n	$1,3,5,\ldots$	$2,4,6,\ldots$	B_n	$1,3,5,\ldots$	$2,4,6,\ldots$
D_n	$1,2,4,6,\ldots$	$3,5,7,\ldots$	$I_2(m)$	1	2
F_4	$1,3$	$2,4$	E_6	$1,4,6$	$2,3,5$
E_7	$1,4,6$	$2,3,5,7$	E_8	$1,4,6,8$	$2,3,5,7$
H_3	$1,3$	2	H_4	$1,3$	$2,4$

□

We keep the assumptions of Lemma 1.5.6 and fix a partition $S = S_1 \amalg S_2$ as above. Since all elements in S_1 commute with each other, we obtain a well-defined involution $w_1 \in W$ by multiplying them together. Similarly, we get a well-defined involution $w_2 \in W$ by multiplying together the generators in S_2. Let $w_c := w_1w_2 \in W$ and $h \geq 2$ be the order of w_c. Note that if we interchange the roles of S_1, S_2 then, since w_1, w_2 are involutions, w_c is replaced by w_c^{-1}. Hence the order of this element is the same number h as before; it is called the *Coxeter number* of W.

The element w_c is obtained by multiplying together the generators in S; any element of this kind is called a *Coxeter element* of W. It can be shown that they are all conjugate in W; we will come back to such questions in Chapter 3.

Since w_1, w_2 are involutions, the subgroup $W' = \langle w_1, w_2 \rangle \subseteq W$ is a dihedral group of order $2h$. We know from Exercise 1.2(b) that

$$w_0' := \mathrm{Prod}(w_1, w_2; h) = \mathrm{Prod}(w_2, w_1; h)$$

is the unique longest (in terms of the generators w_1, w_2) element in W'. Our aim is to show that $w_0' = w_0$ is also the longest element in W, and that the above expressions are also reduced in terms of the generators S.

If we order the set S so that the elements from S_1 precede all elements from S_2 then the Cartan matrix $C = (c_{st})_{s,t \in S}$ has a block shape as follows:

$$C = \left[\begin{array}{ccc|ccc} 2 & & 0 & & & \\ & \ddots & & & c_{st} & \\ 0 & & 2 & & & \\ \hline & & & 2 & & 0 \\ & c_{ts} & & & \ddots & \\ & & & 0 & & 2 \end{array}\right] \qquad (s \in S_1, \quad t \in S_2).$$

We may assume that the Cartan matrix C giving rise to (W, S) is symmetric (see (1.2.8)). Then, by (1.3.2), we have a W-invariant positive-definite scalar product

$(\ ,\ )$ on V such that $(\alpha_s, \alpha_t) = -\cos(\pi/m_{st})$ for all $s, t \in S$. Let $\{\beta_s \mid s \in S\}$ be the dual basis of V, defined by the condition that $(\alpha_s, \beta_t) = \delta_{st}$ for all $s, t \in S$.

The above matrix looks like a blown-up version of the 2×2 Cartan matrices considered in the introduction to this chapter. This can be made more precise, as follows:

1.5.7. Lemma. *There exist positive real numbers c_s $(s \in S)$ such that if we let $\beta_1 := \sum_{s \in S_1} c_s\beta_s$ and $\beta_2 := \sum_{s \in S_2} c_s\beta_s$ then the subspace $V' = \mathbb{R}\beta_1 \oplus \mathbb{R}\beta_2 \subseteq V$ is invariant under W' and the action of w_1, w_2 on V' with respect to the basis $\{\beta_1, \beta_2\}$ is given by the matrices:*

$$\begin{bmatrix} -1 & -c \\ 0 & 1 \end{bmatrix} \quad \text{and} \quad \begin{bmatrix} 1 & 0 \\ -c & -1 \end{bmatrix}, \quad \textit{respectively, where } c \in \mathbb{R}.$$

Proof. Recall that C is a symmetric, indecomposable, positive-definite matrix all of whose off-diagonal entries are non-positive. It is a standard fact about such matrices, due to Frobenius–Perron (see [Bourbaki 1968, Chap. V, §3, No. 5, Lemma 4] or [Carter 1972b, Prop. 10.5.1]), that all eigenvalues are positive real numbers and the smallest of these has an eigenvector all of whose coefficients are positive. Let $\lambda > 0$ denote the smallest eigenvalue and $(c_s)_{s\in S}$ a corresponding eigenvector with $c_s > 0$ for all $s \in S$. We shall show that the numbers $c := \lambda - 2$ and c_s $(s \in S)$ have the required properties.

Let us first determine the action of w_1 and w_2 on α_s for $s \in S$. Since $c_{ss'} = 2\delta_{ss'}$ for all $s, s' \in S_1$, we have $\alpha_s.s' = \alpha_s$ for all $s, s' \in S_1$, $s \neq s'$, and hence

$$\alpha_s.w_1 = \alpha_s.s = -\alpha_s \qquad \text{for all } s \in S_1.$$

A similar argument also shows that $\alpha_t.w_2 = -\alpha_t$ for all $t \in S_2$.

Now take some $t \in S_2$ and consider $\alpha_t.w_1$. For $s \in S_1$ we have $\alpha_t.s = \alpha_t - c_{ts}\alpha_s$. Since $\alpha_s.s' = \alpha_s$ for $s' \in S_1$ with $s \neq s'$, we conclude that

$$\alpha_t.w_1 = \alpha_t.\Big(\prod_{s\in S_1} s\Big) = \alpha_t - \sum_{s\in S_1} c_{st}\alpha_s \qquad \text{for all } t \in S_2.$$

A similar relation also holds for $\alpha_s.w_2$ with $s \in S_1$.

By duality and W-invariance, the above formulas imply formulas for the action of w_1, w_2 with respect to the basis $\{\beta_s \mid s \in S\}$: since w_1, w_2 are involutions, the matrix of w_1 or w_2 with respect to the basis $\{\beta_s \mid s \in S\}$ is the transpose of the matrix with respect to the basis $\{\alpha_s \mid s \in S\}$. Thus, we have

$$\beta_s.w_1 = -\beta_s - \sum_{t\in S_2} c_{st}\beta_t \qquad \text{for } s \in S_1,$$

$$\beta_t.w_1 = \beta_t \qquad \text{for } t \in S_2.$$

Similar relations also hold for the action of w_2. Using these formulas, we obtain

$$\beta_2.w_1 = \Big(\sum_{t\in S_2} c_t\beta_t\Big).w_1 = \sum_{t\in S_2} c_t\beta_t.w_1 = \sum_{t\in S_2} c_t\beta_t = \beta_2.$$

Next, we compute that

$$\begin{aligned}\beta_1.w_1 &= \Big(\sum_{s\in S_1} c_s\beta_s\Big).w_1 = \sum_{s\in S_1} c_s\beta_s.w_1 = \sum_{s\in S_1} c_s\Big(-\beta_s - \sum_{t\in S_2} c_{st}\beta_t\Big)\\ &= -\sum_{s\in S_1} c_s\beta_s - \sum_{t\in S_2}\Big(\sum_{s\in S_1} c_s c_{st}\Big)\beta_t = -\beta_1 - \sum_{t\in S_2}\Big(\sum_{s\in S_1} c_s c_{st}\Big)\beta_t.\end{aligned}$$

To proceed, we use the relation $c_{tt'} = 2\delta_{tt'}$ (for $t, t' \in S_2$) and the fact that $(c_s)_{s\in S}$ is an eigenvector of C with eigenvalue λ, which allow us to write

$$\lambda c_t = \sum_{s\in S} c_s c_{st} = 2c_t + \sum_{s\in S_1} c_s c_{st} \qquad \text{for } t \in S_2.$$

Inserting this into the above expression yields that

$$\beta_1.w_1 = -\beta_1 - \sum_{t\in S_2} (\lambda - 2)c_t\beta_t = -\beta_1 - (\lambda - 2)\beta_2,$$

as desired. A completely similar computation yields that w_2 also acts as desired. Thus, the subspace V' indeed is invariant under the action of W' and the action of the generators is as claimed. □

1.5.8. Theorem. *Assume that (W, S) is irreducible and W is finite. Let $S = S_1 \amalg S_2$ be as in Lemma 1.5.6 and w_1, w_2 be the corresponding involutions. Let $h \geq 2$ be the order of the Coxeter element $w_c = w_1w_2$. Then the following hold.*

(a) *Let $w = \mathrm{Prod}(w_1, w_2; k) \in W$ where $0 \leq k \leq h$. Then this gives a reduced expression for w, i.e., we have $l(w) = k_1|S_1| + k_2|S_2|$ where $k = k_1 + k_2$ and k_i is the number of factors w_i in the above product.*

(b) *We have $w_0 = \mathrm{Prod}(w_1, w_2; h) = \mathrm{Prod}(w_2, w_1; h)$ and $2l(w_0) = |S|h$.*

Proof. We use the set-up of Lemma 1.5.7 and first prove the following statement. Every $y \in \langle w_1, w_2\rangle$ can be written as $y = \mathrm{Prod}(w_1, w_2; m)$ or $y = \mathrm{Prod}(w_2, w_1; m)$, where $0 \leq m \leq h$ and where this expression is reduced. We proceed by induction on $l(y)$. If $l(y) = 0$ then $y = 1$ and there is nothing to prove. Now let $l(y) > 0$ and choose $s \in S$ with $l(sy) < l(y)$. Assume, for example, that $s \in S_1$. The fact that $l(sy) < l(y)$ implies that $\alpha_s.y \leq 0$. Since all coefficients in β_1, β_2 are positive, we deduce that

$$0 \geq (\beta_1, \alpha_s.y) = (\beta_1.y^{-1}, \alpha_s) \quad \text{and} \quad 0 \geq (\beta_2, \alpha_s.y) = (\beta_2.y^{-1}, \alpha_s),$$

and at least one of these inequalities is strict. Thus, there is an $i \in \{1, 2\}$ such that $(\beta_i.y^{-1}, \alpha_s) < 0$. By Lemma 1.5.7, we can write $\beta_i.y^{-1} = x_1\beta_1 + x_2\beta_2$

with $x_1, x_2 \in \mathbb{R}$. Since the scalar product of this vector with α_s is negative, we must have $x_1 < 0$. Now we take any other $s' \in S_1$ and conclude that

$$(\beta_i, \alpha_{s'}.y) = (\beta_i.y^{-1}, \alpha_{s'}) = x_1(\beta_1, \alpha_{s'}) = x_1 c_{s'} < 0.$$

This means that if we express $\alpha_{s'}.y$ as a linear combination of simple roots, then there is at least one negative coefficient. It follows that $\alpha_{s'}.y \leq 0$ and hence $l(s'y) < l(w)$ for all $s' \in S_1$. We can now apply Lemma 1.5.2 with $J = S_1$ to conclude that $y = w_J y'$ with $l(y) = l(w_J) + l(y')$. But all generators in J commute with each other and hence $w_J = w_1$. So we have written $y = w_1 y'$ with $y' \in \langle w_1, w_2 \rangle$ and $l(y) = |S_1| + l(y')$. We can now apply induction to y'.

Now we can prove the two assertions of the theorem. Let w be as in (a). We have just seen that we can also express w as $w = \mathrm{Prod}(w_1, w_2; m)$ or $w = \mathrm{Prod}(w_2, w_1; m)$, where $0 \leq m \leq h$ and where this expression is reduced. Now Exercise 1.2(b) shows that we must have $k = m$. Furthermore, if $k < h$, then there is a unique expression for w as a product of factors w_1 and w_2, i.e., $w = \mathrm{Prod}(w_1, w_2; m)$ is reduced as claimed.

It remains to consider the case where $m = h$. Then $w = \mathrm{Prod}(w_1, w_2; h) = \mathrm{Prod}(w_2, w_1; h)$ and the above argument only shows that at least one of these two expressions is reduced. In order to see that this is in fact the case for both of them, we argue as follows.

If h is even, then the number of generators in S appearing in the two products $\mathrm{Prod}(w_1, w_2; h)$ and $\mathrm{Prod}(w_2, w_1; h)$ is the same. Hence we are done in this case. If h is odd then the elements w_1, w_2 are conjugate by Exercise 1.2(b). Consequently, they must have the same eigenvalues with the same multiplicities. But w_1, w_2 are diagonalizable and the cardinalities of S_1 and S_2 are respectively given by the number of eigenvalues equal to -1. It follows that $|S_1| = |S_2|$ and so both $\mathrm{Prod}(w_1, w_2; h)$ and $\mathrm{Prod}(w_2, w_1; h)$ again have the same length as products in the generators in S.

Finally, note that since $w = \mathrm{Prod}(w_1, w_2; h) = \mathrm{Prod}(w_2, w_1; h)$ are two reduced expressions and all elements in S_1 (or S_2) commute with each other, we have $l(sw) < l(w)$ for all $s \in S_1 \cup S_2 = S$ and so $w_0 = w$ by Proposition 1.5.1. This also yields the equation $2l(w_0) = h|S|$. □

1.5.9. Example. Assume (W, S) is of type H_4, where the Cartan matrix is the third matrix in Example 1.1.4:

$$C := \begin{bmatrix} 2 & -\lambda & 0 & 0 \\ -\lambda & 2 & -1 & 0 \\ 0 & -1 & 2 & -1 \\ 0 & 0 & -1 & 2 \end{bmatrix}, \qquad \lambda = (1 + \sqrt{5})/2.$$

The generators $S = \{s_1, s_2, s_3, s_4\}$ are given by the following matrices (with respect to the basis $\{\alpha_{s_i} \mid 1 \leq i \leq 4\}$), respectively:

$$\begin{bmatrix} -1 & 0 & 0 & 0 \\ \lambda & 1 & 0 & 0 \\ 0 & 0 & 1 & 0 \\ 0 & 0 & 0 & 1 \end{bmatrix}, \begin{bmatrix} 1 & \lambda & 0 & 0 \\ 0 & -1 & 0 & 0 \\ 0 & 1 & 1 & 0 \\ 0 & 0 & 0 & 1 \end{bmatrix}, \begin{bmatrix} 1 & 0 & 0 & 0 \\ 0 & 1 & 1 & 0 \\ 0 & 0 & -1 & 0 \\ 0 & 0 & 1 & 1 \end{bmatrix}, \begin{bmatrix} 1 & 0 & 0 & 0 \\ 0 & 1 & 0 & 0 \\ 0 & 0 & 1 & 1 \\ 0 & 0 & 0 & -1 \end{bmatrix}.$$

We consider the Coxeter element $w_c = (s_1 s_3)(s_2 s_4)$, given by the matrix

$$w_c = \begin{bmatrix} -1 & -\lambda & 0 & 0 \\ \lambda & 1+\lambda & 1 & 1 \\ 0 & -1 & -1 & -1 \\ 0 & 1 & 1 & 0 \end{bmatrix}.$$

The characteristic polynomial of w_c is $f = (X-\zeta)(X-\zeta^{11})(X-\zeta^{19})(X-\zeta^{29}) \in \mathbb{C}[X]$, where $\zeta = \exp(2\pi\sqrt{-1}/30) \in \mathbb{C}$. In particular, this shows that w_c has order 30, and so w_c^{15} has order 2. Since $11 \cdot 15$, $19 \cdot 15$, $29 \cdot 15$ are all congruent to 15 modulo 30, it follows that w_c^{15} has four eigenvalues -1, and hence we must have $w_c^{15} = -\mathrm{id}_V = w_0$. Furthermore, we obtain $|\Phi^+| = l(w_0) = 4 \cdot 15 = 60$.

Similar arguments can be applied to all other irreducible Coxeter systems of finite type. If we already know a formula for $|\Phi|$ (for example, in the case of classical types, see Section 1.4), then the formula $|\Phi| = 2l(w_0) = |S|h$ yields the Coxeter number h. On the other hand, if we can compute directly the Coxeter number h (for example, in the exceptional types as in Example 1.5.9), the same formula yields the number of positive roots. The results are summarized in Table 1.3. Moreover, the positive roots can actually be computed by using a reduced expression of w_0 as given by Theorem 1.5.8 and the fact that $N(w_0) = \Phi^+$ (see Propositions 1.3.5 and 1.5.1).

Table 1.3 Coxeter numbers and positive roots

Type	A_n	B_n	D_n	$I_2(m)$	H_3	H_4	F_4	E_6	E_7	E_8
h	$n+1$	$2n$	$2(n-1)$	m	10	30	12	12	18	30
$\lvert\Phi^+\rvert$	$n(n+1)/2$	n^2	$n(n-1)$	m	15	60	24	36	63	120

For more detailed information about the construction of root systems and finite Coxeter groups we refer the reader to [Bourbaki 1968, Chap. VI, §4].

1.6 BIBLIOGRAPHICAL REMARKS AND EXERCISES

The whole approach to this chapter, i.e., the idea of starting with a Cartan matrix C, is inspired by [Kac 1985]. (Note, however, that there, instead of (C2), it is assumed that the entries of C are in $\mathbb{Z}$.) Otherwise, all arguments already appear, in some form, in [Bourbaki 1968], [Humphreys 1990] or [Curtis and Reiner 1981, §64]. The reader familiar with the theory of Coxeter groups will find that some results appear here in a slightly different order, or are proved in a different way than in the textbooks. Our approach is also well-adapted to an algorithmic point of view. In fact, the logical order of the theory as presented here corresponds exactly to the steps in the actual implementation of finite Coxeter groups in the CHEVIE system. The only question left open by this approach is whether also each finite reflection group can be constructed from a Cartan matrix. This is indeed so, but for a proof we refer to one of the standard texts mentioned above.

In the case of finite Weyl groups, Matsumoto's Theorem 1.2.2 already appears in [Iwahori 1964, Theorem 2.6], but its proof involves a case-by-case argument according to $m_{st} \in \{2,3,4,6\}$. An argument in general is given in [Matsumoto 1964]. Our proof in Section 1.2 is taken from [Tits 1984, Cor. II.1.12]. For the origins of the theory of Coxeter groups see the historical remarks in [Bourbaki 1968].

In Chapter 5, we will briefly discuss complex reflection groups, which are a natural generalization of finite Coxeter groups. One problem with complex reflection groups is that important results like Matsumoto's theorem do not carry over to them in any straightforward way. In an attempt to remedy this situation, general types of length functions and root systems have been introduced and studied by Bremke–Malle and Rampetas–Shoji; see [Bremke and Malle 1997], [Shoji 2000] and the references there.

The discussion of subexpressions in (1.2.4) and that of the Bruhat–Chevalley order in Exercise 1.7 below appears in [Dyer 1987, Remark 5.2]. Lemma 1.2.6 appears as a remark in [Bourbaki 1968, Chap. IV, §2, Exercise 23]; the first explicit reference is probably [Deligne and Lusztig 1976, Lemma 1.6.4]. The results about Cartan matrices of finite type and finite Coxeter groups are standard, but we have put more emphasis than usual on the question of whether the root system is reduced or not. The results about expressing the longest elements in terms of suitably defined Coxeter elements are contained in [Bourbaki 1968, Chap. V, §6].

The automorphisms in Exercise 1.13(b) and a full classification of all automorphisms of finite Coxeter groups can be found in [Bleher et al. 1997].

Exercise 1.1. Consider the Cartan matrix $C := \begin{bmatrix} 2 & -1 & -1 \\ -1 & 2 & -1 \\ -1 & -1 & 2 \end{bmatrix}$. Show that the corresponding root system $\Phi(C)$ is infinite.

Exercise 1.2. Let G be a group generated by two involutions $s \neq t$. Let $m \geq 2$ be the order of st. Prove the following statements:

(a) If $m = \infty$ then each $1 \neq g \in G$ can be expressed uniquely either as $g = \mathrm{Prod}(s,t;k)$ with $k \geq 1$ or as $g = \mathrm{Prod}(t,s;k)$ with $k \geq 1$.

(b) Let $m < \infty$. Then every $1 \neq g \in G$ can be written uniquely either as $g = \mathrm{Prod}(s,t;k)$ with $1 \leq k < m$ or as $g = \mathrm{Prod}(t,s;k)$ with $1 \leq k < m$ or as $g = \mathrm{Prod}(s,t;m) = \mathrm{Prod}(t,s;m)$. Moreover, s and t are conjugate if and only if m is odd.

(c) If $m < \infty$ then G is finite of order $2m$, and we have a presentation $G = \langle s,t \mid s^2 = t^2 = (st)^m = 1\rangle$.

Exercise 1.3. Let G be a finite group and $\rho\colon G \to GL(V)$ be a representation of G on some finite-dimensional $\mathbb{R}$-vector space V. Let $(\,,\,)\colon V \times V \to \mathbb{R}$ be any positive-definite scalar product on V (e.g., choose a basis $(v_1, \ldots, v_n)$ of V and define $(v_i, v_j) = \delta_{ij}$ for $1 \leq i, j \leq n$). Let $(\,,\,)'\colon V \times V \to \mathbb{R}$ be defined by

$$(v, w)' := \frac{1}{|G|} \sum_{g \in G} (v.g, w.g) \qquad \text{for } v, w \in V,$$

where $g \in G$ acts on V via $\rho(g)$. Show that $(\,,\,)'$ is a positive-definite scalar product such that $(v.g, w.g)' = (v, w)'$ for all $v, w \in V$ and $g \in G$.

Exercise 1.4. Let $c, c' \in \mathbb{R}$, $c, c' \leq 0$ be such that $cc' = 4\cos^2(\theta)$ where $\theta = 2\pi/m$ with $2 \leq m < \infty$. Let

$$S := \begin{bmatrix} -1 & 0 \\ -c & 1 \end{bmatrix} \quad \text{and} \quad T := \begin{bmatrix} 1 & -c' \\ 0 & -1 \end{bmatrix}.$$

Show that if we let $P = \begin{bmatrix} 1 & 0 \\ y & x \end{bmatrix}$ (for suitable real numbers $x > 0$, $y \geq 0$), then

$$PSP^{-1} = \begin{bmatrix} -1 & 0 \\ 0 & 1 \end{bmatrix} \quad \text{and} \quad PTP^{-1} = \begin{bmatrix} -\cos(\theta) & \sin(\theta) \\ \sin(\theta) & \cos(\theta) \end{bmatrix}.$$

Show that

$$P(ST)^k P^{-1} = \begin{bmatrix} \cos(k\theta) & -\sin(k\theta) \\ \sin(k\theta) & \cos(k\theta) \end{bmatrix} \qquad \text{for all } k \in \mathbb{N}.$$

Exercise 1.5. Recall that a given Coxeter system (W, S) may arise from different Cartan matrices. The purpose of this exercise is to show that the corresponding reflection representations of W are equivalent. More precisely, assume that W is finite and let $C = (c_{st})_{s,t \in S}$ and $C' = (c'_{st})_{s,t \in S}$ be Cartan matrices which both give rise to (W, S). Then we have embeddings $W = \langle S \rangle \subset GL(V)$ and $W = \langle S \rangle \subset GL(V')$, where V has a basis $\{\alpha_s\}_{s \in S}$ and V' has a basis $\{\alpha'_s\}_{s \in S}$ such that $s \in S$ acts via $\alpha_t \mapsto \alpha_t - c_{st}\alpha_s$ and $\alpha'_t \mapsto \alpha'_t - c'_{st}\alpha'_s$, respectively.

Using the classification in Theorem 1.3.3, show that there exist real numbers $d_s > 0$ $(s \in S)$ such that $d_s c_{st} = c'_{st} d_t$ for all $s, t \in S$. Then consider the linear map $\delta\colon V \to V'$, $\alpha_s \mapsto d_s \alpha'_s$ and check that δ is a W-module isomorphism, i.e., we have $\delta(\alpha_t).s = \delta(\alpha_t.s)$ for all $s, t \in S$.

Exercise 1.6. Let (W, S) be a Coxeter system. Show that the cancellation law implies the *exchange condition*: let $s, s_1, \ldots, s_k \in S$ such that $l(ss_1 \cdots s_k) < l(s_1 \cdots s_k)$. Then $ss_1 \cdots s_{m-1} = s_1 \cdots s_m$ for some $1 \leq m \leq k$.

Exercise 1.7. Let (W, S) be a Coxeter system and F be the map from W to the set of all subsets of W as in (1.2.4). Use the cancellation law to show that $v \leq w$ if and only if $F(v) \subseteq F(w)$. This result shows that the relation $\leq$ is a partial order, called the *Bruhat–Chevalley order* on W.

Exercise 1.8. This exercise is about parabolic subgroups as in (1.2.9). Show that the map $J \mapsto W_J$ is an inclusion-preserving bijection from the set of all subsets of S to the set of all parabolic subgroups of W, and that we have $W_{J\cap K} = W_J \cap W_K$ and $W_{J\cup K} = \langle W_J, W_K\rangle$ for all subsets $J, K \subseteq S$.

Exercise 1.9. Let $m \in \mathbb{Z}$, $m \geq 1$. Show that $\cos(2\pi/m) \in \mathbb{Q}$ if and only if $m \in \{1,2,3,4,6\}$.

Exercise 1.10. Let $W = \langle S\rangle \subset \mathsf{GL}(V)$ be a Coxeter group.

(a) Show that $s, t \in S$ are conjugate in W if and only if there exists a sequence $s = s_0, s_1, \ldots, s_n = t$ with $s_i \in S$ and such that the order of $s_{i-1}s_i$ is finite and odd for all i.

(b) Show that $s, t \in S$ are conjugate if and only if α_s lies in the same W-orbit as some multiple of α_t.

[*Hint*. (a) Use Exercise 1.2(b) to show that s, t are conjugate if a sequence as above exists. For the converse, see [Bourbaki 1968, Chap. IV, §1, No. 3, Prop. 3].

(b) Let $w \in W$ be such that $s = wtw^{-1}$. Then check that $\alpha_s.w$ is an eigenvector of t with eigenvalue -1. Conversely, assume that $x\alpha_s$ and α_t lie in the same W-orbit, for some $x \in \mathbb{R}$. Then use Theorem 1.1.9 to check that the assumptions of Lemma 1.2.6 are satisfied, i.e., $l(sw) = l(wt)$ and $l(swt) = l(w)$. Distinguish the cases where $x > 0$ and $x < 0$.]

Exercise 1.11. Let $n \geq 2$ and $W_n = \langle S\rangle \subset \mathsf{GL}_n(\mathbb{R})$ as in (1.4.1). Now define a matrix $C = (c_{ij})_{0\leq i,j\leq n-1}$ by $c_{ij} = 2(\alpha_i, \alpha_j)/(\alpha_i, \alpha_i)$, where

$$\alpha_0 := 2e_1, \quad \alpha_1 := e_2 - e_1, \quad \ldots, \quad \alpha_{n-1} := e_n - e_{n-1},$$

and $(\,,\,)$ is the standard scalar product on $\mathbb{R}^n$, i.e., we have $(e_i, e_j) = \delta_{ij}$ for all i, j. Show that C is the Cartan matrix given by the Dynkin diagram

C_n
$n\geq 2$
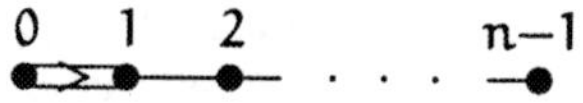

and the elements $t, s_1, \ldots, s_{n-1}$ are the reflections with roots $\alpha_0, \alpha_1, \ldots, \alpha_{n-1}$, respectively. Also determine the corresponding root system.

Exercise 1.12. Let C be a Cartan matrix and $W = \langle S\rangle \subset \mathsf{GL}(V)$ the corresponding reflection group. Show that if C is indecomposable then every endomorphism $\varphi \in \mathsf{End}(V)$ commuting with the action of W is a scalar multiple of the identity.

If W is finite then $\mathbb{R}[W]$ is a semisimple algebra (by Maschke's theorem), and so the above result implies that V is an absolutely simple $\mathbb{R}[W]$-module.

[*Hint*. The equation $\varphi(\alpha_s).s = \varphi(\alpha_s.s) = \varphi(-\alpha_s) = -\varphi(\alpha_s)$ shows that either $\varphi(\alpha_s) = 0$ or $\varphi(\alpha_s)$ is an eigenvector of s with eigenvalue -1. So there exist constants c_s ($s \in S$) such that $\varphi(\alpha_s) = c_s\alpha_s$ for all $s \in S$. Now let $s, t \in S$, $s \neq t$, such that $c_{st} \neq 0$. Evaluating both sides of the identity $\varphi(\alpha_t).s = \varphi(\alpha_t.s)$, we see that $c_s = c_t$. Since C is indecomposable, this implies that all c_s are equal.]

Exercise 1.13. Let $W = \langle S \rangle \subset \mathsf{GL}(V)$ be a finite Coxeter group and $w_0 \in W$ be the longest element.

(a) Assume that (W, S) is irreducible. Show that w_0 is central in W if and only if $w_0 = -\mathrm{id}_V$. Show that this happens exactly in the following cases: B_n ($n \geq 2$), D_n ($n \geq 4$ even), $I_2(m)$ ($m \geq 4$ even), H_3, H_4, F_4, E_7, E_8.

(b) Assume that w_0 is central and let $\mu\colon W \to \mathsf{GL}(V)$ be a group homomorphism with $\mu(w) = \pm\mathrm{id}_V$ for all $w \in W$. Show that $\alpha_\mu\colon W \to W$, $w \mapsto \mu(w)w$, is a group homomorphism which is bijective if and only if $\mu(w_0) = \mathrm{id}_V$.

[*Hint.* (a) For the classical types, work with the realizations in Section 1.4. For the exceptional types, use some explicit computations as in Example 1.5.9. (b) Check the defining relations for W.]

Exercise 1.14. Assume that $W = \langle S \rangle \subset \mathsf{GL}(V)$ is a finite Coxeter group and let $(\,,\,)$ be a W-invariant scalar product on V. For each $0 \neq \alpha \in V$ define

$$w_\alpha\colon V \to V, \quad v \mapsto v - 2\frac{(v, \alpha)}{(\alpha, \alpha)}\alpha,$$

the *reflection with root* α. Note that the formula in (1.3.2) shows that $w_{\alpha_s} = s$ for $s \in S$. Show that if $x \in W$ then $x^{-1}w_\alpha x = w_{\alpha.x}$, and hence $\{w_\alpha \mid \alpha \in \Phi\} \subseteq W$. Show that if Φ is reduced then $|\Phi^+| = |\{w^{-1}sw \mid s \in S, w \in W\}|$.

Exercise 1.15. Let (W, S) be a Coxeter system and fix $s \in S$.

(a) Show that there is a well-defined function $l_s\colon W \to \mathbb{N}_0$ with the following property. If $w = s_1 \cdots s_m$ with $s_i \in S$ is a reduced expression then $l_s(w)$ is the number of $i \in \{1, \ldots, m\}$ such that s_i is conjugate to s in W.

(b) Assume that W is finite and set $N_s = l_s(w_0)$, where $w_0 \in W$ is the longest element. Show that N_s is the size of the conjugacy class of s in W.

[*Hint.* (a) Define $f_s\colon S \to \mathbb{N}_0$ by $f_s(t) = 1$ if s, t are conjugate and $f_s(t) = 0$ otherwise. Consider $\mathbb{N}_0$ as a monoid under addition and check that the assumptions of Matsumoto's Theorem 1.2.2 are satisfied. (b) Let $S' \subseteq S$ be a set of representatives of the elements of S under conjugacy in W. We can assume that $s \in S'$. We must show that $N_s = N_s'$ where N_s' denotes the size of the conjugacy class of s. By definition, we have $l(w_0) = \sum_{s \in S'} N_s$. On the other hand, Exercise 1.14 yields $|\Phi^+| = \sum_{s \in S'} N_s'$. Thus, since $l(w_0) = |\Phi^+|$, it is sufficient to show that $N_s \leq N_s'$ for all $s \in S'$. To see this, take a reduced expression $w_0 = s_1 \cdots s_N$, let $s \in S'$ and $i \in \{1, \ldots, N\}$ be such that s_i, s are conjugate in W. By Proposition 1.3.5, $\alpha := \alpha_i.s_{i-1} \cdots s_1$ is a positive root, and Exercise 1.14 implies that w_α is conjugate to s_i. Deduce from this that $N_s \leq N_s'$.]

2
Parabolic subgroups

Subgroups that deserve special attention are those which share defining properties of their parent groups. In the case of a Coxeter group W with its distinguished generating set S, a subgroup generated by a subset of S is a Coxeter group in its own right. These 2^n subgroups of W (where $n = |S|$) are the parabolic subgroups of W and this chapter is devoted to their structure and properties—both individually as well as seen as a collection.

Parabolic subgroups are also special as they allow for their cosets to be represented by canonical elements of the parent group. This chapter introduces these distinguished coset representatives and describes their properties. The basis of our investigation of the structure of the set of distinguished coset representatives is Deodhar's Lemma 2.1.2. Implicitly, this lemma defines a normal form for the elements of W together with an algorithm to construct it. Using this, the results of this chapter can be derived in a purely combinatorial way, i.e., without reference to the matrix representation that was used in Chapter 1 to define W.

In Section 2.1 we describe the sets of coset representatives and develop algorithms to compute right and double coset representatives. We also obtain factorization and decomposition results. And, in the case where W is finite, we show how distinguished coset representatives give rise to a subalgebra of the group algebra of W, Solomon's descent algebra.

A graphical representation of the set of coset representatives is used in Section 2.2 to illustrate various properties, including the multiplication in the descent algebra. Here the Coxeter groups of classical types are treated among lots of other worked out examples.

In Section 2.3 the parabolic subgroups of W are sorted into equivalence classes with respect to inner automorphisms which we call Coxeter classes. This can be seen as a first step towards sorting the elements of W into conjugacy classes (see Chapter 3). An algorithm for the construction of a Coxeter class is developed. We introduce labels for the Coxeter classes of parabolic subgroups of the Coxeter groups of classical types and determine the orders of the normalizers.

In Section 2.4 we consider actions of parabolic subgroups on the cosets of other parabolic subgroups. Numbers of fixed points of these actions are collected in the parabolic table of marks, a square matrix with rows and columns labelled by the Coxeter classes of parabolic subgroups. We determine properties of marks and use a method of induction to calculate the parabolic table of marks of W from the tables of its parabolic subgroups.

2.1 DISTINGUISHED COSET REPRESENTATIVES

Let (W, S) be a Coxeter system. Recall from (1.2.9) that for each $J \subseteq S$ the subgroup $W_J = \langle s \mid s \in J \rangle$ is called a parabolic subgroup of W, and that such a subgroup is of particular interest since (W_J, J) is a Coxeter system in its own right. With respect to the subgroup W_J the Coxeter group W is partitioned into right cosets $W_J w = \{vw \mid v \in W_J\}$ for certain $w \in W$. Our first aim in this section is to identify a unique set of representatives of the cosets of W_J.

2.1.1. Proposition. *Let* $J \subseteq S$ *and define*

$$X_J = \{w \in W \mid l(sw) > l(w) \textit{ for all } s \in J\}.$$

(a) *For each* $w \in W$ *there exist unique* $v \in W_J$ *and* $x \in X_J$ *such that* $w = vx$. *Moreover,* $l(w) = l(v) + l(x)$.

(b) *For any* $x \in W$ *the following are equivalent:*

  (i) $x \in X_J$;

  (ii) $l(vx) = l(v) + l(x)$ *for all* $v \in W_J$;

  (iii) x *is the unique element of minimal length in* $W_J x$.

In particular, X_J *is a complete set of right coset representatives of* W_J *in* W.

The set X_J is called the set of *distinguished right coset representatives* of W_J in W. We write

$$W = W_J \cdot X_J$$

where the explicit dot indicates that the map from the Cartesian product $W_J \times X_J$ into the group W, which sends a pair (v, x) to the product $w = vx$, is a bijection with the additional property that $l(w) = l(v) + l(x)$.

Proof. (a) Let $w \in W$ and let x be of minimal length in $W_J w$. Write $w = vx$ where $v \in W_J$. Then $l(w) \leq l(v) + l(x)$. Suppose the inequality was strict. Then, by the cancellation law (Theorem 1.2.5), there are v' and x' obtained from reduced expressions for v and x, respectively, by cancelling (at least two) factors such that $w = v'x'$. By Proposition 1.2.10 all reduced expressions for $v \in W_J$ have their factors in J whence $v' \in W_J$. If $x' \neq x$ then x' is an element in $W_J w$ with $l(x') < l(x)$, contradicting the minimal choice of x. Therefore $x' = x$, whence $v' = v$, and the inequality is in fact an equality.

In order to show the uniqueness of x let $y \in W_J x \cap X_J$. Write $y = vx$ with $v \in W_J$. Hence, as we have just seen, $l(y) = l(v) + l(x)$. Now if $y \neq x$ then $v \neq 1$ and there exists an $s \in J$ such that $l(sv) < l(v)$ where $sv \in W_J$. But then $l(sy) = l(svx) < l(v) + l(x) = l(y)$, contradicting $y \in X_J$.

(b) We first show that (i) implies (ii). For, if $x \in X_J$ and $v \in W_J$ then $w = vx$ is the unique product decomposition according to (a), whence $l(vx) = l(v) + l(x)$.

Next, (ii) implies (iii). So let $x \in W$ with $l(vx) = l(v) + l(x)$ for all $v \in W_J$. Then we have $l(vx) \geq l(x)$ for all $vx \in W_J x$, with equality only if $l(v) = 0$, or

$v = 1$. Thus x has minimal length in the coset $W_J x$ and it is unique with that property.

Finally, (iii) implies (i). Let $w \in W$ and let x be the unique element of minimal length in $W_J w$. Then $l(vx) > l(x)$ for all $v \in W_J$, $v \neq 1$. In particular, $l(sx) > l(x)$ for all $s \in J$, whence $x \in X_J$. □

For any coset representative $x \in X_J$ and any $s \in S$ the product xs can be written as $xs = vx'$ for some $v \in W_J$ and some $x' \in X_J$. In fact, only a very limited choice of pairs (v, x') actually occurs in this situation. This is described by the following result which is of basic importance for the study of the structure of the set of distinguished coset representatives.

2.1.2. Lemma (Deodhar's lemma). *Let* $J \subseteq S$, $x \in X_J$ *and let* $s \in S$. *Then either* $xs \in X_J$ *or* $xs = ux$ *for some* $u \in J$.

Proof. If $xs \notin X_J$ there exists, by Proposition 2.1.1, an element $u \in J$, such that $l(uxs) < l(xs)$. On the other hand $l(ux) > l(x)$ since $x \in X_J$. But $l(ux) = l(uxs) \pm 1$ and therefore we must have $l(xs) > l(x)$. By Lemma 1.2.6, this implies $xs = ux$. □

We always have $l(x^{-1}w) \geq l(w) - l(x)$, for $x, w \in W$. The element x is called a *prefix* of w if equality holds, i.e., if $l(x^{-1}w) = l(w) - l(x)$. In that case there is a $y \in W$ such that $w = xy$ and $l(w) = l(x) + l(y)$. Moreover, we call x a *suffix* of w if $l(wx^{-1}) = l(w) - l(x)$.

By Deodhar's lemma, we have, in particular, $xs \in X_J$ whenever $l(xs) < l(x)$. Thus every prefix of a coset representative is a coset representative, a property that makes X_J a *Schreier transversal* (see [Johnson 1990]). This also yields a straightforward algorithm for the explicit computation of the set X_J of distinguished right coset representatives. Using the fact that every non-trivial coset representative x is of the form $x = x's$ for a coset representative x' with $l(x') < l(x)$ and a generator $s \in S$, the elements of X_J can be constructed step by step. In step k the elements of length k are constructed out of those of length $k - 1$.

Algorithm B (Right coset representatives). Given a Coxeter system (W, S) and a subset $J \subseteq S$, the set X_J of distinguished right coset representatives is constructed.

B1. [Initialize.] Set $k \leftarrow 0$, $Y_0 \leftarrow \{1\}$ and $X \leftarrow Y_0$. (Clearly, $1 \in X_J$ and 1 is the only representative of length 0.)

B2. [Next k.] Set $k \leftarrow k + 1$ and

$$Y_k \leftarrow \{xs \mid x \in Y_{k-1},\ s \in S,\ l(xs) > l(x) \text{ and } l(txs) > l(xs) \text{ for all } t \in J\}.$$

Further set $X \leftarrow X \cup Y_k$. (At this stage the set X contains all $x \in X_J$ with $l(x) \leq k$ and Y_k contains those $x \in X_J$ with $l(x) = k$.)

B3. [Loop.] Repeat step B2 until $Y_k = \varnothing$. Then set $X_J \leftarrow X$ and stop. (If there is no coset representative of length k then there is none of any bigger length.)

Here the length conditions like $l(xs) > l(x)$ can be tested in a similar way to Algorithm A (p. 9) by searching in the column of s in the matrix of x (or x^{-1}) for negative entries. Naturally, this algorithm only terminates if X_J is finite, i.e., if W_J has finite index in W.

2.1.3. Example. Suppose that $(W, \{s_1, s_2, s_3\})$ is of type H_3 and let $J = \{s_1, s_2\}$ be such that W_J generates a parabolic subgroup of type $I_2(5)$. The following table illustrates the action of Algorithm B in this situation. Starting with $k = 0$ and $Y_0 = \{1\}$, the table lists for each $x \in X_k$ and each generator $s \in S$ whether $l(xs) < l(x)$ (denoted by an entry $-$ in the column of s) or, if this is not the case, whether $l(txs) < l(xs)$ for some $t \in J$ (denoted by ♣) or whether xs passes all these tests (denoted by ♡). In the latter case xs becomes a member of Y_{k+1}.

k	$x \in Y_k$	xs_1	xs_2	xs_3
0	1	♣	♣	♡
1	s_3	♣	♡	$-$
2	s_3s_2	♡	$-$	♣
3	$s_3s_2s_1$	$-$	♡	♣
4	$s_3s_2s_1s_2$	♡	$-$	♡
5	$s_3s_2s_1s_2s_1$	$-$	♣	♡
	$s_3s_2s_1s_2s_3$	♡	♣	$-$
6	$s_3s_2s_1s_2s_1s_3$	$-$	♡	$-$
7	$s_3s_2s_1s_2s_1s_3s_2$	♡	$-$	♣
8	$s_3s_2s_1s_2s_1s_3s_2s_1$	$-$	♡	♣
9	$s_3s_2s_1s_2s_1s_3s_2s_1s_2$	♣	$-$	♡
10	$s_3s_2s_1s_2s_1s_3s_2s_1s_2s_3$	♣	♣	$-$

The algorithm stops at $k = 11$ since $Y_{11} = \varnothing$ and returns the set of elements listed in the second column as X_J.

2.1.4. Lemma. *Let $J \subseteq S$, $x \in X_J$ and let $w \in W$ such that $xw \in X_J$. Then, if $w = s_1 \cdots s_r$ is a reduced expression for w with $s_j \in S$, $j = 1, \ldots, r$, we have $xs_1 \cdots s_i \in X_J$ for all $i \leq r$.*

Proof. Set $y_i = xs_1 \cdots s_i$ and write $y_i = v_i x_i$ for unique $v_i \in W_J$ and $x_i \in X_J$. By Deodhar's Lemma 2.1.2 we have either $x_{i-1}s_i \in X_J$, or $x_{i-1}s_i = ux_{i-1}$ for some $u \in J$. Let $t_i = x_{i-1}^{-1}x_i$ so that $t_i = s_i$ if $x_{i-1}s_i \in X_J$, and $t_i = 1$ otherwise. But then $xt_1 \cdots t_r = xw$ and $t_1 \cdots t_r = w$. From $l(w) = r$ we now conclude that $t_i = s_i$ in all cases, and therefore $x_i \in X_J$ for all i. □

2.1.5. Transitivity. When considering chains of parabolic subgroups it is useful to have the following notation. For subsets $J \subseteq K \subseteq S$ we denote by X_J^K the set of distinguished right coset representatives of W_J in W_K. We then can factorize X_J as

$$X_J = X_J^K \cdot X_K$$

since $W_J \cdot X_J = W = W_K \cdot X_K = W_J \cdot X_J^K \cdot X_K$.

2.1.6. Remark. The anti-automorphism $w \mapsto w^{-1}$ of W maps the set X_J to $X_J^{-1} = \{x^{-1} \mid x \in X_J\}$ which is a set of left coset representatives for W_J in W. Since $l(w^{-1}) = l(w)$ for all $w \in W$ and inverting products is compatible with the explicit dot, the representatives in X_J^{-1} are also *distinguished*: for each $x \in X_J$ the element x^{-1} is the unique element of minimal length in the left coset $x^{-1}W_J$.

We now determine representatives for the double cosets

$$W_J w W_K = \{uwv \mid u \in W_J,\ v \in W_K\}$$

of W_J and W_K in W where $J, K \subseteq S$ and $w \in W$.

2.1.7. Proposition. *Let* $J, K \subseteq S$ *and define* $X_{JK} = X_J \cap X_K^{-1}$. *Then, for each* $w \in W$, *there are* $u \in W_J$, $v \in W_K$ *and a unique* $d \in X_{JK}$ *such that* $w = udv$ *and* $l(w) = l(u) + l(d) + l(v)$. *Moreover, for each* $d \in W$, *the following are equivalent:*

(i) $d \in X_{JK}$;

(ii) d *is the unique element of minimal length in the double coset* $W_J d W_K$.

In particular, X_{JK} *is a complete set of double coset representatives of* W_J *and* W_K *in* W.

The elements of $X_{JK} = X_J \cap X_K^{-1}$ are called *distinguished double coset representatives* of W_J and W_K in

$$W = \coprod_{d \in X_{JK}} W_J d W_K.$$

Note that the elements u and v in the decomposition $w = udv$ are not necessarily unique, but see Lemma 2.1.9.

Proof. Let $w \in W$ and write $w = ux$ where $u \in W_J$, $x \in X_J$ and $l(w) = l(u) + l(x)$ according to Proposition 2.1.1. Then write $x = dv$ where $d \in X_K^{-1}$, $v \in W_K$ and $l(x) = l(d) + l(v)$. Then $l(w) = l(u) + l(d) + l(v)$. And since d is a prefix of x we have $d \in X_J$, whence $d \in X_J \cap X_K^{-1} = X_{JK}$.

Applying this argument to an element y of minimal length in the double coset $W_J w W_K$ shows $y \in X_{JK}$, whence (ii) implies (i).

In order to prove that (i) implies (ii) we show that X_{JK} contains exactly one element in each double coset. So let $y \in X_{JK} \cap W_J d W_K$ and write $y = udv$ according to the above procedure. But if $y \neq d$ then either $u \neq 1$ (contradicting $y \in X_J$) or $v \neq 1$ (contradicting $y \in X_K^{-1}$). □

2.1.8. Lemma. *Let* $J, K \subseteq S$. *Then the following hold.*

(i) *For all* $x \in X_J$ *and all* $s \in S$ *we have* $xs \notin X_J$ *if and only if* $s \in J^x$.

(ii) *For all* $x \in X_J$ *we have* $x^{-1}X_J \subseteq X_L$ *where* $L = J^x \cap S$.

(iii) *For all* $d \in X_{JK}$ *we have* $dX_{J^d \cap K} = X_{J \cap {}^d K}$.

Proof. (i) follows immediately from Lemma 2.1.2.

(ii) Let $y \in X_J$ and denote $w = x^{-1}y$. We have to show that $w \in X_L$, i.e., that $l(sw) > l(w)$ for all $s \in J^x \cap S$. But, suppose $l(sw) < l(w)$ for some $s \in S$. Then there is a reduced expression of w that starts with s. Now $x \in X_J$ and $xw = y \in X_J$. By Lemma 2.1.4 we then have $xs \in X_J$, whence $s \notin L$ by (i).

(iii) Obviously $J^d \cap K$ and $J \cap {}^dK$ are conjugate subsets of S. Thus $W_{J^d \cap K}$ and $W_{J \cap {}^dK}$ are conjugate in W, whence $|X_{J^d \cap K}| = |X_{J \cap {}^dK}|$. Hence it suffices to show that $dX_{J^d \cap K} \subseteq X_{J \cap {}^dK}$. But this follows from $d^{-1} \in X_K \subseteq X_{J^d \cap K}$ and (ii). □

A consequence of Proposition 2.1.7 is the following partition of the set X_J of coset representatives.

2.1.9. Lemma (Mackey decomposition). *Let* $J, K \subseteq S$. *Then*

$$X_J = \coprod_{d \in X_{JK}} d \cdot X^K_{J^d \cap K}.$$

In particular, each $w \in W$ *has a unique decomposition* $w = udv$ *where* $u \in W_J$, $d \in X_{JK}$, $v \in X^K_{J^d \cap K}$ *and* $l(w) = l(u) + l(d) + l(v)$.

Proof. Let $x \in X_J$ and write $x = udv$ for $u \in W_J$, $v \in W_K$ and a unique $d \in X_{JK}$ such that $l(x) = l(u) + l(d) + l(v)$ according to Proposition 2.1.7. Since $x \in X_J$, we must have $u = 1$ and $x = dv$, whence v is unique. It remains to show that $v \in X^K_{J^d \cap K}$. For that write $v = v_1 y$ such that $v_1 \in W_{J^d \cap K}$, $y \in X^K_{J^d \cap K}$ and $l(v) = l(v_1) + l(y)$. Now suppose $v_1 \neq 1$. Then $l(sv) < l(v)$ for some $s \in J^d \cap K$. But $ds = td$ where $t = dsd^{-1} \in J \cap {}^dK$. Hence

$$l(tx) = l(tdv) = l(dsv) = l(d) + l(sv) < l(d) + l(v) = l(dv) = l(x)$$

contradicting $x \in X_J$. □

2.1.10. Theorem (Solomon). *Assume that* W *is finite. For each* $J \subseteq S$ *define*

$$x_J = \sum_{x \in X_J} x^{-1} \in \mathbb{Q}[W]$$

as the sum over all distinguished left coset representatives of W_J *in* W *in the group algebra* $\mathbb{Q}[W]$ *of* W. *Then, for any* $J, K \subseteq S$,

$$x_J x_K = \sum_{L \subseteq S} a_{JKL} x_L$$

where, for $J, K, L \subseteq S$,

$$X_{JKL} = \{d \in X_{JK} \mid J^d \cap K = L\} \quad \textit{and} \quad a_{JKL} = |X_{JKL}|.$$

Proof. Let $J, K \subseteq S$, and for $L \subseteq J$ denote by x^J_L the sum $\sum_{x \in X^J_L} x^{-1}$. Then

$$x_J x_K = x_J \sum_{d \in X_{JK}} x^J_{J \cap {}^d K} d = \sum_{d \in X_{JK}} x_{J \cap {}^d K} d = \sum_{d \in X_{JK}} x_{J^d \cap K} = \sum_{L \subseteq S} a_{JKL} x_L.$$

Here the first equality follows from Lemma 2.1.9, the second from (2.1.5) and the next from Lemma 2.1.8. □

2.1.11. Corollary. *Let $J, K, L \subseteq S$ and let $x \in X_{JKL}$. Then $x^{-1}X_J \subseteq X_L$. In particular, X_{JJJ} is a subgroup of W.*

Proof. By Lemma 2.1.8(ii) we have $x^{-1}X_J \subseteq X_M$ where $M = J^x \cap S$. From $L = J^x \cap K$ it now follows that $L \subseteq M$, whence $x^{-1}X_J \subseteq X_L$.

Now suppose that $J = K = L$ and let $x, y \in X_{JJJ}$ where $X_{JJJ} = \{d \in X_{JJ} \mid J^d = J\}$. We show that $x^{-1}y \in X_{JJJ}$. From $x^{-1}X_J \subseteq X_J$ we see that $x^{-1}y \in X_J$. Similarly, from $y^{-1}X_J \subseteq X_J$ we have $y^{-1}x \in X_J$, whence $x^{-1}y \in X_J \cap X_J^{-1} = X_{JJ}$. Finally, since $J^x = J$ and $J^y = J$, it follows that J is invariant under conjugation by $x^{-1}y$, whence $x^{-1}y \in X_{JJJ}$. □

2.1.12. Theorem (Kilmoyer, Solomon). *Let $J, K \subseteq S$ and let $d \in X_{JK}$. Then $W_J^d \cap W_K = W_L$ where $L = J^d \cap K$.*

Proof. We certainly have $W_L \subseteq W_J^d \cap W_K$. In order to prove the opposite inclusion we show that $W_J d \cap dW_K = dW_L$.

So let $w \in W_J d \cap dW_K$. Then $w = ud$ for some $u \in W_J$ and $w = dv$ for some $v \in W_K$. Hence $ud = dv$, and $l(u) = l(v)$ since $d \in X_{JK} = X_J \cap X_K^{-1}$.

Now write $v = v_1 \cdots v_r$ as a reduced expression with $v_i \in K$. Let $d_0 = d$ and, for $i = 1, \ldots, r$, let $d_i \in X_J$ such that $d_{i-1}v_i = u_i d_i$ for some $u_i \in W_J$. Then $u = u_1 \cdots u_r$. Moreover, by Deodhar's Lemma 2.1.2, $l(u_i) \le 1$ and $l(u_i) = 1$ if and only if $d_{i-1} = d_i$. But $l(u) = l(v)$ implies $l(u_i) = 1$ and hence $d_i = d$ for all i. Thus $dv_i = u_i d$ with $u_i \in J$ for all i. This means $v_i = u_i^d \in K \cap J^d = L$ from which we conclude $v \in W_L$ and $w = dv \in dW_L$, as required. □

It follows from this last result that the intersection of two conjugates of parabolic subgroups of W is again a conjugate of a parabolic subgroup of W (see Exercise 2.2).

We can also conclude that if two parabolic subgroups of W are conjugate in W, then the conjugation can be performed in such a way that even their generating sets are conjugate.

2.1.13. Corollary. *Let $J, K \subseteq S$. Then J is conjugate to K in W if and only if W_J and W_K are conjugate in W. In particular, $|J| = |K|$ whenever W_J and W_K are conjugate in W.*

Proof. Clearly W_K and W_J are conjugate, if their generating sets are. Conversely, let $w \in W$ be such that $W_J^w = W_K$. Now $w = w_1 d w_2$ for some $w_1 \in W_J$,

$w_2 \in W_K$ and $d \in X_{JK}$. Thus $W_J^d = W_K$. By Theorem 2.1.12 we have $K = J^d \cap K$, whence $K \subseteq J^d$. Analogously, from $d^{-1} \in X_{KJ}$ we get $J \subseteq {}^dK$, hence $J^d = K$. □

We return to the problem of conjugacy among parabolic subgroups of W in Section 2.3. We conclude the present section with a description of the normalizer of W_J in W and a basic lemma about conjugation by coset representatives.

2.1.14. Lemma. *Let* $J \subseteq S$ *and let* $x \in X_J$. *Then* $l(w^x) \geq l(w)$ *for all* $w \in W_J$.

Proof. By the choice of w and x we have $l(wx) = l(w) + l(x)$. And from $xw^x = wx$ we get $l(x) + l(w) = l(wx) = l(xw^x) \leq l(x) + l(w^x)$. □

2.1.15. Proposition. *Let* (W, S) *be a Coxeter system. Then for each* $J \subseteq S$:

(a) *If* $x \in X_J$ *such that* $W_J^x = W_K$ *for some* $K \subseteq S$ *then* $J^x = K$, $x \in X_{JK}$ *and* $l(w^x) = l(w)$ *for all* $w \in W_J$.

(b) $N_W(W_J) = W_J \rtimes X_{JJJ}$.

Proof. (a) By Lemma 2.1.14, we have $l({}^xw) \leq l(w)$ for all $w \in W_J^x = W_K$. This implies $l({}^xs) \leq l(s) = 1$ for all $s \in K$. Since ${}^xs \neq 1$, we have ${}^xK = J$, whence conjugation by x is a length-preserving isomorphism from W_J to W_K. This implies $l(xw^x) = l(x) + l(w^x)$ for all $x \in X_J$, $w \in W_J$, whence $x \in X_J \cap X_K^{-1} = X_{JK}$.

(b) Clearly, $W_J \cap X_{JJJ} = \{1\}$, since $X_{JJJ} \subseteq X_J$, and $N_W(W_J)$ is a union of cosets of W_J in W. Furthermore, by Corollary 2.1.11, X_{JJJ} is a subgroup of W and, by its definition, normalizes W_J. On the other hand, it follows from (a) that every coset representative $x \in X_J$ with $W_J^x = W_J$ is an element of X_{JJJ}. □

2.2 LONGEST REPRESENTATIVES AND THE COSET GRAPH

In this section we associate a graph to the set X_J of distinguished coset representatives that illustrates the structure of this set. We construct various examples of these graphs and derive some basic properties. Moreover, operations for coset graphs are defined such as a semidirect product, restriction and a multiplication.

Throughout the section we assume that W is finite. Using prefixes, X_J can then be described by just one of its elements, the *longest coset representative* d_J. As in Section 1.5, we denote by w_0 the longest element of W and by w_J the longest element of W_J, for $J \subseteq S$.

2.2.1. Lemma. *Let* $J \subseteq S$ *and define* $d_J = w_J w_0$. *Then*

(a) d_J *is the unique element of maximal length in* X_J;

(b) $X_J = \{w \in W \mid w$ *is a prefix of* $d_J\}$;

(c) $d_J^{-1} = d_J^{w_0} = d_{J'}$, *where* $J' = J^{w_0}$.

Proof. (a) Suppose that $x \in X_J$ is an element of maximal length in X_J. Then $w_J x$ must be an element of maximal length in W. Hence $w_J x = w_0$ and $x = w_J w_0$ is the unique element of its length in X_J.

(b) Let $x \in X_J$. Then Deodhar's Lemma 2.1.2 implies $xs \in X_J$ whenever $l(xs) < l(x)$. By repeated application of this fact every prefix of x, and in particular every prefix of d_J, lies in X_J.

Conversely, we have to show that each coset representative is a prefix of d_J.

Suppose first that $x \in X_J$ is such that for all $s \in S$ either $xs \notin X_J$ or $l(xs) < l(x)$. We then can show that $x = w_J w_0 = d_J$. Indeed, if $xs \notin X_J$ then, by (2.1.2), $xs = ux$ for some $u \in J$, whence $l(w_J u) < l(w_J)$. So $l(w_J xs) = l(w_J u) + l(x) = l(w_J) - 1 + l(x)$. And if $l(xs) < l(x)$ then $l(w_J xs) = l(w_J) + l(xs) = l(w_J) + l(x) - 1$. In total we get $l(w_J xs) < l(w_J x)$ for all $s \in S$, and therefore, by (1.5.1), $w_J x = w_0$.

Otherwise, if x is such that $l(xs) > l(x)$ and $xs \in X_J$ for some $s \in S$ then x is a prefix of d_J by downward induction on $l(x)$.

(c) $d_J^{-1} = w_0 w_J = w_J^{w_0} w_0$ and $w_J^{w_0} = w_{J'}$. □

Analogously, the set X_J^{-1} of left coset representatives is the set of all suffixes of $(d_J)^{-1} = w_0 w_J$. By Lemma 1.5.3, $l(ww_0) = l(w_0) - l(w)$ for every $w \in W$, whence every element of W is a prefix of the longest element. This is now seen as a special case of Lemma 2.2.1 with w_0 as the longest coset representative of $W_\varnothing$ in W.

Note that for any $s \in S$ the product ws is a prefix of w if and only if $l(ws) < l(w)$. Moreover, every prefix of w arises by repeatedly constructing prefixes that are one shorter in length. This gives the following algorithm for the explicit construction of the set of all prefixes of a given $w \in W$. As in Algorithm B (p. 40), this set is constructed step by step. Again, length conditions $l(xs) < l(x)$ can be checked by inspecting the matrix of x^{-1} (see Algorithm A, p. 9).

Algorithm C (Prefixes of w). Given a finite Coxeter system (W, S) and an element $w \in W$, the set $P(w)$ of all prefixes of w is determined.

C1. [Initialize.] Set $k \leftarrow l(w)$, $Y_k \leftarrow \{w\}$ and $X \leftarrow Y_k$. (Clearly, $w \in P(w)$ and w is the only prefix of length $l(w)$.)

C2. [Next k.] Set $k \leftarrow k - 1$ and

$$Y_k \leftarrow \{xs \mid x \in Y_{k+1},\ s \in S,\ l(xs) < l(x)\}.$$

Further set $X \leftarrow X \cup Y_k$. (At this stage the set X contains all $x \in P(w)$ with $l(x) \geq k$, and Y_k contains those $x \in P(w)$ with $l(x) = k$.)

C3. [Loop.] If $k > 0$ then repeat step C2. Otherwise set $P(w) \leftarrow X$ and stop. (Now X contains all prefixes of w.)

2.2.2. Example. As before in (2.1.3), let (W, S) be of type H_3 where $S = \{s_1, s_2, s_3\}$ and let $w = s_2 s_1 s_3$. We determine all the prefixes of w by Algorithm C. First, we have $l(w) = 3$ and $Y_3 = \{w\}$. We then determine

$$Y_2 = \{ws \mid s \in S, l(ws) < l(w)\} = \{s_2 s_1, s_2 s_3\}.$$

From this we get $Y_1 = \{s_2\}$ and $Y_0 = \{1\}$, whence $P(w) = \{1, s_2, s_2 s_1, s_2 s_3, w\}$.

In practice, we can assume that the longest element is known for W and W_J by the construction of Theorem 1.5.8 and Table 1.3 on p. 33. Thus the longest coset representative d_J is known and Algorithm C can be used to construct X_J as the set of all prefixes of d_J.

2.2.3. Definition. The *coset graph* Γ_J, for $J \subseteq S$, is the labelled directed graph with vertex set X_J where, for a vertex $x \in X_J$ and any $s \in S$, we draw an edge with label s from x to xs if $xs \in X_J$ and $l(xs) > l(x)$. We also write $x \xrightarrow{s} xs$ for this edge.

Given the set X_J it is clearly possible to determine the edges of the graph Γ_J with vertex set X_J by testing the condition $l(xs) > l(x)$. One might as well keep track of the edges while the set X_J is constructed. The following algorithm is a modified version of Algorithm C that determines X_J together with its graph structure Γ_J.

Algorithm D (Coset graph). Given a finite Coxeter system (W, S) and a subset $J \subseteq S$ the coset graph Γ_J with vertex set X_J and edge set E is determined.

D1. [Initialize.] Set $w \leftarrow d_J$, the longest coset representative of W_J in W, $k \leftarrow l(w)$, $Y_k \leftarrow \{w\}$, $X \leftarrow Y_k$ and $E \leftarrow \varnothing$.

D2. [Next k.] From the set Y_k of coset representatives of length k construct the sets E_{k-1} of new edges and Y_{k-1} of new vertices as follows. Set

$$E_{k-1} \leftarrow \{xs \xrightarrow{s} x \mid x \in Y_k,\ s \in S,\ l(xs) < l(x)\}$$

and set $Y_{k-1} \leftarrow \{y \mid y \xrightarrow{s} x \in E_{k-1}\}$. Further set $X \leftarrow X \cup Y_{k-1}$ and $E \leftarrow E \cup E_{k-1}$.

D3. [Loop.] If $k > 0$ then set $k = k - 1$ and repeat step D2. Otherwise set $\Gamma_J \leftarrow (X, E)$ and stop.

2.2.4. Example. Let $n \geq 1$ and let $(W, \{s_1, s_2, \ldots, s_n\})$ be a Coxeter system of type A_n as in Example 1.5.4, and let $J = \{s_1, s_2, \ldots, s_{n-1}\}$. Then $d_J = w_J w_0 = s_n s_{n-1} \cdots s_1$. In the initialization step of Algorithm C we thus have $Y_n = \{s_n s_{n-1} \cdots s_1\}$. Induction on the length shows that s_j is the only $s \in S$ with the property that $l(s_n s_{n-1} \cdots s_j s) < l(s_n s_{n-1} \cdots s_j)$ (e.g., from the fact that the matrix of $s_j s_{j+1} \cdots s_n$ has negative entries only in its jth row). Therefore $Y_k = \{s_n s_{n-1} \cdots s_{n-k+1}\}$ for $k = 0, \ldots, n$. (The algorithm stops at $k = 0$.) In total we get

$$X_J = \{1,\ s_n,\ s_n s_{n-1},\ \ldots,\ s_n s_{n-1} \cdots s_1\}.$$

If we keep track of the elements $s \in S$ which lead from one coset representative to another we get the following coset graph Γ_J:

Here and in the following examples of coset graphs the edges are drawn as simple lines and understood to point from left to right. Moreover, just write the index i for the label s_i.

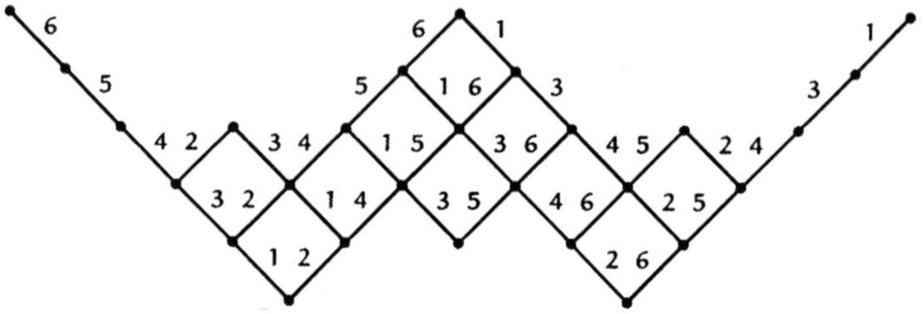

Fig. 2.1. The coset graph of D_5 in E_6.

2.2.5. Example. Let $n \geq 2$ and let $W_n = \langle t, s_1, \ldots, s_{n-1} \rangle$ be the Coxeter group of type B_n, as in Example 1.5.5. Furthermore, let $J = \{t, s_1, \ldots, s_{n-2}\}$, whence (W_J, J) is a Coxeter system of type B_{n-1}. Then $d_J = w_J w_0 = t_{n-1} = s_{n-1} \cdots s_1 t s_1 \cdots s_{n-1}$. Applying Algorithm C to this word $w = d_J$ gives

$$\begin{aligned} X_J = \{1,\ & s_{n-1}, \ldots, s_{n-1} \cdots s_1, \\ & s_{n-1} \cdots s_1 t,\ s_{n-1} \cdots s_1 t s_1, \ldots, s_{n-1} \cdots s_1 t s_1 \cdots s_{n-1}\} \end{aligned}$$

with exactly one coset representative of length k $(0 \leq k \leq 2n-1)$.

Constructing X_J as the set of prefixes of d_J and keeping track of the generators gives the following coset graph Γ_J, where we write 0 for the label t:

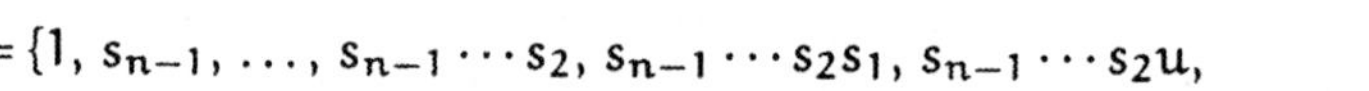

2.2.6. Example. A similar discussion applies to the Coxeter system (W'_n, S') of type D_n $(n \geq 4)$ where $S' = \{u, s_1, \ldots, s_{n-1}\}$ as in (1.4.8). Here, if we choose $J = \{u, s_1, \ldots, s_{n-2}\}$ then (W_J, J) is a Coxeter system of type D_{n-1} and $d_J = w_J w_0 = u_{n-1} = s_{n-1} \cdots s_2 s_1 u s_2 \cdots s_{n-1}$. Algorithm C applied to $w = d_J$ then yields

$$\begin{aligned} X_J = \{1,\ & s_{n-1}, \ldots, s_{n-1} \cdots s_2,\ s_{n-1} \cdots s_2 s_1,\ s_{n-1} \cdots s_2 u, \\ & s_{n-1} \cdots s_2 s_1 u,\ s_{n-1} \cdots s_2 s_1 u s_2, \ldots, s_{n-1} \cdots s_2 s_1 u s_2 \cdots s_{n-1}\} \end{aligned}$$

with two coset representatives of length $n-1$ and otherwise exactly one coset representative of length k $(0 \leq k \leq 2n-2)$. Then Γ_J is as follows with 0 written for u.

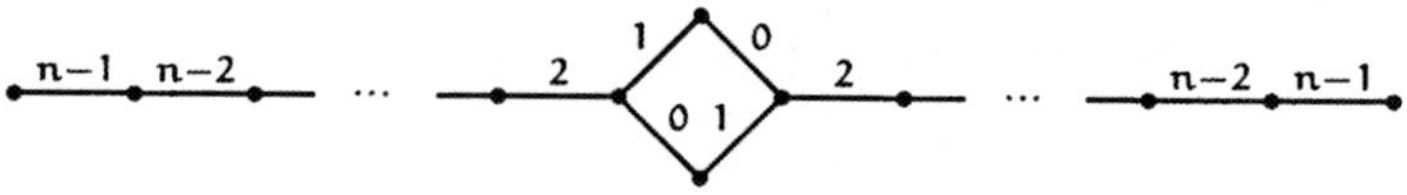

2.2.7. Example. Let (W, S) be of type E_6 with $S = \{s_1, \ldots, s_6\}$ and Coxeter diagram 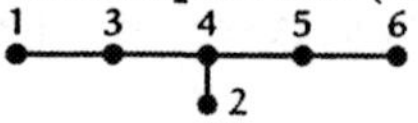

. Let $J = \{s_1, \ldots, s_5\}$. Then Γ_J is as shown in Figure 2.1.

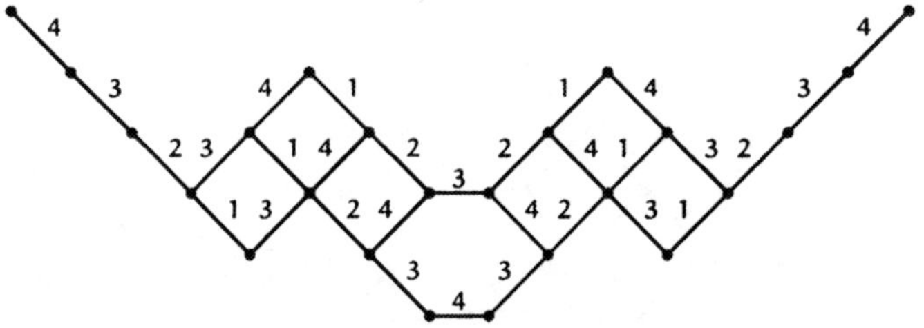

Fig. 2.2. The coset graph of B_3 in F_4.

Figure 2.2 gives the coset graph of B_3 in F_4. In Figure 2.3 on p. 50, a portion of the coset graph of H_3 in H_4 is shown. Here the vertices are those coset representatives x with $l(x) \leq 45/2 = \frac{1}{2}l(d_J)$, where d_J is the longest coset representative. Since w_0 is central in W, the complete coset graph consists of the portion shown, together with its mirror image. In a similar way, Figure 2.4 on p. 51 displays one half of the coset graph of E_7 in E_8. Here most of the labels are left out. Since every (directed) path from the initial vertex to the terminal vertex describes a reduced expression for the longest coset representative, all the labels can be recovered by applying the braid relations to the one reduced expression that is shown.

The coset graph has a number of properties some of which we list here briefly. More properties are studied in the exercises (e.g., 2.7 and 2.10).

We adopt the following language for directed graphs. An edge $v_1 \xrightarrow{s} v_2$ of a directed graph is called a *loop* if $v_1 = v_2$. A *path* is a sequence of vertices $v_0, v_1, \ldots, v_r$, (for some $r > 0$), such that $v_{i-1} \longrightarrow v_i$ for all $i = 1, \ldots, r$. A path $v_0, v_1, \ldots, v_r$ is called a (directed) *cycle* if $v_0 = v_r$.

A directed graph is *acyclic* if it contains no directed cycle. A vertex v is called an *initial vertex* if $v \longrightarrow v'$ for every vertex v' adjacent to v; it is called a *terminal vertex* if $v' \longrightarrow v$ for every such v'. A finite acyclic directed graph contains at least one initial and one terminal vertex.

Now let $J \subseteq S$ and consider the coset graph Γ_J. For any two vertices of Γ_J there is at most one edge between them, and whenever there is an edge going from vertex v_1 to vertex v_2 there is no edge going from v_2 to v_1. In particular, Γ_J has no loops. In fact, Γ_J is an acyclic directed graph without multiple edges and hence it defines a partial order on the set of vertices, i.e., on X_J.

Any vertex of a finite acyclic directed graph belongs to some path from an initial to a terminal vertex. The graph Γ_J has a unique *initial vertex* v_0 corresponding to the coset representative $1 \in X_J$ and (from Exercise 2.10) a unique *terminal vertex* v_∞ corresponding to the longest coset representative d_J. Thus, any vertex in Γ_J belongs to a path from v_0 to v_∞. In particular, Γ_J is connected.

Every path in Γ_J from the initial vertex to the terminal vertex corresponds to a reduced expression of d_J, and every reduced expression of d_J occurs in this way. In fact, Γ_J shows all reduced expressions for any $x \in X_J$ by means of all

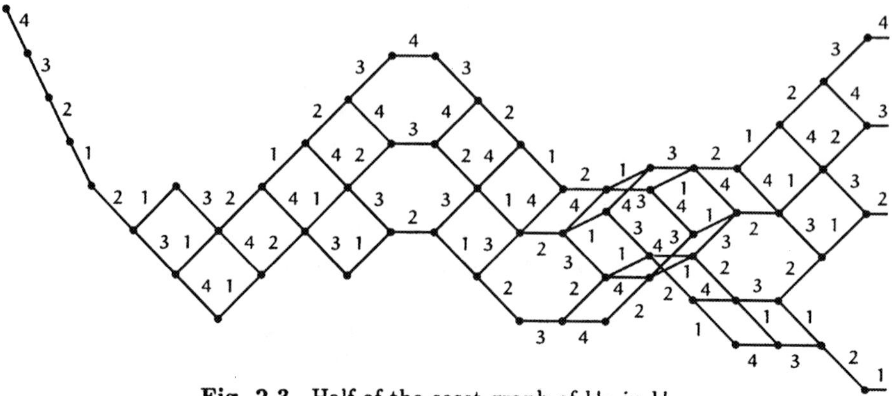

Fig. 2.3. Half of the coset graph of H_3 in H_4.

paths from v_0 to the vertex corresponding to x.

The edge $v_0 \xrightarrow{s} v$ occurs in Γ_J for some $s \in S$ and a vertex v if and only if $s \notin J$. This follows from the fact that $s \in X_J$ if and only if $s \notin J$, which is a direct consequence of the definition of X_J. Thus the subset $J \subseteq S$ is uniquely determined by the graph Γ_J.

The vertices v, v' are *adjacent* if $v \longrightarrow v'$ or $v' \longrightarrow v$ is an edge. An *undirected path* through a directed graph is a sequence of vertices $v_0, v_1, \ldots, v_r$ (for some $r > 0$), such that v_{i-1} is adjacent to v_i for all $i = 1, \ldots, r$. Another useful property of Γ_J is that it is *convex* as subgraph of $\Gamma_\varnothing$: every shortest undirected path in $\Gamma_\varnothing$ between two vertices of Γ_J lies completely in Γ_J.

2.2.8. Transitivity. We denote by Γ_J^K the coset graph of W_J in W_K. The coset graph Γ_J has a semidirect product decomposition

$$\Gamma_J = \Gamma_J^K \ltimes \Gamma_K$$

corresponding to the decomposition of X_J. In this semidirect product the edges correspond to the Cartesian product of the edges of the factors, but the product graph is not a direct product of graphs in the usual sense.

2.2.9. Lemma. *Let $J \subseteq K \subseteq S$. Define a graph $\Gamma_J^K \ltimes \Gamma_K$ as follows. The vertices of $\Gamma_J^K \ltimes \Gamma_K$ are pairs $(x_1, x_2) \in X_J^K \times X_K$; the edges are*

$$(x_1, x_2) \xrightarrow{s} (x_1, x_2 s)$$

for each $x_1 \in \Gamma_J^K$ and every edge $x_2 \xrightarrow{s} x_2 s$ in Γ_K; and

$$(x_1, x_2) \xrightarrow{s} (x_1 u, x_2)$$

whenever $x_2 s = u x_2$ for some $u \in K$ and Γ_J^K contains the edge $x_1 \xrightarrow{u} x_1 u$. Then $\Gamma_J^K \ltimes \Gamma_K = \Gamma_J$.

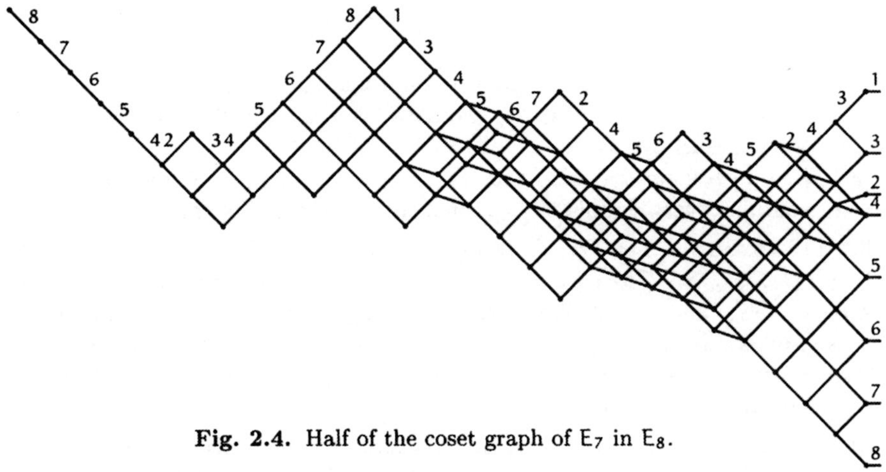

Fig. 2.4. Half of the coset graph of E_7 in E_8.

Proof. If $x_2 \xrightarrow{s} y_2$ is an edge in Γ_K or $x_1 \xrightarrow{u} y_1$ is an edge in Γ_J^K and $ux_2 = x_2 s$ then it is clear that $x_1x_2 \xrightarrow{s} y_1y_2$ is an edge in Γ_J.

Conversely, if $x_1x_2 \xrightarrow{s} y_1y_2$ is an edge in Γ_J then $x_1x_2s = y_1y_2$ and either $x_2s \in X_K$ (whence $x_1 = y_1$, $x_2s = y_2$ and $x_2 \xrightarrow{s} y_2$ is an edge in Γ_K) or $x_2s \notin X_K$. In the latter case $x_2s = ux_2$ for some $u \in K$ and we have $x_1u = y_1$, $x_2 = y_2$ and $x_1 \xrightarrow{s} y_1$ is an edge in Γ_J^K. □

2.2.10. Example. Suppose $(W, \{s_1, s_2, s_3\})$ is of type A_3 and let $J = \{s_1\}$ and $K = \{s_1, s_2\}$. Then the coset graph of W_K in W is given by the horizontal line in the illustration opposite (with arrows pointing from left to right) and the coset graph of W_J in W_K by the vertical line (top to bottom). Their product gives the coset graph of W_J in W as a graph with twelve vertices.

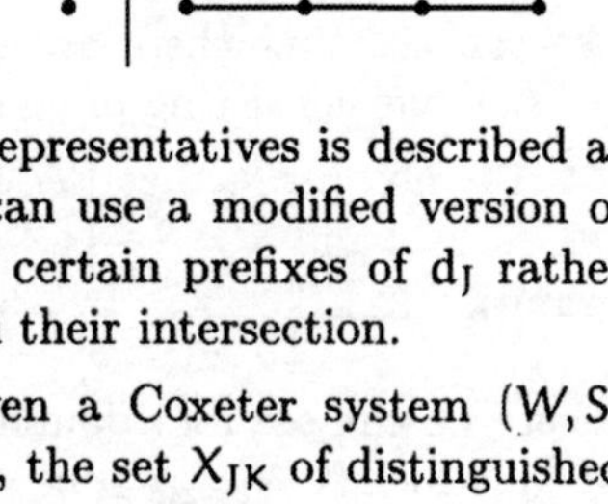

As a subset of X_J the set X_{JK} of double coset representatives is described as $X_{JK} = \{x \in X_J \mid l(xs) > l(x) \text{ for all } s \in K\}$. We can use a modified version of Algorithm C (p. 46) to construct X_{JK} as a set of certain prefixes of d_J rather than explicitly listing X_J and X_K^{-1} in order to form their intersection.

Algorithm E (Double coset representatives). Given a Coxeter system (W, S) and subsets $J, K \subseteq S$ such that W is a finite group, the set X_{JK} of distinguished double coset representatives of W_J and W_K in W is determined.

E1. [Initialize.] Set $X \leftarrow \varnothing$ and $Z \leftarrow \{d_J\}$. (Clearly every $x \in X_{JK}$ is a prefix of d_J.)

E2. [Find a representative.] Choose some $w \in Z$ and set $Z \leftarrow Z - \{w\}$. Write $w = xw'$ for $x \in X_K^{-1}$ and $w' \in W_K$ such that $l(w) = l(x) + l(w')$. Then set $X \leftarrow X \cup \{x\}$ and

$$Z \leftarrow Z \cup \{xs \mid s \in S,\ l(xs) < l(x)\}.$$

(The element x is in X_{JK} since it is obtained as a prefix of d_J and $x \in X_K^{-1}$. Moreover, for every $y \in X_{JK}$ we have that either $y \in X$ or y is a prefix of some $z \in Z$)

E3. [Loop.] Repeat step E2 until $Z = \varnothing$. Then set $X_{JK} \leftarrow X$ and stop. (Now X contains every double coset representative since Z is empty.)

2.2.11. Example. Let (W, S) be of type A_n where $S = \{s_1, \ldots, s_n\}$ and let $J = \{s_1, \ldots, s_{n-1}\}$. Then $d_J = s_n \cdots s_1$. We want to determine X_{JJ}. So we set $Z = \{s_n \cdots s_1\}$ and $X = \varnothing$. We let $w = s_n \cdots s_1$, the only element in Z, and write it as $s_n w'$ where $w' = s_{n-1} \cdots s_1 \in W_J$ and $s_n \in X_J^{-1}$. Thus we obtain a double coset representative and set $X = \{s_n\}$ and $Z = \{1\}$. Finally, 1 is a double coset representative and we get $X_{JJ} = \{1, s_n\}$.

2.2.12. Restricted coset graphs. In terms of the coset graph the Mackey decomposition (Lemma 2.1.9) can be made visible by omitting certain edges. For that purpose we introduce here the *restricted coset graph* $\Gamma_J|_K$ which is obtained from Γ_J by omitting the edges which are not labelled by elements of K.

Let $J, K \subseteq S$, let $\Gamma_J = (V, E)$ and define $\Gamma_J|_K$ as (V, E') where $E' = \{v \xrightarrow{s} v' \in E \mid s \in K\}$. Then

$$\Gamma_J|_K = \coprod_{d \in X_{JK}} \Gamma^K_{J^d \cap K}.$$

2.2.13. Example. Let $n > 0$ and let $(W, \{s_1, s_2, \ldots, s_n\})$ be of type A_n with diagram 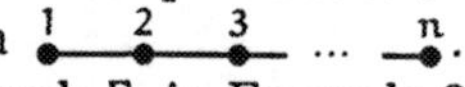. For $J = \{s_1, s_2, \ldots, s_{n-1}\}$ we have determined the coset graph Γ_J in Example 2.2.4. The restricted coset graph $\Gamma_J|_J$ corresponding to the double cosets of A_{n-1} in A_n looks as follows.

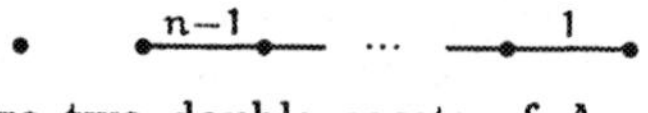

We can see that there are two double cosets of A_{n-1} in A_n, represented by $\{\mathrm{id}, s_n\}$, one consisting of one right coset, the other of $n-1$.

2.2.14. Example. Let (W, S) be of type E_6 as in Example 2.2.7 and let $J = \{s_1, \ldots, s_5\}$ and $K = S - \{s_5\}$. Then W_K has two irreducible components and diagram 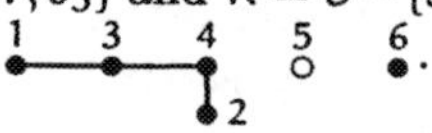. The restricted coset graph $\Gamma_J|_K$ is shown in Figure 2.5.

There we can see four double cosets corresponding to the coset graph of $W_\varnothing$ in $W_{\{6\}}$ (which is $W_{\{1,2,3,4\}}$ in W_K), the coset graph of $W_{\{1,2,3\}}$ in $W_{\{1,2,3,4\}}$ (which is $W_{\{1,2,3,6\}}$ in W_K), the coset graph of $W_{\{2,3,4\}}$ in W_K and the coset graph of $W_{\{1,3,4\}}$ in $W_{\{1,2,3,4\}}$ (which is $W_{\{1,3,4,6\}}$ in W_K).

2.2.15. Products of coset graphs. We now define a multiplication of coset graphs as follows. Let $J, K \subseteq S$ and define

$$\Gamma_J \cdot \Gamma_K = \Gamma_J|_K \ltimes \Gamma_K.$$

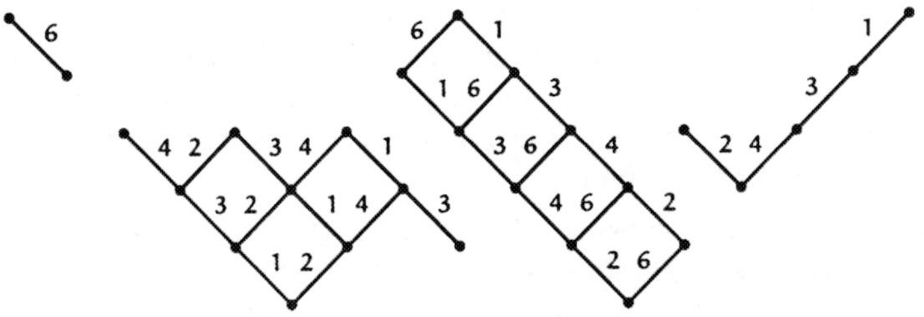

Fig. 2.5. A restricted coset graph.

Then

$$\Gamma_J \cdot \Gamma_K = \Big(\coprod_{d \in X_{JK}} \Gamma^K_{J^d \cap K} \Big) \ltimes \Gamma_K = \coprod_{d \in X_{JK}} \big(\Gamma^K_{J^d \cap K} \ltimes \Gamma_K\big) = \coprod_{d \in X_{JK}} \Gamma_{J^d \cap K} = \coprod_{L \subseteq S} a_{JKL} \Gamma_L$$

by the Mackey decomposition (Lemma 2.1.9) and the transitivity (2.1.5) of coset graphs. (Here we also assume that the product distributes over the disjoint union.)

2.2.16. Example. Let $(W, \{s_1, s_2\})$ be of type A_2. We multiply Γ_J by Γ_K, where $J = \{s_2\}$ and $K = \{s_1\}$. The result is determined as follows. We start with a collection of copies of Γ_K, one copy for each vertex of Γ_J. Next, the initial vertices of these graphs are connected by those edges of Γ_K which are not labelled by an initial label of Γ_K, i.e., an element $s \in S - K$. Finally, the connected components of the resulting graph are completed by further edges to form coset graphs.

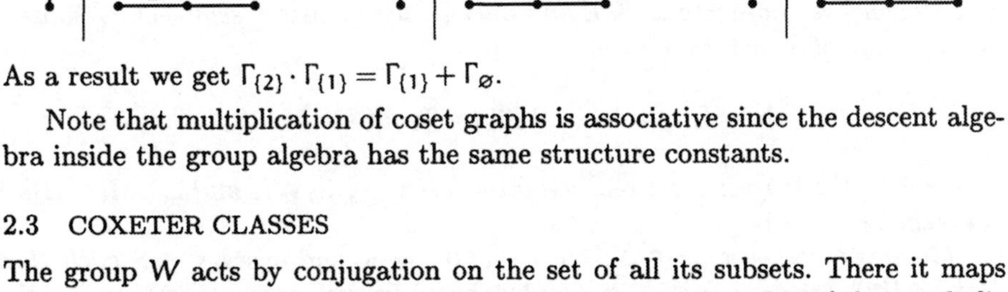

As a result we get $\Gamma_{\{2\}} \cdot \Gamma_{\{1\}} = \Gamma_{\{1\}} + \Gamma_{\varnothing}$.

Note that multiplication of coset graphs is associative since the descent algebra inside the group algebra has the same structure constants.

2.3 COXETER CLASSES

The group W acts by conjugation on the set of all its subsets. There it maps groups to groups and for $J \subseteq S$ we will be interested in the orbit of the parabolic subgroup W_J under this action, in particular in those conjugates of W_J which are again parabolic subgroups, i.e., $W_J^w = W_K$ for some $K \subseteq S$. By Corollary 2.1.13

this amounts to studying the orbit of J under this action, and there in particular those conjugates of J which are again subsets of S. The stabilizer of W_J in this action is the normalizer of W_J in W. In this section we develop an algorithmic solution of the problem and give general descriptions of the orbits in the classical cases. We assume W to be finite.

2.3.1. Definition. Two subsets $J, K \subseteq S$ lie in the same *Coxeter class* of W if and only if there is an element $w \in W$ such that $J^w = K$. In that case we write $J \sim_W K$, or simply $J \sim K$. Note that in general a Coxeter class is not a full orbit under the conjugation action of W.

2.3.2. Proposition. *Let $J \subset S$ be maximal and let $x \in X_J$. Then $J^x \subseteq S$ if and only if $x = 1$ or $x = d_J$. More precisely,*

(i) *the only subsets of S conjugate to J in W are J and J^{w_0}, and*
(ii) *if $J \neq J^{w_0}$ then $N_W(W_J) = W_J$, if $J = J^{w_0}$ then $|N_W(W_J) : W_J| = 2$.*

Proof. (i) If W_K is conjugate to W_J then, by Corollary 2.1.13, $|K| = |J|$, and W_K has a fixed point in the action of W on the (right) cosets of W_J. The coset represented by 1 is fixed by W_J. Now suppose that a non-trivial element $x \in X_J$ represents a coset fixed by W_K. Then $xs \notin X_J$ for all $s \in K$. Since $x \neq 1$ there is an $s \in S$ such that $l(xs) < x$ and $xs \in X_J$. But K is maximal in S so, by Lemma 2.1.2, there remains no $s \in S$ such that $l(xs) > l(x)$ and $xs \in X_J$. By Lemma 2.2.1 then $x = d_J$ and $K = J^{w_0}$.

(ii) By Proposition 2.1.15(b), the index of W_J in its normalizer in W is given as the size of X_{JJJ}. By the definition of the subset X_{JKL} in Theorem 2.1.10 and by Lemma 2.1.8(i), this is the number of coset representatives $x \in X_J$ with the property that $xs \in X_J$ only if s is the unique generator in $S - J$. By (i), this number is 2 if $J^{w_0} = J$, and 1 otherwise. □

Next we show that conjugation of parabolic subgroups of W can always be described by a sequence of longest coset representatives of maximal parabolic subgroups.

2.3.3. Theorem (Lusztig–Spaltenstein, Howlett, Deodhar). *Let J, K be subsets of S. Then J is conjugate to K if and only if there exists a sequence of subsets $J_0, J_1, \ldots, J_r$ of S such that*

$$J_0 = J,\ J_r = K,\ \text{and } J_{i-1}^{d_i} = J_i \text{ for } i = 1, \ldots r,$$

where d_i is the longest coset representative of $W_{J_{i-1}}$ in W_{L_i} and $L_i = J_{i-1} \cup \{s_i\}$ for some $s_i \in S - J_{i-1}$.

Moreover, for every $x \in X_J$ such that $J^x = K$ and every $s \in S$ such that $l(sx) < l(x)$, the d_i, $i = 1, \ldots, r$, can be chosen in such a way that $d_1 = d_J^L$, the longest coset representative of W_J in W_L where $L = J \cup \{s\}$, and that $x = d_1 \cdots d_r$ and $l(x) = l(d_1) + \cdots + l(d_r)$.

Proof. If the sequence exists then clearly $J \sim K$. Conversely, suppose that J and K are conjugate. Let $w \in W$ be such that $J^w = K$ and write $w = w' \cdot x$ for some $w' \in W_J$ and $x \in X_J$. If $l(s^{w'}) > 1$ for some $s \in J$ then also, by Lemma 2.1.14, $l(s^w) > 1$, in contradiction to $s^w \in K$. Hence $s^{w'} \in J$ for all $s \in J$ and thus $J^{w'} = J$. So we may assume $w = x \in X_J$.

If $x = 1$ there is nothing to prove. Suppose $x \neq 1$ and let $u \in S$ such that $l(ux) < l(x)$. Then $u \notin J$ since $x \in X_J$. Let $L = J \cup \{u\}$ and write $x = x_1 x_2$, where $x_1 \in X_J^L$ and $x_2 \in X_L$. By Lemma 2.1.14, we must have $l(s^{x_1}) = 1$ for all $s \in J$, whence $J^{x_1} \subseteq L$.

But J is maximal in L and $x_1 \neq 1$, so by Proposition 2.3.2 $x_1 = d_J^L$, the unique non-trivial coset representative of W_J in W_L which maps J into L. Moreover, by Lemma 2.1.8(i), $x_2 \in X_{J'}$ where $J' = J^{x_1}$. The assertion thus follows by induction on $l(x)$. □

There is an algorithm, based on the last result, for the explicit construction of the Coxeter class of $J \subseteq S$ in W together with the conjugating longest coset representatives.

Algorithm F (Coxeter class graph). Given a finite Coxeter group (W, S) and a subset $J \subseteq S$, the graph $\mathcal{G}_J = (V, E)$ associated with the Coxeter class of J is determined.

F1. [Initialize.] Set $V \leftarrow (J)$, $E \leftarrow \varnothing$, $i \leftarrow 1$ and $l \leftarrow 1$. (The list V of vertices initially contains $V_1 = J$, there are no edges in the edge set E. The pointer i points to the vertex that is to be dealt with next, while l points to the last entry in $V = (V_1, \ldots, V_l)$.)

F2. [Expand a vertex.] Set $K \leftarrow V_i$. For each $s \in S - K$ do the following. Set $M \leftarrow K \cup \{s\}$ and $d \leftarrow d_K^M$. Also set $L \leftarrow K^d$ and $E \leftarrow E \cup \{K \xrightarrow{d} L\}$. If $L \notin V$ then append L to the list V: set $l \leftarrow l + 1$ and $V_l \leftarrow L$.

F3. [Loop]. Set $i \leftarrow i + 1$. If $i \leq l$ go to step F2. Otherwise return $\mathcal{G}_J = (V, E)$.

2.3.4. Example. Suppose (W, S) is of type E_6 with $S = \{s_1, \ldots, s_6\}$, as in Example 2.2.7, and let $J = \{s_1, s_3\}$. Then W_J is of type A_2 and the Coxeter class graph $\mathcal{G}_J$ is determined by Algorithm F as shown in Figure 2.6. For brevity here $K \xrightarrow{s} L$ stands for $K \xrightarrow{d} L$, i.e., $K^d = L$ where $d = d_K^M$ and $M = K \cup \{s\}$.

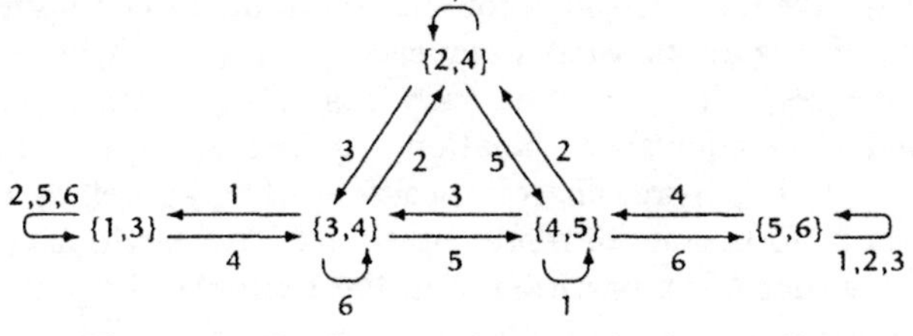

Fig. 2.6. The Coxeter class graph of A_2 in E_6.

2.3.5. Remark. A subset $J \subseteq S$ is called *self-opposed* if $J^{w_M} = J$ for all $M \subseteq S$ with $J \subseteq M$. Non-trivial examples of self-opposed subsets are given by parabolic subgroups of type A_{m-1}^k in a Coxeter group of type A_{n-1} where $n = km$, parabolic subgroups of type D_4, and all maximal parabolic subgroups which are invariant under w_0. In such a case the Coxeter class graph $\mathcal{G}_J$ has a single vertex J and a loop for each $s \in \hat{J} = S - J$. Moreover, the parabolic subgroup $W_{\hat{J}}$ coincides with the group X_{JJJ}.

In general, for any $J \subseteq S$, a generating set for X_{JJJ} can be derived from the graph $\mathcal{G}_J$, together with a set of defining relations. For this the graph $\mathcal{G}_J$ is used to define a groupoid

$$G_J = \{(K, x) \mid K \sim J,\ x \in X_K\}$$

with multiplication $(K, x)(L, y) = (K, xy)$ if $K^x = L$, and undefined otherwise. The groupoid is generated by those (K, d) that correspond to the edges $K \xrightarrow{d} L$ of the graph $\mathcal{G}_J$. In this setting the group X_{JJJ} is described as the set of all $(J, x) \in G_J$ with $J^x = J$. The details of this procedure are beyond the scope of this book. The results of a case-by-case analysis of all irreducible finite Coxeter groups are listed in [Howlett 1980]. It turns out that X_{JJJ} very much resembles a Coxeter group, and in most cases is one. A presentation of the groupoid with respect to the generators (K, d) is given in [Brink and Howlett 1999] and used to derive a presentation for X_{JJJ}. Also in [Borcherds 1998] the groupoid G_J is used to describe X_{JJJ}.

We conclude this section by determining the Coxeter classes of parabolic subgroups and the order of X_{JJJ} for the parabolic subgroups of the Coxeter groups of classical types. By p_J we denote the size of the Coxeter class of J, i.e., the number of parabolic subgroups of W which are conjugate to W_J. The combinatorial objects which will be used as labels are introduced next.

Complete lists of Coxeter classes of parabolic subgroups for the exceptional types are given in Tables A.1 and A.2.

2.3.6. Compositions and partitions. A *composition* of n is an ordered sequence $\mu = (\mu_1, \dots, \mu_k)$ of integers $\mu_i > 0$ such that $|\mu| = \mu_1 + \dots + \mu_k = n$. The composition $\mu = (\mu_1, \dots, \mu_k)$ is called *increasing* if it is an increasing sequence, i.e., if $\mu_1 \leq \dots \leq \mu_k$, and *decreasing* if it is a decreasing sequence, i.e., if $\mu_1 \geq \dots \geq \mu_k$. We write $l(\mu) = k$ for the length of the composition μ.

A *partition* of n is an unordered sequence $\lambda = [\lambda_1, \dots, \lambda_r]$ of integers $\lambda_i > 0$ such that $|\lambda| = \lambda_1 + \dots + \lambda_r = n$. As such it is an equivalence class of compositions of n under the equivalence relation $\sim$ generated by $(\dots, \mu_i, \mu_{i+1}, \dots) \sim (\dots, \mu_{i+1}, \mu_i, \dots)$. The increasing compositions of n, as well as the decreasing compositions of n, form a set of representatives of the partitions of n. We write $\lambda \vdash n$ to indicate that λ is a partition of n. For example, the partitions of $n = 4$ can be written as

$$[1,1,1,1],\ [2,1,1],\ [2,2],\ [3,1],\ [4], \quad \text{or} \quad [1^4],\ [1^2,2^1],\ [2^2],\ [1^1,3^1],\ [4^1],$$

where an exponential notation is used for repeated entries. The number of compositions of n in the equivalence class $\lambda = [1^{l_1}, \ldots, n^{l_n}]$ is $c(\lambda) = \binom{l_1+\cdots+l_n}{l_1,\ldots,l_n}$ (see Exercise 2.12). The unique partition of $n = 0$ is called the *empty partition* and is denoted by $\varnothing$.

2.3.7. Parabolic subgroups in type A_{n-1}. Suppose that $(W, \{s_1, \ldots, s_{n-1}\})$ is of type A_{n-1} as in Proposition 1.4.7, i.e., with the following Dynkin diagram:

A_{n-1}
$(n \geq 2)$

Then there is a bijection between the set of compositions of n and the set of parabolic subgroups of W as follows. For a composition μ of n set $M = M(\mu) = \{\sum_{i=1}^{j} \mu_i \mid j = 1, \ldots, l\}$. Then $W_\mu = \langle s_i \in S \mid i \notin M \rangle$ is a parabolic subgroup of W which is isomorphic to a direct product of symmetric groups $W_\mu \cong \mathfrak{S}_{\mu_1} \times \cdots \times \mathfrak{S}_{\mu_l}$. Every parabolic subgroup of W occurs this way and gets assigned a unique composition of n as label. For example, the diagram

belongs to the parabolic subgroup W_J of (W, S) of type A_{13} where $W \cong \mathfrak{S}_{14}$ and $S = \{s_1, \ldots, s_{13}\}$ and where $J = \{s_2, s_3, s_4, s_5, s_7, s_9, s_{12}, s_{13}\}$. The composition of 14 associated to W_J is $(1, 5, 2, 2, 1, 3)$

2.3.8. Proposition (Coxeter classes in type A_{n-1}). *Let W be of type A_{n-1}. Then the Coxeter classes of parabolic subgroups of W are in bijection with the set of partitions $\{\lambda \mid \lambda \vdash n\}$.*

More precisely, if $\lambda = [1^{l_1}, \ldots, n^{l_n}]$ is a partition of n and W_J is a subgroup in the class with label λ then the type of W_J is $A_1^{l_2} \times \cdots \times A_{n-1}^{l_n}$ and

$$|W_J| = \prod_{i=1}^{n} (i!)^{l_i}, \qquad p_J = \binom{l_1 + \cdots + l_n}{l_1, \ldots, l_n}, \qquad |X_{JJJ}| = l_1! \cdots l_n!.$$

Proof. First note that parabolic subgroups with labels that give rise to different partitions are not isomorphic and therefore cannot be conjugate.

On the other hand, if W_J has label $\mu = (\ldots, \mu_i, \mu_{i+1}, \ldots)$ and $W_{J'}$ has label $\mu' = (\ldots, \mu_{i+1}, \mu_i, \ldots)$ then W_J is conjugate to $W_{J'}$ in W. In order to see this it is sufficient to consider a maximal parabolic subgroup W_J of $W \cong \mathfrak{S}_n$. Then $J = S - \{s_i\}$ for some $i \in \{1, \ldots, n-1\}$ and $W_J \cong \mathfrak{S}_i \times \mathfrak{S}_k$ has as its label the composition (i, k) of n. Now w_0 acts as a permutation on S mapping s_j to s_{n-j}. Similarly, w_J acts simultaneously on $\{s_1, \ldots, s_{i-1}\}$ (mapping s_j to s_{i-j}) and $\{s_{i+1}, \ldots, s_{n-1}\}$ (mapping s_{i+j} to s_{n-j}). Thus, the longest coset representative $d_J = w_J w_0$ maps s_j to s_{j+k} if $j < i$ and to s_{j-i} if $j > i$. The resulting parabolic subgroup $W_J^{d_J}$ is isomorphic to $\mathfrak{S}_k \times \mathfrak{S}_i$ and has as its label the composition (k, i).

It remains to calculate the order of the normalizer. This is determined by the action of $\mathfrak{S}_n$ on the partitions of the set $\{1,\dots,n\}$, where each partition corresponds to a different conjugate of a parabolic subgroup, and the fact that the stabilizer of a partition with parts of sizes $\lambda = [1^{l_1},\dots,n^{l_n}]$ has order $\prod_{i=1}^{n}(i!)^{l_i} l_i!$. □

2.3.9. Parabolic subgroups in type B_n. Consider the Coxeter group $W_n = \langle t, s_1,\dots,s_{n-1}\rangle$ of type B_n, as in (1.4.1), with the following Dynkin diagram:

B_n
$(n\geq 2)$

Every parabolic subgroup W_J of W_n has a direct product decomposition $W_J = W_1 \times W_2$ where W_1 is the irreducible component of W_J containing t or, if $t \notin J$, $W_1 = 1$. Thus $W_1 = \langle t, s_1,\dots,s_{n_J-1}\rangle$ is a Coxeter group of type B_{n_J} for some $n_J \leq n$. Further, W_2 is a parabolic subgroup of $\langle s_{n_J+1},\dots,s_{n-1}\rangle \cong \mathfrak{S}_{n-n_J}$, and as such has as label a composition of $n - n_J$. In this way, the parabolic subgroups of W_n are labelled by the compositions μ with $|\mu| \leq n$.

2.3.10. Proposition (Coxeter classes in type B_n). *Let (W_n, S) be a Coxeter system of type B_n. Then the Coxeter classes of parabolic subgroups of W_n are in bijection with the set $\{\lambda \vdash m \mid 0 \leq m \leq n\}$.*

More precisely, if $m \leq n$, $\lambda = [1^{l_1},\dots,m^{l_m}]$ is a partition of m and W_J is a subgroup in the class with label λ, then the type of W_J is $B_{n-m} \times A_1^{l_2} \times \cdots \times A_{m-1}^{l_m}$ and, if we set $k = l_1 + \cdots + l_m$,

$$|W_J| = 2^{n-m}(n-m)!\prod_{i=1}^{m}(i!)^{l_i}, \quad p_J = \binom{k}{l_1,\dots,l_m}, \quad |X_{JJJ}| = 2^k l_1!\cdots l_m!.$$

Proof. By Proposition 2.3.8, the parabolic subgroups (of a parabolic subgroup of type $B_{n-m} \times A_{m-1}$ of W_n) which give rise to the same partition (of m) are conjugate in W_n. On the other hand, parabolic subgroups corresponding to different partitions (of any $m \leq n$) cannot be conjugate (for $m = n-1$, note that the generator t is not conjugate to any of the s_i).

It follows from the matrix description of W_n in (1.4.1) that if W_J lies in the class with label $\lambda = [1^{l_1},\dots,m^{l_m}]$ then X_{JJJ} is isomorphic to a Coxeter group of type $B_{l_1} \times \cdots \times B_{l_m}$. □

2.3.11. Parabolic subgroups in type D_n. Let $W_n' = \langle u, s_1,\dots,s_{n-1}\rangle$ be a Coxeter group of type D_n as in (1.4.8), i.e., with the following Dynkin diagram:

D_n
$(n\geq 2)$

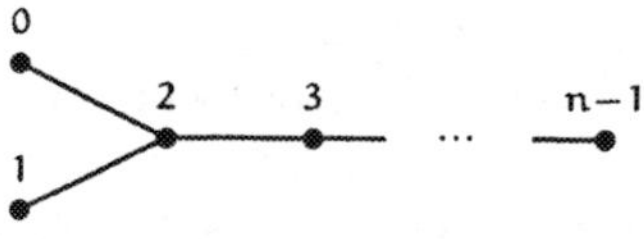

We distinguish four different types of parabolic subgroups W_J of W_n' according to whether u and s_1 belong to J or not. We write $\mu \models m$ to indicate that μ is a composition of m. The parabolic subgroups of W_n' are labelled by the set

$$\{\mu \models m \mid m \le n-2\} \cup \{\mu \models n \mid \mu_1 = 1\} \cup \{(\mu, \pm) \mid \mu \models n,\ \mu_1 > 1\}$$

as follows. If $u \notin J$ then W_J is a parabolic subgroup of $\langle s_1, \ldots, s_{n-1}\rangle \cong \mathfrak{S}_n$ and W_J has as label a composition μ of n. If $s_1 \in J$ we write the label as $(\mu, +)$ in order to indicate that J contains s_1 and not u.

If $s_1 \notin J$ and $u \in J$ then W_J is a parabolic subgroup of $\langle u, s_2, \ldots, s_{n-1}\rangle \cong \mathfrak{S}_n$. Again, W_J has a composition $\mu = (\mu_1, \ldots, \mu_r)$ of n as its label which we write as $(\mu, -)$ to indicate that J contains u and not s_1.

A parabolic subgroup W_J of W'_n which contains both u and s_1 is a direct product $W_1 \times W_2$ of a component $W_1 = \langle u, s_1, \ldots, s_{n_J-1}\rangle$ of type D_{n_J} for some $n_J \le n$ and a parabolic subgroup W_2 of $\langle s_{n_J+1}, \ldots, s_{n-1}\rangle \cong \mathfrak{S}_{n-n_J}$ which has a composition of $n - n_J$ as its label. Here we regard $\langle u, s_1\rangle$ as a group of type D_2 and $\langle u, s_1, s_2\rangle$ as a group of type D_3.

2.3.12. Proposition. *Suppose that $W_n = \langle S\rangle$ is of type B_n and that $W'_n = \langle S'\rangle$, where $S' = \{ts_1t, s_1, \ldots, s_{n-1}\}$ is the subgroup of type D_n as in (1.4.8). For each $J \subseteq S$ the subgroup $W_J \cap W'_n$ is a parabolic subgroup of W'_n. Furthermore, $W_J \cap W'_n = W_J$ unless $t \in J$; in that case $W_J \cap W'_n$ has index 2 in W_J.*

Moreover, if W_J has the composition $\mu = (\mu_1, \mu_2, \ldots, \mu_r)$ of $m \le n$ as its label then $W_J \cap W'_n$ has label μ unless μ is a composition of $n-1$. In that case, the label of $W_J \cap W'_n$ is the composition $(1, \mu_1, \mu_2, \ldots, \mu_r)$ of n.

Proof. If $\{t, s_1\} \subseteq J$ then $W_J \cap W'_n$ has ts_1t instead of t as a generator, which yields a subgroup of index 2. If $t \in J$ but $s_1 \notin J$ then $W_J \cap W'_n$ is generated by $J - \{t\}$ and again of index 2 in W_J. Finally, if $t \notin J$ then $W_J \subseteq W'_n$. □

2.3.13. Proposition (Coxeter classes in type D_n). *Let $W = W'_n$ be the Coxeter group of type D_n. We say that a partition $\lambda = [\lambda_1, \ldots, \lambda_r]$ of n is all-even, if all parts λ_i, $i = 1, \ldots, r$, are even. Then the Coxeter classes of parabolic subgroups of W'_n are labelled by the set*

$$\{\lambda \vdash m \mid m \le n-2\} \cup \{\lambda \vdash n \mid \lambda \text{ has an odd part}\} \cup \{(\lambda, \pm) \mid \lambda \vdash n \text{ all-even}\}.$$

More precisely, suppose that $\lambda = [1^{l_1}, \ldots, m^{l_m}]$ is a partition of $m \le n-2$ or of $m = n$ and set $k = l_1 + \cdots + l_m$. If $m \le n-2$ and W_J is a subgroup in the class with label λ then the type of W_J is $D_{n-m} \times A_1^{l_2} \times \cdots \times A_{m-1}^{l_m}$ and

$$|W_J| = 2^{n-m-1}(n-m)! \prod_{i=1}^{m} (i!)^{l_i}, \quad p_J = \binom{k}{l_1, \ldots, l_m}, \quad |X_{JJJ}| = 2^k l_1! \cdots l_m!.$$

If λ is an all-even partition of n and W_J is a subgroup in the class with label $(\lambda, +)$ or $(\lambda, -)$ then the type of W_J is $A_1^{l_2} \times \cdots \times A_{nm-1}^{l_m}$ and

$$|W_J| = \prod_{i=1}^{m} (i!)^{l_i}, \qquad p_J = \binom{k}{l_1, \ldots, l_m}, \qquad |X_{JJJ}| = 2^k l_1! \cdots l_m!.$$

Otherwise, if λ *is a partition of* n *which has an odd part and* W_J *is a subgroup in the class with label* λ *then the type of* W_J *is* $A_1^{l_2} \times \cdots \times A_{m-1}^{l_m}$ *and*

$$|W_J| = \prod_{i=1}^{m} (i!)^{l_i}, \qquad p_J = \left(2 - \frac{l_1}{k}\right) \binom{k}{l_1, \ldots, l_m}, \qquad |X_{JJJ}| = 2^{k-1} l_1! \cdots l_m!.$$

Proof. Two subsets $J, K \subseteq S$ are conjugate in W_n' if their labelling compositions give rise to the same partition. It remains to decide, for compositions $\mu = (\mu_1, \ldots, \mu_r)$ of n, whether the parabolic subgroup with label $(\mu, +)$ is conjugate to that labelled by $(\mu, -)$. By Theorem 2.3.3, the only way to map a parabolic subgroup W_J of $\langle s_1, s_2, \ldots, s_{n-1}\rangle$ to a parabolic subgroup of $\langle u, s_2, \ldots, s_{n-1}\rangle$ is by conjugation with the longest coset representative $d_J^{J\cup\{u\}}$ of W_J in $W_{J\cup\{u\}}$. Let $\lambda = [\mu_1, \ldots, \mu_r]$ be the partition of n arising from μ. If μ has an odd part l we may assume that $l = \mu_1$ and apply the longest element of D_l to map $\{s_1, \ldots, s_{l-1}\}$ to $\{u, s_2, \ldots, s_{l-1}\}$. In this case λ acts as the label of the class. Otherwise, if all parts of λ are even, they are not conjugate since the longest element in D_l is central for even l. In this case we use the labels $(\lambda, +)$ and $(\lambda, -)$ for the two resulting classes, where $(\lambda, +)$ is the label of the class of subsets which contain t. The remaining details are left to the reader. □

2.4 THE PARABOLIC TABLE OF MARKS

In this section we introduce the parabolic table of marks of W. We first recall the definition and basic properties of the Burnside ring (cf. [Curtis and Reiner 1987, Chap. 11]) and the table of marks of a finite group. In the case of a finite Coxeter group the *parabolic* table of marks is defined as a certain submatrix of the table of marks.

2.4.1. The Burnside ring. Let G be a finite group and let X be a finite set. Then X is called a G-*set* if G acts on X, i.e., if there is a map $X \times G \to X$, $(x, g) \mapsto x.g$, such that $x.1 = x$ and $x.(gh) = (x.g).h$ for all $x \in X$ and all $g, h \in G$. A *homomorphism* from a G-set X to a G-set Y is a map $f\colon X \to Y$ such that $f(x.g) = f(x).g$ for all $x \in X$ and all $g \in G$. The G-sets X and Y are *isomorphic* if there is a bijective homomorphism $f\colon X \to Y$. We denote by $[X]$ the isomorphism type of the G-set X. If X and Y are G-sets then their disjoint union $X \amalg Y$ is a G-set (via $z.g = x.g$ if $z = x \in X$ and $z.g = y.g$ if $z = y \in Y$).

A G-set X is called *transitive* if, for every $x, y \in X$ there is an element $g \in G$ such that $x.g = y$. The *orbit* of $x \in X$ under G is the set $x^G = \{y \in X \mid y = x.g,\ g \in G\}$. Every orbit x^G is a transitive G-set. Every G-set X is a disjoint union of transitive G-sets, the orbits of G on X.

The stabilizer of $x \in X$ in G is the subgroup $\mathrm{Stab}_G(x) = \{g \in G \mid x.g = x\}$. If $y = x.g$ for some $g \in G$ then the stabilizer of y is a conjugate of the stabilizer of x: $\mathrm{Stab}_G(y) = \mathrm{Stab}_G(x)^g$. A transitive G-set X is isomorphic to the G-set $G/\mathrm{Stab}_G(x)$ for any $x \in X$ (via the map that associates with the point $x.g \in X$

the coset $\mathrm{Stab}_G(x)g$ of the stabilizer). If A and B are subgroups of G then the G-sets G/A and G/B are isomorphic if and only if A is conjugate to B in G. (If $f\colon G/A \to G/B$ is an isomorphism of G-sets and $f(A) = Bg$ for some $g \in G$ then, for all $a \in A$,

$$Bg = f(A) = f(Aa) = f(A)a = Bga,$$

whence $A \subseteq B^g$. We also get $B^g \subseteq A$ from $f^{-1}(B) = Ag^{-1}$, hence $A = B^g$. Conversely, if $A = B^g$ for some $g \in G$ then $f(A) = Bg$ defines an isomorphism of G-sets.)

The *Burnside ring* $\Omega(G)$ of G is the free abelian group generated by the isomorphism types $[X]$ of G-sets subject to the relations

$$[X] + [Y] = [X \amalg Y]$$

for all isomorphism types $[X]$ and $[Y]$ of G-sets.

Let $G_1(=1), G_2, \ldots, G_r(=G)$ be representatives of the conjugacy classes of subgroups of G. Then each transitive G-set is isomorphic to G/G_i for some $i \leq r$ and the classes of the transitive G-sets form a $\mathbb{Z}$-basis of $\Omega(G)$. Thus

$$\Omega(G) = \left\{ \sum_{i=1}^{r} a_i [G/G_i] \;\middle|\; a_i \in \mathbb{Z} \right\},$$

the free abelian group generated by the isomorphism types of transitive G-sets.

We define a multiplication in $\Omega(G)$ by the Cartesian product of G-sets. If X and Y are G-sets then so is $X \times Y$ (via $(x, y).g = (x.g, y.g)$) and we set

$$[X] \cdot [Y] = [X \times Y]$$

for all isomorphism types of G-sets in $\Omega(G)$. Then this multiplication is commutative, with the class of the trivial G-set $[G/G]$ as identity and it distributes over the disjoint union of G-sets, thus turning $\Omega(G)$ into a commutative ring. Multiplication in $\Omega(G)$ is described by the formula

$$[G/A \times G/B] = \coprod_d [G/(A^d \cap B)], \qquad \text{for } A, B \leq G, \tag{M}$$

where the union is over the double coset representatives d such that $G = \coprod_d AdB$ (since, for $x_i, y_i \in G$ $(i = 1, 2)$, the pairs (Ax_1, By_1) and (Ax_2, By_2) lie in the same orbit under G if and only if $Ax_1y_1^{-1}B = Ax_2y_2^{-1}B$).

By restriction, any subgroup U of G acts on every G-set X. The *mark* of U on X is the number

$$\beta_X(U) = |\mathrm{Fix}_U(X)| \quad \text{where } \mathrm{Fix}_U(X) = \{x \in X \mid xu = x \text{ for all } u \in U\},$$

the set of fixed points in the G-set X under the action of U. The *table of marks* of G is the $r \times r$ matrix

$$M(G) = \left(\beta_{G/G_i}(G_j)\right)_{i,j=1,\dots,r}$$

where both G_i and G_j run through the system of representatives of conjugacy classes of subgroups of G.

It is clear that a subgroup A of G can only have fixed points on the cosets of a subgroup B of G if it lies in a point stabilizer, i.e., in a conjugate of B. On the other hand, the coset B is a fixed point in the action of the subgroup B on its cosets. Thus, with respect to an ordering of the representatives G_i that is compatible with inclusion of conjugates, the table of marks $M(G)$ is a lower triangular matrix with non-zero entries on the diagonal, and therefore invertible over $\mathbb{Q}$.

Every finite G-set X has an associated vector of fixed point numbers

$$\beta_X = (\beta_X(G_j))_{j=1}^{r}$$

and if $[X] = \sum_{i=1}^{r} a_i[G/G_i]$ in $\Omega(G)$ then $\beta_X = (a_1, \dots, a_r) \cdot M(G)$. We have, for G-sets X and Y,

$$\mathrm{Fix}_U(X \amalg Y) = \mathrm{Fix}_U(X) \amalg \mathrm{Fix}_U(Y) \quad \text{and} \quad \mathrm{Fix}_U(X \times Y) = \mathrm{Fix}_U(X) \times \mathrm{Fix}_U(Y).$$

This implies $\beta_{X \amalg Y} = \beta_X + \beta_Y$ and $\beta_{X \times Y} = \beta_X \cdot \beta_Y$, where addition and multiplication of fixed point vectors is componentwise. Moreover, numbers of fixed points are invariant under conjugation. This yields Burnside's theorem: the map $\beta \colon \Omega(G) \to \mathbb{Z}^r$, $[X] \mapsto \beta_X$ is a well-defined injective homomorphism of rings. In particular, X and Y are isomorphic as G-sets if and only if $\beta_X = \beta_Y$.

We conclude this discussion with an important characterization of marks in terms of the subgroup lattice of G.

2.4.2. Lemma. *Let* $A, B \le G$. *Then*

$$\beta_{G/A}(B) = |N_G(A) : A| \cdot |\{A^g \mid g \in G,\ B \le A^g\}|.$$

Proof. By the definition of a mark, $\beta_{G/A}(B)$ is the number of cosets of A in G which are fixed by the subgroup B. That is,

$$\beta_{G/A}(B) = |\{Ag \mid g \in G, Agb = Ag \text{ for all } b \in B\}|.$$

Now for any given $g \in G$ we have $Agb = Ag$ for all $b \in B$ if and only if $B \le A^g$. And since there are exactly $|A|$ elements g that yield the same coset Ag this means

$$\beta_{G/A}(B) = |\{g \in G \mid B \le A^g\}| / |A|.$$

The claim finally follows from the fact that there are exactly $|N_G(A)|$ elements $g \in G$ which yield the same conjugate A^g of A. □

2.4.3. The parabolic Burnside ring. We now consider the case of a finite Coxeter group (W, S). Let us call a W-set X a "parabolic W-set" if $[X] = [W/W_J]$ for some $J \subseteq S$ in the Burnside ring $\Omega(W)$. If $J, K \subseteq S$, then by formula (M) in (2.4.1) and Theorem 2.1.12,

$$[W/W_J \times W/W_K] = \coprod_{d \in X_{JK}} [W/W_{J^d \cap K}].$$

So the product of two parabolic W-sets in $\Omega(W)$ decomposes as a sum of transitive parabolic W-sets! And the set $\{[W/W_J] \mid J \subseteq S\}$, forms the $\mathbb{Z}$-basis of a subring $\Pi(W)$ of $\Omega(W)$ called the *parabolic Burnside ring* of (W, S).

Let us fix a set $\mathcal{R}$ of representatives of subsets of S under conjugation in W. Then

$$\Pi(W) = \langle [W/W_J] \mid J \subseteq S \rangle = \Big\{ \sum_{J \in \mathcal{R}} \alpha_J [W/W_J] \;\Big|\; \alpha_J \in \mathbb{Z} \Big\}.$$

Define, for $J, K \subseteq S$, the *mark* β_{JK} and the *parabolic table of marks* of W as

$$\beta_{JK} = |\mathrm{Fix}_{W_K}(W/W_J)| \quad \text{and} \quad M(W) = (\beta_{JK})_{J,K \in \mathcal{R}}\,.$$

Then β_{JK} is the mark of W_K on the cosets of W_J in W and $M(W)$ is the submatrix of the ordinary table of marks of W consisting of the rows and columns corresponding to classes of parabolic subgroups.

The mark β_{JK} can be read off the coset graph Γ_J as the number of vertices v which are not involved in a vertex labelled by an element of K,

$$\beta_{JK} = |\{x \in X_J \mid xs \notin X_J \text{ for all } s \in K\}|.$$

2.4.4. Proposition.

(i) *Let* $J, K \subseteq S$. *Then*

$$\beta_{JK} = a_{JKK} = |N_W(W_J) : W_J| \cdot \big|\{W_J^w \mid w \in W,\, W_K \leq W_J^w\}\big|.$$

In particular, $\beta_{JJ} = |N_W(W_J) : W_J| \neq 0$ *and* β_{JJ} *divides* β_{JK} *for every* $K \subseteq S$. *Moreover,* $\beta_{J\varnothing} = |W : W_J|$ *and* $\beta_{SK} = 1$.

(ii) *The parabolic table of marks* $M(W)$ *is invertible over* $\mathbb{Q}$.

(iii) *The linear map* $\beta \colon \Pi(W) \to \mathbb{Z}^{\mathcal{R}}$ *which maps* $[W/W_J]$ *to* $\beta_J = (\beta_{JK})_{K \in \mathcal{R}}$ *is a well-defined injective ring homomorphism.*

Proof. (i) It is easily verified that

$$|N_W(W_J) : W_J| \cdot \big|\{W_J^w \mid w \in W,\, W_K \leq W_J^w\}\big| = |\{x \in X_{JK} \mid J^x \cap K = K\}|.$$

By Lemma 2.4.2, this quantity equals β_{JK} and, by definition, also equals a_{JKK}.

(ii) For an ordering of the representatives in $\mathcal{R}$ which respects subconjugate inclusion, the matrix $M(W)$ is lower triangular with non-zero diagonal entries.

(iii) The map is given by $\sum_{J \in \mathcal{R}} a_J\, [W/W_J] \mapsto (\dots, a_J, \dots) \cdot M(W)$. Thus the claim follows from (ii) and the arithmetical properties of marks. □

1	2m			
A_1'	m	2		
A_1''	m	.	2	
$I_2(m)$	1	1	1	1

(m even)

1	2m		
A_1	m	1	
$I_2(m)$	1	1	1

(m odd)

Fig. 2.7. The parabolic table of marks for $I_2(m)$.

2.4.5. Example. The parabolic tables of marks of the Coxeter groups of dihedral type are shown in Figure 2.7. Such a group has two non-trivial proper parabolic subgroups. Depending on whether m is odd or even they are conjugate or not. Figure 2.8 shows the tables of marks of the Coxeter groups of type B_2 and B_3 and those of type $A_1 \times A_1$ (here called D_2) and A_3 (called D_3).

1	8			
B_1	4	2		
A_1	4	.	2	
B_2	1	1	1	1

1	4			
A_1'	2	2		
A_1''	2	.	2	
D_2	1	1	1	1

1	48						
B_1	24	8					
A_1	24	.	4				
$B_1 \times A_1$	12	4	2	2			
A_2	8	.	4	.	2		
B_2	6	4	2	.	.	2	
B_3	1	1	1	1	1	1	1

1	24				
A_1	12	2			
D_2	6	2	2		
A_2	4	2	.	1	
D_3	1	1	1	1	1

Fig. 2.8. The parabolic tables of marks for B_2, B_3, D_2 and D_3.

2.4.6. Corollary. *Let $K, K' \subseteq S$. Then $K \sim K'$ if and only if $\beta_{JK} = \beta_{JK'}$ for all $J \subseteq S$. Also $J, J' \subseteq S$ are conjugate if and only if $\beta_{JK} = \beta_{J'K}$ for all $K \subseteq S$.*

Proof. If $K \sim K'$ then clearly $\beta_{JK} = \beta_{JK'}$ for all $J \subseteq S$. Conversely, if this happens, then K and K' must lie in the same Coxeter class since the columns of $M(W)$ are linearly independent. The argument for the rows of $M(W)$ is completely analogous. □

For practical purposes, one can use the following induction formula to compute a mark β_{JK}. This is a special case of the general induction formula for marks of subgroups (see [Pfeiffer 1997], Theorem 2.2). For the sake of completeness the proof is presented here.

2.4.7. Proposition. *Let $J, K \subseteq M \subseteq S$ and denote by β^M_{JK} the mark of W_K on W_J in W_M. Then*

$$\beta_{JK} = |N_W(W_K)| \sum_{K' \sim K} \frac{1}{|N_{W_M}(W_{K'})|} \cdot \beta^M_{JK'},$$

where the sum ranges over all representatives K' *of Coxeter classes of* W_M *which are conjugate to* K *in* W.

Note that $|N_W(W_{K'}) : N_{W_M}(W_{K'})| = \beta_{KK}/\beta^M_{K'K'}$.

Proof. We have, by Proposition 2.4.4, $\beta_{JK} = \beta_{JJ} \cdot \left|\{W_J^w \mid W_J^w \geq W_K,\ w \in W\}\right|$ and $\beta^M_{JK'} = \beta^M_{JJ} \cdot \left|\{W_J^w \mid W_J^w \geq W_{K'},\ w \in W_M\}\right|$.

Now we express the total number of inclusions between the conjugacy class of W_K and the conjugacy class of W_J in two different ways. First we note that the number of conjugates of W_K which are contained in a conjugate of W_J is the same for all these conjugates. Likewise the number of conjugates of W_J which do contain a fixed conjugate of W_K does not depend on the choice of this conjugate. Therefore the number of pairs (W_1, W_2) where $W_1 \leq W_2$, and where W_1 is a conjugate of W_K and W_2 is a conjugate of W_J, equals the product of the size of the conjugacy class of W_K and the number of conjugates of W_J that contain W_K. It also equals the product of the size of the conjugacy class of W_J and the number of conjugates of W_K that W_J contains, i.e.,

$$\frac{|W|}{|N_W(W_K)|} \cdot \left|\{W_J^w \geq W_K \mid w \in W\}\right| = \frac{|W|}{|N_W(W_J)|} \cdot |\{W_K^w \leq W_J \mid w \in W\}|.$$

The W-conjugacy class $\{W_K^w \leq W_J \mid w \in W\}$ of subgroups of W_J is a union of W_M-conjugacy classes of subgroups of W_J. Thus

$$|\{W_K^w \leq W_J \mid w \in W\}| = \sum_{K' \sim K} |\{W_{K'}^w \leq W_J \mid w \in W_M\}|$$

where the sum ranges over all representatives K' of Coxeter classes of W_M which are conjugate to K in W and where

$$\frac{|W_M|}{|N_{W_M}(W_J)|} |\{W_{K'}^w \leq W_J \mid w \in W_M\}| = \frac{|W_M|}{|N_{W_M}(W_{K'})|} \left|\{W_J^w \geq W_{K'} \mid w \in W_M\}\right|.$$

Putting these together yields, with Proposition 2.4.4,

$$\beta_{JK} = \beta_{JJ} \cdot \frac{|N_W(W_K)|}{|N_W(W_J)|} \cdot \sum_{K' \sim K} \frac{|N_{W_M}(W_J)|}{|N_{W_M}(W_{K'})|} \cdot \frac{\beta^M_{JK'}}{\beta^M_{JJ}} = \sum_{K' \sim K} \frac{|N_W(W_K)|}{|N_{W_M}(W_{K'})|} \cdot \beta^M_{JK'},$$

as desired. □

This result enables the explicit calculation of the parabolic table of marks $M(W)$ from the knowledge of the index $\beta_{KK} = |N(W_K) : W_K|$ of W_K in its normalizer in W and in W_M and from the fusion $K' \sim K$ of parabolic subgroups, i.e., the question which representatives of parabolic subgroups in W_M are conjugate in W. Both ingredients are known from the preceding section. The fusion problem can be decided by Theorem 2.3.3 and the sizes of normalizers are found in Propositions 2.3.8, 2.3.10 and 2.3.13 and in Tables A.1 and A.2.

Consider the case $M = J$ in Proposition 2.4.7. Then $\beta^M_{JK'} = 1$ for all $K' \subseteq J$ and the mark β_{JK} is given by the formula

$$\beta_{JK} = \sum_{K' \sim K} \left| N_W(W_{K'}) : N_{W_J}(W_{K'}) \right|. \qquad (*)$$

We will use this approach here to derive formulas for the marks of the Coxeter groups of classical type. The parabolic tables of marks of the Coxeter groups of exceptional type are listed in Appendix A.

2.4.8. Let (W, S) be of type A_{n-1}. Let $J \subseteq S$ have $\lambda = [\lambda_1, \dots, \lambda_n] \vdash n$ and let $K \subseteq S$ have $\mu = [1^{m_1}, \dots, n^{m_n}] \vdash n$ as associated labels. Moreover, let K' represent a Coxeter class of W_J which lies in the Coxeter class of K in W. Then K' is a direct product of factors labelled ρ^p, $p = 1, \dots, n$, where each $\rho^p = [1^{r_{p1}}, 2^{r_{p2}}, \dots, n^{r_{pn}}]$ is a partition of λ_p such that the total number of parts q equals m_q for $q = 1, \dots, n$. For the index in the normalizer we have, by Proposition 2.3.8,

$$|N_W(W_K) : W_K| = m_1!\, m_2! \cdots m_n! \quad \text{and} \quad \left|N_{W_J}(W_{K'})\right| = \prod_{p,q} r_{pq}!$$

Thus, by $(*)$, if J has label $\lambda = [\lambda_1, \dots, \lambda_n]$ and K has label $\mu = [1^{m_1}, \dots, n^{m_n}]$ then $\beta_{JK} = \beta_1(\lambda, \mu)$ if we set

$$\beta_1(\lambda, \mu) = \sum_{(r_{pq})} \prod_{q=1}^{n} \frac{m_q!}{r_{1q}! \cdots r_{nq}!} = \sum_{(r_{pq})} \prod_{q=1}^{n} \binom{m_q}{r_{1q}, \dots, r_{nq}},$$

vhere the sum is over all $n \times n$ matrices (r_{pq}) with integer entries $r_{pq} \geq 0$ such hat

$$\sum_{j=1}^{n} j r_{pj} = \lambda_p \quad (p = 1, \dots, n), \quad \text{and} \quad \sum_{i=1}^{n} r_{iq} = m_q \quad (q = 1, \dots, n).$$

4.9. Consider the Coxeter group $W = W_n = \langle S \rangle$ of type B_n and let $J, K \subseteq S$. sume that the Coxeter class of J is labelled by the partition $\lambda = [\lambda_1, \dots, \lambda_n]$ $n_J \leq n$, and the class of K by the partition $\mu = [1^{m_1}, \dots, n^{m_n}]$ of $n_K \leq n$. The class of a subset $K' \subseteq J$ is then labelled by a sequence $(\rho^0, \rho^1, \dots, \rho^n)$ of titions ρ^p, where $\rho^0 = [1^{r_{01}}, 2^{r_{02}}, \dots, n^{r_{0n}}]$ is a partition of some $\lambda_0 \leq n - n_J$ $\rho^p = [1^{r_{p1}}, 2^{r_{p2}}, \dots, n^{r_{pn}}]$ is a partition of λ_p for $p = 1, \dots, n$. The subset s conjugate to K in W if, for each $q = 1, \dots, n$, the total number of parts q q.

For the index in the normalizer we have, by Propositions 2.3.8 and 2.3.10,

$$\frac{|N_W(W_K)|}{|W_K|} = \prod_{q=1}^{n} 2^{m_q} m_q! \quad \text{and} \quad \frac{|N_{W_J}(W_{K'})|}{|W_{K'}|} = \prod_{q=1}^{n} 2^{r_{0q}} r_{0q}!\, r_{1q}! \cdots r_{nq}!$$

Thus, by $(*)$, if J has label $\lambda = [\lambda_1, \ldots, \lambda_n]$ and K has label $\mu = [1^{m_1}, \ldots, n^{m_n}]$ then $\beta_{JK} = \beta_2(\lambda, \mu)$ if we set

$$\beta_2(\lambda,\mu) = \sum_{(r_{pq})} \prod_{q=1}^{n} \frac{2^{m_q} m_q!}{2^{r_{0q}} r_{0q}!\, r_{1q}! \cdots r_{nq}!} = \sum_{(r_{pq})} \prod_{q=1}^{n} 2^{m_q - r_{0q}} \binom{m_q}{r_{0q}, \ldots, r_{nq}},$$

where the sum is over all $(n+1) \times n$ matrices (r_{pq}) with integer entries $r_{pq} \geq 0$ such that

$$\sum_{j=1}^{n} j r_{pj} = \lambda_p \quad (p = 0, \ldots, n) \quad \text{and} \quad \sum_{i=0}^{n} r_{iq} = m_q \quad (q = 1, \ldots, n).$$

2.4.10. Consider the Coxeter group $W = W_n' = \langle S \rangle$ of type D_n and let $J, K \subseteq S'$. Assume that J has label λ or (λ, ϵ) and that K has label μ or (μ, δ). As before, we consider W_n' as a subgroup of the Coxeter group $W_n = \langle S \rangle$ of type B_n where $S = \{t, s_1, \ldots, s_{n-1}\}$ and $S' = \{u, s_1, \ldots, s_{n-1}\}$ with $u = t s_1 t$. Then t acts on S' as the transposition (u, s_1) and $J^t = J$ unless J contains exactly one of $\{u, s_1\}$. It follows from Exercise 2.14 that the mark β_{JK} in this case is determined by the following conditions:

$$\begin{aligned} \beta_{JK} &= \beta_2(\lambda,\mu) \quad \text{if } \lambda \vdash m < n-1, \\ \beta_{JK} + \beta_{J^tK} &= \beta_2(\lambda,\mu) \quad \text{if } \lambda \vdash n, \end{aligned}$$

and $\beta_{J^tK} = \beta_{JK}$ unless the class of J has label (λ, ϵ) and the class of K has label (μ, δ), where $\beta_{J^tK} = 0$ if $\epsilon = \delta$.

2.5 BIBLIOGRAPHICAL REMARKS AND EXERCISES

Deodhar's lemma is Lemma 3.2 in [Deodhar 1977]. Coxeter groups of finite rank are shown in [Brink and Howlett 1993] to be automatic groups (in the sense of [Epstein et al. 1992]). This means that tasks like recognizing a normal form and multiplication of elements can in principle be handled by finite-state machines. Examples of such automata for finite Coxeter groups are constructed in [du Cloux 1990] and [du Cloux 1999] and also in [Casselman 1994] for affine Coxeter groups. Exercises 2.4, 2.5 and 2.6 are concerned with the normal form of elements; in particular, Exercise 2.6 illustrates the role played by Deodhar's lemma. See [Epstein et al. 1992] and also [Sims 1994] for descriptions of the Knuth–Bendix procedure and how it is used to produce confluent rewriting systems which solve the word problem in more general situations than that of a finite Coxeter group.

The descent algebra was first discovered in [Solomon 1976]. There also appears a proof of Theorem 2.1.12. The proof we give of Solomon's Theorem 2.1.10 originates from [Bergeron et al. 1992], where the parabolic Burnside ring is also introduced. The descent algebra associated with the symmetric group has been intensively studied in [Garsia and Reutenauer 1989], and other articles including [Atkinson 1986] and [Atkinson 1992]. The elements of this algebra are also known under different names, like "signature classes" or "zigzags". The subject of [Atkinson and van Willigenburg 1997] is the descent algebra over fields of positive characteristic.

Theorem 2.3.3 appears as Lemma 2.12 in [Lusztig and Spaltenstein 1979]. A proof in terms of root systems is contained in [Howlett 1980]. The result is extended to infinite Coxeter groups in [Deodhar 1982]. Coxeter classes are also discussed in [Richardson 1982]. Our proof of Theorem 2.3.3 here is purely combinatorial and does not depend on a case-by-case analysis.

A variant of our parabolic table of marks appears in the context of hyperplane arrangements in [Orlik and Solomon 1983].

In a completely different way, parabolic subgroups can also be characterized as the subgroups of W which occur as the (pointwise) stabilizer of a subset in the natural reflection representation of W; see [Steinberg 1968, (1.20)].

Parabolic subgroups of W are examples of subgroups which are generated by reflections, i.e., conjugates of the generators S of W. These *reflection subgroups* are Coxeter groups as well. In [Deodhar 1989], [Dyer 1987] and [Dyer 1990] a canonical way to identify a generating system in the sense of the Coxeter presentation is described. In this book reflection subgroups which are not parabolic subgroups occur in Section 5.5. In general, however, they do not give rise to parabolic subalgebras in the Iwahori–Hecke algebra in the sense of Section 9.1.

Throughout the following exercises we assume that (W, S) is a Coxeter system such that W is finite.

Exercise 2.1. For $J, K \subseteq S$ show that $J \subseteq K$ if and only if $X_K \subseteq X_J$.

Exercise 2.2. A set of subgroups of a group G that is closed under conjugation and intersection is called a *Mackey system*. Show that

$$\mathcal{M} = \{W_J^w \mid J \subseteq S, w \in W\},$$

the set of all conjugates of the parabolic subgroups of W, is closed under intersection of subgroups.

Exercise 2.3 ([Solomon 1976]). For each $K \subseteq S$ define

$$Y_K = \{w \in W \mid K = \{s \in S \mid l(sw) > l(w)\}\} \quad \text{and} \quad y_K = \sum_{x \in Y_K} x^{-1}.$$

Show that

(i) $X_J = \coprod_{K \supseteq J} Y_K$,
(ii) $y_K = \sum_{J \supseteq K} (-1)^{|J-K|} x_J$,
(iii) $w_0 = \sum_{J \subseteq S} (-1)^{|J|} x_J$,
(iv) $Y_K w_0 = Y_{\hat{K}}$ and $w_0 y_K = y_{\hat{K}}$, where $\hat{K} = S - K$.

Exercise 2.4. This exercise and the next two are concerned with a normal form for the elements of W. Suppose $S = \{s_1, \dots, s_n\}$ and let $S_i = \{s_1, \dots, s_i\}$ for $i = 0, \dots, n$. Consider the chain

$$1 = W_0 \leq W_1 \leq \cdots \leq W_n = W$$

of parabolic subgroups $W_i = \langle S_i \rangle$ and denote by $X_i = X_{S_{i-1}}^{S_i}$ the set of distinguished right coset representatives of W_{i-1} in W_i.

(i) Show that each $w \in W$ has a unique expression of the form $w = x_1 \cdots x_n$ where $x_i \in X_i$ $(i = 1, \dots, n)$. In other words, W decomposes as

$$W = X_1 \cdot X_2 \cdots X_n.$$

(ii) Let $w \in W$ with $w = x_1 \dots x_n$ for $x_i \in X_i$ and let $s \in S$. Then $ws = x_1 \dots x_{i-1} x_i' x_{i+1} \dots x_n$ for some i, where $x_i' = x_i u \in X_i$ for some $u \in S_i$.

Exercise 2.5. We define a total ordering on the set of all words over $S = \{s_1, \dots, s_n\}$ by ordering the words by length, and words of equal length lexicographically with respect to the ordering $s_1 < s_2 < \cdots < s_n$ on S. This ordering is usually called the *ShortLex* ordering (see [Epstein et al. 1992]). Let S_i, W_i and X_i be as in Exercise 2.4.

(i) Show that the following procedure yields a spanning tree for Γ_J. Start with a copy of the graph Γ_J. Then, while there is a vertex v that is the end point of more than one edge, delete an edge as follows. Choose two edges leading to v labelled by s_i and s_j such that $i < j$. If m_{ij} is odd then delete the edge labelled s_j, otherwise the edge labelled s_i.
(ii) Show that, for each vertex v, the path from the initial vertex v_0 to v in the spanning tree of Γ_J corresponds to the smallest expression with respect to the ShortLex ordering for the coset representative $x \in X_J$ corresponding to v.
(iii) Define the *normal form* of $w \in W$ inductively as follows. The normal form of $1 \in W$ is 1, the empty word. If $w = w_{n-1} x_n$ for $w_{n-1} \in W_{n-1}$ and $x_n \in X_n$ then the normal form of w is the normal form of w_{n-1} (known by induction), followed by the normal form of x_n which we define to be the word corresponding to x_n in the spanning tree of Γ_{J_n}. Show that the normal form of w is the smallest expression for w with respect to the ShortLex ordering.

[*Hint.* If both s_i and s_j are suffixes of $x \in X_J$ then the longest word w_{s_i,s_j} is a suffix of x. Of the two reduced expressions for w_{s_i,s_j} the lexicographically smaller one is the one starting with s_i. It ends in s_i if m_{ij} is odd and in s_j otherwise.]

Exercise 2.6. Let S_i, W_i and X_i be as in Exercise 2.4. As before, for each $w \in W$, let the normal form of w be the smallest expression for w in the ShortLex ordering. With respect to this normal form for the elements of W, the rules

$$s_j^2 \longrightarrow 1$$
$$\mathsf{Prod}(s_j, s_i; m_{ij}) \longrightarrow \mathsf{Prod}(s_i, s_j; m_{ij})$$

for $j > i$, plus for $i \le n$ the tables

$$xs \longrightarrow ux \quad x \in X_i,\ s \in S_i,\ u = xsx^{-1} \in S_{i-1}$$

whenever $xs \notin X_i$, form a *confluent rewriting system* (see [Sims 1994]) for W.

[*Hint.* An element $w \in W$ is in normal form if and only if none of the above rules applies. On the other hand, given any word w in S, its normal form can be obtained by systematically applying the rules: given w in normal form, then $w = x_1 x_2 \dots x_n$, and the normal form is compatible with the coset decomposition. Now let $s \in S$. Then the normal form of ws is found as follows. If $x_n s$ is in the tables, then we are finished by induction on n. Otherwise $x_n s \in X_n$ and it only remains to find the normal form of $x_n s$. This is done by applications of the rules $s_j s_i \cdots \longrightarrow s_i s_j \cdots$ and $s_i^2 \longrightarrow 1$.]

Exercise 2.7. Let $J \subseteq S$ and define, for any $s \in S$, a permutation $\sigma_J(s)$ of the vertices X_J of Γ_J as

$$v.\sigma_J(s) = \begin{cases} vs & \text{if } vs \in X_J, \\ v & \text{otherwise.} \end{cases}$$

Show that this is a permutation representation of W, and that it is equivalent to the action of W on the (right) cosets of W_J.

Exercise 2.8. For $J \subseteq S$ show that $d_J^2 = 1$ if and only if $J^{d_J} = J$.

Exercise 2.9. Let $J \subseteq M \subseteq S$ and denote by $d_J^M = w_J w_M$ the longest coset representative of W_J in W_M. Moreover, let $w \in W$. Show that the following are equivalent:

(i) d_J^M is a prefix of w.
(ii) $w \in X_J$ and w_M is a prefix of $w_J w$.
(iii) $d_J^{J\cup\{s\}}$ is a prefix of w for all $s \in M - J$.

[*Hint.* First, show that if d_J^M is a prefix of w then $l(sw) > l(w)$ for each $s \in J$, whence $w \in X_J$. It then follows that $l(w_J w) = l(w_J) + l(w)$. There exists an element $w' \in W$ such that $w = d_J^M w'$ and $l(w) = l(d_J^M) + l(w')$. Thus $w_J w = w_J d_J^M w' = w_M w'$ where $l(w_M w') = l(w_M) + l(w')$. From this follows the equivalence of (i) and (ii). Now the fact that w_M is a prefix of $w_J w$ is equivalent to the fact that each $s \in M$ is a prefix of $w_J w$. As seen before, $d_J^{J\cup\{s\}}$ is a prefix of w if and only if $w \in X_J$ and t is a prefix of w_J for all $t \in J$ and $t = s$. From this follows the equivalence of (ii) and (iii).]

Exercise 2.10. Given a directed labelled graph Γ with vertices V and edges E, let the *reversed graph* $\operatorname{rev}\Gamma$ be the graph which is obtained from Γ by reversing the edges. That is, $\operatorname{rev}\Gamma$ has vertex set V and edge set $\{v_2 \xrightarrow{s} v_1 \mid v_1 \xrightarrow{s} v_2 \in E\}$.

Show that for $J \subseteq S$ the reversed graph $\operatorname{rev}\Gamma_J$ is isomorphic to the coset graph $\Gamma_{J'}$ where $J' = J^{w_0}$.

Exercise 2.11. Let $J \subseteq S$ and for $x \in X_J$ define the *valency* $\operatorname{val}_J(x)$ of the vertex x in Γ_J as the number of edges of the coset graph Γ_J that involve the vertex x by setting $\operatorname{val}_J(x) = |\{s \in S \mid xs \in X_J\}|$. Furthermore, for a vertex v of Γ_J define $\Gamma_J(v)$ to be the *shifted coset graph* with respect to v, i.e., $\Gamma_J(v)$ is identical to Γ_J except that the direction of every edge is chosen to point away from v: if $x \xrightarrow{s} xs$ in Γ_J then $x \xrightarrow{s} xs$ in $\Gamma_J(v)$ if $l(v^{-1}x) < l(v^{-1}xs)$ and $xs \xrightarrow{s} x$ otherwise. Examples are $\Gamma_J = \Gamma_J(1)$ and $\operatorname{rev}\Gamma_J = \Gamma_J(d_J)$ by Exercise 2.10.

(i) Show that each vertex v of Γ_J has valency $\operatorname{val}_J(v) \geq \operatorname{val}_J(1) = |S - J|$.

(ii) For each vertex v of Γ_J with valency $\operatorname{val}_J(v)$ equal to $\operatorname{val}_J(1)$ show that the shifted graph $\Gamma_J(v)$ is isomorphic to the coset graph Γ_K, where $K = \{s \in S \mid vs \notin X_J\}$. Moreover, K is conjugate to J and every conjugate of J in S arises in this way.

[*Hint*. For $K \subseteq S$ consider the action of W_K on the cosets of W_J. Every fixed point of that action corresponds to a connected component of size 1 in $\Gamma_J|_K$. If W_K has fixed points on Γ_J then W_K is conjugate to a subgroup of W_J.]

Exercise 2.12. Show that the number of compositions of the set $\{1, \ldots, n\}$ that give rise to the partition $\lambda = [1^{l_1}, \ldots, n^{l_n}]$ of n is given by the multinomial coefficient $\binom{l_1+\cdots+l_n}{l_1,\ldots,l_n}$. Conclude that, if (W, S) is of type B_n and $J \subseteq S$ then the product of the size p_J of the Coxeter class of J and the index $|X_{JJJ}|$ of W_J in its normalizer in W is $2^r r!$, where $r = |S - J|$.

Exercise 2.13. Suppose that (W, S) is of type A_{n-1} and let $J \subseteq S$. Show that the Coxeter class of J is uniquely determined by the values $\beta_{JK}/\beta_{J\emptyset}$ where K runs through the subsets $\{s_1, \ldots, s_k\}$ of S, $k = 0, \ldots, n-2$.

[*Hint*. The partition corresponding to $K = \{s_1, \ldots, s_k\}$ is $(1^{n-k}, k)$. Let $\lambda = (\lambda_1, \ldots, \lambda_n) = (1^{l_1}, \ldots, n^{l_n})$ correspond to J. Then, by (2.4.8), we have

$$\frac{\beta_{JK}}{\beta_{J\emptyset}} = \frac{\lambda_1! \cdots \lambda_n!}{n!} \sum_{j \geq k} l_j \frac{(n-k)!}{\lambda_1! \cdots \lambda_n!} \frac{j!}{(j-k)!} = \binom{n}{k}^{-1} \sum_{j \geq k} l_j \binom{j}{k}$$

and the number l_k of parts k in λ is determined by $l_{k+1}, \ldots, l_n$.]

Exercise 2.14. Let G be a finite group, let $H \leq G$ be a normal subgroup of index 2 and let X be a transitive G-set. Then, by restriction, X is an H-set. Show that, if X is not a transitive H-set, then $X = X_1 \amalg X_2$, where X_1 and X_2 are transitive H-sets which are in bijection via $X_2 = X_1.g = \{x.g \mid x \in X_1\}$ for any $g \in G - H$. Moreover, in that case, $X_1 \not\cong X_2$ as H-sets if and only if the normalizer in G of a point stabilizer is a subgroup of H.

[*Hint.* X as an H-set is transitive if and only if the point stabilizer is not contained in H.]

Exercise 2.15. Let W'_n be a Coxeter group of type D_n considered as a subgroup of a Coxeter group W_n of type B_n. Let $Z = \{s_{n-1}s_{n-2}\cdots s_i \mid n \geq i \geq 1\}$. Then

$$X = Z \amalg \{s_{n-1}\cdots s_2 u s_1 s_2 \cdots s_j \mid 0 \leq j \leq n-1\}$$

is the set of distinguished right coset representatives of W'_{n-1} in W'_n and

$$Y = Z \amalg \{s_{n-1}\cdots s_2 s_1 t s_1 s_2 \cdots s_j \mid 0 \leq j \leq n-1\}$$

is the set of distinguished right coset representatives of W_{n-1} in W_n. Let $x \in X$ and show that $W'_{n-1}x \subseteq W_{n-1}y$ for some $y \in Y$ where

$$y = \begin{cases} x & \text{if } x \in Z, \\ tx & \text{otherwise.} \end{cases}$$

[*Hint.* If $x \in Z$ then $x \in Y$. Otherwise, $x = ty$, for some $y \in Y$.]

3
Conjugacy classes and special elements

In this chapter we study the distribution of the elements of a finite Coxeter group W into conjugacy classes. We derive structural properties of the conjugacy classes with respect to the length function of W and achieve a full classification of the conjugacy classes of W. Our point of view is inductive with respect to parabolic subgroups. If everything is known about the conjugacy classes of the proper parabolic subgroups of W then the interesting conjugacy classes are those which do not contain an element of a proper parabolic subgroup. We call them the *cuspidal* classes of W. The two other important concepts in this chapter are *minimal length* (in a conjugacy class) and *cyclic shift* (in the sense of cyclically permuting the factors in a reduced expression). One of our main results shows that the elements of minimal length in a cuspidal class form a single cyclic shift class. Another one shows that cuspidal classes are "rigid" in the sense that elements of different cuspidal classes of W will not be conjugate in any Coxeter group which contains W as a parabolic subgroup.

More generally, various length considerations suggest studying yet another graph associated with W, a graph defined with respect to conjugation by single generators and the length function on the vertex set W. This graph is introduced in Section 3.2 as the *conjugacy graph* of (W, S). The main result about that graph is formulated in Theorem 3.2.7. It has important consequences for the classification of conjugacy classes and reveals a number of fundamental properties of the elements of minimal length in the conjugacy classes of W. These properties lay the foundations for the definition of the character table of the Iwahori–Hecke algebra in Chapter 8. A complete description of the conjugacy classes of W in terms of cuspidal classes and parabolic subgroups is given in Theorem 3.2.12. This result allows us to formulate an algorithm which, given an element w of W, determines the conjugacy class of w in W.

In Section 3.1 the notion of a cuspidal class is formally defined. The Coxeter elements of W are a first example of a cyclic shift class of minimal elements in a cuspidal class. Furthermore, permutation characters on cosets of parabolic subgroups are identified with rows of the parabolic table of marks from Section 2.4.

The proof of Theorem 3.2.7 will be carried out case by case according to the classification of finite irreducible Coxeter groups in Sections 3.3 and 3.4. Section 3.3 provides the tools for a particular Coxeter group W which are then applied to the Coxeter groups of exceptional type and in Section 3.4 to the Coxeter groups of classical type.

3.1 CUSPIDAL CLASSES

Throughout this chapter we assume W to be a finite Coxeter group with generating set S. We begin with the definition that lies at the heart of our approach to the conjugacy classes of W.

3.1.1. Definition. A conjugacy class C of W is called a *cuspidal class* if

$$C \cap W_J = \varnothing \quad \text{for all proper subsets } J \subset S.$$

Furthermore, for every conjugacy class C of W, we denote by

$$C_{\min} = \{w \in C \mid l(w) \leq l(v) \text{ for all } v \in C\}$$

the set of elements of minimal length in C.

The aim of this section is to derive some properties and characterizations of cuspidal classes, and to describe the particular role that is played by the elements of minimal length in a conjugacy class.

Every Coxeter group W contains elements which can be formed as a product over all the generators in S, in some order or other. We have seen such elements already in Section 1.5 where they were used to obtain a reduced expression for the longest element of W. This set of elements turns out to be a typical set of elements of minimal length in a cuspidal conjugacy class of W.

3.1.2. Coxeter elements. A *Coxeter element* of (W, S) is a product of all $s \in S$ in any given order. Thus if $S = \{s_1, \ldots, s_n\}$ then $w_c = s_1 \cdots s_n$ is a Coxeter element of W and any permutation of the s_i in the word w_c gives a Coxeter element, although not necessarily a different one (cf. Exercise 3.2, where the total number of Coxeter elements of W is determined). If (W, S) is of type A_{n-1} then $W \cong \mathfrak{S}_n$ and every Coxeter element of W is an n-cycle. The n-cycles of $\mathfrak{S}_n$ form a conjugacy class in W, although in general not every n-cycle is a Coxeter element. Note that every element $w \in W$ which is a product of, say, k distinct generators in S has length $l(w) = k$. This follows from the fact that such an element can be seen as a product of distinguished coset representatives of length 1 in a suitable chain of parabolic subgroups. Thus, in particular, $l(w_c) = |S|$ for a Coxeter element w_c of (W, S).

An important fact about Coxeter elements is that they are all conjugate in W. But we can go further: any two Coxeter elements are conjugate in a very special way, which is defined as follows.

3.1.3. Definition (Broué–Michel). We say that $w, w' \in W$ are conjugate by *cyclic shift* if there exist a sequence $w = v_0, v_1, \ldots, v_m = w'$ of elements of W and elements x_i, y_i in W such that

$$v_{i-1} = x_i y_i, \quad v_i = y_i x_i, \quad \text{and} \quad l(v_{i-1}) = l(x_i) + l(y_i) = l(v_i)$$

for all $i = 1, \ldots, m$.

Note that every element $w \in W$ which is conjugate to a Coxeter element of W by cyclic shift is itself a Coxeter element. We will see an alternative characterization of cyclic shift in Section 3.2.

3.1.4. Theorem. *Any two Coxeter elements of* (W, S) *are conjugate in* W *by cyclic shift.*

Proof. If S is empty or a singleton set then W has only one Coxeter element and there is nothing to show.

So suppose $|S| = n \geq 2$ and write $S = \{s_1, \dots, s_n\}$ where the labelling is chosen in such a way that s_n is linked to at most one other generator in the Coxeter graph of (W, S). (This is possible since the Coxeter graph of any finite Coxeter group does not contain loops; see Table 1.1 on p. 16.) Hence s_n does not commute with at most one generator in S.

Next, let w be an arbitrary Coxeter element of W, choose a reduced expression of w and write $w = as_nb$, where a is the product of the factors to the left of s_n in the reduced expression and b is the product of the factors to the right. By our choice of s_n, it commutes either with a, whence $w = s_nab$, or with b, whence $w = abs_n$. Clearly, s_nab and abs_n are conjugate by cyclic shift.

Now ab is a Coxeter element of the parabolic subgroup W_J where $J = S - \{s_n\}$. By induction on $|S|$, the particular Coxeter element $s_1 \cdots s_{n-1}$ of W_J and ab are conjugate by cyclic shift. Hence there exist a sequence $v_0, v_1, \dots, v_m$ of elements of W_J and elements x_i, y_i in W_J such that $ab = v_0$, $s_1 \cdots s_{n-1} = v_m$,

$$v_{i-1} = x_i y_i, \quad v_i = y_i x_i, \quad \text{and} \quad l(v_{i-1}) = l(x_i) + l(y_i) = l(v_i)$$

for all $i = 1, \dots, m$. Note that all the elements v_i are then Coxeter elements of W_J. It follows that w and $s_1 \cdots s_n$ are conjugate by cyclic shift if we can show that, for each i, the elements $x_i y_i s_n$ and $y_i x_i s_n$ are conjugate by cyclic shift.

So fix an index $i \in \{1, \dots, m\}$. Since $x_i y_i s_n$ is a Coxeter element of W where every $s_i \in S$ occurs exactly once as a factor, we have that $x_i y_i s_n$ and $y_i s_n x_i$ are conjugate by cyclic shift. Now s_n commutes either with x_i, whence $y_i s_n x_i = y_i x_i s_n$, or with y_i, whence $y_i s_n x_i = s_n y_i x_i$. Clearly, $s_n y_i x_i$ and $y_i x_i s_n$ are conjugate by cyclic shift. □

The conjugacy class of W that contains the Coxeter elements is a cuspidal class. Indeed this follows from counting the fixed points of a Coxeter element in the action of W on the cosets of a parabolic subgroup W_J.

3.1.5. Lemma. *Let* $J \subseteq S$ *and let* w_c *be a Coxeter element of* W. *Then* w_c *fixes a point in the action of* W *on the cosets of* W_J *if and only if* $J = S$.

Proof. Let $x \in X_J$. We have to show that $W_J x w_c = W_J x$ if and only if $J = S$. If $J = S$ then $W_J = W$ and of course $Ww = W$ for every $w \in W$.

Conversely, suppose $W_J x w_c = W_J x$ and write $w_c = s_1 \cdots s_n$ where $S = \{s_1, \dots, s_n\}$. Let $x_0 = x$ and let $x_i \in X_J$, $u_i \in W_J$ such that $x_{i-1} s_i = u_i x_i$, for $i = 1, \dots, n$. Then

$$xw_c = x_0 s_1 \cdots s_n = u_1 \cdots u_i x_i s_{i+1} \cdots s_n = u_1 \cdots u_n x_n.$$

By Deodhar's Lemma 2.1.2 we have either $u_i \in J$ and $x_i = x_{i-1}$ or $u_i = 1$ and $x_i = x_{i-1}s_i$. According to this we set either $t_i = 1$ or $t_i = s_i$ such that $x_i = x_{i-1}t_i$, for all i. Then $t_1 \cdots t_n$ is a Coxeter element of some parabolic subgroup W_K of W and $x_0 t_1 \cdots t_n = x_n$. By our hypothesis $W_J x w_c = W_J x$ we must have $x = x_0 = x_n$ and therefore $t_1 \cdots t_n = 1$. But the only parabolic subgroup of W which has a Coxeter element equal to 1 is W_K for $K = \varnothing$. Therefore $t_i = 1$ and $x_i = x$ for all i, and $u_i = x s_i x^{-1} \in J$ for all i. It follows that $J = xSx^{-1} = S$. □

We conclude the treatment of the special case of Coxeter elements by summarizing the results as follows.

3.1.6. Proposition. *Let* C *be the conjugacy class of* W *containing the Coxeter elements. Then*

(i) C *is a cuspidal class;*
(ii) $C_{\min} = \{w \in W \mid w$ *is a Coxeter element of* $W\}$.

Proof. (i) As an immediate consequence of Lemma 3.1.5 we find that no conjugate of a Coxeter element of W has a fixed point in the action of W on the cosets of a proper parabolic subgroup W_J of W, and therefore $C \cap W_J = \varnothing$ (since any $w \in W_J$ fixes the coset W_J).

(ii) By (i), any reduced expression for any element of C involves all generators of W, whence $l(w) \geq |S|$ for all $w \in C$. Consequently, $C_{\min}$ coincides with the set of Coxeter elements. □

3.1.7. Permutation characters. For each $J \subseteq S$ we define a map π_J as

$$\pi_J : W \to \mathbb{N}_0, \quad w \mapsto \pi_J(w) = |\mathrm{Fix}_w(W/W_J)|$$

and call π_J the *permutation character* of W on the cosets of W_J.

Recall from Section 2.4 that the entry β_{JK} in the parabolic table of marks of W is the number of fixed points of the parabolic subgroup W_K of W in the action of W on the cosets of W_J. We can now identify β_{JK} as value of the permutation character π_J. (See Exercise 3.3 for a different solution of this problem.) Recall also from Corollary 1.2.3 that, for any $w \in W$ we define $J(w) = \{s_1, \ldots, s_k\}$ if $w = s_1 \cdots s_k$ is a reduced expression.

3.1.8. Proposition. *Let* $w \in W$ *and let* C *be the conjugacy class of* w *in* W. *Let* $w' \in C_{\min}$ *and* $K = J(w')$. *Then* $\pi_J(w) = \beta_{JK}$ *for all* $J \subseteq S$. *In particular,* $\pi_J(c_K) = \beta_{JK}$ *if* c_K *is a Coxeter element of* W_K, $K \subseteq S$.

Proof. We may assume that $w \in C_{\min}$ and that $K = J(w)$ since π_J is constant on C. Let $J \subseteq S$ and let $\Omega = W/W_J$. Then $\mathrm{Fix}_{W_K}(\Omega) = \beta_{JK}$. Clearly, $\mathrm{Fix}_{W_K}(\Omega) \subseteq \mathrm{Fix}_w(\Omega)$ since $w \in W_K$. We will show that $\mathrm{Fix}_w(\Omega) \subseteq \mathrm{Fix}_s(\Omega)$ for all $s \in K$. It then follows that $\mathrm{Fix}_w(\Omega) = \mathrm{Fix}_{W_K}(\Omega)$ (see Exercise 3.5), whence $\pi_J(w) = \beta_{JK}$.

So let $x \in X_J$ such that $W_J x \in \Omega$ is fixed by w. Then $xw \in W_J x$ and we have to show that $xs \in W_J x$ for every $s \in K$. To that end we write $w = s_1 \cdots s_r$ as a reduced expression for certain $s_i \in S$. Then $\{s_1, \ldots, s_r\} = J(w) = K$. Define $x_0 = x$ and, for $i = 1, \ldots, r$, let $x_i \in X_J$ such that $x_{i-1}s_i = u_i x_i$, for some $u_i \in W_J$. Then $x_r = x_0 = x$. By Deodhar's Lemma 2.1.2, either $u_i \in J$ and $x_i = x_{i-1}$, or $u_i = 1$ and $x_i = x_{i-1}s_i$. Let $u = u_1 \cdots u_r$. Then $xw = ux_r = ux$ and we see that $u^x = w$. Now $l(u) \geq l(w) = r$ since $w \in C_{\min}$. Therefore, we must have $u_i \in J$ and $x_i = x$ for all i. But then $xs_i \in W_J x$ for all $s_i \in J(w) = K$, as required. □

This leads us to a first characterization of cuspidal classes in terms of permutation character values.

3.1.9. Corollary. *Let $w \in W$. Then the conjugacy class C of w in W is cuspidal if and only if $\pi_J(w) = 0$ for all proper subsets $J \subset S$.*

Proof. If $\pi_J(w) = 0$ for all $J \subset S$ then w is not contained in any conjugate of W_J, for all $J \subset S$. Hence the conjugacy class of w in W contains no element of W_J and is by definition cuspidal.

Conversely, assume that the class of w is cuspidal. Then $J(w) = J(w') = S$ for each element w' of minimal length in the conjugacy class of w. Furthermore, we have $J(w_c) = S$ as well, for a Coxeter element w_c of W. By Proposition 3.1.8 $\pi_J(w) = \pi_J(w_c)$, and $\pi_J(w_c) = 0$ for all $J \subset S$ by Lemma 3.1.5. □

A similar result shows that if we consider a realization of W as a subgroup of $\mathsf{GL}(V)$ where the elements of S are reflections (see Section 1.1 and (1.2.8)), then the fixed point space of an element in a cuspidal class is trivial.

3.1.10. Lemma. *Assume that $W \subset \mathsf{GL}(V)$ as in Section 1.1. Let $w \in W$ and let $p_w(q) = \det(q\,\mathrm{id}_V - w) \in \mathbb{R}[q]$ be the characteristic polynomial of w on V. Then the conjugacy class of w in W is cuspidal if $p_w(1) \neq 0$.*

Proof. If $w \in W_J$ for some proper subset $J \subset S$ and if $p^J_w(q)$ is its characteristic polynomial on the natural module V_J of W_J then, by Exercise 3.11, $p_w(q) = (q-1)^{|S-J|}p^J_w(q)$. Therefore $p_w(1) = 0$ for the characteristic polynomial $p_w(q)$ of w and any of its conjugates in W. □

The converse of this lemma is also valid, $p_w(1) \neq 0$ for all w in cuspidal classes of W; see Exercise 3.16.

Whether or not a conjugacy class C is cuspidal can be decided by looking at a single element of minimal length in C by virtue of the following two results.

3.1.11. Corollary (Howlett). *Let C be a conjugacy class of W and let $w, w' \in C_{\min}$. Then $J(w)$ is conjugate to $J(w')$.*

Proof. Let $K = J(w)$ and $K' = J(w')$. Then, by Proposition 3.1.8, we have $\beta_{JK} = \beta_{JK'}$ for all $J \subseteq S$. And by Corollary 2.4.6, this is the case if and only if K is conjugate to K' in W. □

3.1.12. Proposition. *Let C be a conjugacy class of W. Then the following are equivalent:*

(i) *C is a cuspidal conjugacy class.*
(ii) *$J(w) = S$ for all $w \in C_{\min}$.*
(iii) *There exists an element $w \in C_{\min}$ such that $J(w) = S$.*

Proof. (i) $\Longrightarrow$ (ii): if C is a cuspidal class then, by definition, $J(w) = S$ for all $w \in C$. (ii) trivially implies (iii) since $C \neq \varnothing$. It remains to show (iii) $\Longrightarrow$ (i).

Let $w' \in C \cap W_J$ for a subset $J \subseteq S$. Let $u \in W$ such that $(w')^u = w$ and write $u = vx$ with $v \in W_J$ and $x \in X_J$. Then we also have $(w')^v \in C \cap W_J$ and we may assume that $v = 1$. By Lemma 2.1.14, we then have $l(w) = l((w')^x) \geq l(w')$. Since $w \in C_{\min}$ we conclude that $l(w) = l(w')$ and hence that $w' \in C_{\min}$. Now Corollary 3.1.11 implies $J(w') = S$. □

3.1.13. Corollary. *Let $w \in C_{\min}$, set $J = J(w)$ and let $v \in W_J$. Then $J(w^v) = J$.*

Proof. Since $w \in C_{\min}$ it also has minimal length in its conjugacy class C^J in W_J. By Proposition 3.1.12, the class C^J is cuspidal in W_J, whence $J(w^v) = J$ for all $v \in W_J$. □

We record the fact that the concept of elements of minimal length in W is compatible with the embedding of parabolic subgroups in W.

3.1.14. Lemma. *Let $J \subseteq S$ and let $w \in W_J$ be of minimal length in its conjugacy class in W_J. Then w has minimal length in its conjugacy class in W.*

Proof. Let $u \in W$ and $w' = w^u$. Write $u = vx$ for some $v \in W_J$ and $x \in X_J$. Then $w^v \in W_J$ and $l(w) \leq l(w^v)$ since w has minimal length in its class in W_J. Also $l(w^v) \leq l(w^{vx})$ by Lemma 2.1.14, since $x \in X_J$. Thus $l(w) \leq l(w')$ for all $w' \in C$. □

It follows from Lemma 3.1.14 that a Coxeter element c_K of W_K for any $K \subseteq S$ has minimal length in its conjugacy class in W, since c_K has minimal length in its class in W_K. From $J(c_K) = K$ we now get a criterion for when Coxeter elements of parabolic subgroups are conjugate. Recall from Corollary 2.1.13 that the subset J of S is conjugate to K if and only if W_J is conjugate to W_K.

3.1.15. Proposition. *Let $J, K \subseteq S$ and let c_J, c_K be Coxeter elements of W_J, W_K, respectively. Then c_J and c_K are conjugate in W if and only if J and K are conjugate in W.*

Proof. If J and K are conjugate in W, then c_J is certainly conjugate to a Coxeter element c of W_K, and c is conjugate to c_K by Theorem 3.1.4.

Conversely, if c_K and c_J are conjugate in W, they both are elements of minimal length in the same conjugacy class of W and thus, by Corollary 3.1.11, we have that $J = J(c_J)$ is conjugate to $K = J(c_K)$. □

3.1.16. Example. Suppose that (W, S) is of type A_{n-1}. Then W is isomorphic to the symmetric group $\mathfrak{S}_n$. We will illustrate the notions of "minimal length" and "cuspidal class" in this example. The *cycle structure* of an element $w \in \mathfrak{S}_n$ is the partition of n that is given by the lengths of the orbits of w in its action on $\{1, \ldots, n\}$. It is easy to see that conjugation in $\mathfrak{S}_n$ preserves the cycle structure and that any two elements with the same cycle structure are conjugate in $\mathfrak{S}_n$.

Moreover, let $J \subseteq S$ and let c_J be a Coxeter element of the parabolic subgroup W_J of W. Then c_J has cycle structure λ, where λ is the label of the Coxeter class of J according to Proposition 2.3.8. By Proposition 3.1.6, c_J has minimal length in its conjugacy class in W_J and by Lemma 3.1.14 it also has minimal length in its conjugacy class in W. It follows that, for (W, S) of type A_{n-1},

- an element of W has minimal length in its class if and only if it is a Coxeter element in some parabolic subgroup of W; and
- the conjugacy class containing the Coxeter elements of W (which are n-cycles, whence the class has label $[n]$), is the *only* cuspidal class of W.

Here, in the case of type A, the concepts of Coxeter elements and parabolic subgroups are sufficient to completely describe cuspidal classes and the elements of minimal length. In general, a Coxeter group will have more than one cuspidal class and not only Coxeter elements will occur as elements of minimal length.

3.2 THE CONJUGACY GRAPH

In the preceding section we have seen that elements of minimal length play a special role among all elements of a conjugacy class. In this section we study conjugacy classes with respect to the length function in more detail.

3.2.1. Definition. The *conjugacy graph* of (W, S) is the directed graph defined as follows. Its vertices are the elements of W. Given $w, w' \in W$ and $s \in S$ we have a labelled edge $w \xrightarrow{s} w'$, if $w' = sws$ and $l(w') \leq l(w)$.

As a consequence of Remark 1.1.5, we have for $w \in W$ and $s \in S$ that either $l(sws) = l(w)$ or $l(sws) = l(w) \pm 2$. Thus, by considering the edges of the conjugacy graph as directed edges we effectively exclude the case $l(sws) = l(w) + 2$ from the relation "being conjugate by a generator" on W.

3.2.2. Example. Let (W, S) be of type A_2 where $S = \{s_1, s_2\}$. Then $|W| = 6$ and the conjugacy graph of W is easily determined as in Figure 3.1. Here the

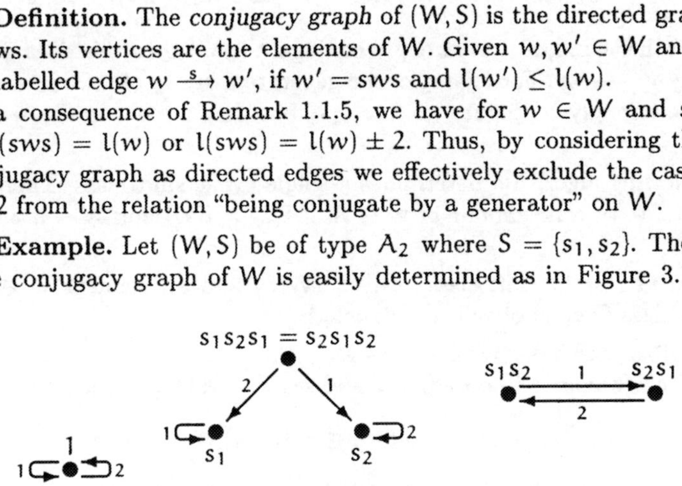

Fig. 3.1. The conjugacy graph of type A_2.

vertices of the graph are written as words in the generators S, while for the edge labels we write $1,2$ as abbreviations for s_1, s_2.

3.2.3. Definition. We denote by $\longrightarrow$ the pre-order relation induced on W by the conjugacy graph: we write $w \longrightarrow w'$ for $w, w' \in W$ if there are sequences $s_1, s_2, \ldots, s_r \in S$ and $w = w_0, w_1, \ldots, w_r = w'$ such that $w_i \in W$ and $w_{i-1} \xrightarrow{s_i} w_i$ for all $i = 1, \ldots, r$, i.e., if $w_{i-1}^{s_i} = w_i$ and $l(w_i) \leq l(w_{i-1})$. We also define

$$\mathsf{Cyc}(w) = \{v \in W \mid w \longrightarrow v \text{ and } v \longrightarrow w\}.$$

In Exercise 3.7 it is shown that $\mathsf{Cyc}(w)$ is precisely the cyclic shift class of w, in the sense of Definition 3.1.3.

3.2.4. Definition. We call $w, w' \in W$ *elementarily strongly conjugate* in W, if $l(w) = l(w')$ and there is an $x \in W$ such that $wx = xw'$ and $l(wx) = l(w) + l(x)$ or $xw = w'x$ and $l(xw) = l(x) + l(w)$. We call w and w' *strongly conjugate* if there is a sequence $w = w_0, w_1, \ldots, w_r = w'$ of elements $w_i \in W$ such that w_{i-1} is elementarily strongly conjugate to w_i.

3.2.5. Remark. We point out some useful relations. Note that if $w, w' \in W$ are such that $l(w) = l(w')$ and $w \longrightarrow w'$, then w, w' are strongly conjugate. Indeed, it is sufficient to consider the case where $w \neq w'$ and $w' = sws$ for some $s \in S$. We must show that $l(sw) > l(w)$ or $l(ws) > l(w)$. But this follows from Lemma 1.2.6 which shows that if $l(sw) = l(ws) < l(w)$ then necessarily $sws = w$.

In particular, this shows that elements in the same cyclic shift class are strongly conjugate. We also see that

$$\mathsf{Cyc}(w) = \{v \in W \mid w \longrightarrow v \text{ and } l(w) = l(v)\}. \qquad (*)$$

On the other hand, if w is strongly conjugate to w' then this does not necessarily imply $w \longrightarrow w'$. For example, the generators s_1, s_2 of a Coxeter group of type A_2 are strongly conjugate via $x = s_2 s_1$ but they are not in the same cyclic shift class.

The following algorithm determines a single cyclic shift class. There we call an element $z \in W$ a *neighbour* of $w \in W$ if $l(z) = l(w)$ and $w \xrightarrow{s} z$ for some $s \in S$.

Algorithm G (Cyclic shift class). Given (W, S) and an element $w \in W$, the cyclic shift class $\mathsf{Cyc}(w)$ of w is constructed.

G1. [Initialize.] Set $Y \leftarrow \{w\}$ and $X \leftarrow \varnothing$.

G2. [Find neighbours.] Choose $y \in Y$ and set $X \leftarrow X \cup \{y\}$. Then set

$$Z \leftarrow \{sys \mid s \in S,\, y \in Y,\, l(sys) = l(y)\} - X.$$

Finally set $Y \leftarrow (Y - \{y\}) \cup Z$. (We have $w \in X \cup Y \subseteq \mathsf{Cyc}(w)$. Moreover, $X \cup Y$ contains all neighbours of any $x \in X$.)

G3. [Loop.] Repeat step G2 until $Y = \varnothing$. Then return $\mathrm{Cyc}(w) = X$.

3.2.6. Definition. Let $w \in W$. The cyclic shift class $\mathrm{Cyc}(w)$ is called *terminal* if $w \longrightarrow w'$ implies $w' \in \mathrm{Cyc}(w)$ for all $w' \in W$. This does not depend on the choice of w as the representative. Indeed, let $v \in \mathrm{Cyc}(w)$ and suppose that $v \longrightarrow w'$ for some $w' \in W$. Then, by $(*)$, we have $w \longrightarrow v$, whence $w \longrightarrow w'$ and thus $w' \in \mathrm{Cyc}(w)$.

Clearly, if w is an element of minimal length in some conjugacy class C of W then $\mathrm{Cyc}(w)$ is a terminal cyclic shift class. Note that, for $w, w' \in W$ such that $w \longrightarrow w'$, we either have $w' \in \mathrm{Cyc}(w)$ or $l(w') < l(w)$. Consequently, for each $w \in W$ there exists some $w' \in W$ such that $w \longrightarrow w'$ and $\mathrm{Cyc}(w')$ is terminal.

It will turn out (Theorem 3.2.9(a)) that there are no "hidden" terminal cyclic shift classes in the conjugacy graph of (W, S): the graph has the strong property that every terminal cyclic shift class is a subset of C_{min} for some conjugacy class C of W.

We can now state the main theorem of this chapter which reveals important properties of cuspidal classes. For $s \in S$ and $w \in W$ we define, as in Exercise 1.15, the length function $l_s(w)$ as the number of generators in S conjugate to s occurring in a reduced expression of w.

3.2.7. Theorem. *For a finite irreducible Coxeter group $W = \langle S \rangle$ the following hold:*

(P1) *Let $w \in W$ be such that $J(w) = S$ and that $\mathrm{Cyc}(w)$ is terminal. Then the conjugacy class of w in W is a cuspidal class.*

(P2) *Let C be a cuspidal class of W. Then $C_{\mathrm{min}} = \mathrm{Cyc}(w)$ for any $w \in C_{\mathrm{min}}$.*

(P3) *Let C, C' be cuspidal classes of W and let $w \in C_{\mathrm{min}}$, $w' \in C'_{\mathrm{min}}$. Then $C = C'$ if and only if $p_w(q) = p_{w'}(q)$ and $l_s(w) = l_s(w')$ for all $s \in S$.*

We will prove Theorem 3.2.7 case by case for every type of irreducible finite Coxeter group. The general strategy follows three steps.

Step 1. We start with an irreducible Coxeter group $W = \langle S \rangle$ and determine a complete set of representatives w of terminal cyclic shift classes $\mathrm{Cyc}(w)$ such that $J(w) = S$. How this can be organized in an efficient way is the subject of the next section.

Step 2. Then we show that the conjugacy class of every such representative is indeed a cuspidal class of W, thus verifying property (P1). This will be achieved using the characteristic polynomial and Lemma 3.1.10.

Step 3. Next we show that any two different representatives are not conjugate in W. It then follows that a cuspidal class C contains only one terminal cyclic shift class which must then coincide with C_{min}. This establishes property (P2) for W. In order to distinguish the representatives, we simply check that (P3) holds for them. Note that this automatically establishes (P3) for any choice of representatives, by Exercise 3.8.

3.2.8. Example. Let (W, S) be of type $I_2(m)$, for $m \geq 5$, i.e., with the following Coxeter graph:

$I_2(m)$
$(m \geq 5)$

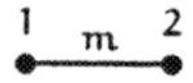

Then the cuspidal classes of W are represented by $\{(s_1 s_2)^i \mid i = 1, \ldots, \lfloor m/2 \rfloor\}$ (where $\lfloor x \rfloor$ is the largest integer $n \leq x$). Indeed, this can be seen as follows.

If $l(w)$ is odd, i.e., if $w = s_1(s_2 s_1)^i$ or $w = s_2(s_1 s_2)^i$ for some i, then $w \longrightarrow s_1$ or $w \longrightarrow s_2$. If $w \neq 1$ and $l(w)$ is even, i.e., $w = (s_1 s_2)^i$ or $w = (s_2 s_1)^i$ for some i, then $\mathrm{Cyc}(w) = \{w, sws\}$ where $s \in S$ is such that $l(sw) < l(w)$. Clearly, the conjugacy classes containing 1, s_1 or s_2 are not cuspidal.

So we are left with $(s_1 s_2)^i$ for $i = 1, \ldots, \lfloor m/2 \rfloor$. That these elements are pairwise not conjugate in W and that their classes are indeed cuspidal follows from the fact that each such element $(s_1 s_2)^i$ has a different characteristic polynomial in the reflection representation. Indeed, these polynomials are easily determined using Exercise 1.4: in the standard reflection representation we have, for $a = s_1 s_2$,

$$a^i = \begin{bmatrix} \cos(i\theta) & -\sin(i\theta) \\ \sin(i\theta) & \cos(i\theta) \end{bmatrix} \qquad \text{with } \theta = 2\pi/m,$$

and so $p_{a^i}(q) = q^2 - 2\cos(2i\pi/m)q + 1$, for $i = 1, \ldots, \lfloor m/2 \rfloor$. In particular, we see that (P3) is valid in this case.

The proof of Theorem 3.2.7 is carried out case by case in the next two sections. Section 3.3 provides the tools needed for the proof in a particular example; this will be used for the exceptional types. In Section 3.4 these ideas are put into a general framework that allows us to prove the theorem for the classical types A_n, B_n and D_n.

For the remainder of this section we take Theorem 3.2.7 as a fact and work out some of its consequences.

3.2.9. Theorem (Geck–Pfeiffer). *Let (W, S) be a Coxeter system such that W is finite and let C be a conjugacy class of W.*

(a) *For each $w \in C$ there exists an element $w' \in C_{\min}$ such that $w \longrightarrow w'$.*

(b) *Let $w, v \in C_{\min}$. Then there exists an element $w' \in \mathrm{Cyc}(w)$ and an element $x \in W$ such that w' is elementarily strongly conjugate to v via x. In particular, any two elements in $C_{\min}$ are strongly conjugate.*

Proof. Using Exercise 3.10, it is sufficient to prove the theorem for irreducible Coxeter systems (W, S).

(a) Let $w \in C$ and suppose first that there exists an element $w' \in W$ such that $J(w')$ is a proper subset of S and $w \longrightarrow w'$. Then, by induction on $|S|$, we have $w' \longrightarrow w''$ for some $w'' \in W_J$ which is of minimal length in its class in W_J. By Lemma 3.1.14, w'' also has minimal length in its class in W and therefore transitivity of the relation $\longrightarrow$ yields $w \longrightarrow w'' \in C_{\min}$.

Otherwise we have $J(w') = S$ for all $w' \in W$ with $w \longrightarrow w'$. Now let $w' \in W$ be such that $\mathrm{Cyc}(w')$ is terminal and $w \longrightarrow w'$. By (P1), $J(w') = S$ then implies that the conjugacy class C, which contains both w' and w, is cuspidal. This in turn, by (P2), implies that the terminal cyclic shift class $\mathrm{Cyc}(w')$ coincides with $C_{\min}$, whence in particular w' has minimal length in C and the claim is proved.

(b) From $v, w \in C_{\min}$ we have in particular that $l(v) = l(w)$ and $w^a = v$ for some $a \in W$. Let $J = J(w)$ and write $a = bx$ for some $b \in W_J$ and $x \in X_J$. Set $w' := w^b$. Then $w' \in W_J$ and $v = (w')^x$, whence $l(w') \leq l(v)$ by Lemma 2.1.14. But we also have $l(w') \geq l(w) = l(v)$ since w has minimal length in its class in W_J. Therefore, we have $l(w') = l(v)$. Since x is a distinguished coset representative we also have $l(w'x) = l(w') + l(x)$ which shows that w' is elementarily strongly conjugate to v.

Since w has minimal length in its conjugacy class in W_J where $J = J(w)$, its cyclic shift class $\mathrm{Cyc}(w)$ is terminal. This implies, by (P1), that the conjugacy class of w in W_J is cuspidal. Finally, since $l(w') = l(w)$, it follows that $w' \in \mathrm{Cyc}(w)$ by (P2). □

The following algorithm performs the conjugation of an arbitrary element $w \in W$ to an element of minimal length in the conjugacy class. It is based on Theorem 3.2.9(a).

Algorithm H (Minimal length element). Given (W, S) and an element $w \in W$, an element of minimal length in the conjugacy class of w in W is determined.

H1. [Cyclic shift class.] Determine $Z \leftarrow \mathrm{Cyc}(w)$ by Algorithm G (p. 80).

H2. [Loop over Z.] Check if $l(svs) < l(v)$ for some $v \in Z$ and some $s \in S$. If such v, s exist set $w \leftarrow svs$ and go back to step H1.

H3. [Terminate.] Otherwise return w.

Straightforward proofs for the two parts of Theorem 3.2.9 exist in particular cases. For (b), we have seen in Theorem 3.1.4 that the Coxeter elements of W form a single cyclic shift class, and in Proposition 3.1.6 that they exhaust all of $C_{\min}$ in their conjugacy class C. For (a), the assertion can be shown to hold for involutions, as follows. As before, we denote, for $K \subseteq S$, by w_K the longest element of W_K.

3.2.10. Proposition (Richardson, Howlett). *Let $w \in W$ be an involution, i.e., $w^2 = 1$. Then $w \longrightarrow w_K$ for some $K \subseteq S$. Moreover, w_K is central in W_K and has minimal length in its class in W.*

Proof. First, suppose that w_K is central in W_K for some $K \subseteq S$. Let $v \in W$ be arbitrary and write $v = v'y$ for some $v' \in W_K$ and $y \in X_K$. Then $(w_K)^v = (w_K)^{v'y} = w_K^y$ and $l(w_K^v) = l(w_K^y) \geq l(w_K)$ by Lemma 2.1.14. Therefore w_K has minimal length in its conjugacy class in W.

It remains to show, for any $w \in W$, that $w \longrightarrow w_K$ for such a central longest element w_K. For $w = 1$ the assertion holds with $K = \varnothing$. So suppose $l(w) \geq 1$.

Note that, since $w^{-1} = w$, the right descent set $\mathcal{R}(w) = \{s \in S \mid l(ws) < l(w)\}$ and the left descent set $\mathcal{L}(w) = \{s \in S \mid l(sw) < l(w)\}$ coincide. Set

$$K = \{s \in \mathcal{R}(w) \mid sw = ws\}.$$

If $w = w_K$ then clearly w_K is central in W_K.

Otherwise we can, by Lemma 1.5.2, write $w = w_K w'$ for some $w' \in W$ in such a way that $l(w) = l(w_K) + l(w')$. This implies that $w' \in X_K$. It follows that $(w')^2 = 1$ since w_K commutes with w and both w and w_K have order 2.

Since we have $w' \neq 1$ we can now choose an element $s \in \mathcal{R}(w') = \mathcal{L}(w')$. From $w' \in X_K$ it follows that $\mathcal{R}(w') \cap K = \emptyset$. Thus $s \in \mathcal{R}(w) - K$, whence $sw \neq ws$. But then $l(sws) < l(w)$ by Lemma 1.2.6, and by induction on the length we have that $sws \longrightarrow w_L$ for some $L \subseteq S$ with w_L central in W_L. Therefore also $w \longrightarrow w_L$. □

A further consequence of Theorem 3.2.7 is that cuspidal classes never fuse!

3.2.11. Theorem. *Let W be a finite Coxeter group with generating set S. Let $J \subseteq S$ and let $w \in W_J$ such that the conjugacy class C^J of w in W_J is cuspidal in W_J. Then*

$$C^J = C \cap W_J,$$

where C is the conjugacy class of w in W.

Proof. It is sufficient to prove the theorem in the case where W is irreducible (see Exercise 3.10). Let $w' \in W_J$ and $x \in W$ such that $(w')^x = w$. We may assume that $x \in X_{JJ}$. For otherwise, write $x = udv$ for some $u, v \in W_J$ and $d \in X_{JJ}$; then $(w')^u \in W_J$ and vwv^{-1} lies in the cuspidal class of w in W_J, whence we can replace w' by $(w')^u$ and w by vwv^{-1}.

Now $(w')^x = w \in W_J \cap W_{J^x}$, where $W_J \cap W_{J^x} = W_{J \cap J^x}$ by Theorem 2.1.12. But the smallest parabolic subgroup of W containing w is W_J. Thus $J^x = J$.

We consider W_J as a direct product of irreducible components and write $W_J = W_1 \times W_2$ where W_1 is the direct product of all components of type A and W_2 is the direct product of all irreducible components not of type A. By inspection of the list of Coxeter graphs in Table 1.1 (p. 16) we see that either $W_2 = 1$ or W_2 is irreducible.

Since $J^x = J$, it follows that conjugation by x is a type-preserving permutation of the irreducible components of W_J. (Note that the order of the product of two generators is preserved under conjugation.) Hence we have $W_i^x = W_i$ for $i = 1, 2$. Now we can write $C^J = C_1 \times C_2$ where C_i is a cuspidal class in W_i for $i = 1, 2$. It remains to show that $C_i^x = C_i$ for $i = 1, 2$. For $i = 1$, this is clear since a group of type A only has one cuspidal class (see Example 3.1.16).

Now consider C_2 and let $w \in C_2$. Then it suffices to check the conditions in (P3) for w and $w' = w^x$ (with respect to W_2). Now, w and $w' = w^x$ certainly have the same characteristic polynomial in the reflection representation of W.

So, using Exercise 3.11, we see that w and w' also have the same characteristic polynomial in the reflection representation of W_2.

Finally, we have to show that $l_s(w) = l_s(w')$ for all $s \in J_2 := W_2 \cap J$, where l_s is defined with respect to W_2. This follows from the fact that if $s, t \in J_2$ are conjugate in W then s, t are already conjugate in W_2.

Indeed, since W_2 is irreducible the Coxeter graph of W_2 is connected. Thus there exists a sequence of generators $s_0, s_1, \ldots, s_l \in J_2 \subseteq S$ connecting s and t in such a way that

$$s = s_0,\ t = s_l, \text{ and the order of } s_{j-1}s_j \text{ is larger than 2 for all } j = 1, \ldots, l.$$

Now, since W is finite, whence the Coxeter graph of W is a tree (see the diagrams in 1.1), this sequence is unique. And since s and t are conjugate in W, it follows from Exercise 1.10 that the order of $s_{j-1}s_j$ is odd for all j. But this in turn implies that s and t are conjugate in W_2, as desired. □

A complete description of the conjugacy classes of W in terms of cuspidal classes of parabolic subgroups is our next result.

3.2.12. Theorem. *Consider the set $\mathcal{P}$ of pairs (J, D) where $J \subseteq S$ and D is a cuspidal class of W_J together with the equivalence relation induced on $\mathcal{P}$ by the conjugation action of W, i.e., where (J, D) and (J', D') in $\mathcal{P}$ are related if there is an element $x \in W$ such that $J^x = J'$ and $D^x = D'$. The map*

$$\Lambda \colon C \mapsto \{(J, C \cap W_J) \mid J = J(w) \text{ for some } w \in C_{\min}\}$$

is a bijection from the set of conjugacy classes of elements of W to the set of equivalence classes of pairs in $\mathcal{P}$. In particular, for every conjugacy class C of W, the set $\{J(w) \mid w \in C_{\min}\}$ is a Coxeter class of subsets of S.

In other words, the conjugacy classes of W are parametrized by the pairs (J, D) where J runs through a system of representatives of the Coxeter classes of subsets of S, and, for each such J, D runs through the cuspidal classes of W_J.

Proof. For a given conjugacy class C of W let us first show that the label $\Lambda(C) = \{(J, C \cap W_J) \mid J = J(w) \text{ for some } w \in C_{\min}\}$ is indeed a set of pairs (J, D) in $\mathcal{P}$. For this let $w \in C_{\min}$, let $J = J(w)$ and let D be the class of w in W_J. Clearly, w has minimal length in D since $D \subseteq C$. By property (P1), D is then a cuspidal class of W_J. From Theorem 3.2.11 it then follows that $D = C \cap W_J$ and thus $(J, C \cap W_J) \in \mathcal{P}$.

Next, let $w, w' \in C_{\min}$ and let us show that the corresponding pairs are equivalent. Denote $J = J(w)$ and $J' = J(w')$. By Corollary 3.1.11, there exists an element $x \in W$ such that $J^x = J'$, whence $W_J^x = W_{J'}$. If D is the class of w in W_J then, by Theorem 3.2.11, $D = C \cap W_J$, and similarly, if D' is the class of w' in $W_{J'}$ then $D' = C \cap W_{J'}$. But then, $D^x = C^x \cap W_J^x = D'$, whence the pairs (J, D) and (J', D') are indeed equivalent.

Further, let $(J, D) \in \Lambda(C)$, let $x \in W$ such that $(J^x, D^x) \in \mathcal{P}$ is equivalent to (J, D) and let us show that $(J^x, D^x) \in \Lambda(C)$. For this choose w of minimal length in D. Then, by Lemma 3.1.14, w has minimal length in its class C in W as well. Moreover, since $J^x \subseteq S$, we have $l(w^x) = l(w)$ and therefore $w^x \in C_{\min}$. Also $J(w) = J$ and the fact that $J^x \subseteq S$ imply $J(w^x) = J(w)^x = J^x$. Furthermore, $C \cap W_{J^x} = C^x \cap W_J^x = D^x$ and therefore $(J^x, D^x) \in \Lambda(C)$.

We have thus seen that $\Lambda(C)$ is an equivalence class of pairs in $\mathcal{P}$. It remains to show that the map Λ is injective and surjective.

Suppose $\Lambda(C) = \Lambda(C')$ for conjugacy classes C and C′ of W. Then $(J, D) \in \Lambda(C')$ for any $(J, D) \in \Lambda(C)$. But then $w \in D = C \cap W_J = C' \cap W_J$ implies $C \cap C' \neq \varnothing$, whence $C = C'$. Therefore Λ is injective.

Finally, given any pair $(J, D) \in \mathcal{P}$, let w be of minimal length in D and let us show that $(J, D) \in \Lambda(C)$ where C is the conjugacy class of w in W. Since D is a cuspidal class of W_J, Lemma 3.1.14 shows that $w \in C_{\min}$. By Proposition 3.1.12, $J(w) = J$ since D is a cuspidal class of W_J. Moreover, by Theorem 3.2.11, we have $C \cap W_J = D$. Therefore Λ is surjective and hence a bijection. □

A mechanical procedure which captures parts of this proof in order to identify the conjugacy class of w in W is given by the following algorithm.

Algorithm I (Class identification). We assume that a set of representatives of the Coxeter classes of parabolic subgroups of W and, for each such representative, a set of representatives of its cuspidal classes are given. Then, given $w \in W$, the class label (J, D) of w in W is determined.

- **I1.** [Minimal length.] Find an element $w_1 \in C_{\min}$ such that $w \longrightarrow w_1$ by Algorithm H (p. 83) and set $J \leftarrow J(w_1)$. (By (P1), then the conjugacy class of w_1 in W_J is a cuspidal class.)
- **I2.** [Coxeter class.] Let $K \subseteq S$ be the conjugate of J which represents the Coxeter class of J in W, and use Algorithm F (p. 55) to find an element $x \in X_J$ such that $J^x = K$. Set $w_2 = w_1^x$. (Then $w_2 \in W_K$ and w_2 has minimal length it its conjugacy class in W_K.)
- **I3.** [Conjugacy class.] By Algorithm G (p. 80), construct the cyclic shift class $\mathrm{Cyc}(w_2)$ of w_2. Then let w_3 be the unique representative of the cuspidal classes of W_K which is contained in $\mathrm{Cyc}(w_2)$ and return the class label (K, w_3). (This is the label which will be returned for all elements in the conjugacy class of w in W and for those elements only.)

3.2.13. Remark. We have a parametrization of the conjugacy classes of involutions in W in terms of parabolic subgroups: the conjugacy classes of involutions of W are in bijection to the set

$$\{[J] \mid J \subseteq S,\ w_J \text{ central in } W_J\},$$

where [J] denotes the Coxeter class of the subset $J \subseteq S$ in W. Indeed, this is a consequence of the description of the conjugacy classes of W in Theorem 3.2.12

and the minimal length property of classes of involutions described in Proposition 3.2.10.

We conclude the list of consequences of Theorem 3.2.7 with a general property of the elements of W.

3.2.14. Corollary. *Every element of W is conjugate to its inverse.*

Proof. We can assume that W is irreducible and that w lies in a cuspidal class of W. Let $\varepsilon_1, \ldots, \varepsilon_n$ (where $n = |S|$) be the (complex) eigenvalues of w in the natural reflection representation. Then $\varepsilon_1^{-1}, \ldots, \varepsilon_n^{-1}$ are the eigenvalues of w^{-1} and so

$$p_{w^{-1}}(q) = \prod_{i=1}^{n}(q - \varepsilon_i^{-1}) = \prod_{i=1}^{n} \overline{(q - \varepsilon_i)} = \overline{p_w(q)},$$

where the bar denotes complex conjugation. But, since the reflection representation is defined over $\mathbb{R}$, we have $\overline{p_w(q)} = p_w(q)$. Now (P3) implies that w and w^{-1} are conjugate (note that w^{-1} also lies in a cuspidal class and that clearly $l_s(w) = l_s(w^{-1})$ for all $s \in S$.). □

In [Carter 1972a, Theorem C] it is shown, case by case, that every element $w \in W$ is a product of two involutions. This of course implies that w is conjugate to its inverse in W. It would be interesting to see a proof of this factorization property in the current framework supplied by the conjugacy graph.

3.3 ALGORITHMS AND REDUCTION THEOREMS

The proof of Theorem 3.2.7 for a particular Coxeter group W depends on a series of lemmas and algorithms which we develop in this section.

The first move towards step 1 above is suggested by the following observation. For $A, B, C \subseteq W$, we write "$A \longrightarrow B$ via C" if for each $w \in A$ there exists an element w' in B such that $w \longrightarrow w'$ via a conjugating element in C.

3.3.1. Lemma. *Let $J \subseteq S$. Then $W \longrightarrow W_J \cdot X_{JJ}$ via W_J.*

Proof. Write $w \in W$ as uxv where $u, v \in W_J$ and $x \in X_{JJ}$. Also write $v = v_1 \cdots v_l$ as a reduced expression, $v_i \in J$. If $l = 0$ there is nothing to show. So suppose $l > 0$. We then have $w \longrightarrow v_l u x v_1 \cdots v_{l-1}$ and by induction on l we have $v_l u x v_1 \cdots v_{l-1} \longrightarrow vux$. Hence $w \longrightarrow vux \in W_J X_{JJ}$ via $v \in W_J$. □

The following algorithm continues this sort of reduction along a chain of parabolic subgroups.

Algorithm J (Short list). Given (W, S) and a chain $S = J_0 \supseteq J_1 \supseteq \cdots \supseteq J_r = \varnothing$ of subsets of S (for some $r > 0$), a (small) set M is constructed, which for each $w \in W$ contains an element $m \in M$ such that $w \longrightarrow m$, or $W \longrightarrow M$ for short.

J1. [Initialize.] Set $i \leftarrow 0$, $W_0 \leftarrow W$ and $Y_0 \leftarrow \{1\}$. (Clearly, $W \longrightarrow W = W_0 \cdot Y_0$.)

J2. [Next i.] Increase i by 1 and set $W_i \leftarrow W_{J_i}$ and

$$Y_i \leftarrow \{x \cdot y \mid x \in X_{J_i}^{J_{i-1}},\ y \in Y_{i-1}\},\ l(xys) > l(xy) \text{ for all } s \in J_i\}$$

(Now $W \longrightarrow W_{i-1} \cdot Y_{i-1}$ implies $W \longrightarrow W_i \cdot Y_i$: let $w \in W_{i-1}$, $y \in Y_{i-1}$ and write $w = vx$ for $v \in W_i$, $x \in X_{J_i}^{J_{i-1}}$. Then $xy \in X_{J_i}$ and we can write $xy = y'u$ where $y' \in X_{J_i}^{-1}$, $u \in W_{J_i}$ and $l(xy) = l(y') + l(u)$. But then $wy = wy'u \longrightarrow uwy'$ if this is performed along a reduced expression for $u = u_1 \cdots u_s$, $u_j \in J_i$, since $l(uwy') \leq l(u) + l(w) + l(y') = l(wy'u)$.)

J3. [Loop.] Repeat step J2 until $i = r$ and $J_i = \varnothing$. Then return $M = Y_r$. (Finally, $W \longrightarrow Y_r = W_r \cdot Y_r$.)

Note that the sets Y_i $(i = 0, \ldots, r)$ are sets of double coset representatives: we always have $Y_i \subseteq X_{J_i J_i}$. In particular, for $i = 1$, we have $Y_1 = X_{J_1 J_1}$.

The following example shows Algorithm J in action.

3.3.2. Example. Let (W, S) be of type H_3 and $S = \{s_1, s_2, s_3\}$ such that the Coxeter graph of (W, S) is given by 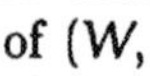. Then W has 120 elements. We consider the chain $S \supset J \supset K \supset L = \varnothing$ where $J = \{s_1, s_2\}$ and $K = \{s_1\}$. First we determine the set of coset representatives of W_J in W in the form of the coset graph Γ_J, as follows:

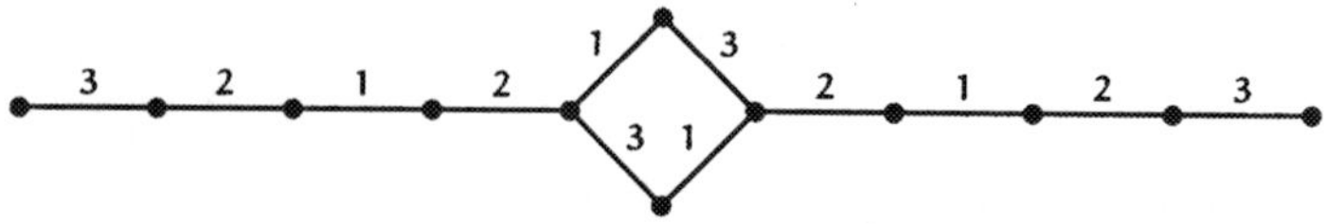

(Again, we leave out the arrows and understand them to point from left to right.) We determine

$$Y_1 = X_{JJ} = \{1,\ s_3,\ s_3s_2s_1s_2s_3,\ s_3s_2s_1s_2s_1s_3s_2s_1s_2s_3\},$$

which means that we only have to consider 4 cosets out of 12. The cosets of W_K in W_J are given by the following coset graph:

In order to determine Y_2, we have to consider 20 elements of the form $x \cdot y$ where

$$x \in \{1,\ s_2,\ s_2s_1,\ s_2s_1s_2,\ s_2s_1s_2s_1\} \quad \text{and} \quad y \in Y_1.$$

We have to select those xy with $l(xys) > l(xy)$ for all $s \in K = \{s_1\}$ and we find that this is always the case, except for

$$\begin{array}{ll} x = s_2, & y = s_3s_2s_1s_2s_1s_3s_2s_1s_2s_3; \\ x = s_2s_1, & y = 1 \quad \text{or} \quad s_3; \\ x = s_2s_1s_2, & y = s_3s_2s_1s_2s_1s_3s_2s_1s_2s_3; \\ x = s_2s_1s_2s_1, & y = 1 \quad \text{or} \quad s_3. \end{array}$$

This leaves 14 elements in Y_2. Continuing with $L = \emptyset$ and $X_L^K = \{1, s_1\}$, we get $Y_3 = Y_2 \cup s_1 Y_2$, a total of 28 elements out of 120. Note that in this last step the condition on the elements in $X_{J_3}^{J_2} \cdot Y_2$ is empty, whence $M = Y_3 = X_L^K \cdot Y_2$.

The following lemma suggests how to proceed further.

3.3.3. Lemma. *Let $M \subseteq W$ such that $W \longrightarrow M$. Then the following hold.*

(i) *M contains a set of representatives of cyclic shift classes of elements of minimal length.*

(ii) *We have $W \longrightarrow M'$ for*

$$M' = \{m \in M \mid l(sxs) \geq l(x) \text{ for all } x \in \mathrm{Cyc}(m),\ s \in S\}.$$

Proof. (i) For each $w \in W$ there is an element $m \in M$ such that $w \longrightarrow m$. Now let C be a conjugacy class of W and let $w \in C_{\min}$. Moreover, let $m \in M$ such that $w \longrightarrow m$. Then $l(m) = l(w)$ since w has minimal length in its class, and therefore m represents the cyclic shift class of w in W.

(ii) Let $m \in M$ and assume $l(sxs) < l(x)$ for some $x \in \mathrm{Cyc}(m)$ and $s \in S$. Then $x \longrightarrow sxs$ by definition, and $m \longrightarrow x$ since $x \in \mathrm{Cyc}(m)$. Now $sxs \longrightarrow m'$ for some $m' \in M$ since $W \longrightarrow M$. And $m' \neq m$ since $l(m') < l(m)$. Thus $M \longrightarrow M - \{m\}$ and we can replace M by $M - \{m\}$. Continuing this way until $M = M'$ completes the proof. □

Algorithm K (Terminal cyclic shift representatives). Given (W, S) and a set M such that $W \longrightarrow M$, a set of representatives of the terminal cyclic shift classes of W is constructed.

K1. [Initialize.] Set $Y \leftarrow M$ and $X \leftarrow \emptyset$.

K2. [Choose a representative.] Choose $y \in Y$ and set $Z \leftarrow \mathrm{Cyc}(y)$. Then set $Y \leftarrow Y - Z$.

K3. [Terminal?] If $l(szs) \geq l(z)$ for all $s \in S$ and $z \in Z$ then set $X \leftarrow X \cup \{y\}$. (We have $M \longrightarrow X \cup Y$ and $X \cap \mathrm{Cyc}(x) = \{x\}$ for all $x \in X$.)

K4. [Loop.] Repeat steps K2 and K3 until $Y = \emptyset$. Then return X.

3.3.4. Example. Let us continue with the example of H_3. Since we are only interested in cuspidal classes we just have to consider 19 out of the 28 elements of the list M constructed earlier in Example 3.3.2.

It turns out that, for all but five elements w in the list, there exists an element $w' \in W$ such that $w \longrightarrow w'$ and $l(w') < l(w)$. The five remaining elements are

$$s_1s_2s_1s_2s_1s_3s_2s_1s_2s_1s_3s_2s_1s_2s_3,$$

$$s_2s_1s_2s_1s_3s_2s_1s_2s_3 \quad \text{and} \quad s_1s_2s_1s_2s_3s_2s_1s_2s_3,$$

$$s_1s_2s_1s_2s_3,$$

$$s_1s_2s_3,$$

for which we observe that the two elements of length 9 are conjugate by cyclic shift via s_2. We have thus found four terminal cyclic shift classes.

At this stage, Algorithms J and K have provided us with the complete list of representatives of the terminal cyclic shift classes of W, none of which contains an element of a proper parabolic subgroup of W. In order to see that this list constitutes a complete list of representatives of the cuspidal classes of W we have to show that (i) they do indeed represent cuspidal classes and (ii) no two elements from the list are conjugate in W.

For this twofold purpose we simply calculate the characteristic polynomial $p_w(q) \in \mathbb{C}[q]$ of each representative w:

$$\begin{aligned}
w &= s_1s_2s_3, & p_w(q) &= (q+1)(q^2-\alpha q+1).\\
w &= s_1s_2s_1s_2s_3, & p_w(q) &= (q+1)(q^2-q+1).\\
w &= s_1s_2s_1s_2s_3s_2s_1s_2s_3, & p_w(q) &= (q+1)(q^2-\bar{\alpha} q+1).\\
w &= s_1s_2s_1s_2s_1s_3s_2s_1s_2s_1s_3s_2s_1s_2s_3, & p_w(q) &= (q+1)^3.
\end{aligned}$$

Here $\alpha = (-1+\sqrt{5})/2$ and $\bar{\alpha} = (-1-\sqrt{5})/2$ (cf. Table B.7).

None of the polynomials $p_w(q)$ is divisible by $(q-1)$. This implies, by Lemma 3.1.10, that all these elements do represent cuspidal classes. Moreover, the $p_w(q)$ are all distinct. Therefore no two representatives are conjugate in W.

Proof of Theorem 3.2.7. For any irreducible Coxeter group $W = \langle S \rangle$ the theorem can now be checked as follows. Using Algorithms J and K, a set of representatives of terminal cyclic shift classes is constructed. Then it remains to show three things in order to establish Theorem 3.2.7 for W:

(i) that each representative lies in a cuspidal class, verifying (P1);
(ii) that any two distinct representatives are not conjugate in W, whence each cuspidal class contains just one terminal class, verifying (P2);
(iii) that the characteristic polynomial and the length functions l_s are sufficient to distinguish the cuspidal classes of W, verifying (P3).

This is done for the classical types in the next section. Type $I_2(m)$ has already been settled in Example 3.2.8.

For the exceptional types, we proceed by the same strategy as that outlined above for type H_3. Using an implementation of Algorithms J and K in GAP, we find for each irreducible Coxeter system (W, S) representatives of terminal cyclic shift classes. The numbers of terminal cyclic shift classes in each case are listed in the following table:

Type of (W, S)	H_3	H_4	F_4	E_6	E_7	E_8
terminal classes	4	20	9	5	12	30

Reduced expressions and the characteristic polynomials for these representatives are printed in the tables in Appendix B. From these, we see that (P1), (P2) and (P3) are satisfied. Note that in all but one case it is already sufficient to use the characteristic polynomials in order to distinguish the cuspidal classes. The exception occurs in type F_4 for the two cuspidal classes with characteristic polynomial $(q+1)(q^3+1)$. □

3.4 MINIMAL ELEMENTS IN CLASSICAL TYPES

In this section we verify Theorem 3.2.7 for Coxeter groups of classical types, i.e., for the types A_n, B_n and D_n.

Throughout the section we will use the notation and the general set-up from Section 1.4. There we defined, for $n \geq 2$, a Coxeter group W_n of type B_n generated by a set $S = \{t, s_1, \ldots, s_{n-1}\}$ satisfying the relations encoded in the Coxeter diagram

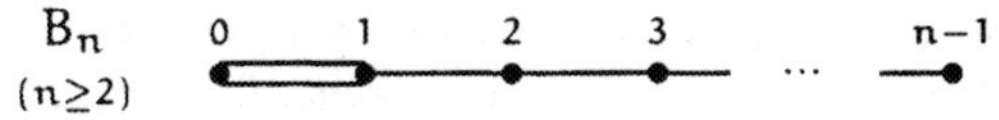

where we write i short for the generator s_i, $i = 1, \ldots, n-1$, and 0 for t. We also defined elements

$$t_0 = t \quad \text{and} \quad t_i = s_i t_{i-1} s_i \quad \text{for } i = 1, \ldots, n-1.$$

Then we have $t_i t_j = t_j t_i$ for all $i, j = 0, \ldots, n-1$, and

$$t_i s_j = \begin{cases} s_i t_{i-1} & \text{if } j = i, \\ s_{i+1} t_{i+1} & \text{if } j = i+1, \\ s_j t_i & \text{otherwise.} \end{cases} \tag{t}$$

The normal subgroup W'_n of W_n is a Coxeter group of type D_n generated by the set $S' = \{u, s_1, \ldots, s_{n-1}\}$ of reflections, where $u = ts_1t$. The relations satisfied by the generators in S' are encoded in the Coxeter diagram

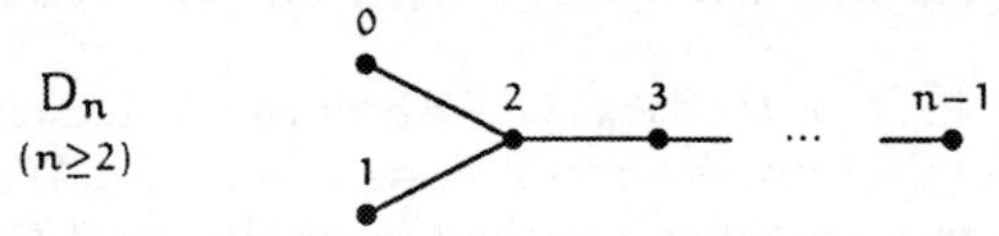

where we write i for the generator s_i, $i = 1, \ldots, n-1$, and 0 for u. In (1.4.8), we also defined elements

$$u_1 = us_1 = s_1 u \quad \text{and} \quad u_i = s_i u_{i-1} s_i \quad \text{for } i = 2, \ldots, n-1.$$

We have $u_i = tt_i$ and $u_i u_j = u_j u_i$ for all $i, j = 1, \ldots, n-1$. Moreover,

$$u_i s_j = \begin{cases} u & \text{if } j = i = 1, \\ uu_i & \text{if } j = 1 \text{ and } i \geq 2, \\ s_i u_{i-1} & \text{if } j = i \geq 2, \\ s_{i+1} u_{i+1} & \text{if } j = i+1, \\ s_j u_i & \text{otherwise.} \end{cases} \tag{u}$$

Both W_n and W'_n contain as a parabolic subgroup the group $\mathfrak{S}_n$ of type A_{n-1} which is generated by $\{s_1, \ldots, s_{n-1}\}$ subject to the relations in the following Coxeter diagram:

A_{n-1} $(n \geq 2)$

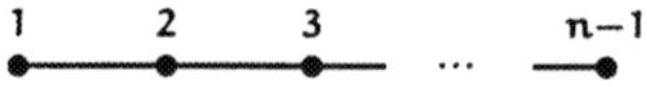

The parametrization of the conjugacy classes of the symmetric group by partitions is well known and the fact that $\mathfrak{S}_n$ has only one cuspidal class is established in Example 3.1.16. The proof of Theorem 3.2.7 can in this case be completed by the following result.

3.4.1. Proposition. *Let $n \geq 2$ and let $W = \mathfrak{S}_n$ be a Coxeter group of type A_{n-1} with generating set $\{s_1, \ldots, s_{n-1}\}$ as above. Let C be a cuspidal class of W and let $w \in C$. Then $w \longrightarrow s_1 \cdots s_{n-1}$ via elements of $\mathfrak{S}_{n-1} = \langle s_1, \ldots, s_{n-2} \rangle$.*

Proof. First, consider the case $n = 2$. The group $\mathfrak{S}_2$ has only two elements: 1 lies in a proper parabolic subgroup and therefore not in a cuspidal class, and s_1 already has the required form.

Next, let $n > 2$ and let $J = S - \{s_{n-1}\}$. Then, by Example 2.2.11, the set of double coset representatives of W_J in W is $X_{JJ} = \{1, s_{n-1}\}$. By Lemma 3.3.1, there is an element $w' \in W_J$ such that either $w \longrightarrow w'$ or $w \longrightarrow w's_{n-1}$ via elements of W_J. In the first case w lies in a proper parabolic subgroup and hence not in a cuspidal class and we are done.

In the second case let $K = J - \{s_{n-2}\}$. By induction on n we have that either $w' \longrightarrow w''$, where $J(w'')$ is a proper subset of J, or $w' \longrightarrow s_1 \cdots s_{n-2}$ via elements of W_K. Since s_{n-1} commutes with all generators of W_K, we have for $w's_{n-1}$ that either $w's_{n-1} \longrightarrow w''s_{n-1}$, in which case w does not lie in a cuspidal class, or that $w \longrightarrow w's_{n-1} \longrightarrow s_1 \cdots s_{n-2}s_{n-1}$, as required. □

Proof of Theorem 3.2.7 for type A_{n-1}. We have seen in Example 3.1.16 that the conjugacy class C of the Coxeter element $w = s_1 \cdots s_{n-1}$ in W is a cuspidal class. Since there is only one representative, there is nothing more to show. We note in passing that the characteristic polynomial of w in the reflection representation is $p_w(q) = 1 + q + q^2 + \cdots + q^{n-1}$. □

3.4.2. Blocks and signed block forms. Given $m \geq 0$ and $e \geq 1$ such that $m + e \leq n$, we define a *positive block* $b^+_{m,e}$ of length e and starting at m by

$$b^+_{m,e} = s_{m+1}s_{m+2} \cdots s_{m+e-1} \in W_n$$

and a *negative block* $b^-_{m,e}$ of length e and starting at m by

$$b^-_{m,e} = t_m s_{m+1}s_{m+2} \cdots s_{m+e-1} \in W_n.$$

The *support* of a block $b^\pm_{m,e}$ is defined as the set $\{m+1, m+2, \ldots, m+e\}$.

The following multiplicative properties of negative blocks are easily verified. Let $m \geq 0$, $e \geq 1$ and $i \leq n-1$. Then

$$b^-_{m,e}s_i = \begin{cases} s_i b^-_{m,e} & \text{if } i < m \text{ or } i \geq m+e+1, \\ s_{i+1} b^-_{m,e} & \text{if } m+1 \leq i \leq m+e-2, \\ b^-_{m,e-1} & \text{if } e > 1 \text{ and } i = m+e-1, \\ b^-_{m,e+1} & \text{if } i = m+e. \end{cases}$$

This also shows that blocks with disjoint support sets commute.

An element $w \in W_n$ is in *signed block form* if it is written as a product of positive and negative blocks whose support sets form a partition of the set $\{1,\dots,n\}$, i.e., if

$$w = b^{\epsilon_1}_{m_1,e_1} \cdots b^{\epsilon_r}_{m_r,e_r} \tag{1}$$

for $\epsilon_i = \pm$, m_i, e_i $(i = 1,\dots,r)$, such that $m_1 = 0$, $m_{i+1} = m_i + e_i$ for $i = 1,\dots,r-1$, and $m_r + e_r = n$. Note that in this case the block lengths e_i form a composition $(e_1,\dots,e_r)$ of n, i.e., $e_1 + \cdots + e_r = n$.

If in (1) all the negative blocks come before the positive blocks, i.e., if there exists an index $l \leq r$ such that ϵ_i is "$-$" for all $i \leq l$ and "$+$" for all $i > l$ then we set $\beta = (e_1,\dots,e_l)$, the composition corresponding to the negative blocks, and $\alpha = (e_{l+1},\dots,e_r)$, the composition corresponding to the positive blocks, and denote

$$w_{(\alpha,\beta)} = b^-_{m_1,e_1} \cdots b^-_{m_l,e_l} \cdot b^+_{m_{l+1},e_{l+1}} \cdots b^+_{m_r,e_r}. \tag{2}$$

Note that the defining expression of any element w as in (1) is reduced, since it is of form $d_0 \cdots d_{n-1}$ where each d_i is a distinguished coset representative of the parabolic subgroup W_i in W_{i+1}; see Example 2.2.5. For the extreme cases where $l = 0$ or $l = r$, i.e., where all blocks are positive or all are negative, we introduce the following notation, for any composition α of n:

$$w^+_\alpha = w_{(\alpha,\varnothing)} \qquad \text{and} \qquad w^-_\alpha = w_{(\varnothing,\alpha)}. \tag{3}$$

3.4.3. Characteristic polynomials for classical types. Let (α,β) be a pair of compositions such that $|\alpha| + |\beta| = n$. Then, using the matrix description in (1.4.1), we have (note that α corresponds to positive and β to negative blocks):

$$\det(q\,\mathrm{id}_{\mathbb{R}^n} - w_{(\alpha,\beta)}) = \prod_{i=1}^{r}(q^{\alpha_i} - 1)\prod_{j=1}^{s}(q^{\beta_j} + 1),$$

where $\alpha = (\alpha_1,\dots,\alpha_r)$ and $\beta = (\beta_1,\dots,\beta_s)$. Finally, if β is empty then α is a composition of n and $w_{(\alpha,\varnothing)}$ lies in the parabolic subgroup of W_n isomorphic to $\mathfrak{S}_n$. In that case, we have to divide the above expression by $q-1$ in order to obtain the characteristic polynomial with respect to the natural representation of $\mathfrak{S}_n$ (see Exercise 3.11).

3.4.4. Lemma. *Let $W_n = \langle S \rangle$ be a Coxeter group of type B_n generated by $S = \{t, s_1, \ldots, s_{n-1}\}$ as above. Let C be a cuspidal class of W_n and let $w \in C$. Then $w \longrightarrow w_\alpha^-$ for a composition $\alpha = (\alpha_1, \ldots, \alpha_r)$ of n.*

Proof. It will be convenient to set $s_0 = t$. First we consider the case $n = 1$ where $W_n = \langle t \rangle$. Then, for $w \in W_n$, we either have $w = 1$ and w lies in a proper parabolic subgroup of W_n or $w = t$ whence $w = w_\alpha^-$ for the composition $\alpha = (1)$ of 1.

Next let $n \geq 2$ and let $J = S - \{s_{n-1}\}$. Then, using Example 2.2.5, we find that $X_{JJ} = \{1, s_{n-1}, t_{n-1}\}$. By Lemma 3.3.1, there exists an element $w' \in W_J$ such that $w \longrightarrow w'd$ via elements of W_J where $d \in \{1, s_{n-1}, t_{n-1}\}$. We must have $d \neq 1$ since otherwise the conjugate w' of w lies in a proper parabolic subgroup of W_n, contradicting the cuspidality of C.

Further, let $K = J - \{s_{n-2}\}$. By induction on n there is either a composition $\alpha' = (\alpha'_1, \ldots, \alpha'_r)$ of $n-1$ such that $w' \longrightarrow w_{\alpha'}^-$ via elements of W_K, or there is an element w'' of a proper parabolic subgroup of W_J such that $w' \longrightarrow w''$ via elements of W_K.

Since d commutes with W_K we can apply the same sequences of conjugating elements to $w'd$ and obtain that either $w'd \longrightarrow w_{\alpha'}^- d$ or $w'd \longrightarrow w''d$. In the first case we let

$$\alpha = \begin{cases} (\alpha'_1, \ldots, \alpha'_{r-1}, \alpha'_r + 1) & \text{if } d = s_{n-1}, \\ (\alpha'_1, \ldots, \alpha'_{r-1}, \alpha'_r, 1) & \text{if } d = t_{n-1}. \end{cases}$$

Then α is a composition of n and $w_{\alpha'}^- d = w_\alpha^-$, proving the claim.

In the other case suppose $d = s_{n-1}$. Then $w \longrightarrow w''d$ where $w''d$ lies in a proper parabolic subgroup of W, contradicting the cuspidality of C.

Finally, suppose $d = t_{n-1}$ and $w \longrightarrow w''t_{n-1}$ for some element w'' of a proper parabolic subgroup of W_J. Then there exists an index $i \in \{0, \ldots, n-2\}$ such that $w'' = w_1 \cdot w_2$, where $w_1 \in \langle s_0, s_1, \ldots, s_{i-1} \rangle$ and $w_2 \in \langle s_{i+1}, \ldots, s_{n-2} \rangle$.

Let $y = s_{n-1} \cdots s_{i+1}$. Then we can write $t_{n-1} = y \cdot t_i \cdot y^{-1}$ and we have $s_k^y = s_{k+1}$ for $k \in \{i+1, \ldots, n-2\}$. Therefore $w_1 w_2 t_{n-1} \longrightarrow w_1 w_2^y t_i = w_1 t_i w_2^y$ where $w_2^y \in \langle s_{i+2}, \ldots, s_{n-1} \rangle$. Hence $w_1 t_i w_2^y$ lies in the proper parabolic subgroup $\langle S - \{s_{i+1}\} \rangle$, again contradicting the cuspidality of C. □

3.4.5. Lemma (Block exchange). *Let $m > k \geq 1$ and let $p \geq 0$. Moreover, let*

$$y = (s_{p+m} \cdots s_{p+2})(s_{p+m+1} \cdots s_{p+4}) \cdots (s_{p+m+k-1} \cdots s_{p+2k}) \cdot s_{p+2k-1}(s_{p+2k-3} s_{p+2k-2}) \cdots (s_{p+1} \cdots s_{p+k}).$$

Then $b_{p,m}^- b_{p+m,k}^- \longrightarrow b_{p,k}^- b_{p+k,m}^-$ via y.

Note that the conjugating word y here is a particular reduced expression for the longest coset representative of W_L in W_M, where $M = \{s_{p+1}, \ldots, s_{p+m+k-1}\}$ and $L = M - \{s_{p+m}\}$.

Proof. We will only give a proof for $p = 0$. The general formula follows from an analogous calculation with modified indices. We proceed in two major steps. The strategy is to perform a cyclic shift on the first product of blocks until a possible reduction in length via conjugation becomes obvious, and then to perform a cyclic shift on the shorter element obtained this way until it equals the second product of blocks. So we start by defining

$$x_i = s_{m+i-1} \cdots s_{2i} \qquad \text{and} \qquad w_i = (s_{m+i-1} \cdots s_{2i})\, b^-_{0,\,m+i}\, b^-_{0,\,m+k}$$

for $i = 1, \ldots k$. Then

$$b^-_{0,\,m}\, b^-_{m,\,k} = b^-_{0,\,m+1}(s_{m-1} \cdots s_1)\, b^-_{0,\,m+k} = (s_m \cdots s_2)\, b^-_{0,\,m+1}\, b^-_{0,\,m+k} = w_1.$$

Furthermore

$$\begin{aligned} w_i &\xrightarrow{x_i} b^-_{0,\,m+i} b^-_{0,\,m+k}(s_{m+i-1} \cdots s_{2i}) \\ &= b^-_{0,\,m+i+1}(s_{m+i-1} \cdots s_{2i+1})\, b^-_{0,\,m+k} \\ &= (s_{m+i} \cdots s_{2i+2})\, b^-_{0,\,m+i+1}\, b^-_{0,\,m+k} = w_{i+1}, \end{aligned}$$

for $i = 1, \ldots, k-1$. Next we define

$$y_i = s_{2i-1} \cdots s_{k+i-1} \qquad \text{and} \qquad v_i = b^-_{0,\,m+k}\, b^-_{0,\,k+i}(s_{k+i-2} \cdots s_{2i} s_{2i-1})$$

for $i = k, \ldots, 1$. Then

$$w_k = (s_{m+k-1} \cdots s_{2k})\, b^-_{0,\,m+k}\, b^-_{0,\,2k}(s_{2k} \cdots s_{m+k-1}) \xrightarrow{x_k} b^-_{0,\,m+k}\, b^-_{0,\,2k} = v_k.$$

Moreover, for $i = k, \ldots, 2$, we have

$$\begin{aligned} v_i &\xrightarrow{y_i} (s_{k+i-1} \cdots s_{2i-1})\, b^-_{0,\,m+k}\, b^-_{0,\,k+i-1} \\ &= b^-_{0,\,m+k}\, b^-_{0,\,k+i-1}(s_{k+i-3} \cdots s_{2i-3}) = v_{i-1}. \end{aligned}$$

Finally,

$$\begin{aligned} v_1 &= b^-_{0,\,m+k}\, b^-_{0,\,k}(s_k \cdots s_1) \xrightarrow{y_1} (s_k \cdots s_1)\, b^-_{0,\,m+k}\, b^-_{0,\,k} \\ &= b^-_{k,\,m}\, b^-_{0,\,k} = b^-_{0,\,k}\, b^-_{k,\,m}, \end{aligned}$$

as desired. □

We can use the block exchange lemma to get a small set of elements labelled by the partitions of n which contains all representatives of terminal cyclic shift classes in cuspidal conjugacy classes of W. In fact it then turns out that the set $\{w^-_\alpha \mid \alpha \vdash n\}$ is a set of representatives of the cuspidal classes of W.

3.4.6. Proposition. *Let $W_n = \langle S \rangle$ be a Coxeter group of type B_n with $S = \{t, s_1, \ldots, s_{n-1}\}$ as above. Let C be a cuspidal class of W and let $w \in C$. Then $w \longrightarrow w^-_\alpha$ for an increasing composition $\alpha = (\alpha_1, \ldots, \alpha_r)$ of n.*

Proof. By Lemma 3.4.4, we have $w \longrightarrow w_\alpha^-$ for a composition $\alpha = (\alpha_1, \ldots, \alpha_r)$ of n. We can rearrange the blocks of α by increasing length as follows. If $\alpha_1 \leq \cdots \leq \alpha_r$ then there is nothing to do. So suppose $\alpha_i > \alpha_{i+1}$ for some i. We have

$$w_\alpha^- = b_{m_1,\alpha_1}^- \cdots b_{m_i,\alpha_i}^- b_{m_{i+1},\alpha_{i+1}}^- \cdots b_{m_r,\alpha_r}^-.$$

By Lemma 3.4.5, we have

$$b_{m_i,\alpha_i}^- b_{m_i+\alpha_i,\alpha_{i+1}}^- \longrightarrow b_{m_i,\alpha_{i+1}}^- b_{m_i+\alpha_{i+1},\alpha_i}^-$$

via some $y \in \langle s_{m_i+1}, \ldots, s_{m_{i+1}+\alpha_{i+1}-1}\rangle$. Now y commutes with all the other blocks b_{m_j,α_j}^-, $j \neq i, i+1$. Therefore

$$w_\alpha^- \longrightarrow b_{m_1,\alpha_1}^- \cdots b_{m_i,\alpha_{i+1}}^- b_{m_i+\alpha_{i+1},\alpha_i}^- \cdots b_{m_r,\alpha_r}^- = w_{\alpha'}^-$$

for $\alpha' = (\alpha_1, \ldots, \alpha_{i+1}, \alpha_i, \ldots, \alpha_r)$. Continuing this way we finally get that $w \longrightarrow w_\alpha^-$ for an increasing composition α of n. □

Proof of Theorem 3.2.7 for type B_n. By Proposition 3.4.6, we know that $\{w_\alpha^- \mid \alpha \vdash n\}$ contains a complete set of representatives of those terminal cyclic shift classes which lie in cuspidal conjugacy classes of W_n.

By (3.4.3), the characteristic polynomial of w_α^- on the natural module V is $\det(q\,\mathrm{id}_{\mathbb{R}^n} - w_\alpha^-) = \prod_{i=1}^r (q^{\alpha_i} + 1)$ if $\alpha = (\alpha_1, \ldots, \alpha_r)$. Since this is not divisible by $(q-1)$ it follows that the conjugacy class of w_α^- in W_n is a cuspidal class (see Lemma 3.1.10). Since different partitions of n give rise to different polynomials (use Exercise 3.12), it follows that no two elements of $\{w_\alpha^- \mid \alpha \vdash n\}$ are conjugate. Therefore, this set does indeed represent the cuspidal classes of W_n and (P3) holds. □

3.4.7. Proposition (Standard representatives in type B_n). *Let $W_n = \langle S\rangle$ be a Coxeter group of type B_n, as above.*

(i) *The conjugacy classes of W_n are parametrized by the pairs of partitions (α, β) with $|\alpha| + |\beta| = n$.*

(ii) *Let $\mathcal{P}$ be the set of pairs (α, β) consisting of a decreasing composition α and an increasing composition β such that $|\alpha| + |\beta| = n$. Then the set*

$$\{w_{(\alpha,\beta)} \mid (\alpha, \beta) \in \mathcal{P}\},$$

where $w_{(\alpha,\beta)}$ is defined as in (3.4.2), is a complete set of representatives of minimal length of the conjugacy classes of W_n.

Proof. (i) We use Theorem 3.2.12 to show the parametrization of the conjugacy classes of W_n. By Proposition 2.3.10, the Coxeter classes of parabolic subgroups of W_n are parametrized by the partitions α with $|\alpha| = m$ where $0 \leq m \leq n$. A representative W_J of the class whose label α is a partition of m is a direct product of a Coxeter group W_1 of type B_{n-m} and a product W_2 of irreducible components of type A. By Example 3.1.16, the Coxeter group W_2 only has one cuspidal class; and by the above proof of Theorem 3.2.7 for type B_n, the cuspidal classes of W_1 are represented by $\{w_\beta^- \mid \beta \vdash n-m\}$. So we have a bijective correspondence between the equivalence classes of pairs (J, D) as in Theorem 3.2.12 and pairs of partitions (α, β) as above.

(ii) Now fix a pair (J, D) with a corresponding pair of partitions (α, β) and write $W_J = W_1 \times W_2$ as in the proof of (i). Assume that the parts of α are arranged in decreasing order and those of β in increasing order. Then the definition in (3.4.2) shows that $w_{(\alpha,\beta)} = w_\beta^- \cdot w_2$, where $w_\beta^- \in W_1$ lies in a cuspidal class and $w_2 \in W_2$ is a Coxeter element. By Example 3.1.16 and the above proof of Theorem 3.2.7 for type B_n, the elements w_β^- and w_2 have minimal length in their conjugacy classes in W_1 and W_2, respectively. Hence $w_{(\alpha,\beta)}$ has minimal length in its conjugacy class in W_J (by Exercise 3.10) and, finally, also in its conjugacy class in W_n (by Lemma 3.1.14). □

3.4.8. Example. We illustrate Proposition 3.4.7 by listing the standard representatives of the conjugacy classes of W_n in the cases $n = 2$ and $n = 3$:

(α,β)	$w_{(\alpha,\beta)}$
$(11.\varnothing)$	1
(1.1)	t_0
$(\varnothing.11)$	t_0t_1
$(2.\varnothing)$	s_1
$(\varnothing.2)$	t_0s_1

(α,β)	$w_{(\alpha,\beta)}$	(α,β)	$w_{(\alpha,\beta)}$
$(111.\varnothing)$	1	(1.2)	t_0s_1
(11.1)	t_0	(2.1)	t_0s_2
(1.11)	t_0t_1	$(\varnothing.12)$	$t_0t_1s_2$
$(\varnothing.111)$	$t_0t_1t_2$	$(3.\varnothing)$	s_1s_2
$(21.\varnothing)$	s_1	$(\varnothing.3)$	$t_0s_1s_2$

Here we write compositions without commas and parentheses.

Finally, we consider the group W_n' of type D_n. Since W_n' is not a parabolic subgroup of W_n, the validity of $w \longrightarrow w'$ for $w, w' \in W_n'$ may depend on whether w, w' are regarded as elements of (W_n, S) or of (W_n', S'). We can, by Lemma 1.4.12, relate the relation $\longrightarrow$ in W_n' to that in W_n in the following way.

3.4.9. Lemma. *Let $w, w' \in W_n' \subset W_n$. If $w \longrightarrow w'$ in W_n then, in W_n', either $w \longrightarrow w'$ or $twt \longrightarrow w'$.*

Proof. By induction on the length of the conjugating element in type B_n it is sufficient to consider the situation where $w' = sws$ for some $s \in \{t, s_1, \ldots, s_{n-1}\}$. If $s = t$ then $w \longrightarrow twt$ via t in W_n trivially implies $twt \longrightarrow w'$ by 1 in W_n'. By Lemma 1.4.12, conjugation by s_i, for some $i = 1, \ldots, n-1$, does not affect $l_t(w) = l(w) - l'(w)$. Thus if $s = s_i$ and $l(s_iws_i) \leq l(w)$ then also $l'(s_iws_i) \leq l'(w)$. Hence, if $w \longrightarrow w'$ via s_i in W_n then we also have $w \longrightarrow w'$ via s_i in W_n'. □

3.4.10. Lemma. *Let $\alpha = (\alpha_1, \ldots, \alpha_l)$ be a composition of n such that l is even and let $w_\alpha^- \in W_n$ be defined as in equation* (3) *in* (3.4.2). *Then $w_\alpha^- \in W_n'$, and tw_α^-t and w_α^- are conjugate by cyclic shift in W_n'.*

Proof. Set $M_\alpha := \{\alpha_1 + \cdots + \alpha_i \mid i = 0, \ldots, n-1\}$. Then, as an element of W_n, we have

$$w_\alpha^- = a_0 \cdots a_{n-1} \qquad \text{where, for } i = 0, \ldots, n-1,\ a_i = \begin{cases} t_i & \text{if } i \in M_\alpha, \\ s_i & \text{otherwise.} \end{cases}$$

It follows that, as an element of W'_n, we have

$$w_\alpha^- = b_1 \cdots b_{n-1} \qquad \text{where, for } i = 1, \ldots, n-1,\ b_i = \begin{cases} u_i & \text{if } i \in M_\alpha, \\ s_i & \text{otherwise.} \end{cases}$$

Note that both these expressions for w_α^- are reduced expressions. Now $tw_\alpha^- t = w_\alpha$ unless $a_1 = s_1$, since every a_i different from s_1 commutes with t (see relation (t) on p. 91). Suppose $a_1 = s_1$ and let $k = \alpha_1$ (which then is greater than 1). Then $tw_\alpha^- t = ub_2 \cdots b_{n-1}$.

Clearly, $b_1 \cdots b_{n-1}$ and $b_k \cdots b_{n-1} s_1 \cdots s_{k-1}$ are conjugate by cyclic shift. But, since $b_1 \cdots b_{n-1}$ contains an odd number of factors u_i it follows (see relation (u) on p. 91) that

$$b_k \cdots b_{n-1} s_1 \cdots s_{k-1} = ub_2 \cdots b_{n-1},$$

whence w_α^- and $tw_\alpha^- t$ are conjugate by cyclic shift. □

It then follows from Lemma 3.4.9 that we can use the conjugating element from the Coxeter group W_n of type B_n that establishes $w \longrightarrow C_{\min}$ as a model for achieving the same effect in the Coxeter group W'_n of type D_n.

3.4.11. Proposition. *Let $W'_n = \langle S' \rangle$ be a Coxeter group of type D_n with $S' = \{u, s_1, \ldots, s_{n-1}\}$ as above. Let C be a cuspidal class of W'_n and let $w \in C$. Then $w \longrightarrow w_\alpha^-$ for an increasing composition α of n with an even number of parts.*

Proof. Let $w \in W'_n$ and consider it as an element of W_n. Then, by Proposition 3.4.6, there exists an element $w' \in W_n$ such that $w \longrightarrow w'$ in W_n and where either w' is an element of a proper parabolic subgroup of W_n or $w' = w_\alpha^-$ in W_n for some increasing composition α of n. In any case, by Lemma 3.4.9 we then have either $w \longrightarrow w'$ or $w \longrightarrow tw't$ in W'_n.

If w' lies in a proper parabolic subgroup W_J of W_n then it lies in the proper parabolic subgroup $W_J \cap W'_n$ of W'_n and $tw't$ lies in the proper parabolic subgroup $(W_J \cap W'_n)^t$ of W'_n (see Proposition 2.3.12). Therefore, in this case, w does not lie in a cuspidal class.

So suppose that $w' = w_\alpha^-$. This partition α has an even number of parts since the element w we start with lies in W'_n, the normal subgroup of W_n containing the elements with an even number of factors t. The claim now follows since $tw_\alpha^- t$ and w_α^- are conjugate by cyclic shift in W'_n, by Lemma 3.4.10. □

Proof of Theorem 3.2.7 for type D_n. The characteristic polynomial of w_α^- on V is the same, regardless of whether w_α^- is seen as an element of W_n or of W_n'. As we noted in the proof for type B_n, the characteristic polynomial of w_α^- is not divisible by $(q-1)$ and different partitions α give rise to different polynomials. Therefore, the set $\{w_\alpha^- \mid \alpha \vdash n,\ l(\alpha) \text{ even}\}$ is indeed a complete set of representatives of the cuspidal classes of W_n'.

We also see that (P3) is valid in this case. □

3.4.12. Proposition (Standard representatives in type D_n). *Let* $W_n' = \langle S' \rangle$ *be a Coxeter group of type* D_n*, as above.*

(i) *The conjugacy classes of* W *are parametrized by the pairs of partitions* (α, β) *where* $|\alpha| + |\beta| = n$ *and* β *has an even number of parts, and where there are two classes if* $\beta = \varnothing$ *and all parts of* α *are even.*

(ii) *Let* $\mathcal{P}_1$ *be the set of all decreasing compositions* $\alpha = (\alpha_1, \ldots, \alpha_l)$ *of* n *with* α_i *even for all* i. *Let* $\mathcal{P}_2$ *be the set of pairs* (α, β) *consisting of a decreasing composition* α *and an increasing composition* β *such that* $|\alpha| + |\beta| = n$, β *has an even number of parts and* $\alpha \notin \mathcal{P}_1$. *Then the set*

$$\{w_{(\alpha,\beta)} \mid (\alpha, \beta) \in \mathcal{P}_2\} \amalg \{w_\alpha^+ \mid \alpha \in \mathcal{P}_1\} \amalg \{t w_\alpha^+ t \mid \alpha \in \mathcal{P}_1\}$$

is a complete set of representatives of minimal length of the conjugacy classes of W_n'.

Proof. (i) By Proposition 2.3.13, the Coxeter classes of parabolic subgroups of W_n' are parametrized by the partitions of $m \le n-2$ and the partitions of n where every all-even partition α occurs twice as $(\alpha, +)$ and $(\alpha, -)$.

So let α be a partition of m and suppose first that α is not an all-even partition of n. Then a representative W_J of the class whose label α is a partition of m is a direct product of a Coxeter group W_1 of type D_{n-m} and a product W_2 of irreducible components of type A. By Example 3.1.16, the Coxeter group W_2 only has one cuspidal class; and by the above proof of Theorem 3.2.7 for type D_n, the cuspidal classes of W_1 are represented by $\{w_\beta^- \mid \beta \vdash n-m,\ l(\beta) \text{ even}\}$.

Next, let α be an all-even partition of n and let W_J be a representative of the Coxeter class with label $(\alpha, +)$. Then $W_J \le \langle s_1, \ldots, s_{n-1} \rangle$ and tW_Jt is a representative of the Coxeter class with label $(\alpha, -)$. Both W_J and tW_Jt are direct products of Coxeter groups of type A and, by Example 3.1.16, have only one cuspidal class.

Thus we have a bijective correspondence between the equivalence classes of pairs (J, D) as in Theorem 3.2.12 and the pairs of partitions or pairs consisting of an all-even partition of n and a sign, as above.

(ii) Now fix a pair (J, D) with a corresponding pair (α, β) or $(\alpha, \pm)$ and assume that the parts of α are arranged in decreasing order and those of β in increasing order.

In the case of $(\alpha, +)$ and $(\alpha, -)$ the elements w_α^+ and tw_α^+t have, by Example 3.1.16, minimal length in their conjugacy class in W_J, and thus minimal length in their conjugacy class in W, by Lemma 3.1.14.

In all the other cases (α, β) we write $W_J = W_1 \times W_2$ as in the proof of (i) and conclude that $w_{(\alpha,\beta)}$ has minimal length in its conjugacy class in W as in the proof of Proposition 3.4.7. □

Note how the fusion of conjugacy classes of W'_n, if seen as conjugacy classes of W_n, is being dealt with already at the level of Coxeter classes of parabolic subgroups; see Proposition 2.3.13.

3.4.13. Example. We illustrate Proposition 3.4.12 by listing the standard representatives of the conjugacy classes of W'_n in the case $n = 4$:

(α,β)	$w_{(\alpha,\beta)}$	(α,β)	$w_{(\alpha,\beta)}$
$(1111.\varnothing)$	1	$(22.+)$	s_1s_3
(11.11)	$t_0t_1 = u_1$	$(22.-)$	us_3
$(\varnothing.1111)$	$t_0t_1t_2t_3 = u_1u_2u_3$	$(\varnothing.22)$	$t_0s_1t_2s_3 = s_1u_2s_3$
$(211.\varnothing)$	s_1	$(31.\varnothing)$	s_1s_2
(1.12)	$t_0t_1s_2 = u_1s_2$	$(\varnothing.13)$	$t_0t_1s_2s_3 = u_1s_2s_3$
(2.11)	$t_0t_1s_3 = u_1s_3$	$(4.+)$	$s_1s_2s_3$
		$(4.-)$	us_2s_3

Here we write the compositions without commas and parentheses.

3.5 BIBLIOGRAPHICAL REMARKS AND EXERCISES

The conjugacy classes of the symmetric group are easily found to be parametrized by partitions. The conjugacy classes of the Coxeter groups of type B_n and D_n have been known for a long time (see [Specht 1937, Young 1977]). The conjugacy classes of the remaining irreducible Coxeter groups have been studied individually. A systematic description of the conjugacy classes of all finite Coxeter groups of crystallographic type was first given in [Carter 1972a]. This is built on the observation that every element of W is a product of two involutions. Our approach through cuspidal classes and parabolic subgroups is new and completely independent of the earlier results.

Coxeter elements form the subject of [Bourbaki 1968, Chap. V, §6]. The proof of Theorem 3.1.4 given there is formulated in terms of a graph-theoretical argument. Another proof of the theorem is found in [Carter 1972b, Section 10.3]. The relation of conjugacy by cyclic shift (in the form of Definition 3.1.3) appears in [Broué and Michel 1996, Def. 3.16].

The converse of Lemma 3.1.10 can be proved by a general argument, which involves the fact that the stabilizer in W of a vector in V is a parabolic subgroup; see [Steinberg 1968, (1.20)].

Theorem 3.2.9 appears in [Geck and Pfeiffer 1993], where it is proved for the classical types, and the algorithmic solution for the exceptional types is outlined.

For the non-crystallographic types, a proof is given in [Geck et al. 1996]. The intermediate Theorem 3.2.7 which we use here to derive Theorem 3.2.9 and its proof through a case-by-case analysis are new.

The description of the conjugacy graph for classes of involutions in the sense of Proposition 3.2.10 is the theme of [Richardson 1982]. Our proof is a variation of Howlett's proof of the theorem (private communication).

This chapter represents only the beginning of a structure theory of the conjugacy graph. In Exercise 3.2 it is shown that the size of the cyclic shift class containing the Coxeter elements of W is a power of 2. The tables in Appendix B show the sizes of all cyclic shift classes of elements of minimal length in cuspidal classes. Is there any pattern to spot? What can the size of a cyclic shift class be in general, or its structure? Other topics which are not mentioned include conjugacy graphs of non-cuspidal classes (can their structure be captured in an inductive way?) and the graph that is obtained from the conjugacy graph by collapsing cyclic shift classes into single points. Elements of maximal length play the role of elements of minimal length for Coxeter groups twisted by an automorphism and are studied in [Kim 1998] and in [Geck et al. 2000].

Exercise 3.1. Let (W, S) be (i) of type A_4, (ii) of type D_4, (iii) of type F_4. Determine in each case the set of Coxeter elements of W and, for each Coxeter element of W, all its different reduced expressions.

Exercise 3.2 ([Shi 1997]). Let (W, S) be any Coxeter system such that W is finite. Show that, if e is the number of edges of the Coxeter graph of W, the number of Coxeter elements of W is 2^e.

[*Hint.* Suppose that $S = \{s_1, s_2, \ldots, s_n\}$. Take any $s_i, s_j \in S$, where $i < j$, such that s_i and s_j are connected by an edge in the Coxeter graph of (W, S). Then s_i and s_j do not commute. Now let w be a Coxeter element of W. Show that, if in any reduced expression for w the generator s_i occurs to the left of s_j then this is the case for all reduced expressions of w. Associate with w the set of all edges $\{s_i, s_j\}$ $(i < j)$ of the Coxeter graph where s_i occurs to the left of s_j in w. Show that w (as a Coxeter element) is uniquely determined by this set and that each subset of edges of the Coxeter graph occurs as a set associated with a Coxeter element.]

Exercise 3.3. The parabolic marks (the numbers of fixed points on W-sets of the form $[W/W_J]$) of an arbitrary subgroup of W coincide with the marks of a class of parabolic subgroups associated with it: for $U \leq W$ we define the *parabolic closure* $\mathrm{parc}_W(U)$ of U in W as

$$\mathrm{parc}_W(U) = \bigcap \{W_J^w \mid J \subseteq S, w \in W, U \leq W_J^w\},$$

the intersection of all conjugates of parabolic subgroups of W that contain U. Then $\mathrm{parc}_W(U)$ is a conjugate of W_K for some $K \subseteq S$. Show that

(i) $\mathsf{parc}_W(U)$ is conjugate to W_K if and only if $|\mathsf{Fix}_U(W/W_J)| = \beta_{JK}$ for all $J \subseteq S$;

(ii) $\mathsf{parc}_W(U) = \mathsf{Stab}(\mathsf{Fix}_U(V))$ for every subgroup U of W, where V is the natural module of W;

(iii) $\mathsf{parc}_W(\langle c_J \rangle) = W_J$ if $J \subseteq S$ and c_J is a Coxeter element of W_J.

[*Hint*. Let $U' = \mathsf{parc}_W(U)$. We have $\mathsf{Fix}_U(W/W_J) \supseteq \mathsf{Fix}_{U'}(W/W_J)$ since $U \subseteq U'$. Now let $x \in \mathsf{Fix}_U(W/W_J)$. We may assume that $x = W_J$. But then $U \leq W_J$ and, by the definition of U', $U' \leq W_J$, whence $x \in \mathsf{Fix}_{U'}(W/W_J) = \mathsf{Fix}_U(W/W_J)$. The converse follows from the fact that U' is conjugate to some W_K, and that all the columns of the parabolic table of marks are different.]

Exercise 3.4. Let parc_W be defined as in Exercise 3.3. Let $J \subseteq S$ and denote by π_J the permutation character of W on the cosets of W_J as in (3.1.7). For $w \in W$ show that $\pi_J(w) = \beta_{JK}$ if and only if $\mathsf{parc}_W(\langle w \rangle) \sim W_K$.

[*Hint*. Note that a permutation character value coincides with the mark of a cyclic subgroup, both being the same number of fixed points.]

Exercise 3.5. Let G be a group and let Ω be a G-set. Suppose G is generated by $\{g_1, \ldots, g_n\}$ and, for $g \in G$, write $\mathsf{Fix}_g(\Omega)$ for $\mathsf{Fix}_{\langle g \rangle}(\Omega)$. Show that

$$\mathsf{Fix}_G(\Omega) = \bigcap_{i=1}^{n} \mathsf{Fix}_{g_i}(\Omega).$$

Exercise 3.6. Given a finite Coxeter group $W = \langle S \rangle$, we define a monoid M by a presentation with

generators:	$\hat{s}$, $s \in S$	
relations:	$\hat{s}^3 = \hat{s}$	for all $s \in S$
	$\mathsf{Prod}(\hat{s}, \hat{t}; m_{st}) = \mathsf{Prod}(\hat{t}, \hat{s}; m_{st})$	for $s, t \in S$, $s \neq t$, $m_{st} < \infty$,

where m_{st} as usual is the order of the product st in W.

(i) Show that the map $W \times M \to W$ given by

$$(w, \hat{s}) \mapsto \begin{cases} sws & \text{if } l(sws) \leq l(w), \\ w & \text{otherwise,} \end{cases}$$

defines an action of the monoid M on W.

(ii) Let $w \in W$ and describe the role of both the conjugacy class of w in W and the cyclic shift class $\mathsf{Cyc}(w)$ under this action.

Exercise 3.7. In this exercise two characterizations of "cyclic shift" are compared. Recall the definition of $\mathsf{Cyc}(w)$ from Section 3.2 and the notion of cyclic shift from Definition 3.1.3. Show that the cyclic shift class of w in the latter sense is precisely the set $\mathsf{Cyc}(w)$.

[*Hint*. Use Lemma 1.2.6.]

Exercise 3.8. Let $w \in W$ and $s \in S$. Show that $J(v) = J(w)$ and $l_s(v) = l_s(w)$ for all $v \in \mathrm{Cyc}(w)$.

Exercise 3.9. Let $J \subseteq S$, let $w \in W_J$ and suppose $w \longrightarrow w'$ for some $w' \in W$. Show that $w' \in W_J$ and $w \longrightarrow w'$ in W_J (i.e., via elements of W_J).

Exercise 3.10. Given a reducible Coxeter group $W = W_1 \times W_2$, show the following:

(i) If C^i is a conjugacy class of W_i ($i = 1, 2$), then $C = C^1 \times C^2$ is a conjugacy class of W and $C_{\mathrm{min}} = C^1_{\mathrm{min}} \times C^2_{\mathrm{min}}$.
(ii) The conjugacy class $C = C^1 \times C^2$ of W is a cuspidal class if and only if C^1 is a cuspidal class of W_1 and C^2 is a cuspidal class of W_2.
(iii) Suppose $w_i \in W_i$ ($i = 1, 2$). Then $w_1 w_2 \in W$ and $\mathrm{Cyc}(w_1 w_2) = \mathrm{Cyc}(w_1) \times \mathrm{Cyc}(w_2)$.

Exercise 3.11. Let $W = \langle S \rangle \subset \mathrm{GL}(V)$ be a finite Coxeter group, where V denotes the natural module. Let $J \subseteq S$ and $W_J \subseteq W$ be the corresponding parabolic subgroup, with natural module $V_J \subseteq V$ spanned by the simple roots corresponding to the elements of J. Show that $V = V_J \oplus \mathrm{Fix}_{W_J}(V)$ and that

$$\det(q\,\mathrm{id}_V - w) = (q-1)^{|S-J|} \det(q\,\mathrm{id}_{V_J} - w) \qquad \text{for all } w \in W_J.$$

[*Hint.* Choose a W-invariant scalar product on V. Then each $s \in S$ is a reflection as in Exercise 1.14 and so $\mathrm{Fix}_{W_J}(V)$ is the orthogonal complement to V_J.]

Exercise 3.12. Let $\alpha = [\alpha_1, \ldots, \alpha_r]$ be a partition of n and let q be an indeterminate. Show that α is uniquely determined by the polynomial $\prod_{i=1}^{r}(q^{\alpha_i}+1)$.
[*Hint.* Proceed by induction on r; see also [Bleher et al. 1997, (2.7)].]

Exercise 3.13 ([Solomon 1976]). Recall from Section 2.4 the definitions of the Burnside ring $\Omega(W)$ and the parabolic Burnside ring $\Pi(W)$. Let $\Sigma(W) = \mathbb{Z}[x_J \mid J \subseteq S]$ (where $x_J = \sum_{x \in X_J} x^{-1}$) and let $\theta \colon \Sigma(W) \otimes \mathbb{Q} \to \Omega(W) \otimes \mathbb{Q}$ be the linear map defined by $x_J \mapsto [W/W_J]$ for all $J \subseteq S$. Show that:

(i) θ is a ring homomorphism with kernel generated by $\{x_J - x_K \mid J \sim K\}$ and image $\Pi(W) \otimes \mathbb{Q}$;
(ii) $\ker \theta$ is the radical of the algebra $\Sigma(W) \otimes \mathbb{Q}$.

[*Hint.* Note that $[W/W_J] = [W/W_K]$ if and only if $J \sim K$.]

Exercise 3.14. Show that the parabolic Burnside ring $\mathbb{Q} \otimes \Pi(W)$ is isomorphic to the ring of class functions of W with values in $\mathbb{Q}$ (and pointwise addition and multiplication) generated by $\{\pi_J \mid J \subseteq S\}$.

Exercise 3.15 ([Solomon 1976]). Let π_J be the permutation character of W on the cosets of W_J, as in (3.1.7), and let ε be the sign character of W defined by $\varepsilon(s) = -1$ for all $s \in S$, as in Remark 1.1.5. Let θ be as in Exercise 3.13 and let

θ' be the composition of θ and the isomorphism from Exercise 3.14, defined by $x_J \mapsto \pi_J$ for all $J \subseteq S$. Show that $\theta'(w_0) = \varepsilon$ and conclude that

$$\varepsilon = \sum_{J \subseteq S} (-1)^{|J|} \pi_J.$$

[*Hint*. By Exercise 2.3, $w_0 = \sum_{J \subseteq S} (-1)^{|J|} x_J$. Hence

$$(\theta'(w_0), \varepsilon) = \sum_{J \subseteq S} (-1)^{|J|} (\pi_J, \varepsilon) = \sum_{J \subseteq S} (-1)^{|J|} (1_J, \varepsilon_J) = 1$$

since $(1_J, \varepsilon_J) = 0$ unless $J = \varnothing$. Here 1_J is the trivial and ε_J the sign character of W_J. That $(\theta'(w_0), \chi) = 0$ for every irreducible character $\chi \neq \varepsilon$ follows from the fact that $\pm\theta'(w_0)$ is an irreducible character. This in turn follows from $(\theta'(w_0), \theta'(w_0)) = (1_W, \theta'(w_0)^2) = (1_W, 1_W) = 1$. Note that $\theta'(w_0)$ is a rational character and therefore invariant under complex conjugation, and that $\theta'(w_0)^2 = 1$ since $w_0^2 = 1$.]

Exercise 3.16. For $w \in W$ show that if w lies in a cuspidal class of W then $q - 1$ does not divide $p_w(q)$.

[*Hint*. This follows from a case-by-case analysis, using (3.4.3) and the tables of polynomials in Appendix B. For a general proof, use [Steinberg 1968, (1.20)].]

Exercise 3.17. Suppose that $|S| = n$ and let $w \in W$ be an element of a cuspidal class of W. Show that w can be written as a product of n reflections. Conclude that $l(w) \equiv n \pmod 2$.

[*Hint*. This follows from a case-by-case analysis using the tables in Appendix B for the exceptional types. For a general proof, see [Carter 1972a, Lemma 2].]

Exercise 3.18. Suppose that $w \in W$ is an involution. Show that:

(i) the cyclic shift class $\mathrm{Cyc}(w)$ of w has size 1;
(ii) if $w' \in W$ is such that $w \longrightarrow w'$, then there exists an element $x \in W$ such that $w^x = w'$ and $l(w) = 2l(x) + l(w')$.

[*Hint*. The left descent set $\mathcal{L}(w) = \{s \in S \mid l(sw) < l(w)\}$ of an involution w coincides with its right descent set $\mathcal{R}(w)$.]

4
The braid monoid and good elements

In this chapter, we consider the braid group and the braid monoid associated with a Coxeter system (W, S). We introduce the braid group from a purely algebraic point of view, as a group given by a presentation which is similar to that of W, but where the quadratic relations are omitted. Our aim is to prove the basic results due to Deligne and Brieskorn–Saito about a normal form for the elements in the braid group. This will be achieved in Sections 4.1 and 4.2. The subsequent section deals with the application of these results to the problem of finding elements with certain "good" properties in the conjugacy classes of W (assuming W to be finite). The desired property is formulated in Definition 4.3.1; note that it refers in an essential way to the corresponding braid group. We determine, case by case, "good" elements of minimal length in the conjugacy classes of all finite Coxeter groups. These results will play a crucial role in our discussion of rationality properties and the computation of the values of the irreducible characters of Iwahori–Hecke algebras.

At the end of Section 4.1, assuming that W is finite, we also give a brief indication of the topological meaning of the braid group as the fundamental group of a certain orbit space of W in its standard reflection representation. In type A, one obtains in this way the classical braid group studied in [Artin 1947]. Via this topological interpretation, the theory of "algebraic" braid groups is connected with the theory of knots and links.

In Section 4.4, we will introduce Iwahori–Hecke algebras as certain quotients of the group algebra of the braid group. These quotients depend on the choice of a ground ring and a collection of elements from that ground ring which play the role of "parameters". We prove the basic result due to Bourbaki that the algebras obtained in this way have a standard basis indexed by the elements of W.

In Section 4.5, as a first application, we briefly discuss Jones's construction of an invariant for oriented knots and links, via a certain trace function on Iwahori–Hecke algebras of type A. Our approach here follows some suggestions by J. Michel.

Later, in Section 8.4, we will see that if W arises as the Weyl group of a group G with a BN-pair then a corresponding Iwahori–Hecke algebra also appears in a completely different context, namely as the endomorphism algebra of a certain representation of G. In Chapter 8, we will also begin to develop systematically the representation theory of Iwahori–Hecke algebras.

4.1 THE BRAID MONOID AND THE BRAID GROUP

Let (W, S) be a Coxeter system, i.e., W is a group, $S \subseteq W$ is a finite subset, and we have a presentation with

$$\begin{array}{lll} \text{generators:} & s \in S & \\ \text{relations:} & (st)^{m_{st}} = 1 & \text{for all } s, t \in S \text{ with } m_{st} < \infty. \end{array}$$

In fact, there are two types of defining relations involved: the *quadratic relations* $s^2 = 1$ for all $s \in S$, and order relations $(st)^{m_{st}} = 1$ for $s \neq t$ with $m_{st} < \infty$. Now the latter relations can also be rewritten in the form

$$\mathsf{Prod}(s, t; m_{st}) = \mathsf{Prod}(t, s; m_{st}) \quad \text{where } s, t \in S \text{ and } s \neq t.$$

These relations will be called *braid relations*.

Recall from Section 1.2 that Matsumoto's theorem was concerned with maps from W to monoids so that the images of the generators in S satisfy the braid relations. There is a universal monoid for this, which is called the *braid monoid* of (W, S) and is denoted by $B^+ = B^+(W, S)$. It is given by a presentation on generators in bijection with the elements of S, subject to the braid relations. In order to distinguish the generators of W from those of B^+, we denote the latter ones by bold symbols. Thus, we have a presentation of B^+ as a monoid with

$$\begin{array}{lll} \text{generators:} & \mathbf{s},\ s \in S & \\ \text{relations:} & \mathsf{Prod}(\mathbf{s}, \mathbf{t}; m_{st}) = \mathsf{Prod}(\mathbf{t}, \mathbf{s}; m_{st}) & \text{for } s, t \in S,\ s \neq t,\ m_{st} < \infty. \end{array}$$

4.1.1. Remark. By Matsumoto's theorem, there exists a unique map $r\colon W \to B^+$ such that $r(w) = \mathbf{s}_1 \cdots \mathbf{s}_k$ whenever $w \in W$ and $w = s_1 \cdots s_k$ $(s_i \in S)$ is a reduced expression. We shall denote

$$B^+_{\mathbf{red}} := \{r(w) \mid w \in W\}.$$

An intrinsic description of $B^+_{\mathbf{red}}$ is given in Exercise 4.1. Note that the map r has the following multiplicative property:

$$r(w_1 w_2) = r(w_1) r(w_2) \qquad \text{if } w_1, w_2 \in W \text{ with } l(w_1 w_2) = l(w_1) + l(w_2).$$

Indeed, if $l(w_1) + l(w_2) = l(w_1 w_2)$, then a reduced expression for $w_1 w_2$ is obtained by concatenating reduced expressions for w_1 and w_2.

The aim of this section is to discuss some basic results about B^+, including a normal form for its elements. This is based on a theory of divisibility of elements in B^+, whose proof will be given in Section 4.2. The overall strategy will be to establish the results first in W, then to transport them to $B^+_{\mathbf{red}}$ via the map r, and finally to extend them to all of B^+.

4.1.2. Remark. The length function on W can be extended to a length function $l\colon B^+ \to \mathbb{N}_0$ by setting $l(g) = k$ if $g = s_1 \cdots s_k$ with $s_i \in S$. Note that this is well defined since the braid relations do not affect the number of factors in an expression of g as above. Note also that the analogue of Corollary 1.2.3 holds: if $g = s_1 \cdots s_k = t_1 \cdots t_k$ with $s_i, t_j \in S$ then $\{s_1, \ldots, s_k\} = \{t_1, \ldots, t_k\}$.

We can certainly write each element $g \in B^+$ as a product $g = g_1 \cdots g_n$ with $g_i \in B^+_{\mathsf{red}}$ for all i. In other words, B^+ is generated, as a monoid, by its subset B^+_{red}. As in the case of W itself, there may be several expressions of $g \in B^+$ as a product $g = g_1 \cdots g_n$ with $g_i \in B^+_{\mathsf{red}}$. We shall see below, in Definition 4.1.5, that we obtain a unique expression if we require that g_1 has maximum possible length as first factor, g_2 has maximum possible length as second factor and so on.

We have just seen that B^+ is generated by the subset B^+_{red}. These generators obviously satisfy the multiplicative relations in Remark 4.1.1. It may look surprising that these relations form in fact a set of defining relations for B^+:

4.1.3. Proposition. *The monoid B^+ has a presentation with*

generators: $r(w)$, *for* $w \in W$
relations: $r(w_1w_2) = r(w_1)r(w_2)$ *if* $l(w_1w_2) = l(w_1) + l(w_2)$.

Proof. For the sake of separating the notation, we let $\tilde{W} = \{\tilde{w} \mid w \in W\}$ be a set in bijection to W and let M be the monoid defined by a presentation as above, with generators $\tilde{W}$ and relations $\tilde{w} = \tilde{w}_1\tilde{w}_2$ if $w = w_1w_2$ and $l(w) = l(w_1) + l(w_2)$.

Consider the map $f\colon \tilde{W} \to B^+$, $\tilde{w} \mapsto r(w)$. Using the multiplicative property of r in Remark 4.1.1, we conclude that f has a unique extension to a monoid homomorphism $M \to B^+$, which we denote by the same symbol.

Conversely, consider the map $f'\colon \{\mathbf{s} \mid s \in S\} \to M$, $\mathbf{s} \mapsto \tilde{s}$. In order to show that f' extends to a unique monoid homomorphism from B^+ to M, we must check that $\mathsf{Prod}(\tilde{s}, \tilde{t}; m_{st}) = \mathsf{Prod}(\tilde{t}, \tilde{s}; m_{st})$ for all $s, t \in S$ with $s \neq t$ and $m_{st} < \infty$. But, by (1.1.7)(a), m_{st} is the order of st, which implies that $\mathsf{Prod}(s, t; m_{st})$ and $\mathsf{Prod}(t, s; m_{st})$ are reduced expressions. On the other hand, applying repeatedly the defining relations in M yields that $\tilde{w} = \tilde{s}_1 \cdots \tilde{s}_k$ if $w = s_1 \cdots s_k$ with $s_i \in S$ is a reduced expression. Thus, we can conclude that $\mathsf{Prod}(\tilde{s}, \tilde{t}; m_{st}) = \mathsf{Prod}(\tilde{t}, \tilde{s}; m_{st})$, as desired.

So we have homomorphisms $f\colon M \to B^+$ and $f'\colon B^+ \to M$ such that $f \circ f' = \mathsf{id}_{B^+}$, $f' \circ f = \mathsf{id}_M$. It follows that $M \cong B^+$, as desired. □

We define a concept of divisibility in B^+ as follows: let $g, h \in B^+$. We say that g *divides* h, and write $g \prec h$, if $h = gg'$ with $g' \in B^+$. This defines a partial order relation on B^+. We can now state the following basic result:

4.1.4. Theorem (Deligne). *The monoid B^+ is left simplifiable, i.e., if we have $gh = gh'$ with $g, h, h' \in B^+$ then $h = h'$. Moreover, for each $g \in B^+$, there exists a unique element $w \in W$ of maximum possible length such that $r(w) \prec g$.*

The proof will be given in Section 4.2. In Algorithm L (p. 116), the above result is made constructive. What we will do now is to discuss consequences of this result.

Using Theorem 4.1.4 we define a map $\alpha\colon B^+ \to B^+_{\mathrm{red}}$ as follows. Let $g \in B^+$. Then $\alpha(g) = r(w)$ where $w \in W$ is the unique element of maximum possible length such that $r(w) \prec g$. Since B^+ is left simplifiable, we can define a second map $\omega\colon B^+ \to B^+$ by the condition that $g = \alpha(g)\omega(g)$ for all $g \in B^+$. Repeating the above process with $\omega(g)$, we obtain a normal form for the elements of B^+:

4.1.5. Definition. Let $g \in B^+$ and write $g = g_1 \cdots g_n$ with $g_i \in B^+_{\mathrm{red}}$ for all i. We say that this expression is the *Deligne normal form* of g if

$$g_i = \alpha(g_i \cdots g_n) \qquad \text{for } 1 \leq i \leq n-1.$$

We will see lots of examples for normal forms in subsequent sections. Here is a simple example which shows what is going on:

4.1.6. Example. Let (W, S) be of type A_3, with generators s_1, s_2, s_3 and the usual relations. Recall our convention that we denote $r(s_i) = s_i$. Consider the element $g = (s_1s_2s_3)^4 \in B^+$. We leave it to the reader to check that, by rearranging the given expression using the braid relations, we obtain

$$(s_1s_2s_3)(s_1s_2s_3)(s_1s_2s_3)(s_1s_2s_3) = (s_1s_2s_1s_3s_2s_1)(s_1s_2s_1s_3s_2s_1).$$

Now, we have $w_0 = s_1s_2s_1s_3s_2s_1$ (the longest element in W). Thus, setting $\Delta = r(w_0)$, we see that $g = g_1g_2$ with, $g_1 = g_2 = \Delta$, is the Deligne normal form of g. In Section 4.3, we will study in general the question of what happens when we raise $r(w)$ to a suitable power in B^+.

4.1.7. Parabolic submonoids. For $J \subseteq S$ we have a parabolic subgroup $W_J = \langle J \rangle \subseteq W$; see (1.2.9). Similarly, we now define a *parabolic submonoid* B_J^+ of B^+ by

$$B_J^+ := \langle r(s) \mid s \in J \rangle \subseteq B^+.$$

In (1.2.9) we saw that (W_J, J) itself is a Coxeter system. Now it is natural to ask whether B_J^+ is the braid monoid associated with the Coxeter system (W_J, J), i.e., the monoid $B^+(W_J, J)$ given by a presentation with

generators: s, $s \in J$
relations: $\mathrm{Prod}(s, t; m_{st}) = \mathrm{Prod}(t, s; m_{st})$ for $s, t \in J$, $s \neq t$, $m_{st} < \infty$.

This is indeed so: we certainly have a canonical surjective map $B^+(W_J, J) \to B_J^+$, induced by identifying the generators of $B^+(W_J, J)$ with those of B_J^+. In order to construct an inverse of this map, we proceed as follows. Let $g \in B_J^+$ and write $g = g_1 \cdots g_n$ with $g_i \in B^+_{\mathrm{red}}$ in Deligne normal form. Let $g_i = r(w_i)$ with $w_i \in W$ for all i. Since g is a product of generators from J, the same holds for each g_i

(see Remark 4.1.2), and hence $w_i \in W_J$ for all i. Thus, we can define an inverse of the above map by sending g to the product of the images of $w_1, \dots, w_n$ in $B^+_{\mathrm{red}}(W_J, J)$. Hence, indeed, we have a canonical isomorphism $B^+(W_J, J) \cong B^+_J$.

There is a refinement of the Deligne normal form for the elements of B^+. This is based on the existence of a unique longest element in finite Coxeter groups.

For $J \subseteq S$, if W_J is finite, there exists a unique longest element in W_J (see Proposition 1.5.1), which we shall denote by w_J.

The refined normal form for the elements in B^+ is defined as follows. Let $g \in B^+$ and define $J(g) := \{s \in S \mid r(s) \prec g\} \subseteq S$. We claim that $W_{J(g)}$ is finite and that $\Delta_{J(g)} \prec g$, where we denote

$$\Delta_J := r(w_J) \in B^+_{\mathrm{red}} \qquad \text{for each subset } J \subseteq S \text{ with } W_J \text{ finite.}$$

Indeed, write $g = g_1 \cdots g_n$ in Deligne normal form, where $g_1 = \alpha(g)$. Since $r(s) \prec g$ for all $s \in J(g)$ we must have $r(s) \prec \alpha(g)$ for all $s \in J(g)$. This property also holds in W: if we write $\alpha(g) = r(w)$ for $w \in W$, we have $l(sw) < l(w)$ for all $s \in J(g)$. By Lemma 1.5.2, this implies that $W_{J(g)}$ is finite and that $w = w_{J(g)}w'$ with $l(w) = l(w_{J(g)}) + l(w')$. So we also have $\alpha(g) = r(w) = r(w_{J(g)})r(w')$. It follows that $g = \Delta_{J(g)}g'$ for some $g' \in B^+$. The uniqueness of g' follows from the left simplifiability of B^+. Repeating this process with g', we obtain the desired normal form for the elements in B^+:

4.1.8. Definition. Let $g \in B^+$ and write $g = g_1 \cdots g_n$ with $g_i \in B^+_{\mathrm{red}}$ for all i. We say that this expression is the *Brieskorn–Saito normal form* of g if

$$g_i = \Delta_{J_i} \quad \text{where} \quad J_i = J(g_i \cdots g_n) \quad \text{for } 1 \leq i \leq n.$$

The above constructions work without finiteness assumptions on W. Now we assume that W is finite, and let $w_0 \in W$ be the longest element. The element

$$\Delta := r(w_0) \in B^+_{\mathrm{red}}$$

plays a quite important role in the structure theory of B^+, and it is therefore called the *fundamental element* of B^+. The following result contains some basic properties of this element. Recall from Lemma 1.5.3 that conjugation by w_0 defines an automorphism $W \to W$, $w \mapsto \bar{w}$, such that $\overline{S} = S$.

4.1.9. Lemma. *We have $r(w)\Delta = \Delta r(\bar{w})$ for all $w \in W$. In particular, the element Δ^2 commutes with all elements of B^+. Moreover, if $g \in B^+$, then $g \prec \Delta^{2n}$ for some $n \geq 1$.*

Proof. To prove the first statement, it is enough to show that

$$r(s)\Delta = \Delta r(\bar{s}) \qquad \text{for each } s \in S.$$

To see this, let $s \in S$ and set $x := sw_0$. Then $w_0 = sx$, $l(w_0) = l(x)+1$, and hence $\Delta = r(s)r(x)$. On the other hand, we have $x = w_0\bar{s}$ and so, by a similar argument,

$\Delta = r(x)r(\bar{s})$. It follows that $r(s)\Delta = r(s)(r(x)r(\bar{s})) = (r(s)r(x))r(\bar{s}) = \Delta r(\bar{s})$, as desired. Since $w \mapsto \bar{w}$ has order 2, this implies that Δ^2 commutes with all elements of B^+.

Finally, let $g \in B^+$ and write $g = g_1 \cdots g_n$ with $g_i \in B^+_{\text{red}}$. Then $g_i = r(w_i)$ for some $w_i \in W$. By Lemma 1.5.3, we have $w_0 = w_i(w_i^{-1}w_0)$ and $l(w_0) = l(w_i) + l(w_i^{-1}w_0)$. This implies that $g_i = r(w_i) \prec r(w_0) = \Delta$. In particular, we have $g_i \prec \Delta^2$. Let us write $\Delta^2 = g_i h_i$ with $h_i \in B^+$. Since Δ^2 commutes with each h_i, we can compute that

$$\Delta^{2n} = \Delta^{2n-2} g_1 h_1 = g_1 \Delta^{2n-2} h_1 = \cdots = g_1 \cdots g_n h_n \cdots h_1 = g h_n \cdots h_1,$$

and so $g \prec \Delta^{2n}$. □

4.1.10. Proposition. *Assume that W is finite and let $B = B(W, S)$ be the group defined by a presentation with*

generators: s, $s \in S$
relations: $\mathrm{Prod}(s, t; m_{st}) = \mathrm{Prod}(t, s; m_{st})$ *for* $s, t \in S$, $s \neq t$, $m_{st} < \infty$.

Then the canonical map $B^+ \to B$ induced by identifying the generators of B^+ with those of B is injective and, moreover, every element of B can be written in the form g/h where $g, h \in B^+$. (We can even choose h to be a power of Δ.) Thus, the monoid B^+ can be embedded into a group, which is called the Artin–Tits braid group associated with (W, S).

Proof. First, we show that B^+ can be embedded into some group G. Such a group is constructed as follows. The set $\mathbb{N}_0 \times B^+$ is a monoid with multiplication defined by $(m, g) \cdot (m', g) := (m + m', gg')$ and identity element $(0, 1)$. We consider the equivalence relation on $\mathbb{N}_0 \times B^+$ defined by $(m, g) \sim (m', g') \iff \Delta^{2m} g' = \Delta^{2m'} g$. Let us check that this is indeed an equivalence relation: symmetry and reflexivity are clear; assume now that we have $(m, g) \sim (m', g')$ and $(m', g') \sim (m'', g'')$. Since Δ^2 commutes with all elements of B^+, we have

$$\begin{aligned}
\Delta^{2(m'+m'')} g &= \Delta^{2m''} \Delta^{2m'} g = \Delta^{2m''} \Delta^{2m} g' && \text{since } (m, g) \sim (m', g') \\
&= \Delta^{2m} \Delta^{2m''} g' = \Delta^{2m} \Delta^{2m'} g'' && \text{since } (m', g') \sim (m'', g'') \\
&= \Delta^{2m'} \Delta^{2m} g''.
\end{aligned}$$

Since B^+ is left simplifiable (see Theorem 4.1.4), we conclude that $\Delta^{2m''} g = \Delta^{2m} g''$ and hence $(m, g) \sim (m'', g'')$, as desired.

Now let G be the quotient of $\mathbb{N}_0 \times B^+$ by this equivalence relation. Denote the equivalence class of (m, g) by $[m, g]$. Then define a multiplication in G by

$$[m, g] \cdot [m', g'] := [m + m', gg'].$$

(Since Δ^2 commutes with all elements of B^+, it is readily checked that this is well defined.) Thus, G is a monoid with identity element $[0, 1]$. We have a natural monoid homomorphism $B^+ \to G$, $g \mapsto [0, g]$. This is clearly injective.

So it remains to show that each element in G has a left and a right inverse. Let $[m, g] \in G$. By Lemma 4.1.9, there exists some $n \geq 1$ such that $g \prec \Delta^{2n}$. We choose $n \geq m$ and write $\Delta^{2n} = gg'$ with $g' \in B^+$. Then $[m, g] \cdot [n-m, g'] = [n, gg'] = [n, \Delta^{2n}]$. But note that (n, Δ^{2n}) is equivalent to $(0, 1)$, and hence $[n-m, g']$ is a right inverse to $[m, g]$. But it is a general fact about monoids that if every element has a right inverse then it also has a left inverse. (Indeed, given x, we take a right inverse y to x and a right inverse z to y; then we have $x = x(yz) = (xy)z = z$, i.e., $yx = yz = 1$.) Thus, G is a group and we have an embedding of B^+ into G, where every element of G can be expressed as required.

Now the embedding $B^+ \hookrightarrow G$ certainly factors through the canonical map $B^+ \to B$. Hence the latter map is also injective. □

4.1.11. Remark. It is shown in [Brieskorn and Saito 1972, (5.5)] that if B^+ embeds into a group such that each element of that group can be expressed as a quotient of two elements in B^+, then W must be finite.

We only remark briefly that the braid group $B(W, S)$ has a topological meaning. Recall from Chapter 1 that W is naturally a group of linear transformations of a real vector space V, where the elements of S are represented by reflections. Now consider the set $T \subset W$ of all conjugates of the elements in S. Each $t \in T$ is a reflection in W, and so we have a corresponding reflecting hyperplane $H_t \subset V$. Let $V_{\mathbb{C}}$ denote the complexified space and $(H_t)_{\mathbb{C}} \subseteq V_{\mathbb{C}}$ $(t \in T)$ the corresponding hyperplanes. Then W induces an action on $V_{\mathbb{C}}$ and also on

$$\mathcal{M} := V_{\mathbb{C}} - \bigcup_{t \in T} (H_t)_{\mathbb{C}}.$$

It is shown in [Brieskorn 1971] that the fundamental group of the orbit space of $\mathcal{M}$ under the action of W is naturally isomorphic to $B(W, S)$, via a map which describes explicitly the paths corresponding to the generators of $B(W, S)$. In type A, one obtains in this way the classical braid groups studied in [Artin 1947].

4.2 DIVISIBILITY IN BRAID MONOIDS

Let (W, S) be a Coxeter system. The main purpose of this section is to provide a proof of Theorem 4.1.4. After some preparations, this will be accomplished in Theorems 4.2.3 and 4.2.6 below.

We start with a result about W. Let $v, w \in W$. We say that v divides w, and write $v \prec w$, if $w = vx$ with $x \in W$ and $l(w) = l(v) + l(x)$. This defines a partial order relation on W. (Note that this is just the restriction to W via the map r of the relation on B^+ defined earlier; in Chapter 2, we called v a prefix of w if $v \prec w$.)

4.2.1. Proposition. *Let $v, w \in W$. Then the set*

$$\{x \in W \mid x \prec w \text{ and } l(vx) = l(v) + l(x)\}$$

contains a unique maximal element (with respect to the order relation $\prec$).

Proof. Call the above set M and choose $w' \in M$ of maximum possible length. We must show that if $y \in M$ then $y \preceq w'$. Assume this is not the case, and let $y \in M$ such that $y \not\preceq w'$. Let $y = s_1 \cdots s_k$ ($s_i \in S$) be a reduced expression. Then there exists some $m \geq 1$ such that $s_1 \cdots s_{m-1} \preceq w'$ but $s_1 \cdots s_m \not\preceq w'$. Let $z := s_1 \cdots s_{m-1}$ and $s := s_m$. Thus, starting with the above assumption, we have now found an element z having the following properties:

$$z \in M,\ z \preceq w',\ zs \in M,\ zs \not\preceq w', \qquad \text{for some } s \in S \text{ with } l(zs) > l(z).$$

We assume that z has been chosen of maximum possible length among all elements having these properties. Since $z \preceq w'$, we can write $w' = zu$ with $u \in W$ and $l(w') = l(z) + l(u)$. Since $zs \in M$, we have $l(zs) \leq l(w')$ by the maximality of w'. So $l(z) < l(zs) \leq l(w')$, and hence $u \neq 1$. It follows that there exists some $t \in S$ such that $l(tu) < l(u)$. We have $l(zt) > l(z)$, $zt \preceq w'$, and hence also $zt \in M$. By the choice of z, this implies that $s \neq t$.

Now we have $l(zs) > l(z)$, $l(zt) > l(z)$, and $zs, zt \in M$. We claim that then $m_{st} < \infty$ and $z\Delta_{st} \in M$. Indeed, since $zs \preceq w$ and $zt \preceq w$, we can write $w = zsa = ztb$ with $a, b \in W$ and $l(zsa) = l(zs) + l(a) = l(zt) + l(b)$. Letting $c := sa = tb$, this shows that $s, t \preceq c$. But this is exactly a situation where we can apply Lemma 1.2.1, and so $m_{st} < \infty$ and $\Delta_{st} \preceq c$. Writing $c = \Delta_{st}c'$ with $c' \in W$ and $l(c) = l(\Delta_{st}) + l(c')$, we find that $w = zc = z\Delta_{st}c'$ and, hence, $z\Delta_{st} \preceq w$. Furthermore, we have $l(vz\Delta_{st}) = l(vz) + l(\Delta_{st})$ by (a "left" version of) Proposition 2.1.1 and so $z\Delta_{st} \in M$, as desired.

The element z had the property that $zt \preceq w'$. But since $zs \not\preceq w'$ and $zs \preceq z\Delta_{st}$, we cannot have $z\Delta_{st} \preceq w'$. So there exists some l, $1 \leq l < m_{st}$, such that $z\,\mathrm{Prod}(t, s; l) \preceq w'$ but $z\,\mathrm{Prod}(t, s; l+1) \not\preceq w'$. Hence the element $z\,\mathrm{Prod}(t, s; l)$ has all the properties that we required for z. But its length is strictly larger than that of z since $l \geq 1$, and we have reached a contradiction. Hence our assumption was wrong, and we have $y \preceq w'$ for all $y \in M$, as desired. □

We can now define two functions $\alpha_2, \omega_2 \colon W \times W \to W$ as follows. Given $v, w \in W$ let $w' \in W$ be the unique maximal element in the set M as in Proposition 4.2.1. Writing $w = w'z$ with $z \in W$ and $l(w) = l(w') + l(z)$, we let $\alpha_2(v, w) := vw'$ and $\omega_2(v, w) := z$. The functions α_2, ω_2 have the following fundamental properties, which will allow us to extend them to functions on B^+:

4.2.2. Lemma. *Let $u, v \in W$ be such that $l(uv) = l(u) + l(v)$. Then $uv = \alpha_2(u, v)\omega_2(u, v)$. Moreover, for any $w \in W$ we have*

$$\alpha_2(uv, w) = \alpha_2(u, \alpha_2(v, w)) \quad \textit{and} \quad \omega_2(uv, w) = \omega_2(u, \alpha_2(v, w))\omega_2(v, w).$$

Proof. The first equation is clear by definition. Now consider the map α_2. By definition, we have $\alpha_2(v, w) = vw'$ with $w' \preceq w$ and such that $l(vw') = l(v) + l(w')$. So, since $v \preceq vw'$ and $l(uv) = l(u) + l(v)$, we have $\alpha_2(u, \alpha_2(v, w)) = \alpha_2(u, vw') = uvw''$ with $w'' \preceq w' \preceq w$. Hence $uvw'' \preceq \alpha_2(uv, w)$. Conversely, let $\alpha_2(uv, w) = uvz$ with $z \prec w$. Then $w'' \prec z \prec w$. Now $l(vz) = l(v) + l(z)$ and hence $z \prec w'$, since $\alpha_2(v, w) = vw'$. But $l(uvz) = l(uv) + l(z)$ and $z \prec w'$ imply that $uvz \prec \alpha_2(u, vw') = uvw''$, and hence $z \prec w''$. Thus, $z = w''$, and we are done.

Now consider the map ω_2. By definition, we have

$$\alpha_2(uv, w)\omega_2(uv, w) = uvw = u\alpha_2(v, w)\omega_2(v, w)$$

and $\alpha_2(u, \alpha_2(v, w))\omega_2(u, \alpha_2(v, w)) = u\alpha_2(v, w)$. Multiplying the latter equation from the right by $\omega_2(v, w)$ yields that

$$\alpha_2(u, \alpha_2(v, w))\omega_2(u, \alpha_2(v, w))\omega_2(v, w) = u\alpha_2(v, w)\omega_2(v, w).$$

Comparison with the first equation above shows that

$$\alpha_2(u, \alpha_2(v, w))\omega_2(u, \alpha_2(v, w))\omega_2(v, w) = \alpha_2(uv, w)\omega_2(uv, w),$$

and we are done using the formula for α_2. □

We shall regard the functions α_2, ω_2 as functions $B^+_{\text{red}} \times B^+_{\text{red}} \to B^+_{\text{red}}$, via the bijection $r\colon W \to B^+_{\text{red}}$. Now we can establish the existence of a function $\alpha\colon B^+ \to B^+_{\text{red}}$ as in Theorem 4.1.4.

4.2.3. Theorem. *There is a unique function $\alpha\colon B^+ \to B^+_{\text{red}}$ such that*

$$\alpha(gh) = \alpha_2(g, \alpha(h)) \qquad \textit{for all } g \in B^+_{\text{red}} \textit{ and } h \in B^+.$$

For each $g \in B^+$ we have $\alpha(g) \prec g$, and $\alpha(g)$ is the unique element of maximum possible length in B^+_{red} which divides g.

Proof. For each $k \geq 0$ let $B^+_k := \{g \in B^+ \mid l(g) \leq k\}$. We shall define α recursively on each B^+_k. For $k = 0$, we have $B^+_k = \{1\}$ and we set $\alpha(1) = 1$. Now let $k \geq 1$. We assume that α has already been defined on B^+_{k-1} and satisfies $\alpha(gh) = \alpha_2(g, \alpha(h))$ whenever $g \in B^+_{\text{red}}$ and $h, gh \in B^+_{k-1}$. Now let $g \in B^+_k$. Writing $g = g_1 \cdots g_n$ with $g_i \in B^+_{\text{red}}$ and $l(g_2 \cdots g_n) < k$, we set

$$\alpha(g) := \alpha_2(g_1, \alpha(g_2 \cdots g_n)).$$

We must check that this is well defined. Thus, we have to consider another decomposition of g as a product of elements from B^+_{red}, and we must show that we obtain the same result. Now any two expressions of g as products of elements from B^+_{red} can be transformed into each other by applying a finite sequence of the defining relations given in Proposition 4.1.3. Thus, it is sufficient to consider another decomposition of the form $g = g_1 \cdots g_{i-1} h h' g_{i+1} \cdots g_n$ with $hh' = g_i$

and $h, h' \neq 1$. By induction, we can even assume that $i = 1$. So, letting $g' := g_2 \cdots g_n$, we must show that $\alpha_2(h, \alpha(h'g')) = \alpha_2(g_1, \alpha(g'))$. Indeed, we have:

$$\begin{aligned} \alpha_2(h, \alpha(h'g')) &= \alpha_2(h, \alpha_2(h', \alpha(g'))) && \text{by induction, since } l(h'g') < k, \\ &= \alpha_2(hh', \alpha(g')) && \text{by Lemma 4.2.2, since } hh' \in B^+_{\mathrm{red}}, \\ &= \alpha_2(g_1, \alpha(g')). \end{aligned}$$

Thus, α is well defined and, by construction, it satisfies $\alpha(gh) = \alpha_2(g, \alpha(h))$ for all $g \in B^+_{\mathrm{red}}$ and $h \in B^+$.

It remains to show that $\alpha(g) \prec g$ and that $\alpha(g)$ is the unique element of maximum possible length in B^+_{red} which divides g.

First, we show that $\alpha(g) \prec g$ by induction on $l(g)$. If $l(g) = 0$, then $g = \alpha(g) = 1$ and there is nothing to prove. Now let $l(g) > 0$ and write $g = g_1 g'$ with $1 \neq g_1 \in B^+_{\mathrm{red}}$ and $g' \in B^+$. Then $\alpha(g) = \alpha_2(g_1, \alpha(g')) = g_1 h$ for some $h \prec \alpha(g')$, by the definition of α_2. By induction, since $l(g') < l(g)$, we have $\alpha(g') \prec g'$, and hence $\alpha(g) \prec g_1 h \prec g_1 g' = g$.

Now let g' be any element in B^+_{red} such that $g' \prec g$. Then we can write $g = g'h$ for some $h \in B^+$ and we have $\alpha(g) = \alpha(g'h) = \alpha_2(g', \alpha(h)) = g'h'$ for some $h' \prec \alpha(h)$, by the definition of α_2. Thus, $g' \prec \alpha(g)$, and we are done. ☐

We have just seen that for each $g \in B^+$ there exists a unique $g' \in B^+_{\mathrm{red}}$ of maximum possible length such that $g' \prec g$. Thus, we can write $g = g'h$ for some $h \in B^+$, but at this stage it is not clear that h is unique. The proof of this uniqueness assertion will also be achieved using a suitable function $\omega\colon B^+ \to B^+$.

4.2.4. Theorem. *There is a unique function* $\omega\colon B^+ \to B^+$ *such that*

$$\omega(gh) = \omega_2(g, \alpha(h))\omega(h) \qquad \textit{for all } g \in B^+_{\mathrm{red}} \textit{ and } h \in B^+.$$

For each $g \in B^+$, *we have* $g = \alpha(g)\omega(g)$.

Proof. This is proved similarly to Theorem 4.2.3. Using the notation of that proof, we define ω recursively on each B^+_k, for $k \geq 0$. We set $\omega(1) = 1$ for $1 \in B^+_0$. Now let $k \geq 1$ and assume that ω has already been defined on B^+_{k-1} and satisfies $\omega(gh) = \omega_2(g, \alpha(h))\omega(h)$ whenever $g \in B^+_{\mathrm{red}}$ and $h, gh \in B^+_{k-1}$. Now let $g \in B^+_k$. We would then like to define

$$\omega(g) := \omega_2(g_1, \alpha(g'))\omega(g'),$$

where $g = g_1 g'$ with $g_1 \in B^+_{\mathrm{red}}$ and $g' \in B^+$. But we must check that this does not depend on the chosen decomposition of g. As in the proof of Theorem 4.2.3, it is sufficient to consider the case where $g_1 = hh'$ with $1 \neq h, h' \in B^+_{\mathrm{red}}$, and we have to check that

$$\omega_2(h, \alpha(h'g'))\omega(h'g') = \omega_2(hh', \alpha(g'))\omega(g').$$

Indeed, the left-hand side equals $\omega_2(h, \alpha_2(h', \alpha(g')))\omega(h'g')$ by Theorem 4.2.3. Since $l(h'g') < k$, we can apply induction and obtain that

$$\omega_2(h, \alpha_2(h', \alpha(g')))\omega_2(h', \alpha(g'))\omega(g') = \omega_2(hh', \alpha(g'))\omega(g')$$

where the last equality is by Lemma 4.2.2. Thus, ω is well defined and, by construction, it satisfies $\omega(gh) = \omega_2(g, \alpha(h))\omega(h)$ for all $g \in B^+_{\mathrm{red}}$ and $h \in B^+$.

It remains to show that $g = \alpha(g)\omega(g)$ for all $g \in B^+$. We do this by induction on $l(g)$. If $l(g) = 0$ then $g = \alpha(g) = \omega(g) = 1$, and there is nothing to prove. Now let $l(g) > 0$ and write $g = g_1 g'$ with $1 \neq g_1 \in B^+_{\mathrm{red}}$ and $g' \in B^+$. Then

$$\begin{aligned}
\alpha(g)\omega(g) &= \alpha(g_1g')\omega(g_1g') = \alpha_2(g_1, \alpha(g'))\omega(g_1g') && \text{by Theorem 4.2.3,}\\
&= \alpha_2(g_1, \alpha(g'))\omega_2(g_1, \alpha(g'))\omega(g') && \text{by the construction of } \omega,\\
&= g_1\alpha(g')\omega(g') && \text{by the definition of } \alpha_2, \omega_2 \text{ (see Lemma 4.2.2),}\\
&= g_1g' = g && \text{by induction, since } l(g') < k.
\end{aligned}$$

□

4.2.5. Corollary. *Let* $g \in B^+$. *Then there exists a unique* $h \in B^+$ *such that* $g = \alpha(g)h$, *and we have* $h = \omega(g)$.

Proof. We proceed by induction on $l(g)$. If $l(g) = 0$ then $g = \alpha(g) = 1$, and there is nothing to prove. Now let $l(g) > 0$, and write $g = \alpha(g)h$ for some $h \in B^+$.

Let $b := \alpha(g)\alpha(h) \in B^+$. By the definition of α, we can write

$$\alpha(b) = \alpha_2(\alpha(g), \alpha(h)) = \alpha(g)x \qquad \text{for some } x \prec \alpha(h) \prec h.$$

Then $\alpha(b) \prec \alpha(g)h = g$ and hence $\alpha(b) \prec \alpha(g)$ by Theorem 4.2.3. Since also $\alpha(g) \prec \alpha(b)$, we must have $x = 1$. This shows that $\alpha(g) = \alpha_2(\alpha(g), \alpha(h))$. By Proposition 4.2.1, there exists a unique $h' \in B^+_{\mathrm{red}}$ such that $h' \prec \alpha(h)$ and $b = \alpha_2(\alpha(g), \alpha(h))h'$; moreover, we have $h' = \omega_2(\alpha(g), \alpha(h))$. Thus, $\omega_2(\alpha(g), \alpha(h)) = \alpha(h)$. Since $g = \alpha(g)h$ and $\alpha(g) \neq 1$, we certainly have $l(h) < l(g)$ and hence, by induction, $h = \alpha(h)\omega(h)$. We conclude that

$$\omega(g) = \omega(\alpha(g)h) = \omega_2(\alpha(g), \alpha(h))\omega(h) = \alpha(h)\omega(h) = h,$$

using Theorem 4.2.4. □

4.2.6. Theorem. *The monoid* B^+ *is left simplifiable, i.e, if we have* $gh = gh'$ *with* $g, h, h' \in B^+$ *then* $h = h'$.

Proof. We write $g = g_1 \cdots g_n$ with $g_i \in B^+_{\mathrm{red}}$ for all i, and proceed by induction on n. First let $g = g_1 \in B^+_{\mathrm{red}}$. The formula $\alpha(gh) = \alpha_2(g, \alpha(h))$ and the definition of α_2 show that there exists some $h_1 \prec \alpha(h) \prec h$ such that $\alpha(gh) = gh_1$. Similarly, $\alpha(gh') = gh_1'$ for some $h_1' \prec \alpha(h') \prec h'$. We have $gh = gh'$ and hence $gh_1 = gh_1'$. Now note that $g, h_1, h_1' \in B^+_{\mathrm{red}}$ and $gh_1 = gh_1' \in B^+_{\mathrm{red}}$. In other words, we can write $g = r(w)$, $h_1 = r(w_1)$ and $h_1' = r(w_1')$ for some $w, w_1, w_1' \in W$, and we have $l(ww_1) = l(w) + l(w_1) = l(w) + l(w_1')$. This implies that $ww_1 = ww_1'$. Since W is a group, it follows that $w_1 = w_1'$ and hence $h_1 = r(w_1) = r(w_1') = h_1'$. Now we can write $h = h_1h_2$ and $h' = h_1'h_2'$ for some $h_2, h_2' \in B^+$, and so $(gh_1)h_2 = gh = gh' = (gh_1')h_2'$. Since $gh_1 = \alpha(gh) = \alpha(gh') = gh_1'$, Corollary 4.2.5 implies that $h_2 = h_2'$, and hence $h = h_1h_2 = h_1'h_2' = h'$, as desired.

Now assume that $g = g'g''$ with $1 \neq g' \in B^+_{\text{red}}$ and $g'' \in B^+$. Thus, $g'(g''h) = gh = gh' = g'(g''h')$. Since $g' \in B^+_{\text{red}}$, we can apply the previous argument to conclude that $g''h = g''h'$. Then, applying induction to g'' yields $h = h'$. □

Finally, we present an algorithm for computing the Deligne normal form.

Algorithm L (Deligne normal form). Given a Coxeter system (W, S), the corresponding braid monoid $B^+ = B^+(W, S)$ and $g = w_1 \cdots w_k \in B^+$ (where $w_i \in W$ for $1 \leq i \leq k$), the element $w \in W$ such that $w = \alpha(g)$ is found, i.e., $w \in W$ is the unique element of maximum possible length such that $w \prec g$.

L1. [Initialize.] Set $i \leftarrow k$ and $w \leftarrow w_k$.

L2. [Next i.] Find $\alpha_2(w_{i-1}, w)$, as follows. Use Algorithm C (p. 46) to construct the set $P(w) = \{x_1, x_2, \ldots\}$ of all prefixes of w, where $l(x_1) \geq l(x_2) \geq \ldots$. Let $t \geq 1$ be minimal such that $l(w_{i-1}x_t) = l(w_{i-1}) + l(x_t)$. (Since $1 \in P(w)$, such an i exists.) Then set

$$w \leftarrow w_{i-1}x_t = \alpha_2(w_{i-1}, w).$$

L3. [Loop.] Set $i \leftarrow i - 1$ and repeat step L2 until $i = 1$. Then $w = \alpha(g)$.

Once we have an algorithm for computing α, we also have an algorithmic description for ω via the formula in Theorem 4.2.4. Hence we obtain the Deligne normal form of g by computing the sequence $g_1 := \alpha(g), g_2 := \alpha(\omega(g)), g_3 := \alpha(\omega^2(g)), \ldots,$ until we find $n \geq 1$ such that $g = g_1 \cdots g_n$.

4.3 GOOD ELEMENTS IN BRAID MONOIDS

Assume from now on that (W, S) is a Coxeter system where W is a finite group. Let B^+ be the corresponding braid monoid. We have a canonical surjective monoid homomorphism $\pi\colon B^+ \to W$ and an injection $r\colon W \to B^+$, induced by identifying the generators of B^+ with those of W. In this section we define the notion of "good" elements in W, using the embedding of W in B^+ via r.

Recall from Section 4.1 the definition of the fundamental element $\Delta := r(w_0) \in B^+_{\text{red}}$ where $w_0 \in W$ is the longest element of W. Further, for each subset $J \subseteq S$, let $W_J \subseteq W$ be the corresponding parabolic subgroup with $w_J \in W_J$ its longest element and recall that

$$\Delta_J := r(w_J) = w_J \in B^+_{\text{red}}.$$

In particular, we have $\Delta = \Delta_S$. Now our main interest is in elements of W. Let $w \in W$ and $d \geq 1$ denote its order. Then, of course, we have $\mathbf{w}^d = 1$ in W. So the element $\mathbf{w}^d$ lies in the *pure braid monoid*, which is defined as the subset

$$P^+ = P^+(W, S) := \{g \in B^+ \mid \pi(g) = 1\}.$$

We shall say that $w \in W$ is "good" if $\mathbf{w}^d$ is of a particular shape; see the precise definition below. Thus, the notion of good elements in W uses in an essential way the braid monoid B^+.

4.3.1. Definition (Geck–Michel). Let $w \in W$ be of order d. Then w is a *good element* if there exist a strictly decreasing sequence of subsets

$$J_1 \supset J_2 \supset \cdots \supset J_k,$$

of S and a sequence of positive integers $d_1, d_2, \ldots, d_k$ such that

$$\mathbf{w}^d = \Delta_{J_1}^{d_1} \Delta_{J_2}^{d_2} \cdots \Delta_{J_k}^{d_k}.$$

4.3.2. Remark. Let $w \in W$ be a good element of order d such that $\mathbf{w}^d$ is given by an expression as in Definition 4.3.1. Then the following hold.

(a) The d_i are necessarily even, since $\mathbf{w}^d \in P^+$. Indeed, applying π, we have

$$w_{J_1}^{d_1} w_{J_2}^{d_2} \cdots w_{J_k}^{d_k} = 1.$$

If d_1 were odd, then we would find $w_{J_1} = w_{J_1}^{d_1} = w_{J_k}^{-d_k} \cdots w_{J_2}^{-d_2} \in W_{J_2}$, which is impossible since $J_1 \neq J_2$. Thus, d_1 must be even. We now have $w_{J_2}^{d_2} \cdots w_{J_k}^{d_k} = 1$ and repeat the above argument.

(b) All factors $\Delta_{J_i}^{d_i}$ commute with each other, since each d_i is even and, hence, $\Delta_{J_i}^{d_i}$ commutes with all elements $\mathbf{s}$ for $s \in J_j$, $j \geq i$ (see Lemma 4.1.9).

(c) The Deligne and the Brieskorn–Saito normal form for $\mathbf{w}^d$ coincide. (This follows from the fact that the J_i form a decreasing sequence of subsets.)

4.3.3. Lemma. *Assume that $w_1, w_2 \in W$ are such that w_1 is strongly conjugate to w_2, in the sense of Definition 3.2.4. Then $\mathbf{w}_1$ and $\mathbf{w}_2$ are conjugate in* B.

Proof. It is sufficient to show that if w_1 and w_2 are elementarily strongly conjugate (see Definition 3.2.4) then $\mathbf{w}_1$ is conjugate to $\mathbf{w}_2$ in B. So let $x \in W$ such that $w_1 x = x w_2$ and $l(w_1 x) = l(w_1) + l(x)$. But then $\mathbf{w}_1 \mathbf{x} = \mathbf{x} \mathbf{w}_2$ in B and $\mathbf{w}_1$ is conjugate to $\mathbf{w}_2$ via $\mathbf{x}$. □

Here is the standard example of a good element in W:

4.3.4. Proposition. *Let us fix a labelling $S = \{s_1, \ldots, s_n\}$ where $n \geq 1$. Then the Coxeter element $w = s_1 \cdots s_n \in W$ is a good element, and we have*

$$\mathbf{w}^h = \Delta^2,$$

where $h \geq 2$ is the Coxeter number. Moreover, if h is even then $\mathbf{w}^{h/2} = \Delta$.

Proof. Using the compatibility with direct products (see Exercise 4.2), we may assume without loss of generality that (W, S) is irreducible.

Now, first, we consider the case where the ordering of $S = \{s_1, \ldots, s_n\}$ has been chosen so that it is compatible with a partition $S = S_1 \amalg S_2$ as in Lemma 1.5.6. Then $w = w_c = w_1 w_2$ as in Theorem 1.5.8. Now distinguish the cases where h is even or odd. If h is even, we have $w_c^{h/2} = w_0$ and the product on the left-hand side is reduced. It follows that $\mathbf{w}_c^{h/2} = \Delta$ and hence $\mathbf{w}_c^h = \Delta^2$ as desired. Now let h be odd and recall that we have a factorization $w = w_1 w_2$. Then $w_0 = \mathsf{Prod}(w_1, w_2; h) = \mathsf{Prod}(w_2, w_1; h)$, where these products are reduced. It follows that $\Delta = \mathsf{Prod}(\mathbf{w}_1, \mathbf{w}_2; h) = \mathsf{Prod}(\mathbf{w}_2, \mathbf{w}_1; h)$, and hence $\Delta^2 = \mathsf{Prod}(\mathbf{w}_1, \mathbf{w}_2; 2h) = (\mathbf{w}_1\mathbf{w}_2)^h = \mathbf{w}_c^h$, as desired.

Now consider the general case, where we have chosen any ordering of S. By Theorem 3.1.4, all Coxeter elements are conjugate by cyclic shift in W. So Lemma 4.3.3 implies that $\mathbf{w}$ is conjugate in B to an element $\mathbf{w}_c$ as considered above, i.e., there exists some $g \in B$ such that $g\mathbf{w}g^{-1} = \mathbf{w}_c$. We have just seen that $\mathbf{w}_c^h = \Delta^2$. So it follows that $g\mathbf{w}^h g^{-1} = \Delta^2$. Since Δ^2 commutes with all elements of B (see Proposition 4.1.9) we conclude that $\mathbf{w}^h = \Delta^2$. □

Not every element $w \in W$ is good. If, for example, (W, S) is of type B_2 with $S = \{s, t\}$ then $(\mathbf{sts})^2 \in B^+(W, S)$ is not of the required form. However, it turns out that every conjugacy class contains at least one good element. That is the main consequence of the investigation of braid monoids in this chapter for the finite Coxeter group W.

4.3.5. Theorem (Geck–Michel). *Let C be a conjugacy class of W. Then there exists a good element $w \in C_{\min}$.*

It is readily checked that it is sufficient to prove Theorem 4.3.5 for irreducible Coxeter groups, since the concepts of conjugacy, longest elements and braid monoids are compatible with taking direct products. Furthermore, for each irreducible type it is sufficient to verify the theorem for the cuspidal classes.

For any given Coxeter group (W, S) the following strategy can be used to establish Theorem 4.3.5. Given a list of representatives of the cuspidal classes of W, for each such representative its cyclic shift class X is constructed by Algorithm G (p. 80). Let d be the order of the elements in X. Then for each $w \in X$ the Brieskorn–Saito normal form (Definition 4.1.8) of $\mathbf{w}^d$ in B^+ is determined and checked for the required properties.

The theorem thus follows from a case-by-case analysis. The results of this analysis for the exceptional types H_3, H_4, F_4, E_6, E_7 and E_8, are listed in the tables in Appendix B. For the types $I_2(m)$, A_{n-1}, B_n and D_n, see Example 4.3.6, (4.3.7) and Proposition 4.3.11 below.

4.3.6. Example. Let (W, S) be of type $I_2(m)$ $(m \geq 5)$ with the following Coxeter graph:

$I_2(m)$
$(m \geq 5)$

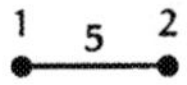

Recall from (Example 3.2.8) that the cuspidal classes of W are the classes of the form $\{(s_1s_2)^i, (s_2s_1)^i\}$ for some $1 \leq i \leq m/2$. The longest element in W is $w_0 = \mathsf{Prod}(s_1, s_2; m) = \mathsf{Prod}(s_2, s_1; m)$. Hence $\Delta^2 = (s_1s_2)^m = (s_2s_1)^m$ and we see that both $(s_1s_2)^i$ and $(s_2s_1)^i$ are good elements.

We are now going to determine good elements in the Coxeter groups of type A_{n-1}, B_n and D_n. Throughout we will work in the set-up of Section 3.4. That is, the Coxeter group of type B_n is generated by $\{t, s_1, s_2, \ldots, s_{n-1}\}$, where $\{s_1, s_2, \ldots, s_{n-1}\}$ generate a Coxeter group of type A_{n-1}. Moreover, with $u = ts_1t$, the elements $\{u, s_1, s_2, \ldots, s_{n-1}\}$ generate a Coxeter group of type D_n.

4.3.7. Good elements in type A_{n-1}. Let (W, S) be of type A_{n-1} $(n \geq 2)$. Then W is isomorphic to the symmetric group $\mathfrak{S}_n$ and the only cuspidal class of W is the conjugacy class labelled by the partition $[n]$ (see Example 3.1.16 and Proposition 3.4.1). The elements of C_{min} in this case are precisely the Coxeter elements of W which are good by Proposition 4.3.4. Thus, Theorem 4.3.5 holds for type A_{n-1}.

Recall from Section 3.4 the definition of the elements $t_i = s_i \cdots s_1 t s_1 \cdots s_i$ (for $0 \leq i \leq n-1$) in type B_n and $u_i = s_i \cdots s_2 u s_1 s_2 \cdots s_i$ (for $1 \leq i \leq n-1$) together with $u_0 = 1$ in type D_n. Note that under the above embedding of D_n into B_n we have $u_i = tt_i = t_i t$ for all $i = 0, \ldots, n-1$. In type B_n, for $m \geq 0$, the elements $\{t_m, s_{m+1}, \ldots, s_{m+e-1}\}$ generate a subgroup which is a Coxeter group of type B_e. In type D_n, for $m \geq 1$, the elements $\{u_m, s_{m+1}, \ldots, s_{m+e-1}\}$ generate a subgroup which is a Coxeter group of type B_e.

In order to establish Theorem 4.3.5 for the types B_n and D_n we need some more properties of the positive and negative blocks $b^{\pm}_{m,e}$ as defined in (3.4.2). We can then show that the standard representatives $w_{(\alpha,\beta)}$ are good elements. Recall the definition of the negative block

$$b^-_{m,e} = \begin{cases} t_m s_{m+1} \cdots s_{m+e-1} & \text{for } m \geq 0 \text{ in type } B_n, \\ u_m s_{m+1} \cdots s_{m+e-1} & \text{for } m \geq 1 \text{ in type } D_n. \end{cases}$$

With one possible exception, two blocks $b^{\pm}_{m,e}$ and $b^{\pm}_{m',e'}$ commute with each other if their support sets $\{m, m+1, \ldots, m+e-1\}$ and $\{m', m'+1, \ldots, m'+e'-1\}$ are disjoint (see Exercise 4.6). The exception occurs in type D_n if one block is the special block $b^-_{0,e} = s_1 \cdots s_{e-1} = b^+_{0,e}$ with $e > 1$ and the other block is a negative block. This situation is studied in Lemma 4.3.9 below.

We define

$$t_{m,e} = t_m \cdots t_{m+e-1} \quad \text{or} \quad t_{m,e} = u_m \cdots u_{m+e-1}$$

if W is of type B_n or D_n respectively, and denote by $\mathbf{t}_{m,e} = r(t_{m,e})$ the corresponding expression in B^+.

4.3.8. Lemma. *Let W be of type B_n or D_n. Let $e \geq 1$ and $m \geq 0$.*

(i) *The order of the negative block $b^-_{m,e}$ in W is $2e$.*
(ii) *We have $(\mathbf{b}^-_{m,e})^e = \mathbf{t}_{m,e}$ in B^+ (for $m \geq 1$ in type D_n).*

Proof. Assume first that (W, S) is of type B_n. The matrix description in Section 1.4 shows that the subgroup $W' = \langle t_m, s_{m+1}, \ldots, s_{m+e-1}\rangle \subseteq W$ is a Coxeter group of type B_e and that $b^-_{m,e}$ is a Coxeter element in W'. The Coxeter number of W' is $2e$ and the longest element of W' is $t_m \cdots t_{m+e-1}$. Therefore, $b^-_{m,e}$ has order $2e$ and $(b^-_{m,e})^e = t_m \cdots t_{m+e-1}$. This equation holds in B^+ since $t_m \cdots t_{m+e-1}$ is reduced and $e \cdot l(b^-_{m,e}) = (2m+e)e = l(t_m \cdots t_{m+e-1})$, whence $(b^-_{m,e})^e$ also is reduced.

If (W, S) is of type D_n and $m \geq 1$ then, again using the matrix description in Section 1.4, the subgroup $W' = \langle u_m, s_{m+1}, \ldots, s_{m+e-1}\rangle \subseteq W$ is a Coxeter group of type D_e and $b^-_{m,e}$ is a Coxeter element in W'. The longest element in W' is $u_m \cdots u_{m+e-1}$ and the proof is completed as for type B_n by observing that in this case $e \cdot l(b^-_{m,e}) = (2m+e-1)e = l(u_m \cdots u_{m+e-1})$. □

We now consider the block $b^-_{0,e} = b^+_{0,e} = s_1 s_2 \cdots s_{e-1}$ in type D_n.

4.3.9. Lemma. *Let (W, S) be of type D_n. For $e \geq 2$ define $a(e) = s_1 s_2 \cdots s_{e-1}$ and $\hat{a}(e) = u s_2 \cdots s_{e-1}$ and denote by $\mathbf{a}(e)$ and $\hat{\mathbf{a}}(e)$ the corresponding elements in B^+.*

(i) *The order of $a(e)$ and of $\hat{a}(e)$ in W is e.*
(ii) *We have*

$$\mathsf{Prod}(\mathbf{a}(e), \hat{\mathbf{a}}(e); e) = \mathsf{Prod}(\hat{\mathbf{a}}(e), \mathbf{a}(e); e) = \mathbf{t}_{0,e}[= \mathbf{u}_1 \cdots \mathbf{u}_{e-1}]$$

in B^+.

Proof. Both $a(e)$ and $\hat{a}(e)$ are Coxeter elements of parabolic subgroups of type A_{e-1}, whence (i).

In order to prove (ii), we consider the embedding of W into the Coxeter group W' of type B_n given by $u \mapsto t s_1 t$ and $s_i \mapsto s_i$ for $i = 1, \ldots, n-1$. Here $\hat{a}(e) = t a(e) t$. Thus $\mathsf{Prod}(\hat{a}(e), a(e); e) = (t a(e))^e$ if e is even, and $(t a(e))^e t$ if e is odd. Now $t a(e) = b^-_{0,e}$ in W', whence $(t a(e))^e = t_0 t_1 \cdots t_{e-1}$ by Lemma 4.3.8. Finally, from $u_i = t t_i = t_i t$ for $i = 0, \ldots, n-1$, we get in any case that $\mathsf{Prod}(\hat{a}(e), a(e); e) = u_1 \cdots u_{e-1}$. The equation holds in B^+ since the expressions on both sides are reduced. The statement for $\mathsf{Prod}(\mathbf{a}(e), \hat{\mathbf{a}}(e); e)$ follows analogously. □

Let α be a partition of n with parts $1 \leq \alpha_1 \leq \alpha_2 \leq \cdots \leq a_r$. As in (3.4.2), we define $w^-_\alpha = w_{(\varnothing,\alpha)}$, i.e.,

$$w^-_\alpha = b^-_{m_1,\alpha_1} b^-_{m_2,\alpha_2} \cdots b^-_{m_r,\alpha_r},$$

where $m_i = \alpha_1 + \cdots + \alpha_{i-1}$ for all i.

4.3.10. Lemma. *Assume that* (W, S) *is of type* B_n *or* D_n, $n \geq 2$. *Let* α *be a partition of* n *with parts* $1 \leq \alpha_1 \leq \alpha_2 \leq \cdots \leq a_r$ *(with* r *even if* W *is of type* D_n*) and let* d *be the least common multiple of the parts* α_i.

(i) *The order of* w_α^- *in* W *is* $2d$.

(ii) *Let* $e_i = d/\alpha_i$ *for* $i = 1, \ldots, r$. *Then*

$$(w_\alpha^-)^d = t_{m_1,\alpha_1}^{e_1} \cdots t_{m_r,\alpha_r}^{e_r},$$

where $m_i = \alpha_1 + \cdots + \alpha_{i-1}$ *for all* i.

Proof. (i) If W is of type D_n, we consider the embedding of W into the Coxeter group of type B_n where $b_{m,e}^-$ becomes $tb_{m,e}^-$. But, since α in this case has an even number of parts, all factors t cancel and w_α^- in D_n coincides with the corresponding element in B_n. In any case, w_α^- thus is a product of commuting blocks b_{m_i,α_i}^-, where b_{m_i,α_i}^- has order $2\alpha_i$ by Lemma 4.3.8. This implies that the order of w_α^- is indeed $2d$.

(ii) Assume first that $\alpha_1 = 1$ if W is of type D_n. Then all negative blocks involved in w_α^- commute with each other and $(w_\alpha^-)^d$ equals the product of the terms $(b_{m_i,\alpha_i}^-)^d$, $i = 1, \ldots, r$. Lemma 4.3.8 now yields the desired form.

Now suppose that we are in type D_n and $\alpha_1 > 1$. Then $b_{m_1,\alpha_1}^- = a(\alpha_1)$, and we have

$$a(e)b_{m,d}^- = b_{m,d}^-\hat{a}(e) \quad \text{and} \quad \hat{a}(e)b_{m,d}^- = b_{m,d}^-a(e)$$

for $d \geq 0$ and $1 \leq e - 1 \leq m$. Thus we get

$$(w_\alpha^-)^d = \mathsf{Prod}(a(\alpha_i), \hat{a}(\alpha_i); d) \cdot (b_{m_2,\alpha_2}^-)^d \cdots (b_{m_r,\alpha_r}^-)^d,$$

where $\mathsf{Prod}(a(\alpha_i), \hat{a}(\alpha_i); d) = (u_1 \cdots u_{\alpha_1 - 1})^{e_1}$ by Lemma 4.3.9. We finally note that $u_1 \cdots u_{\alpha_1-1} = t_{0,\alpha_1}$, and again Lemma 4.3.8 yields the desired form. □

The following proposition summarizes the results obtained so far and establishes Theorem 4.3.5 for the types B_n and D_n. (Recall that it suffices to consider the cuspidal classes.)

4.3.11. Proposition (Good elements in types B_n and D_n). *Assume the Coxeter system* (W, S) *is of type* B_n *or* D_n $(n \geq 2)$. *Let* α *be a partition of* n *with parts* $1 \leq \alpha_1 \leq \alpha_2 \leq \cdots \leq \alpha_r$ *(with* r *even if* W *is of type* D_n*) and let* $d \geq 1$ *be the least common multiple of the parts* α_i *of* α. *Then*

$$(w_\alpha^-)^d = \Delta_{J_r}^{e_r} \Delta_{J_{r-1}}^{e_{r-1}-e_r} \Delta_{J_{r-2}}^{e_{r-2}-e_{r-1}} \cdots \Delta_{J_1}^{e_1-e_2},$$

where $e_i = d/\alpha_i$ *for all* $i \geq 1$, $J_r = S$ *and* $J_{i-1} = J_i - \{s_{m_i}, \ldots, s_{m_i+\alpha_i-1}\}$ *for all* $i > 1$. *Moreover,*

$$(w_\alpha^-)^{2d} = \Delta_{J_r}^{2e_r} \Delta_{J_{r-1}}^{2(e_{r-1}-e_r)} \Delta_{J_{r-2}}^{2(e_{r-2}-e_{r-1})} \cdots \Delta_{J_1}^{2(e_1-e_2)},$$

whence w_α^- *is a good element.*

Proof. By Lemma 4.3.10 we have that $(w_\alpha^-)^d = t_{m_1,\alpha_1}^{e_1} \cdots t_{m_r,\alpha_r}^{e_r}$, where $e_i = d/\alpha_i$. Since the parts α_i are in ascending order, we have $e_1 \geq e_2 \geq \cdots \geq e_r$. Now,

$$t_{m_1,\alpha_1} t_{m_2,\alpha_2} \cdots t_{m_r,\alpha_r} = t_{0,n} = t_0 \cdots t_{n-1} = w_0,$$

where w_0 is the longest element of W. So we obtain

$$(w_\alpha^-)^d = \Delta^{e_r} t_{m_1,\alpha_1}^{e_1-e_r} \cdots t_{m_{r-1},\alpha_{r-1}}^{e_{r-1}-e_r},$$

Now we proceed in the same way with the first $r-1$ factors in this product and obtain

$$(w_\alpha^-)^d = \Delta^{e_r} \Delta_{J_{r-1}}^{e_{r-1}-e_r} t_{m_1,\alpha_1}^{e_1-e_{r-1}} \cdots t_{m_{r-2},\alpha_{r-2}}^{e_{r-2}-e_{r-1}},$$

where $J_{r-1} = S - \{s_{m_r}, \ldots s_{m_r+\alpha_r-1}\}$. Continuing in this way yields the desired expression.

The expression for $(w_\alpha^-)^{2d}$ results from the expression for $(w_\alpha^-)^d$ and the fact that all terms in that expression commute with each other. □

Note that the number of distinct subsets J_i occurring in that expression for $(w_\alpha^-)^d$ can become arbitrarily large. It is given by the number of distinct parts of α.

4.4 IWAHORI–HECKE ALGEBRAS

Throughout this section, we fix a Coxeter system (W, S) and a commutative ring A with 1. We consider the monoid algebra $A[B^+]$ of the corresponding braid monoid $B^+ = B^+(W, S)$. Thus, $A[B^+]$ consists of all formal linear combinations $\sum_{b \in B^+} r_b b$ where $r_b \in A$ and only a finite number of coefficients r_b are non-zero. The aim of this section is to show that the quotient of $A[B^+]$ by the ideal generated by certain quadratic relations is an A-algebra which is finitely generated and free over A, with a standard basis labelled by the elements of W.

4.4.1. Definition. Let $\{a_s, b_s \mid s \in S\} \subseteq A$ be such that $a_s = a_t$ and $b_s = b_t$ whenever $s, t \in S$ are conjugate in W. Let $\mathcal{J} \subseteq A[B^+]$ be the two-sided ideal generated by all elements of form

$$s^2 - a_s 1_{B^+} - b_s s, \qquad (s \in S).$$

Then the A-algebra $H = H_A(W, S, \{a_s, b_s \mid s \in S\}) := A[B^+]/\mathcal{J}$ is called the *Iwahori–Hecke algebra* of (W, S) over the ring A and with parameters $\{a_s, b_s \mid s \in S\}$. For any $s \in S$, the image of s in H will be denoted by T_s; the image of 1_{B^+} in H is a unit element in H; it will be denoted by 1_H.

Another way of expressing the above definition is that H is the associative A-algebra (with identity 1) which is given by a presentation with

generators: T_s, $s \in S$,
relations: $\mathsf{T}_s^2 = a_s 1_{\mathsf{H}} + b_s \mathsf{T}_s$ for all $s \in S$,
$\mathrm{Prod}(\mathsf{T}_s, \mathsf{T}_t; m_{st}) = \mathrm{Prod}(\mathsf{T}_t, \mathsf{T}_s; m_{st})$ for $s \neq t$, $m_{st} < \infty$,

where, as usual, m_{st} denotes the order of $st \in W$ for $s, t \in S$.

4.4.2. Example. Assume that $a_s = 1$ and $b_s = 0$ for all $s \in S$. Then the quadratic relations in Definition 4.4.1 just read $s^2 - 1_{B^+}$ for all $s \in S$. Hence, in this case, H is canonically isomorphic to the group algebra $\mathsf{A}[W]$.

Our aim is to show that H is always free as an A-module, with a natural basis indexed by the elements of W. Note that this can only work with the above condition on the parameters a_s, b_s; see Exercise 4.7.

4.4.3. Lemma. *Let* $\mathsf{H} = \mathsf{H}_{\mathsf{A}}(W, S, \{a_s, b_s \mid s \in S\})$ *be an Iwahori–Hecke algebra over a commutative ring* A *as above.*

(a) *For each* $w \in W$ *there is a well-defined element* $\mathsf{T}_w \in \mathsf{H}$ *such that* $\mathsf{T}_w = \mathsf{T}_{s_1} \cdots \mathsf{T}_{s_n}$ *whenever* $w = s_1 \cdots s_n$ $(s_i \in S)$ *is a reduced expression. In particular, we have* $\mathsf{T}_1 = 1_{\mathsf{H}}$.

(b) *For* $s \in S$ *and* $w \in W$ *we have*

$$\mathsf{T}_s \mathsf{T}_w = \begin{cases} \mathsf{T}_{sw} & \text{if } l(sw) > l(w), \\ a_s \mathsf{T}_{sw} + b_s \mathsf{T}_w & \text{if } l(sw) < l(w). \end{cases}$$

An analogous relation also holds for the product $\mathsf{T}_w \mathsf{T}_s$.

(c) *As an* A*-module, the algebra* H *is generated by* $\{\mathsf{T}_w \mid w \in W\}$.

Proof. Consider the monoid $\mathcal{M} = \{\mathsf{T}_{s_1} \cdots \mathsf{T}_{s_n} \mid s_1, \ldots, s_n \in S \text{ for some } n \geq 0\}$ where the operation is given by multiplication. The map $S \to \mathcal{M}$, $s \mapsto \mathsf{T}_s$ satisfies the assumptions of Matsumoto's Theorem 1.2.2, which gives (a).

Now let $s \in S$ and $w \in W$. Take a reduced expression $w = s_1 \cdots s_n$ with $s_i \in S$. Then $\mathsf{T}_w = \mathsf{T}_{s_1} \cdots \mathsf{T}_{s_n}$ by (a). If $l(sw) = l(w) + 1$, then $sw = ss_1 \cdots s_n$ is also reduced, hence (a) yields that

$$\mathsf{T}_{sw} = \mathsf{T}_{ss_1 \cdots s_n} = \mathsf{T}_s \mathsf{T}_{s_1} \cdots \mathsf{T}_{s_n} = \mathsf{T}_s \mathsf{T}_w.$$

On the other hand, if $l(sw) = l(w) - 1$, we may assume that the reduced expression for w has been chosen so that $s = s_1$. Using the defining relations for H, we then compute that

$$\begin{aligned} \mathsf{T}_{s_1} \mathsf{T}_w = \mathsf{T}_{s_1}^2 \mathsf{T}_{s_2} \cdots \mathsf{T}_{s_n} &= (a_{s_1} 1_{\mathsf{H}} + b_{s_1} \mathsf{T}_{s_1}) \mathsf{T}_{s_2} \cdots \mathsf{T}_{s_n} \\ &= a_{s_1} \mathsf{T}_{s_2} \cdots \mathsf{T}_{s_n} + b_{s_1} \mathsf{T}_{s_1} \cdots \mathsf{T}_{s_n} \\ &= a_{s_1} \mathsf{T}_{s_2 \cdots s_n} + b_{s_1} \mathsf{T}_{s_1 \cdots s_n} = a_{s_1} \mathsf{T}_{s_1 w} + b_{s_1} \mathsf{T}_w. \end{aligned}$$

This completes the proof of (b).

Finally, in order to prove (c), consider the A-submodule $H' \subseteq H$ generated by $\{T_w \mid w \in W\}$. We must show that $H' = H$. Since H' contains the identity 1_H, it will be sufficient to show that H' is a left ideal in H. For this purpose, we just have to check that H' is invariant under left multiplication by all T_s, $s \in S$. But this is clear by (b). □

4.4.4. Remark. (a) The relations in Lemma 4.4.3 provide rules for multiplying any two elements $T_w, T_{w'}$ for $w, w' \in W$. Indeed, taking a reduced expression $w = s_1 \cdots s_n$ with $s_i \in S$, we can write $T_w = T_{s_1} \cdots T_{s_n}$. Thus, we are reduced to the case where $w = s \in S$, in which case we can apply Lemma 4.4.3.

(b) Assume that a_s is invertible in A, for all $s \in S$. Then the corresponding generator T_s of H is also invertible, with inverse given by

$$T_s^{-1} = a_s^{-1} T_s - a_s^{-1} b_s T_1.$$

Consequently, any T_w ($w \in W$) is invertible in H, since we have $T_w = T_{s_1} \cdots T_{s_n}$ for a reduced expression $w = s_1 \cdots s_n$ with $s_i \in S$ and so $T_w^{-1} = T_{s_n}^{-1} \cdots T_{s_1}^{-1}$. Note also that in this case, the map

$$A[B(W,S)] \to H, \qquad w \mapsto T_w \qquad (w \in W),$$

is an A-algebra homomorphism.

4.4.5. Corollary. *The Iwahori–Hecke algebra* $H = H_A(W, S, \{a_s, b_s \mid s \in S\})$ *in Definition 4.4.1 also has a presentation as an associative A-algebra with 1 and*

generators: T_w, $w \in W$,
relations: $T_s^2 = a_s T_1 + b_s T_s$ *for all* $s \in S$ *and*
$T_w = T_{s_1} \cdots T_{s_n}$ *if* $w = s_1 \cdots s_n$, *with* $s_i \in S$, *is reduced.*

Proof. For the sake of separating the notation, we let $\tilde{H}$ be the A-algebra given by a presentation with

generators: $\tilde{T}_w$, $w \in W$,
relations: $\tilde{T}_s^2 = a_s \tilde{T}_1 + b_s \tilde{T}_s$ for all $s \in S$ and
$\tilde{T}_w = \tilde{T}_{s_1} \cdots \tilde{T}_{s_n}$ if $w = s_1 \cdots s_n$, with $s_i \in S$, is reduced.

By Lemma 4.4.3(a), we have $T_w = T_{s_1} \cdots T_{s_n}$ whenever $w = s_1 \cdots s_n$ ($s_i \in S$) is reduced. So there is a unique A-algebra homomorphism $f \colon \tilde{H} \to H$ such that $f(\tilde{T}_w) = T_w$ for all $w \in W$. Conversely, consider the map $f' \colon \{T_s \mid s \in S\} \mapsto \tilde{H}$, $T_s \mapsto \tilde{T}_s$. We must check that $\mathrm{Prod}(\tilde{T}_s, \tilde{T}_t; m_{st}) = \mathrm{Prod}(\tilde{T}_t, \tilde{T}_s; m_{st})$ for all $s, t \in S$, $s \neq t$, $m_{st} < \infty$. This is done by an argument which is completely analogous to that in the proof of Proposition 4.1.3. Thus, f' extends to a unique A-algebra homomorphism $H \to \tilde{H}$, and we see that f, f' are inverse to each other. □

4.4.6. Theorem (Bourbaki). *Let* $H = H_A(W, S, \{a_s, b_s \mid s \in S\})$ *be an Iwahori–Hecke algebra over the commutative ring A as in Definition 4.4.1. Then H is free as an A-module and the elements* $\{T_w \mid w \in W\}$ *form an A-basis of H.*

Proof. In view of Lemma 4.4.3, it is sufficient to show that the elements T_w ($w \in W$) are linearly independent. For this purpose, we construct an explicit realization of H as an algebra of linear operators on some free A-module V. This is done as follows. Let V be the free A-module with a basis indexed by the elements of W, $\{e_w \mid w \in W\}$ say. For each $s \in S$, we define an A-linear map $\lambda_s \colon V \to V$ by the rule

$$e_w\lambda_s = \begin{cases} e_{sw} & \text{if } l(sw) > l(w), \\ a_s e_{sw} + b_s e_w & \text{if } l(sw) < l(w). \end{cases}$$

(We write λ_s on the right because it will be used to define an action of H on V, and we will usually write such an action on the right.) We claim that the assignment $T_s \mapsto \lambda_s$ ($s \in S$) extends to an A-algebra homomorphism $\varphi\colon H \to \mathrm{End}_A(V)$.

To prove this, we use the defining relations in Corollary 4.4.5. First let us check the quadratic relation. Let $s \in S$ and $w \in W$. Then

$$e_w\lambda_s^2 = \begin{cases} (e_w\lambda_s)\lambda_s = e_{sw}\lambda_s = a_s e_w + b_s e_{sw} & \text{if } l(sw) > l(w), \\ (a_s e_{sw} + b_s e_w)\lambda_s = a_s e_w + b_s a_s e_{sw} + b_s^2 e_w & \text{if } l(sw) < l(w). \end{cases}$$

Hence λ_s^2 acts in the same way as $a_s \mathrm{id}_V + b_s\lambda_s$. Now let $w \in W$ and take two reduced expressions $w = s_1 \cdots s_n = s_1' \cdots s_n'$ with $s_i, s_i' \in S$. We must show that

$$\lambda := \lambda_{s_1} \cdots \lambda_{s_n} - \lambda_{s_1'} \cdots \lambda_{s_n'} = 0.$$

First, we consider the effect of λ on e_1. We have $e_1\lambda_s = e_s$ for all $s \in S$. Hence

$$e_1(\lambda_{s_1} \cdots \lambda_{s_n}) = e_{w^{-1}} = e_1(\lambda_{s_1'} \cdots \lambda_{s_n'}),$$

and so $e_1\lambda = 0$. In order to show that this implies $\lambda = 0$, we introduce A-linear maps $\rho_s \colon V \to V$ (for $s \in S$) by the rule

$$e_w\rho_s = \begin{cases} e_{ws} & \text{if } l(ws) > l(w), \\ a_s e_{ws} + b_s e_w & \text{if } l(ws) < l(w). \end{cases}$$

The crux of the whole proof consists of showing that λ_s and ρ_t commute for all $s, t \in S$. In order to evaluate $(e_w\lambda_s)\rho_t$ and $(e_w\rho_t)\lambda_s$ for $w \in W$, one has to distinguish a number of cases according to the relative length of w, sw, wt, swt.

Case 1. $l(sw) > l(w)$ and $l(swt) > l(sw)$. Then we also have $l(swt) > l(wt) > l(w)$ and hence

$$(e_w\lambda_s)\rho_t = e_{sw}\rho_t = e_{swt} = e_{wt}\lambda_s = (e_w\rho_t)\lambda_s.$$

Case 2. $l(sw) > l(w)$, $l(swt) < l(sw)$ and $l(wt) < l(w)$. Then we also have $l(swt) = l(w) > l(wt)$ and hence

$$(e_w\lambda_s)\rho_t = e_{sw}\rho_t = a_t e_{swt} + b_t e_{sw} = (a_t e_{wt} + b_t e_w)\lambda_s = (e_w\rho_t)\lambda_s.$$

Case 3. $l(sw) > l(w)$, $l(swt) < l(sw)$ and $l(wt) > l(w)$. Then we also have $l(swt) = l(w)$ and hence $sw = wt$ by Lemma 1.2.6. Thus, s, t are conjugate in W and hence, by assumption, $a_s = a_t$, $b_s = b_t$. So the two expressions

$$(e_w\lambda_s)\rho_t = e_{sw}\rho_t = a_t e_{swt} + b_t e_{sw},$$
$$(e_w\rho_t)\lambda_s = e_{wt}\lambda_s = a_s e_{swt} + b_s e_{wt}$$

are equal.

Case 4. $l(sw) < l(w)$, $l(swt) < l(sw)$. Then we also have $l(swt) < l(wt) < l(w)$ and hence

$$(e_w\lambda_s)\rho_t = (a_s e_{sw} + b_s e_w)\rho_t = a_s a_t e_{swt} + a_s b_t e_{sw} + b_s a_t e_{wt} + b_s b_t e_w,$$
$$(e_w\rho_t)\lambda_s = (a_t e_{wt} + b_t e_w)\lambda_s = a_t a_s e_{swt} + a_t b_s e_{wt} + b_t a_s e_{sw} + b_t b_s e_w.$$

Again, we see that these two expressions are equal.

Case 5. $l(sw) < l(w)$, $l(swt) > l(sw)$ and $l(wt) > l(w)$. Then we also have $l(swt) = l(w) < l(wt)$ and hence

$$(e_w\lambda_s)\rho_t = (a_s e_{sw} + b_s e_w)\rho_t = a_s e_{swt} + b_s e_{wt} = e_{wt}\lambda_s = (e_w\rho_t)\lambda_s.$$

Case 6. $l(sw) < l(w)$, $l(swt) > l(sw)$ and $l(wt) < l(w)$. Then we also have $l(swt) = l(w)$ and hence $sw = tw$, as in case 3 above. So we conclude that

$$(e_w\lambda_s)\rho_t = (a_s e_{sw} + b_s e_w)\rho_t = a_s e_{swt} + b_s a_t e_{wt} + b_s b_t e_w,$$
$$(e_w\rho_t)\lambda_s = (a_t e_{wt} + b_t e_w)\lambda_s = a_t e_{swt} + b_t a_s e_{sw} + b_t b_s e_w.$$

Since we have $a_s = b_s$ and $a_t = b_t$, the above two expressions are again equal.

Thus, we see that $\lambda_s\rho_t = \rho_t\lambda_s$ for all $s, t \in S$, as claimed.

Now let us return to the problem of showing that $\lambda = 0$. We prove by induction on $l(w)$ that $e_w\lambda = 0$. If $w = 1$, then we already know that $e_1\lambda = 0$. Now let $w \neq 1$. There exists some $t \in S$ such that $l(wt) < l(w)$. Then $e_{wt}\lambda = 0$ by induction and, hence, using the fact that ρ_t commutes with λ, we compute that $e_w\lambda = e_{(wt)t}\lambda = (e_{wt}\rho_t)\lambda = (e_{wt}\lambda)\rho_t = 0$, as desired.

Thus, we have an A-algebra homomorphism $\varphi\colon H \to \mathrm{End}_A(V)$, $T_s \mapsto \lambda_s$, as desired. Via this map, we can define an action of H on V by $vh := v\varphi(h)$ for $v \in V, h \in H$. We use this action to finish the proof, as follows. Suppose we have a linear relation $\sum_{w\in W} a_w T_w = 0$, where only a finite number of coefficients $a_w \in A$ is non-zero. We have already seen above that if $w = s_1 \cdots s_n$ with $s_i \in S$ is a reduced expression, then $e_1(\lambda_{s_1} \cdots \lambda_{s_n}) = e_{w^{-1}}$ and so $e_1 T_w = e_{w^{-1}}$. Hence, applying the above linear relation to e_1 yields

$$0 = e_1\Big(\sum_{w\in W} a_w T_w\Big) = \sum_{w\in W} a_w e_{w^{-1}}.$$

But $\{e_w \mid w \in W\}$ is linearly independent, whence $a_w = 0$ for all $w \in W$. □

As a first application, we can now show that each parabolic subgroup of W gives rise to a subalgebra of H such that H is a free module over that subalgebra.

4.4.7. Parabolic subalgebras. Let $H = H_A(W, S, \{a_s, b_s \mid s \in S\})$ be an Iwahori–Hecke algebra as in Definition 4.4.1. Let $J \subseteq S$ and $W_J \subseteq W$ be the corresponding parabolic subgroup. Then the multiplication rules in Lemma 4.4.3 show that the A-submodule

$$H_J := \langle T_w \mid w \in W_J \rangle_A \subseteq H$$

is in fact a subalgebra with the same identity element as in H. We call H_J a *parabolic subalgebra* of H. Recall from (1.2.9) that (W_J, J) itself is a Coxeter system. Then we have a canonical surjective map

$$H_A(W_J, J, \{a_s, b_s \mid s \in J\}) \to H_J \qquad \text{(a)}$$

given by identifying corresponding generators. By Theorem 4.4.6, the elements $\{T_w \mid w \in W_J\}$ form A-bases of the above two algebras. Hence the above map is an isomorphism. Thus, $H_J \subseteq H$ can be canonically identified with the Iwahori–Hecke algebra over A associated with (W_J, J) and with parameters $\{a_s, b_s \mid s \in J\}$.

Furthermore, recall from Section 2.1 that each $w \in W$ can be written uniquely $w = w'x$ with $w' \in W_J$ and $x \in X_J$, where $X_J \subseteq W$ is the set of distinguished right coset representatives of W_J in W. Since $l(w) = l(w') + l(x)$, we also have $T_w = T_{w'}T_x$. Hence we obtain a direct sum decomposition of H_J-modules

$$H = \bigoplus_{x \in X_J} H_J T_x. \qquad \text{(b)}$$

In particular, H is free as a (left) H_J-module, with a basis given by $\{T_x \mid x \in X_J\}$.

4.5 INVARIANTS OF KNOTS AND LINKS

To give a first application of Iwahori–Hecke algebras, we briefly discuss Jones's construction of an invariant for oriented *knots and links* via certain linear functionals on Iwahori–Hecke algebras of type A.

4.5.1. If n is a positive integer, an oriented n-*link* is an embedding of n copies of the interval $[0,1] \subset \mathbb{R}$ into $\mathbb{R}^3$ such that 0 and 1 are mapped to the same point (the orientation is induced by the natural ordering of $[0,1]$); a 1-link is also called a *knot*. We are only interested in knots and links modulo isotopy, i.e., homeomorphic transformations which preserve the orientation. We refer to [Crowell and Fox 1977] for precise versions of the above definitions. The problem of classifying oriented links modulo isotopy can be translated into a purely algebraic problem about Artin braid groups. This is based on the following facts.

For $n \geq 1$ let $B(\mathfrak{S}_n)$ be the Artin braid group corresponding to $W = \mathfrak{S}_n$, with respect to the standard generators $s_i = (i, i+1)$ for $1 \leq i \leq n-1$. The generators of $B(\mathfrak{S}_n)$ can be represented by oriented diagrams as indicated below; writing any $g \in B(\mathfrak{S}_n)$ as a product of the generators and their inverses, we also obtain a diagram for g, by concatenating the diagrams for the generators. "Closing" such a diagram by joining the end points, we obtain the plane projection of an oriented link in $\mathbb{R}^3$:

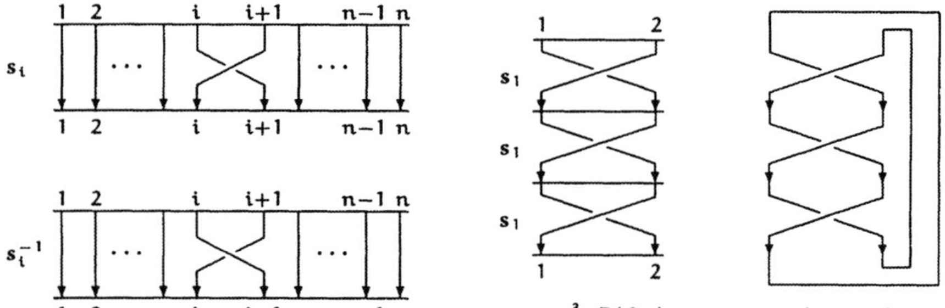

$g = s_1^3 \in B(\mathfrak{S}_2)$ closure of g

By Alexander's theorem (see [Vogel 1990]), every oriented link in $\mathbb{R}^3$ is isotopic to the closure of an element in $B(\mathfrak{S}_n)$, for some $n \geq 1$. The question of when two links in $\mathbb{R}^3$ are isotopic can also be expressed algebraically. For this purpose, we consider the infinite disjoint union

$$B_\infty := \coprod_{n \geq 1} B(\mathfrak{S}_n).$$

Given $g, g' \in B_\infty$, we write $g \sim g'$ if one of the following relations is satisfied:

(I) We have $g, g' \in B(\mathfrak{S}_n)$ and $g' = x^{-1}gx$ for some $x \in B(\mathfrak{S}_n)$.

(II) We have $g \in B(\mathfrak{S}_n)$, $g' \in B(\mathfrak{S}_{n+1})$ and $g' = gs_n$ or $g' = gs_n^{-1}$.

The above two relations are called *Markov relations*. By a classical result due to Markov (see [Traczyk 1998]), two elements of B_∞ are equivalent under the equivalence relation generated by $\sim$ if and only if the corresponding links obtained by closure are isotopic. Thus, to define an invariant of oriented links is the same as to define a map on B_∞ which takes equal values on elements $g, g' \in B_\infty$ satisfying (I) or (II). In order to solve the latter problem, the idea is to look for maps which factor through the corresponding Iwahori–Hecke algebras, for suitable values of the parameters.

Let A be a commutative ring with 1 and fix two invertible elements $u, v \in A$. For any $n \geq 1$, we consider the Iwahori–Hecke algebra $H_A(\mathfrak{S}_n)$ associated with $\mathfrak{S}_n$ over the ring A and with parameters $a_{s_i} = u$, $b_{s_i} = v$ for $1 \leq i \leq n-1$. By (4.4.7), we may regard $H_A(\mathfrak{S}_n)$ as a parabolic subalgebra of $H_A(\mathfrak{S}_{n+1})$.

4.5.2. Theorem (Jones, Ocneanu). *There is a unique family of A-linear maps $\tau_n \colon H_A(\mathfrak{S}_n) \to A$ $(n \geq 1)$ such that the following conditions hold:*

(M1) $\tau_1(T_1) = 1$;

(M2) $\tau_{n+1}(hT_{s_n}) = \tau_{n+1}(hT_{s_n}^{-1}) = \tau_n(h)$ *for all* $n \geq 1$ *and* $h \in H_A(\mathfrak{S}_n)$;

(M3) $\tau_n(hh') = \tau_n(h'h)$ *for all* $n \geq 1$ *and* $h, h' \in H_A(\mathfrak{S}_n)$.

Moreover, we have $\tau_{n+1}(h) = v^{-1}(1-u)\tau_n(h)$ *for all* $n \geq 1$ *and* $h \in H_A(\mathfrak{S}_n)$.

Proof. We begin with some preliminary remarks on the distinguished coset representatives of $\mathfrak{S}_n$ in $\mathfrak{S}_{n+1}$. By Example 2.2.4 (see also Example 2.2.13), these are given by

$$X_n := \{1, s_n, s_n s_{n-1}, \ldots, s_n s_{n-1} \cdots s_1\}.$$

It follows that each $w \in \mathfrak{S}_{n+1}$ can be written as $w = y x_n z$, where $y, z \in \mathfrak{S}_n$ and $x_n \in \{1, s_n\}$, and where $l(w) = l(y) + l(x_n) + l(z)$.

To simplify notation, we will now write $H_n = H_A(\mathfrak{S}_n)$ for all $n \geq 1$. We claim that, for any $n \geq 2$, the map

$$\psi_n \colon H_n \oplus (H_n \otimes_{H_{n-1}} H_n) \to H_{n+1}, \qquad a \oplus (b \otimes c) \mapsto a + bT_{s_n}c,$$

is an isomorphism of A-modules. First of all, note that ψ_n is well defined. Indeed, one has to check that for any $b, c \in H_n$ and $d \in H_{n-1}$, we have $bdT_{s_n}c = bT_{s_n}dc$, which is clear since all elements in $\mathfrak{S}_{n-1}$ commute with s_n. Now let $w \in \mathfrak{S}_{n+1}$. By the above remarks on coset representatives, we can write $w = y x_n z$ with $y, z \in \mathfrak{S}_n$, $x_n \in \{1, s_n\}$ and where $l(w) = l(y) + l(x_n) + l(z)$. It follows that we also have $T_w = T_y T_{x_n} T_z$ in H_{n+1}, where $T_y, T_z \in H_n$. So ψ_n is surjective. On the other hand, the direct sum decomposition in (4.4.7)(b) yields

$$H_n \otimes_{H_{n-1}} H_n \cong_A \bigoplus_{x \in X_{n-1}} H_n \otimes T_x,$$

which means that this is a free A-module of rank equal to $|\mathfrak{S}_n||X_{n-1}|$. Consequently, $H_n \oplus (H_n \otimes_{H_{n-1}} H_n)$ is a free A-module of rank equal to $|\mathfrak{S}_{n+1}| = |\mathfrak{S}_n|(1 + |X_{n-1}|)$. Hence ψ_n is also injective, and the above claim is proved.

We can now define τ_n recursively as follows. For $n = 1$, we set $\tau_1(T_1) = 1$. Now let $n \geq 1$ and assume that τ_n has been already defined. Then we set

$$\tau_{n+1}(a + bT_{s_n}c) := \frac{1-u}{v}\tau_n(a) + \tau_n(bc) \qquad \text{for all } a, b, c \in H_n. \qquad (*)$$

Using the isomorphism ψ_n, we see that τ_{n+1} is well defined. It is clear that (M1) holds. We have to check that (M2) and (M3) are also satisfied. First consider (M2). Let $h \in H_n$. Then we have $\tau_{n+1}(hT_{s_n}) = \tau_n(h)$ by the definition. Furthermore, by Remark 4.4.4(b), we have $T_{s_n}^{-1} = u^{-1}T_{s_n} - u^{-1}vT_1$ and so

$$\begin{aligned} \tau_{n+1}(hT_{s_n}^{-1}) &= u^{-1}\tau_{n+1}(hT_{s_n}) - u^{-1}v\tau_{n+1}(h) \\ &= u^{-1}\tau_n(h) - u^{-1}(1-u)\tau_n(h) = \tau_n(h), \end{aligned}$$

as desired. In order to see that (M3) holds, we proceed again by induction on n. The case $n = 1$ is trivial. Now let $n \geq 1$ and assume that (M3) holds for τ_n. In order to check (M3) for τ_{n+1}, it is enough to consider the case where $h' = T_{s_i}$ for some $1 \leq i \leq n$. Furthermore, using ψ_n, we can assume that $h = T_y T_{x_n} T_z$ with $y, z \in \mathfrak{S}_n$ and $x_n \in \{1, s_n\}$. Now there are several cases to consider, according to whether $i = n$ or $i < n$ or $x_n = 1$ or $x_n = s_n$. If $i < n$ or $x_n = 1$, the desired relation follows trivially from the definition of τ_{n+1}. For example, suppose that $i < n$ and $x_n = s_n$. Then $T_y T_{s_i}, T_{s_i} T_z \in H_n$ and, using induction and the definition of τ_{n+1}, we have $\tau_{n+1}(T_y T_{s_n} T_z T_{s_i}) = \tau_n(T_y T_z T_{s_i}) = \tau_n(T_{s_i} T_y T_z) = \tau_{n+1}(T_{s_i} T_y T_{s_n} T_z)$. Now let us consider the remaining case where $i = n$ and $x_n = s_n$. Then we have to consider the following possibilities:

(a) $y, z \in \mathfrak{S}_{n-1}$.
(b) $y \in \mathfrak{S}_{n-1}$ and $z = cs_{n-1}d$ with $c, d \in \mathfrak{S}_{n-1}$ and $l(z) = l(c) + l(d) + 1$.
(c) The same as in (b), with the roles of y, z interchanged.
(d) $y = as_{n-1}b$ and $z = cs_{n-1}d$ with $a, b, c, d \in \mathfrak{S}_{n-1}$ and $l(y) = l(a) + l(b) + 1$, $l(z) = l(c) + l(d) + 1$.

In case (a), (M3) holds since T_{s_n} commutes with T_y, T_z. In case (b), we obtain:

$$\begin{aligned}
\tau_{n+1}(T_y T_{s_n} T_z T_{s_n}) &= \tau_{n+1}(T_y T_{s_n} T_c T_{s_{n-1}} T_d T_{s_n}) = \tau_{n+1}(T_y T_c T_{s_n} T_{s_{n-1}} T_{s_n} T_d) \\
&= \tau_{n+1}(T_y T_c T_{s_{n-1}} T_{s_n} T_{s_{n-1}} T_d) = \tau_n(T_y T_c T_{s_{n-1}}^2 T_d) \\
&= u\tau_n(T_y T_c T_d) + v\tau_n(T_y T_c T_{s_{n-1}} T_d), \\
\tau_{n+1}(T_{s_n} T_y T_{s_n} T_z) &= \tau_{n+1}(T_{s_n}^2 T_y T_z) = u\tau_{n+1}(T_y T_z) + v\tau_{n+1}(T_{s_n} T_y T_z) \\
&= u\tau_{n+1}(T_y T_c T_{s_{n-1}} T_d) + v\tau_n(T_y T_z) \\
&= v^{-1}u(1-u)\tau_n(T_y T_c T_{s_{n-1}} T_d) + v\tau_n(T_y T_c T_{s_{n-1}} T_d).
\end{aligned}$$

But, using the definition of τ_n, we find that the first summand in each of the above expressions equals $v^{-1}u(1-u)\tau_{n-1}(T_y T_c T_d)$. Hence we have equality. Case (c) is treated similarly. Finally, in case (d), similar computations yield:

$$\begin{aligned}
\tau_{n+1}(T_y T_{s_n} T_z T_{s_n}) &= u\tau_n(T_y T_c T_d) + v\tau_n(T_y T_c T_{s_{n-1}} T_d) \qquad \text{(as before)} \\
&= u\tau_n(T_a T_{s_{n-1}} T_b T_c T_d) + v\tau_n(T_a T_{s_{n-1}} T_b T_c T_{s_{n-1}} T_d), \\
\tau_{n+1}(T_{s_n} T_y T_{s_n} T_z) &= \tau_{n+1}(T_{s_n} T_a T_{s_{n-1}} T_b T_{s_n} T_z) = \tau_n(T_a T_{s_{n-1}}^2 T_b T_z) \\
&= u\tau_n(T_a T_b T_c T_{s_{n-1}} T_d) + v\tau_n(T_a T_{s_{n-1}} T_b T_c T_{s_{n-1}} T_d).
\end{aligned}$$

Again, using the definition of τ_n, we find that the first summand in each of the above expressions equals $u\tau_{n-1}(T_a T_b T_c T_d)$. Hence we have equality. Thus, we have proved the existence of a family of A-linear maps with the required properties. Now assume that $(\tau'_n)_{n \geq 1}$ is another family which also satisfies (M1), (M2) and (M3). Then we must show that $\tau_n(T_w) = \tau'_n(T_w)$ for all $n \geq 1$ and $w \in \mathfrak{S}_n$. We proceed by induction on n. For $n = 1$, we can apply (M1). Now assume that $n \geq 1$ and let $w \in \mathfrak{S}_{n+1}$. We can assume that $w \notin \mathfrak{S}_n$ and so we can write $w = ys_n z$ with $y, z \in \mathfrak{S}_n$ and $l(w) = l(y) + l(z) + 1$. But then we also have $T_w = T_y T_{s_n} T_z$ and so, using (M2) and induction, $\tau_{n+1}(T_w) = \tau_n(T_y T_z) = \tau'_n(T_y T_z) = \tau'_{n+1}(T_w)$. Thus, we have also proved the uniqueness statement. □

4.5.3. The HOMFLY-PT polynomial. We can now construct a two-variable invariant of oriented knots and links as follows. Consider an oriented link L and assume that it is isotopic to the closure of $g \in B(\mathfrak{S}_n)$ for $n \geq 1$. Then we set

$$X_L(u, v) := \tau_n(\bar{g}) \in \mathbb{Z}[u^{\pm 1}, v^{\pm 1}] \qquad \text{with } \tau_n \text{ as in Theorem 4.5.2.}$$

Here, $\bar{g}$ denotes the image of g under the natural map $A[B(\mathfrak{S}_n)] \to H_A(\mathfrak{S}_n)$, $w \mapsto T_w$ ($w \in \mathfrak{S}_n$). The properties (M2) and (M3) make sure that $\tau_n(\bar{g})$ does not depend on the choice of g. If we make the change of variables $u = t^2$ and $v = tx$, we can identify the above invariant* with the HOMFLY-PT polynomial $P_L(t, x)$ in [Jones 1987, (6.2)]. Furthermore, the *Jones polynomial* $J_L(t)$ is obtained by setting $u = t^2$, $v = \sqrt{t}(t-1)$ (see [Jones 1987, §11]). Note that we still obtain a Laurent polynomial in $\sqrt{t}$ since the defining formula (∗) in the proof of Theorem 4.5.2 now reads:

$$\tau_{n+1}(a + bT_{s_n}c) = \frac{1-t^2}{\sqrt{t}(t-1)}\tau_n(a) + \tau_n(bc) = -\frac{t+1}{\sqrt{t}}\tau_n(a) + \tau_n(bc). \qquad \text{(J)}$$

Finally, setting $u = 1$ and $v = \sqrt{t} - 1/\sqrt{t}$, we obtain the classical *Alexander polynomial* $A_L(t)$, where formula (∗) now simply reads:

$$\tau_{n+1}(a + bT_{s_n}c) = \tau_n(bc). \qquad \text{(A)}$$

See Exercise 4.10 for an example of two knots which have the same Alexander polynomial but different Jones polynomials. The Alexander polynomial is originally defined using a suitable matrix representation of the braid group; we will discuss this in more detail in (4.5.5) below.

Formula (∗) in the proof of Theorem 4.5.2 actually provides an algorithm for computing $X_L(u, v)$. This is illustrated by the following examples.

4.5.4. Example. The knot defined by the element $s_1^3 \in B(\mathfrak{S}_2)$ is called the *trefoil knot*; see the picture in (4.5.1). We have $T_{s_1}^3 = uvT_1 + (u+v^2)T_{s_1}$ and so, using formula (∗) in the proof of Theorem 4.5.2:

$$\begin{aligned}\tau_2(T_{s_1}^3) &= uv\tau_2(T_1) + (u+v^2)\tau_2(T_{s_1}) = uv\frac{1-u}{v}\tau_1(T_1) + (u+v^2)\tau_1(T_1)\\ &= u(1-u) + u + v^2 = 2u - u^2 + v^2,\end{aligned}$$

and so $P_{\text{trefoil}}(t, x) = 2t^2 - t^4 + t^2x^2$. For $n \geq 1$, the n-component unlink is defined as the closure of $1 \in B(\mathfrak{S}_n)$. Using formula (∗) again, we obtain

$$\tau_n(T_1) = \left(\frac{1-u}{v}\right)^{n-1} \quad \text{and} \quad P_{n\text{-unlink}}(t, x) = \left(\frac{t^{-1}-t}{x}\right)^{n-1}$$

*The invariant in [Jones 1987, (6.1)] depends on two variables q, λ. These are related with our variables u, v by the formulas $u = q\lambda$ and $v = \sqrt{\lambda}(q-1)$.

We will now establish the promised alternative interpretation of the Alexander polynomial. The proof of this will rely on some general results about trace functions on Iwahori–Hecke algebras which will be developed in Chapter 8.

4.5.5. The Alexander polynomial and the Burau representation. We have defined the Alexander polynomial of a link L by the formula $A_L(t) = \tau_n(\bar{g})$, where L is the closure of $g \in B(\mathfrak{S}_n)$ and where we have $u = 1$, $v = \sqrt{t} - 1/\sqrt{t}$. We now define a representation of $B(\mathfrak{S}_n)$ as follows. For $1 \le i \le n-1$, let

$$M_i := \left[\begin{array}{c|cc|c} I_{i-1} & 0 & 0 & 0 \\ \hline 0 & 1-t & t & 0 \\ 0 & 1 & 0 & 0 \\ \hline 0 & 0 & 0 & I_{n-i-1} \end{array}\right] \in M_n(A),$$

where I_k denotes the identity matrix of size k. It is checked in Exercise 4.11 that we have a representation $\beta_n \colon A[B(\mathfrak{S}_n)] \to M_n(A)$, $s_i \mapsto M_i$. Furthermore, considering $V = A^n$ as an $A[B(\mathfrak{S}_n)]$-module via $xs_i = xM_i$, there exists some $0 \neq x \in V$ such that $xM_i = x$ for all i and such that the quotient space $V' = V/\langle x \rangle$ is A-free of rank $n-1$. Hence we obtain an induced representation $\beta_n' \colon A[B(\mathfrak{S}_n)] \to \mathrm{End}_A(V')$ which is called the *reduced Burau representation*. The following identity shows that $A_L(t)$ can be expressed in terms of β_n':

$$\tau_n(T_w) = (-1/\sqrt{t})^{l(w)-n+1} \frac{1-t}{1-t^n} \det(1 - \beta_n'(w)) \quad \text{for all } w \in \mathfrak{S}_n. \qquad (A')$$

In order to prove (A′), consider the Iwahori–Hecke algebra $H_A(\mathfrak{S}_n)$ as above, where the parameters are $a_{s_i} = u = 1$ and $b_{s_i} = v = \sqrt{t} - 1/\sqrt{t}$ for all i. On the other hand, we also set $T_w' := \sqrt{t}^{\,l(w)} T_w$ for any $w \in \mathfrak{S}_n$. Then we have

$$(T_{s_i}')^2 = tT_{s_i}^2 = tT_1 + t(\sqrt{t} - 1/\sqrt{t})T_{s_i} = tT_1 + (t-1)T_{s_i}'.$$

It is shown in Exercise 4.11 that then the map $T_w' \mapsto (-1)^{l(w)}\beta_n'(w)$ ($w \in \mathfrak{S}_n$) defines a representation of $H_A(\mathfrak{S}_n)$. Consider the A-linear map

$$\tau_n' \colon H_A(\mathfrak{S}_n) \to A, \quad T_w' \mapsto (-1)^{l(w)} \det(1 - \beta_n'(w)) \qquad \text{for } w \in \mathfrak{S}_n.$$

We must show that $\tau_n'(T_w') = (-1)^{l(w)}\sqrt{t}^{\,n-1-l(w)}(1+t+\cdots+t^{n-1})\tau_n(T_w)$ for all $w \in \mathfrak{S}_n$. Our proof of this fact rests on the following two ingredients, which will be established in Chapter 8 (for Iwahori–Hecke algebras of any type):

(a) By (8.1.14), τ_n' is a trace function on $H_A(\mathfrak{S}_n)$, i.e., we have $\tau_n'(h_1h_2) = \tau_n'(h_2h_1)$ for all $h_1, h_2 \in H_A(\mathfrak{S}_n)$.

(b) By Corollary 8.2.6, every trace function on $H_A(\mathfrak{S}_n)$ is uniquely determined by its values on basis elements T_w where w runs over a set of representatives of minimal length in the conjugacy classes of $\mathfrak{S}_n$.

Thus, taking also into account the results in Section 3.4 on standard representatives for $\mathfrak{S}_n$, we see that it is enough to compare the values of τ_n and τ'_n on basis elements corresponding to elements $w \in \mathfrak{S}_n$ of the form $w = s_{i_1} \cdots s_{i_k}$ where $1 \leq i_1 < \cdots < i_k \leq n-1$. Using the defining formula (4.5.3)(A) for τ_n, we immediately find that $\tau_n(T_w) = 0$ if $l(w) < n-1$ and $\tau_n(T_w) = 1$ if $l(w) = n-1$. The values for τ'_n are computed in Exercise 4.12. Comparison of these values yields (A$'$).

4.6 BIBLIOGRAPHICAL REMARKS AND EXERCISES

The presentation for B^+ in Proposition 4.1.3 is well known; it appears, for example, in [Deligne 1997, §1.4]. Normal forms of elements in the braid monoid of type A_n were first investigated in [Garside 1969], and in the general case in [Brieskorn and Saito 1972] and [Deligne 1972]. The solution to the conjugacy problem in Exercise 4.5 can also be found in these articles. Our presentation here follows [Michel 1999]. Exercise 4.1 is due to Tits; see [Brieskorn and Saito 1972, (5.6)]. The fact that Coxeter elements are good is contained, for example, in [Brieskorn and Saito 1972, Lemma 5.8]. The general notion of good elements and Theorem 4.3.5 appeared in [Geck and Michel 1997].

For an introduction to knots and links, see [Crowell and Fox 1977]. The topological interpretation of the braid group as the fundamental group of a certain orbit space (mentioned at the end of Section 4.1) has been generalized to finite complex reflection groups in [Broué et al. 1998].

The origin of the theory of Hecke algebras lies in the paper [Iwahori 1964]; this will be considered in more detail in Section 8.4. The fact that abstract Iwahori–Hecke algebras as in Definition 4.4.1 are free over the ground ring is proved in [Bourbaki 1968, Chap. IV, §2, Exercise 23].

Theorem 4.5.2 and the definition of the two-variable invariant $P_L(t, x)$ is a combination of Theorem 5.1 and Proposition 6.2 in [Jones 1987]; we have followed the argument given there, with some simplifications suggested by J. Michel. (The simplification arises from the fact that we use two different parameters in the quadratic relations in Definition 4.4.1: as a consequence, the "singularities" mentioned in [Jones 1987, p. 349, Notes (1)] simply disappear.) The Jones polynomial also admits an interpretation in terms of suitable representations; see [Jones 1987, §11]. Generalizations of Theorem 4.5.2 to types B_n and D_n have been found in [Geck and Lambropoulou 1997] and [Geck 1998d]. It is explicitly known how these trace functions, also called *Markov traces*, can be expressed as linear combinations of irreducible characters; for type A, this is due originally to Ocneanu (see [Wenzl 1988, §3]); for types B (for special choices of the parameters) and D, see [Orellana 1999]. The case of arbitrary parameters in type B_n is settled in [Iancu 2000]. For a survey about recent developments in the theory of knots and links, see [Birman 1993].

The matrix representation in (4.5.5) is due to Burau; see [Jones 1987, §2]. A slightly different proof for (4.5.5)(A$'$) can be found in [Jones 1987, §7]. It was an open question for many years whether that representation is faithful for $n \geq 4$. For large n, it has been shown recently in [Moody 1993] that this is not the case. Exercise 4.9 is Example 6.8 in [Jones 1987].

Exercise 4.1. This exercise gives an intrinsic description of the set B^+_{red}. Let $g \in B^+$. We say that g contains a square if we can write $g = s_1 \cdots s_k$ with $s_1, \ldots, s_k \in S$ and $s_i = s_{i+1}$ for some $1 \leq i < k$. Show that

$$B^+_{\mathsf{red}} = \{g \in B^+ \mid g \text{ does not contain a square}\}.$$

[*Hint.* If g contains a square, it certainly does not lie in B^+_{red}. Assume now that $g \notin B^+_{\mathsf{red}}$. Thus, if $b = s_1 \cdots s_k$ with $s_i \in S$, then $l(s_1 \cdots s_k) < k$. Use the exchange condition (see Exercise 1.6) to show that there exists some $l \geq 2$ such that $s_1 \cdots s_{l-1}$ is reduced and $s_1 \cdots s_{l-1} = s_2 \cdots s_l$. Then we also have $s_1 \cdots s_{l-1} = s_2 \cdots s_l \in B^+$, and hence g contains a square.]

Exercise 4.2. Let (W, S) be a Coxeter system and assume that we have a partition $S = S_1 \cup S_2$ such that $S_1 \cap S_2 = \varnothing$ and every element of S_1 commutes with every element of S_2. Let $W_i = \langle S_i \rangle$ for $i = 1, 2$, so that we have $W = W_1 \times W_2$. Show that the natural map given by multiplication,

$$B^+(W_1, S_1) \times B^+(W_2, S_2) \to B^+(W, S), \quad (g_1, g_2) \mapsto g_1 g_2,$$

is an isomorphism which also extends to the corresponding braid groups.

Exercise 4.3. Let B^+ be the braid monoid associated with a Coxeter system (W, S). Show that there exists a bijection $\mathsf{rev}\colon B^+ \to B^+$ satisfying $\mathsf{rev} \circ \mathsf{rev} = \mathsf{id}_{B^+}$ and $\mathsf{rev}(gh) = \mathsf{rev}(h)\,\mathsf{rev}(g)$ for all $g, h \in B^+$. Thus, using rev, left divisibility in B^+ translates to right divisibility and vice versa.

Exercise 4.4. Let $\alpha\colon B^+ \to B^+_{\mathsf{red}}$ be the function defined in Theorem 4.2.3. Show that we have $\alpha(gh) = \alpha(g\alpha(h))$ for any $g, h \in B^+$.

[*Hint.* Write $g = g_1 \cdots g_n$ with $g_i \in B^+_{\mathsf{red}}$ and use induction on n.]

Exercise 4.5. Assume that (W, S) is a Coxeter system with W finite. Let B^+ be the braid monoid and B the corresponding braid group.

(a) Let $a, g \in B^+$ such that $a \prec ga$. Show that $\alpha(a) \prec g\alpha(a)$.
(b) Show that $g, h \in B^+$ are conjugate in B if and only if there exist elements $g_1, \ldots, g_n \in B^+_{\mathsf{red}}$ and $h_1, \ldots, h_{n+1} \in B^+$ such that $h_1 = g$, $h_{n+1} = h$ and $h_i g_i = g_i h_{i+1}$ for all i.

The above statements provide an algorithm for checking if two elements of B^+ are conjugate in B. This is based on the following two facts: (1) two elements of B^+ can only be conjugate if they have the same length, and there are only a finite number of elements of B^+ of a given length; (2) in order to find out which elements of B^+ are conjugate to a given $g \in B^+$, it is enough (by (b)) to conjugate g successively with elements from B^+_{red}, which is a finite set.

Use this algorithm to show that, in the case where $W = \mathfrak{S}_5$ with $s_i = (i, i+1)$ for $1 \leq i \leq 4$, $s_1s_3s_2s_1s_4s_3$ and $s_1s_2s_1s_3s_2s_1$ are not conjugate in B.

[*Hint.* (a) Use Exercise 4.4. (b) Multiplying by a suitable power of Δ^2, we may assume that $ga = ah$ for some $a \in B^+$. Let $a = g_1 \cdots g_n$ with $g_i \in B^+_{\text{red}}$ in Deligne normal form. We proceed by induction on n. Since $g_1 = \alpha(a)$, we can use (a) and conclude that $g_1 \prec gg_1$. Let $h_2 = g_1^{-1}gg_1 \in B^+$. Then we have

$$g_1h_2g_2 \cdots g_n = gg_1 \cdots g_n = ga = ah = g_1 \cdots g_nh,$$

and so $h_2g_2 \cdots g_n = g_2 \cdots g_nh$. Now apply induction.]

Exercise 4.6. Let W be of type B_n or D_n, $n \geq 2$, and let $b^-_{m,e}$ as in (3.4.2).

(a) Determine the matrix of the action of $b^-_{m,e}$ on the subspace of $\mathbb{R}^n$ generated by $\{e_{m+1}, \ldots, e_{m+e}\}$, and show that the eth power of this matrix is $-\text{id}$. Determine the length of t_m and of $b^-_{m,e}$ and show that $(b^-_{m,e})^e$ is a reduced expression.

(b) Let $m' \geq m + e$ and suppose $m \geq 1$ in the case where W is of type D_n and $b^{\pm}_{m',d'}$ is a negative block. Show that $b^{\pm}_{m,d}\,b^{\pm}_{m',d'} = b^{\pm}_{m',d'}\,b^{\pm}_{m,d}$.

[*Hint.* By Lemma 4.3.8, $(b^-_{m,e})^e = t_m \cdots t_{m+e-1}$ and the length of this element is $(2m+e)e$.]

Exercise 4.7. Show that the conditions on the parameters in Definition 4.4.1 are necessary in order that the elements $\{T_w \mid w \in W\}$ are linearly independent. Do this by taking suitable $s, t \in S$, $s \neq t$, such that m_{st} is odd (using Exercise 1.10(a)), and let $w := \text{Prod}(t, s; m_{st} - 1)$. Then $sw = wt$ and $l(sw) = l(wt) > l(w)$. Now evaluate $T_s^2T_w = T_s(T_sT_w) = T_s(T_wT_t) = (T_wT_t)T_t = T_wT_t^2$ and compare the resulting expressions.

Exercise 4.8. Let $H = H_A(W, S, \{a_s, b_s \mid s \in S\})$ be an Iwahori–Hecke algebra as in Definition 4.4.1. Show that the map

$$\iota\colon H \to H, \qquad \sum_{w\in W} a_wT_w \mapsto \sum_{w\in W} a_wT_{w^{-1}},$$

is an A-algebra anti-automorphism of order 2.

Exercise 4.9. Let L be an oriented link and L' be the oriented link obtained from L by reversing the orientation in all components of L. Show that if L is the closure of $g = s_{i_1} \cdots s_{i_k} \in B(\mathfrak{S}_n)$, then L' is the closure of $g' = s_{i_k} \cdots s_{i_1}$. Use this and Exercise 4.8 to show that $X_L(u, v) = X_{L'}(u, v)$.

Exercise 4.10. The purpose of this exercise is to give an example of two knots which have the same Alexander polynomial but different Jones polynomials. The two knots we will consider are obtained by closing the braids $g = s_1^{-3}s_2^{-1}s_1^{-3}s_2$ and $h = s_1^{-3}s_2^{-1}s_1^{3}s_2$ in $B(\mathfrak{S}_3)$. Let $\hat{g}$ and $\hat{h}$ denote their closures. Show that

$$X_{\hat{g}}(u, v) = u^{-4}(v^2-2u-1)^2 \quad \text{and} \quad X_{\hat{h}}(u, v) = -u^{-2}(v^2+2u-1)(u^2-v-2).$$

Now specialize the variables as indicated in (4.5.3). Further, show that $\hat{\mathfrak{g}}$ and $\hat{\mathfrak{h}}$ are isotopic to the "granny knot" and the "square knot", respectively, which are printed in [Crowell and Fox 1977, Fig. 55, p. 131]. (We thank S. Lambropoulou for this remark.)

Exercise 4.11. Consider the Burau matrices M_i $(1 \leq i \leq n-1)$ defined as in (4.5.5). Prove the following statements.

(a) We have a representation $\beta_n \colon A[B(\mathfrak{S}_n)] \to M_n(A)$, $s_i \mapsto M_i$. Consider $V = A^n$ as an $A[B(\mathfrak{S}_n)]$-module via $xs_i = xM_i$.

(b) There exists some $0 \neq x \in V$ such that $xM_i = x$ for all i and that the quotient space $V' = V/\langle x \rangle$ is A-free of rank $n-1$. Determine explicitly matrices for the induced action of s_i on V'.

(c) Let $H_A(\mathfrak{S}_n)$ be the Iwahori–Hecke algebra of $\mathfrak{S}_n$ over A and with parameters $a_{s_i} = t$ and $b_{s_i} = t-1$ for all i. Show that the assignment $T_{s_i} \mapsto -M_i$ defines a representation of $H_A(\mathfrak{S}_n)$.

By (a) and (b), we obtain an induced representation $\beta'_n \colon A[B(\mathfrak{S}_n)] \to \mathrm{End}_A(V')$.

Exercise 4.12. Let $\tau'_n \colon H_A(\mathfrak{S}_n) \to A$ be defined as in (4.5.5):

$$\tau'_n(T'_w) = (-1)^{l(w)} \det(1 - \beta'_n(w)) \qquad (w \in \mathfrak{S}_n).$$

The purpose of this exercise is to compute the above value for $w = s_{i_1} \cdots s_{i_k} \in \mathfrak{S}_n$, where $1 \leq i_1 < \cdots < i_k \leq n-1$. Proceed as follows.

(a) Use Exercise 4.11(b) to show that $\det(X - \beta_n(g)) = (X-1)\det(X - \beta'_n(g))$ for all $g \in B(\mathfrak{S}_n)$. Deduce from this that if $l(w) < n-1$, then $\det(X - \beta_n(w))$ will be divisible by $(X-1)^2$ and so $\det(1 - \beta'_n(w)) = 0$.

(b) Let $w = s_1 \cdots s_{n-1}$. Show that then $\beta_n(w)$ has the following entries: the first row consists of $1-t, t(1-t), \ldots, t^{n-2}(1-t), t^{n-1}$; we have an entry 1 at position $(i, i-1)$ for $i > 1$ and all other entries are 0. Show that its characteristic polynomial is $X^n + (t-1)X^{n-1} + \cdots + t^{n-2}(t-1)X - t^{n-1}$. Deduce from this that $\tau'_n(T'_w) = (-1)^{n-1}(1 + t + \cdots + t^{n-1})$.

5

Irreducible characters of finite Coxeter groups

Let W be a finite Coxeter group. In this chapter we begin the systematic study of the irreducible characters of W. This will be based on two basic constructions, involving the natural module V of W. Recall from Chapter 1 that W is defined in terms of a Cartan matrix and this gives rise to a natural embedding $W \subset \mathsf{GL}(V)$, where V is an $\mathbb{R}$-vector space on which the generators of W act as reflections. The main theme of this chapter is to study the induced actions of W on the exterior and on the symmetric powers of V.

In Section 5.1 we will show that if W is irreducible, then the exterior powers $\bigwedge^d V$ $(0 \leq d \leq \dim_{\mathbb{R}} V)$ are pairwise non-isomorphic simple modules for W. Thus, this gives a method for constructing at least some irreducible characters of W. Also note that the corresponding character values can be explicitly computed from the characteristic polynomials of the elements of W acting on V. (In Chapter 8, we will extend this to the associated Iwahori–Hecke algebra.)

On the other hand, the study of the action of W on the symmetric powers of V is in fact part of the general theory of invariants for finite groups. Here, we will only concentrate on those aspects of that theory which will allow us to define the "b-invariants" of the irreducible characters of W and the "j-induction". These will be important tools for constructing and labelling the irreducible characters. For example, we will see in Section 5.3 that the irreducible characters of groups of type E_6, E_7 and E_8 have a unique labelling by pairs (n, b), where n denotes the degree and b denotes the "b-invariant". In Sections 5.4, 5.5 and 5.6 we will construct the irreducible characters of Coxeter groups of classical type.

Our approach also yields a quite straightforward proof of the fact that if $K \supseteq \mathbb{Q}$ is an extension field such that the character of V can be realized over K, then K is already a splitting field for *every* irreducible character of W; see Theorem 5.3.8 for the exceptional types.

In this chapter, we will not say much about the problem of computing the values of the irreducible characters in classical types (except for the special cases in Proposition 5.4.11 and Exercises 5.8–5.9). That problem will be treated on the level of Iwahori–Hecke algebras in Chapter 10.

We will only need some standard results from the character theory of finite groups. Our references are [Isaacs 1976] or [Goldschmidt 1980], which are largely sufficient. Throughout, we shall work with representations over $\mathbb{R}$ and $\mathbb{C}$.

5.1 EXTERIOR POWERS OF THE REFLECTION REPRESENTATION

Throughout this section, we assume that $W = \langle S \rangle \subset \mathsf{GL}(V)$ is a finite Coxeter group as in Definition 1.1.3, defined in terms of a Cartan matrix $C = (c_{st})_{s,t \in S}$. Recall that we have a basis $\{\alpha_s \mid s \in S\}$ of V such that the action of $s \in S$ is given by $\alpha_t.s = \alpha_t - c_{st}\alpha_s$, for all $t \in S$.

The aim of this section is to show that the induced action of W on the exterior powers of V yields absolutely irreducible and pairwise non-isomorphic W-modules if (W, S) is irreducible. (For $V = \bigwedge^1 V$ itself, see Exercise 1.12.)

5.1.1. For $0 \geq 1$, let $\bigwedge^d V$ be the dth *exterior power* of V. Let us briefly recall some basic properties. We have a canonical map

$$\bigotimes^d V \to \bigwedge^d V, \qquad v_1 \otimes \cdots \otimes v_d \mapsto v_1 \wedge \cdots \wedge v_d,$$

which is universal with respect to the property that $v_1 \otimes \cdots \otimes v_d$ is mapped to 0 whenever $v_i = v_j$ for some $i \neq j$. Any map $\varphi \in \mathsf{End}_{\mathbb{R}}(V)$ induces a linear map

$$\varphi^{\wedge d} \colon \bigwedge^d V \to \bigwedge^d V, \qquad v_1 \wedge \cdots \wedge v_d \mapsto (v_1\varphi) \wedge \cdots \wedge (v_d\varphi).$$

If $\{v_1, \ldots, v_n\}$ is any basis of V, then the elements

$$\{v_{i_1} \wedge \cdots \wedge v_{i_d} \mid 1 \leq i_1 < \cdots < i_d \leq n\}$$

form a basis of $\bigwedge^d V$. In particular, the dimension of $\bigwedge^d V$ is $\binom{n}{d}$. (For $d = 0$, we have $\bigwedge^0 V = \mathbb{R}$ and $\varphi^{\wedge 0} = \mathsf{id}_{\mathbb{R}}$.)

5.1.2. Lemma. *The vector space $\bigwedge^d V$ is a W-module where the action of $s \in S$ is given by the formula:*

$$(v_1 \wedge \cdots \wedge v_d).s = (v_1.s) \wedge \cdots \wedge (v_d.s) \qquad (v_i \in V).$$

The (-1)-eigenspace and the 1-eigenspace of the action of s on $\bigwedge^d V$ have dimension $\binom{n-1}{d-1}$ and $\binom{n-1}{d}$, respectively, where $n = \mathsf{dim}_{\mathbb{R}} V = |S|$.

Proof. For any $s \in S$, denote by $\rho_s \in \mathsf{End}_{\mathbb{R}}(V)$ the map given by the action of s. Then we have a corresponding map $\rho_s^{\wedge d} \in \mathsf{End}_{\mathbb{R}}(\bigwedge^d V)$. It is clear that the assignment $s \mapsto \rho_s^{\wedge d}$ defines a representation of W on $\bigwedge^d V$. So it remains to prove the assertion about the dimensions of the eigenspaces. We can find a basis of V consisting of eigenvectors for ρ_s. Then the corresponding basis vectors of $\bigwedge^d V$ are also eigenvectors for $\rho_s^{\wedge d}$. Let $\lambda_1, \ldots, \lambda_n$ be the eigenvalues of s on V. Then the eigenvalues of $\rho_s^{\wedge d}$ are $\{\lambda_{i_1} \cdots \lambda_{i_d} \mid 1 \leq i_1 < \cdots < i_d \leq n\}$. By Lemma 1.1.2, precisely one λ_i equals -1 and all the other λ_i are equal to 1. This yields the desired formulas. □

Let $J \subseteq S$ be any subset and $W_J \subseteq W$ the corresponding parabolic subgroup. Then, by (1.2.9), we have $W_J = \langle J \rangle \subset \mathrm{GL}(V_J)$, where $V_J \subseteq V$ is spanned by $\{\alpha_s \mid s \in J\}$, and this is a Coxeter group with respect to the Cartan matrix $C_J := (c_{ss'})_{s,s' \in J}$. The following result shows how $\bigwedge^d V$ decomposes when we restrict the action from W to W_J. This will be useful in induction arguments.

5.1.3. Lemma. *Assume that* $|S| \geq 2$ *and let* $J = S - \{t\}$ *for some* $t \in S$. *Then, for any* $d > 0$, *we have a direct sum decomposition of* W_J*-modules,*

$$\bigwedge^d V = V' \oplus \bigwedge^d V_J \qquad \textit{where} \qquad V' \cong \bigwedge^{d-1} V_J.$$

Proof. By Exercise 3.11, there exists some $0 \neq v \in V$ such that $V = V_J \oplus \langle v \rangle$ and $v.s = v$ for all $s \in J$. We consider the basis $\{\alpha_s \mid s \in J\} \cup \{v\}$ of V. Now write $S = \{s_1, \ldots, s_n\}$ and assume the labelling is such that $t = s_n$ and $J = \{s_1, \ldots, s_{n-1}\}$. We set $\alpha_i := \alpha_{s_i}$ for $1 \leq i \leq n-1$. Then we have a direct sum decomposition $\bigwedge^d V = (\bigwedge^d V_J) \oplus V'$ where $\bigwedge^d V_J$ and V' are spanned by

$$\begin{aligned} &\{\alpha_{i_1} \wedge \cdots \wedge \alpha_{i_{d-1}} \wedge \alpha_{i_d} \mid 1 \leq i_1 < \cdots < i_d \leq n-1\} \\ \text{and} \quad &\{\alpha_{i_1} \wedge \cdots \wedge \alpha_{i_{d-1}} \wedge v \mid 1 \leq i_1 < \cdots < i_{d-1} \leq n-1\}, \end{aligned}$$

respectively. Since both V_J and $\langle v \rangle$ are stable under the action of W_J, it follows that the above decomposition of $\bigwedge^d V$ is also stable under this action. The map

$$\alpha_{i_1} \wedge \cdots \wedge \alpha_{i_{d-1}} \mapsto \alpha_{i_1} \wedge \cdots \wedge \alpha_{i_{d-1}} \wedge v \qquad (1 \leq i_1 < \cdots < i_{d-1} \leq n-1)$$

certainly defines a vector space isomorphism between $\bigwedge^{d-1} V_J$ and V'. Since each $s \in J$ fixes v, that isomorphism is compatible with the action of W_J. □

5.1.4. Theorem. *Assume that* (W, S) *is irreducible. Then, for* $0 \leq d \leq |S|$, *the* W*-modules* $\bigwedge^d V$ *are absolutely irreducible and pairwise non-isomorphic.*

Proof. We proceed by induction on $|S|$. There is nothing to prove if $S = \{s\}$ is a singleton set. Now assume that $|S| \geq 2$. We first prove that each $\bigwedge^d V$ is irreducible. If $d = 0$ or $d = |S|$, then $\bigwedge^d V$ is one-dimensional and we are done. Now assume that $0 < d < |S|$. The Coxeter graph of (W, S) is one of the graphs in Theorem 1.3.3. We see that we can find an element $t \in S$ such that the graph of the subsystem $J = S - \{t\}$ is also connected. Then, by Lemma 5.1.3, we have a direct sum decomposition $\bigwedge^d V = V' \oplus \bigwedge^d V_J$ where $V' \cong \bigwedge^{d-1} V_J$ as W_J-modules. Since (W_J, J) is irreducible, we can apply induction and conclude that V' and $\bigwedge^d V_J$ are absolutely irreducible and non-isomorphic W_J-modules. Hence V' and $\bigwedge^d V_J$ are the only subspaces of $\bigwedge^d V$ which are invariant under the action of W_J. It follows that either $\bigwedge^d V$ is absolutely irreducible as a W-module, or $\bigwedge^d V_J$ is a non-zero invariant subspace. We check that the last possibility cannot occur. Let $S = \{t = s_1, \ldots, s_n\}$. The generator t is connected to exactly one generator $s_i \in J$, i.e., we have $c_{ts_i} \neq 0$ and $c_{ts_j} = 0$ for all $j \neq i, 1$. Assume that the labelling is such that $i = 2$. With these conventions, we consider the vector $v := \alpha_{s_2} \wedge \alpha_{s_3} \wedge \cdots \wedge \alpha_{s_{d+1}} \in \bigwedge^d V_J$. We have

$$v.t = (\alpha_{s_2} - c_{ts_2}\alpha_t) \wedge \alpha_{s_3} \wedge \cdots \wedge \alpha_{s_{d+1}}.$$

Since $c_{ts_2} \neq 0$, this new vector does not lie in $\bigwedge^d V_J$. So $\bigwedge^d V$ must be irreducible.

The fact that $\bigwedge^d V$ and $\bigwedge^{d'} V$ are non-isomorphic for $d \neq d'$ can be seen by looking at the eigenvalues of $s \in S$. By Lemma 5.1.2, the (-1)-eigenspace and the 1-eigenspace of s on $\bigwedge^d V$ have dimensions $\binom{|S|-1}{d-1}$ and $\binom{|S|-1}{d}$, respectively. These two numbers together uniquely determine d. □

How can we recognize that a given character of W is the character afforded by an exterior power $\bigwedge^d V$? For this purpose, the following result will be useful, namely in connection with the uniqueness results in Chapter 6 below.

For any W-module M, denote by $\mathrm{Fix}_W(M)$ the space of *fixed points* of W on M. Note that $\dim_{\mathbb{R}} \mathrm{Fix}_W(M)$ is equal to the scalar product of the character afforded by M and the trivial character of W.

5.1.5. Proposition. *Let* $J \subseteq S$ *and* $0 \leq d \leq |S|$. *Then*

$$\dim_{\mathbb{R}} \mathrm{Fix}_{W_J}\Big(\bigwedge^d V\Big) = \binom{|S - J|}{d}.$$

Proof. First we prove by induction on $|S|$ that $\mathrm{Fix}_W(\bigwedge^d V) = \{0\}$ if $d > 0$. This is seen as follows. If $S = \{s\}$ is a singleton set then $d = 1$ and $V^{\wedge 1} = V$ is the one-dimensional sign representation. Hence W has no fixed points on V and we are done in this case. Now let $|S| > 1$ and $J := S - \{t\}$ for some $t \in S$. Assume, if possible, that there exists a non-zero vector $v \in \bigwedge^d V$ which is fixed by all $s \in S$. By Lemma 5.1.3, we have a direct sum decomposition $\bigwedge^d V = V' \oplus \bigwedge^d V_J$ with $V' \cong \bigwedge^{d-1} V_J$ as W_J-modules. We can write $v = v_1 + v_2$ with $v_1 \in V'$ and $v_2 \in \bigwedge^d V_J$, and, since these two subspaces are invariant under the action of W_J, we conclude that v_1, v_2 are also fixed by all elements of W_J. By induction, this implies that $v_2 = 0$ and $v_1 = 0$ or $d = 1$. Since $v \neq 0$, we conclude that $d = 1$. Thus, v is a non-zero fixed point of W in V, which is impossible by Exercise 1.1. Hence our assumption was wrong and we must have $\mathrm{Fix}_W(\bigwedge^d V) = \{0\}$ if $d > 0$.

Now we can prove the assertion of the proposition by induction on $|S - J|$. If this is zero, then $J = S$ and $\binom{|S-J|}{d} = 0$ unless $d = 0$, in which case the result is 1. If $d = 0$, then $V^{\wedge 0}$ is the trivial W-module and so all vectors are fixed. The fact that there are no fixed points on $\bigwedge^d V$ if $d > 0$ is proved above. So we are done in this case. Now let $|S - J| \geq 1$. Then we can find some $t \in S$ such that $J \subseteq J' := S - \{t\}$. Trivially, the fixed points of W_J on $\bigwedge^d V$ are the fixed points of W_J on the restriction of $\bigwedge^d V$ to $W_{J'}$. By Lemma 5.1.3 this restriction is isomorphic to $\bigwedge^{d-1} V_{J'} \oplus \bigwedge^d V_{J'}$. If $d = 0$, the first summand is 0 and we are done by induction. Now assume that $d > 0$. Since taking fixed points commutes with direct sums we can now compute

$$\begin{aligned}
\dim_{\mathbb{R}} \mathrm{Fix}_{W_J}\Big(\bigwedge^{d} V\Big) &= \dim_{\mathbb{R}} \mathrm{Fix}_{W_J}\Big(\bigwedge^{d-1} V_{J'}\Big) + \dim_{\mathbb{R}} \mathrm{Fix}_{W_J}\Big(\bigwedge^{d} V_{J'}\Big) \\
&= \binom{|J' - J|}{d-1} + \binom{|J' - J|}{d} \qquad \text{by induction} \\
&= \binom{|J' - J| + 1}{d} = \binom{|S - J|}{d},
\end{aligned}$$

as desired. □

5.1.6. Remark. How can we compute the values of the character of W afforded by $\bigwedge^d V$? For this purpose, it is not just sufficient to know the values of the character afforded by V itself, but we also need to know the eigenvalues of each element of W on V. Indeed, let $\varphi\colon V \to V$ be an endomorphism and

$$\det(q\,\mathrm{id}_V - \varphi) = q^n + a_1 q^{n-1} + \cdots + a_{n-1} q + a_n$$

be the characteristic polynomial of φ, where q is an indeterminate, $n = |S|$, and $a_1, \ldots, a_n \in \mathbb{R}$. Then $(-1)^d a_d$ is the dth elementary symmetric function in the eigenvalues of φ (in $\mathbb{C}$). So, by the definition of $\varphi^{\wedge d}\colon \bigwedge^d V \to \bigwedge^d V$, we have that

$$(-1)^d a_d = \mathrm{Tr}(\varphi^{\wedge d}) \qquad \text{for } 1 \leq d \leq n.$$

Thus, the values of the characters of the W-modules $\bigwedge^d V$ are given in terms of the characteristic polynomials of the elements of W acting on V. These polynomials are presented in (3.4.3) and in the tables in Appendix B.

5.2 MACDONALD–LUSZTIG–SPALTENSTEIN INDUCTION

In the previous section, we considered the action of a finite Coxeter group $W \subset \mathrm{GL}(V)$ on the exterior powers of V. In a different direction, we will now consider the induced action of W on the symmetric powers of V. In fact, the basic result about this action (see Theorem 5.2.6) can be developed in a completely general setting, which we will now describe.

5.2.1. Symmetric powers. Let V be a finite-dimensional vector space over $\mathbb{C}$. For any $d \geq 0$, let $\mathcal{S}^d(V)$ be the dth symmetric power of V. Recall that $\mathcal{S}^d(V)$ is defined as the quotient of $\bigotimes^d V$ by the subspace generated by all elements of the form $v_{1.\sigma} \otimes \cdots \otimes v_{d.\sigma} - v_1 \otimes \cdots \otimes v_d$ where $v_i \in V$ and $\sigma \in \mathfrak{S}_d$. We denote the image of $v_1 \otimes \cdots \otimes v_d$ in $\mathcal{S}^d(V)$ by $v_1 \cdots v_d$. (For $d = 0$, we have $\mathcal{S}^0 = \mathbb{C}$.) Given also $d' \geq 0$, we have a well-defined product map

$$\mathcal{S}^d(V) \times \mathcal{S}^{d'}(V) \mapsto \mathcal{S}^{d+d'}(V), \qquad (v_1 \cdots v_d, v_1' \cdots v_{d'}') \mapsto v_1 \cdots v_d v_1' \cdots v_{d'}'.$$

This gives $\mathcal{S}(V) := \bigoplus_{d \geq 0} \mathcal{S}^d(V)$ the structure of a commutative algebra over $\mathbb{C}$. The algebra $\mathcal{S}(V)$ is called the *symmetric algebra* of V.

Assume now that $\{v_1, \ldots, v_n\}$ is a basis of V, where $\dim_{\mathbb{C}} V = n$. Then

$$\{v_1^{i_1} \cdots v_n^{i_n} \mid i_1, \ldots, i_n \geq 0, \quad i_1 + \cdots + i_n = d\}$$

is a $\mathbb{C}$-basis of $\mathcal{S}^d(V)$; in particular, the dimension of $\mathcal{S}^d(V)$ is $\binom{n+d-1}{d}$. Furthermore, if $X_1, \ldots, X_n$ are indeterminates over $\mathbb{C}$, then

$$\mathbb{C}[X_1, \ldots, X_n] \to \mathcal{S}(V), \qquad X_1^{i_1} \cdots X_n^{i_n} \mapsto v_1^{i_1} \cdots v_n^{i_n},$$

is an isomorphism of $\mathbb{C}$-algebras. We see that, under this isomorphism, $\mathcal{S}^d(V)$ corresponds to the space of homogeneous polynomials of degree d.

Any linear map $\varphi \colon V \to V$ naturally induces a map

$$\varphi^{(d)} \colon \mathcal{S}^d(V) \to \mathcal{S}^d(V), \qquad v_1 \cdots v_d \mapsto (v_1\varphi) \cdots (v_d\varphi).$$

If φ is diagonalizable with eigenvalues $\varepsilon_1, \ldots, \varepsilon_n$, then $\varphi^{(d)}$ is also diagonalizable, with eigenvalues given by $\varepsilon_1^{i_1} \cdots \varepsilon_{i_n}^{i_n}$ where $(i_1, \ldots, i_n)$ runs over all sequences of non-negative integers with $i_1 + \cdots + i_n = d$. Thus, we have

$$\mathrm{Tr}(\varphi^{(d)}) = \sum_{i_1+\cdots+i_n=d} \varepsilon_1^{i_1} \cdots \varepsilon_n^{i_n}.$$

5.2.2. The b-invariants. Now let G be a finite group and V be a $\mathbb{C}G$-module with $\dim_{\mathbb{C}} V < \infty$. Then, for any $d \geq 0$, the space $\mathcal{S}^d(V)$ is naturally a module for G, where the action of $g \in G$ on $v_1 \cdots v_d \in \mathcal{S}^d(V)$ (where $v_i \in V$) is given by $(v_1 \cdots v_d).g = (v_1.g) \cdots (v_d.g)$. Let $\rho_V^{(d)}$ be the character afforded by $\mathcal{S}^d(V)$. Note that $\rho_V^{(1)}$ is just the character afforded by V. For any $\chi \in \mathrm{Irr}(G)$, we set

$$n_d(\chi) = (\rho_V^{(d)}, \chi)_G = \frac{1}{|G|} \sum_{g \in G} \rho_V^{(d)}(g)\chi(g^{-1}),$$

i.e., $n_d(\chi)$ is the multiplicity of χ in $\rho_V^{(d)}$. Let q be an indeterminate over $\mathbb{C}$ and consider the formal power series, also called *Molien series*,

$$P_\chi(q) := \sum_{d \geq 0} n_d(\chi) q^d \in \mathbb{C}[[q]].$$

Assume now that $n_d(\chi) \neq 0$ for some $d \geq 0$. Then we can write

$$P_\chi(q) = \gamma_\chi q^{b_\chi} + \text{combination of higher powers of } q,$$

where $b_\chi \geq 0$ and $\gamma_\chi \neq 0$ is a positive integer. In analogy to [Lusztig 1979a], we call b_χ the b-*invariant* of χ. If it is necessary to indicate the underlying module V, we will write $P_\chi^V(q)$, b_χ^V and γ_χ^V for $P_\chi(q)$, b_χ and γ_χ, respectively. If $n_d(\chi) = 0$ for all $d \geq 0$, we set $b_\chi = \infty$ and $\gamma_\chi = 0$.

5.2.3. Example. Let $\chi \in \mathrm{Irr}(G)$. Then we have $b_\chi > 0$ unless $\chi = 1_G$ (the trivial character). Moreover, we have $P_{1_G}(q) = 1+$ combination of higher powers of q. This follows directly from the defining formula of $P_{1_G}(q)$ in (5.2.2).

In the following, we will address the following problems:

(1) Determine b_χ for all $\chi \in \mathrm{Irr}(G)$; in particular, determine when $b_\chi < \infty$.
(2) For any $\chi \in \mathrm{Irr}(G)$ with $b_\chi < \infty$, determine the coefficient γ_χ in $P_\chi(q)$.

By definition, the b-invariants can be computed by working out explicitly the characters afforded by $S^d(V)$. The following result yields an alternative and, in general, more efficient method for computing $P_\chi(q)$.

5.2.4. Proposition (Molien). *Let V be a $\mathbb{C}G$-module. Then we have*

$$P_\chi(q) = P_\chi^V(q) = \frac{1}{|G|}\sum_{g\in G}\frac{\chi(g^{-1})}{\det(\mathrm{id}_V - q\cdot\varphi_g)} \qquad \textit{for all } \chi \in \mathrm{Irr}(G),$$

where $\varphi_g \in \mathrm{End}_{\mathbb{C}}(V)$ denotes the map defined by the action of g. In particular, $P_\chi(q)$ is a rational function in q.

Proof. Using the formula in (5.2.2), we obtain

$$P_\chi(q) = \sum_{d\geq 0} n_d(\chi)q^d = \frac{1}{|G|}\sum_{g\in G}\Big(\sum_{d\geq 0}\rho_V^{(d)}(g)q^d\Big)\chi(g^{-1}).$$

Hence, it remains to evaluate the sum $\sum_{d\geq 0}\rho_V^{(d)}(g)q^d$. To do this, let us fix an element $g \in G$. We may assume that g acts diagonally on V, with eigenvalues $\varepsilon_1, \ldots, \varepsilon_n$ say, where $n = \dim_{\mathbb{C}} V$. Using the formula in (5.2.1), we have

$$\begin{aligned}\sum_{d\geq 0}\rho_V^{(d)}(g)q^d &= \sum_{d\geq 0}\Big(\sum_{i_1+\cdots+i_n=d}\varepsilon_1^{i_1}\cdots\varepsilon_n^{i_n}\Big)q^d \\ &= \prod_{j=1}^{n}\Big(1+\varepsilon_j q + (\varepsilon_j q)^2 + \cdots\Big) = \prod_{j=1}^{n}\frac{1}{1-\varepsilon_j q},\end{aligned}$$

and the term on the right-hand side equals $1/\det(\mathrm{id}_V - q\cdot\varphi_g)$, as required. □

Next we consider the question of how the b-invariants behave with respect to the induction of characters from subgroups.

5.2.5. Lemma. *Let V be a $\mathbb{C}G$-module and $H \subseteq G$ be a subgroup. Then $U := V/\mathrm{Fix}_H(V)$ can be naturally regarded as a $\mathbb{C}H$-module. Let $d \geq 0$. Then we have*

$$\mathrm{Res}_H^G(\rho_V^{(d)}) = \sum_{a=0}^{d} m_a\rho_U^{(a)} \qquad \textit{where } m_a := \dim_{\mathbb{C}} S^{d-a}(\mathrm{Fix}_H(V)).$$

Here, $\rho_U^{(a)}$ *denotes the character of* H *afforded by* $S^a(U)$*. In particular, if* $\psi \in \mathrm{Irr}(H)$ *and* $\chi \in \mathrm{Irr}(G)$ *are such that* χ *has non-zero multiplicity in* $\mathrm{Ind}_H^G(\psi)$*, then we must have* $b_\chi^V \geq b_\psi^U$*.*

Proof. Set $n = \dim_{\mathbb{C}} V$ and $m = \dim_{\mathbb{C}} U$. Let $g \in H$ and let $\varepsilon_1, \ldots, \varepsilon_m$ be the eigenvalues of g on U. Now, as $\mathbb{C}H$-modules, we have $V \cong U \oplus \mathrm{Fix}_H(V)$. Consequently, the eigenvalues of g on V are $\varepsilon_1, \ldots, \varepsilon_m$ plus $n-m$ eigenvalues equal to 1. It follows that the eigenvalues of g on $S^d(V)$ are given by $\varepsilon_1^{i_1} \cdots \varepsilon_m^{i_m} 1^{i_{m+1}} \cdots 1^{i_n}$ where $(i_1, \ldots, i_n)$ runs over all sequences of non-negative integers such that $i_1 + \cdots + i_n = d$. Using the formula in (5.2.1), we obtain:

$$\begin{aligned}
\rho_V^{(d)}(g) &= \sum_{i_1+\cdots+i_n=d} \varepsilon_1^{i_1} \cdots \varepsilon_m^{i_m} 1^{i_{m+1}} \cdots 1^{i_n} \\
&= \sum_{a=0}^{d} \Big(\sum_{i_1+\cdots+i_m=a} \varepsilon_1^{i_1} \cdots \varepsilon_m^{i_m} \Big) \Big(\sum_{j_1+\cdots+j_{n-m}=d-a} 1 \Big) \\
&= \sum_{a=0}^{d} \Big(\sum_{i_1+\cdots+i_m=a} \varepsilon_1^{i_1} \cdots \varepsilon_m^{i_m} \Big) \dim_{\mathbb{C}} S^{d-a}(\mathrm{Fix}_H(V)) = \sum_{a=0}^{d} m_a \rho_U^{(a)}(g),
\end{aligned}$$

as desired. Now let $\chi \in \mathrm{Irr}(G)$ and $\psi \in \mathrm{Irr}(H)$ be such that χ occurs with non-zero multiplicity in $\mathrm{Ind}_H^G(\psi)$. Assume that $b_\chi^V = d$. Using Frobenius reciprocity, we see that ψ occurs with non-zero multiplicity in $\mathrm{Res}_H^G(\chi)$ and, hence, also in $\mathrm{Res}_H^G(\rho_V^{(d)})$. The above formula then shows that $b_\psi^U \leq b_\chi^V$. □

5.2.6. Theorem (Macdonald–Lusztig–Spaltenstein). *Let* G *be a finite group and* V *be a* $\mathbb{C}G$*-module. Let* $H \subseteq G$ *be a subgroup and set* $U := V/\mathrm{Fix}_H(V)$*. Then we may naturally regard* U *as a* $\mathbb{C}H$*-module. Let* $\psi \in \mathrm{Irr}(H)$ *and* $0 \leq d < \infty$*. Assume that* $b_\psi^U = d$ *and* $\gamma_\psi^U = 1$*, i.e., we have*

$$P_\psi^U(q) = q^d + \textit{combination of higher powers of } q.$$

Then $\mathrm{Ind}_H^G(\psi)$ *has a unique constituent* $\chi \in \mathrm{Irr}(G)$ *with* $b_\chi^V = d$ *and* $\gamma_\chi^V = 1$*, i.e.,*

$$P_\chi^V(q) = q^d + \textit{combination of higher powers of } q.$$

Proof. By Lemma 5.2.5, we already know that all irreducible constituents of $\mathrm{Ind}_H^G(\psi)$ have b-invariants greater than or equal to d. Furthermore, the assumption on $P_\psi^U(q)$ means that we have $(\rho_U^{(a)}, \psi)_H = 0$ for $a < d$ and $(\rho_U^{(d)}, \psi)_H = 1$. Using Frobenius reciprocity and Lemma 5.2.5, we now compute

$$\begin{aligned}
(\rho_V^{(d)}, \mathrm{Ind}_H^G(\psi))_G = (\mathrm{Res}_H^G(\rho_V^{(d)}), \psi)_H &= \sum_{a=0}^{d} m_a (\rho_U^{(a)}, \psi)_H \\
&= m_d = \dim_{\mathbb{R}} S^0(\mathrm{Fix}_H(V)) = 1.
\end{aligned}$$

Thus, $\mathrm{Ind}_H^G(\psi)$ and $\rho_V^{(d)}$ have precisely one irreducible constituent in common, which yields a character χ as required. □

5.2.7. Remark. We keep the assumptions of Theorem 5.2.6. Then the uniqueness statement in that result can also be used to obtain some information about splitting fields. Let $K \subseteq \mathbb{C}$ be a subfield such that

(a) we have $\det(\mathrm{id}_V - q\varphi_g) \in K[q]$ for all $g \in G$, and
(b) the character $\psi \in \mathrm{Irr}(H)$ is already afforded by a KH-module.

Then the unique irreducible constituent $\chi \in \mathrm{Irr}(G)$ of $\mathrm{Ind}_H^G(\psi)$ with $b_\chi^V = b_\psi^U$ is also afforded by a KG-module.

Proof. First we check that the values of χ lie in K. Indeed, let σ be an automorphism of a sufficiently large algebraic extension of K which contains all character values of χ. Let χ^σ be the character obtained by algebraic conjugation (see [Goldschmidt 1980, (2.2)]). Now Molien's formula shows that $P_{\chi^\sigma}(q) = P_\chi(q)^\sigma = P_\chi(q)$, since $\det(\mathrm{id}_V - q\varphi_g) \in K[q]$ is fixed under σ for all $g \in G$ by condition (a). In particular, this means that $b_{\chi^\sigma} = b_\chi$. But, since ψ has values in K (by condition (b)), we have $\psi^\sigma = \psi$. Hence, by Frobenius reciprocity, χ^σ also occurs in $\mathrm{Ind}_H^G(\psi)$. The uniqueness statement in Theorem 5.2.6 therefore implies that $\chi = \chi^\sigma$. Since this holds for all σ as above, the values of χ must lie in K. But, furthermore, χ occurs with multiplicity 1 in $\mathrm{Ind}_H^G(\psi)$ and the latter module is afforded by a KG-module (again by condition (b) and [Goldschmidt 1980, (4.8)]). These two conditions imply that χ is also afforded by a KG-module; see [Goldschmidt 1980, (4.6) and (4.7)]. □

5.2.8. The j-induction. As above, let V be a $\mathbb{C}G$-module and $H \subseteq G$ be a subgroup. For any $d \geq 0$, let $\mathrm{Irr}(H, d)$ be the set of all $\psi \in \mathrm{Irr}(H)$ such that $P_\psi^U(q) = q^d +$ combination of higher powers of q, where $U = V/\mathrm{Fix}_H(V)$ as before. Then Theorem 5.2.6 shows that we have a map

$$j_H^G\colon \mathrm{Irr}(H, d) \to \mathrm{Irr}(G, d), \qquad \text{called the } j\text{-}induction,$$

which is defined by the condition that $j_H^G(\psi)$, for $\psi \in \mathrm{Irr}(H, d)$, is the unique irreducible constituent of $\mathrm{Ind}_H^G(\psi)$ whose b-invariant is equal to d.

In Exercise 5.2, it is shown that the j-induction is compatible with direct products. The j-induction is also transitive in the following sense. Let $H' \subseteq G$ be a subgroup with $H \subseteq H'$. Then we have

$$j_H^G(\psi) = j_{H'}^G(j_H^{H'}(\psi)) \qquad \text{for all } \psi \in \mathrm{Irr}(H, d).$$

Indeed, by Theorem 5.2.6, we can write $\mathrm{Ind}_H^{H'}(\psi) = \psi' + \theta_1$, where $\psi' = j_H^{H'}(\psi)$ and θ_1 is a sum of characters whose b-invariants are strictly bigger than d. We can again apply Theorem 5.2.6 to ψ' and write $\mathrm{Ind}_{H'}^G(\psi') = \chi + \theta_2$, where $\chi = j_{H'}^G(\psi')$ and θ_2 is a sum of characters whose b-invariants are strictly bigger than d. Using Lemma 5.2.5 and the transitivity of induction, we conclude that $\mathrm{Ind}_H^G(\psi) = \chi + \theta_3$, where θ_3 also is a sum of characters whose b-invariants are strictly bigger than d. Hence we must have $\chi = j_H^G(\psi)$.

Note that there are characters for which the assumptions of Theorem 5.2.6 are not satisfied; see Remarks 5.3.7 and 5.6.7. On the other hand, we have not yet seen any "real" examples of characters where those assumptions are satisfied! We will now see that a large variety of such examples is provided by complex reflection groups.

5.2.9. Complex reflection groups. Let V be a finite-dimensional vector space over $\mathbb{C}$. Given $0 \neq v \in V$ and an element $g \in GL(V)$ of finite order, we say that g is a *complex reflection* with root v if v is an eigenvector for g and we have $V = \langle v \rangle \oplus \ker(\mathrm{id}_V - g)$. Note that then v is an eigenvector of g whose eigenvalue is not equal to 1.

Now let $G \subset GL(V)$ be a finite group and $T \subseteq G$ be a subset consisting of complex reflections. We say that G is a *complex reflection group* with respect to T if G is generated by T and T is closed under conjugation.

Further note that if $g \in T$ and $m \geq 1$, then the elements g, g^m can only be conjugate in G if $g^m = g$ (since, with respect to some basis of V, the element g is represented by a diagonal matrix with one eigenvalue not equal to 1 and all other eigenvalues equal to 1). Hence, we can always "normalize" T such that

$$|\langle g \rangle \cap T| = 1 \qquad \text{for all } g \in T.$$

See also Exercise 5.3, where the example of cyclic groups is discussed.

5.2.10. Theorem. *Let $G \subseteq GL(V)$ be a complex reflection group with respect to $T \subseteq G$ and $\delta_G \colon G \to \mathbb{C}^\times$ be the linear character given by $\delta_G(g) = \det(g)$ for $g \in G$. Then, if T is "normalized" as above, we have*

$$P_{\delta_G}^V(q) = q^{|T|} + \textit{combination of higher powers of } q.$$

Proof. For each $g \in T$ we fix a root $v_g \in V$. (Note that v_g is only unique up to non-zero scalar multiples.) Now, for any $x \in G$, we have $x^{-1}gx \in T$ and $v_g.x$ is a root of $x^{-1}gx$. Hence we must have $v_g.x = \delta_g(x) v_{x^{-1}gx}$ for some $\delta_g(x) \in \mathbb{C}^\times$. It is straightforward to check that we have

$$\delta_g(1) = 1 \quad \text{and} \quad \delta_g(x^m) = \prod_{i=0}^{m-1} \delta_{x^{-i}gx^i}(x) \quad \text{for all } x \in G \text{ and } m \geq 1. \qquad (1)$$

We now proceed in three steps.

Step 1. First we claim that if we set $\delta'(x) := \prod_{g \in T} \delta_g(x)$, then we have

$$\delta_G(x) = \delta'(x) \qquad \text{for all } x \in T.$$

This is proved as follows. Fix $x \in T$. Now T decomposes into orbits under x and we can compute $\delta'(x)$ by taking the product over the various orbits. Let $g \in T$

and $m \geq 1$ be such that $\{x^{-i}gx^{i} \mid 0 \leq i \leq m-1\}$ is the orbit of g. If $x^m = 1$ then relation (1) directly shows that $1 = \delta_g(x^m) = \prod_{i=0}^{m-1} \delta_{x^{-i}gx^{i}}(x)$. So the orbit does not give a non-trivial contribution to $\delta'(x)$.

Now assume that $x^m \neq 1$. But then we still have $x^{-m}gx^m = g$ and so

$$v_g.x^m = \delta_g(x^m)v_g.$$

Thus, v_g is an eigenvector for x^m. Now x^m is a complex reflection with root v_x. If v_g is not a multiple of v_x then v_g must be an eigenvector with eigenvalue 1 and so relation (1) shows again that $1 = \delta_g(x^m) = \prod_{i=0}^{m-1} \delta_{x^{-i}gx^{i}}(x)$, i.e., the orbit does not give a non-trivial contribution to $\delta'(x)$.

The only remaining possibility is that v_g is a multiple of v_x. But then g is equal to some power of x. Our assumption that T is "normalized" therefore implies that $g = x$ and so $\delta'(x) = \delta_x(x)$. But $\delta_x(x)$ is the non-trivial eigenvalue of x, and so $\delta'(x) = \det(x) = \delta_G(x)$. Thus, our claim is proved.

Step 2. Note that $v_g \in \mathcal{S}^1(V)$. We claim that the subspace of $\mathcal{S}(V)$ spanned by

$$v_T := \Big(\prod_{g\in T} v_g\Big) \in \mathcal{S}^{|T|}(V) \tag{2}$$

is a $\mathbb{C}G$-module which affords δ_G; in particular, $b_{\delta_G} \leq |T|$. To prove this, note that $v_g.x = \delta_g(x)v_{x^{-1}gx}$ for any $x, g \in T$. Hence, since T is closed under conjugation, we conclude that $v_T.x = \delta'(x)v_T$. So the assertion follows from step 1.

Step 3. Assume that δ_G occurs with non-zero multiplicity in $\mathcal{S}^d(V)$ for some $d \geq 0$. Let $0 \neq f \in \mathcal{S}^d(V)$ be such that $f.g = \delta_G(g)f$ for all $g \in G$. We claim that then v_g divides f in $\mathcal{S}(V)$, for every $g \in T$. To see this, fix $g \in T$ and set $H_g = \ker(\mathrm{id}_V - g)$. Recall from (5.2.1) that if $B = (v_1, \ldots, v_n)$ is any basis of V, then we obtain an isomorphism $\mathbb{C}[X_1, \ldots, X_n] \cong \mathcal{S}(V)$, which sends X_i to v_i. Choosing B to be adapted to the eigenspace decomposition of g, we can arrange notation such that H_g is spanned by $v_1, \ldots, v_{n-1}$ and $v_n = v_g$. Then we can write $f = f_0 + f_1v_n + f_2v_n + \cdots$ where each f_i is a polynomial in $v_1, \ldots, v_{n-1}$. Then we have $f_i.g = f_i$ for $i = 0, 1, 2, \ldots$ and $v_n.g = \delta_G(g)v_n$. This yields the identity

$$\begin{aligned}\delta_G(g)f = f.g &= f_0.g + (f_1.g)(v_n.g) + (f_2.g)(v_n.g) + \cdots \\ &= f_0 + \delta_G(g)f_1v_n + \delta_G(g)f_2v_n + \cdots = (1 - \delta_G(g))f_0 + \delta_G(g)f.\end{aligned}$$

Since $\delta_G(g) \neq 1$, we conclude that $f_0 = 0$, and so $v_g = v_n$ divides f in $\mathcal{S}(V)$.

Finally, note that $\mathcal{S}(V) \cong \mathbb{C}[X_1, \ldots, X_n]$ is a unique factorization domain and that v_g and v_h are coprime for $g \neq h$ (they correspond to different homogeneous polynomials of degree 1). Hence we conclude that $v_T = \prod_{g\in T} v_g$ divides f in $\mathcal{S}(V)$. This means, first of all, that $d \geq |T|$ and so (using step 2) the b-invariant of δ_G must be $|T|$. Moreover, if $d = |T|$, then f is a multiple of v_T of the same homogeneous degree, which means that f must be a scalar multiple of v_T. This shows that δ_G has multiplicity 1 in $\mathcal{S}^{|T|}(V)$. □

5.2.11. Corollary (Macdonald). *Keep the assumptions of Theorem 5.2.10 and let* $H \subseteq G$ *be a subgroup such that* $H = \langle T \cap H \rangle$. *Let* δ_H *be the restriction of* δ_G *to* H. *Then there exists a unique* $\chi \in \mathrm{Irr}(G)$ *such that* $b_\chi^V = |T \cap H|$ *and*

$$\mathrm{Ind}_H^G(\delta_H) = \chi + \textit{combination of } \chi' \in \mathrm{Irr}(G) \textit{ with } b_{\chi'}^V > b_\chi^V.$$

With the notation in (5.2.8), we have $\chi = j_H^G(\delta_H)$.

Proof. Let $U = V/\mathrm{Fix}_H(V)$. Then we can regard H as a subgroup of $GL(U)$. The elements of $T \cap H$ are still represented by complex reflections on U and we still have $|\langle g \rangle \cap T \cap H| = 1$ for all $g \in T \cap H$. Moreover, δ_H is the character given by the determinant on $GL(U)$. Thus, all the assumptions of Theorem 5.2.10 are satisfied and so $P_{\delta_H}^U(q) = q^{|T \cap H|} +$ combination of higher powers of q. All that remains is to apply Theorem 5.2.6. □

5.3 APPLICATIONS TO FINITE COXETER GROUPS

Let $W = \langle S \rangle \subset GL(V)$ be a finite Coxeter group as in Definition 1.1.3, arising from a Cartan matrix such that the corresponding root system $\Phi \subset V$ is reduced. We now apply the methods developed so far to the exterior and the symmetric powers of the natural module V. We will see that "most" irreducible characters of W are determined in this way. In this section, we deal with the exceptional types; the classical types will be considered in Sections 5.4, 5.5 and 5.6.

5.3.1. A finite Coxeter group W as above is an example of a complex reflection group. More precisely, we may naturally regard W as a subgroup of $GL(V_\mathbb{C})$, where $V_\mathbb{C}$ is the complexified space. Let $T = \{wsw^{-1} \mid s \in S, w \in W\}$. Since S consists of (complex) reflections by Lemma 1.1.2, the same holds for T.

We conclude that W is a complex reflection group with respect to T, since T is closed under conjugation and W is already generated by the subset $S \subseteq T$. Moreover, the condition that $|\langle w \rangle \cap T| = 1$ for all $w \in T$ is trivially satisfied since the elements of T have order 2. Finally, the character $\delta_{V_\mathbb{C}}$ is nothing but the sign character ε of W, since $\det(w) = -1$ for all $w \in T$. Theorem 5.2.10, Exercise 1.14 and Proposition 1.5.1 show that

$$b_\varepsilon = |T| = |\Phi^+| = l(w_0) \qquad \text{where } w_0 \in W \text{ is the longest element.} \qquad \text{(a)}$$

Applying Corollary 5.2.11, we see that for each subgroup $W' \subseteq W$ such that $W' = \langle T \cap W' \rangle$, there exists a unique $\chi \in \mathrm{Irr}(W)$ such that $b_\chi = |T \cap W'|$ and

$$\mathrm{Ind}_{W'}^W(\varepsilon_{W'}) = \chi + \text{combination of } \chi' \in \mathrm{Irr}(W) \text{ with } b_{\chi'} > b_\chi. \qquad \text{(b)}$$

Here, $\varepsilon_{W'}$ denotes the restriction of the sign character ε of W to W'. With the notation of (5.2.8), we have $\chi = j_{W'}^W(\varepsilon_{W'})$. We will see in Section 5.5 that this construction yields all irreducible characters of Coxeter groups of type B_n.

5.3.2. Remark. Assume that $W' = W_J \subset \mathsf{GL}(V_J)$ is a parabolic subgroup for some $J \subseteq S$, where V_J is spanned by the simple roots corresponding to the generators in J; see (1.2.9). First note that, setting $U = V/\mathsf{Fix}_{W_J}(V)$ as above, we have $U \cong V_J$ as W_J-modules (see Exercise 3.11), and so

$$P_\psi^{U^c}(q) = P_\psi^{(V_J)^c}(q) \qquad \text{for all } \psi \in \mathsf{Irr}(W_J).$$

Now the formula in Exercise 5.4 shows that $|T \cap W_J| = l(w_J)$, where $w_J \in W_J$ is the longest element. Hence, applying the j-induction, we obtain

$$\chi_J := j_{W_J}^{W}(\varepsilon_{W_J}) \in \mathsf{Irr}(W) \qquad \text{with} \qquad b_{\chi_J} = l(w_J).$$

For example, we have $\chi_\varnothing = 1_W$ and $\chi_S = \varepsilon$; moreover, for any $s \in S$, $\chi_{\{s\}}$ is the character afforded by V. We will see in Section 5.4 that this construction yields all irreducible characters of the symmetric groups.

5.3.3. Fake degrees. Let $W = \langle S \rangle \subset \mathsf{GL}(V)$ be a finite Coxeter group as above. Let $\chi \in \mathsf{Irr}(W)$ and consider the corresponding Molien series $P_\chi(q) = P_\chi^{V^c}(q)$; see (5.2.2). We have already remarked in Proposition 5.2.4 that $P_\chi(q)$ is in fact a rational function in q. Can we find an explicit representation of $P_\chi(q)$ as a quotient of two polynomials in q? The answer is yes, but the proof requires some additional results from the theory of polynomial invariants of finite reflection groups. Since, in any case, we do not need these results in this book, let us just briefly describe the answer. We set

$$R_\chi := P_\chi(q) \prod_{i=1}^{|S|} (1 - q^{d_i}),$$

where $d_1, \ldots, d_{|S|}$ are the so-called *degrees* of W. Then R_χ is a polynomial in q, where the coefficient of q^i is the multiplicity of χ in the ith piece of a natural grading of the regular representation of W; see [Carter 1985, (11.1.11)]. Furthermore, we have a factorization

$$\prod_{i=1}^{|S|} \frac{1 - q^{d_i}}{1 - q} = \sum_{w \in W} q^{l(w)};$$

see [Carter 1972b, (9.4.9)] and [Bourbaki 1968, Chap. V, §5].

Lusztig has called the polynomials R_χ the *fake degrees* of W. They play an important role in the representation theory of finite groups of Lie type; see [Carter 1985, §11.3]. The R_χ have been computed in all cases; see [Lusztig 1979a], [Lusztig 1977], [Beynon and Lusztig 1978] and the references therein. In CHEVIE, they can all be obtained via the function **FakeDegree**.

We will now apply the above techniques to the irreducible characters of finite Coxeter groups of exceptional type. We begin with the dihedral groups.

5.3.4. Coxeter groups of type $I_2(m)$. Let $m \geq 3$ and $W = \langle s, t \rangle$ be the dihedral group of order $2m$, where s, t are elements of order 2 and st has order m. We set $a = st$. Then $W = \{1, a, a^2, \ldots, a^{m-1}, s, sa, sa^2, \ldots, sa^{m-1}\}$. Let 1_W and ε denote the trivial and the sign character, respectively. If m is even, we have two more linear characters ε_1 and ε_2 which are given by $\varepsilon_1(s) = \varepsilon_2(t) = 1$ and $\varepsilon_1(t) = \varepsilon_2(s) = -1$. All the remaining irreducible characters of W have degree 2, and they are afforded by the following representations:

$$\varphi_j \colon a \mapsto \begin{bmatrix} \zeta^j & 0 \\ 0 & \zeta^{-j} \end{bmatrix}, \qquad s \mapsto \begin{bmatrix} 0 & 1 \\ 1 & 0 \end{bmatrix},$$

where $\zeta = \exp(2\pi\sqrt{-1}/m) \in \mathbb{C}$ and $1 \leq j \leq (m-1)/2$ for m odd, $1 \leq j \leq (m-2)/2$ for m even, respectively. (This is proved in a similar way to the corresponding statement on the level of the associated Iwahori–Hecke algebra; see Theorem 8.3.1 below.) Let V_j be a $\mathbb{C}W$-module affording φ_j and χ_j be its character. We have

$$\chi_j(a^i) = \zeta^{ij} + \zeta^{-ij} = 2\cos(2\pi ij/m) \qquad \text{and} \qquad \chi_j(sa^i) = 0.$$

Let V be the natural module for W defined in terms of a Cartan matrix C as in Definition 1.1.3. We will now compute the b-invariants of the irreducible characters of W with respect to the embedding $W \subset \mathsf{GL}(V_{\mathbb{C}})$, where $V_{\mathbb{C}}$ denotes the complexified space as in (5.3.1). The character afforded by V is identified as follows. The defining equation for the action of s, t on V in terms of the entries of C shows that the trace of $a = st$ on V must be $2\cos(2\pi/m)$. A comparison with the above matrices shows that $V_1 \cong V_{\mathbb{C}}$.

By Example 5.2.3 and (5.3.1) we already know that

$$P_{1_W}^{V_{\mathbb{C}}}(q) = 1 \quad + \text{ combination of higher powers of } q,$$
$$P_{\varepsilon}^{V_{\mathbb{C}}}(q) = q^m + \text{ combination of higher powers of } q.$$

In order to compute the remaining b-invariants, we use Molien's formula in Proposition 5.2.4. The characteristic polynomials occurring in the denominators of that formula are given by

$$\det(\mathrm{id}_{V_{\mathbb{C}}} - qa^i) = (1 - \zeta^i q)(1 - \zeta^{-i} q) \quad \text{and} \quad \det(\mathrm{id}_{V_{\mathbb{C}}} - qsa^i) = 1 - q^2.$$

Let us now consider an irreducible character χ_j of degree 2. We have $\chi_j(a^i) = \zeta^{ij} + \zeta^{-ij}$ and $\chi_j(sa^i) = 0$ for all $1 \leq i \leq m$. Using Exercise 5.5 (note that $2j < m$), we can evaluate Molien's formula and obtain

$$P_{\chi_j}^{V_{\mathbb{C}}}(q) = \frac{1}{2m} \sum_{i=1}^{m} \frac{\zeta^{ij} + \zeta^{-ij}}{(1-\zeta^i q)(1-\zeta^{-i} q)} = q^j + \text{comb. of higher powers of } q.$$

Finally, assume that m is even and consider the two remaining linear characters of W. Let $j \in \{1,2\}$. We have $\varepsilon_j(a^i) = (-1)^i$ and $\varepsilon_j(sa^i) = \varepsilon_j(s)(-1)^i$. Inserting this into Molien's formula, we obtain

$$P_{\varepsilon_j}^{V_c}(q) = \frac{1}{2m}\Big(\sum_{i=1}^{m} \frac{(-1)^i}{(1-\zeta^i q)(1-\zeta^{-i} q)}\Big) + \varepsilon_j(s)\frac{1}{2m}\Big(\sum_{i=1}^{m} \frac{(-1)^i}{1-q^2}\Big).$$

Now, the second sum certainly is 0 and the first sum can be evaluated using Exercise 5.5 (note that $\zeta^{m/2} = -1$). Hence we have

$$P_{\varepsilon_1}^{V_c}(q) = P_{\varepsilon_2}^{V_c}(q) = q^{m/2} + \text{combination of higher powers of } q.$$

Thus, the b-invariants of all irreducible characters of W are determined.

The irreducible characters for the remaining exceptional types are constructed in an algorithmic fashion, which we shall now explain. The results, together with some additional information, are presented in the tables in Appendix C.

5.3.5. Algorithms. Assume we are given a Dynkin diagram as in Table 1.2 (p. 20) and let (W, S) be the corresponding Coxeter system. We now describe how to compute—using the GAP implementation of the Dixon–Schneider algorithm [Schneider 1990]—the character table of W and the b-invariants of all irreducible characters of W.

Step 1. From the Dynkin diagram, we can write down explicit matrices for each generator $s \in S$. Thus, we obtain a realization of W as a subgroup of $GL_n(\mathbb{R})$, where $n = |S|$; see Section 1.1. In GAP, we create the corresponding matrix group record using the function `Group`. Applying the function `ConjugacyClasses` to that record yields a complete set of representatives for the conjugacy classes of W. (One could also use the representatives of minimal length which we have already obtained in Chapter 3.)

Step 2. We apply the function `PermGroup` to the matrix group record of W and obtain the record of a permutation group. (It will be more efficient to work with permutations than with matrices.) Then the function `CharTable` returns the ordinary character table of W. The actual computations in GAP take just a few minutes; only type E_8 requires something close to half an hour.

Step 3. Computing the traces and the characteristic polynomials of the matrix representatives of the conjugacy classes from step 1, we explicitly obtain the character of the standard reflection representation of W. Thus, we can identify it in the character table of W which was computed in step 2.

Step 4. Finally, we use the GAP function `MolienSeries` to compute the Molien series of every $\chi \in \mathrm{Irr}(W)$. (GAP actually returns a rational function in q describing that series.) Note that, by Proposition 5.2.4, the computation only requires the character table of W and the characteristic polynomials of step 3.

5.3.6. Let (W, S) be a Coxeter system of type H_3, H_4, F_4, E_6, E_7 or E_8. We choose a labelling $S = \{s_1, s_2, \ldots\}$ as in the Dynkin diagram in Table 1.2. Applying the algorithm described in (5.3.5), we obtain the character table of W and the b-invariants of all irreducible characters of W.

The complete lists of characters, together with their b-invariants and some additional information concerning characters of parabolic type (see (6.3.6)) and families (see (6.5.10)), are given in Tables C.1–C.6 in Appendix C. These tables also define the notation for the irreducible characters which is used below.

By inspection, we find the smallest field (denoted $\mathbb{Q}(W)$) in which all character values $\chi(w)$ ($\chi \in \mathrm{Irr}(W)$, $w \in W$) lie. Moreover, we can identify the character (denoted ρ) of the standard reflection representation as described in step 2 of (5.3.5), find $\rho(w_c)$ where $w_c \in W$ is a Coxeter element (simply by multiplying the matrices corresponding to the generators in S and taking the trace), and compute the exterior powers $\rho^{\wedge d}$. The results are as follows:

Type	$\mathbb{Q}(W)$	ρ	$\rho(w_c)$	$\rho^{\wedge 0}, \rho^{\wedge 1}, \ldots, \rho^{\wedge \lvert S\rvert}$
H_3	$\mathbb{Q}(\sqrt{5})$	$3_s'$	$(-1+\sqrt{5})/2$	$1_r, 3_s', 3_s, 1_r'$
H_4	$\mathbb{Q}(\sqrt{5})$	4_t	$(-1+\sqrt{5})/2$	$1_r, 4_t, 6_s, 4_t', 1_r'$
F_4	$\mathbb{Q}$	4_2	0	$1_1, 4_2, 6_2, 4_5, 1_4$
E_6	$\mathbb{Q}$	6_p	-1	$1_p, 6_p, 15_p, 20_s, 15_p', 6_p', 1_p'$
E_7	$\mathbb{Q}$	$7_a'$	-1	$1_a, 7_a', 21_a, 35_a', 35_a, 21_a', 7_a, 1_a'$
E_8	$\mathbb{Q}$	8_z	-1	$1_x, 8_z, 28_x, 56_z, 70_y, 56_z', 28_x', 8_z', 1_x'$

In types F_4, E_6, E_7 and E_8, the reflection character is the unique faithful irreducible character of degree $|S|$ whose value on $s \in S$ is $|S| - 2$. In types H_3 and H_4, we obtain a uniqueness statement when we also take into account the value on a Coxeter element.

It turns out that, with very few exceptions, each $\chi \in \mathrm{Irr}(W)$ is uniquely determined by the pair $(\chi(1), b_\chi)$. The exceptions only occur in types F_4 and H_4:

Type H_4. The two irreducible characters denoted 30_s and $\overline{30}_s$ in Table C.2 have the same b-invariant. They can be distinguished from each other by the fact that their values on a Coxeter element are given by $(-1+\sqrt{5})/2$ and $(-1-\sqrt{5})/2$, respectively. They also have different Molien series:

$$P_{30_s}(q) = q^{10} + 2q^{12} + 3q^{14} + \text{combination of higher powers of } q,$$

$$P_{\overline{30}_s}(q) = q^{10} + q^{12} + 2q^{14} + \text{combination of higher powers of } q.$$

Type F_4. There are several characters which have the same degree and the same b-invariant. We use the notation in Table C.3. The two characters of degree 6 are distinguished by the fact that $6_2 = \bigwedge^2 4_2$. For the remaining characters, the tables show that they are distinguished by their values on the generators s_1 and s_3. We have followed the convention in [Carter 1985, p. 413] which says that, if

$s \in S$ corresponds to a long root, then 1_2, 2_1, 8_1, 8_2, and 9_2 have values $+1$, $+2$, 4, -4 and 3 on s, respectively. (See also [Lusztig 1984, (4.10)] for an explicit construction of all irreducible characters.)

5.3.7. Remark. Let (W, S) be as in (5.3.6). The computation of the Molien series of the irreducible characters of W (using the algorithms described in (5.3.5)) yields the following results. First of all, we always have $P_\chi(q) = \gamma_\chi q^{b_\chi} +$ combination of higher powers of q, where $0 \le b_\chi < \infty$ and $\gamma_\chi \neq 0$. Moreover, with a few exceptions listed below, we always have $\gamma_\chi = 1$. The exceptions only occur in types E_7 and E_8; they are as follows, where we use the notation for the characters as in Tables C.5 and C.6:

$$E_7 \quad \begin{cases} P_{280_a}(q) &= 2q^{18} + 5q^{20} + \text{higher powers of } q, \\ P_{280'_a}(q) &= 2q^{9} + 5q^{11} + \text{higher powers of } q, \\ P_{512_a}(q) &= 2q^{12} + 4q^{14} + 13q^{16} + \text{higher powers of } q. \end{cases}$$

$$E_8 \quad \begin{cases} P_{2688_y}(q) &= 2q^{20} + 3q^{22} + \text{higher powers of } q, \\ P_{2100_x}(q) &= 2q^{16} + 4q^{18} + \text{higher powers of } q, \\ P_{2100'_x}(q) &= 2q^{28} + 4q^{30} + \text{higher powers of } q, \\ P_{5600_w}(q) &= 2q^{19} + 5q^{21} + \text{higher powers of } q. \end{cases}$$

The determination of γ_χ for all types of (W, S) will be completed in Remark 5.6.7.

5.3.8. Theorem (Benard, Benson–Grove, Kondo). *Let (W, S) be a Coxeter system of exceptional type $I_2(m)$ $(m \ge 3)$, H_3, H_4, F_4, E_6, E_7 or E_8. Let*

$$K := \mathbb{Q}(\cos(2\pi/m_{st}) \mid s, t \in S) \subset \mathbb{R}$$

where, as usual, m_{st} denotes the order of $st \in W$. Then the following hold.

(a) *For any $\chi \in \mathrm{Irr}(W)$, we have $\mathbb{Q}(\chi(w) \mid w \in W) \subseteq K$, where equality holds for the character of the reflection representation of W.*

(b) *Every $\chi \in \mathrm{Irr}(W)$ has Schur index 1 over K, i.e., χ can be realized over K.*

Proof. Let ρ be the character of the standard reflection representation. Recall that ρ is the character afforded by an embedding $W = \langle S \rangle \subset GL(V)$ as in Definition 1.1.3, where the action on V is defined in terms of a Cartan matrix $C = (c_{st})_{s,t \in S}$. (Note that ρ is independent of the choice of C; see Exercise 1.5.) Using Lemma 1.1.6 and Definition 1.1.1(C2) we see that for $s, t \in S$, the trace of st on V equals $2\cos(2\pi/m_{st})$. Thus, we have $K \subseteq \mathbb{Q}(\rho(w) \mid w \in W)$. Moreover, ρ can be realized over K. To see this, we choose C to be "standard", in the sense of (1.3.7). Then we have in fact $K = \mathbb{Q}(c_{st} \mid s, t \in S)$. So, using the defining formula for the action of $s \in S$ on V, we see that the matrix of $s \in S$ with respect to the basis $\{\alpha_s\}_{s \in S}$ of V has coefficients in K.

Now consider any $\chi \in \mathrm{Irr}(W)$. First note that the results in (5.3.4) and the table in (5.3.6) show that the values of χ lie in K. We have already seen above that ρ can be realized over K. So, by construction, the same holds for all symmetric powers of ρ. Hence, by general properties of the Schur index (see, for example, [Goldschmidt 1980, §4]), χ can be realized over an extension field $K' \supseteq K$ such that the degree $[K' : K]$ divides the greatest common divisor of the coefficients of $P_\chi(q)$. In particular, if that greatest common divisor is 1, then we have $K' = K$. The results in (5.3.4) and Remark 5.3.7 show that the latter is always the case, except if we are in type E_8 and $\chi \in \{2100_x, 2100'_x\}$, in which case the greatest common divisor is at most 2. Since 2100_x and $2100'_x$ are obtained from each other by tensoring with the sign character, it is enough to show that 2100_x can be realized over K. This can be seen as follows. Since we already know that 2100_x can be realized over an extension field of degree at most 2 over K, it is sufficient to show that there exists some character of W which can be realized over K and in which 2100_x occurs with odd multiplicity. Now, from Table C.6 we see that 2100_x has value 210 on a generator $s \in S$. Hence, by Frobenius reciprocity, the multiplicity of 2100_x in the character obtained by inducing the trivial character of $\langle s \rangle$ to W equals $(2100 + 210)/2 = 1155$, which is odd, as required. □

Note that it may actually happen that the character values of $\chi \in \mathrm{Irr}(W)$ lie in a field which is properly contained in K. Then the question arises whether χ can be realized over that subfield or not. (The answer to this question is not always affirmative!) We will consider this in more detail in Section 6.3.

5.4 THE CHARACTERS OF THE SYMMETRIC GROUP

Let $n \geq 1$ and $\mathfrak{S}_n$ be the symmetric group on $\{1, \ldots, n\}$. The purpose of this section is to prove some basic results about the irreducible characters of $\mathfrak{S}_n$, using the methods developed in Section 5.2.

For $1 \leq i \leq n-1$, denote by $s_i \in \mathfrak{S}_n$ the basic transposition $(i, i+1)$. Then, by Proposition 1.4.7, the pair $(\mathfrak{S}_n, \{s_1, \ldots, s_{n-1}\})$ is a Coxeter system associated with the Dynkin diagram

A_{n-1} 

(By convention, we also set $\mathfrak{S}_0 = \{1\}$.) Throughout, we denote by ε the sign character of $\mathfrak{S}_n$. (If $n = 0, 1$, then ε is the trivial character of the trivial group.) As a first and crucial step, we will show in Theorem 5.4.5 that the method of Remark 5.3.2 yields all irreducible characters of $\mathfrak{S}_n$.

Recall from Section 3.4 that the classes of $\mathfrak{S}_n$ are parametrized by the partitions of n. Hence so are the irreducible characters of $\mathfrak{S}_n$. We begin by introducing some basic combinatorial notions concerning compositions and partitions.

5.4.1. The a^*-invariant of a composition. Let λ be a *composition* of n. This means that $\lambda = (\lambda_1, \ldots, \lambda_r)$ is an *ordered* sequence of positive integers such that $|\lambda| := \sum_i \lambda_i = n$. We define a corresponding integer $a^*(\lambda)$ by

$$a^*(\lambda) := \frac{1}{2}\sum_{i=1}^{r} \lambda_i(\lambda_i - 1) = \sum_{i=1}^{r} \binom{\lambda_i}{2}.$$

An interpretation of $a^*(\lambda)$ will be given in (5.4.3) below. Note that we certainly have $\sum_i \lambda_i^2 \leq (\sum_i \lambda_i)^2 = n^2$ and so

$$0 \leq a^*(\lambda) \leq \frac{1}{2}n(n-1) = \binom{n}{2}, \qquad \text{with equality only for } \lambda = [n].$$

This property will also play a role in step 1 of the proof of Theorem 5.4.5.

5.4.2. The a-invariant of a partition. Now let $\lambda \vdash n$ be a partition of n. This means that $\lambda = [\lambda_1, \ldots, \lambda_r]$ is an *unordered* sequence of positive integers with $|\lambda| := \sum_i \lambda_i = n$. We define a corresponding integer $a(\lambda)$ by the formula

$$a(\lambda) := \sum_{1 \leq j < i \leq r} \min\{\lambda_i, \lambda_j\}.$$

Note that if the parts of λ are ordered such that $\lambda_1 \geq \lambda_2 \geq \cdots \geq \lambda_r > 0$, then we have $a(\lambda) = \sum_{i=1}^{r}(i-1)\lambda_i$. (As a convention, we also call the empty sequence $\varnothing = []$ a partition of 0 and let its a-invariant be 0.)

It will sometimes be convenient to visualize a partition λ in the usual way by a *Young diagram*, i.e., an array of boxes in left-justified rows, with λ_1 boxes in the first row, λ_2 boxes in the second row, and so on. Here, we will always assume that the parts of λ are ordered as a weakly decreasing sequence. Flipping such a diagram over its main diagonal (from upper left to lower right), we obtain the Young diagram of another partition $\lambda^* \vdash n$, which is called the *conjugate partition*. Here is an example, where $\lambda = [4, 2, 1]$ and $\lambda^* = [3, 2, 1, 1]$:

λ : 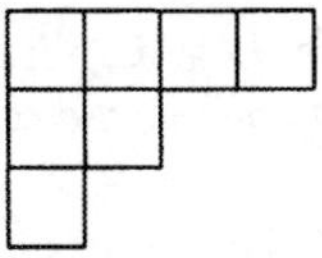

λ^* : 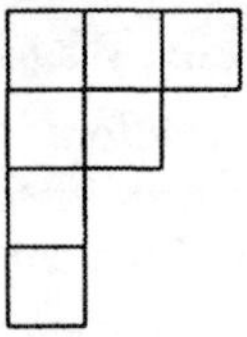

Thus, formally we have $\lambda^* = [\lambda_1^*, \ldots, \lambda_s^*]$, where $s = \lambda_1$ and

$$\lambda_j^* = |\{1 \leq i \leq r \mid \lambda_i \geq j\}| \qquad \text{for } 1 \leq j \leq s.$$

It is straightforward to check that we have $a^*(\lambda^*) = a(\lambda)$ for all $\lambda \vdash n$.

5.4.3. Young subgroups. Assume that $\lambda = (\lambda_1, \ldots, \lambda_r)$ is a composition of n. Then we have a corresponding *Young subgroup* $\mathfrak{S}_\lambda \subseteq \mathfrak{S}_n$ consisting of all those permutations of $1, \ldots, n$ which permute among themselves the first λ_1 numbers, the next λ_2 numbers, and so on. Any Young subgroup $\mathfrak{S}_\lambda$ is a parabolic subgroup of $\mathfrak{S}_n$ in the sense of (1.2.9): the corresponding subset $J \subseteq \{s_1, \ldots, s_{n-1}\}$ is obtained by deleting the generators indexed by $\lambda_1 + \cdots + \lambda_i$ for $1 \leq i < r$ from S. Let $w_0^\lambda \in \mathfrak{S}_\lambda$ be the longest element. Using Example 1.5.4, we have

$$l(w_0^\lambda) = a^*(\lambda).$$

More generally, the conjugates of a Young subgroup are described as follows. Let $\{1, \ldots, n\} = A_1 \amalg \cdots \amalg A_r$ be any decomposition into subsets such that

$|A_i| = \lambda_i$ for all i. We obtain a permutation $w \in \mathfrak{S}_n$ by first listing the elements of A_1, then those of A_2 and so on. Then $w\mathfrak{S}_\lambda w^{-1}$ is the stabilizer of the above decomposition. In particular, we see that $\mathfrak{S}_\lambda$ and $\mathfrak{S}_\mu$ are conjugate in $\mathfrak{S}_n$ if λ and μ are compositions which give rise to the same partition of n.

5.4.4. Definition. Let λ be a composition of n. Let 1_λ be the trivial character of $\mathfrak{S}_\lambda$ and ε_λ be the sign character of $\mathfrak{S}_\lambda$. We set

$$\pi_\lambda := \mathrm{Ind}_{\mathfrak{S}_\lambda}^{\mathfrak{S}_n}(1_\lambda) \qquad \text{and} \qquad \theta_\lambda := \mathrm{Ind}_{\mathfrak{S}_\lambda}^{\mathfrak{S}_n}(\varepsilon_\lambda) = \varepsilon \otimes \pi_\lambda.$$

(The formula $\varepsilon \otimes \pi_\lambda = \theta_\lambda$ comes from a general compatibility between induction and tensor products; see, for example, [Goldschmidt 1980, (3.4)].) Then, using Macdonald's Corollary 5.2.11 and Remark 5.3.2, together with the formulas in (5.4.3), (5.4.2), we can uniquely define an irreducible character χ_λ of $\mathfrak{S}_n$ by the condition that

$$b_{\chi_\lambda} = a(\lambda) \quad \text{and} \quad \theta_{\lambda^*} = \chi_\lambda + \text{combination of } \chi \in \mathrm{Irr}(\mathfrak{S}_n) \text{ with } b_\chi > a(\lambda),$$

where the b-invariants are taken with respect to the natural representation of $\mathfrak{S}_n$ given by the Dynkin diagram on p. 154. Note that the above characters are independent of the ordering of the parts of λ. Thus, when we write π_λ, θ_λ or χ_λ, we may tacitly assume that λ is a partition of n. For example, $\chi_{[1^n]}$ is the sign character (with b-invariant $n(n-1)/2$) and $\chi_{[n]}$ is the trivial character (with b-invariant 0). If $n = 0$, then $\chi_\varnothing$ is the trivial character of the trivial group.

5.4.5. Theorem. *We have* $\mathrm{Irr}(\mathfrak{S}_n) = \{\chi_\lambda \mid \lambda \vdash n\}$. *Moreover, the following hold.*

(a) *Each* χ_λ *is afforded by a* $\mathbb{Q}\mathfrak{S}_n$*-module. Thus,* $\mathbb{Q}$ *is a splitting field for* $\mathfrak{S}_n$.
(b) *Each* χ_λ *is an integral linear combination of the characters* π_μ *(for* $\mu \vdash n$*) and also of the characters* θ_μ *(for* $\mu \vdash n$*).*

Proof. Let $\lambda^1, \ldots, \lambda^m$ be all partitions of n. We shall write $a_i = a(\lambda^i)$, $a_i^* = a^*(\lambda^i)$, $\chi_i = \chi_{\lambda^i}$, $\pi_i = \pi_{\lambda^i}$ and $\theta_i = \theta_{\lambda^i}$. We now proceed in four steps.

Step 1. Assume that the ordering of the λ^i is such that $a_1^* \leq a_2^* \leq \cdots \leq a_m^*$. For each $j \in \{1, \ldots, m\}$, let $w_j \in \mathfrak{S}_n$ be an element of cycle type λ^j. We claim that the matrix of all values $(\pi_i(w_j))_{i,j}$ is lower triangular with non-zero elements on the diagonal.

This immediately follows by combining Proposition 2.4.4(ii) and Proposition 3.1.8. A direct proof goes as follows. First note that the diagonal entries are non-zero since $w_i \in \mathfrak{S}_{\lambda^i}$ for all i. Now assume that i, j are such that $\pi_i(w_j) \neq 0$. Using the formula for the values of the induced character (see [Isaacs 1976, (5.1)]), we see that then w_j must be conjugate to some element of $\mathfrak{S}_{\lambda^i}$. But this means that λ^j is a refinement of λ^i, i.e., if λ^i has parts λ_k^i (for $1 \leq k \leq r_i$), then there exist partitions $\mu^{ik} \vdash \lambda_k^i$ such that λ^j is obtained by concatenating the μ^{ik} and reordering. Then $a_j^* = \sum_k a^*(\mu^{ik})$ and so, using the inequality in (5.4.1), we obtain $a_j^* \leq \sum_k \lambda_k^i(\lambda_k^i - 1)/2 = a_i^*$. If $a_j^* < a_i^*$, we have $j \leq i$ by the chosen ordering. Now assume that $a_j^* = a_i^*$. Then we must have $a^*(\mu^{ik}) = \lambda_k^i(\lambda_k^i - 1)/2$ for all i and so $\mu^{ik} = [\lambda_k^i]$, again by (5.4.1). This means that $\lambda^j = \lambda^i$, i.e., $i = j$.

Step 2. Now we claim that the matrix of scalar products

$$((\chi_i, \theta_j)_{\mathfrak{S}_n})_{i,j}$$

is non-singular. Indeed, as we already remarked in (5.4.3), we have $\varepsilon \otimes \pi_i = \theta_i$ for all i. Hence, the non-singularity of the matrix in step 1 implies that the matrix of all values $(\theta_i(w_j))_{i,j}$ is also non-singular. But that matrix is the product of the above matrix with the character table of $\mathfrak{S}_n$.

Step 3. By step 2, we already know that $\chi_i \neq \chi_j$ for $i \neq j$. Since the total number of irreducible characters equals the number of conjugacy classes and, hence, equals the number of partitions of n, we conclude that $\mathrm{Irr}(\mathfrak{S}_n) = \{\chi_i \mid 1 \leq i \leq m\}$. Moreover, the assertion in (a) follows from Remark 5.2.7: we have $\det(\mathrm{id}_V - qw_j) \in \mathbb{Q}[q]$ by (3.4.3) and each ε_λ is certainly afforded over $\mathbb{Q}$.

Step 4. It remains to prove (b). Assume that the ordering of the λ_i is such that $a_1 \leq a_2 \leq \cdots \leq a_m$, and consider the matrix of scalar products

$$Z = ((\chi_i, \theta_{j^*})_{\mathfrak{S}_n})_{i,j},$$

where j^* is such that $(\lambda^j)^* = \lambda^{j^*}$. We define an equivalence relation $\sim$ on $\{1, \ldots, m\}$ by $i \sim j$ if and only if $a_i = a_j$. Then Corollary 5.2.11 shows that Z is a block lower triangular matrix with blocks corresponding to the equivalence classes under $\sim$. Moreover, each diagonal block has precisely one non-zero entry in each column, and that non-zero entry is 1. But, by step 2, Z is non-singular. Hence each diagonal block must be non-singular. This is only possible if the diagonal blocks are permutation matrices. It follows that Z can be inverted over $\mathbb{Z}$, and so every χ_i is an integral linear combination of the θ_j. Tensoring with ε and using the formula $\varepsilon \otimes \pi_j = \theta_j$ also shows that each χ_i is an integral linear combination of the π_j. □

Now we can prove a characterization of the characters χ_λ without reference to the Macdonald–Lusztig–Spaltenstein induction. First, we need the following:

5.4.6. Definition. The *dominance order*, denoted $\trianglelefteq$, on the set of partitions of n is defined as follows. Let $\lambda = [\lambda_1, \ldots, \lambda_r] \vdash n$ and $\mu = [\mu_1, \ldots, \mu_s] \vdash n$ be such that $\lambda_1 \geq \lambda_2 \geq \cdots \geq \lambda_r > 0$ and $\mu_1 \geq \mu_2 \geq \cdots \geq \mu_s > 0$. Then we have

$$\lambda \trianglelefteq \mu \quad \text{if and only if} \quad \sum_{i=1}^{k} \lambda_i \leq \sum_{i=1}^{k} \mu_i \quad \text{for } 1 \leq k \leq \min\{r, s\}.$$

This is a partial order which is compatible with the a-invariants, i.e., we have $\lambda \trianglelefteq \mu \Rightarrow a(\lambda) \geq a(\mu)$, with equality only for $\lambda = \mu$; see Exercise 5.6(a).

5.4.7. Theorem. *Let $\lambda \vdash n$. Then π_λ and θ_{λ^*} have a unique irreducible constituent in common, and that common constituent is χ_λ. In fact, we have*

$$\pi_\lambda = \chi_\lambda + \text{linear combination of } \chi_\mu \text{ with } \lambda \trianglelefteq \mu,\ \mu \neq \lambda,$$
$$\theta_{\lambda^*} = \chi_\lambda + \text{linear combination of } \chi_\mu \text{ with } \mu \trianglelefteq \lambda,\ \mu \neq \lambda.$$

Proof. We proceed in four steps.

Step 1. Assume that $\lambda, \mu \vdash n$ are such that $(\pi_\lambda, \theta_\mu)_{\mathfrak{S}_n} \neq 0$. Then we claim that there exists some $x \in \mathfrak{S}_n$ such that $x^{-1}\mathfrak{S}_\lambda x \cap \mathfrak{S}_\mu = \{1\}$. Indeed, by the Mackey formula and Frobenius reciprocity, we have

$$0 \neq (\pi_\lambda, \theta_\mu)_{\mathfrak{S}_n} = \sum_{d \in D_{\lambda\mu}} (1_{H_d}, \varepsilon_{H_d})_{H_d},$$

where $H_d := d^{-1}\mathfrak{S}_\lambda d \cap \mathfrak{S}_\mu$ and $D_{\lambda\mu}$ is the set of distinguished double coset representatives with respect to the parabolic subgroups $\mathfrak{S}_\lambda$ and $\mathfrak{S}_\mu$; see Proposition 2.1.7. By the Kilmoyer–Solomon Theorem 2.1.12, each H_d is again a parabolic subgroup of $\mathfrak{S}_n$. Hence, if $H_d \neq \{1\}$ for some $d \in D_{\lambda\mu}$, then H_d contains a generator $s_i \in S$. But then 1_{H_d} and ε_{H_d} are certainly different characters of H_d, hence their scalar product is 0. Therefore, we must have $H_{d_0} = \{1\}$ for some $d_0 \in D_{\lambda\mu}$. Thus, $x = d_0 \in D_{\lambda\mu}$ is an element as required.

Step 2. Let $\lambda, \mu \vdash n$ again be such that $(\pi_\lambda, \theta_\mu)_{\mathfrak{S}_n} \neq 0$. Now we claim that then we must have $\lambda \trianglelefteq \mu^*$.

By step 1, we already know that we have $x^{-1}\mathfrak{S}_\lambda x \cap \mathfrak{S}_\mu = \{1\}$ for some $x \in \mathfrak{S}_n$. In order to show that this implies $\lambda \trianglelefteq \mu^*$, it will be convenient to reformulate the above condition as follows. As in (5.4.3), we can regard $x^{-1}\mathfrak{S}_\lambda x$ as the stabilizer of a decomposition $\{1, \ldots, n\} = A_1 \amalg \cdots \amalg A_r$ where the cardinalities of the sets A_i are the parts of λ. Similarly, we have a decomposition $\{1, \ldots, n\} = B_1 \amalg \cdots \amalg B_s$ corresponding to $\mathfrak{S}_\mu$, where the cardinalities of the sets B_j are the parts of μ. Now, the condition $x^{-1}\mathfrak{S}_\lambda x \cap \mathfrak{S}_\mu = \{1\}$ means that

$$|A_i \cap B_j| \leq 1 \qquad \text{for all } i, j. \tag{$*$}$$

Let us now consider the Young diagram of λ, where the boxes in the ith row are filled by the elements of A_i. Similarly, we consider the Young diagram of μ, where the boxes in the jth row are filled by the elements of B_j. We now argue by the *pigeon hole principle*. Consider the first k rows of the λ-diagram. Let $x_k := \sum_{i=1}^{k} |A_i|$ be the total number of boxes in those rows. By condition $(*)$, no two numbers from the same row of the λ-diagram are in the same row of the μ-diagram. Hence, the first k columns of the latter diagram must contain at least x_k boxes. Recalling the definition of μ^*, this means precisely that $\lambda \trianglelefteq \mu^*$.

Step 3. Now we show that π_λ and θ_{λ^*} always have a unique irreducible constituent in common, and that this constituent has multiplicity 1 in π_λ and in θ_{λ^*}. For this purpose, let $\lambda^1, \dots, \lambda^m$ be all the partitions of n. We again write $a_i = a(\lambda^i)$, $\chi_i = \chi_{\lambda^i}$, $\pi_i = \pi_{\lambda^i}$ and $\theta_i = \theta_{\lambda^i}$. Assume that the ordering is such that $a_1 \leq \cdots \leq a_m$. By step 2 and Exercise 5.6(a), we see that the matrix of scalar products

$$((\pi_i, \theta_{j^*})_{\mathfrak{S}_n})_{i,j}$$

is lower triangular, where j^* is such that $(\lambda^j)^* = \lambda^{j^*}$. Moreover, by Theorem 5.4.5, each irreducible character of $\mathfrak{S}_n$ is an integral linear combination of the π_i and also an integral linear combination of the θ_{j^*}. It follows that the above matrix has determinant ± 1. Being also lower triangular, we conclude that the diagonal entries must be equal to 1. This means precisely that π_i and θ_{i^*} have a unique irreducible constituent in common, with multiplicity 1.

Step 4. It remains to prove the assertions about the decompositions of π_λ and θ_{λ^*}. For this purpose, we use the following two relations.

(a) Assume that χ_μ occurs in π_λ. Since χ_μ has also multiplicity 1 in θ_{μ^*}, we deduce that $(\pi_\lambda, \theta_{\mu^*})_{\mathfrak{S}_n} \neq 0$ and so $\lambda \trianglelefteq \mu$ by step 2.

(b) If χ_μ occurs in θ_{λ^*}, then we have $a(\mu) \geq a(\lambda)$; see Definition 5.4.4.

First of all, we can now identify the unique common constituent of π_λ and θ_{λ^*}, which exists by step 3. If this common constituent is χ_μ, then (a) and (b) imply that $\lambda \trianglelefteq \mu$ and $a(\mu) \geq a(\lambda)$. Now Exercise 5.6(a) shows that $\lambda = \mu$, as desired.

The assertion about the decomposition of π_λ is already contained in (a). Finally, assume that χ_μ occurs in θ_{λ^*}. Since χ_μ also occurs in π_μ, we have $(\pi_\mu, \theta_{\lambda^*})_{\mathfrak{S}_n} \neq 0$ and so $\mu \trianglelefteq \lambda$. □

The multiplicities $\kappa_{\mu\lambda} = (\pi_\lambda, \chi_\mu)_{\mathfrak{S}_n}$ are called *Kostka numbers*; they can also be characterized in purely combinatorial terms; see Exercise 6.5. The above result shows that the matrix $(\kappa_{\mu\lambda})$ of all Kostka numbers is triangular with 1 along the diagonal, if the partitions are ordered according to the dominance order. This immediately implies the following result, which is in fact a model for what will happen for finite Coxeter groups in general.

5.4.8. Corollary. *Let $\chi \in \mathrm{Irr}(\mathfrak{S}_n)$. Then there exists some $\lambda \vdash n$ such that χ has multiplicity 1 in $\mathrm{Ind}_{\mathfrak{S}_\lambda}^{\mathfrak{S}_n}(1_\lambda)$. Moreover, χ is uniquely determined by its multiplicities in $\mathrm{Ind}_{\mathfrak{S}_\lambda}^{\mathfrak{S}_n}(1_\lambda)$, for all $\lambda \vdash n$.*

5.4.9. Corollary. *For each $\lambda \vdash n$, we have $\varepsilon \otimes \chi_\lambda = \chi_{\lambda^*}$.*

Proof. Using the formula $\varepsilon \otimes \pi_\lambda = \theta_\lambda$ (see (5.4.3)), we obtain:

$$\begin{aligned}
(\varepsilon \otimes \chi_\lambda, \pi_{\lambda^*})_{\mathfrak{S}_n} &= (\chi_\lambda, \varepsilon \otimes \pi_{\lambda^*})_{\mathfrak{S}_n} = (\chi_\lambda, \theta_{\lambda^*})_{\mathfrak{S}_n} = 1, \\
(\varepsilon \otimes \chi_\lambda, \theta_\lambda)_{\mathfrak{S}_n} &= (\chi_\lambda, \varepsilon \otimes \theta_\lambda)_{\mathfrak{S}_n} = (\chi_\lambda, \pi_\lambda)_{\mathfrak{S}_n} = 1.
\end{aligned}$$

Thus, since $(\lambda^*)^* = \lambda$, we see that $\varepsilon \otimes \chi_\lambda$ occurs both in π_{λ^*} and in $\theta_{(\lambda^*)^*}$. It remains to use the uniqueness statement in Theorem 5.4.7. □

The following result yields an explicit realization of χ_λ.

5.4.10. Proposition. *Let* $n \geq 1$ *and* $\lambda \vdash n$. *Consider the Young diagram of* λ, *where the boxes are filled with the numbers* $\{1, \ldots, n\}$ *(in some ordering). Let* $R \cong \mathfrak{S}_\lambda$ *be the subgroup of* $\mathfrak{S}_n$ *which stabilizes each row of that diagram and* $C \cong \mathfrak{S}_{\lambda^*}$ *be the subgroup which stabilizes each column. Let*

$$e_\lambda := \sum_{x \in R} \sum_{y \in C} \varepsilon(y) yx \in \mathbb{C}[\mathfrak{S}_n].$$

Then $\chi_\lambda \in \mathrm{Irr}(\mathfrak{S}_n)$ *is afforded by* $e_\lambda \mathbb{C}[\mathfrak{S}_n]$, *and we have* $e_\lambda^2 = n!^{-1} \chi_\lambda(1)\, e_\lambda$.

Proof. Consider the trivial character of R and the sign character of C. By [Goldschmidt 1980, (2.8)], the corresponding central idempotents are given by

$$f := |R|^{-1} \sum_{x \in R} x \qquad \text{and} \qquad g := |C|^{-1} \sum_{y \in C} \varepsilon(y) y.$$

Then we have $e_\lambda = |R||C|\, gf$. Now we proceed in three steps.

Step 1. We claim that if we write e_λ as a linear combination of the elements of $\mathfrak{S}_n$, then the coefficient of $1 \in \mathfrak{S}_n$ is 1; in particular, this means that $e_\lambda \neq 0$.

To prove this, first note that $R \cap C = \{1\}$. (Indeed, if some permutation does not move a given $i \in \{1, \ldots, n\}$ either out of its row or out of its column in the diagram of λ, it must fix i.) Hence the products yx ($x \in R$, $y \in C$) are all distinct. Thus, e_λ is a linear combination of $|R||C|$ elements of $\mathfrak{S}_n$, and the coefficient of $1 \in \mathfrak{S}_n$ is 1.

Step 2. Let $A := \mathbb{C}[\mathfrak{S}_n]$. We claim that $e_\lambda A$ affords χ_λ.

To prove this, first note that $e_\lambda A \neq 0$ by step 1. Now, we have $e_\lambda A = g(fA) \subseteq gA$ and so $e_\lambda A$ is a submodule of gA. On the other hand, the map $fA \to e_\lambda A$, $fa \mapsto gfa$, is a surjective A-algebra homomorphism and so $e_\lambda A$ is a factor module of fA. But, since R is conjugate to $\mathfrak{S}_\lambda$ and C is conjugate to $\mathfrak{S}_{\lambda^*}$, the modules fA and gA afford the characters π_λ and θ_{λ^*}. Using Theorem 5.4.7, we conclude that $\dim_\mathbb{C} \mathrm{Hom}_A(fA, gA) = 1$ and that $e_\lambda A$ affords χ_λ.

Step 3. It remains to show that $e_\lambda^2 = n!^{-1}\chi_\lambda(1)\, e_\lambda$. For this purpose, let $\varphi \in \mathrm{End}_A(A)$ be given by left multiplication with e_λ. Then the image of $e_\lambda A$ under φ is $e_\lambda^2 A$ and, hence, contained in $e_\lambda A$. So, by restriction, φ defines an endomorphism of $e_\lambda A$. By step 2 and Schur's lemma, φ acts as a scalar on $e_\lambda A$. So we have $e_\lambda^2 = e_\lambda.\varphi = \xi e_\lambda$ for some $\xi \in \mathbb{C}$. It follows that $a\varphi^2 = (e_\lambda a)\varphi = e_\lambda^2 a = \xi e_\lambda a = \xi(a\varphi)$ for all $a \in A$ and so $\varphi^2 = \xi\varphi$. Hence φ only has eigenvalues 0 and ξ, where ξ has multiplicity equal to $\dim_\mathbb{C}(e_\lambda A)$. It follows that

$$\mathrm{Tr}(\varphi) = \xi \dim_\mathbb{C}(e_\lambda A) = \xi \chi_\lambda(1).$$

On the other hand, let us consider the matrix of φ with respect to the basis $\mathfrak{S}_n$ of A. Let $w \in \mathfrak{S}_n$ and consider $w\varphi = e_\lambda w$. If we write this as a linear combination of elements of $\mathfrak{S}_n$, then the coefficient of w is the same as that of $1 \in \mathfrak{S}_n$ in e_λ, which is 1 by step 1. So we have $\mathrm{Tr}(\varphi) = n!$. A comparison with the above expression for $\mathrm{Tr}(\varphi)$ shows that ξ has the desired form. □

We now use the above result to obtain the following (slightly surprising) relation between the value of $\chi_\lambda \in \mathrm{Irr}(\mathfrak{S}_n)$ on a generator s_i and the a-invariants of λ and λ^*. (In Section 6.5, this will be generalized to all finite Coxeter groups.)

5.4.11. Proposition (Beynon–Lusztig). *Let* $n \geq 2$ *and* $\lambda \vdash n$. *Then we have*

$$\frac{|T|\chi_\lambda(s)}{\chi_\lambda(1)} = a(\lambda^*) - a(\lambda), \qquad \textit{where } s \in T \textit{ and } T \subseteq \mathfrak{S}_n \textit{ as in (5.3.1)}.$$

Proof. First note that T is the conjugacy class of $s_1 = (1,2)$. Let $\hat{T}$ be the sum of the elements in T. Then $\hat{T}$ belongs to the centre of the group algebra $\mathbb{C}[\mathfrak{S}_n]$ and the left-hand side of the desired identity equals $\omega_\lambda(\hat{T})$, where ω_λ denotes the central character corresponding to χ_λ; cf. (6.2.1). Now consider two subgroups $R, C \subseteq \mathfrak{S}_n$ associated with λ and the corresponding element $e_\lambda \in \mathbb{C}[\mathfrak{S}_n]$ as in Proposition 5.4.10. First we show that

$$e_\lambda \hat{T} = \omega_\lambda(\hat{T}) e_\lambda \qquad \text{and} \qquad \omega_\lambda(\hat{T}) = \sum_{\substack{x \in R,\, y \in C \\ yx \in T}} \varepsilon(y). \tag{1}$$

This is seen as follows. Since $\hat{T}$ lies in the centre of $\mathbb{C}[\mathfrak{S}_n]$, it acts as a scalar on $e_\lambda \mathbb{C}[\mathfrak{S}_n]$. By definition, that scalar is $\omega_\lambda(\hat{T})$. Using the explicit formula in Proposition 5.4.10, a comparison of the coefficient of $1 \in \mathfrak{S}_n$ on both sides of $e_\lambda \hat{T} = \omega_\lambda(\hat{T}) e_\lambda$ yields (1). Next we claim that

$$\text{if } x \in R \text{ and } y \in C \text{ are such that } yx \in T, \text{ then we have } x = 1 \text{ or } y = 1. \tag{2}$$

Indeed, suppose that $yx \in T$. Then we have $yx = (i,j)$, where $1 \leq i < j \leq n$. There are two cases: either i, j belong to the same row or to the same column of the Young diagram of λ. Where i, j belong to the same row, the image of any $k \in \{1, \ldots, n\}$ under y belongs to the same row as k. (If $k \notin \{i, j\}$, this is clear since $(ky).x = k(i,j) = k$; furthermore, since $(iy)x = j$ and $(jy)x = i$, we have that i, j, iy, jy all belong to the same row.) Thus, $y \in R \cap C = \{1\}$. Where i, j belong to the same column, a similar argument shows that $x \in R \cap C = \{1\}$. This completes the proof of (2).

Using (1) and (2), we now obtain that $\omega_\lambda(\hat{C}) = |T \cap R| - |T \cap C|$. By Exercise 5.4, the sets T, $T \cap R$ and $T \cap C$ consist precisely of the reflections in $\mathfrak{S}_n$, $R \cong \mathfrak{S}_\lambda$ and $C \cong \mathfrak{S}_{\lambda^*}$, respectively. Using (5.3.1)(a), this yields the desired identity. □

Finally, we consider the exterior powers $\bigwedge^d V$ of the natural module for $\mathfrak{S}_n$ defined in terms of the above Cartan matrix. Recall from Theorem 5.1.4 that the modules $\bigwedge^d V$, for $0 \leq d \leq n-1$, are simple and pairwise non-isomorphic.

5.4.12. Proposition. *Let* $\rho^{\wedge d} \in \mathrm{Irr}(\mathfrak{S}_n)$ *be afforded by* $\bigwedge^d V$. *Then we have*

$$\rho^{\wedge d} = \chi_{[n-d,1^d]} \qquad \text{and} \qquad b_{\rho^{\wedge d}} = \frac{1}{2}d(d+1) \qquad \text{for } 0 \leq d \leq n-1.$$

In particular, the character of the natural module V *itself is* $\chi_{[n-1,1]}$.

Proof. Fix d and let $\mu = [\mu_1, \ldots, \mu_s]$ be any partition of n. Then $\mathfrak{S}_\mu$ is the parabolic subgroup corresponding to a subset J whose cardinality is $n - s$; see (5.4.3). Hence, using Proposition 5.1.5, we have

$$(\rho^{\wedge d}, \pi_\mu)_{\mathfrak{S}_n} = \dim_{\mathbb{R}} \mathrm{Fix}_{\mathfrak{S}_\mu}\Big(\bigwedge^d V\Big) = \binom{|S-J|}{d} = \binom{s-1}{d}.$$

Now let $\lambda = [\lambda_1, \ldots, \lambda_r] \vdash n$ be such that $\rho^{\wedge d} = \chi_\lambda$ and $\lambda_1 \geq \cdots \geq \lambda_r > 0$. Then, by Theorem 5.4.7, $\rho^{\wedge d}$ occurs with multiplicity 1 in π_λ. So the above formula shows that λ must have at least $d+1$ parts. It follows that

$$a(\lambda) = \sum_{i=1}^{r}(i-1)\lambda_i \geq \sum_{i=1}^{d+1}(i-1) = \frac{1}{2}d(d+1) = a([n-d, 1^d]).$$

On the other hand, the partition $[n-d, 1^d]$ does have $d+1$ parts. Thus, applying the above scalar product formula to that partition, we see that χ_λ occurs with multiplicity 1 in $\pi_{[n-d,1^d]}$. Hence, by Theorem 5.4.7, we must have $[n-d, 1^d] \trianglelefteq \lambda$. Now Exercise 5.6(a) implies that $\lambda = [n-d, 1^d]$, as desired. □

In Exercise 5.8, the value of $\rho^{\wedge d}$ on an n-cycle of $\mathfrak{S}_n$ is determined. Further important results about the characters of $\mathfrak{S}_n$, including the branching rule and the Littlewood–Richardson rule, will be discussed in Section 6.1. These results rely on the connections between the character theory of $\mathfrak{S}_n$ and symmetric functions. We close this section by stating the classical result due to Frobenius which lies at the origin of these connections.

Let $x_1, \ldots, x_N$ be indeterminates (for some $N \geq n$). For any $\lambda \vdash n$ we set

$$s_\lambda := \det(x_i^{\lambda_j+N-j})/\det(x_i^{N-j}) \in \mathbb{Q}(x_1, \ldots, x_N),$$

where λ has parts $\lambda_1 \geq \cdots \geq \lambda_r > 0$ and where we set $\lambda_j = 0$ for $j > r$. It is easily checked that s_λ is not only a rational function but a symmetric polynomial in $x_1, \ldots, x_N$. It is called the *Schur function* associated with λ; see [Macdonald 1995, I, §3]. Furthermore, for any $\mu \vdash n$, let p_μ be the corresponding *Newton power sum*, as defined in Exercise 5.10. Then we have the following fundamental result.

5.4.13. Theorem (Frobenius). *Let* $w_\mu \in \mathfrak{S}_n$ *have cycle type* μ. *Then*

$$p_\mu = \sum_{\lambda \vdash n} \chi_\lambda(w_\mu)\, s_\lambda.$$

For an easily readable and self-contained proof, see [Ledermann 1987, §4.8].

5.5 THE CHARACTERS OF COXETER GROUPS OF TYPE B

Let $n \geq 1$ and $W_n = \langle t, s_1, \ldots, s_{n-1} \rangle \subset \mathsf{GL}_n(\mathbb{R})$ be defined as in (1.4.1). By Proposition 1.4.3, this is a finite Coxeter group with Dynkin diagram

B_n

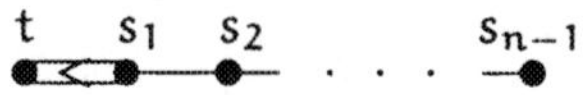

Furthermore, consider the linear character $\varepsilon' \colon W_n \to \{\pm 1\}$ given by $\varepsilon'(t) = -1$ and $\varepsilon'(s_i) = 1$ for $1 \leq i \leq n-1$. Let $W'_n \subseteq W_n$ be the kernel of ε', as in (1.4.8). For $n \geq 2$, we set $u = ts_1t$. Then, by Proposition 1.4.10, $W'_n = \langle u, s_1, \ldots, s_{n-1} \rangle$ is a Coxeter group with Dynkin diagram

D_n

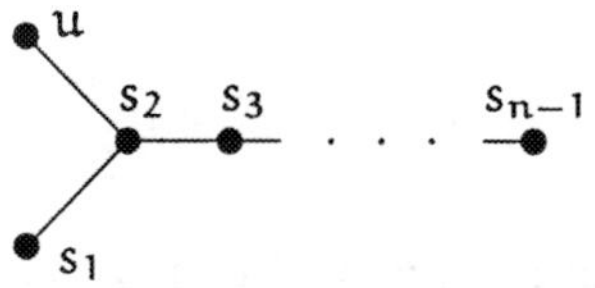

(We set $u = 1$ if $n = 1$; note that $W'_1 = \{1\}$.) The purpose of this and the following section is to construct the irreducible characters of W_n and of W'_n. Throughout, we denote by ε the sign character of W_n. We will see in Theorem 5.5.6 that every $\chi \in \mathsf{Irr}(W_n)$ is obtained by applying the j-induction to the restriction of ε to a suitable subgroup of W_n. The characters of W'_n are then obtained by restriction and Clifford's theorem; see (5.6.1).

Recall from Section 3.4 that the classes of W_n are parametrized by pairs of partitions of n. Hence so are the irreducible characters. We begin by defining suitable reflection subgroups of W_n, to which we will then apply the j-induction.

5.5.1. Reflection subgroups of W_n. Let λ and μ be partitions such that $|\lambda| + |\mu| = n$. We wish to define a corresponding subgroup $W'_{(\lambda,\mu)} \subseteq W_n$.

Let $\lambda_1, \ldots, \lambda_r$ be the parts of λ and $\mu_1, \ldots, \mu_s$ be the parts of μ, taken in some fixed order. For each i, we consider the group $W'_{\lambda_i} \subseteq \mathsf{GL}_{\lambda_i}(\mathbb{R})$, and for each j, we consider the group $W_{\mu_j} \subseteq \mathsf{GL}_{\mu_j}(\mathbb{R})$. (As a convention, we set $W_0 = W'_0 = \{1\}$.) Taking block diagonal matrices corresponding to the parts of λ and of μ, we can naturally embed the direct product of the above groups into W_n. Using that embedding, we now set

$$W'_{(\lambda,\mu)} := W'_{\lambda_1} \times \cdots \times W'_{\lambda_r} \times W_{\mu_1} \times \cdots \times W_{\mu_s} \subseteq W_n.$$

We obtain generators for $W'_{(\lambda,\mu)}$ as follows. Similarly to (1.4.1), we define:

$$t_0 := t \quad \text{and} \quad t_i := s_i t_{i-1} s_i \quad \text{for } 1 \leq i \leq n-1,$$
$$t'_0 := u \quad \text{and} \quad t'_i := s_i t'_{i-1} s_i \quad \text{for } 1 \leq i \leq n-1.$$

For $1 \leq i \leq r$ and $1 \leq j \leq s$, we set $n_i = \lambda_1 + \cdots + \lambda_{i-1}$ and $m_j = |\lambda| + \mu_1 + \cdots + \mu_{j-1}$. (In particular, we have $n_1 = 0$ and $m_1 = |\lambda|$.) Then the matrix description of W_n immediately shows that

$$W'_{\lambda_i} = \langle t'_{n_i}, s_{n_i+1}, \ldots, s_{n_i+\lambda_i-1} \rangle \quad \text{and} \quad W_{\mu_j} = \langle t_{m_j}, s_{m_j+1}, \ldots, s_{m_j+\mu_j-1} \rangle.$$

In fact, the above generators for W'_{λ_i} and W_{μ_j} satisfy the defining relations for Coxeter systems of type D_{λ_i} and B_{μ_j}, respectively. It follows that

$$\begin{aligned} &(W'_{\lambda_i}, \{t'_{n_i}, s_{n_i+1}, \ldots, s_{n_i+\lambda_i-1}\}) && \text{is a Coxeter system of type } D_{\lambda_i}, \\ &(W_{\mu_j}, \{t_{m_j}, s_{m_j+1}, \ldots, s_{m_j+\mu_j-1}\}) && \text{is a Coxeter system of type } B_{\mu_j}. \end{aligned}$$

(Here, we regard D_0, D_1 and B_0 as the empty systems.)

5.5.2. Lemma. *Let* $T \subseteq W_n$ *be the set of all conjugates of the generators* $t, s_1, \ldots, s_{n-1}$, *as in (5.3.1). Then we have* $W'_{(\lambda,\mu)} = \langle T \cap W'_{(\lambda,\mu)} \rangle$ *and*

$$|T \cap W'_{(\lambda,\mu)}| = \Big(\sum_{i=1}^{r} \lambda_i(\lambda_i - 1)\Big) + \Big(\sum_{j=1}^{s} \mu_j^2\Big) = 2a^*(\lambda) + 2a^*(\mu) + |\mu|.$$

Proof. First note that all t_i and all t'_i lie in T. Hence the first statement is clear since $W'_{(\lambda,\mu)}$ is generated by a subset, S' say, of T. Even more, S' can be chosen such that $(W'_{(\lambda,\mu)}, S')$ itself is a Coxeter system. Taking all conjugates of S' inside $W'_{(\lambda,\mu)}$, we obtain a subset $T' \subseteq W'_{(\lambda,\mu)}$.

Now we regard W_n as a subgroup of $GL_n(\mathbb{C})$. Then T consists of complex reflections in the sense of (5.2.9). So $T \cap W'_{(\lambda,\mu)}$ consists of complex reflections in the Coxeter group $W'_{(\lambda,\mu)}$. Hence Exercise 5.4 shows that $T \cap W'_{(\lambda,\mu)} = T'$. Now we are reduced to a computation entirely inside $W'_{(\lambda,\mu)}$. In fact, we can even reduce to the various direct factors of $W'_{(\lambda,\mu)}$. By Exercise 1.14, we have to compute the number of positive roots for each factor. For a factor of type D_{λ_i}, this number is given by $\lambda_i(\lambda_i - 1)$ (see Proposition 1.4.11); for a factor of type B_{μ_j}, the corresponding number is μ_j^2 (see Proposition 1.4.7). It remains to add up these values and to recall the definition of the a^*-invariant from (5.4.1). □

5.5.3. Definition. Let λ, μ be partitions with $|\lambda| + |\mu| = n$. Fix an ordering of the parts of the dual partitions λ^* and μ^* and let $W'_{(\lambda^*,\mu^*)} \subseteq W_n$ be the corresponding subgroup, as defined in (5.5.1). We set

$$b(\lambda, \mu) := 2a^*(\lambda^*) + 2a^*(\mu^*) + |\mu^*| = 2a(\lambda) + 2a(\mu) + |\mu|.$$

Using (5.4.1) and (5.4.2), it follows that $0 \leq b(\lambda, \mu) \leq n^2$, with $b(\lambda, \mu) = 0$ only if $(\lambda, \mu) = ([n], \varnothing)$ and $b(\lambda, \mu) = n^2$ only if $(\lambda, \mu) = (\varnothing, [1^n])$. Let $\varepsilon_{(\lambda^*,\mu^*)}$ be the restriction of the sign character ε to $W'_{(\lambda^*,\mu^*)}$ and set

$$\theta_{(\lambda^*,\mu^*)} := \mathrm{Ind}_{W'_{(\lambda^*,\mu^*)}}^{W_n}(\varepsilon_{(\lambda^*,\mu^*)}).$$

Then, using Macdonald's Corollary 5.2.11 and (5.3.1), together with the formula in Lemma 5.5.2, we can uniquely define an irreducible character $\chi_{(\lambda,\mu)}$ of W_n by the condition that $b_{\chi_{(\lambda,\mu)}} = b(\lambda, \mu)$ and

$$\theta_{(\lambda^*,\mu^*)} = \chi_{(\lambda,\mu)} + \text{combination of } \chi \in \mathrm{Irr}(W_n) \text{ with } b_\chi > b(\lambda,\mu).$$

Here, the b-invariants are taken with respect to the natural representation of W_n given by the Dynkin diagram on p. 163. As in Definition 5.4.4, note that the above definitions are independent of the chosen ordering of the parts of λ^* and μ^*. For example, we have $W_{([1^n],\varnothing)} = \{1\}$ and $W_{(\varnothing,[n])} = W_n$; so $\chi_{([n],\varnothing)}$ is the trivial character and $\chi_{(\varnothing,[1^n])}$ is the sign character. They have b-invariants 0 and n^2, respectively.

5.5.4. We shall also need the following full construction of the irreducible characters of W_n, based on Clifford theory. First recall from (1.4.1) that $N_n := \langle t_0, t_1, \ldots, t_{n-1}\rangle$ is an elementary abelian normal subgroup of order 2^n and that W_n/N_n is canonically isomorphic to $\mathfrak{S}_n$; in fact, we have $W_n \cong N_n \rtimes \mathfrak{S}_n$.

For any $a, b \geq 0$ with $a+b = n$, we define a linear character $\eta_{a,b}\colon N_n \to \{\pm 1\}$ by $\eta_{a,b}(t_i) = 1$ for $i < a$ and $\eta_{a,b}(t_j) = -1$ for $j \geq a$. Then we let $\mathrm{Irr}(W_n \mid \eta_{a,b})$ be the set of all $\chi \in \mathrm{Irr}(W_n)$ whose restriction to N_n contains $\eta_{a,b}$.

Since W_n acts on N_n by permutation of the t_i, the characters $\eta_{a,b}$ are a complete set of representatives for the orbits of $\mathrm{Irr}(N_n)$ under the induced action of W_n. Hence, Clifford's theorem (see [Isaacs 1976, (6.5)]) shows that

$$\mathrm{Irr}(W_n) = \coprod_{a+b=n} \mathrm{Irr}(W_n \mid \eta_{a,b}). \tag{1}$$

Now fix $a, b \geq 0$ with $a + b = n$. Again, since W_n acts on N_n by permutation of the t_i, we see that the stabilizer of $\eta_{a,b}$ in W_n is the subgroup $W_{a,b} := W_a \times W_b \subseteq W_n$, where the embedding comes from the diagonal embedding $GL_a(\mathbb{R}) \times GL_b(\mathbb{R}) \subseteq GL_n(\mathbb{R})$. We let $\mathrm{Irr}(W_{a,b} \mid \eta_{a,b})$ be the set of all $\psi \in \mathrm{Irr}(W_{a,b})$ whose restriction to N_n contains $\eta_{a,b}$.

Now $\eta_{a,b}$ can be extended to $W_{a,b}$, namely to the character $1_{W_a} \boxtimes \varepsilon'_b$, where $\boxtimes$ denotes outer tensor product and ε'_b denotes the restriction of ε' to W_b. Hence, [Isaacs 1976, (6.17)] yields the following description of $\mathrm{Irr}(W_{a,b} \mid \eta_{a,b})$.

Let $H_a \subseteq W_a$ and $H_b \subseteq W_b$ be the natural subgroups isomorphic to $\mathfrak{S}_a$ and $\mathfrak{S}_b$, respectively. Consider a pair of partitions (λ, μ) such that $|\lambda| = a$ and $|\mu| = b$. Then we have corresponding irreducible characters $\chi_\lambda \in \mathrm{Irr}(\mathfrak{S}_a)$ and $\chi_\mu \in \mathrm{Irr}(\mathfrak{S}_b)$, as in Definition 5.4.4. Let $\tilde{\chi}_\lambda \in \mathrm{Irr}(W_a)$ be the pull-back of χ_λ under the natural projection $W_a \to H_a \cong \mathfrak{S}_a$. Similarly, we define $\tilde{\chi}_\mu \in \mathrm{Irr}(W_b)$. Then the desired description is:

$$\mathrm{Irr}(W_{a,b} \mid \eta_{a,b}) = \{\tilde{\chi}_\lambda \boxtimes (\varepsilon'_b \otimes \tilde{\chi}_\mu) \mid \lambda \vdash a, \mu \vdash b\}. \tag{2}$$

Finally, by [Isaacs 1976, (6.11)], the induction $\mathrm{Ind}_{W_a \times W_b}^{W_n}$ defines a bijection

$$\mathrm{Irr}(W_{a,b} \mid \eta_{a,b}) \xrightarrow{\sim} \mathrm{Irr}(W_n \mid \eta_{a,b}), \qquad \tilde{\chi}_\lambda \boxtimes (\varepsilon'_b \otimes \tilde{\chi}_\mu) \mapsto \tilde{\chi}_{(\lambda,\mu)}. \tag{3}$$

Thus, combining (1) and (3), we obtain a natural parametrization of $\mathrm{Irr}(W_n)$ by pairs of partitions (λ, μ) such that $|\lambda|+|\mu| = n$. The above construction shows, for example, that $\tilde{\chi}_{([n],\varnothing)}$ is the trivial character and $\tilde{\chi}_{(\varnothing,[1^n])}$ is the sign character.

For the identification of $\chi_{(\lambda,\mu)}$ and $\tilde{\chi}_{(\lambda,\mu)}$, we shall need the following result.

5.5.5. Lemma. *Recall that we have $W_n = N_n \rtimes H_n$, where $H_n \cong \mathfrak{S}_n$ and N_n is elementary abelian of order 2^n. Then, for any $\nu \vdash n$, we have*

$$b_{\tilde{\chi}_\nu} = 2a(\nu) \qquad \text{and} \qquad b_{\varepsilon'\otimes\tilde{\chi}_\nu} = 2a(\nu) + n,$$

where $\tilde{\chi}_\nu \in \mathrm{Irr}(W_n)$ is the pull-back of $\chi_\nu \in \mathrm{Irr}(\mathfrak{S}_n)$ under $W_n \to H_n \cong \mathfrak{S}_n$.

Proof. Each $w \in W_n$ can be written uniquely as $w = yh$ where $y \in N_n$ and $h \in H_n$. Using Molien's formula in Proposition 5.2.4, we obtain

$$\begin{aligned} P_{\tilde{\chi}_\nu}(q) &= \frac{1}{|H_n|}\sum_{h\in H_n} \chi_\nu(h)\frac{1}{2^n}\sum_{y\in N_n} \frac{1}{\det(\mathrm{id}_{\mathbb{R}^n} - qhy)}, \quad \text{since } \tilde{\chi}_\nu(yh) = \chi_\nu(h),\\ &= \frac{1}{|H_n|}\sum_{h\in H_n} \frac{\chi_\nu(h)}{\det(\mathrm{id}_{\mathbb{R}^n} - q^2h)}, \qquad \text{using Exercise 5.7.} \end{aligned}$$

Now, for any $h \in H_n$, the denominator in the above formula is $(1-q^2)$ times the analogous polynomial (evaluated at q^2) for h in the $(n-1)$-dimensional natural representation of $H_n \cong \mathfrak{S}_n$. It follows that $P_{\tilde{\chi}_\nu} = P_{\chi_\nu}(q^2)/(1-q^2)$ and so $b_{\tilde{\chi}_\nu} = 2b_{\chi_\nu}$. A similar computation, using the second formula in Exercise 5.7, also shows that $P_{\varepsilon'\otimes\tilde{\chi}_\nu}(q) = q^n P_{\tilde{\chi}_\nu}(q)$ and, hence, $b_{\varepsilon'\otimes\tilde{\chi}_\nu} = b_{\tilde{\chi}_\nu} + n$. It remains to use the formula $b_{\chi_\nu} = a(\nu)$; see Definition 5.4.4. □

5.5.6. Theorem. *The characters $\chi_{(\lambda,\mu)}$, as constructed in Definition 5.5.3, exhaust all of $\mathrm{Irr}(W_n)$. Moreover, the following hold.*

(a) *Each character $\chi_{(\lambda,\mu)}$ equals the character $\tilde{\chi}_{(\lambda,\mu)}$, as constructed in (5.5.4).*
(b) *Each $\chi_{(\lambda,\mu)}$ is afforded by a $\mathbb{Q}W_n$-module. So $\mathbb{Q}$ is a splitting field for W_n.*
(c) *We have $\varepsilon \otimes \chi_{(\lambda,\mu)} = \chi_{(\mu^*,\lambda^*)}$ and $\varepsilon' \otimes \chi_{(\lambda,\mu)} = \chi_{(\mu,\lambda)}$.*

Proof. First note that the fact that the characters $\chi_{(\lambda,\mu)}$ exhaust all of $\mathrm{Irr}(W_n)$ is clear by (5.5.4), once (a) is proved. In order to prove (a), we proceed as follows. We can write $W'_{(\lambda^*,\mu^*)} = W'_{\lambda^*} \times W_{\mu^*} \subseteq W_a \times W_b$, where W'_{λ^*} is the direct product of all factors $W'_{\lambda_i^*}$ and W_{μ^*} is the direct product of all factors $W_{\mu_j^*}$. Then we also have $\varepsilon_{(\lambda^*,\mu^*)} = \varepsilon_{\lambda^*} \boxtimes \varepsilon_{\mu^*}$, where ε_{λ^*} and ε_{μ^*} denote the sign characters of W'_{λ^*} and W_{μ^*}, respectively. Since the j-induction is compatible with direct products (see Exercise 5.2), we obtain

$$j_{W_{(\lambda^*,\mu^*)}}^{W_a\times W_b}(\varepsilon_{(\lambda^*,\mu^*)}) = j_{W'_{\lambda^*}}^{W_a}(\varepsilon_{\lambda^*}) \boxtimes j_{W_{\mu^*}}^{W_b}(\varepsilon_{\mu^*}). \tag{1}$$

The second factor in the right-hand side of (1) can be rewritten as follows. We have a natural projection $W_b \to \mathfrak{S}_b$ with kernel $N_b = \langle t_a, \ldots, t_{n-1}\rangle$. We also have $W_{\mu^\bullet} \twoheadrightarrow \mathfrak{S}_{\mu^\bullet}$ under this projection. By looking at the values of the elements t_i, s_i, t_i', we see that

$$\varepsilon_{\mu^\bullet} = \mathrm{Res}^{W_b}_{W_{\mu^\bullet}}(\varepsilon_b') \otimes \tilde{\varepsilon}_{\mu^\bullet},$$

where $\tilde{\varepsilon}_{\mu^\bullet}$ denotes the pull-back of the sign character of $\mathfrak{S}_{\mu^\bullet}$. So we have

$$\mathrm{Ind}^{W_b}_{W_{\mu^\bullet}}(\varepsilon_{\mu^\bullet}) = \mathrm{Ind}^{W_b}_{W_{\mu^\bullet}}\Big(\mathrm{Res}^{W_b}_{W_{\mu^\bullet}}(\varepsilon_b') \otimes \tilde{\varepsilon}_{\mu^\bullet}\Big) = \varepsilon_b' \otimes \mathrm{Ind}^{W_b}_{W_{\mu^\bullet}}(\tilde{\varepsilon}_{\mu^\bullet}).$$

Since induction and pulling back via $W_b \twoheadrightarrow \mathfrak{S}_b$ commute with each other, we obtain the following identity, using Theorem 5.4.7:

$$\mathrm{Ind}^{W_b}_{W_{\mu^\bullet}}(\varepsilon_{\mu^\bullet}) = \varepsilon_b' \otimes \Big(\tilde{\chi}_\mu + \text{combination of } \tilde{\chi}_{\mu_1} \text{ with } \mu_1 \trianglelefteq \mu,\ \mu_1 \neq \mu\Big).$$

Using the formula in Lemma 5.5.5, we see that $\varepsilon_b' \otimes \tilde{\chi}_{\mu_1}$ (for any $\mu_1 \vdash b$) has b-invariant equal to $2a(\mu_1)+b$. Moreover, by Exercise 5.6(a), we have $a(\mu_1) > a(\mu)$ for $\mu_1 \trianglelefteq \mu$, $\mu_1 \neq \mu$. Thus, we conclude that

$$j^{W_b}_{W_{\mu^\bullet}}(\varepsilon_{\mu^\bullet}) = \varepsilon_b' \otimes \tilde{\chi}_\mu. \tag{2}$$

Now let us rewrite the first factor on the right-hand side of (1). By (1.4.8), we have a natural projection $W_a' \twoheadrightarrow \mathfrak{S}_a$ with kernel $N_a' = \langle tt_1, \ldots, tt_{a-1}\rangle$. We also have $W_{\lambda^\bullet}' \twoheadrightarrow \mathfrak{S}_{\lambda^\bullet}$ under this projection. Looking again at the values of the elements s_i, t_i, we see that $\varepsilon_{\lambda^\bullet}$ is the pull-back of the sign character of $\mathfrak{S}_{\lambda^\bullet}$. Again using Theorem 5.4.7 and the fact that induction and pulling back via $W_a' \twoheadrightarrow \mathfrak{S}_a$ commute with each other, we obtain that

$$\begin{aligned}\mathrm{Ind}^{W_a}_{W_{\lambda^\bullet}^\circ}(\varepsilon_{\lambda^\bullet}) = \mathrm{Ind}^{W_a}_{W_a^\circ}\Big(\mathrm{Ind}^{W_a'}_{W_{\lambda^\bullet}^\circ}(\varepsilon_{\lambda^\bullet})\Big) &= \mathrm{Ind}^{W_a}_{W_a^\circ}(\tilde{\chi}_\lambda') \\ &+ \text{combination of } \mathrm{Ind}^{W_a}_{W_a^\circ}(\tilde{\chi}_{\lambda_1}') \text{ with } \lambda_1 \trianglelefteq \lambda,\ \lambda_1 \neq \lambda,\end{aligned}$$

where $\tilde{\chi}_{\lambda_1}' \in \mathrm{Irr}(W_a')$, for any $\lambda_1 \vdash a$, denotes the pull-back of $\chi_{\lambda_1} \in \mathrm{Irr}(\mathfrak{S}_a)$. But we also have the pull-back $\tilde{\chi}_{\lambda_1} \in \mathrm{Irr}(W_a)$ under the projection $W_a \twoheadrightarrow \mathfrak{S}_a$; its restriction to W_a' clearly equals $\tilde{\chi}_{\lambda_1}'$. Hence we have

$$\mathrm{Ind}^{W_a}_{W_a^\circ}(\tilde{\chi}_{\lambda_1}') = \tilde{\chi}_{\lambda_1} \otimes \mathrm{Ind}^{W_a}_{W_a^\circ}(1_{W_a'}) = \tilde{\chi}_{\lambda_1} + (\varepsilon_a' \otimes \tilde{\chi}_{\lambda_1}) \quad \text{for all } \lambda_1 \vdash a.$$

So, by an argument similar to that for the proof of (2), we deduce that

$$j^{W_a}_{W_{\lambda^\bullet}^\circ}(\varepsilon_{\lambda^\bullet}) = \tilde{\chi}_\lambda. \tag{3}$$

Putting things together, we obtain the following conclusion, proving (a):

$$\begin{aligned}
\chi_{(\lambda,\mu)} &= j_{W_a\times W_b}^{W_n}\left(j_{W'_{\lambda^*,\mu^*}}^{W_a\times W_b}(\varepsilon_{(\lambda^*,\mu^*)})\right) && \text{by transitivity}\\
&= j_{W_a\times W_b}^{W_n}\left(j_{W'_{\lambda^*}}^{W_a}(\varepsilon_{\lambda^*})\boxtimes j_{W_{\mu^*}}^{W_b}(\varepsilon_{\mu^*})\right) && \text{by (1)}\\
&= j_{W_a\times W_b}^{W_n}\left(\tilde{\chi}_\mu\boxtimes(\varepsilon'_b\otimes\tilde{\chi}_\mu)\right) && \text{by (2) and (3)}\\
&= \mathrm{Ind}_{W_a\times W_b}^{W_n}\left(\tilde{\chi}_\lambda\boxtimes(\varepsilon'_b\otimes\tilde{\chi}_\mu)\right) = \tilde{\chi}_{(\lambda,\mu)} && \text{by (5.5.4)(3).}
\end{aligned}$$

Now Remark 5.2.7 shows that all characters $\chi_{(\lambda,\mu)}$ are afforded by $\mathbb{Q}W_n$-modules. So (a) implies (b). Finally, the construction in (5.5.4) easily shows that the statements in (c) hold for the characters $\tilde{\chi}_{(\lambda,\mu)}$. So (a) also implies (c). ☐

In Exercise 5.9, the value of $\chi_{(\lambda,\mu)}$ on a Coxeter element of W_n is determined.

We can now also identify the characters afforded by the exterior powers of the natural module V for W_n given by the Dynkin diagram of type B_n on p. 163. Recall from Proposition 1.4.3 that we have $V = \mathbb{R}^n$ via the embedding $W_n \subset \mathrm{GL}_n(\mathbb{R})$.

5.5.7. Proposition. *Let $\rho^{\wedge d}$ be the character afforded by $\bigwedge^d V$. Then we have*

$$\rho^{\wedge d} = \chi_{([n-d],[1^d])} \qquad \textit{and} \qquad b_{\rho^{\wedge d}} = d^2 \qquad \textit{for } 0 \leq d \leq n.$$

In particular, the character of the natural module V itself is $\chi_{([n-1],[1])}$.

Proof. If $d = 0$ or $d = n$, then $\rho^{\wedge d}$ is the trivial or the sign character, respectively. So the assertion holds in this case. Now let us assume that $0 < d < n$ and consider the parabolic subgroup $W_{n-d} = \langle t, s_1, \ldots, s_{n-d-1}\rangle \subseteq W_n$. We have

$$\begin{aligned}
\mathrm{Ind}_{W_{n-d}}^{W_n}(1_{W_{n-d}}) &= \mathrm{Ind}_{W_{n-d}\times W_d}^{W_n}\left(\mathrm{Ind}_{W_{n-d}}^{W_{n-d}\times W_d}(1_{W_{n-d}})\right)\\
&= \mathrm{Ind}_{W_{n-d}\times W_d}^{W_n}\left(1_{W_{n-d}}\boxtimes\chi_{\mathrm{reg}}\right),
\end{aligned}$$

where χ_{reg} denotes the character of the regular representation of W_d and the embedding $W_{n-d}\times W_d \subseteq W_n$ comes from the diagonal embedding $\mathrm{GL}_{n-d}(\mathbb{R})\times \mathrm{GL}_d(\mathbb{R}) \subseteq \mathrm{GL}_n(\mathbb{R})$. On the other hand, by Proposition 5.1.5, we know that $\rho^{\wedge d}$ occurs with multiplicity 1 in the above induced character. So there exists some $\psi \in \mathrm{Irr}(W_d)$ of degree 1 such that $\rho^{\wedge d}$ occurs in $\mathrm{Ind}_{W_{n-d}\times W_d}^{W_n}(1_{W_{n-d}}\boxtimes\psi)$. But the index of $W_{n-d}\times W_d$ in W_n equals $\binom{n}{d} = \rho^{\wedge d}(1)$ and we have $\psi(1) = 1$. So, by looking at the character degrees, we conclude that $\rho^{\wedge d}$ not only occurs in but also must be equal to the above induced character. In particular, this implies that the restriction of $1_{W_{n-d}}\boxtimes\psi$ to N_n cannot be the trivial character. Consequently, ψ must be of the form $\psi = \varepsilon'_d\otimes\tilde{\eta}$, where $\tilde{\eta}$ is pulled back from a linear character $\eta \in \mathrm{Irr}(\mathfrak{S}_d)$ via $B_d \to H_d \cong \mathfrak{S}_d$. Thus, we have found that

$$\chi_n^d = \mathrm{Ind}_{W_{n-d}\times W_d}^{W_n}\left(1_{W_{n-d}}\boxtimes(\varepsilon'_d\otimes\tilde{\eta})\right).$$

A comparison with the construction in (5.5.4) shows that $\rho^{\wedge d} = \chi_{([n-d],\mu)}$, where $\mu \vdash d$ is such that $\eta = \chi_\mu$. The formula for the b-invariants in Definition 5.5.3 and Exercise 5.6 now yields that

$$b_{\rho^{\wedge d}} = 2a([n-d]) + 2a(\mu) + d \leq 2a([1^d]) + d = d^2,$$

with equality only if $\mu = [1^d]$. Hence it remains to show that $b_{\rho^{\wedge d}} \geq d^2$.

To prove this, consider the parabolic subgroup $W_d = \langle t, s_1, \ldots, s_{d-1} \rangle \subseteq W_n$. By Exercise 5.1, we have $\rho^{\wedge d} = \varepsilon \otimes \rho^{\wedge n-d}$. So, using Proposition 5.1.5, we obtain

$$\left(\rho^{\wedge d}, \mathrm{Ind}_{W_d}^{W_n}(\varepsilon_{W_d})\right)_{W_n} = \left(\rho^{\wedge n-d}, \mathrm{Ind}_{W_d}^{W_n}(1_{W_d})\right)_{W_n} = \binom{n-d}{n-d} = 1.$$

Now ε_{W_d} has b-invariant equal to the length of the longest element in W_d (see Remark 5.3.2), which is d^2. All that remains is to use Lemma 5.2.5. □

5.6 THE CHARACTERS OF COXETER GROUPS OF TYPE D

We keep the set-up and notation of the previous section. Recall that we have an embedding $W'_n \subseteq W_n$, where W'_n is normal of index 2. The aim of this section is to construct the irreducible characters of W'_n, by using the restriction of characters from W_n (see (5.6.1)) and the j-induction from characters of subgroups of W'_n which are isomorphic to $\mathfrak{S}_n$ (see Proposition 5.6.3).

5.6.1. Let (λ, μ) be a pair of partitions with $|\lambda| + |\mu| = n$. Let $\chi_{(\lambda,\mu)} \in \mathrm{Irr}(W_n)$ be as in Theorem 5.5.6 and denote by $\chi'_{(\lambda,\mu)}$ the restriction of $\chi_{(\lambda,\mu)}$ to W'_n. Since W'_n is a normal subgroup of index 2 in W_n, the situation is controlled as in [Isaacs 1976, (6.19)]: either $\chi'_{(\lambda,\mu)}$ is irreducible or the sum of two different irreducible characters of W'_n; furthermore, all irreducible characters of W'_n are obtained in this way. More precisely, we know by Theorem 5.5.6 that $\varepsilon' \otimes \chi_{(\lambda,\mu)} = \chi_{(\mu,\lambda)}$, where $\varepsilon' \colon W_n \to \{\pm 1\}$ is the linear character given by $\varepsilon'(t) = -1$ and $\varepsilon'(s_i) = 1$ for all i. Hence we obtain the following:

(a) If $\lambda \neq \mu$, then we have $\chi'_{(\lambda,\mu)} = \chi'_{(\mu,\lambda)} \in \mathrm{Irr}(W'_n)$.

(b) If $\lambda = \mu$, then n is even and we have $\chi'_{(\lambda,\lambda)} = \chi_{(\lambda,+)} + \chi_{(\lambda,-)}$ with two distinct characters $\chi_{(\lambda,+)}, \chi_{(\lambda,-)} \in \mathrm{Irr}(W'_n)$.

So, $\mathrm{Irr}(W'_n)$ is obtained by taking the characters $\chi'_{(\lambda,\mu)}$ for unordered pairs of partitions $\lambda \neq \mu$ and, in the case where n is even, also the characters $\chi_{(\lambda,\pm)}$ for $\lambda \vdash n/2$. Each irreducible character of W'_n is obtained exactly once in this way. For an example of this labelling, see Tables 6.1 (p. 191) and 10.4 (p. 346).

5.6.2. The fact that $V = \mathbb{R}^n$ is also the natural module for W'_n via the embedding $W'_n \subseteq W_n \subset \mathrm{GL}(\mathbb{R})$ (see Proposition 1.4.10) allows us to relate the Molien series (as in (5.2.2)) for the irreducible characters of W'_n to those for the irreducible characters of W_n. In fact, we have:

$$P_{\chi_{(\lambda,\mu)}}(q) + P_{\chi_{(\mu,\lambda)}}(q) = P_{\chi'_{(\lambda,\mu)}}(q) \qquad \text{if } \lambda \neq \mu,$$
$$P_{\chi_{(\lambda,\lambda)}}(q) = P_{\chi_{(\lambda,+)}}(q) = P_{\chi_{(\lambda,-)}}(q) \qquad \text{if } n \text{ is even and } \lambda \vdash n/2.$$

Indeed, using the above-mentioned fact about the natural module, Frobenius reciprocity and the defining formula in (5.2.2), the above relations are an immediate consequence of the two identities in (5.6.1)(a) and (b).

Combining this with the knowledge of the b-invariants for the characters of W_n from Definition 5.5.3, we obtain the following formulas:

$$b_{\chi'_{(\lambda,\mu)}} = b_{\chi_{(\lambda,\mu)}} = 2a(\lambda) + 2a(\mu) + |\mu| \qquad \text{if } \lambda \neq \mu \text{ are such that } |\lambda| \geq |\mu|,$$
$$b_{\chi_{(\lambda,\pm)}} = b_{\chi_{(\lambda,\lambda)}} = 4a(\lambda) + n/2 \qquad \text{if } n \text{ is even and } \lambda \vdash n/2.$$

Now, the only missing piece in the above description is a construction of the two characters $\chi_{(\lambda,\pm)}$ occurring in (5.6.1)(b). This is provided by the following result.

5.6.3. Proposition. *Assume that $n \geq 2$ is even and let $\lambda \vdash n/2$. Now let*

$$H_n^+ := \langle s_1, \ldots, s_{n-1} \rangle \quad \textit{and} \quad H_n^- := tH_n^+t = \langle u, s_2, \ldots, s_{n-1} \rangle.$$

Then $H_n^\pm \subseteq W'_n$ is a parabolic subgroup canonically isomorphic to $\mathfrak{S}_n$. If $\chi_{2\lambda}^\pm \in \mathrm{Irr}(H_n^\pm)$ corresponds to $\chi_{2\lambda} \in \mathrm{Irr}(\mathfrak{S}_n)$ under $H_n^\pm \cong \mathfrak{S}_n$, we have

$$\chi_{(\lambda,+)} = j_{H_n^+}^{W'_n}(\chi_{2\lambda}^+) \qquad \textit{and} \qquad \chi_{(\lambda,-)} = j_{H_n^-}^{W'_n}(\chi_{2\lambda}^-),$$

where 2λ is the partition of n obtained by multiplying the parts of λ by 2.

Proof. First note that all irreducible characters of $\mathfrak{S}_n$ satisfy the assumptions for applying the j-induction (by the construction in Definition 5.4.4). So we must prove that $\chi'_{(\lambda,\lambda)} = \psi_\lambda^+ + \psi_\lambda^-$, where we define

$$\psi_\lambda^+ := j_{H_n^+}^{W'_n}(\chi_{2\lambda}^+) \in \mathrm{Irr}(W'_n) \qquad \text{and} \qquad \psi_\lambda^- := j_{H_n^-}^{W'_n}(\chi_{2\lambda}^-) \in \mathrm{Irr}(W'_n).$$

Next note that $\psi_\lambda^\pm$ have the correct b-invariants, by using the formulas in (5.6.2) and Definition 5.4.4. Furthermore, since conjugation with t sends H_n^+ to H_n^- and preserves the b-invariants (by (5.6.2)), we conclude that conjugation with t in fact transforms ψ_λ^+ into ψ_λ^-. Hence, it will be sufficient to prove that

$$\left(\mathrm{Ind}_{\mathfrak{S}_n}^{W_n}(\chi_{2\lambda}^+), \chi_{(\lambda,\lambda)}\right)_{W_n} \neq 0, \quad \text{where we identify } \mathfrak{S}_n = H_n^+ \subseteq W_n.$$

Now, using the notation of (5.5.4), we see that $\chi_{(\lambda,\lambda)}$ is j-induced from the character $\tilde{\chi}_\lambda \boxtimes (\varepsilon'_{n/2} \otimes \tilde{\chi}_\lambda)$ of $W_{n/2} \times W_{n/2} \subseteq W_n$. Furthermore, we have

$$W_n = (W_{n/2} \times W_{n/2}) \cdot \mathfrak{S}_n, \qquad (W_{n/2} \times W_{n/2}) \cap \mathfrak{S}_n = \mathfrak{S}_{n/2} \times \mathfrak{S}_{n/2} \subseteq \mathfrak{S}_n.$$

The Mackey formula therefore implies that the above scalar product is equal to

$$\left(\chi_{2\lambda}^{+}, \mathrm{Ind}_{\mathfrak{S}_{n/2}\times\mathfrak{S}_{n/2}}^{\mathfrak{S}_n}(\chi_\lambda \boxtimes \chi_\lambda)\right)_{\mathfrak{S}_n}.$$

But this is non-zero (even equal to 1) by the construction of the characters of $\mathfrak{S}_n$ in Definition 5.4.4. Indeed, first recall that χ_λ is j-induced from the sign character of the Young subgroup $\mathfrak{S}_{\lambda^*}$. Hence, by Exercise 5.2, $\chi_\lambda \boxtimes \chi_\lambda$ is j-induced from the sign character of $\mathfrak{S}_{\lambda^*} \times \mathfrak{S}_{\lambda^*} \subseteq \mathfrak{S}_n$. But, concatenating λ^* with itself and reordering the parts, we see that the latter Young subgroup is conjugate (in $\mathfrak{S}_n$) to the Young subgroup corresponding to $(2\lambda)^*$. Hence we conclude that

$$\chi_{2\lambda}^{+} = j_{\mathfrak{S}_{n/2}\times\mathfrak{S}_{n/2}}^{\mathfrak{S}_n}(\varepsilon_{\lambda^*} \boxtimes \varepsilon_{\lambda^*}),$$

and so $\chi_{2\lambda}^{+}$ occurs in the above induced character, as claimed. □

5.6.4. Corollary. *Every irreducible character of W'_n is afforded by a $\mathbb{Q}W'_n$-module and so $\mathbb{Q}$ is a splitting field for W'_n.*

Proof. Using Theorem 5.5.6, we already know that each restriction $\chi'_{(\lambda,\mu)}$ is afforded by a $\mathbb{Q}W'_n$-module. Hence, the only problem is to show that the characters $\chi_{(\lambda,\pm)}$ in (5.6.1)(b) are afforded by $\mathbb{Q}W'_n$-modules. Using Remark 5.2.7, this follows from Proposition 5.6.3 and Theorem 5.4.5. □

5.6.5. Remark. Let us define a linear character $\varepsilon''\colon W_n \to \{\pm 1\}$ by $\varepsilon''(t) = 1$ and $\varepsilon''(s_i) = -1$ for all i. We have $\varepsilon'' = \varepsilon \otimes \varepsilon'$ where ε is the sign character and ε' is as in (5.6.1). So, using Theorem 5.5.6(c), we conclude that

$$\varepsilon'' \otimes \chi_{(\lambda,\mu)} = \chi_{(\lambda^*,\mu^*)}.$$

Now, by looking at the values of the elements s_i, t, u, we see that the restriction of ε'' to W'_n is the sign character of W'_n. Let us denote that restriction again by the same symbol. Then we have

$$\varepsilon'' \otimes \chi'_{(\lambda,\mu)} = \chi'_{(\lambda^*,\mu^*)} \qquad \text{if } \lambda \neq \mu,$$
$$\varepsilon'' \otimes \chi_{(\lambda,\pm)} = \chi_{(\lambda^*,\pm)} \qquad \text{if } n \text{ is even and } \lambda \vdash n/2,$$

The first statement is clear by (5.6.1)(a), while the second easily follows from Proposition 5.6.3 and Corollary 5.4.9.

5.6.6. Remark. Recall that the natural module for W_n is $V = \mathbb{R}^n$, where W_n acts via the embedding $W_n \subset \mathrm{GL}_n(\mathbb{R})$. Now Proposition 1.4.10 shows that the restriction of V to W'_n is the natural module given by the Dynkin diagram of type D_n on p. 163. Using Proposition 5.5.7, it follows that, for $0 \leq d \leq n-1$, the character $\chi'_{([n-d],[1^d])}$ is afforded by $\bigwedge^d V$.

5.6.7. Remark. Recall from Remark 5.3.7 that, in exceptional groups, the constant γ_χ in (5.2.2) is always equal to 1 or 2 (where the value 2 only occurs for very few irreducible characters). Furthermore, we always have $\gamma_\chi = 1$ for the irreducible characters of groups of type $I_2(m)$, A_n and B_n; see (5.3.4), Definition 5.4.4 and Definition 5.5.3. The formulas in (5.6.2) show that, in type D_n, we have

$$\gamma_{\chi'_{(\lambda,\mu)}} = \begin{cases} 1 & \text{if } \lambda \neq \mu,\ |\lambda| \neq |\mu|, \\ 2 & \text{if } \lambda \neq \mu,\ |\lambda| = |\mu|, \end{cases} \quad \text{and} \quad \gamma_{\chi_{(\lambda,+)}} = \gamma_{\chi_{(\lambda,-)}} = 1 \text{ for } \lambda = \mu.$$

Combining the above results and using the compatibility with direct products in Exercise 5.2, we can now conclude that, for any finite Coxeter group $W \subset \mathsf{GL}(V)$ and any $\chi \in \mathsf{Irr}(W)$, we have $\gamma_\chi = 2^d$ for some $d \geq 0$.

5.7 BIBLIOGRAPHICAL REMARKS AND EXERCISES

A first introduction to invariant theory can be found in [Benson and Grove 1985, Chap. 7]. For more on that subject, with a particular emphasis on algorithmic aspects, see [Sturmfels 1993].

The fact that the exterior powers of the reflection representation of W are irreducible and pairwise non-isomorphic is due to R. Steinberg, see [Bourbaki 1968, Chap. V, §2, Exercise 3(d)]. The remaining statements on exterior powers are taken from [Curtis et al. 1971, §9], where they are given in the Hecke algebra context. See also [Steinberg 1968, §14].

The j-induction in the completely general setting of Theorem 5.2.6 appears in [Lusztig and Spaltenstein 1979, (3.2)], generalizing an earlier construction in [Macdonald 1972]; see also [Lusztig 1979a, §3]. The proof of Theorem 5.2.10 follows [Cohen 1976, (1.8)]; it is a generalization of that in [Macdonald 1972]. See also [Bourbaki 1968, Chap. V, no. 5.4] and [Springer 1974, §2]. The important role that the j-induction plays in finite Weyl groups, especially in connection with the theory of so-called special characters, was discovered by [Lusztig 1979a]; see also Theorem 6.5.13.

Finite complex reflection groups are a natural generalization of finite Coxeter groups. They have been classified by [Shephard and Todd 1954]; see also [Springer 1974] and [Cohen 1976]. The subject has recently seen a revival in connection with Broué's conjectures about modular representations of finite Chevalley groups; see [Broué and Malle 1993], [Broué et al. 1998], [Broué et al. 1999].

The irreducible characters of finite Coxeter groups of type F_4, E_6, E_7, E_8 and H_4 were first determined in [Kondo 1965], [Frame 1951], [Frame 1970], and [Grove 1974]. Here, we have obtained the character tables of these groups by a straightforward application of the Dixon–Schneider algorithm [Schneider 1990].

The results in Section 5.4 go back to [Frobenius 1900]; for modern treatments of this work, see [Macdonald 1995], [Fulton 1997], and also the chapter on $\mathfrak{S}_n$ in [Ledermann 1987]. Our construction of the irreducible characters of

$\mathfrak{S}_n$ uses the j-induction in an essential way. This is closer to the approach followed in [Specht 1935]; see also [Macdonald 1972]. In step 2 of the proof of Theorem 5.4.5 we show one implication of the so-called Gale–Ryser theorem, following [Goldschmidt 1993, (6.3)]. The description of the idempotents in Proposition 5.4.10 is contained in [Macdonald 1995, I.7, Example 15]. The argument for proving Proposition 5.4.11 is taken from [Beynon and Lusztig 1978, p. 419], where it is attributed to the referee of that paper. A different argument, using fake degrees and applicable to complex reflection groups, is given in [Malle 2000].

A comprehensive treatment of the character theory of $\mathfrak{S}_n$ can be found in [James and Kerber 1981]. For an application of the character theory of $\mathfrak{S}_n$ to a general result about characters of finite groups, see [Alperin 1987].

The construction of the irreducible characters of W_n in (5.5.4) is just a special case of a more general construction which works for wreath products $G \wr \mathfrak{S}_n$, where G is any finite group; see Appendix I.A in [Macdonald 1995]. The description of $\mathrm{Irr}(W_n)$ in terms of the j-induction of sign characters of suitable subgroups as in (5.5.1) goes back to [Specht 1937]. The results on type D_n go back to Young; see [Young 1977]. The statement about $\chi_{(\lambda,\pm)}$ in Proposition 5.6.3 appears in [Lusztig 1984, (4.6.2)].

The results on splitting fields are due to Young, Specht (for the classical types) and Kondo, Benard and Benson–Grove (for the exceptional types); see [James and Kerber 1981], [Benard 1971] and [Benson and Grove 1973] and the references therein. These results have also been generalized to complex reflection groups; see [Bessis 1997] and the references therein.

Finally, see [Pahlings 1976] (and the references therein), where it is shown that a finite Coxeter group W is almost always determined up to isomorphism by its character table; see also [Bleher et al. 1997] for the classification of automorphisms of W and its character table.

Exercise 5.1. Let W be a finite Coxeter group with natural module V. Let ε be the sign character of W. Show that

$$\varepsilon \otimes \bigwedge^{d} V \cong \bigwedge^{|S|-d} V \qquad \text{for } 0 \leq d \leq |S|.$$

[*Hint.* Show that the character values are the same, using Remark 5.1.6 and the identity $\det(q\,\mathrm{id}_V - w) = (-q)^{|S|}\varepsilon(w)\det(q^{-1}\,\mathrm{id}_V - w^{-1})$.]

Exercise 5.2. The purpose of this exercise is to show that the j-induction of (5.2.8) is compatible with direct products. For $i = 1, 2$, let V_i be a $\mathbb{C}$-vector space with $\dim_{\mathbb{C}} V_i < \infty$ and $G_i \subset \mathrm{GL}(V_i)$ be a finite group. Let $H_i \subseteq G_i$ be a subgroup and set $U_i = V_i / \mathrm{Fix}_{H_i}(V_i)$. Let $\psi_i \in \mathrm{Irr}(H_i)$ and assume that

$$P^{U_i}_{\psi_i}(q) = \gamma_i q^{d_i} + \text{combination of higher powers of } q,$$

where $\gamma_i \neq 0$. Now set $G = G_1 \times G_2$, $H = H_1 \times H_2$, $V = V_1 \oplus V_2$, and $U = U_1 \oplus U_2$. We can naturally regard $\mathrm{GL}(V_1) \times \mathrm{GL}(V_2)$ as a subgroup of $\mathrm{GL}(V)$, by a diagonal embedding. Show that we have $U = \mathrm{Fix}_H(V)$, and that the following holds:

(a) We have $P^U_{\psi_1 \boxtimes \psi_2}(q) = \gamma_1\gamma_2 q^{d_1+d_2}$ + combination of higher powers of q.

(b) If $\gamma_1 = \gamma_2 = 1$, then we have $j_H^G(\psi_1 \boxtimes \psi_2) = j_{H_1}^{G_1}(\psi) \boxtimes j_{H_2}^{G_2}(\psi_2)$.

Exercise 5.3. Let $G = \langle g \rangle$ be a cyclic group of order $m \geq 2$ say. Let $\zeta \in \mathbb{C}$ be a primitive mth root of unity and $\chi \in \mathrm{Irr}(G)$ be defined by $\chi(g) = \zeta$. Then we have $\mathrm{Irr}(G) = \{\chi^j \mid 1 \leq j \leq m\}$. Show that

$$P^V_{\chi^j}(q) = q^j + \text{combination of higher powers of } q, \quad \text{for } 1 \leq j \leq m,$$

where V is a $\mathbb{C}G$-module affording χ. Check that $G \subset \mathrm{GL}(V)$ is a complex reflection group with respect to $T := \{g\}$ and that T is "normalized" as in (5.2.9).

Exercise 5.4. Let $W = \langle S \rangle \subset \mathrm{GL}(V)$ be a finite Coxeter group as in Definition 1.1.3. Let $T = \{wsw^{-1} \mid s \in S, w \in W\}$. We may consider W as a subgroup of $\mathrm{GL}(V_\mathbb{C})$, where $V_\mathbb{C}$ denotes the complexified space. We have already seen in (5.3.1) that the elements of T are complex reflections. Let $J \subseteq S$.

(a) Show that T contains all complex reflections in W.

(b) Show that $|T \cap W_J| = l(w_J)$ where $w_J \in W_J$ is the longest element.

[*Hint.* (a) Let $g \in W$ be any complex reflection. First note that g must have order 2 (since the character of $V_\mathbb{C}$ is real-valued). Assume, if possible, that $g \notin T$ and define $T' \subseteq W$ to be the union of T and all conjugates of g. Then W is also a complex reflection group with respect to T' and the assumptions of Theorem 5.2.10 are still satisfied. Hence we also have $b_{\delta_G} = |T'|$. (b) The elements of $T \cap W_J$ are complex reflections in W_J; hence we must have $T \cap W_J = \{wsw^{-1} \mid s \in J, w \in W_J\}$, by (a). By Exercise 1.14, $|T \cap W'|$ is the number of positive roots of W_J. It remains to use Proposition 1.5.1.]

Exercise 5.5. Let $\zeta \in \mathbb{C}$ be a primitive mth root of unity and let $j \geq 1$ be an integer such that $2j \leq m$. Show that we have

$$\sum_{i=1}^{m} \frac{\zeta^{ij}}{(1-\zeta^i q)(1-\zeta^{-i}q)} = mq^j + \text{combination of higher powers of } q.$$

[*Hint.* The sum on the left-hand side equals

$$\zeta^{ij}\Big(\sum_{a\geq 0} \zeta^{ia} q^a\Big)\Big(\sum_{b \geq 0} \zeta^{-ib} q^b\Big) = \sum_{d \geq 0}\Big(\sum_{a=0}^{d} \zeta^{i(2a-d+j)}\Big) q^d.$$

Fix some $d \leq j$. Then our assumption on j implies $0 \leq 2a-d+j \leq m$. Moreover, we actually have $2a-d+j < m$ if $2j < m$. Check that $\sum_{i=1}^m \zeta^{i(2a-d+j)} = 0$ unless we have either $d = j$ and $a = 0$ or $a = d = j = m/2$.]

Exercise 5.6. This exercise is about the dominance order on partitions (see Definition 5.4.6) and the a-invariants (see (5.4.2)). Let $\lambda, \mu \vdash n$.

(a) Show that if $\lambda \trianglelefteq \mu$, then $a(\lambda) \geq a(\mu)$, with equality only for $\lambda = \mu$.
(b) Show that we have $\lambda \trianglelefteq \mu$ if and only if $\mu^* \trianglelefteq \lambda^*$.

[*Hint.* For (b), see [Macdonald 1995, (I.1.11)]. To prove (a), let $\lambda = [\lambda_1, \ldots, \lambda_r]$ and $\mu = [\mu_1, \ldots, \mu_s]$. To simplify notation, assume that $r = s$ by adding to λ or μ some parts equal to 0, if necessary. Define $r \times r$ matrices $A = (a_{ij})$ and $B = (b_{ij})$ with $a_{ij} = \lambda_j$ and $b_{ij} = \mu_j$. Then, by definition, we have $a(\lambda) = \sum_{i<j} a_{ij}$ and $a(\mu) = \sum_{i<j} b_{ij}$. Moreover, the assumption means that, for each i, we have $\sum_{j \leq i} a_{ij} \leq \sum_{j \leq i} b_{ij}$. Then use the fact that $\sum_j a_{ij} = \sum_j b_{ij} = n$.]

Exercise 5.7. This exercise is used in Lemma 5.5.5; it is due to [Lusztig 1977, (2.4.2)]. We keep the notation of Section 5.5. Recall that $W_n = N_n \rtimes H_n$ where $N_n = \langle t_0, t_1, \ldots, t_{n-1} \rangle$ is elementary abelian of order 2^n and $H_n \cong \mathfrak{S}_n$. Fix $h \in H_n$ and show that

$$\frac{1}{2^n} \sum_{y \in N_n} \frac{1}{\det(\mathrm{id}_{\mathbb{R}^n} - qhy)} = \frac{1}{\det(\mathrm{id}_{\mathbb{R}^n} - q^2h)},$$

$$\frac{1}{2^n} \sum_{y \in N_n} \frac{\varepsilon'(y)}{\det(\mathrm{id}_{\mathbb{R}^n} - qhy)} = \frac{q^n}{\det(\mathrm{id}_{\mathbb{R}^n} - q^2h)}.$$

[*Hint.* Reduce to the case where h is an n-cycle in $\mathfrak{S}_n$, in which case a direct computation shows that $\det(\mathrm{id}_{\mathbb{R}^n} - qhy) = 1 - \varepsilon'(y)q^n$.]

Exercise 5.8. Let $\lambda \vdash n$ and $\chi_\lambda \in \mathrm{Irr}(\mathfrak{S}_n)$ be as in Definition 5.4.4. Let $w_{[n]} \in \mathfrak{S}_n$ be an element of cycle type $[n]$. Show that

$$\chi_\lambda(w_{[n]}) = \begin{cases} (-1)^d & \text{if } \lambda = [n-d, 1^d] \text{ for some } 0 \leq d \leq n-1, \\ 0 & \text{otherwise.} \end{cases}$$

[*Hint.* Regard $\mathfrak{S}_n$ as a Coxeter group in the usual way. Then, by Proposition 5.4.12, $\chi_{[n-d,1^d]}$ is the character of the dth exterior power of the natural module for $\mathfrak{S}_n$. Now use Remark 5.1.6 and the formulas in (3.4.3) to compute the value of that character on $w_{[n]}$. The assertion about the remaining characters then follows from the second orthogonality relations for the characters of $\mathfrak{S}_n$ and the fact that the centralizer of $w_{[n]}$ in $\mathfrak{S}_n$ has order n.]

Exercise 5.9. Let $n \geq 1$ and $W_n \subseteq GL_n(\mathbb{R})$ be a Coxeter group of type B_n as in Section 5.5. Let λ, μ be a pair of partitions with $|\lambda|+|\mu| = n$ and $\chi_{(\lambda,\mu)} \in \mathrm{Irr}(W_n)$ be the corresponding irreducible character. Let $w_{\mathrm{cox}} = ts_1 \cdots s_{n-1} \in W_n$ be a Coxeter element. Show that

$$\chi_{(\lambda,\mu)}(w_{\mathrm{cox}}) = \begin{cases} -(-1)^d & \text{if } \lambda = \varnothing,\ \mu = [n-d, 1^d] \text{ for some } 0 \leq d \leq n-1, \\ (-1)^d & \text{if } \mu = \varnothing,\ \lambda = [n-d, 1^d] \text{ for some } 0 \leq d \leq n-1, \\ 0 & \text{otherwise.} \end{cases}$$

[*Hint.* If $\mu = \varnothing$ or $\lambda = \varnothing$ then $\chi_{(\lambda,\mu)}$ or $\varepsilon' \otimes \chi_{(\lambda,\mu)}$ is pulled back from an irreducible character of $\mathfrak{S}_n$ under the projection $W_n \to H_n \cong \mathfrak{S}_n$. Then use the results in Exercise 5.8 and the fact that $C_{W_n}(w_{\text{cox}})$ has order $2n$.]

Exercise 5.10. Let $\mu \vdash n$ and $x_1, \ldots, x_N$ be indeterminates (where $N \geq n$). The corresponding *Newton power sum* is defined by

$$p_\mu := \prod_{i=1}^{N} (x_1^i + \cdots + x_N^i)^{m_i} \in \mathbb{Z}[x_1, \ldots, x_N],$$

where m_i denotes the number of parts of μ which are equal to i. Furthermore, let λ be a composition of n, with parts $\lambda_1, \ldots, \lambda_r$. Then we define $x^\lambda := x_1^{\lambda_1} \cdots x_N^{\lambda_N}$, where we set $\lambda_j = 0$ for $j > r$. Show that we have

$$p_\mu = \sum_\lambda \pi_\lambda(w_\mu)\, x^\lambda,$$

where the sum is over all compositions λ of n, π_λ is the corresponding permutation character as in Definition 5.4.4, and $w_\mu \in \mathfrak{S}_n$ has cycle type μ.

The above "generating function" is a first step towards Frobenius's classical approach to the character theory of $\mathfrak{S}_n$; see [Ledermann 1987, Chap. 4].

[*Hint.* An explicit formula for $\pi_\lambda(w_\mu)$ is provided by (2.4.8) in combination with Proposition 3.1.8 (we may assume that w_μ has minimal length in its class). Then check that we obtain the same formula for the coefficients in the expansion of p_μ as a linear combination of monomials. See also [Ledermann 1987, §4.6].]

Exercise 5.11. Let λ be a composition of n and π_λ be the corresponding permutation character of $\mathfrak{S}_n$, as in Definition 5.4.4. Let $f \geq 1$ and denote by $f\lambda$ the composition of fn obtained by multiplying all parts of λ by f. (A similar notation will be used for partitions of n.) Show that we have

$$\pi_{f\lambda}(w_{f\mu}) = \pi_\lambda(w_\mu) \qquad \text{for any } \mu \vdash n,$$

where $w_\mu \in \mathfrak{S}_n$ has cycle type μ and $w_{f\mu} \in \mathfrak{S}_{fn}$ has cycle type $f\mu$.

[*Hint.* Use the set-up and the notation of Exercise 5.10, where $N \geq fn$. First note that $p_{f\mu} = \prod_{i=1}^m (x_1^{fi} + \cdots + x_N^{fi})^{m_i} = p_\mu(x_1^f, \ldots, x_N^f)$. Furthermore, write $f\lambda = (f\lambda_1, \ldots, f\lambda_r, 0, \ldots, 0)$. Then $x^{f\lambda} = x_1^{f\lambda_1} \cdots x_r^{f\lambda_r}$ and so the coefficient of $x^{f\lambda}$ in $p_{f\mu}$ is the same as that of x^λ in p_μ.]

6
Parabolic subgroups and induced characters

In Chapter 5, we determined the irreducible characters for each finite Coxeter group W. For the exceptional types, this was done using some explicit computations, which yield the complete character tables. For the classical types, we described explicit constructions which yield parametrizations of the irreducible characters in terms of partitions or pairs of partitions.

Now we are interested in finding systematic ways (valid for all types) of describing the irreducible characters of W in terms of the induction of characters from parabolic subgroups. Of course, we cannot expect to obtain a nice theory when we just look at all parabolic subgroups and all constituents of all induced characters. However, it turns out that we do obtain interesting results when we impose restrictions either on the choice of the characters that are induced or on the irreducible constituents that we consider in a given induced character.

On the one hand, this leads to the concept of characters of "parabolic type" which was developed by Benson, Curtis, Iwahori and Kilmoyer. Here, we only induce the trivial characters of the various parabolic subgroups. It turns out that, with very few and explicitly known exceptions, every irreducible character of W is uniquely determined by its multiplicities in these induced characters. Following [Benson and Curtis 1972], these results will be established in Sections 6.2 and 6.3.

On the other hand, we follow Lusztig's idea of attaching "$\mathfrak{a}$-invariants" to the irreducible characters and considering the corresponding "truncated induction" of characters. For the classical types, this relies on a labelling of the irreducible characters in terms of Lusztig's "symbols", which will be introduced in Section 6.4. Finally, in Section 6.5, we consider the $\mathfrak{a}$-invariants, the truncated induction and the corresponding notion of "families" in general, thereby giving a purely axiomatic treatment of a number of results in [Lusztig 1984, Chap. 4].

The above results lead to convenient characterizations of the irreducible characters of W, which is particularly important for applications to the character theory of a finite reductive group G with Weyl group W. For example, in [Lusztig 1984, Prop. 12.6], the results of Benson and Curtis are used to establish a property of the parametrization of the unipotent characters of G. On the other hand, the $\mathfrak{a}$-invariants occur in the degree formulas for the irreducible characters of G and in connection with the so-called "unipotent support" for these characters; see [Geck and Malle 2000] and the references therein.

6.1 INDUCTION OF CHARACTERS IN CLASSICAL TYPES

In this section we study the induction of characters of Coxeter groups of classical type. We will see that this can be reduced to statements about the symmetric group and then, ultimately, to the combinatorics of partitions. The following definition introduces the most important notion in this context.

6.1.1. Definition. Let $n \geq 0$ and consider the symmetric group $\mathfrak{S}_n$. (We set $\mathfrak{S}_0 = \{1\}$.) Let $k, l \geq 0$ be such that $n = k + l$. Then we have an embedding $\mathfrak{S}_k \times \mathfrak{S}_l \subseteq \mathfrak{S}_n$, where the first factor is regarded as a permutation group on $\{1, \ldots, k\}$ and the second factor is regarded as a permutation group on $\{k+1, \ldots, n\}$. Given $\lambda \vdash k$ and $\mu \vdash l$, we have corresponding irreducible characters $\chi_\lambda \in \mathrm{Irr}(\mathfrak{S}_k)$ and $\chi_\mu \in \mathrm{Irr}(\mathfrak{S}_l)$. Denoting by $\boxtimes$ the outer tensor product of characters, we have

$$\mathrm{Ind}_{\mathfrak{S}_k \times \mathfrak{S}_l}^{\mathfrak{S}_n}\left(\chi_\lambda \boxtimes \chi_\mu\right) = \sum_{\nu \vdash n} c_{\lambda\mu}^{\nu} \chi_\nu \qquad \text{where } c_{\lambda\mu}^{\nu} \in \mathbb{N}_0.$$

The $c_{\lambda\mu}^{\nu}$ are called *Littlewood–Richardson (LR) coefficients.*

A combinatorial rule for computing the LR coefficients will be explained without proof in (6.1.6). For the moment, we do not need to know precisely how this works. The following result establishes a basic property of the LR coefficients.

6.1.2. Lemma. *Let $n, k, l \geq 0$ be such that $n = k + l$. Let $\lambda \vdash k$, $\mu \vdash l$ and set $\nu_0 := \lambda + \mu$; see Exercise 6.4. Then we have $c_{\lambda\mu}^{\nu_0} = 1$ and $c_{\lambda\mu}^{\nu} = 0$ unless $\nu \trianglelefteq \nu_0$.*

Proof. By Definition 5.4.4, χ_λ and χ_μ are obtained by applying the j-induction to the sign characters of the Young subgroups $\mathfrak{S}_{\lambda^*} \subseteq \mathfrak{S}_k$ and $\mathfrak{S}_{\mu^*} \subseteq \mathfrak{S}_l$, respectively. Hence, using Exercise 5.2, we also obtain that

$$\mathrm{Ind}_{\mathfrak{S}_{\lambda^*} \times \mathfrak{S}_{\mu^*}}^{\mathfrak{S}_n}\left(\varepsilon_{\mathfrak{S}_{\lambda^*}} \boxtimes \varepsilon_{\mathfrak{S}_{\mu^*}}\right) = \chi_{\nu_0} + \text{combination of } \chi_\nu \text{ with } a(\nu) > a(\nu_0),$$

where $\nu_0 \vdash n$ is determined by the condition that $\mathfrak{S}_{\nu_0^*}$ is conjugate to $\mathfrak{S}_{\lambda^*} \times \mathfrak{S}_{\mu^*}$ inside $\mathfrak{S}_n$. This means that the parts of ν_0^* are given by taking all the parts of λ^* and μ^* and arranging them in decreasing order. So we have $\nu_0 = \lambda + \mu$ by Exercise 6.4(b). Finally, by Theorem 5.4.7, if χ_ν occurs with non-zero multiplicity in the above induced character, then we have $\nu \trianglelefteq \nu_0$, as desired. ☐

We now extend the above formulas to Coxeter groups of type B_n. We will work with the group W_n and the generators $\{t, s_1, \ldots, s_{n-1}\}$ as defined in Section 5.5; we set $W_0 = \{1\}$ by convention. Recall from Section 5.5 that every irreducible character of W_n is of the form $\chi_{(\lambda,\mu)}$, where λ, μ are partitions with $|\lambda| + |\mu| = n$.

6.1.3. Lemma. *Let $n \geq 1$ and $k, l \geq 0$ be integers such that $n = k + l$. Let (λ_1, λ_2) and (μ_1, μ_2) be pairs of partitions with $|\lambda_1| + |\lambda_2| = k$ and $|\mu_1| + |\mu_2| = l$. Then, using the diagonal embedding $W_k \times W_l \subseteq W_n$ (as in Section 5.5), we have*

$$\mathrm{Ind}_{W_k \times W_l}^{W_n}\left(\chi_{(\lambda_1,\lambda_2)} \boxtimes \chi_{(\mu_1,\mu_2)}\right) = \sum_{(\nu_1,\nu_2)} c_{\lambda_1\mu_1}^{\nu_1} c_{\lambda_2\mu_2}^{\nu_2} \chi_{(\nu_1,\nu_2)},$$

where the sum runs over all pairs of partitions (ν_1, ν_2) *with* $|\nu_i| = |\lambda_i| + |\mu_i|$ *for* $i = 1, 2$.

Proof. With the notation of (5.5.4), we have

$$\chi_{(\lambda_1,\lambda_2)} = \mathrm{Ind}_{W_{a_1}\times W_{a_2}}^{W_k}\Big(\tilde{\chi}_{\lambda_1} \boxtimes (\varepsilon'_{a_2} \otimes \tilde{\chi}_{\lambda_2})\Big), \quad \text{where } k = a_1 + a_2,\ \lambda_i \vdash a_i,$$

$$\chi_{(\mu_1,\mu_2)} = \mathrm{Ind}_{W_{b_1}\times W_{b_2}}^{W_l}\Big(\tilde{\chi}_{\mu_1} \boxtimes (\varepsilon'_{b_2} \otimes \tilde{\chi}_{\mu_2})\Big), \quad \text{where } l = b_1 + b_2,\ \mu_i \vdash b_i.$$

Thus, using the transitivity of induction and its compatibility with direct products, we see that the induced character that we want to compute is equal to

$$\mathrm{Ind}_{W_{a_1}\times W_{a_2}\times W_{b_1}\times W_{b_2}}^{W_n}\Big(\tilde{\chi}_{\lambda_1} \boxtimes (\varepsilon'_{a_2} \otimes \tilde{\chi}_{\lambda_2}) \boxtimes \tilde{\chi}_{\mu_1} \boxtimes (\varepsilon'_{b_2} \otimes \tilde{\chi}_{\mu_2})\Big).$$

Now the subgroups $W_{a_1} \times W_{a_2} \times W_{b_1} \times W_{b_2}$ and $W_{a_1} \times W_{b_1} \times W_{a_2} \times W_{b_2}$ are conjugate in W_n. Hence we can also express the above character as

$$\begin{aligned}
&\mathrm{Ind}_{W_{a_1}\times W_{b_1}\times W_{a_2}\times W_{b_2}}^{W_n}\Big(\tilde{\chi}_{\lambda_1} \boxtimes \tilde{\chi}_{\mu_1} \boxtimes (\varepsilon'_{a_2} \otimes \tilde{\chi}_{\lambda_2}) \boxtimes (\varepsilon'_{b_2} \otimes \tilde{\chi}_{\mu_2})\Big)\\
&\quad = \mathrm{Ind}_{W_{a_1}\times W_{b_1}\times W_{a_2}\times W_{b_2}}^{W_n}\Big(\tilde{\chi}_{\lambda_1} \boxtimes \tilde{\chi}_{\mu_1} \boxtimes (\varepsilon'_{a_2+b_2} \otimes (\tilde{\chi}_{\lambda_2} \boxtimes \tilde{\chi}_{\mu_2}))\Big),
\end{aligned}$$

where we use the fact that $\varepsilon'_{a_2} \boxtimes \varepsilon'_{b_2} = \varepsilon'_{a_2+b_2}$. Now, setting $n_1 = a_1 + b_1$ and $n_2 = a_2 + b_2$ and using again the transitivity of induction and its compatibility with direct products, we can rewrite the above expression as

$$\mathrm{Ind}_{W_{n_1}\times W_{n_2}}^{W_n}\Big(\tilde{\psi}_{(\lambda_1,\mu_1)} \boxtimes (\varepsilon_{n_2} \otimes \tilde{\psi}_{(\lambda_2,\mu_2)})\Big),$$

where the tilde means (as usual) pulling back a character of $\mathfrak{S}_{n_i}$ via the natural map $W_{n_i} \to \mathfrak{S}_{n_i}$ (for $i = 1, 2$) and where we set

$$\psi_{(\lambda_i,\mu_i)} = \mathrm{Ind}_{\mathfrak{S}_{a_i}\times\mathfrak{S}_{b_i}}^{\mathfrak{S}_{n_i}}\Big(\chi_{\lambda_i} \boxtimes \chi_{\mu_i}\Big) \qquad \text{for } i = 1, 2.$$

All that remains is to use the defining formula in Definition 6.1.1. □

6.1.4. Lemma. *Let* $n \geq 1$ *and consider the subgroup* $\mathfrak{S}_n \cong H_n \subseteq W_n$. *Let* $\nu \vdash n$ *and* $\chi_\nu \in \mathrm{Irr}(H_n)$ *be the corresponding irreducible character. Then*

$$\mathrm{Ind}_{H_n}^{W_n}(\chi_\nu) = \sum_{\lambda,\mu} c_{\lambda\mu}^{\nu}\, \chi_{(\lambda,\mu)}.$$

Proof. Let $\chi_{(\lambda,\mu)} \in \mathrm{Irr}(W_n)$. By Frobenius reciprocity, it is enough to compute the multiplicity of χ_ν in the restriction of $\chi_{(\lambda,\mu)}$ to H_n. Let us compute that restriction using Mackey's formula. With the notation of (5.5.4), we have

$$\chi_{(\lambda,\mu)} = \mathrm{Ind}_{W_a\times W_b}^{W_n}\Big(\tilde{\chi}_\lambda \boxtimes (\varepsilon'_b \otimes \tilde{\chi}_\mu)\Big).$$

Now recall that we have a semidirect product decomposition $W_n = N_n \rtimes H_n$. Similarly, we have $W_a = N_a \rtimes H_a$ and $W_b = N_b \rtimes H_b$. Now note that $N_n \subseteq$

$W_a \times W_b$. So we have $W_n = H_n(W_a \times W_b)$ and $H_n \cap (W_a \times W_b) = H_a \times H_b$. Hence Mackey's formula shows that

$$\mathrm{Res}^{W_n}_{H_n}(\chi_{(\lambda,\mu)}) = \mathrm{Ind}^{H_n}_{H_a \times H_b}\left(\chi_\lambda \boxtimes \chi_\mu\right).$$

So the desired multiplicity is equal to the LR coefficient $c^{\nu}_{\lambda\mu}$. □

For type D_n, we can use the results in Section 5.6 and some Clifford theory in order to obtain formulas for the induction of characters from parabolic subgroups. We will consider this in more detail in Section 6.4.

As a first application of the above results, we obtain the following "multiplicity 1" statement for the types B_n and D_n. (For $\mathfrak{S}_n$, see Theorem 5.4.7.) We denote the restriction of $\chi_{(\lambda,\mu)} \in \mathrm{Irr}(W_n)$ to W'_n by $\chi'_{(\lambda,\mu)}$. We have $\chi'_{(\lambda,\mu)} \in \mathrm{Irr}(W'_n)$ if $\lambda \neq \mu$ and $\chi'_{(\lambda,\lambda)} = \chi_{(\lambda,+)} + \chi_{(\lambda,-)}$ with $\chi_{(\lambda,\pm)} \in \mathrm{Irr}(W'_n)$ otherwise.

To simplify notation, for any finite group G and a subgroup $H \subseteq G$, we denote by 1_H^G the character obtained by inducing the trivial character from H to G.

6.1.5. Proposition (Benson–Curtis). *Let $n \geq 1$ and (λ, μ) be a pair of partitions with $|\lambda| + |\mu| = n$. Let $\nu_0 := \lambda + \mu$; see Exercise 6.4. Then we have*

$$\left(\chi_{(\lambda,\mu)}, 1^{W_n}_{H_\nu}\right)_{W_n} = \left(\chi'_{(\lambda,\mu)}, 1^{W'_n}_{H_\nu}\right)_{W'_n} = \begin{cases} 1 & \text{if } \nu = \nu_0, \\ 0 & \text{if } \nu \not\trianglelefteq \nu_0, \end{cases} \tag{a}$$

where $\nu \vdash n$ and H_ν is the corresponding parabolic subgroup in $H_n \subseteq W'_n \subseteq W_n$. Moreover, if $\lambda = \mu$, set $H_+ := H_{\nu_0}$ and $H_- := tH_+t \subseteq W'_n$. Then we have

$$\left(\chi_{(\lambda,\pm)}, 1^{W'_n}_{H_\pm}\right)_{W'_n} = 1 \qquad \text{and} \qquad \left(\chi_{(\lambda,\pm)}, 1^{W'_n}_{H_\mp}\right)_{W'_n} = 0. \tag{b}$$

In particular, every irreducible character of W_n and of W'_n is of "parabolic type", in the sense of Definition 6.3.1 below.

Proof. Let ν be any partition of n. By Theorem 5.4.7, we can write

$$1^{\mathfrak{S}_n}_{\mathfrak{S}_\nu} = \sum_{\nu \trianglelefteq \nu'} \kappa_{\nu'\nu}\chi_{\nu'}, \qquad \text{where } \kappa_{\nu'\nu} \in \mathbb{N}_0, \quad \kappa_{\nu\nu} = 1.$$

Using the transitivity of induction and Lemma 6.1.4, we obtain

$$\left(\chi_{(\lambda,\mu)}, 1^{W_n}_{H_\nu}\right)_{W_n} = \sum_{\nu \trianglelefteq \nu'} \kappa_{\nu'\nu}\left(\chi_{(\lambda,\mu)}, \mathrm{Ind}^{W_n}_{H_n}(\chi_{\nu'})\right)_{W_n} = \sum_{\nu \trianglelefteq \nu'} \kappa_{\nu'\nu}\, c^{\nu'}_{\lambda\mu}.$$

Now, by Lemma 6.1.2, we also have $c^{\nu'}_{\lambda\mu} = 0$ unless $\nu' \trianglelefteq \nu_0$. Hence, the above sum is 0 unless $\nu \trianglelefteq \nu_0$. Using Lemma 6.1.2 once more, we obtain

$$\left(\chi_{(\lambda,\mu)}, 1^{W_n}_{H_\nu}\right)_{W_n} = \sum_{\nu \trianglelefteq \nu' \trianglelefteq \nu_0} k_{\nu'\nu} c^{\nu'}_{\lambda\mu} = \begin{cases} 1 & \text{if } \nu = \nu_0, \\ 0 & \text{if } \nu \not\trianglelefteq \nu_0. \end{cases}$$

Now consider the subgroup $W'_n \subseteq W_n$ of type D_n. Since $\chi'_{(\lambda,\mu)}$ is the restriction of $\chi_{(\lambda,\mu)}$, the transitivity of induction and Frobenius reciprocity imply that

$$\left(\chi'_{(\lambda,\mu)}, 1_{H_\nu}^{W'_n}\right)_{W'_n} = \left(\chi_{(\lambda,\mu)}, 1_{H_\nu}^{W_n}\right)_{W_n},$$

as desired. Now assume that $\lambda = \mu$ and consider the subgroups $H_\pm$. Since $\chi'_{(\lambda,\lambda)}$ is invariant under conjugation with t, the above equations already show that

$$\left(\chi_{(\lambda,+)} + \chi_{(\lambda,-)}, 1_{H_\pm}^{W'_n}\right)_{W'_n} = \left(\chi'_{(\lambda,\lambda)}, 1_{H_\pm}^{W'_n}\right)_{W'_n} = \left(\chi'_{(\lambda,\lambda)}, 1_{H_{\nu_0}}^{W'_n}\right)_{W'_n} = 1.$$

Now $\chi_{(\lambda,-)}$ and $\chi_{(\lambda,+)}$ are obtained from each other via conjugation with t. Hence, in order to complete the proof, it is enough to show that

$$\left(\chi_{(\lambda,+)}, 1_{H_+}^{W'_n}\right)_{W'_n} \neq 0.$$

This is seen as follows. By Proposition 5.6.3 (note that $\nu_0 = 2\lambda$), we have

$$\mathrm{Ind}_{H_n}^{W'_n}(\chi_{\nu_0}) = \chi_{(\lambda,+)} + \text{combination of } \chi \in \mathrm{Irr}(W'_n) \text{ with } b_\chi > b_{(\lambda,\lambda)}.$$

So all that remains is to use the transitivity of induction and Theorem 5.4.7, which states that χ_{ν_0} occurs with multiplicity 1 in $1_{H_+}^{H_n}$. □

In order to obtain more precise results about the decomposition of induced characters into irreducible characters, we need an explicit combinatorial rule for computing the LR coefficients. Using the transition between irreducible characters of $\mathfrak{S}_n$ and symmetric functions provided by Theorem 5.4.13, such a rule is equivalent to a multiplication rule for Schur functions. Then the results in [Macdonald 1995, I, §9] yield:

6.1.6. The Littlewood–Richardson rule. *Let λ, μ, ν be partitions such that $|\nu| = |\lambda| + |\mu|$. Then $c_{\lambda\mu}^{\nu}$ equals the number of tableaux $\mathbb{T}$ of shape $\nu - \lambda$ and weight μ such that $w(\mathbb{T})$ is a lattice permutation.*

We have to define the notions involved in the above statement. First, if λ, ν are any partitions, we write $\lambda \subseteq \nu$ if the following holds: assuming that λ has parts $\lambda_1 \geq \cdots \geq \lambda_r > 0$ and ν has parts $\nu_1 \geq \cdots \geq \nu_s > 0$, we require that $r \leq s$ and $\lambda_i \leq \nu_i$ for $1 \leq i \leq r$. Assume now that this is the case. If we identify λ and ν in the usual way with their Young diagrams (see (5.4.2)), this means that the diagram of λ is contained in the diagram of ν. Removing the boxes of ν which belong to λ, we obtain a *skew diagram* which we denote by $\nu - \lambda$. For example:

$\nu - \lambda$:

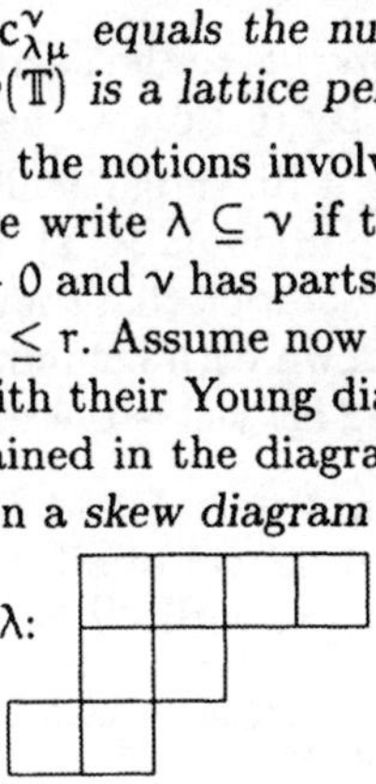

$\nu = [5, 3, 2]$

$\lambda = [1, 1]$

(Note that, in general, a skew diagram need not be connected.) By definition, a *tableau** $\mathbb{T}$ of shape $\nu - \lambda$ is a filling of the boxes of $\nu - \lambda$ with positive integers such that the entries are weakly increasing from the left to the right along each row and strictly increasing down each column. (For a more formal definition of tableaux, see [Macdonald 1995, p. 5].) The *weight* of such a tableau $\mathbb{T}$ is the sequence $(a_1, a_2, \ldots)$, where a_i denotes the number of occurrences of i in $\mathbb{T}$. When we say that the weight is equal to a partition, we implicitly mean that $a_1 \geq a_2 \geq \ldots$. The *word* $w(\mathbb{T})$ is the sequence obtained by reading the entries of $\mathbb{T}$ from right to left in successive rows, starting with the top row. For example:

$\mathbb{T}$:

	1	1	2	3
	2	3		
1	4			

weight of $\mathbb{T}$: $(3, 2, 2, 1)$

$w(\mathbb{T}) = (3, 2, 1, 1, 3, 2, 4, 1)$

Finally, any sequence $(a_1, a_2, \ldots, a_N)$ with $a_i \in \{1, \ldots, n\}$ is called a *lattice permutation* if, for $1 \leq r \leq N$ and $1 \leq i \leq n-1$, the number of occurrences of i in $(a_1, \ldots, a_r)$ is not less than the number of occurrences of $i+1$. In the above example, $w(\mathbb{T})$ is not a lattice permutation, since the above condition fails for $r = 1$ and $i = 2$.

The Littlewood–Richardson rule is an extremely powerful tool. For example, we can use it to obtain a combinatorial formula for computing the Kostka numbers; see Exercise 6.5. We now discuss a number of further consequences.

6.1.7. Corollary (Pieri's Rule). *Let $n, k, l \geq 0$ be such that $n = k + l$. Then, for any $\lambda \vdash k$, we have*

$$\mathrm{Ind}_{\mathfrak{S}_k \times \mathfrak{S}_l}^{\mathfrak{S}_n}\left(\chi_\lambda \boxtimes 1_{\mathfrak{S}_l}\right) = \sum_{\nu} \chi_\nu,$$

where the sum is over all partitions $\nu \vdash n$ whose Young diagram is obtained from that of λ by adding l boxes, with no two boxes in the same column.

Proof. The trivial character of $\mathfrak{S}_l$ corresponds to the partition $\mu = [l]$. Hence we must compute the LR coefficients $c_{\lambda[l]}^{\nu}$. By (6.1.6), we have to consider certain skew tableaux $\mathbb{T}$ of shape $\nu - \lambda$ and weight $[l]$. In particular, this means that all entries of $\mathbb{T}$ are equal to 1 and so the condition on $w(\mathbb{T})$ is empty. On the other hand, the entries of $\mathbb{T}$ have to satisfy the condition that they are weakly increasing across each row and strictly increasing down each column. It follows that there is at most one box in each column of $\mathbb{T}$. This yields the desired statement. $\square$

6.1.8. The branching rule for $\mathfrak{S}_n$. Let $n \geq 1$ and $\lambda \vdash n$. Assume that λ has parts $\lambda_1 \geq \cdots \geq \lambda_r > 0$. Let $I(\lambda)$ be the set of all $d \in \{1, \ldots, r\}$ such that $d = r$ or $d < r$ and $\lambda_d > \lambda_{d+1}$. For $d \in I(\lambda)$, let $\lambda^{(d)}$ be the partition of $n - 1$

*Here we follow the conventions in [Fulton 1997]. Thus, a tableau is what is often called a semistandard tableau, for example, in [Macdonald 1995].

which is obtained from λ by decreasing λ_d by 1. (If $d = r$ and $\lambda_r = 1$, we simply delete the part λ_r from λ.) Then the partitions $\lambda^{(d)}$ are precisely the partitions of $n-1$ whose Young diagram can be obtained from that of λ by removing one box. Hence, by Frobenius reciprocity and Pieri's rule (with $k = n-1$), we have

$$\mathrm{Res}^{\mathfrak{S}_n}_{\mathfrak{S}_{n-1}}(\chi_\lambda) = \sum_{d \in I(\lambda)} \chi_{\lambda^{(d)}}.$$

Using the definition in (5.4.2), we have $a(\lambda) = a(\lambda^{(d)}) + d - 1$ for all $d \in I(\lambda)$.

6.1.9. Pieri's rule for type B_n. Let $n = k+l \geq 1$ where $k, l \geq 0$. Consider the parabolic subgroup $W_k \times H_l \subseteq W_n$, where W_k is generated by $\{t, s_1, \ldots, s_{k-1}\}$ and H_l is generated by $\{s_{k+1}, \ldots, s_{n-1}\}$. (We have $W_k = \{1\}$ or $H_l = \{1\}$ if $k = 0$ or $l = 0$, respectively.) Let $\chi_{(\lambda',\mu')} \in \mathrm{Irr}(W_k)$. Now, by Definition 6.1.1, we have

$$c^{[n]}_{\lambda\mu} = \begin{cases} 1 & \text{if } \lambda = [d] \text{ and } \mu = [n-d] \text{ for some } 0 \leq d \leq n, \\ 0 & \text{otherwise.} \end{cases}$$

Using the transitivity of induction, Lemma 6.1.4 and Lemma 6.1.3, we obtain

$$\begin{aligned} \mathrm{Ind}^{W_n}_{W_k \times H_l}\left(\chi_{(\lambda',\mu')} \boxtimes 1_{H_l}\right) &= \sum_{d=0}^{l} \mathrm{Ind}^{W_n}_{W_k \times W_l}\left(\chi_{(\lambda',\mu')} \boxtimes \chi_{([d],[n-d])}\right) \\ &= \sum_{(\lambda,\mu)} \Big(\sum_{d=0}^{l} c^{\lambda}_{\lambda'[d]}\, c^{\mu}_{\mu'[l-d]}\Big)\chi_{(\lambda,\mu)} = \sum_{(\lambda,\mu)} \chi_{(\lambda,\mu)}, \end{aligned}$$

where (λ, μ) runs over all pairs of partitions such that, for some $0 \leq d \leq l$, the Young diagram of λ (or of μ) can be obtained from that of λ' (or of μ') by adding d (or $l-d$) boxes, with no two in the same column. In particular, if $l = 1$, we obtain an analogue for the *branching rule*:

$$\mathrm{Res}^{W_n}_{W_{n-1}}(\chi_{(\lambda,\mu)}) = \sum_{d \in I(\lambda)} \chi_{(\lambda^{(d)},\mu)} + \sum_{d \in I(\mu)} \chi_{(\lambda,\mu^{(d)})},$$

where we use Frobenius reciprocity and the notation of (6.1.8).

6.2 CENTRAL CHARACTERS IN CLASSICAL TYPES

The aim of this section is to prove a result which allows us to identify the irreducible characters of a Coxeter group of type B_n or D_n in terms of their restrictions to parabolic subgroups of type A; see Theorem 6.2.9. The whole discussion relies on the description of a suitable generating set for the centre of the group algebra of $\mathfrak{S}_n$. We use the following notation.

6.2.1. Let G be any finite group and $\mathrm{Cl}(G)$ be the set of conjugacy classes of G. For $C \in \mathrm{Cl}(G)$, we write $\hat{C} := \sum_{g\in C} g \in \mathbb{C}[G]$. Then, by Example 7.1.8,

$$\{\hat{C} \mid C \in \mathrm{Cl}(G)\} \text{ is a basis of the centre of } \mathbb{C}[G]. \qquad \text{(a)}$$

If $g \in C$, we sometimes write $\hat{g}$ for $\hat{C}$. Each $\chi \in \mathrm{Irr}(G)$ determines a corresponding *central character* ω_χ, which is an algebra homomorphism from the centre of $\mathbb{C}[G]$ to $\mathbb{C}$. The values of ω_χ are given by the following formula (see Example 7.5.4):

$$\omega_\chi(\hat{C}) = \frac{|C|\chi(g)}{\chi(1)} \qquad \text{where } g \in C. \qquad \text{(b)}$$

Now we can state the following result about the centre of $\mathbb{C}[\mathfrak{S}_n]$. To simplify notation, for any $\lambda \vdash n$, we let $C_\lambda \subseteq \mathfrak{S}_n$ be the conjugacy class of elements of cycle type λ and denote by ω_λ the central character corresponding to $\chi_\lambda \in \mathrm{Irr}(\mathfrak{S}_n)$.

6.2.2. Proposition. *Let $n \geq 3$. Then the elements $\hat{C}_{[k]}$ $(1 \leq k \leq n-1)$ generate the centre of $\mathbb{Q}[\mathfrak{S}_n]$ (as a $\mathbb{Q}$-algebra).*

Proof. Consider any $\lambda \vdash n$, with parts $\lambda_1 \geq \cdots \geq \lambda_r > 1$. Then we can write:

$$\hat{C}(\lambda) := \hat{C}_{[\lambda_1]} \cdots \hat{C}_{[\lambda_r]} = \sum_{\mu\vdash n} m_{\lambda\mu}\hat{C}_\mu, \qquad (1)$$

where $m_{\lambda\mu} = |\{(w_1,\ldots,w_r) \in C_{[\lambda_1]} \times \cdots \times C_{[\lambda_r]} \mid w_1\cdots w_r = w\}|$, for a fixed $w \in C_\mu$. Let $n_1(\lambda)$ be the number of parts λ_i which equal 1. Then we claim:

$$m_{\lambda\mu} \neq 0 \quad \Rightarrow \quad \lambda = \mu \quad \text{or} \quad n_1(\mu) > n_1(\lambda). \qquad (2)$$

To prove this, for each part λ_i of λ, let $w_i \in \mathfrak{S}_n$ be a cycle of length λ_i. Let $\mu \vdash n$ be the cycle type of $w := w_1\cdots w_r$. If all cycles w_i are disjoint, then we clearly have $\mu = \lambda$. Now assume that this is not the case. For each i, denote by I_i the set of points which are moved by w_i. Then, noting that $n_1(\mu)$ is the number of fixed points of w, we have $n - n_1(\mu) \leq |\bigcup_i I_i| \leq n - n_1(\lambda)$. But the second equality is strict if not all cycles w_i are disjoint. Thus, $n_1(\mu) > n_1(\lambda)$ and (2) is proved.

Since we certainly have $m_{\lambda\lambda} \neq 0$ for all $\lambda \vdash n$, the system of equations given by (1) can be inverted recursively using (2), by starting with the partition $\lambda = [1^n]$ and proceeding by decreasing value of $n_1(\lambda)$. Thus, for any $\lambda \vdash n$, we can express $\hat{C}_\lambda$ as a linear combination of $\hat{C}(\lambda)$ and terms $\hat{C}(\mu)$ with $n_1(\mu) > n_1(\lambda)$. Hence, the centre of $\mathbb{Q}[\mathfrak{S}_n]$ is generated by $\hat{C}_{[k]}$ for $1 \leq k \leq n$. But, since $n \geq 3$, $\hat{C}_{[n]}$ can be omitted from the generating set. Indeed, we have an expression

$$\underbrace{\hat{C}_{[2]}\cdots\hat{C}_{[2]}}_{n-1 \text{ factors}} = \sum_{\mu\vdash n} m'_\mu \hat{C}_\mu, \qquad \text{where } m'_\mu \in \mathbb{N}_0.$$

Since an n-cycle can be expressed as a product of $n-1$ elements of $C_{[2]}$, we have $m'_{[n]} \neq 0$. Furthermore, we have seen above that all $\hat{C}_\mu$ with $\mu \neq [n]$ can be expressed as a linear combination of products of terms $\hat{C}_{[k]}$ where $1 \leq k \leq n-1$. Hence the same also holds for $\hat{C}_{[n]}$. □

6.2.3. Corollary (Benson–Curtis). *Let $n \geq 3$. Then, if $\chi, \chi' \in \mathrm{Irr}(\mathfrak{S}_n)$ are such that $\omega_\chi(\hat{C}_{[k]}) = \omega_{\chi'}(\hat{C}_{[k]})$ for all $2 \leq k \leq n-1$, we have $\chi = \chi'$.*

Proof. Assume that the above hypothesis is satisfied. Trivially, it also holds for $k = 1$. So, by Proposition 6.2.2, ω_χ and $\omega_{\chi'}$ have the same value on a set of algebra generators for the centre of $\mathbb{C}[\mathfrak{S}_n]$. Since the central characters are algebra homomorphisms, we have $\omega_\chi = \omega_{\chi'}$ and so $\chi = \chi'$, as claimed. □

Now we turn to the problem of characterizing the irreducible characters of groups of type B_n and D_n by their restrictions to suitable parabolic subgroups. As usual, we work with the groups $W_n = \langle t, s_1, \ldots, s_{n-1}\rangle$ and $W_n' = \langle u, s_1, \ldots, s_{n-1}\rangle$, where $u = 1$ for $n = 1$ and $u = ts_1t$ for $n \geq 2$; see Section 5.5. In both cases, we have a parabolic subgroup $H_n = \langle s_1, \ldots, s_{n-1}\rangle \cong \mathfrak{S}_n$.

6.2.4. Identifying characters of W_n and W_n': Part 1. Let $n \geq 1$ and χ be an irreducible character of W_n or of W_n'. Then χ is labelled by a pair of partitions (λ, μ). The partition $\nu_0 \vdash n$ such that $\nu_0 = \lambda + \mu$ is determined as follows.

For any $\nu \vdash n$, let $J_\nu \subseteq \{s_1, \ldots, s_{n-1}\}$ be the subset which generates the parabolic subgroup $H_\nu \subseteq H_n$. Note that the Coxeter graph of H_ν is a disjoint union of graphs of type A. Denote by $n(J_\nu, \chi)$ the multiplicity of the trivial character of H_ν in the restriction of χ to H_ν and set

$$a_0 := \min\{a(\nu) \mid \nu \vdash n \text{ such that } n(J_\nu, \chi) \neq 0\}.$$

Then, by Frobenius reciprocity and Proposition 6.1.5(a) (see also Exercise 5.6 for the relation with the a-invariants), $\nu_0 \vdash n$ is the unique partition of n such that $a(\nu_0) = a_0$ and $n(J_{\nu_0}, \chi) \neq 0$. Thus, given the restrictions of χ to all parabolic subgroups of the form H_ν ($\nu \vdash n$), we can determine $\nu_0 = \lambda + \mu$.

Moreover, in type D_n, there are two types of characters: those which are invariant under conjugation with t (under the embedding $W_n' \subseteq W_n$) and those which are not. The latter are precisely the characters χ for which we have $\lambda = \mu$ (see Proposition 5.6.3). By Proposition 6.1.5(b), this is the case if and only if

$$n(J, \chi) \neq n(tJt, \chi) \qquad \text{for some } J \subseteq \{s_1, \ldots, s_{n-1}\}.$$

Furthermore, in this case, $\lambda = \mu$ is determined by $\nu_0 = 2\lambda$. Thus, given all the restrictions of χ to parabolic subgroups of the form H_ν or $tH_\nu t$ ($\nu \vdash n$), we can determine whether or not χ is invariant under conjugation with t and, if this is not so, we can find λ and the sign $\pm$ such that $\chi = \chi_{(\lambda,\pm)}$.

Next, we wish to express the values of the central character corresponding to $\chi_{(\lambda,\mu)} \in \mathrm{Irr}(W_n)$ in terms of the values of the central characters ω_λ and ω_μ.

6.2.5. Central characters for type B_n. Let $n \geq 1$ and (λ, μ) be a pair of partitions with $n = |\lambda| + |\mu|$. With the notation of (5.5.4), the corresponding irreducible character of W_n is given by

$$\chi_{(\lambda,\mu)} = \mathrm{Ind}_{W_a \times W_b}^{W_n}\left(\tilde{\chi}_\lambda \boxtimes (\varepsilon_b' \otimes \tilde{\chi}_\mu)\right) \qquad \text{where } a = |\lambda| \text{ and } b = |\mu|.$$

Now let $w \in W_n$ and let $w_1, \ldots, w_m \in W_a \times W_b$ be representatives of the conjugacy classes of $W_a \times W_b$ which are contained in the W_n-conjugacy class of w. We write each w_i in the form $w_i = w_i^{(a)} w_i^{(b)}$ where $w_i^{(a)} \in W_a$ and $w_i^{(b)} \in W_b$. Furthermore, we also have semidirect product decompositions $W_a = N_a \rtimes H_a$ and $W_b = N_b \rtimes H_b$. The corresponding factorizations of $w_i^{(a)}$ and $w_i^{(b)}$ will be written as $w_i^{(a)} = n_i^{(a)} h_i^{(a)}$ and $w_i^{(b)} = n_i^{(b)} h_i^{(b)}$. Then the explicit induction formula in [Isaacs 1976, p. 64] yields

$$\chi_{(\lambda,\mu)}(w) = \sum_{i=1}^{m} \frac{|C_{W_n}(w)|}{|C_{W_a \times W_b}(w_i)|} \chi_\lambda(h_i^{(a)})\, \varepsilon_b'(n_i^{(b)})\, \chi_\mu(h_i^{(b)}). \qquad \text{(a)}$$

Now denote the central character corresponding to $\chi_{(\lambda,\mu)}$ by $\omega_{(\lambda,\mu)}$. Inserting the identity $\chi_{(\lambda,\mu)}(1) = [W_n : W_a \times W_b]\, \chi_\lambda(1)\, \chi_\mu(1)$ into the above expression, a straightforward computation yields

$$\omega_{(\lambda,\mu)}(\hat{w}) = 2^n \sum_{i=1}^{m} \frac{|C_{H_a}(h_i^{(a)})|\,|C_{H_b}(h_i^{(b)})|}{|C_{W_a}(w_i^{(a)})|\,|C_{W_b}(w_i^{(b)})|} \varepsilon_b'(n_i^{(b)}) \omega_\lambda(\hat{h}_i^{(a)}) \omega_\mu(\hat{h}_i^{(b)}), \quad \text{(b)}$$

where ω_λ and ω_μ are the central characters corresponding to $\chi_\lambda \in \mathrm{Irr}(\mathfrak{S}_a)$ and $\chi_\mu \in \mathrm{Irr}(\mathfrak{S}_b)$, respectively. We now apply the above formula to a number of cases.

6.2.6. Lemma. *Let* $n \geq 1$ *and* $\chi_{(\lambda,\mu)} \in \mathrm{Irr}(W_n)$. *Let* $u = ts_1t$. *Then we have*

$$\omega_{(\lambda,\mu)}(\hat{t}) = |\lambda| - |\mu| \quad \textit{and} \quad \omega_{(\lambda,\mu)}(\widehat{us_1}) = \frac{1}{2}\left((|\lambda| - |\mu|)^2 - n\right) \quad (n \geq 2).$$

Proof. In the setting of (6.2.5), assume that $w \in N_n$. Then all w_i also lie in N_n, since N_n is a normal subgroup. So we have $h_i^{(a)} = 1$ and $h_i^{(b)} = 1$ for all i and the formula in (6.2.5)(a) simplifies to

$$\omega_{(\lambda,\mu)}(\hat{w}) = \sum_{i=1}^{m} [W_a : C_{W_a}(w_i^{(a)})]\, [W_b : C_{W_b}(w_i^{(b)})]\, \varepsilon_b'(w_i^{(b)}).$$

Now, for $w = t$, the W_n-class of w consists of all t_i. The intersection of that class with $W_a \times W_b$ consists of the class of t (of length a) and that of t_a (of length b). This yields the first formula. Now let $w = tt_1$. Then the W_n-class of w consists of all t_it_j for $i < j$. The intersection of that class with $W_a \times W_b$ consists of the class of tt_1 (of length $a(a-1)/2$), that of t_at_{a+1} (of length $b(b-1)/2$) and, for $a, b \neq 0$, that of tt_a (of length $a + b$). This yields the second formula. □

6.2.7. Identifying characters of W_n and W'_n: Part 2. Let $n \geq 1$ and χ be an irreducible character of W_n or of W'_n. Then χ is labelled by a pair of partitions (λ, μ). We have already seen in (6.2.4) that the restrictions of χ to certain parabolic subgroups determine the sum $\lambda + \mu$. Now we can show that the integers $a, b \geq 0$ such that $\lambda \vdash a$ and $\mu \vdash b$ are also determined.

First assume that we are in type B_n. Considering the restriction of χ to the parabolic subgroups $\{1\}$ and $\langle t \rangle$, we see that the values $\chi(1)$ and $\chi(t)$ are determined and, hence, so are a and b by the first formula in Lemma 6.2.6.

Now assume that $n \geq 2$ and that we are in type D_n. Considering the restriction of χ to the parabolic subgroups $\{1\}$ and $\langle u, s_1 \rangle$, we see that the values $\chi(1)$ and $\chi(us_1)$ are determined. Now, by Remark 6.2.4, it only remains to consider the case where $\lambda \neq \mu$. Then χ is the restriction of $\chi_{(\lambda,\mu)} \in \mathrm{Irr}(W_n)$ to W'_n and so we have $\chi(us_1) = \chi_{(\lambda,\mu)}(us_1)$. Hence the second formula in Lemma 6.2.6 determines $(a-b)^2$ and, therefore, $\{a, b\}$. Note that this is sufficient since $\chi_{(\lambda,\mu)}$ and $\chi_{(\mu,\lambda)}$ have the same restriction to W'_n.

6.2.8. Lemma. *Let $n \geq 2$ and $\chi_{(\lambda,\mu)} \in \mathrm{Irr}(W_n)$, where $a = |\lambda|$ and $b = |\mu|$.*

(a) *Let $2 \leq k \leq n$ and consider the parabolic subgroup $W_{n-k} \times H_k \subseteq W_n$. Let $w_k \in H_k$ be a k-cycle and $v = 1$, $v = t$ or $v = us_1$. Then we have*

$$\omega_{(\lambda,\mu)}(\widehat{vw_k}) = 2^{k-1}\Big(\xi(v)\,\omega_\lambda(\hat{w}_k) + \eta(v)\,\omega_\mu(\hat{w}_k)\Big),$$

where the values of $\xi(v)$, $\eta(v)$ are given by the following table:

	$v=1$	$v=t,\ k\leq n-1$	$v=us_1,\ k\leq n-2$
$\xi(v)$	1	$a-b-k$	$\binom{a-k}{2}+\binom{b}{2}-(a-k)b$
$\eta(v)$	1	$a-b+k$	$\binom{b-k}{2}+\binom{a}{2}-(b-k)a$

(b) *Assume that $a = |\lambda| = |\mu|$. Let $k \geq l \geq 2$ be such that $k + l \leq n$ and consider the parabolic subgroup $W_{n-k-l} \times H_{k+l} \subseteq W_n$. Let $w_{[k,l]} \in H_{k+l}$ be an element of cycle type $[k, l]$ and $v = 1$ or $v = us_1$. Then we have*

$$\begin{aligned}\omega_{(\lambda,\mu)}(\widehat{vw_{[k,l]}}) = 2^{k+l-2}\xi(v)\Big(\omega_\lambda(\hat{w}_k)\omega_\mu(\hat{w}_l) + \omega_\lambda(\hat{w}_l)\omega_\mu(\hat{w}_k)\Big)\\ + 2^{k+l-2}\eta(v)\Big(\omega_\lambda(\hat{w}_{[k,l]}) + \omega_\mu(\hat{w}_{[k,l]})\Big),\end{aligned}$$

where the values of $\xi(v)$, $\eta(v)$ are given by the following table:

	$v=1$	$v=us_1,\ k+l\leq n-2$
$\xi(v)$	1	$\binom{a-k}{2}+\binom{a-l}{2}-(a-k)(a-l)$
$\eta(v)$	1	$\binom{a-k-l}{2}+\binom{a}{2}-(a-k-l)a$

In the above formulas, it is understood that $\omega_\lambda(\hat{w})$ or $\omega_\mu(\hat{w})$ are set equal to 0, if $w \in W_n$ is not conjugate to an element in $H_a \cong \mathfrak{S}_a$ or $H_b \cong \mathfrak{S}_b$, respectively.

Proof. Since this is similar to that of Lemma 6.2.6, we just give a sketch. In (a), we have to consider the intersection of the W_n-class of vw_k with $W_a \times W_b$. If $v = 1$, then w_k is conjugate to an element either in W_a or in W_b. Hence that intersection consists of the class of a k-cycle $w_k^{(a)} \in W_a$ and the class of a k-cycle $w_k^{(b)} \in W_b$. We have

$$|C_{W_a}(w_k^{(a)})| = 2^{a-k+1}k(a-k)! \quad \text{and} \quad |C_{H_a}(w_k^{(a)})| = k(a-k)!;$$

analogous formulas also hold with respect to W_b. Thus, we have $\xi(1) = \eta(1) = 1$ as claimed. Now let $v = t$. Then the intersection of the W_n-conjugacy class of tw_k with $W_a \times W_b$ has representatives which we denote by $w_i = w_i^{(a)}w_i^{(b)}$, where $w_i^{(a)} = n_i^{(a)}h_i^{(a)}$ and $w_i^{(b)} = n_i^{(b)}h_i^{(b)}$, as in (6.2.5). We have

i	$n_i^{(a)}$	$h_i^{(a)}$	$n_i^{(b)}$	$h_i^{(b)}$	$c_i^{(a)}$	$c_i^{(b)}$
1	t	$w_k^{(a)}$	1	1	$2^{k-a-1}(a-k)$	2^{-b}
2	1	1	t_a	$w_k^{(b)}$	2^{-a}	$2^{k-b-1}(b-k)$
3	t	1	1	$w_k^{(b)}$	$2^{-a}a$	2^{k-b-1}
4	1	$w_k^{(a)}$	t_a	1	2^{k-a-1}	$2^{-b}b$

where $c_i^{(a)} = |C_{H_a}(h_i^{(a)})|/|C_{W_a}(w_i^{(a)})|$ and $c_i^{(b)} = |C_{H_b}(h_i^{(b)})|/|C_{W_b}(w_i^{(b)})|$. This yields the desired values for $\xi(t)$ and $\eta(t)$. Now let $v = us_1 = ts_1ts_1$. Since $t, ts_1t \in N_n$, we can argue as above, except that there are some more cases to consider, according to which of the factors t, ts_1t and w_k is conjugate to an element in W_a or in W_b. A similar discussion also applies to (b). □

6.2.9. Theorem (Benson–Curtis). *Let $n \geq 2$. Then every irreducible character of W_n or of W'_n is uniquely determined by its restrictions to all parabolic subgroups whose Coxeter graph is a disjoint union of graphs of type A.*

Proof. First consider the group W_n of type B_n. Let $\chi \in \text{Irr}(W_n)$. Then we have $\chi = \chi_{(\lambda,\mu)}$ for a pair of partitions (λ, μ). We must show that λ and μ are determined by the restrictions of χ to parabolic subgroups as above. By (6.2.4) and (6.2.7), we already know that the sum $\lambda + \mu$ and the integers $a, b \geq 0$ such that $a = |\lambda|$ and $b = |\mu|$ are determined as required. In particular, this is already sufficient to determine λ and μ if $n = 2$. Now assume that $n \geq 3$. Let $2 \leq k \leq n-1$ and consider the parabolic subgroup $W_{n-k} \times H_k \subseteq W_n$. Let $w_k \in H_k$ be a k-cycle. Then we can express $\omega_\chi(\hat{w}_k)$ and $\omega_\chi(\widehat{tw_k})$ as linear combinations of $\omega_\lambda(\hat{w}_k)$ and $\omega_\mu(\hat{w}_k)$, where the matrix of coefficients is given by the first two columns in the table in Lemma 6.2.8(a). The determinant of that matrix is $2k \neq 0$. So the values of $\omega_\lambda(\hat{w}_k)$ and $\omega_\mu(\hat{w}_k)$ are determined for all $2 \leq k \leq n-1$. Trivially, this also holds for $k = 1$. But then λ and μ are determined; see Corollary 6.2.3.

The argument for type D_n is essentially the same but technically slightly more complicated. We can assume that $n \geq 4$. Consider the group $W_n' \subseteq W_n$ and let $\chi \in \mathrm{Irr}(W_n')$. Then χ is labelled by an unordered pair of partitions (λ, μ). By (6.2.4), the restrictions of χ to parabolic subgroups of the required type already determine $\lambda + \mu$ and whether $\lambda = \mu$ or $\lambda \neq \mu$; in particular, if $\lambda = \mu$, then λ and the sign $\pm$ such that $\chi = \chi_{(\lambda,\pm)}$ are completely determined. Hence, it remains to consider the case where $\lambda \neq \mu$. Then χ is the restriction of $\chi_{(\lambda,\mu)} \in \mathrm{Irr}(W_n)$. Using (6.2.4) again, the integers $a, b \geq 0$ such that $a = |\lambda|$ and $b = |\mu|$ are determined as required. We now consider again suitable values of ω_χ. Since χ is the restriction of $\chi_{(\lambda,\mu)}$, the values of ω_χ also determine the values for $\omega_{(\lambda,\mu)}$ on classes which lie in W_n'. So we can again use Lemma 6.2.8. Now there are two cases to consider.

Case 1. We have $a \neq b$. Then consider a k-cycle w_k as before, for $2 \leq k \leq n-2$. We can express $\omega_\chi(\hat{w}_k)$ and $\omega_\chi(\widehat{us_1 w_k})$ as linear combinations of $\omega_\lambda(\hat{w}_k)$ and $\omega_\mu(\hat{w}_k)$, where the coefficient matrix is given by the first and third columns in the table of Lemma 6.2.8(a). The determinant of that matrix is invertible since $a \neq b$. So we conclude again that the values of $\omega_\lambda(\hat{w}_k)$ and $\omega_\mu(\hat{w}_k)$ are determined for all $2 \leq k \leq n-2$. Trivially, this also holds for $k = 1$. If both a and b are non-zero, we can apply Corollary 6.2.3 and so λ and μ are determined. But, if $a = n$ and $b = 0$, then we have $\omega_{(\lambda,\mu)}(\hat{w}_k) = 2^{k-1}\omega_\lambda(\hat{w}_k)$ for all $2 \leq k \leq n$ and so λ is again determined by Corollary 6.2.3.

Case 2. We have $a = b$. Let $2 \leq k, l \leq a-1$ and consider an element $w_{[k,l]}$ as in Lemma 6.2.8(b). Since $\eta(us_1) - \xi(us_1) = 2(a-k)(a-l) \neq 0$ but $\eta(1) = \xi(1)$, a suitable linear combination of $\omega_{(\lambda,\mu)}(\hat{w}_{[k,l]})$ and $\omega_{(\lambda,\mu)}(\widehat{us_1 w_{[k,l]}})$ is equal to

$$A_{kl} := \omega_\lambda(\hat{w}_k)\omega_\mu(\hat{w}_l) + \omega_\lambda(\hat{w}_l)\omega_\mu(\hat{w}_k).$$

Hence, A_{kl} is determined by the restrictions of χ to parabolic subgroups of the required type. On the other hand, we can still use Lemma 6.2.8(a) and obtain that $\omega_\lambda(\hat{w}_k) + \omega_\mu(\hat{w}_k)$ is also determined, for $2 \leq k \leq a-1$. It remains to show that the above conditions determine the unordered pair $\{\lambda, \mu\}$ uniquely. So suppose that we also have $\lambda', \mu' \vdash a$ such that, for all $2 \leq k, l \leq a-1$:

$$\omega_\lambda(\hat{w}_k) + \omega_\mu(\hat{w}_k) = \omega_{\lambda'}(\hat{w}_k) + \omega_{\mu'}(\hat{w}_k),$$
$$A_{kl} = A'_{kl} := \omega_{\lambda'}(\hat{w}_k)\omega_{\mu'}(\hat{w}_l) + \omega_{\lambda'}(\hat{w}_l)\omega_{\mu'}(\hat{w}_k).$$

But then Exercise 6.6 shows that either $\omega_\lambda(\hat{w}_k) = \omega_{\lambda'}(\hat{w}_k)$ and $\omega_\mu(\hat{w}_k) = \omega_{\mu'}(\hat{w}_k)$ for all $2 \leq k \leq a-1$, or $\omega_\lambda(\hat{w}_k) = \omega_{\mu'}(\hat{w}_k)$ and $\omega_\mu(\hat{w}_k) = \omega_{\lambda'}(\hat{w}_k)$ for all $2 \leq k \leq a-1$. Using Corollary 6.2.3, it follows that either $\lambda = \lambda'$ and $\mu = \mu'$ or $\lambda = \mu'$ and $\mu = \lambda'$, as desired. □

Note that the above result is not true in general for finite Coxeter groups. For example, if (W, S) is of type $I_2(m)$ with $m \geq 5$, then all $\chi \in \mathrm{Irr}(W)$ of degree 2 have the same restriction to all proper parabolic subgroups of W (i.e., have the same value on 1 and on $s \in S$), and there exist at least two such characters.

6.3 CHARACTERS OF PARABOLIC TYPE

Let (W, S) be a Coxeter system where W is finite. In Section 2.4 we studied the parabolic Burnside ring of W, i.e., the permutation representations of W on the cosets of the various parabolic subgroups W_J, $J \subseteq S$. We will now study in more detail in which way the irreducible characters of W occur as constituents of the corresponding permutation representations. To simplify notation, we shall denote the characters of these permutation representations by $1_{W_J}^W$, for $J \subseteq S$. Furthermore, for any $J \subseteq S$, let $n(J, \chi)$ be the multiplicity of χ in $1_{W_J}^W$.

6.3.1. Definition (Curtis–Iwahori–Kilmoyer). Let $\chi \in \mathrm{Irr}(W)$. We say that χ is of *parabolic type* if $n(J, \chi) = 1$ for some $J \subseteq S$.

One important aspect of the above notion is that any $\chi \in \mathrm{Irr}(W)$ of parabolic type can be realized over the field $\mathbb{Q}(\chi) := \mathbb{Q}(\chi(w) \mid w \in W)$. This is an immediate consequence of general results on the Schur index; see, for example, [Goldschmidt 1980, Chap. 4]. Indeed, if $\chi \in \mathrm{Irr}(W)$ is of parabolic type, then χ occurs with multiplicity 1 in a character which is obviously realized over $\mathbb{Q}$ (namely $1_{W_J}^W$), and so χ has Schur index 1.

Another important property is that, in most cases (with known exceptions to be described below), a given irreducible character $\chi \in \mathrm{Irr}(W)$ will be uniquely determined by the multiplicities $n(J, \chi)$ for all $J \subseteq S$. In Corollary 5.4.8, we have already seen that this is true for groups of type A. This yields the following result which will allow us also to handle the types B_n and D_n.

6.3.2. Lemma. *Let $I \subseteq S$ be such that the Coxeter graph of $W_I \subseteq W$ is a disjoint union of graphs of type A. Let $\chi, \chi' \in \mathrm{Irr}(W)$ be such that $n(J, \chi) = n(J, \chi')$ for all subsets $J \subseteq I$. Then we have $\chi(w) = \chi'(w)$ for all $w \in W_I$.*

Proof. Let us denote the restriction of χ to W_I by ψ. Then, by the transitivity of induction and Frobenius reciprocity, we have

$$n(J, \psi) = (\psi, 1_{W_J}^{W_I})_{W_I} = (\chi, 1_{W_J}^W)_W = n(J, \chi) \qquad \text{for all } J \subseteq I.$$

A similar statement also holds for the restriction ψ' of χ'. Hence the assumption implies that $n(J, \psi) = n(J, \psi')$ for all $J \subseteq I$. Now, since W_I is a direct product of groups of type A, the irreducible characters of W_I can be expressed as integral linear combinations of the characters $1_{W_J}^{W_I}$ for all $J \subseteq I$. Indeed, in the case where W_I itself is irreducible, this is contained in Theorem 5.4.5(b); the general case is a direct consequence of this, using the compatibility of the induction of characters with direct products. Hence, all irreducible characters of W_I have the same multiplicity in ψ and in ψ', and so we have $\psi = \psi'$ as claimed. □

6.3.3. Theorem (Benson–Curtis). *Consider a Coxeter system (W, S) of type A_{n-1}, B_n or D_n ($n \geq 2$). Then each $\chi \in \mathrm{Irr}(W)$ is of parabolic type and is uniquely determined by the multiplicities $n(J, \chi)$ for all $J \subseteq S$.*

Proof. For type A_{n-1}, see Corollary 5.4.8. For groups of type B_n and D_n, all irreducible characters are of parabolic type by Proposition 6.1.5. The uniqueness statement follows from Lemma 6.3.2 and Theorem 6.2.9. □

6.3.4. Example. Consider the decomposition of the induced characters $1_{W_J}^W$ for the groups W_3 and W_4'. We label the generators in the usual way, as in the diagrams in Section 5.5. The results are given in Table 6.1. To determine this table, one first computes the character tables of W_3 and W_4' by the method described in (5.3.5). The labelling of the irreducible characters in terms of pairs of partitions is determined using the b-invariants and the rules in (6.2.4), (6.2.7). (See also Tables 10.3 and 10.4 on pp. 336 and 346.)

Table 6.1 Characters of parabolic type in types B_3 and D_4

B_3	∅	0	1	01	02	12	012
$\chi_{(3.\varnothing)}$	1	1	1	1	1	1	1
$\chi_{(2.1)}$	3	2	2	1	1	1	.
$\chi_{(21.\varnothing)}$	2	2	1	1	1	.	.
$\chi_{(\varnothing.3)}$	1	.	1	.	.	1	.
$\chi_{(1.2)}$	3	1	2	.	1	1	.
$\chi_{(11.1)}$	3	2	1	.	1	.	.
$\chi_{(1.11)}$	3	1	1	.	.	.	.
$\chi_{(\varnothing.21)}$	2	.	1	.	.	.	.
$\chi_{(111.\varnothing)}$	1	1	.	.	.	.	.
$\chi_{(\varnothing.111)}$	1	.	.	.	.	.	.

D_4	∅	0	01	02	03	13	012	013	023	123	0123
$\chi_{(\varnothing.4)}$	1	1	1	1	1	1	1	1	1	1	1
$\chi_{(1.3)}$	4	3	2	2	2	2	1	1	1	1	.
$\chi_{(\varnothing.31)}$	3	2	2	1	1	1	1	1	.	.	.
$\chi_{(2.-)}$	3	2	1	1	1	2	.	1	.	1	.
$\chi_{(2.+)}$	3	2	1	1	2	1	.	1	1	.	.
$\chi_{(1.21)}$	8	4	2	1	2	2	.	1	.	.	.
$\chi_{(\varnothing.22)}$	2	1	1	.	1	1	.	1	.	.	.
$\chi_{(11.2)}$	6	3	1	1	1	1	.	.	.	.	.
$\chi_{(\varnothing.211)}$	3	1	1	.	.	.	.	.	.	.	.
$\chi_{(11.-)}$	3	1	.	.	.	1	.	.	.	.	.
$\chi_{(11.+)}$	3	1	.	.	1	.	.	.	.	.	.
$\chi_{(1.111)}$	4	1	.	.	.	.	.	.	.	.	.
$\chi_{(\varnothing.1111)}$	1	.	.	.	.	.	.	.	.	.	.

The subsets J are specified in the first row, where 1, 2, 3 stand for s_1, s_2, s_3, respectively, and 0 stands for t (in type B_3) or for u (in type D_4). For example, 013 is $\{u, s_1, s_3\}$ in type D_4.

6.3.5. Characters of parabolic type for dihedral groups. Let $m \geq 3$ and $W = \langle s, t\rangle$ be a Coxeter group of type $I_2(m)$, where s, t have order 2 and st has order m. The irreducible characters of W are described in (5.3.4). We have the following induced characters:

$$1_{W_\varnothing}^W = 1_W + \varepsilon + \varepsilon_1 + \varepsilon_2 + 2\sum_j \chi_j, \qquad 1_W^W = 1_W,$$

$$1_{W_{\{s\}}}^W = 1_W + \varepsilon_1 + \sum_j \chi_j \quad \text{and} \quad 1_{W_{\{t\}}}^W = 1_W + \varepsilon_2 + \sum_j \chi_j,$$

where it is understood that $\varepsilon_1, \varepsilon_2$ are omitted if m is odd. Thus, every $\chi \in \mathrm{Irr}(W)$ is of parabolic type. Note that, for $m \geq 2$, there are at least two characters χ_j, and all of them have the same multiplicities in the above induced characters.

6.3.6. Characters of parabolic type in exceptional groups. Let (W, S) be of exceptional type H_3, H_4, F_4, E_6, E_7 or E_8, as in (5.3.6). Having computed the character table of W, we can also determine the multiplicities $n(J, \chi)$ for all $\chi \in \mathrm{Irr}(W)$ and $J \subseteq S$. Note that the values of those induced characters are determined using Proposition 3.1.8 and the tables in Appendix A.

(a) With one exception, every $\chi \in \mathrm{Irr}(W)$ is of parabolic type. In the tables in Appendix C we give the *maximal* subset $J \subseteq S$ such that $n(J, \chi) = 1$, where we use the following total order $\preceq$ on the set of all subsets of S. First, we assume that a numbering $S = \{s_1, s_2, \dots\}$ is chosen as in the Dynkin diagram in Table 1.2 (p. 20). Now, given $J, J' \subseteq S$, we write $J \preceq J'$ if either $|J| < |J'|$ or $|J| = |J'|$ and the set of indices labelling the generators in J is lexicographically smaller than the analogous set for J'. The exception is the character 48_{rr} in type H_4; the multiplicities of 48_{rr} in $1_{W_J}^W$ are printed in Table C.2.

(b) Let $\chi \in \mathrm{Irr}(W)$ be such that all character values are rational. Then, with some exceptions, χ is uniquely determined by the multiplicities $\{n(J, \chi)\}_{J \subseteq S}$. The exceptions only occur in types H_3, H_4, E_7 and E_8; in these cases, there exist pairs of characters with the same multiplicities:

Type	Exceptional rational characters
H_3	$\{4_r, 4_r'\}$
H_4	$\{8_r, 8_{rr}\}$, $\{16_r, 16_{rr}\}$, $\{16_r', 16_{rr}'\}$
E_7	$\{512_a, 512_a'\}$
E_8	$\{4096_x, 4096_z\}$, $\{4096_x', 4096_z'\}$

(Note that the degree of an exceptional character turns out to be a power of 2.)

The above uniqueness statement is checked as follows. First note that we have $n(\varnothing, \chi) = \chi(1)$ and $n(\{s\}, \chi) = (\chi(1) + \chi(s))/2$ for $s \in S$. Thus, $\chi(1)$ and $\chi(s)$ (for $s \in S$) are uniquely determined by the multiplicities $n(J, \chi)$. Looking through the tables in Appendix C, we see that χ itself is uniquely determined by $\chi(1)$ and the values $\chi(s)$ (for $s \in S$) unless χ is non-rational or χ occurs in the above table, or we are in type E_6 and χ has degree 15 or we are in type E_7 and χ has degree 280. But, in the latter cases, the subsets J printed in the last column of the tables are different.

6.3.7. Remark. Let W be an irreducible finite Coxeter group with generating set S. Let V be the standard module for W and $\rho^{\wedge d} \in \mathrm{Irr}(W)$ be afforded by the dth exterior power of V, for $0 \leq d \leq |S|$. By Proposition 5.1.5, we have

$$(\rho^{\wedge d}, 1_{W_J}^W)_W = \binom{|S - J|}{d} \qquad \text{for all } J \subseteq S.$$

The above results imply that $\rho^{\wedge d}$ is the only irreducible character of W whose multiplicities in $1_{W_J}^W$ are given by the above formula, unless (W, S) is of type $I_2(m)$ $(m \geq 5)$, H_3 or H_4. Note that, in these exceptional cases, even ρ itself is not uniquely determined by its multiplicities in $1_{W_J}^W$.

As an application, we can now completely solve the problem of determining the Schur indices of the irreducible characters of finite Coxeter groups.

6.3.8. Theorem (Benard, Benson–Curtis, Benson–Grove). *Let W be a finite Coxeter group, with generating set S. Set $K = \mathbb{Q}(\cos(2\pi/m_{st}) \mid s,t \in S) \subset \mathbb{R}$ where, as usual, m_{st} denotes the order of $st \in W$. Then the following hold:*

(a) *Every $\chi \in \mathrm{Irr}(W)$ can be realized over K and we have $K = \bigcup_{\chi \in \mathrm{Irr}(W)} \mathbb{Q}(\chi)$ where $\mathbb{Q}(\chi) := \mathbb{Q}(\chi(w) \mid w \in W)$.*

(b) *If (W, S) is irreducible then, with one exception, $\chi \in \mathrm{Irr}(W)$ has Schur index 1, i.e., χ can be realized over $\mathbb{Q}(\chi)$. The exception is $\chi = 48_{rr}$ in type H_4, all of whose values are in $\mathbb{Q}$, but which cannot be realized over $\mathbb{Q}$.*

Proof. If W is not irreducible, then we can write $W = W_1 \times W_2$ where W_1, W_2 are Coxeter groups. The irreducible characters of W are obtained as outer tensor products of the irreducible characters of W_1 and W_2. We have corresponding fields K_1 and K_2 defined by analogy with the definition of K. Then we see that K is generated by K_1 and K_2. Thus, the proof of (a) can be reduced to the case where (W, S) is irreducible. Assume now that this is the case.

Now (a) holds by the results in Chapter 5; see Theorem 5.3.8 for the exceptional types, Theorem 5.4.5 for type A_n, Theorem 5.5.6 for type B_n and Corollary 5.6.4 for type D_n. The latter results actually imply that (b) holds for groups of type A_n, B_n and D_n. It remains to prove (b) for the exceptional types. The remarks following Definition 6.3.1 in combination with the results in (6.3.5) and (6.3.6) are used to settle all the remaining cases except for the character of degree 48 in type H_4. So, it only remains to show that the latter character really has Schur index 2 over $\mathbb{Q}$. To see this, it seems to be necessary to perform some explicit computations inside W; see Exercise 6.10 for the details. □

6.4 SYMBOLS FOR THE CLASSICAL TYPES

In this section, we introduce in an *ad hoc* way the a-invariants for the irreducible characters in groups of classical type. (A more conceptual framework will be described in Section 6.5.) Here, we will use these invariants as a principle for organizing the set of irreducible characters in such a way that the formulas for induced characters in Section 6.1 take a particularly simple and elegant form; see Proposition 6.4.7 for type B_n and Proposition 6.4.9 for type D_n.

We begin by introducing β-sets and symbols, which provide alternative ways of labelling the irreducible characters of $\mathfrak{S}_n$, W_n and W_n'.

6.4.1. A finite subset $X \subset \mathbb{N}_0$ is called a *β-set*. We usually write the entries of a β-set in the form $\beta_1 > \beta_2 > \cdots > \beta_s$, where $|X| := s \geq 0$. The rank and the a-invariant of X are defined by $\mathrm{rk}(X) = a(X) = 0$ if $X = \varnothing$ and, otherwise, by:

$$\mathrm{rk}(X) = \sum_{i=1}^{s} \beta_i - \binom{s}{2} \quad \text{and} \quad a(X) = \sum_{i=1}^{s} (i-1)\beta_i - \sum_{i=1}^{s-1} \binom{i}{2}.$$

We let Φ_n be the set of β-sets of rank n, taken modulo the equivalence relation generated by the following *shift operation*:

$$X \sim X^{+t} := \{\beta_1+t, \beta_2+t, \ldots, \beta_s+t, t-1, t-2, \ldots, 1, 0\} \quad (t \geq 0).$$

It is verified checked that β-sets which are equivalent under shift have the same rank and the same a-invariant. Now let $X \in \Phi_n$, with entries $\beta_1 > \cdots > \beta_s$. Using shift, we can assume that $0 \notin X$. Then we associate with X the partition

$$\lambda_X := [\beta_1-(s-1), \beta_2-(s-2), \ldots, \beta_s] \vdash n.$$

We have $a(\lambda_X) = a(X)$, where the a-invariant of a partition is defined as in (5.4.2). The assignment $X \mapsto \lambda_X$ (where $0 \notin X$) defines a bijection between Φ_n and the set of all partitions of n. To simplify notation, we will sometimes write $[X] := \chi_{\lambda_X} \in \mathrm{Irr}(\mathfrak{S}_n)$ if this is convenient.

6.4.2. Lusztig's symbols. By definition, a *symbol* of defect $d \geq 0$ is an unordered pair of two β-sets $X, Y \subset \mathbb{N}_0$ such that $|X|-|Y| = \pm d$; we shall write such a symbol in the form $\Lambda = \binom{X}{Y}$. The rank and the b-invariant of Λ are defined as

$$\mathrm{rk}(\Lambda) := \mathrm{rk}(X) + \mathrm{rk}(Y) + \left\lfloor \left(\frac{d}{2}\right)^2 \right\rfloor, \quad \text{where } \lfloor x \rfloor = \text{largest integer} \leq x,$$
$$b(\Lambda) := 2a(X) + 2a(Y) + \mathrm{rk}(Y), \quad \text{assuming } |X| > |Y| \text{ or } \mathrm{rk}(X) \geq \mathrm{rk}(Y).$$

The shift operation on β-sets induces a shift operation on symbols by

$$\Lambda \sim \Lambda^{+t} = \binom{X^{+t}}{Y^{+t}} \qquad (t \geq 0).$$

Equivalent symbols have the same rank and the same b-invariant. We denote by $\Phi_{n,d}$ the set of symbols of rank n and defect d, taken modulo shift.

Now let (λ, μ) be a pair of partitions such that $|\lambda| + |\mu| = n$. We have a corresponding symbol $\Lambda = \binom{X}{Y} \in \Phi_{n,1}$, where X and Y are β-sets associated with λ and μ, respectively, such that $|X| = |Y| + 1$. We can now also label the irreducible characters of W_n as

$$\mathrm{Irr}(W_n) = \{\chi_\Lambda \mid \Lambda \in \Phi_{n,1}\}.$$

Then $b(\Lambda)$, as defined above, is also the b-invariant of χ_Λ; see Definition 5.5.3. Whenever it is convenient, we will just write $[\Lambda]$ instead of χ_Λ.

We can also associate with (λ, μ) a symbol $\Lambda' = \binom{X}{Y} \in \Phi_{n,0}$, where X, Y are β-sets associated with λ and μ, respectively, such that $|X| = |Y|$ and $\mathrm{rk}(X) \geq \mathrm{rk}(Y)$. In this case, $\Lambda = \binom{X}{Y}$ is called *degenerate* if $X = Y$; otherwise, Λ is called *non-degenerate*. Then the irreducible characters of $W_n' \subset W_n$ are labelled as

$$\mathrm{Irr}(W_n') = \{\chi_\Lambda \mid \Lambda \in \Phi_{n,0},\ \Lambda \text{ non-degenerate}\}$$
$$\cup \{\chi_{(\Lambda,\pm)} \mid \Lambda \in \Phi_{n,0},\ \Lambda \text{ degenerate}\}.$$

Again, the b-invariant of a symbol is also the b-invariant of the corresponding irreducible character of W_n'; see (5.6.1). Whenever it is convenient, we write $[\Lambda]$ instead of χ_Λ and $[\Lambda]_\pm$ instead of $\chi_{(\Lambda,\pm)}$.

6.4.3. Definition. Let $\Lambda = \binom{X}{Y}$ be a symbol, where $|X| = r$ and $|Y| = s$. Consider the collection of all elements of X and Y and write them in decreasing order: $\gamma_1 \geq \gamma_2 \geq \cdots \geq \gamma_{r+s} \geq 0$. Then the *a-invariant* of Λ is defined by

$$a(\Lambda) := \sum_{i=1}^{r+s} (i-1)\gamma_i - \sum_{i=1}^{\lfloor \frac{r+s}{2} \rfloor} \binom{r+s-2i}{2}.$$

It is readily verified that equivalent symbols have the same a-invariant. Also note that we have $a(\Lambda) = a(\Lambda')$ if Λ and Λ' are symbols which contain the same entries, arranged perhaps differently in rows.

The above definition of $a(\Lambda)$ can be regarded as a formal generalization of the definition of the a-invariant of a β-set; see (6.4.1). In Section 6.5, we will see that the above definition of the a-invariants is the only possible one, subject to a certain set of conditions; see also Remark 6.5.14.

6.4.4. Special symbols. Now let us consider the two cases which are of interest to us, namely when Λ is a symbol of defect 1 or 0. In these cases, Lusztig has observed the following relation between the a-invariants and the b-invariants. First assume that $\Lambda = \binom{X}{Y}$ has defect 1, where X has entries $\alpha_1 > \cdots > \alpha_r$ and Y has entries $\beta_1 > \cdots > \beta_s$, $r = s+1$. Using Exercise 6.7, we find that

$$\begin{aligned} b(\Lambda) - a(\Lambda) &= \sum_{i=1}^{s+1}(i-1)\alpha_i + \sum_{i=1}^{s}(i-1)\beta_i + \sum_{i=1}^{s}\beta_i - \sum_{i=1}^{s+1}\sum_{j=1}^{s}\min\{\alpha_i,\beta_j\} \\ &= \sum_{i=1}^{s}\Big(i\alpha_{i+1} - \sum_{j\leq i}\min\{\alpha_{i+1},\beta_j\}\Big) + \sum_{i=1}^{s}\Big(i\beta_i - \sum_{j\leq i}\min\{\alpha_j,\beta_i\}\Big) \\ &= \sum_{1\leq j\leq i\leq s}\Big(\alpha_{i+1} - \min\{\alpha_{i+1},\beta_j\}\Big) + \sum_{1\leq j\leq i\leq s}\Big(\beta_i - \min\{\alpha_j,\beta_i\}\Big). \end{aligned}$$

Thus, we always have $b(\Lambda) - a(\Lambda) \geq 0$, with equality if and only if

$$\alpha_1 \geq \beta_1 \geq \alpha_2 \geq \beta_2 \geq \cdots \geq \beta_s \geq \alpha_{s+1}.$$

If this condition holds, Λ is called a *special symbol*. Likewise, if $\Lambda = \binom{X}{Y}$ is a symbol of defect 0, where $|X| = |Y| = s$ and $\text{rk}(X) \geq \text{rk}(Y)$, then we find that

$$b(\Lambda) - a(\Lambda) = \sum_{1\leq j<i\leq s}\Big(\alpha_i - \min\{\alpha_i,\beta_j\}\Big) + \sum_{1\leq j\leq i\leq s}\Big(\beta_i - \min\{\alpha_j,\beta_i\}\Big) \geq 0,$$

with equality if and only if $\alpha_1 \geq \beta_1 \geq \alpha_2 \geq \beta_2 \geq \cdots \geq \alpha_s \geq \beta_s$. Again, if this is the case, Λ is called *special*. Note that a degenerate symbol is always special.

6.4.5. Dual β-sets and symbols. Let B be a β-set of rank n with entries $\beta_1 > \beta_2 > \cdots > \beta_s$. Assume that B corresponds to $\lambda \vdash n$; see (6.4.1). Let $t \geq 0$ be such that $\beta_i \leq t$ for all i. Then a β-set corresponding to λ^* is given by

$$B^* := \{t, t-1, \ldots, 1, 0\} - \{t-\beta_s, t-\beta_{s-1}, \ldots, t-\beta_1\}.$$

(For a simple proof, see [Macdonald 1995, (I.1.7)].) Now let $\Lambda = \binom{X}{Y}$ be a symbol of rank n and defect d. Then we define the *dual symbol* Λ^* as follows. Let $t := 0$ if $X = Y = \varnothing$ and $t := \max\{\alpha \in X, \beta \in Y\}$ otherwise. Then we set

$$\Lambda^* = \begin{pmatrix} Y^* \\ X^* \end{pmatrix}.$$

Note that Λ^* is again a symbol of rank n and defect d.

If $d = 1$ (or $d = 0$) and Λ labels an irreducible character χ of W_n (or W'_n) then Λ^* labels $\varepsilon \otimes \chi$, where ε is the sign character. This follows from Corollary 5.4.9, Theorem 5.5.6 and Remark 5.6.5.

6.4.6. Proposition. *Let Λ be a symbol of rank n and defect 1 (or 0), which labels the irreducible character χ of W_n (or W'_n). Then we have*

$$\omega_\chi(\hat{T}) = b(\Lambda^*) - b(\Lambda) = a(\Lambda^*) - a(\Lambda),$$

where ω_χ denotes the central character corresponding to χ, T denotes the set of reflections in W_n (or W'_n) and $\hat{T}$ is the sum of the elements in T.

Proof. We work with the groups $W'_n \subseteq W_n$. For any element $w \in W_n$ (or $w \in W'_n$), we denote by $\hat{w}$ (or $\hat{w}'$) the sum of the elements in the corresponding conjugacy class in W_n (or W'_n). Now we proceed in several steps.

Step 1. We show that $\omega_\chi(\hat{T}) = b(\Lambda^*) - b(\Lambda)$. First, consider W_n. We denote ω_χ by $\omega_{(\lambda,\mu)}$, where (λ, μ) is the pair of partitions associated with Λ. Since T is the union of the classes of t and s_1, we have

$$\omega_{(\lambda,\mu)}(\hat{T}) = \omega_{(\lambda,\mu)}(\hat{t}) + \omega_{(\lambda,\mu)}(\hat{s}_1).$$

The second term can be evaluated using Proposition 5.4.11 and Lemma 6.2.8:

$$\omega_{(\lambda,\mu)}(\hat{s}_1) = 2(a(\lambda^*) - a(\lambda) + a(\mu^*) - a(\mu)). \tag{1}$$

Combining this with the first formula in Lemma 6.2.6 and recalling the formula for the b-invariant in (6.4.2), we see that $\omega_\chi(\hat{T}) = b(\Lambda^*) - b(\Lambda)$, as desired.

Now consider the group $W'_n \subseteq W_n$. If $n \leq 3$, then W'_n is of type A_1, $A_1 \times A_1$ or A_3, so the desired assertion is contained in Proposition 5.4.11. Now assume that $n \geq 4$. Then T equals the class of s_2. The formula in (1) already yields the corresponding value of the central character for the group W_n. But, since $t \in C_{W_n}(s_2)$, we have $[W'_n : C_{W'_n}(s_2)] = [W_n : C_{W_n}(s_2)]$. So, denoting by $\omega'_{(\lambda,\mu)}$

(or $\omega_{(\lambda,\pm)}$) the central character corresponding to $\chi'_{(\lambda,\mu)} \in \mathrm{Irr}(W'_n)$ if $\lambda \neq \mu$ (or $\chi_{(\lambda,\pm)}$ if $\lambda = \mu$), the expression in (1) also equals

$$\frac{[W'_n : C_{W'_n}(s_2)]\chi_{(\lambda,\mu)}(s_2)}{\chi_{(\lambda,\mu)}(1)} = \begin{cases} \omega'_{(\lambda,\mu)}(\hat{s}'_2) & \text{if } \lambda \neq \mu, \\ \frac{1}{2}(\omega_{(\lambda,+)}(\hat{s}'_2) + \omega_{(\lambda,-)}(\hat{s}'_2)) & \text{if } \lambda = \mu. \end{cases}$$

Now, since $\chi_{(\lambda,-)}$ is obtained from $\chi_{(\lambda,+)}$ via conjugation with t but the W'_n-conjugacy class of s_2 is invariant under conjugation with t, we have $\omega_{(\lambda,+)}(\hat{s}'_2) = \omega_{(\lambda,-)}(\hat{s}'_2)$. It remains to recall the formulas for the b-invariants from (5.6.2).

Step 2. Next we show that if $\tilde{\Lambda}$ is a symbol which contains the same entries as Λ (arranged perhaps differently in rows), then we have

$$b(\Lambda^*) - b(\Lambda) = b(\tilde{\Lambda}^*) - b(\tilde{\Lambda}). \tag{2}$$

To see this, let q be an indeterminate. We write $\Lambda = \binom{X}{Y}$ where $|X| = r$, $|Y| = s$ and $r = s + d$ with $d \in \{0, 1\}$. As before, let (λ, μ) be the pair of partitions associated with Λ. Assume that λ has parts $\lambda_1 \geq \lambda_2 \geq \cdots$. By adding to λ some parts equal to 0 if necessary, we have $X = \{\alpha_1 > \cdots > \alpha_r\}$ where $\alpha_i = \lambda_i + r - i$ for all i. We associate with X the polynomial

$$\psi_X(q) := \prod_{i=1}^{r}\prod_{j=1}^{\lambda_i}(1 - q^{r-i+j}) = \prod_{i=1}^{r}\frac{\varphi_{\alpha_i}(q)}{\varphi_{r-i}(q)},$$

where, for any integer $t \geq 0$, we set $\varphi_t(q) := (1-q)(1-q^2)\cdots(1-q^t)$. Let d_X be the degree of ψ_X; we have

$$\begin{aligned} d_X &= \sum_{i=1}^{r}\sum_{j=1}^{\lambda_i}(r - i + j) = r|\lambda| + \Big(\sum_{i=1}^{r}\lambda_i(\lambda_i+1)/2\Big) - \Big(\sum_{i=1}^{r} i\lambda_i\Big) \\ &= r|\lambda| + \Big(\sum_{i=1}^{r}\lambda_i(\lambda_i-1)/2\Big) - \Big(\sum_{i=1}^{r}(i-1)\lambda_i\Big) = r|\lambda| + a(\lambda^*) - a(\lambda). \end{aligned}$$

Similarly, for the β-set Y associated with μ, we have a polynomial $\psi_Y(q)$ of degree $d_Y = s|\mu| + a(\mu^*) - a(\mu)$. Now $\psi_X(q)\psi_Y(q)$ has degree

$$d_X + d_Y = sn + d|\lambda| + a(\lambda^*) - a(\lambda) + a(\mu^*) - a(\mu).$$

Using the formula in (6.4.2), we see that for $d = 0$ or $d = 1$, the above expression equals $sn + (b(\Lambda^*) - b(\Lambda))/2$. On the other hand, note that $\psi_X(q)\psi_Y(q)$ clearly only depends on d and on the collection of all entries of X and Y, but not on the particular arrangement of these entries in the rows of Λ. Thus, (2) is proved.

Step 3. Now consider the identity $b(\Lambda^*) - b(\Lambda) = a(\Lambda^*) - a(\Lambda)$. By step 2, the expression on the right-hand side only depends on the collection of all entries of Λ, but not on the particular arrangement of these entries in the rows of Λ. By Definition 6.4.3, the analogous property also holds for the left-hand side of the above identity. Hence, we can assume that Λ is a special symbol. But then Λ^* is also special, i.e., we have $a(\Lambda) = b(\Lambda)$ and $a(\Lambda^*) = b(\Lambda^*)$. $\square$

We will now study the behaviour of the a-invariants with respect to the induction of characters from parabolic subgroups. We begin with the groups W_n.

6.4.7. Proposition (Induction and a-invariants in type B_n). *Let $n, k, l \geq 0$ be such that $n = k + l$. Let $\chi' \in \mathrm{Irr}(W_k)$ be labelled by $\Lambda' \in \Phi_{k,1}$. Then we have*

$$\mathrm{Ind}_{W_k \times H_l}^{W_n}\left(\chi' \boxtimes \varepsilon_{H_l}\right) = \chi_{(=)} + \chi_{(>)},$$

where $\chi_{(=)}$ (or $\chi_{(>)}$) is a combination of characters $\chi \in \mathrm{Irr}(W_n)$ which are labelled by symbols Λ with $a(\Lambda) = a(\Lambda') + l(l-1)/2$ (or $a(\Lambda) > a(\Lambda') + l(l-1)/2$). Furthermore, using shift, we can assume that Λ' has at least l entries. Then

$$\chi_{(=)} = [\Lambda] \quad \text{or} \quad [\Lambda_1] + [\Lambda_2],$$

where either Λ is the unique symbol of rank n and defect 1 which can be obtained from Λ' by increasing the l largest entries by 1, or Λ_1, Λ_2 are the two unique symbols which can be obtained in that way.

Proof. Let (λ', μ') be the pair of partitions associated with Λ'. Using the "dual" version of Pieri's rule in Exercise 6.3 and arguing as in (6.1.9), we obtain

$$\mathrm{Ind}_{W_k \times H_l}^{W_n}\left(\chi_{(\lambda', \mu')} \boxtimes \varepsilon_{W_l}\right) = \sum_{\lambda, \mu} \chi_{(\lambda, \mu)},$$

where the sum is over all pairs of partitions (λ, μ) such that, for some $0 \leq d \leq l$, the Young diagram λ (or μ) is obtained from that of λ' (or μ') by adding d (or $l - d$) boxes, with no two in the same *row*. Reformulating this in terms of symbols, we find that any (λ, μ) occuring in the above sum corresponds to a symbol which can be obtained from Λ' by increasing l entries by 1.

Let us now consider the corresponding a-invariants. Let $\Lambda' = \binom{X}{Y}$ where $|X| = s + 1$, $|Y| = s$ and $l \leq 2s + 1$. Consider all the entries of X and Y and write them in decreasing order: $\gamma_1 \geq \gamma_2 \geq \cdots \geq \gamma_{2s+1}$. Now, if Λ is obtained from Λ' by increasing l entries by 1, we obtain, using the formula in Definition 6.4.3,

$$a(\Lambda) - a(\Lambda') = \sum_{j=1}^{l} (i_j - 1), \qquad \text{where } 1 \leq i_1 < \cdots < i_l \leq 2s + 1.$$

We have $\sum_{j=1}^{l}(i_j - 1) \geq \sum_{j=1}^{l}(j - 1) = l(l-1)/2$, with equality if and only if $i_j = j$ for all j, i.e., if and only if the l largest entries in Λ' are increased by 1.

Finally, since all elements of X (or Y) are pairwise distinct, we are in one of the following two cases (where we set $\gamma_{l+1} := -\infty$ if $l = 2s+1$):

(1) If $\gamma_{l+1} < \gamma_l$, then there is precisely one symbol Λ of rank n and defect 1 which can be obtained from Λ' by increasing the l largest entries by 1.

(2) If $\gamma_{l+1} = \gamma_l$, then there are exactly two symbols Λ_1, Λ_2 of rank n and defect 1 which can be obtained from Λ' by increasing the l largest entries by 1.

Thus, $\chi_{(=)}$ has either one or two irreducible constituents, as claimed. □

6.4.8. Example. Consider the induction of characters from $W_2 = W_2 \times H_1$ to W_3. The results are printed in Table 6.2. (By Lemmas 6.1.3 and 6.1.4, the computation of that table can be reduced to a similar problem for $\mathfrak{S}_3$.) We can also see explicit examples for the two cases in Proposition 6.4.7:

Case 1. Consider $\psi_{(11,\varnothing)} \in \mathrm{Irr}(W_2)$. Then there is a unique way of increasing the largest entry (which is 2) of the corresponding symbol by 1.

Case 2. Consider $\psi_{(\varnothing,11)} \in \mathrm{Irr}(W_2)$. Then there are two possible ways of increasing the largest entry (which is 2) of the corresponding symbol by 1.

Table 6.2 Induction of characters from $W(B_2)$ to $W(B_3)$

			$\psi \in \mathrm{Irr}(W_2)$	$\psi_{(2.\varnothing)}$	$\psi_{(1.1)}$	$\psi_{(11.\varnothing)}$	$\psi_{(\varnothing.2)}$	$\psi_{(\varnothing.11)}$
			Symbol	$\binom{2}{\varnothing}$	$\binom{2,0}{1}$	$\binom{2,1}{0}$	$\binom{1,0}{2}$	$\binom{2,1,0}{2,1}$
			a_ψ	0	1	1	1	4
			b_ψ	0	1	2	2	4
$\chi \in \mathrm{Irr}(W_3)$	Symbol	a_χ	b_χ	Multiplicity of χ in $\mathrm{Ind}_{W_2}^{W_3}(\psi)$				
$\chi_{(3.\varnothing)}$	$\binom{3}{\varnothing}$	0	0	1	.	.	.	.
$\chi_{(2.1)}$	$\binom{3,0}{1}$	1	1	1	1	.	.	.
$\chi_{(21.\varnothing)}$	$\binom{3,1}{0}$	1	2	1	.	1	.	.
$\chi_{(\varnothing.3)}$	$\binom{1,0}{3}$	1	3	.	.	.	1	.
$\chi_{(1.2)}$	$\binom{2,0}{2}$	2	2	.	1	.	1	.
$\chi_{(11.1)}$	$\binom{2,1}{1}$	3	3	.	1	1	.	.
$\chi_{(1.11)}$	$\binom{3,1,0}{2,1}$	4	4	.	1	.	.	1
$\chi_{(\varnothing.21)}$	$\binom{2,1,0}{3,1}$	4	5	.	.	.	1	1
$\chi_{(111.\varnothing)}$	$\binom{3,2,1}{1,0}$	4	6	.	.	1	.	.
$\chi_{(\varnothing.111)}$	$\binom{3,2,1,0}{3,2,1}$	9	9	.	.	.	.	1

Here, we use a simplified notation for partitions, where $[\lambda_1, \ldots, \lambda_r]$ is simply written without the square brackets and without separating commas between the parts λ_i. The symbol $\varnothing$ stands for the empty partition $[\,]$.

We now turn to groups of type D_n. Recall from Section 5.6 that the irreducible characters are obtained via restriction from a group of type B_n.

6.4.9. Proposition (Induction and a-invariants in type D_n). *Let $n \geq 2$ and $k, l \geq 0$ be such that $n = k + l$. Let Λ' be a symbol of rank k and defect 0. If $\chi' \in \mathrm{Irr}(W'_k)$ is labelled by Λ', we have*

$$\mathrm{Ind}_{W'_k \times H_l}^{W'_n}\left(\chi' \boxtimes \varepsilon_{H_l}\right) = \chi_{(=)} + \chi_{(>)},$$

where $\chi_{(=)}$ (or $\chi_{(>)}$) is a combination of characters $\chi \in \mathrm{Irr}(W'_n)$ which are labelled by symbols Λ with $a(\Lambda) = a(\Lambda') + l(l-1)/2$ (or $a(\Lambda) > a(\Lambda') + l(l-1)/2$). Furthermore, using shift, assume that Λ' has at least l entries. Then:

(a) *If Λ' is non-degenerate, we have $\chi_{(=)} = [\Lambda]$ or $[\Lambda_1] + [\Lambda_2]$, where Λ (or Λ_1, Λ_2) are obtained from Λ' exactly as in Proposition 6.4.7, by increasing the l largest entries of Λ' by 1. Note that Λ (or Λ_1, Λ_2) are non-degenerate.*

(b) *If k is even and $\chi' = [\Lambda']_\pm$ with Λ' degenerate, we have $\chi_{(=)} = [\Lambda]_{(\pm)}$, where Λ is the unique symbol obtained from Λ' by increasing the l largest entries by 1. Here, $[\Lambda]_{(\pm)}$ stands for $[\Lambda]_\pm$ if l is even, and for $[\Lambda]$ if l is odd.*

Finally, if $n \geq 2$ is even and $H_n^\pm \subseteq W'_n$ are as in Proposition 5.6.3, we have

$$\mathrm{Ind}_{H_n^\pm}^{W'_n}(\varepsilon_{H_n^\pm}) = \left[\begin{pmatrix} n/2, \ldots, 2, 1 \\ n/2, \ldots, 2, 1 \end{pmatrix}\right]_\pm + \tau,$$

where τ is a linear combination of characters $\chi \in \mathrm{Irr}(W'_n)$ which are labelled by symbols Λ with $a(\Lambda) > n(n-1)/2$.

Proof. Let (λ', μ') be the pair of partitions associated with Λ' and define

$$\psi' := \mathrm{Ind}_{W'_k \times H_l}^{W'_n}\left(\chi' \boxtimes \varepsilon_{H_l}\right).$$

Given $W'_k \times H_l$ in W'_n, we also have a parabolic subgroup $W_k \times H_l$ in W_n, where $W'_k \subseteq W_k$. Using the transitivity of induction, it follows that

$$\mathrm{Ind}_{W'_n}^{W_n}(\psi') = \mathrm{Ind}_{W'_k \times H_l}^{W_n}\left(\chi' \boxtimes \varepsilon_{H_l}\right) = \mathrm{Ind}_{W_k \times H_l}^{W_n}\left(\mathrm{Ind}_{W'_k}^{W_k}(\chi') \boxtimes \varepsilon_{H_l}\right). \quad (1)$$

Now assume we are in case (a). Then we have $\lambda' \neq \mu'$ and

$$\mathrm{Ind}_{W'_k}^{W_k}(\chi') = \chi_{(\lambda',\mu')} + \chi_{(\mu',\lambda')} \qquad \text{where } \chi_{(\lambda',\mu')}, \chi_{(\mu',\lambda')} \in \mathrm{Irr}(W_k).$$

Using (1) and Pieri's rule as in the proof of Proposition 6.4.7, we obtain

$$\mathrm{Ind}_{W'_n}^{W_n}(\psi') = \sum_{\lambda,\mu}\left(\chi_{(\lambda,\mu)} + \chi_{(\mu,\lambda)}\right), \quad (2)$$

where the sum is over all pairs of partitions (λ, μ) such that, for some $0 \leq d \leq l$, the Young diagram of λ (or μ) is obtained from that of λ' (or μ') by adding d (or

$l-d$) boxes, with no two in the same row. We now restrict both sides of (2) to W_n'. Since χ' is invariant under conjugation with t, the induced character ψ' also has that property. So, by Mackey's formula, the restriction of the left-hand side of (2) equals $2\psi'$. On the other hand, $\chi_{(\lambda,\mu)}$ and $\chi_{(\mu,\lambda)}$ have the same restriction to W_n'. Hence, we obtain the following identity from (2):

$$\psi' = \sum_{\lambda,\mu} \chi'_{(\lambda,\mu)} \qquad \text{where } \chi'_{(\lambda,\mu)} = \text{Res}_{W_n'}^{W_n}(\chi_{(\lambda,\mu)}).$$

We can now continue as in the proof of Proposition 6.4.7. Any (λ,μ) occurring in the above sum corresponds to a symbol which can be obtained from Λ' by increasing l entries by 1, and the condition on the a-invariant is satisfied. Furthermore, such a symbol is necessarily non-degenerate if the l *largest* entries are increased. Hence we have $\psi' = \chi_{(=)} + \chi_{(>)}$, where $\chi_{(=)}$ and $\chi_{(>)}$ have the required form. Thus, (a) is proved.

Now assume we are in case (b). Then we have $\lambda' = \mu'$ and

$$\text{Ind}_{W_k'}^{W_k}(\chi') = \chi_{(\lambda',\lambda')} \in \text{Irr}(W_k).$$

Using (1) and Pieri's rule for the characters of W_n, we now obtain that

$$\text{Ind}_{W_n'}^{W_n}(\psi') = \text{Ind}_{W_k \times H_l}^{W_n}\left(\chi_{(\lambda',\mu')} \boxtimes \varepsilon_{H_l}\right) = \sum_{\lambda,\mu} \chi_{(\lambda,\mu)}, \tag{3}$$

where the sum is over all pairs of partitions (λ,μ) such that, for some $0 \leq d \leq l$, the Young diagram of λ (or μ) is obtained from that of λ' by adding d (or $l-d$) boxes, with no two in the same row. We now restrict both sides of (3) to W_n'. By Mackey's formula, the left-hand side becomes $\psi' + (\psi')^t$, where $(\psi')^t$ denotes the composition of ψ' with the automorphism of W_n' given by conjugation with t. On the other hand, the restriction of $\chi_{(\lambda,\mu)}$ to W_n' is irreducible if $\lambda \neq \mu$ and splits into two irreducible components if $\lambda = \mu$. So we obtain the identity

$$\psi + (\psi')^t = \sum_{\lambda=\mu}\left(\chi_{(\lambda,+)} + \chi_{(\lambda,-)}\right) + \sum_{\lambda\neq\mu} \chi'_{(\lambda,\mu)}. \tag{3'}$$

Now we reformulate this again in terms of symbols and consider the corresponding a-invariants. Any pair (λ,μ) occurring on the right-hand side of (3′) corresponds to a symbol Λ which can be obtained from Λ' by increasing l entries by 1. This already shows (as in the proof of Proposition 6.4.7) that the required condition on the a-invariants is satisfied. Thus, we have $\psi' = \chi_{(=)} + \chi_{(>)}$, but it remains to determine which irreducible characters occurring on the right-hand side of (3′) are the constituents of $\chi_{(=)}$. Any such constituent is labelled by a symbol which can be obtained from Λ' by increasing the l largest entries by 1.

Since we are dealing with symbols of defect 0, there is a unique such symbol, Λ say; moreover, Λ is certainly degenerate if n is also even. Hence we have

$$\chi_{(=)} + \chi_{(=)}^t = \begin{cases} 2[\Lambda] & \text{if } n \text{ is odd,} \\ [\Lambda]_+ + [\Lambda]_- & \text{if } n \text{ is even.} \end{cases} \tag{4}$$

But this certainly implies that $\chi_{(=)} = [\Lambda]$ (if n is odd) and $\chi_{(=)} = [\Lambda]_+$ or $[\Lambda]_-$ (if n is even). The final step consists of identifying the sign $\pm$ in the case where n is even. Assume first that $\chi' = [\Lambda']_+$. Then, by Proposition 5.6.3, we have

$$\chi' = j_{H_{(2\lambda)^*}}^{W'_k}(\varepsilon_{H_{(2\lambda)^*}}).$$

On the other hand, by (6.4.4), the b-invariant of a symbol is always equal to or greater than the a-invariant; moreover, we have $a(\Lambda') = b_{\chi'}$ (since Λ' is special). Hence $\chi_{(>)}$ is a linear combination of irreducible characters of W'_n with b-invariant strictly bigger than that of $\chi' \boxtimes \varepsilon_{H_l}$. Consequently, since $\chi_{(=)}$ is irreducible, we must have

$$\chi_{(=)} = j_{W'_k \times H_l}^{W'_n}(\chi' \boxtimes \varepsilon_{H_l}).$$

Using the transitivity of the j-induction (see (5.2.8)), it follows that

$$\chi_{(=)} = j_{H_{(2\lambda)^*} \times H_l}^{W'_n}\left(\varepsilon_{H_{(2\lambda)^*}} \boxtimes \varepsilon_{H_l}\right).$$

Hence, using Proposition 5.6.3 once more, we conclude that $\chi_{(=)} = [\Lambda]_+$. Finally, since we know the sum of $\chi_{(=)}$ and $\chi_{(=)}^t$ by (4) and since

$$\mathrm{Ind}_{W'_k \times H_l}^{W'_n}\left([\Lambda']_- \boxtimes \varepsilon_{H_l}\right) = (\psi')^t = \chi_{(=)}^t + \chi_{(>)}^t,$$

it is clear that $\chi_{(=)} = [\Lambda]_-$ if $\chi' = [\Lambda']_-$. This completes the proof of (b). The above discussion also applies to the case $k = 0$, where it yields the desired formula for the induction of the sign character from H_n. Then, applying conjugation with t and using Proposition 5.6.3, we also obtain that

$$\mathrm{Ind}_{H_n^-}^{W'_n}(\varepsilon_{H_n^-}) = \left[\begin{pmatrix} n/2, \ldots, 2, 1 \\ n/2, \ldots, 2, 1 \end{pmatrix}\right]_- + \text{further terms},$$

where "further terms" means a combination of characters $\chi \in \mathrm{Irr}(W'_n)$ which are labelled by symbols with a-invariant strictly bigger than $n(n-1)/2$. □

6.5 FAMILIES AND a-FUNCTIONS

Let W be a finite Coxeter group with generating set S. We now formally introduce a-invariants and a corresponding truncated induction of characters from parabolic subgroups. This will be done in a setting which is an axiomatic version of the theory as developed originally by Lusztig; see also Remark 6.5.14 below.

6.5.1. Let $\mathcal{A} = \{a_{W'} \colon \mathrm{Irr}(W') \to \mathbb{N}_0 \mid W' \subseteq W \text{ parabolic}\}$ be a collection of functions. To simplify the notation, we shall often write a_ψ for $a_{W'}(\psi)$, if the parabolic subgroup W' under consideration is clear from the context. We require that the functions in $\mathcal{A}$ satisfy the following condition. For any parabolic subgroups $W'' \subseteq W' \subseteq W$ and any $\psi \in \mathrm{Irr}(W'')$, $\chi \in \mathrm{Irr}(W')$, we have that

$$m(\chi, \psi) \neq 0 \quad \Rightarrow \quad a_\psi \leq a_\chi, \tag{a}$$

where $m(\chi, \psi)$ denotes the multiplicity of χ in $\mathrm{Ind}_{W''}^{W'}(\psi)$. Using the a-functions, we now define an operation called J-*induction*, as follows. Let $W'' \subseteq W' \subseteq W$ be parabolic subgroups and $\psi \in \mathrm{Irr}(W'')$. Then we set

$$J_{W''}^{W'}(\psi) := \sum_{\substack{\chi \in \mathrm{Irr}(W') \\ a_\chi = a_\psi}} m(\chi, \psi)\chi.$$

Extending by linearity, we obtain a map $J_{W''}^{W'} \colon \mathbb{Z}\mathrm{Irr}(W'') \to \mathbb{Z}\mathrm{Irr}(W')$. By (a), it follows that the J-induction is transitive: we have

$$J_{W''}^{W} = J_{W'}^{W} \circ J_{W''}^{W'} \colon \mathbb{Z}\mathrm{Irr}(W'') \to \mathbb{Z}\mathrm{Irr}(W). \tag{b}$$

For example, if $a_{W'} = 0$ for all W', then (a) is satisfied and the corresponding J-induction is just the ordinary induction of characters from subgroups.

6.5.2. Definition (Lusztig). Let $\mathcal{A}$ be a collection of a-functions as in (6.5.1). For any parabolic subgroup $W' \subseteq W$, we define a partition on $\mathrm{Irr}(W')$ into disjoint pieces which we call $\mathcal{A}$-*families*. This is done recursively as follows.

If $W' = \{1\}$, there is only one $\mathcal{A}$-family; it consists of the trivial character. Assume now that $W' \neq \{1\}$ and that $\mathcal{A}$-families have already been defined for all parabolic subgroups $W'' \subseteq W'$, $W'' \neq W'$. Then $\chi, \chi' \in \mathrm{Irr}(W')$ $(\chi \neq \chi')$ are said to be in the same $\mathcal{A}$-family if there exists a sequence $\chi = \chi_0, \chi_1, \ldots, \chi_m = \chi'$ with $\chi_k \in \mathrm{Irr}(W')$ for all k, such that the following holds. For $0 \leq k \leq m-1$, there exist parabolic subgroups $W_k'' \subseteq W'$, $W_k'' \neq W'$, and $\psi_k, \psi_k' \in \mathrm{Irr}(W_k'')$ in the same $\mathcal{A}$-family for W_k'', such that

either χ_k occurs in $J_{W_k''}^{W'}(\psi_k)$ and χ_{k+1} occurs in $J_{W_k''}^{W'}(\psi_k')$
or χ_k^* occurs in $J_{W_k''}^{W'}(\psi_k)$ and χ_{k+1}^* occurs in $J_{W_k''}^{W'}(\psi_k')$,

where, for any $\chi \in W'$, we set $\chi^* := \varepsilon_{W'} \otimes \chi$ and $\varepsilon_{W'}$ is the sign character of W'.

The following two remarks show that families are compatible with tensoring by the sign character ε and with taking direct products.

6.5.3. Remark. Let $\mathcal{A}$ be a collection of a-functions as in (6.5.1). Let $W' \subseteq W$ be a parabolic subgroup and $\mathcal{F} \subseteq \mathrm{Irr}(W')$. Then the following holds:

$\mathcal{F}$ is an $\mathcal{A}$-family if and only if $\{\chi^* = \varepsilon \otimes \chi \mid \chi \in \mathcal{F}\}$ is an $\mathcal{A}$-family.

Proof. We proceed by induction on $|W'|$. If $W' = \{1\}$, there is nothing to prove. Now assume that $W' \neq \{1\}$ and that $\mathcal{F} \subseteq \mathrm{Irr}(W')$ is an $\mathcal{A}$-family. Let $\chi, \chi' \in \mathcal{F}$. We must show that χ^* and $(\chi')^*$ lie in the same family. By the definition of families, it is enough to consider the case where $\chi, \chi' \in \mathcal{F}$ are such that there exist a parabolic subgroup $W'' \subseteq W'$, $W'' \neq W'$ and characters $\psi, \psi' \in \mathrm{Irr}(W'')$ in the same $\mathcal{A}$-family of $\mathrm{Irr}(W'')$ such that

$$\begin{aligned}&\text{either } \chi \text{ occurs in } J_{W''}^{W'}(\psi) \text{ and } \chi' \text{ occurs in } J_{W''}^{W'}(\psi')\\ &\text{or } \chi^* \text{ occurs in } J_{W''}^{W'}(\psi) \text{ and } (\chi')^* \text{ occurs in } J_{W''}^{W'}(\psi').\end{aligned}$$

Then the definition of families yields the desired statement, using the fact that $\varepsilon_{W'} \otimes \chi^* = \chi$ for all $\chi \in \mathrm{Irr}(W')$. □

6.5.4. Direct products. Let W be a finite Coxeter group and assume that we have $W = W_1 \times W_2$ where $W_1, W_2 \subseteq W$ are parabolic subgroups. Furthermore, assume that for $i = 1, 2$ we are given a collection $\mathcal{A}_i$ of a-functions for W_i as in (6.5.1). Then we define a collection of a-functions for W as follows. Let $W' \subseteq W$ be a parabolic subgroup. Then we can write $W' = W_1' \times W_2'$ where $W_i' \subseteq W_i$ are parabolic for $i = 1, 2$. Moreover, every irreducible character of W' is of the form $\chi_1 \boxtimes \chi_2$ where $\chi_i \in \mathrm{Irr}(W_i')$ for $i = 1, 2$. So we can set

$$a_{\chi_1 \boxtimes \chi_2} := a_{\chi_1} + a_{\chi_2}. \tag{1}$$

This defines a function $a_{W'}\colon \mathrm{Irr}(W') \to \mathbb{N}_0$. The resulting collection of a-functions will be denoted by $\mathcal{A}_1 + \mathcal{A}_2$. Since the induction of characters is compatible with direct products, (6.5.1)(a) also holds for $\mathcal{A}_1 + \mathcal{A}_2$. Furthermore, for parabolic subgroups $W_i' \subseteq W_i$ and $\chi_i \in \mathrm{Irr}(W_i')$ $(i = 1, 2)$, we have

$$J_{W_1' \times W_2'}^{W_1 \times W_2}(\chi_1 \boxtimes \chi_2) = J_{W_1'}^{W_1}(\chi_1) \boxtimes J_{W_2'}^{W_2}(\chi_2). \tag{2}$$

Finally, since $(\chi_1 \boxtimes \chi_2)^* = \chi_1^* \boxtimes \chi_2^*$, the definition immediately shows that if $\mathcal{F}_i$ is an $\mathcal{A}_i$-family of $\mathrm{Irr}(W_i')$ for $i = 1, 2$, then

$$\{\chi_1 \boxtimes \chi_2 \mid \chi_1 \in \mathcal{F}_1, \chi_2 \in \mathcal{F}_2\} \tag{3}$$

is an $(\mathcal{A}_1 + \mathcal{A}_2)$-family of $\mathrm{Irr}(W')$, and all such families arise in this way.

Our aim will be to find a "natural" set of conditions such that there exists a unique collection of a-functions which satisfies that set of conditions. The following definition suggests such a set of conditions. To give some idea of where condition (b) comes from, note that we have already encountered such a property for type A_{n-1} in Proposition 5.4.11.

6.5.5. Definition. A collection $\mathcal{A}$ as in (6.5.1) is called *normalized* if, for any parabolic subgroup $W' \subseteq W$, we have $a_{1_{W'}} = 0$ where $1_{W'}$ denotes the trivial character of W'. Furthermore, we say that $\mathcal{A}$ is *saturated* if the following two conditions hold for all parabolic subgroups $\{1\} \neq W' \subseteq W$ and all $\chi \in \mathrm{Irr}(W')$:

(a) There exists a proper parabolic subgroup $W'' \subset W'$ and $\psi \in \mathrm{Irr}(W'')$ such that χ or $\chi^* = \varepsilon_{W'} \otimes \chi$ occurs in $J_{W''}^{W'}(\psi)$.

(b) We have $a_{\chi^*} - a_\chi = \omega_\chi(\hat{T}')$, with ω_χ defined as in (6.2.1).

Here, we let $T' := \{wsw^{-1} \mid s \in S', w \in W'\}$, where W' is generated by $S' \subseteq S$; moreover, $\hat{T}'$ denotes the sum of the elements in T', which is an element of the centre of $\mathbb{C}[W']$. By Exercise 5.4, T' is the set of all reflections in W', and that set is the intersection of the corresponding set for W with W'.

Note that the b-invariants define a collection of a-functions as in (6.5.1) but which is not saturated: for example, condition (b) in Definition 6.5.5 fails for the characters of degree 512 in type E_7 and the characters of degree 4096 in type E_8. Moreover, condition (a) fails in type $I_2(m)$ for $m \geq 4$; see (5.3.4).

6.5.6. Lemma. *For any finite Coxeter group W, there exists at most one collection of a-functions as in (6.5.1) which is normalized and saturated.*

Proof. We proceed by induction on $|W|$. If $W = \{1\}$, then $\mathrm{Irr}(W) = \{1_W\}$. So the normalization forces $a_{1_W} = 0$. Now assume that $W \neq \{1\}$ and let $\mathcal{A}$ be a normalized and saturated collection of a-functions for W as in (6.5.1). Then, for any parabolic subgroup $W' \subseteq W$, the functions

$$\mathcal{A}' := \{a_{W''} \colon \mathrm{Irr}(W'') \to \mathbb{N}_0 \mid W'' \subseteq W' \text{ parabolic}\}$$

form a normalized and saturated collection of a-functions for W'. So, if $W' \neq W$, then $\mathcal{A}'$ is unique by induction. It remains to show that the function $a \colon \mathrm{Irr}(W) \to \mathbb{N}_0$ is also uniquely determined. For any $\chi \in \mathrm{Irr}(W)$, we define

$$a'_\chi := \max\{a_\psi\}_\psi,$$

where ψ runs over all irreducible characters of all proper parabolic subgroups $W' \subset W$ such that $m(\chi, \psi) \neq 0$. Then, by (6.5.1)(a), we have

$$a'_\chi \leq a_\chi \qquad \text{for all } \chi \in \mathrm{Irr}(W). \tag{$*$}$$

Moreover, the condition in Definition 6.5.5 shows that

$$a_\chi = a'_\chi \quad \text{or} \quad a_{\chi^*} = a'_{\chi^*}, \qquad \text{for all } \chi \in \mathrm{Irr}(W). \tag{$**$}$$

Now there are two cases to consider.

Case 1. Assume that $a'_{\chi^*} - a'_\chi \leq \omega_\chi(\hat{T})$. We claim that then $a_\chi = a'_\chi$. Indeed, if we had $a'_\chi \neq a_\chi$, then $a_{\chi^*} = a'_{\chi^*}$ by ($**$) and $a'_\chi < a_\chi$ by ($*$). So we would obtain the following contradiction to Definition 6.5.5(b):

$$a_{\chi^*} - a_\chi < a'_{\chi^*} - a'_\chi \leq \omega_\chi(\hat{T}).$$

Hence we have $a_{\chi^*} = a'_{\chi^*}$ and so $a_\chi = a'_{\chi^*} - \omega_\chi(\hat{T})$.

Case 2. Assume that $a'_{\chi^*} - a'_\chi \geq \omega_\chi(\hat{T})$. We claim that then $a_{\chi^*} = a'_{\chi^*}$. Indeed, if we had $a'_{\chi^*} \neq a_{\chi^*}$ then $a'_\chi = a_\chi$ by (**) and $a'_{\chi^*} < a_{\chi^*}$ by (*). So, again, we would obtain a contradiction to Definition 6.5.5(b):

$$a_{\chi^*} - a_\chi > a'_{\chi^*} - a'_\chi \geq \omega_\chi(\hat{T}).$$

Hence we have $a_{\chi^*} = a'_{\chi^*}$ and so $a_\chi = a'_{\chi^*} - \omega_\chi(\hat{T})$ is again determined. □

Now we have to show that there does indeed exist a collection of a-functions as in (6.5.1) which is normalized and saturated.

6.5.7. Definition (Lusztig's a-function; cf. also Remark 6.5.14). Let W be a finite Coxeter group. We define a function $\chi \mapsto a_\chi$ on $\mathrm{Irr}(W)$ as follows. If $W = \{1\}$, then $\mathrm{Irr}(W) = \{1_W\}$ and we set $a_{1_W} = 0$. Now assume that $W \neq \{1\}$.

If W is irreducible, we define a_χ case by case. If W is of type A_{n-1} $(n \geq 1)$, then $W \cong \mathfrak{S}_n$ and every $\chi \in \mathrm{Irr}(W)$ is labelled by a partition $\lambda \vdash n$. In this case, we set $a_\chi := b_\chi = a(\lambda)$, where $a(\lambda)$ is defined in (5.4.2) and b_χ is the b-invariant. If W is of type B_n or D_n, then every $\chi \in \mathrm{Irr}(W)$ is labelled by a symbol Λ of rank n and defect 1 or 0. In this case, we set $a_\chi := a(\Lambda)$; see Definition 6.4.3. For type $I_2(m)$, we let the trivial and the sign character have a-invariants 0 and m, respectively; all the remaining characters have a-invariant 1. Finally, if W is of type H_3, H_4, F_4, E_6, E_7 or E_8, the definition of a_χ is contained in the tables in Appendix C.

If W is not irreducible, we write $W = W_1 \times \cdots \times W_m$ where $W_i \subset W$ are parabolic subgroups which are themselves irreducible. Each $\chi \in \mathrm{Irr}(W)$ is of the form $\chi = \chi_1 \boxtimes \cdots \boxtimes \chi_m$ where $\chi_i \in \mathrm{Irr}(W_i)$ for all i. Then we set

$$a_\chi = a_{\chi_1 \boxtimes \cdots \boxtimes \chi_m} := a_{\chi_1} + \cdots + a_{\chi_m}. \tag{1}$$

The corresponding collection of a-functions will be denoted by $\mathcal{A}_0$.

Note that, at this stage, it is not even clear that the conditions in (6.5.1) hold for $\mathcal{A}_0$. But we will see that this is indeed the case and that, moreover, $\mathcal{A}_0$ is normalized and saturated.

6.5.8. Families for type A_{n-1}. Let $W = \mathfrak{S}_n$ $(n \geq 1)$ with generators $s_i = (i, i+1)$ for $1 \leq i \leq n-1$. Every proper parabolic subgroup of W is again a direct product of symmetric groups. So, by Definition 6.5.7, we have

$$a_\chi = b_\chi \qquad \text{for all } \chi \in \mathrm{Irr}(W'),\ W' \subseteq W \text{ parabolic}.$$

The condition in (6.5.1)(a) holds by Lemma 5.2.5. In fact, the J-induction is nothing but the j-induction; see (5.2.8). Furthermore, we have seen in Section 5.4 that each $\chi \in \mathrm{Irr}(W)$ is obtained by applying the j-induction to the sign character of a suitable parabolic subgroup. This shows that the condition in Definition 6.5.5(a) is satisfied. Finally, the condition in Definition 6.5.5(b) holds by Proposition 5.4.11. Since the j-induction takes irreducible characters to irreducible characters, we can now conclude that each $\chi \in \mathrm{Irr}(\mathfrak{S}_n)$ forms a family by itself.

We now turn to the problem of showing that the conditions in (6.5.1)(a) and Definition 6.5.5 hold in general for all types.

6.5.9. Lemma. *The conditions in (6.5.1) are satisfied for Lusztig's a-functions.*

Proof. We argue by induction on $|W|$. Then it is enough to show that for any proper parabolic $W' \subset W$, we have $m(\chi, \psi) \neq 0 \Rightarrow a_\chi \geq a_\psi$ for all $\psi \in \mathrm{Irr}(W')$ and $\chi \in \mathrm{Irr}(W)$. (Note that this implies, in particular, that $a_\chi \geq 0$.) If $W = \{1\}$, there is nothing to prove, since $a_{1_W} = 0$. Now assume that $W \neq \{1\}$. By (6.5.4), we can also assume that W is irreducible. By the transitivity of induction, we can further assume that $W' \subset W$ is maximal. We now proceed case by case.

For type A_{n-1}, see (6.5.8). If W is of type $I_2(m)$, the desired condition is checked using the induction formulas in (6.3.5). If W is of exceptional type H_3, H_4, F_4, E_6, E_7 or E_8, we check the desired property by explicit computation. For this purpose, we have to compute the complete table describing the induction of characters from W' to W. That table is determined using the ordinary character tables of W' and W (computed by the algorithm in (5.3.5)) and the fusion map from $\mathrm{Cl}(W')$ to $\mathrm{Cl}(W)$ (computed using Algorithm I, p. 86). These computations can be easily performed in GAP, for example.

Now assume that $W = W_n$ is of type B_n. Let $\chi \in \mathrm{Irr}(W_n)$ and $\psi \in \mathrm{Irr}(W')$ be such that $m(\chi, \psi) \neq 0$. With the notation of Section 6.1, we have $W' = W_k \times H_l$ where $n = k + l$ and $l \geq 1$. Using also (6.5.8), we can write

$$\psi = \psi_1 \boxtimes J_{H_l'}^{H_l}(\varepsilon_{H_l'}) = J_{W_k \times H_l'}^{W_k \times H_l}\left(\psi_1 \boxtimes \varepsilon_{H_l'}\right)$$

for some $\psi_1 \in \mathrm{Irr}(W_k)$ and some parabolic subgroup $H_l' \subseteq H_l$. Consequently, we have $m(\chi, \psi_1 \boxtimes \varepsilon_{H_l'}) \neq 0$, where $a(\psi) = a(\psi_1 \boxtimes \varepsilon_{H_l'})$. Now H_l' itself is a direct product of symmetric groups. Let us write $H_l' = \tilde{H}_l \times H_r$, where H_r $(r \geq 1)$ is one such factor and $\tilde{H}_l$ is the product of all the other factors. Then $W_k \times \tilde{H}_l' \subseteq W_{n-r}$ and so, by the transitivity of induction,

$$\mathrm{Ind}_{W_k \times H_l'}^{W_n}\left(\psi_1 \boxtimes \varepsilon_{H_l'}\right) = \mathrm{Ind}_{W_{n-r} \times H_r}^{W_n}\left(\mathrm{Ind}_{W_k \times \tilde{H}_l}^{W_{n-r}}\left(\psi_1 \boxtimes \varepsilon_{\tilde{H}_l}\right) \boxtimes \varepsilon_{H_r}\right).$$

By induction, we have $a(\psi_2) \geq a(\psi_1 \boxtimes \varepsilon_{\tilde{H}_l}) = a(\psi) - a(\varepsilon_{H_r}) = a(\psi) - r(r-1)/2$ for all $\psi_2 \in \mathrm{Irr}(W_{n-r})$ which occur in the character

$$\mathrm{Ind}_{W_k \times \tilde{H}_l}^{W_{n-r}}\left(\psi_1 \boxtimes \varepsilon_{\tilde{H}_l}\right).$$

Hence, we have $m(\chi, \psi_2 \boxtimes \varepsilon_{H_r}) \neq 0$ for some $\psi_2 \in \mathrm{Irr}(W_{n-r})$ with $a(\psi_2) \geq a(\psi) - r(r-1)/2$. But now we have reached a situation where we can apply Proposition 6.4.7, to yield $a(\chi) \geq a(\psi)$, as desired.

Finally, assume that $W = W_n'$ is of type D_n, $n \geq 2$. Then we can argue as above, now using Proposition 6.4.9. Note that this also works without any further problem for the parabolic subgroups $W' = H_n^{\pm}$. □

6.5.10. Families for the exceptional types. Let W be a finite Coxeter group of type H_3, H_4, F_4, E_6, E_7 or E_8. Then the tables in Appendix C contain the information concerning the a-function and the corresponding families. The families are computed using the tables for the J-induction from all maximal parabolic subgroups in Appendix D, which in turn are computed from complete induce/restrict matrices for the relevant subgroups.

Let $W = \langle s, t\rangle$ be a dihedral group where st has order $m \geq 3$. The only proper parabolic subgroups are the trivial subgroup and the two subgroups of type A_1 generated by s and by t. For these subgroups, we already know the a-functions by (6.5.8). Using the notation in (6.3.5), we have the following table:

χ	1_W	χ_j	ε_1	ε_2	ε
a_χ	0	1	1	1	m
b_χ	0	j	$m/2$	$m/2$	m

Families:
$\{1_W\}$, $\{\varepsilon\}$,
$\{\varepsilon_1, \varepsilon_2, \chi_1, \chi_2, \ldots\}$

The families are determined using the induction formulas in (6.3.5). Note that $\varepsilon_1^* = \varepsilon_2$ and $\chi_j^* = \chi_j$ for all j. The a-invariants can also be determined by following the steps in the proof of Lemma 6.5.6. For the b-invariants, see (5.3.4).

6.5.11. Proposition (Families in type B_n). *Let $n \geq 1$ and $\Lambda, \Lambda' \in \Phi_{n,1}$. Using shift, we can assume that Λ and Λ' have the same total number of entries. Then $[\Lambda]$ and $[\Lambda']$ lie in the same family of $\mathrm{Irr}(W_n)$ if and only if Λ and Λ' contain the same entries (arranged perhaps differently in rows).*

(For examples, see Table 6.2, p. 199.)

Proof. We proceed by induction on n. If $n = 1$, we only have the trivial and the sign character, which form families by themselves. Hence the assertion holds in this case. Now let $n \geq 2$. Assume first that $[\Lambda], [\Lambda']$ lie in the same family. We must show that Λ and Λ' contain the same entries. By the definition of families, it is enough to consider the case where there exist a proper parabolic subgroup $W' \subset W_n$ and $\psi, \psi' \in \mathrm{Irr}(W')$ in the same family of $\mathrm{Irr}(W')$ such that

$$\text{either } [\Lambda] \text{ occurs in } J_{W'}^{W_n}(\psi) \text{ and } [\Lambda'] \text{ occurs in } J_{W'}^{W_n}(\psi')$$
$$\text{or } [\Lambda^*] = \varepsilon \otimes [\Lambda] \text{ occurs in } J_{W'}^{W_n}(\psi) \text{ and } [(\Lambda')^*] = \varepsilon \otimes [\Lambda'] \text{ occurs in } J_{W'}^{W_n}(\psi').$$

Now, two symbols contain the same entries if and only if the corresponding dual symbols contain the same entries (see (6.4.5)). Hence, it is enough to consider the first of the above two possibilities. Furthermore, since the definition of families is recursive and the J-induction is transitive, we can actually assume that W' is a maximal parabolic subgroup. So we can write $W' = W_k \times H_l$ where $n = k + l$ and $l \geq 1$. Correspondingly, we have $\psi = \psi_1 \boxtimes \chi_\lambda$ and $\psi' = \psi_1' \boxtimes \chi_{\lambda'}$, where $\psi_1, \psi_1' \in \mathrm{Irr}(W_k)$ and $\lambda, \lambda' \vdash l$. Now, by assumption, ψ and ψ' lie in the same family of W'. Since families are compatible with direct products, we conclude that ψ_1, ψ_1' must lie in the same family of W_k and $\chi_\lambda, \chi_{\lambda'}$ must lie in the same family of H_l. Using (6.5.8), we see that $\lambda = \lambda'$. Moreover, $\chi_\lambda = \chi_{\lambda'}$ is obtained by applying the J-induction to the sign character of a parabolic subgroup $H_l' \subseteq H_l$.

We can write $H'_l = H_{l_1} \times \cdots \times H_{l_r}$ where $l_1 + \cdots + l_r = l$ and $H_{l_i} \cong \mathfrak{S}_{l_i}$ for all i. Then we have

$$J_{W'}^{W_n}(\psi) = J_{W_k \times H_l}^{W_n}\Big(\psi \boxtimes \chi_\lambda\Big) = J_{W_k \times H_{l_1} \times \cdots \times H_{l_r}}^{W_n}\Big(\psi_1 \boxtimes \varepsilon_{H_{l_1}} \boxtimes \cdots \boxtimes \varepsilon_{H_{l_r}}\Big).$$

Using the transitivity of the J-induction, we can compute the above character in a sequence of steps, by first inducing from $W_k \times H_{l_1}$ to W_{k+l_1}, then from $W_{k+l_1} \times H_{l_2}$ to $W_{k+l_1+l_2}$, and so on. A similar discussion also applies to ψ'. In this way, we see that we may actually assume without loss of generality that $\chi_\lambda = \chi_{\lambda'}$ is the sign character of H_l. Now we can proceed as follows. By induction, we know that ψ_1, ψ_2 are labelled by symbols which contain the same entries. Hence, using Proposition 6.4.7, we see that $J_{W'}^{W_n}(\psi)$ and $J_{W'}^{W_n}(\psi')$ are linear combinations of irreducible characters which are labelled by symbols which all contain the same entries. Hence, Λ and Λ' contain the same entries, as required.

Conversely, assume that Λ and Λ' contain the same entries. We must show that $[\Lambda]$ and $[\Lambda']$ lie in the same family. Now, if Λ satisfies one of the conditions (a), (b) or (c) in Exercise 6.8, then so does Λ'. In this case, Exercise 6.8, Proposition 6.4.7 and induction show that $[\Lambda']$ and $[\Lambda']$ lie in the same family. Furthermore, if Λ does not satisfy the above conditions, then Λ^* does. Now we can repeat the above argument with the dual symbols and conclude that $[\Lambda^*]$ and $[(\Lambda')^*]$ lie in the same family. But then $[\Lambda]$ and $[\Lambda']$ also lie in the same family, thanks to the compatibility with tensoring by the sign character. □

6.5.12. Proposition (Families in type D_n). *Let $n \geq 3$. If Λ is a degenerate symbol of rank n and defect 0, then each of the two characters $[\Lambda]_\pm \in \mathrm{Irr}(W'_n)$ forms a family by itself. Now let Λ, Λ' be non-degenerate symbols of rank n and defect 0, such that Λ and Λ' have the same total number of entries. Then $[\Lambda]$ and $[\Lambda']$ lie in the same family of $\mathrm{Irr}(W'_n)$ if and only if Λ and Λ' contain the same entries (arranged perhaps differently in rows).*

For example, the following are families in D_4:

$$\left\{\left[\begin{pmatrix}2,0\\3,1\end{pmatrix}\right], \left[\begin{pmatrix}1,0\\3,2\end{pmatrix}\right], \left[\begin{pmatrix}2,1\\3,0\end{pmatrix}\right]\right\}, \left\{\left[\begin{pmatrix}2,1\\2,1\end{pmatrix}\right]_+\right\}, \left\{\left[\begin{pmatrix}2,1\\2,1\end{pmatrix}\right]_-\right\}.$$

Proof. If Λ, Λ' are non-degenerate symbols, the assertion is proved by a similar argument to that in Proposition 6.5.11 (using Proposition 6.4.9 instead of Proposition 6.4.7 and an analogous version of Exercise 6.8). Now assume that Λ is degenerate. We proceed by induction on n to show that each of the two characters $[\Lambda]_\pm$ forms a family by itself. If $n = 2$, then we are in type $A_1 \times A_1$ and the assertion is clear by (6.5.8). Now let $n \geq 4$. Assume that $\chi \in \mathrm{Irr}(W'_n)$ belongs to the same family as $[\Lambda]_\pm$. Up to replacing $[\Lambda]_\pm$ and χ by their duals if necessary (note that Λ^* is again degenerate), this means that there exist a proper parabolic subgroup $W' \subset W'_n$ and $\psi, \psi' \in \mathrm{Irr}(W')$ in the same family such that $[\Lambda]_\pm$ occurs in $J_{W'}^{W'_n}(\psi)$ and χ occurs in $J_{W'}^{W'_n}(\psi')$. By similar reduction arguments to that in the proof of Proposition 6.5.11, we may assume without loss of generality that W' is maximal. Then W' is of the form $W' = W'_k \times H_l$ (where $n = k + l$ and $l \geq 1$) or of the form $W' = H_n^{\pm}$. Moreover, we can further assume that ψ is of the form $\psi = \psi_1 \boxtimes \varepsilon_{H_l}$ in the first case or of the form $\psi = \varepsilon_{H_n^{\pm}}$ in the second case. In both cases, the formulas in Proposition 6.4.9 show that if ψ_1 is labelled by a non-degenerate symbol, then all constituents of $J_{W'}^{W'_n}(\psi)$ are also labelled by non-degenerate symbols. Hence, ψ_1 must be labelled by a degenerate symbol. But then we have $\psi = \psi'$ (by induction) and $J_{W'}^{W'_n}(\psi) \in \mathrm{Irr}(W'_n)$ by Proposition 6.4.9. Hence we have $\chi = [\Lambda]_{\pm}$, as desired. □

Now we can state the following fundamental result.

6.5.13. Theorem (Lusztig). *Let W be a finite Coxeter group. Then Lusztig's a-functions (see Definition 6.5.7) define a collection of a-functions which is normalized and saturated (see Definition 6.5.5). Conversely, this is the unique collection of a-functions satisfying these conditions. Moreover, the corresponding families have the following properties, for any parabolic subgroup $W' \subseteq W$:*

(a) *The function $a_{W'}\colon \mathrm{Irr}(W') \to \mathbb{N}_0$ is constant on the families of $\mathrm{Irr}(W')$.*
(b) *We have $a_\chi \leq b_\chi$ for all $\chi \in \mathrm{Irr}(W')$, with equality for exactly one member of each family of $\mathrm{Irr}(W')$. If $a_\chi = b_\chi$, we say that χ is a special character.*
(c) *For any special $\chi \in \mathrm{Irr}(W')$, we have $\gamma_\chi = 1$, i.e., $P_\chi(q) = q^{b_\chi} +$ combination of higher powers of q; for the definition of P_χ, γ_χ see (5.2.2).*

Proof. By Lemma 6.5.6, we already know that there exists at most one collection of a-functions which is normalized and saturated. So it is enough to show that the collection $\mathcal{A}_0$ defined in terms of Lusztig's a-functions in Definition 6.5.7 satisfies the required conditions. By Lemma 6.5.9, the condition in (6.5.1)(a) is satisfied. By definition, $\mathcal{A}_0$ is also normalized. Hence it remains to show that $\mathcal{A}_0$ is saturated. This will be done by induction on $|W|$. It is enough to check that conditions (a) and (b) in Definition 6.5.5 hold for the characters of W itself. If $W = \{1\}$, there is nothing to prove. Now assume that $W \neq \{1\}$.

If W is irreducible, we proceed case by case, as follows. For type A_{n-1}, we settled these points in (6.5.8). For type $I_2(m)$, we simply use (5.3.4) and (6.5.10). For the remaining exceptional types, the tables in Appendix D show that the condition in Definition 6.5.5(a) holds. The information concerning the character values on reflections in the tables in Appendix C can be used to check that the condition in Definition 6.5.5(b) is also satisfied. Note that these tables also describe the effect of tensoring with the sign character.

It remains to consider groups of type B_n and D_n. Condition (b) in Definition 6.5.5 is established in Proposition 6.4.6. For type B_n, condition (a) in Definition 6.5.5 holds by Exercise 6.8. For type D_n, we can argue as follows. If we have an irreducible character which is labelled by a non-degenerate symbol, then we can use Exercise 6.8 again. If that symbol is degenerate, then the corresponding character is obtained by applying the J-induction to the sign character of a suitable parabolic subgroup. (This follows from Proposition 6.4.9 and (6.5.8).)

Thus, we have shown that $\mathcal{A}_0$ satisfies all the required conditions in the case where W is irreducible. If W is not irreducible, we can write $W = W_1 \times W_2$ where $W_1, W_2 \subset W$ are proper parabolic subgroups. Every $\chi \in \mathrm{Irr}(W)$ is of the form $\chi = \chi_1 \boxtimes \chi_2$ where $\chi_i \in \mathrm{Irr}(W_i)$ for $i = 1, 2$. By induction, χ_1 or χ_1^* occurs in $J_{W_1'}^{W_1}(\psi_1)$ for some proper parabolic $W_1' \subseteq W_1$ and $\psi_1 \in \mathrm{Irr}(W_1')$. Hence $\chi_1 \boxtimes \chi_2$ occurs in $J_{W_1'}^{W_1}(\psi_1) \boxtimes \chi_2$ or $(\chi_1 \boxtimes \chi_2)^* = \chi_1^* \boxtimes \chi_2^*$ occurs in $J_{W_1'}^{W_1}(\psi_1) \boxtimes \chi_2^*$. So, using (6.5.4)(2), we see that part (a) of Definition 6.5.5 is satisfied.

Now consider the set of reflections $T_i \subseteq W_i$ $(i = 1, 2)$. Then we have $T = T_1 \cup T_2$. Moreover, if z_i is in the centre of $\mathbb{C}[W_i]$ for $i = 1, 2$, we have $\omega_{\chi_1 \boxtimes \chi_2}(z_1 z_2) = \omega_{\chi_1}(z_1)\omega_{\chi_2}(z_2)$. It follows that

$$\omega_{\chi_1 \boxtimes \chi_2}(\widehat{T}) = \omega_{\chi_1 \boxtimes \chi_2}(\widehat{T}_1 + \widehat{T}_2) = \omega_{\chi_1}(\widehat{T}_1) + \omega_{\chi_2}(\widehat{T}_2).$$

By induction and the additivity of the a-functions, this yields that part (b) of Definition 6.5.5 is also satisfied. Thus, $\mathcal{A}_0$ satisfies all the required conditions.

Finally, let us show that the properties in (a), (b) and (c) hold. First note that the b-invariants and the Molien series are compatible with direct products; see Exercise 5.2. Hence, since the same also holds for the a-invariants and the families, it is enough to consider the case where W is irreducible. Then (a) and (b) are verified using the explicit description of the families in all cases: for type A_{n-1}, there is nothing to prove, see (6.5.8); for types B_n and D_n, see (6.4.4), Proposition 6.5.11 and Proposition 6.5.12; for type $I_2(m)$ and all the remaining exceptional types, just look through the tables in Appendix C.

Now consider the coefficient γ_χ of q^{b_χ} in P_χ. By Remark 5.6.7, we can have $\gamma_\chi \neq 1$ only in types D_n, E_7 or E_8. For types E_7 and E_8, the tables in Appendix C show that if $\gamma_\chi \neq 1$ then we also have $a_\chi \neq b_\chi$. Finally, let $\chi \in \mathrm{Irr}(W_n')$ be such that $\gamma_\chi \neq 1$. Then χ is labelled by a pair of partitions (λ, μ) with $\lambda \neq \mu$ and $|\lambda| = |\mu|$. This means that a corresponding symbol $\Lambda = \binom{X}{Y}$ satisfies $X \neq Y$, $\mathrm{rk}(X) = \mathrm{rk}(Y)$. But then the conditions in (6.4.4) show that $a_\chi \neq b_\chi$. □

6.5.14. Remark. Lusztig has defined a-functions in terms of the "generic degrees" of the irreducible characters of W (see [Lusztig 1979a] and Remark 9.4.10 below). Then the corresponding collection of a-invariants is normalized and it can be shown by general arguments that it satisfies the conditions in (6.5.1)(a) and Definition 6.5.7(b). The condition in Definition 6.5.7(a) only follows by inspection, after explicitly describing the J-induction in all cases; see [Lusztig 1984, (5.30)]. The point about the approach that we followed in this section is that it is independent of the theory of Iwahori–Hecke algebras. Thus, conversely, the explicit formulas in Definition 6.5.7 (which are taken from [Lusztig 1984, Chap. 4]) yield a piece of information concerning the generic degrees for W.

Alternatively, one can also define a-invariants in terms of the "leading coefficients" of character values of Iwahori–Hecke algebras (see [Lusztig 1981b]) or in terms of the a-function on W itself (see [Lusztig 1985]). In the case where W is a finite Weyl group, all these definitions coincide (see [Lusztig 1987b, §3]).

6.6 BIBLIOGRAPHICAL REMARKS AND EXERCISES

The classical reference for symmetric functions is [Macdonald 1995]. For further approaches to the Littlewood–Richardson rule, see [Fulton 1997, Chap. 5], [James and Kerber 1981, (2.8.13)] and [Gasharov 1998]. Pieri's rule also appears as the "2nd version of Young's rule" in [James and Kerber 1981, (2.8.2)]. For a direct proof in terms of Schur functions, see [Macdonald 1995, (I.5.16)]. For the other classical types, see also [Zelevinsky 1981].

Definition 6.3.1 appears in [Curtis et al. 1971, (4.1)]. The fact that all irreducible characters of Weyl groups are of parabolic type and the results in Section 6.2 are proved in [Benson and Curtis 1972]. Our exposition closely follows that paper, except that we have a completely elementary proof of Corollary 6.2.3. For extensions of these results to complex reflection groups, see [Malle 2000]. We addressed the question of fields of definition for the irreducible characters in Chapter 5 (see also the bibliographic remarks there). Historically, the last open case concerning Schur indices in Theorem 6.3.8 was the character of degree 48 in type H_4, which is solved in [Benson and Grove 1973].

Symbols were introduced in [Lusztig 1977] and used to label the unipotent characters of finite classical groups. For many purposes, they are more convenient than the labelling by pairs of partitions. See [Malle 1995] for a generalization to e-symbols. We have followed the exposition in [Olsson 1986]. The argument in step 2 of the proof of Proposition 6.4.6 uses the polynomial identities concerning the *content* of a partition in [Macdonald 1995, (I.1), Example 3]; a general argument (using fake degrees) for the first equality in that proposition is given in [Malle 2000, §6]. All the ideas and results concerning a-functions and families in Section 6.5—except for the uniqueness statement in Lemma 6.5.6—are due to Lusztig; see [Lusztig 1979a], [Lusztig 1982] and [Lusztig 1984, Chap. 4].

Exercise 6.1. Let $d \geq 0$ and λ, μ be partitions such that $\mu \vdash |\lambda| + d$. Assume that λ has parts $\lambda_1 \geq \cdots \geq \lambda_r$ and μ has parts $\mu_1 \geq \cdots \geq \mu_s$. Show that λ, μ satisfy the conditions in Pieri's rule if and only if $\mu_i \geq \lambda_i \geq \mu_{i+1}$ for all i, where we set $\lambda_i := 0$ for $i > r$ and $\mu_i := 0$ for $i > s$.

Exercise 6.2. Show that we have the following implications for $n \geq 3$:

(a) If $\chi, \chi' \in \mathrm{Irr}(\mathfrak{S}_n)$ have the same restriction to $\mathfrak{S}_{n-1}$, then $\chi = \chi'$.
(b) If $\chi, \chi' \in \mathrm{Irr}(W_n)$ have the same restriction to W_{n-1}, then $\chi = \chi'$.

Statement (b) can be found in [Shoji 1979, (3.5)].

[*Hint.* (a) is clear by Corollary 6.2.3. To prove (b), we use the notation of (6.1.9). Let $\chi = \chi_{(\lambda,\mu)} \in \mathrm{Irr}(W_n)$. If both $I(\lambda)$ and $I(\mu)$ are non-empty, then λ and μ are determined by the pairs $(\lambda, \mu^{(d)})$ (for $d \in I(\mu)$) and $(\lambda^{(d')}, \mu)$ (for $d' \in I(\lambda)$). If $I(\lambda)$ is empty, then $\lambda = \varnothing$ and we can use (a). Argue similarly if $I(\mu) = \varnothing$.]

Exercise 6.3. Let $n, k, l \geq 0$ be integers such that $n = k + l$. Let $\lambda \vdash k$. Show that we have the following "dual" version of Pieri's rule:

$$\mathrm{Ind}_{\mathfrak{S}_a \times \mathfrak{S}_b}^{\mathfrak{S}_n}\Big(\chi_\lambda \boxtimes \varepsilon_b\Big) = \sum_\mu \chi_\mu, \qquad \varepsilon_b = \text{sign character of } \mathfrak{S}_b,$$

where the sum is over all partitions $\mu \vdash n$ whose Young diagram is obtained from that of λ by adding l boxes, with no two in the same *row*.

[*Hint.* Tensor Pieri's rule in Corollary 6.1.7 with ε_n and use Corollary 5.4.9.]

Exercise 6.4. Let $n, k, l \geq 0$ be integers such that $n = k+l$. Let $\lambda \vdash k$ and $\mu \vdash l$. We define a partition $\nu \vdash n$ as follows. Assume that λ has parts $\lambda_1 \geq \cdots \geq \lambda_r$ and μ has parts $\mu_1 \geq \cdots \geq \mu_s$. We may assume that $r = s$, by adding to λ or to μ some parts equal to 0 if necessary. Then ν is the partition with parts $\lambda_1 + \mu_1 \geq \cdots \geq \lambda_r + \mu_r$. We call ν the sum of λ and μ, and write $\nu = \lambda + \mu$.

(a) Show that $a(\lambda + \mu) = a(\lambda) + a(\mu)$, with the a-invariant as in (5.4.2).
(b) Show that ν^* is obtained by taking all parts of λ^* and μ^* and arranging them in decreasing order.

[*Hint.* For (b), see [Macdonald 1995, (I.1.8)].]

Exercise 6.5. This exercise is concerned with some properties of *Kostka numbers*, which we encountered in the remarks following Theorem 5.4.7. For the proof of (a), we rely on Pieri's rule. The relation in (b) goes back to Frobenius; its proof requires some further reading in [Fulton 1997] or [Macdonald 1995]. Let $n \geq 0$ and μ be any composition of n. Then we have

$$\mathrm{Ind}_{\mathfrak{S}_\mu}^{\mathfrak{S}_n}(1_{\mathfrak{S}_\mu}) = \sum_{\lambda \vdash n} \kappa_{\lambda\mu}\, \chi_\lambda, \qquad \text{where } \kappa_{\lambda\mu} \text{ are the Kostka numbers.}$$

It will be convenient to use the convention that $\kappa_{\lambda\nu} = 0$ if ν is any sequence of integers which is not a composition of n.

(a) Use Pieri's rule to prove the following rule for computing $\kappa_{\lambda\mu}$. If μ has (non-zero) parts $\mu_1, \mu_2, \ldots, \mu_r$, then $\kappa_{\lambda\mu}$ is the number of λ-tableaux $\mathbb{T}$ of weight μ, i.e., for $1 \leq i \leq r$, the number of occurrences of i in $\mathbb{T}$ equals μ_i.
(b) For any $w \in \mathfrak{S}_n$, we set $\mu_w := (\mu_1 - 1 + 1.w, \ldots, \mu_n - n + n.w)$ (where we use the convention that $\mu_i = 0$ for $i > r$). Then show that

$$\sum_{w \in \mathfrak{S}_n} \varepsilon(w)\, \kappa_{\lambda\mu_w} = \begin{cases} 1 & \text{if } \lambda = \mu, \\ 0 & \text{otherwise.} \end{cases}$$

[*Hint.* To prove (a), proceed by induction on r. The result is clear for $r = 1$. If $r > 1$, let $k = n - \mu_r$ and $\mu' := [\mu_1, \ldots, \mu_{r-1}] \vdash k$. Using the transitivity of induction, we have

$$\mathrm{Ind}_{\mathfrak{S}_\mu}^{\mathfrak{S}_n}(1_{\mathfrak{S}_\mu}) = \mathrm{Ind}_{\mathfrak{S}_k \times \mathfrak{S}_{\mu_r}}^{\mathfrak{S}_n}\left(\mathrm{Ind}_{\mathfrak{S}_{\mu'}}^{\mathfrak{S}_k}(1_{\mathfrak{S}_{\mu'}}) \boxtimes 1_{\mathfrak{S}_{\mu_r}}\right).$$

Now use induction and Pieri's rule to obtain a recursive formula for $\kappa_{\lambda\mu}$. For another proof, using symmetric functions, see [Macdonald 1995, (I.6.4)]. Once (a) is established, (b) can be proved directly by formal manipulations using generating functions; see [Fulton 1997, §6.1, Exercise 4].]

Exercise 6.6. This exercise is used in the proof of Theorem 6.2.9. Let $m \geq 1$ and x_k, x_k', y_k, y_k' $(1 \leq k \leq m)$ be complex numbers such that

$$\begin{aligned} x_k + y_k &= x_k' + y_k' && \text{for } 1 \leq k \leq m, \\ x_k y_l + x_l y_k &= x_k' y_l' + x_l' y_k' && \text{for } 1 \leq k, l \leq m. \end{aligned}$$

Show that then either $x_k = x_k'$, $y_k = y_k'$ for all k, or $x_k = y_k'$, $y_k = x_k'$ for all k.

Exercise 6.7. Let Λ be a symbol. Show that the formulas for $b(\Lambda)$ and $a(\Lambda)$ (see (6.4.4)) can be rewritten in the following form (see [Lusztig 1979a]):

$$b(\Lambda) = 2\sum_{i=1}^{r}(i-1)\alpha_i + 2\sum_{i=1}^{s}(i-1)\beta_i + \sum_{i=1}^{s}\beta_i - \sum_{i=1}^{\lfloor\frac{r+s}{2}\rfloor}\binom{r+s-2i}{2},$$

$$a(\Lambda) = \sum_{i=1}^{r}(i-1)\alpha_i + \sum_{i=1}^{s}(i-1)\beta_i + \sum_{i=1}^{r}\sum_{j=1}^{s}\min\{\alpha_i, \beta_j\} - \sum_{i=1}^{\lfloor\frac{r+s}{2}\rfloor}\binom{r+s-2i}{2}.$$

[*Hint.* You will need the identity: $\binom{k}{2} + \binom{l}{2} = \lfloor(\frac{k+l-1}{2})^2\rfloor + \lfloor(\frac{k-l}{2})^2\rfloor$ for $k, l \geq 0$.]

Exercise 6.8. Let $n \geq 1$ and Λ be a symbol of rank n and defect 1. Consider the collection of all entries of Λ and write them in decreasing order: $\gamma_1 \geq \gamma_2 \geq \cdots \geq \gamma_{2s+1}$. Using shift, we may assume that 0 occurs at most once in this list. Show that one of the following cases occurs:

Case 1. Λ satisfies one of the following conditions: (a) $\gamma_i \neq 0$ for all i, (b) $\gamma_i - \gamma_{i+1} \geq 2$ for some i, or (c) $\gamma_{i-1} > \gamma_i = \gamma_{i+1} + 1$ or $\gamma_i - 1 = \gamma_{i+1} > \gamma_{i+2}$ for some i. Show that then $\chi_{[\Lambda]}$ is a component of $J_{W_k \times H_l}^{W_n}(\chi_{[\Lambda']} \boxtimes \varepsilon_{H_l})$, where $l \geq 1$ and Λ' is a symbol of rank k and defect 1.

Case 2. Λ^* satisfies one of the above conditions (a), (b) or (c), where Λ^* is the dual symbol defined in (6.4.5). Note that 0 occurs at most once in Λ^*.

Show that an analogous result holds for non-degenerate symbols of defect 0.

[*Hint.* Check that if (a), (b) or (c) holds, there exists some $l \geq 1$ such that the l largest entries of Λ can be decreased by 1 and the result is a symbol of rank $n-l$ and defect 1. Now use Proposition 6.4.7. Check that, if this is not the case, then Λ^* satisfies (a), (b) or (c); see also [Lusztig 1984, (4.5.6)].]

Exercise 6.9. Assume that (W, S) is irreducible and let $\gamma\colon W \to W$ be a group automorphism such that $\gamma(S) = S$. Then γ also induces a permutation of $\mathrm{Irr}(W)$. Show the following assertion, which can be found in [Lusztig 1984, (4.17)]: the families of $\mathrm{Irr}(W)$ (as given by Theorem 6.5.13) are permuted by γ and such a family is stable under γ if and only if each character in that family is fixed by γ.

[*Hint.* It is enough to consider the case where W is irreducible and γ is not inner. Then, by Theorem 1.3.3, either γ has order 2 and we are in type $I_2(m)$ (m even), D_n (n even) or F_4; or γ has order 3 and we are in type D_4. Now describe explicitly the action of γ on $\mathrm{Irr}(W)$, using the information provided in (5.3.4), Section 5.6, Table C.3 and Table 6.1.]

Exercise 6.10. The purpose of this exercise is to provide a proof for the fact that the unique irreducible character χ of degree 48 of a Coxeter group W of type H_4 cannot be realized over $\mathbb{Q}$, although all its values are rational. We follow the method of [Benson and Grove 1973], but we replace some theoretical arguments by explicit computations in GAP. The idea is to proceed as follows:

Step 1. Try to find a subgroup $H \subsetneq W$ such that there exists some $\psi \in \mathrm{Irr}(H)$ which satisfies the following conditions:

(a) ψ is the restriction of the standard reflection representation of W;
(b) ψ is rational-valued;
(c) ψ occurs with odd multiplicity in the restriction of χ.

By [Benson and Grove 1973], it is possible to find such a subgroup H of order 48. In GAP, we can do this as follows. Let C be a Cartan matrix of type H_4 and $S = \{s_1, s_2, s_3, s_4\}$. Then, writing out the defining equation for the action of $s \in S$ on $V = \mathbb{R}^4$, we obtain an explicit realization of W as a subgroup of $GL_4(\mathbb{R})$, where s_1, s_2, s_3, s_4 are given as in Example 1.5.9. We create the corresponding matrix group record in GAP (using the command `Group`) and compute the conjugacy class C of $s_1 s_3 \in W$ (using the command `ConjugacyClass`). Now the idea is simply to choose randomly three elements $a, b, c \in C$ (using the command `Random`), consider the subgroup $H = \langle a, b, c\rangle$, check if its order is 48 (using `Size`), compute its character table (using `CharTable`) and check by inspection if H has a rational-valued irreducible character of degree 4. After several such random tries, one finds elements a, b, c satisfying these properties. For example:

$$a = \begin{bmatrix} 1 & 0 & 0 & 0 \\ 0 & 1 & 1 & 1 \\ -1-2\lambda & -2-2\lambda & -2-\lambda & -1-\lambda \\ 1+2\lambda & 2+2\lambda & 1+\lambda & \lambda \end{bmatrix}, \quad b = \begin{bmatrix} 0 & 1 & 0 & 0 \\ 1 & 0 & 0 & 0 \\ -1-\lambda & -1-\lambda & -1 & -1 \\ 0 & 0 & 0 & 1 \end{bmatrix},$$

$$c = \begin{bmatrix} 1 & \lambda & \lambda & 0 \\ 1+\lambda & 2\lambda & \lambda & \lambda \\ -1-\lambda & -1-2\lambda & -1-\lambda & -\lambda \\ -1-2\lambda & -1-2\lambda & -2\lambda & -\lambda \end{bmatrix}, \qquad \text{where } \lambda = (1+\sqrt{5})/2.$$

(Using Algorithm A (p. 9), it can also be verified directly that $a, b, c \in W$.) In fact, one can now also see that H has a centre of order 2, that H has a unique irreducible character of degree 4 (which we denote by ψ), and that the value of ψ on the non-trivial central element is -4. Using the fact that $(ab)^6$ has trace -4 and $(ab)^2$ has trace 2, it immediately follows from the character table that ψ is afforded by the embedding $H \subseteq \mathsf{GL}_4(\mathbb{R})$. Thus, (a) and (b) are fulfilled. In order to find the fusion of the conjugacy classes of H into those of W, we use the command `SubgroupFusions`. We can now restrict χ to H and compute that ψ occurs with multiplicity 7 in that restriction. Thus, (c) is also fulfilled.

Step 2. Try to show that the character ψ of H cannot be realized over $\mathbb{Q}$. If this is true then, since ψ can be realized over $\mathbb{Q}(\lambda)$ and $\mathbb{Q}(\lambda)$ has degree 2 over $\mathbb{Q}$, ψ has Schur index 2. This would imply, by condition (c) and [Goldschmidt 1980, §4], that the Schur index of χ is divisible by 2, i.e., χ cannot be realized over $\mathbb{Q}$.

For this purpose, we proceed as follows. Using the embedding

$$\mathbb{Q}(\lambda) \hookrightarrow M_2(\mathbb{Q}), \qquad \lambda \mapsto \begin{bmatrix} 0 & 1 \\ 1 & 1 \end{bmatrix},$$

we can "blow up" the matrices in step 1 to obtain a representation of H by 8×8 matrices with rational coefficients. The character of the "blown-up" representation of H is 2ψ (by construction). Let $D \subseteq M_8(\mathbb{Q})$ be its endomorphism algebra; we have $\dim_{\mathbb{Q}} D = 4$. Now, if ψ could be realized over $\mathbb{Q}$, then D would be the endomorphism algebra of a direct sum of two copies of the same absolutely irreducible representation and, hence, would be isomorphic to $M_2(\mathbb{Q})$. Thus, it remains to show that $D \not\cong M_2(\mathbb{Q})$.

This will be done by the following method (which we learned from G. Nebe). Consider the map $\tau\colon D \to \mathbb{Q}$ given by taking the trace of matrices in $M_8(\mathbb{Q})$. If D were isomorphic to $M_2(\mathbb{Q})$ then τ would be a non-zero scalar multiple of the usual trace on $M_2(\mathbb{Q})$ (see Exercise 7.1). Now we know that there exists some $0 \neq A \in M_2(\mathbb{Q})$ such that A and A^2 have trace 0 (check this!). Hence it will be enough to prove that there exists no $0 \neq m \in D$ with $\tau(m) = \tau(m^2) = 0$.

For this purpose, we compute a basis $\{m_i \mid i = 0, 1, 2, 3\}$ for D. (In GAP, this can be done by computing the centralizer of the subalgebra generated by the "blow-ups" of a, b, c inside the full matrix algebra.) Assume that $m_0 = \mathrm{id}_8$ and $\tau(m_i) = 0$ for $i = 1, 2, 3$. Then we must show, for any $\xi_1, \xi_2, \xi_3 \in \mathbb{Q}$:

$$m := \xi_1 m_1 + \xi_2 m_2 + \xi_3 m_3 \neq 0 \quad \Rightarrow \quad \tau(m^2) \neq 0. \qquad (*)$$

This is done as follows. We compute an orthogonal basis $\{m_1', m_2', m_3'\}$ in the subspace $\langle m_1, m_2, m_3 \rangle \subseteq D$, with respect to the bilinear form $(x, y) \mapsto \tau(xy)$. Up to renumbering and multiplying by squares in $\mathbb{Q}$, we find that the scalar products (m_i', m_i') for $i = 1, 2, 3$ are given by $-2, 6, 6$, respectively. Thus, condition $(*)$ translates to the condition that the equation $-2\,\xi_1^2 + 6\,\xi_2^2 + 6\,\xi_3^2 = 0$ has no non-zero solutions in $\mathbb{Q}$.

7
Representation theory of symmetric algebras

In the previous chapters, we have studied the irreducible characters of finite Coxeter groups. For this purpose, we have only required some basic knowledge about the character theory of finite groups, as can be found, for example, in [Isaacs 1976]. In later chapters, we will want to study in a similar way the irreducible representations and characters of Iwahori–Hecke algebras. For example, we would like to have analogous versions of the usual orthogonality relations for the irreducible characters of a finite group. The purpose of this chapter is to lay the foundations for such a more general theory.

The common feature of group algebras of finite groups and Iwahori–Hecke algebras associated with finite Coxeter groups is that they are both examples of *symmetric* algebras. An (associative) algebra H is called symmetric if it carries a symmetric non-degenerate bilinear form $(\ ,\ )$ such that $(\mathsf{ab}, \mathsf{c}) = (\mathsf{a}, \mathsf{bc})$ for all $\mathsf{a}, \mathsf{b}, \mathsf{c} \in \mathsf{H}$. In this chapter we develop the basic aspects of the representation theory of this class of algebras. This includes both "ordinary" character theory, i.e., the theory of representations of a split semisimple algebra, and "modular" representation theory, i.e., the theory of decomposition maps.

After some general remarks about trace functions, the main results on Schur relations and orthogonality relations will be proved in Section 7.2. The subsequent section contains a discussion of integrality properties. These can be seen as generalizations of known facts about characters of finite groups, e.g., the fact that the degrees of the irreducible characters of a finite group divide the group order. We also introduce the notion of a *character table* for H; see (7.3.11).

Then Section 7.4 deals with decomposition maps. Usually, this is done for algebras over complete discrete valuation rings. But in view of our applications to Iwahori–Hecke algebras, we have to make sure that such a theory also works for more general ground rings. The main applications are Tits's Theorem 7.4.6 and the semisimplicity criterion in Theorem 7.4.7.

The modular representation theory will be developed further in Section 7.5, where we study properties of the decomposition map in more detail. These results will play a role in Chapter 11, where we describe algorithms for computing character values of Iwahori–Hecke algebras.

We make a number of general assumptions. All of our rings have identity elements, and ring homomorphisms preserve these identity elements. In particular, an algebra over a commutative ring is unital. All of our modules are right modules but endomorphisms are written on the left, unless explicitly stated otherwise.

We shall need *Wedderburn's theorem* (see [Curtis and Reiner 1981, §3B]), in the following form. Let H be a finite-dimensional algebra over a field K. The radical $\mathrm{rad}(H)$ is the nilpotent ideal consisting of all elements of H which act as 0 on each simple H-module. The algebra H is semisimple if and only if $\mathrm{rad}(H) = \{0\}$. If this is the case, there is a direct sum decomposition $H = \bigoplus_V H(V)$ where V runs over the simple H-modules (up to isomorphism), and each $H(V)$ is a simple K-algebra. This decomposition is such that an element $h \in H(V)$ acts on a simple module V' as 0 unless V and V' are isomorphic. Since $H(V)$ is a simple algebra, we have an isomorphism $\rho_V \colon H(V) \to M_{n_V}(D_V)$ onto a full matrix algebra, where D_V is a division algebra over K and n_V is the multiplicity of V as a composition factor of H regarded as a module over itself. Moreover, we have $D_V \cong \mathrm{End}_H(V)$ and $\dim_K V = n_V \dim_K D_V$.

We shall say that a simple module V is *split simple* if $\dim_K D_V = 1$, and that H is *split* if all simple modules are split simple.

In Section 7.5 we will also require some basic familiarity with non-semisimple algebras, for which we refer to Chapters 5 and 6 in [Curtis and Reiner 1981].

7.1 TRACE FUNCTIONS

We start with an associative A-algebra H where A is any commutative ring. We assume that H is finitely generated and free over A. The purpose of this section is to present the basic results about trace functions and the Gaschütz–Ikeda lemma for symmetric algebras.

7.1.1. Definition. A *trace function* on H is an A-linear map $\tau \colon H \to A$ such that $\tau(hh') = \tau(h'h)$ for all $h, h' \in H$. The set of trace functions on H is an A-module, with pointwise defined operations. We say that a trace function τ is a *symmetrizing trace* or that H is a *symmetric algebra* if the bilinear form

$$H \times H \to A, \quad (h, h') \mapsto \tau(hh'),$$

is non-degenerate, i.e., if the determinant of the matrix $(\tau(bb'))_{b,b' \in \mathcal{B}}$ is a unit in A for some (and hence every) A-basis $\mathcal{B}$ of H.

If τ is a symmetrizing trace on H and $\mathcal{B}$ is a basis for H, we denote by $\mathcal{B}^\vee = \{b^\vee \mid b \in \mathcal{B}\}$ the dual basis; it is uniquely determined by the requirement that $\tau(b^\vee b') = \delta_{bb'}$ for all $b, b' \in \mathcal{B}$.

7.1.2. Example. Let $n \geq 1$ and $H = M_n(A)$, the algebra of $n \times n$ matrices with entries in A. By Exercise 7.1, *every* trace function on H is a scalar multiple of the usual matrix trace, which we denote by Tr.

For $1 \leq a, b \leq n$ let E_{ab} be the $n \times n$ matrix with (a, b)th entry 1 and 0 otherwise. Then $\mathcal{B} = \{E_{ab} \mid 1 \leq a, b \leq n\}$ is an A-basis for H. We have

$$E_{ab} E_{a'b'} = \delta_{ba'} E_{ab'} \qquad \text{for all } 1 \leq a, b, a', b' \leq n.$$

It follows that Tr is a symmetrizing trace and the basis dual to $\mathcal{B}$ is given by $E_{ab}^{\vee} = E_{ba}$ for all a, b. Note that the sets $\mathcal{B}$ and $\mathcal{B}^{\vee}$ are equal.

7.1.3. Example. Let $H = \bigoplus_{i \in I} M_{n_i}(A)$ where I is a finite index set and n_i are positive integers for all i. Composing the projection on the ith factor with the usual matrix trace on that factor, we obtain a trace function τ_i on H. Let $d_i \in A$ $(i \in I)$ and set $\tau := \sum_{i \in I} d_i \tau_i$. Then, clearly, τ is a trace function on H.

For each $i \in I$ let $\mathcal{B}_i$ be a basis for $M_{n_i}(A) \subseteq H$ as in Example 7.1.2, with dual basis $\mathcal{B}_i^{\vee}$ taken with respect to the usual matrix trace, i.e., the restriction of τ_i to $M_{n_i}(A)$. Then $\mathcal{B} = \coprod_i \mathcal{B}_i$ is a basis for H. We have

$$\tau(b'b^{\vee}) = d_i \delta_{ii'} \delta_{bb'} \qquad \text{for } b \in \mathcal{B}_i \text{ and } b' \in \mathcal{B}_{i'}.$$

It follows that τ is non-degenerate if and only if all d_i are units in A. If this is the case, the set $\bigcup_i \{d_i^{-1} b^{\vee} \mid b \in \mathcal{B}_i\}$ is a basis dual to $\mathcal{B}$ with respect to τ.

By using Wedderburn's theorem, this example shows that all split semisimple algebras over a field are symmetric.

7.1.4. Example. Let V be an H-module which is finitely generated and free over A. Such a module will be called an H-*lattice*. The action of H on V will be written in the form $v \mapsto vh$ $(v \in V, h \in H)$. We obtain a corresponding algebra homomorphism

$$\rho_V \colon H \to \mathsf{End}_A(V), \quad \text{where } v\rho_V(h) = vh \text{ for } v \in V, h \in H.$$

We call ρ_V the *representation afforded by* V. The corresponding *character* is the A-linear map defined by $\chi_V \colon H \to A$, $h \mapsto \mathsf{Tr}(\rho_V(h))$, where Tr denotes the usual matrix trace. Then χ_V is indeed a trace function.

7.1.5. Example. Let G be a finite group and $H = A[G]$ the group algebra of G over A. If $h \in H$ we write $h = \sum_{g \in G} a(h)_g g$ with $a(h)_g \in A$. Then the map $\tau \colon H \to A$, $h \mapsto a(h)_1$, is a trace function. Let $\mathcal{B} := \{g \mid g \in G\}$ be the standard basis of H. Then, for all $g, h \in G$, we have $\tau(gh) = 1$ if $h = g^{-1}$ and $\tau(gh) = 0$ otherwise. It follows that τ is a symmetrizing trace, and that the dual basis is given by $\mathcal{B}^{\vee} = \{g^{-1} \mid g \in G\}$. Again, $\mathcal{B} = \mathcal{B}^{\vee}$ as sets.

7.1.6. The centre and the space of trace functions. Let τ be a symmetrizing trace on H. Then we have a duality between the centre of H and the space of trace functions on H, in the following sense.

For any $\lambda \in H^* := \mathsf{Hom}_A(H, A)$ we can uniquely define an element $\lambda^* \in H$ by the condition that $\tau(\lambda^* h) = \lambda(h)$ for all $h \in H$. The definition shows that, for example, $\tau^* = 1_H$. More generally, we have

$$\lambda^* = \sum_{b \in \mathcal{B}} \lambda(b) b^{\vee}, \qquad \text{where } \mathcal{B} \text{ is any } A\text{-basis of } H.$$

In order to prove such a relation, we just have to multiply both sides by an arbitrary basis element and compute the trace on the resulting elements. The above relations show that the map $H^* \to H$, $\lambda \mapsto \lambda^*$ is an A-linear isomorphism.

For any $h, h' \in H$, define their *commutator* by $[h, h'] := hh' - h'h$. Let $[H, H] \subseteq H$ be the A-submodule spanned by all commutators. By definition, an A-linear map $\lambda: H \to A$ is a trace function if and only if $[H, H] \subseteq \ker(\lambda)$. Thus, the A-module of trace functions on H is canonically isomorphic to the dual space of $H/[H, H]$, i.e., to $\mathrm{Hom}_A(H/[H, H], A)$.

7.1.7. Lemma. *Let $\lambda \in H^*$ and λ^* be the corresponding element in H. Then λ is a trace function if and only if λ^* lies in the centre of H. Moreover, the centre of H is a free A-submodule of H if and only if $\mathrm{Hom}_A(H/[H, H], A)$ is a free A-module.*

Note that, if $H/[H, H]$ is a free A-module then so is $\mathrm{Hom}_A(H/[H, H], A)$, but the converse is not true in general.

Proof. Let $\lambda \in H^*$. Then λ is a trace function if and only if $\lambda(hh') = \lambda(h'h)$ for all $h, h' \in H$. Using the definition of λ^*, this is equivalent to the condition that $\tau(\lambda^*(hh' - h'h)) = 0$ for all $h, h' \in H$. On the other hand, we have $\lambda^* \in Z(H)$, i.e., $\lambda^* h = h\lambda^*$ for all $h \in H$, if and only if $\tau((\lambda^* h - h\lambda^*)h') = 0$ for all $h' \in H$. Comparing these two conditions, we see that they are equivalent if and only if

$$\tau(\lambda^* h' h) = \tau(h\lambda^* h') \qquad \text{for all } h, h' \in H.$$

But this certainly holds since τ is a trace function. The map $\lambda \mapsto \lambda^*$ therefore defines an A-module isomorphism between $\mathrm{Hom}_A(H/[H, H], A)$ and the centre of H. Hence the centre is free if and only if $\mathrm{Hom}_A(H/[H, H], A)$ is free. □

7.1.8. Example. Let $H = A[G]$ be the group algebra over A of a finite group G, with symmetrizing trace as before. Let $\mathrm{Cl}(G)$ be the set of conjugacy classes of G. For each $C \in \mathrm{Cl}(G)$, we let τ_C be the indicator function on the conjugacy class C^{-1} of G. Then $\{\tau_C \mid C \in \mathrm{Cl}(G)\}$ certainly is an A-basis for the space of trace functions on H. For each C let $\hat{C}$ be the unique element in the centre of $A[G]$ such that $\hat{C} = \tau_C^*$. Writing $\hat{C} = \sum_{g \in G} a(C)_g g$ with $a(C)_g \in A$, we have

$$a(C)_g = \tau(\hat{C} \cdot g^{-1}) = \hat{C}^*(g^{-1}) = \tau_C(g^{-1}) = \begin{cases} 1 & \text{if } g \in C, \\ 0 & \text{otherwise.} \end{cases}$$

Thus, $\hat{C}$ is the sum of the elements in C. Using Lemma 7.1.7, we have thus recovered the well-known result that these class sums form an A-basis for the centre of H. (We will generalize this later to construct a basis for the centre of an Iwahori–Hecke algebra; see Corollary 8.2.4.)

For the remainder of this section, we assume that H is symmetric, with symmetrizing trace τ. The existence of τ allows certain representation-theoretic constructions which we will now explain.

7.1.9. Definition. Let V, V' be (right) H-modules and $\mathcal{B}$ be a basis of H, with dual basis $\mathcal{B}^\vee$. For any $\varphi \in \mathrm{Hom}_A(V, V')$ we define $I(\varphi) \in \mathrm{Hom}_A(V, V')$ by

$$I(\varphi)(v) := \sum_{b \in \mathcal{B}} \varphi(vb)b^\vee \qquad (v \in V).$$

The definition immediately implies the following rules: if V, V', V'' are H-modules and $\varphi \in \mathrm{Hom}_H(V, V')$, $\psi \in \mathrm{Hom}_A(V', V'')$, $\pi \in \mathrm{Hom}_A(V'', V)$, then

$$I(\psi \circ \varphi) = I(\psi) \circ \varphi \quad \text{and} \quad I(\varphi \circ \pi) = \varphi \circ I(\pi).$$

The following result will show that the operator $I(\varphi)$ provides a tool that turns any A-homomorphism into an H-homomorphism.

7.1.10. Lemma. *Let* $\varphi \in \mathrm{Hom}_A(V, V')$. *Then the homomorphism* $I(\varphi)$ *lies in* $\mathrm{Hom}_H(V, V')$ *and does not depend on the choice of the basis* $\mathcal{B}$.

Proof. First we check the independence of the basis. Let $\mathcal{C}$ be another basis of H, with dual basis $\{c^\vee \mid c \in \mathcal{C}\}$. We have equations $c = \sum_{b \in \mathcal{B}} \lambda_{bc} b$ where (λ_{bc}) is an invertible matrix of elements of A. For $b \in \mathcal{B}$ we have $\tau(b^\vee c) = \sum_{b'} \lambda_{b'c} \tau(b^\vee b') = \lambda_{bc}$. Hence $b^\vee = \sum_{c \in \mathcal{C}} \lambda_{bc} c^\vee$. We can now compute that

$$\sum_c \varphi(vc)c^\vee = \sum_{c,b} \lambda_{bc} \varphi(vb)c^\vee = \sum_b \varphi(vb)\Big(\sum_c \lambda_{bc} c^\vee\Big) = \sum_b \varphi(vb)b^\vee$$

for all $v \in V$, as desired. Next we check that $I(\varphi) \in \mathrm{Hom}_H(V, V')$. For any $h \in H$ and $b \in \mathcal{B}$ we write $hb = \sum_{b'} \rho(h)_{b'b} b'$ with $\rho(h)_{b'b} \in A$. Then

$$\tau(b'^\vee hb) = \sum_{b''} \rho(h)_{b''b} \tau(b'^\vee b'') = \rho(h)_{b'b},$$

and so $b'^\vee h = \sum_b \rho(h)_{b'b} b^\vee$. For any $v \in V$, we can now compute that

$$\begin{aligned} I(\varphi)(vh) &= \sum_b \varphi(vhb)b^\vee = \sum_{b,b'} \rho(h)_{b'b} \varphi(vb')b^\vee \\ &= \sum_{b'} \varphi(vb')\Big(\sum_b \rho(h)_{b'b} b^\vee\Big) = \sum_{b'} \varphi(vb')b'^\vee h = (I(\varphi)(v))h, \end{aligned}$$

and so $I(\varphi)$ commutes with the action of H. □

One of the important properties of the construction in Definition 7.1.9 lies in the fact that it provides a criterion for an H-module to be projective. Recall that an H-module V is *projective* if every epimorphism $M \xrightarrow{\pi} V$ of H-modules splits, that is, if there exists some $\iota \in \mathrm{Hom}_H(V, M)$ such that $\pi \circ \iota = \mathrm{id}_V$. For example, free H-modules are projective.

7.1.11. Lemma (Gaschütz–Ikeda). *Let* V *be an* H*-module which is projective as an* A*-module. Then* V *is projective (as an* H*-module) if and only if there exists some* $\varphi \in \mathrm{End}_A(V)$ *such that* $I(\varphi) = \mathrm{id}_V$.

Proof. Suppose first that there exists some $\varphi \in \mathrm{End}_A(V)$ such that $I(\varphi) = \mathrm{id}_V$. Let $M \xrightarrow{\pi} V$ be an epimorphism of H-modules. Since V is projective over A, we can find an A-linear map $\mu \colon V \to M$ such that $\pi \circ \mu = \mathrm{id}_V$. We compose this equation with φ and apply $I(\cdot)$ to both sides. The right-hand side just gives id_V. Since π is an H-module homomorphism, the left-hand side yields

$$I(\pi \circ \mu \circ \varphi) = \pi \circ I(\mu \circ \varphi).$$

Hence $\iota := I(\mu \circ \varphi) \in \mathrm{Hom}_H(V, M)$ is the desired map.

Conversely, assume that every epimorphism $M \xrightarrow{\pi} V$ splits. Let $M := V \otimes_A H$ where H is regarded as a right module over itself, and M is a right module in a natural way. Choose a basis $\mathcal{B}$ of H. Then every element $m \in M$ can be written uniquely in the form $m = \sum_{b \in \mathcal{B}} v_b \otimes b$ where $v_b \in V$. We write $1_H = \sum_{b \in \mathcal{B}} \xi_b b^\vee$ where $\xi_b \in A$, and define $\varphi \in \mathrm{End}_A(M)$ by

$$\varphi\Big(\sum_{b \in \mathcal{B}} v_b \otimes b\Big) := \sum_{b \in \mathcal{B}} \xi_b v_b \otimes 1.$$

Using relations $hb = \sum_{b'} \rho(h)_{b'b} b'$ as in the proof of Lemma 7.1.10, a straightforward computation shows that $I(\varphi) = \mathrm{id}_M$.

Now define $\pi \colon M \to V$ by $\pi(v \otimes h) := vh$. Then $\pi \in \mathrm{Hom}_H(M, V)$ and π is surjective. Since V is projective there exists some $\iota \in \mathrm{Hom}_H(V, M)$ such that $\pi \circ \iota = \mathrm{id}_V$. We set $\varphi' := \pi \circ \varphi \circ \iota \in \mathrm{End}_A(V)$. Since π and ι commute with the action of H we can now conclude

$$I(\varphi') = I(\pi \circ \varphi) \circ \iota = \pi \circ I(\varphi) \circ \iota = \pi \circ \mathrm{id}_M \circ \iota = \pi \circ \iota = \mathrm{id}_V.$$

Hence $\varphi' \in \mathrm{End}_A(V)$ has the desired property. □

7.1.12. Example. Let $H = M_n(A)$ for some $n \geq 1$ and τ be the usual trace, as in Example 7.1.2. Let $\mathcal{B} = \{E_{ab} \mid 1 \leq a, b \leq n\}$ be the standard basis of H and recall that $E_{ab}^\vee = E_{ba}$. Consider the H-lattice $V = A^{1 \times n}$ (row vectors), with H acting by matrix multiplication. Let $(e_1, \ldots, e_n)$ be the standard basis of V. Let $\varphi \in \mathrm{End}_A(V)$ be such that $\varphi(e_b) = \sum_c \varphi_{cb} e_c$ where $\varphi_{cb} \in A$. Now we can compute that

$$\begin{aligned} I(\varphi)(e_a) &= \sum_{a',b'} \varphi(e_a E_{a'b'}) E_{b'a'} = \sum_{a',b'} \varphi(\delta_{aa'} e_{b'}) E_{b'a'} \\ &= \sum_{b'} \Big(\sum_c \varphi_{cb'} e_c\Big) E_{b'a} = \sum_{b',c} \varphi_{cb'} e_c E_{b'a} = \sum_{b',c} \varphi_{cb'} \delta_{b'c} e_a \\ &= \mathrm{Tr}(\varphi) e_a. \end{aligned}$$

Hence, choosing any $\varphi \in \mathrm{End}_A(V)$ with $\mathrm{Tr}(\varphi) = 1$, we see that $I(\varphi) = \mathrm{id}_V$. In particular, V is projective.

7.1.13. Example. Let G be a finite group and $H = A[G]$ the group algebra over A, with symmetrizing trace τ as in Example 7.1.5. Recall that the basis dual to $\mathcal{B} = \{g \mid g \in G\}$ is $\mathcal{B}^\vee = \{g^{-1} \mid g \in G\}$. For any H-lattice V, we have

$$I(\mathrm{id}_V)(v) = \sum_{g \in G} \mathrm{id}_V(vg)g^{-1} = \sum_{g \in G} v = |G|\, v \qquad (v \in V).$$

Hence, if $|G|1_A \in A$ is a unit, then every H-lattice is projective. In particular, if A is a field whose characteristic does not divide $|G|$ then $A[G]$ is a semisimple algebra. Thus, we have recovered *Maschke's theorem* (see [Isaacs 1976, (1.9)]).

7.2 SCHUR RELATIONS AND SCHUR ELEMENTS

Let H be a finite-dimensional algebra over a field K and assume that we are given a symmetrizing trace τ on H. We will now see that the existence of τ implies that we have orthogonality relations for the irreducible characters of H. The main tool will be the operator $I(\cdot)$ of Definition 7.1.9.

Any H-module will be tacitly assumed to be finite-dimensional. Lemma 7.1.10 shows that if V, V' are H-modules such that $\mathrm{Hom}_H(V, V') = 0$ then $I(\varphi) = 0$ for all $\varphi \in \mathrm{Hom}_K(V, V')$. On the other hand, if $V = V'$ is a simple module, we know by *Schur's lemma* that $\mathrm{End}_H(V)$ is a division algebra over K. Recall that V is split simple if $\mathrm{End}_H(V) = K \cdot \mathrm{id}_V$.

7.2.1. Theorem. *Let V be a split simple H-module. Then there is a unique element $c_V \in K$ such that*

$$I(\varphi) = c_V \,\mathrm{Tr}(\varphi)\, \mathrm{id}_V \qquad \textit{for all } \varphi \in \mathrm{End}_K(V).$$

Furthermore, the constant c_V only depends on the isomorphism class of V.

The element $c_V \in K$ will be called the *Schur element* associated with V.

Proof. The uniqueness of c_V is clear since there certainly exists some endomorphism of V with trace 1. Now let $\varphi \in \mathrm{End}_K(V)$. By Lemma 7.1.10 we have $I(\varphi) \in \mathrm{End}_H(V)$. Since V is split simple, we have $I(\varphi) = c_{V,\varphi}\, \mathrm{id}_V$ for some $c_{V,\varphi} \in K$. We first consider special choices for φ for which we can explicitly compute these constants.

Assume that $\dim_K V = n$ and choose a basis $(v_1, \ldots, v_n)$. For $1 \leq i, j \leq n$ let $\varphi_{ij} \in \mathrm{End}_K(V)$ such that $\varphi_{ij}(v_l) = \delta_{il} v_j$. To simplify notation, we write c_{ij} instead of $c_{V,\varphi_{ij}}$. For $h \in H$ let also $v_l h = \sum_k \rho(h)_{kl} v_k$ where $\rho(h)_{kl} \in K$. Using this notation, a straightforward computation yields that

$$I(\varphi_{ij})(v_l) = \sum_{b,k} \rho(b)_{il} \rho(b^\vee)_{kj} v_k.$$

Since $I(\varphi_{ij}) = c_{ij}\, \mathrm{id}_V$ we conclude that

$$\sum_b \rho(b)_{il}\rho(b^\vee)_{kj} = \delta_{kl}c_{ij} \qquad \text{for all } i,j,k,l.$$

The dual basis of $\mathcal{B}^\vee$ is again $\mathcal{B}$. By Lemma 7.1.10, we can therefore interchange the roles of $\mathcal{B}$ and its dual basis in the above computation without affecting the result. Thus, we also obtain

$$\sum_b \rho(b^\vee)_{i'l'}\rho(b)_{k'j'} = \delta_{k'l'}c_{i'j'} \qquad \text{for all } i',j',k',l'.$$

Now we set $i' = k$, $l' = j$, $k' = i$ and $j' = l$. Then the left-hand sides of the last two equations are the same. Hence so are the right-hand sides and we have reached the conclusion that

$$\delta_{kl}c_{ij} = \delta_{ij}c_{kl} \qquad \text{for all } i,j,k,l.$$

We set $k = l = 1$ and $c_V := c_{11}$. If $i \neq j$ then $c_{ij} = \delta_{ij}c_V = 0$, and if $i = j$ then $c_{ii} = c_{11} = c_V$. This implies that $I(\varphi_{ij}) = \delta_{ij}c_V\,\mathrm{id}_V$. Since $\mathrm{Tr}(\varphi_{ij}) = \delta_{ij}$, this proves the assertion for all endomorphisms φ_{ij}. The defining formula in Definition 7.1.9 shows that the assignment $\varphi \mapsto c_{V,\varphi}$ is linear in φ. Since the φ_{ij} form a basis of $\mathrm{End}_K(V)$, we conclude that for any $\varphi \in \mathrm{End}_K(V)$ we have $I(\varphi) = c_V\,\mathrm{Tr}(\varphi)\,\mathrm{id}_V$.

It remains to show that $c_V = c_{V'}$ if V' is an H-module isomorphic to V. Let $\sigma\colon V' \to V$ be an H-module isomorphism and $\varphi' := \sigma^{-1}\circ\varphi\circ\sigma \in \mathrm{End}_K(V')$. Since σ commutes with the action of H we can compute that

$$c_{V'}\,\mathrm{Tr}(\varphi')\,\mathrm{id}_{V'} = I(\varphi') = \sigma^{-1}\circ I(\varphi)\circ\sigma = c_V\,\mathrm{Tr}(\varphi)\,\mathrm{id}_V.$$

Since certainly $\mathrm{Tr}(\varphi') = \mathrm{Tr}(\varphi)$, this implies $c_{V'} = c_V$. □

For some purposes, it may be convenient to reformulate these relations in terms of matrix representations and characters.

7.2.2. Corollary (Schur relations). *Let V, V' be split simple* H*-modules affording matrix representations $\rho\colon H \to M_n(K)$ and $\rho'\colon H \to M_m(K)$, respectively, where $n = \dim_K V$ and $m = \dim_K V'$. Then we have*

$$\sum_{b\in\mathcal{B}} \rho(b)_{il}\rho'(b^\vee)_{kj} = \begin{cases} \delta_{ij}\delta_{kl}c_V & \text{if } V = V' \text{ and } \rho = \rho', \\ 0 & \text{if } V \not\cong V', \end{cases}$$

for all $1 \leq i,l \leq n$, $1 \leq j,k \leq m$.

Proof. Choose bases $(v_1,\ldots,v_n)$ and $(v'_1,\ldots,v'_m)$ of V and V' defining the representations ρ and ρ', respectively. For $1 \leq i \leq n$ and $1 \leq j \leq m$ let $\varphi_{ij} \in \mathrm{Hom}_K(V,V')$ be defined by $\varphi_{ij}(v_l) = \delta_{il}v'_j$. The above relations are then obtained by writing out the equations $I(\varphi_{ij}) = 0$ (if $V \not\cong V'$) and $I(\varphi_{ij}) = \delta_{ij}c_V\,\mathrm{id}_V$ (if $V = V'$ and $\rho = \rho'$). □

We can now prove a kind of converse to Theorem 7.2.1:

7.2.3. Remark. Let V be an H-module and $0 \neq c \in K$ be such that $I(\varphi) = c\,\mathsf{Tr}(\varphi)\,\mathsf{id}_V$ for all $\varphi \in \mathsf{End}_K(V)$. Then V is split simple and $c = c_V$.

Proof. Assume, if possible, that V is not simple. Choosing a basis adapted to a non-trivial proper invariant subspace of V, we obtain a matrix representation $\rho\colon H \to M_n(K)$ afforded by V such that $\rho(h)_{n1} = 0$ for all $h \in H$. Writing out the equation $I(\varphi) = c\,\mathsf{Tr}(\varphi)\,\mathsf{id}_V$ as in the proof of Corollary 7.2.2 and choosing $k = l = n$ and $i = j = 1$ yields that $0 = c$, contrary to our assumption. Hence V is simple. Now consider any $\varphi \in \mathsf{End}_H(V)$. Take $\varphi' \in \mathsf{End}_K(V)$ with $\mathsf{Tr}(\varphi') = 1$. Then we compute that

$$c\,\mathsf{Tr}(\varphi \circ \varphi')\,\mathsf{id}_V = I(\varphi \circ \varphi') = \varphi \circ I(\varphi') = c\,\mathsf{Tr}(\varphi')\varphi = c\varphi.$$

We conclude that φ is a scalar multiple of id_V. Hence V is split simple. □

7.2.4. Corollary (Orthogonality relations). *Let V and V' be split simple H-modules and denote by χ_V and $\chi_{V'}$ their characters, respectively. Then*

$$\sum_{b \in \mathcal{B}} \chi_V(b)\chi_{V'}(b^\vee) = \begin{cases} c_V \dim_K V & \text{if } \chi_V = \chi_{V'}, \\ 0 & \text{otherwise.} \end{cases}$$

Proof. First note that we have $\chi_V = \chi_{V'}$ if and only if $V \cong V'$. (This follows easily from Exercise 7.4.) Using the notation of the proof of Corollary 7.2.2, we have $\chi_V(h) = \sum_i \rho(h)_{ii}$ and $\chi_{V'}(h) = \sum_j \rho(h)'_{jj}$ for all $h \in H$. It remains to insert this into the above expression and to evaluate using the Schur relations. □

7.2.5. Example. Let $H = K[G]$ be a group algebra with symmetrizing trace τ as in Example 7.1.5. Assume that K is a field of characteristic 0. Let V be a split simple H-module. Then $I(\mathsf{id}_V) = c_V(\dim_K V)\,\mathsf{id}_V$. On the other hand, we have seen in Example 7.1.13 that $I(\mathsf{id}_V) = |G|\,\mathsf{id}_V$. Hence we have $c_V = |G|/\dim_K V$. Combining this with Corollary 7.2.4, we obtain the well-known orthogonality relations for the (absolutely) irreducible characters of G.

Combining the Gaschütz–Ikeda Lemma 7.1.11 with the above results on the Schur relations, we obtain the following semisimplicity criterion.

7.2.6. Theorem. *A split simple H-module V is projective if and only if $c_V \neq 0$. In particular, assuming that H is split, H is a semisimple algebra if and only if all Schur elements are non-zero. If this is the case, we have*

$$\tau = \sum_V \frac{1}{c_V}\chi_V$$

where V runs over the simple H-modules (up to isomorphism).

Proof. Let V be a split simple H-module. Assume first that $c_V \neq 0$. Then we can take any $\varphi \in \mathrm{End}_K(V)$ with $\mathrm{Tr}(\varphi) = c_V^{-1}$ (which certainly exists) and obtain $I(\varphi) = c_V \,\mathrm{Tr}(\varphi)\,\mathrm{id}_V = \mathrm{id}_V$. Hence Lemma 7.1.11 shows that V is projective. Conversely, assume that V is projective. Again by using Lemma 7.1.11, there exists some $\varphi \in \mathrm{End}_K(V)$ such that $I(\varphi) = \mathrm{id}_V$. Hence $\mathrm{Tr}(\varphi)c_V = 1$, and so $c_V \neq 0$. Thus, assuming that H is split, all Schur elements are non-zero if and only if all simple H-modules are projective, which is certainly equivalent to H being semisimple.

It remains to prove the assertion about the expression for τ. Assume that H is split and semisimple. Let $H = \bigoplus_V H(V)$ be the Wedderburn decomposition of H (see the introduction to this chapter), where V runs over the simple H-modules (up to isomorphism). For each V there is an isomorphism $\rho_V \colon H(V) \to M_{n_V}(K)$, where $n_V = \dim_K V$.

By Example 7.1.2, the restriction of τ to $H(V)$ is of the form $d_V \mathrm{Tr}_V \circ \rho_V$ for some $d_V \in K$, where Tr_V denotes the usual trace on $M_{n_V}(K)$. Now note that the composition of the projection $H \to H(V)$ with $\mathrm{Tr}_V \circ \rho_V$ just equals the character χ_V. Thus we have $\tau = \sum_V d_V \chi_V$, and we must show that $d_V c_V = 1$ for all V. Since τ is non-degenerate, we have $d_V \neq 0$ for all V; see Example 7.1.3.

Let $\mathcal{B} = \coprod_V \mathcal{B}(V)$ be a basis of H which is adapted to the decomposition $H = \bigoplus_V H(V)$. For each V, let $\tilde{\mathcal{B}}(V)$ be the basis dual to $\mathcal{B}(V)$ taken with respect to the trace function $\mathrm{Tr}_V \circ \rho_V$ on $H(V)$. Then $\coprod_V \{d_V^{-1}\tilde{b} \mid \tilde{b} \in \tilde{\mathcal{B}}(V)\}$ is the basis of H which is dual to $\mathcal{B}$ with respect to the form τ; see again Example 7.1.3.

Now fix a simple module V, an endomorphism $\varphi \in \mathrm{End}_K(V)$ and a vector $v \in V$. We consider the defining equation for c_V:

$$c_V \,\mathrm{Tr}(\varphi) v = I(\varphi)(v) = \sum_{b \in \mathcal{B}} \varphi(vb)b^\vee = \sum_{V'} \sum_{b \in \mathcal{B}(V')} \varphi(vb)b^\vee.$$

Since $h \in H(V')$ acts as 0 on V unless V and V' are isomorphic, we can restrict the above sum to $\mathcal{B}(V)$ and obtain

$$c_V \,\mathrm{Tr}(\varphi) v = \sum_{b \in \mathcal{B}(V)} \varphi(vb)b^\vee = d_V^{-1} \sum_{b \in \mathcal{B}(V)} \varphi(vb)\tilde{b}.$$

The summation on the right-hand side is the defining formula for the Schur element of V regarded as a module for $H(V)$ with symmetrizing trace $\mathrm{Tr}_V \circ \rho_V$. Example 7.1.12 shows that this Schur element equals 1, and so the right-hand side evaluates to $d_V^{-1}\,\mathrm{Tr}(\varphi)v$. This holds for all $v \in V$ and all $\varphi \in \mathrm{End}_K(V)$. Hence we conclude that $c_V = d_V^{-1}$. □

The Schur relations can also be used to obtain explicit formulas for idempotents in H. For this purpose, assume that H is split semisimple and let $H = \bigoplus_V H(V)$ be the Wedderburn decomposition of H, where V runs over the simple

H-modules (up to isomorphism). For each simple H-module V, we have an isomorphism $\rho_V\colon H(V) \to M_{n_V}(K)$ where $n_V = \dim_K V$. Let $\mathcal{B}$ be any basis of H. By Theorem 7.2.6, the Schur element c_V is non-zero, and we set

$$e_{ij}^V := \frac{1}{c_V}\sum_{b\in\mathcal{B}} \rho_V(b)_{ji} b^\vee, \qquad 1 \le i,j \le n_V.$$

7.2.7. Proposition (Formulas for idempotents). *Recall that* H *is assumed split and semisimple. Then, in the above set-up, the following hold.*

(a) *We have* $\rho_{V'}(e_{ij}^V) = 0$ *if* $V \not\cong V'$ *and* $\rho_V(e_{ij}^V)$ *is the matrix with* (i,j)*th coefficient* 1 *and coefficient* 0 *otherwise. Hence,* $\{e_{ij}^V\}$ *is a basis of* H.

(b) *We have* $1_H = \sum_V \sum_{i=1}^{n_V} e_V^{ii}$ *and this is a decomposition of* 1_H *into orthogonal primitive idempotents. Moreover, for fixed* V*, the idempotents* e_{ii}^V $(1 \le i \le n_V)$ *are all conjugate by units in* H.

(c) *Let* $1_H = \sum_V e_V$ *with* $e_V \in H(V)$*. Then each* e_V *is a central primitive idempotent and, denoting by* χ_V *the character of* V*, we have*

$$e_V = \sum_{i=1}^{n_V} e_{ii}^V = \frac{1}{c_V}\sum_{b\in\mathcal{B}} \chi_V(b) b^\vee.$$

Proof. The statement in (a) immediately follows using the Schur relations in Corollary 7.2.2. We can now assume that $H = \bigoplus_V M_{n_V}(K)$ and that, in the direct summand corresponding to V, the element e_{ij}^V is the matrix with (i,j)th coefficient 1 and coefficient 0 otherwise. From this description, it is obvious that the e_{ii}^V are mutually orthogonal idempotents and that $1_H = \sum_V \sum_{i=1}^{n_V} e_{ii}^V$. Moreover, for fixed V, two matrices e_{ii}^V and e_{jj}^V are certainly conjugate by an invertible matrix in $M_{n_V}(K)$. Finally, each e_{ii}^V is primitive, since $\dim_K(e_{ii}^V M_{n_V}(K) e_{ii}^V) = 1$. Thus, all the statements in (b) are proved. Now consider (c). The element $\sum_{i=1}^{n_V} e_{ii}^V$ is the identity matrix in $M_{n_V}(K)$ and so must be equal to e_V (since the sum of all these elements is 1_H). The primitivity just expresses the fact that $H(V)$ cannot be decomposed further as a direct sum of ideals in H. The last equality in the formula for e_V holds since $\chi_V(h) = \sum_{i=1}^{n_V} \rho_V(h)_{ii}$ for any $h \in H$. □

Finally, we obtain yet another characterization of the Schur elements. Consider a split simple H-module V, with character χ_V. If $z \in H$ lies in the centre of H then z commutes with the action of any element of H on V. Hence Schur's lemma implies that z acts as a scalar on V; we denote this scalar by $\omega_V(z)$. Thus, we have defined a K-algebra homomorphism $\omega_V\colon Z(H) \to K$, which we call the *central character* associated with V.

7.2.8. Proposition. *Assume that* H *is split semisimple. For any simple* H*-module* V*, let* $\chi_V^* \in Z(H)$ *be the unique element such that* $\tau(\chi_V^* h) = \chi_V(h)$ *for all* $h \in H$ *(as in Lemma 7.1.7). Then the following hold:*

(a) *We have* $\omega_{V'}(\chi_V^*) = 0$ *if* V, V' *are not isomorphic, and* $\omega_V(\chi_V^*) = c_V$.

(b) *The elements* $\{\chi_V^*\}$, *where* V *runs over the simple* H-*modules (up to isomorphism), form a* K-*basis of* $Z(H)$.

(c) *We have* $\chi_V^* = c_V e_V$, *where* $e_V \in H$ *is the central idempotent associated with* V *as in Proposition 7.2.7.*

Proof. Let $\mathcal{B}$ be any K-basis of H. Then the definition of χ_V^* shows that

$$\chi_V^* = \sum_{b \in \mathcal{B}} \chi_V(b) b^\vee.$$

Comparison with the formula in Proposition 7.2.7 already proves (c). Since e_V acts as 0 on V' if V, V' are not isomorphic, this proves $\omega_{V'}(e_V) = 0$, and hence also $\omega_{V'}(\chi_V^*) = 0$, unless V, V' are isomorphic. Now consider the formula $\tau = \sum_V c_V^{-1} \chi_V$ of Corollary 7.2.6. Taking "duals", it is equivalent to the formula $1_H = \sum_V c_V^{-1} \chi_V^*$. Applying $\omega_{V'}$ yields that $1 = \omega_{V'}(1_H) = \omega_{V'}(\chi_{V'}^*) c_{V'}^{-1}$. This completes the proof of (a).

Now consider (b). By Lemma 7.1.7, the correspondence $\lambda \mapsto \lambda^*$ is a K-linear isomorphism between the space of trace functions on H and the centre of H. Hence, (b) holds if and only if the set $\{\chi_V\}$ is a basis of the space of trace functions on H, which in turn holds since H is split and semisimple (see Exercise 7.4). □

7.3 GROTHENDIECK GROUPS AND INTEGRALITY

We now combine the settings of the previous two sections. Assume that A is an integral domain and K is a field containing A. Let H be an A-algebra, finitely generated and free over A.

We denote by $KH = H \otimes_A K$ the K-algebra obtained by extension of scalars from A to K. We may naturally consider H as a subset of KH. A similar convention will be applied to any ring B such that $A \subseteq B \subseteq K$, where we have natural inclusions $H \subseteq BH \subseteq KH$.

In what follows, it will be convenient to use the language of *Grothendieck groups* (see, for example, [Curtis and Reiner 1981, Section 16B]). Let $R_0(KH)$ be the Grothendieck group of the category of finite-dimensional KH-modules. Thus, $R_0(KH)$ is generated by expressions $[V]$, one for each KH-module V (up to isomorphism), with relations $[V] = [V'] + [V'']$ for each short exact sequence $0 \to V' \to V \to V'' \to 0$ of KH-modules. Two KH-modules V, V' give rise to the same element in $R_0(KH)$ if and only if V and V' have the same composition factors, counting multiplicities. It follows that $R_0(KH)$ is free abelian, with basis given by the isomorphism classes of simple modules. Finally, let $R_0^+(KH)$ be the subset of $R_0(KH)$ consisting of elements $[V]$ where V is a KH-module. This is a monoid whose identity element is the class of the 0-module.

7.3.1. Definition. Let X be an indeterminate over K and $\mathsf{Maps}(H, K[X])$ the K-algebra of maps from H to $K[X]$ (with pointwise multiplication of maps as algebra multiplication). We define a map

$$p_K \colon R_0^+(KH) \to \mathsf{Maps}(H, K[X]),$$

by associating with the class of a KH-module V the map which assigns to each $h \in H$ the characteristic polynomial of $\rho_V(h)$, where ρ_V is the representation afforded by V (see Example 7.1.4). Considering $\mathsf{Maps}(H, K[X])$ as a semigroup with respect to multiplication, the map p_K is a semigroup homomorphism. (We leave it to the reader to check that p_K is well defined.)

Let $\mathrm{Irr}(KH)$ be the set of all characters χ_V where V is a simple KH-module.

7.3.2. Lemma (Brauer–Nesbitt). *Assume that* $\mathrm{Irr}(KH)$ *is a linearly independent subset of* $\mathrm{Hom}_K(KH, K)$. *Then* $p_K \colon R_0^+(KH) \to \mathsf{Maps}(H, K[X])$ *is injective.*

Proof. Let V, V' be KH-modules such that $p_K([V]) = p_K([V'])$. Since $[V]$, $[V']$ only depend on the composition factors of V, V', we may assume that V, V' are semisimple modules. Let

$$V = \bigoplus_{i=1}^{n} a_i V_i \quad \text{and} \quad V' = \bigoplus_{i=1}^{n} b_i V_i$$

where the V_i are pairwise non-isomorphic simple KH-modules and $a_i, b_i \geq 0$ for all i. We must show that $a_i = b_i$ for all i.

If, for some i, we have both $a_i > 0$ and $b_i > 0$, then we can write $V = V_i \oplus \tilde{V}$ and $V' = V_i \oplus \tilde{V}'$. Since p_K is a semigroup homomorphism we compute that

$$p_K([V_i]) \cdot p_K([\tilde{V}]) = p_K([V]) = p_K([V']) = p_K([V_i]) \cdot p_K([\tilde{V}']).$$

We can divide by $p_K([V_i])$ (check this!) and conclude that $p_K([\tilde{V}]) = p_K([\tilde{V}'])$. Continuing in this way, we arrive at a situation where we have two semisimple modules such that V_i occurs either in one or the other, but not both. This discussion shows that we may assume, without loss of generality, that for all i, we have $a_i = 0$ or $b_i = 0$. Taking characters yields that

$$\chi_V = \sum_i a_i \chi_{V_i} \quad \text{and} \quad \chi_{V'} = \sum_i b_i \chi_{V_i}.$$

For each $h \in H$, the character values $\chi_V(h)$ and $\chi_{V'}(h)$ appear as coefficients in the polynomials $p_K([V])(h)$ and $p_K([V'])(h)$, respectively. Hence we have that

$$\sum_i (a_i - b_i)\chi_{V_i} = 0.$$

By assumption, the characters χ_{V_i} are linearly independent. So we conclude that $(a_i - b_i)1_K = 0$ for all i. Since also $a_i = 0$ or $b_i = 0$, we conclude that $a_i 1_K = 0$ and $b_i 1_K = 0$ for all i. If K has characteristic 0, this immediately shows that $a_i = b_i = 0$ for all i, and we are done.

Now assume that K has characteristic $p > 0$. We have just seen that then p divides all a_i and all b_i, and so $\frac{1}{p}[V]$ and $\frac{1}{p}[V']$ exist in $R_0^+(KH)$. Consequently, we also have $p_K(\frac{1}{p}[V]) = p_K(\frac{1}{p}[V'])$. Repeating the above argument with $\frac{1}{p}[V]$ and $\frac{1}{p}[V']$ yields that the multiplicity of V_i in each of these modules is still divisible by p. This can be repeated again and again, and hence the multiplicities of V_i in V and V' should be divisible by arbitrary powers of p. This forces $a_i = b_i = 0$ for all i, as desired. □

7.3.3. Remark. The assumption of the Brauer–Nesbitt lemma is certainly satisfied if KH is split. This follows immediately from Wedderburn's theorem (see Exercise 7.4). In general, it holds if there exists a field extension $K' \supseteq K$ such that $V \otimes_K K'$ is a direct sum of split simple $K'H$-modules, for every simple KH-module V. (This follows from [Curtis and Reiner 1981, Theorem 7.9].) This condition in turn holds, for example, whenever K is a perfect field. For more details see [Curtis and Reiner 1981, Chap. 7].

7.3.4. Lemma (Compatibility with field extensions). *Let $K \subseteq K'$ be a field extension. Then there is a canonical map $d_K^{K'}: R_0(KH) \to R_0(K'H)$ given by $[V] \mapsto [V \otimes_K K']$. Furthermore, we have a commutative diagram*

$$\begin{array}{ccc} R_0^+(KH) & \xrightarrow{p_K} & \mathsf{Maps}(H, K[X]) \\ {\scriptstyle d_K^{K'}}\downarrow & & \downarrow{\scriptstyle t_K^{K'}} \\ R_0^+(K'H) & \xrightarrow{p_{K'}} & \mathsf{Maps}(H, K'[X]) \end{array}$$

where $t_K^{K'}: \mathsf{Maps}(H, K[X]) \hookrightarrow \mathsf{Maps}(H, K'[X])$ is the canonical embedding. If, moreover, KH is split, then $d_K^{K'}$ is an isomorphism which preserves isomorphism classes of simple modules.

Proof. To simplify notation, we write $K'V$ instead of $V \otimes_K K'$ for any KH-module V. Then the map $V \mapsto K'V$ induces a map $R_0(KH) \to R_0(K'H)$. To check this, recall once more the precise definition of Grothendieck groups and use the fact that tensoring over K preserves short exact sequences, which holds since we are only dealing with finite-dimensional vector spaces over a field. Now take a KH-module V. Then $p_K([V])(h) = p_{K'}([K'V])(h) = p_{K'}(d_K^{K'}([V]))(h)$ for any $h \in H$. Hence $(t_K^{K'} \circ p_K)([V])$ is the restriction of $(p_{K'} \circ d_K^{K'})([V])$ to KH. This shows the commutativity of the above diagram.

Now assume that KH is split. Let V be a simple KH-module. By Wedderburn's theorem the representation $KH \to \mathrm{End}_K(V)$ afforded by V is surjective. Tensoring with K' shows that the representation $K'H \to \mathrm{End}_K(V) \otimes_K K' \cong \mathrm{End}_{K'}(K'V)$ afforded by $K'V$ is also surjective. This certainly implies that $K'V$ is split simple.

We always have $\mathrm{Hom}_{K'H}(K'V_1, K'V_2) \cong \mathrm{Hom}_{KH}(V_1, V_2) \otimes_K K'$ for any KH-modules V_1, V_2 (see [Curtis and Reiner 1962, (29.5)]). Hence, if V_1, V_2 are simple and not isomorphic, then $K'V_1$, $K'V_2$ cannot be isomorphic either.

Finally, every simple $K'H$-module occurs as a composition factor of $K'H$ regarded as a module over itself. Since the latter is just obtained from KH by tensoring with K' and since tensoring preserves short exact sequences, we conclude that each simple $K'H$-module is of the form $K'V$ for some simple KH-module V. □

The above results did not really make use of the fact that KH is obtained from H by scalar extension from the ground ring A to K. We will now bring into play this additional information. Ideally, the ring A should have the property that every finitely generated and torsion-free A-module is free. The lack of this property will be overcome by the use of valuation rings.

7.3.5. Valuation rings. A subring $\mathcal{O} \subseteq K$ is called a *valuation ring* if, for each non-zero element $x \in K$, we have $x \in \mathcal{O}$ or $x^{-1} \in \mathcal{O}$. Such a valuation ring is a local ring whose maximal ideal we denote by $J(\mathcal{O})$. We summarize the results on valuation rings that we need as follows:

(V1) If $I \subseteq A$ is a prime ideal there exists a valuation ring $\mathcal{O} \subseteq K$ such that $A \subseteq \mathcal{O}$ and $J(\mathcal{O}) \cap A = I$.

(V2) Every finitely generated torsion-free module over a valuation ring in K is free.

(V3) The intersection of all valuation rings $\mathcal{O} \subseteq K$ with $A \subseteq \mathcal{O}$ is the integral closure of A in K; each valuation ring itself is integrally closed in K.

Proofs of (V1) and (V2) can be found in [Goldschmidt 1980, Section 5]; for (V3) see [Matsumura 1986, Theorem 10.4]. Usually, we are not interested in knowing exactly what kind of valuation rings we can find in K. Note that, for example, they need not be Noetherian even if A is. (In fact, it is a general result that a valuation ring is Noetherian if and only if it is a principal ideal domain and, hence, a discrete valuation ring; see [Matsumura 1986, Theorem 11.1].) The main point for us is (V1), i.e., the existence of suitable valuation rings.

7.3.6. Example. Assume that A is a Noetherian ring with field of fractions K. Moreover, let $\theta\colon A \to L$ be a homomorphism into a field L which is the field of fractions of $\theta(A)$ and whose kernel is a principal ideal generated by a non-zero (and non-invertible) element $\Phi \in A$, say. Let

$$\mathcal{O} = \{a/b \in K \mid a, b \in A, \theta(b) \neq 0\} = \{a/b \in K \mid b \text{ not divisible by } \Phi\}$$

be the localization of A in the prime ideal $\ker(\theta)$. Then every non-zero element of K can be written in the form $\Phi^n a/b$ where $n \in \mathbb{Z}$ and $a, b \in A$ are such that $\theta(a) \neq 0$, $\theta(b) \neq 0$; here, the integer n is uniquely determined. This already shows that $\mathcal{O}$ is a valuation ring in K. Furthermore, we obtain a map $\nu\colon K \to \mathbb{Z} \cup \{\infty\}$ by sending $\Phi^n a/b$ as above to n, and setting $\nu(0) := \infty$ by convention. It is readily checked that ν is in fact a *discrete valuation*, i.e., we have

$$\nu(xy) = \nu(x) + \nu(y) \quad \text{and} \quad \nu(x+y) \geq \min(\nu(x), \nu(y)) \quad \text{for } x, y \in K.$$

Furthermore, $\mathcal{O} = \{x \in K \mid \nu(x) \geq 0\}$, $J(\mathcal{O}) = \{x \in K \mid \nu(x) > 0\} = (\Phi)$, and θ canonically extends to a map $\mathcal{O} \to L$, $a/b \mapsto \theta(a)/\theta(b)$. For more about discrete valuation rings see, for example, [Curtis and Reiner 1981, §4C] or [Matsumura 1986, Chap. 11]. A far-reaching generalization of the above construction is discussed in Exercise 7.8.

An example of a non-discrete valuation ring is as follows. Let K be an algebraic closure of $\mathbb{Q}$ and p be any rational prime number. By (V1) there exists a valuation ring $\mathcal{O} \subseteq K$ with maximal ideal $J(\mathcal{O})$ such that $J(\mathcal{O}) \cap \mathbb{Z} = p\mathbb{Z}$. It is easily checked that then $J(\mathcal{O})$ cannot be a principal ideal.

The importance of valuation rings lies in the following construction.

7.3.7. Realizing modules over $\mathcal{O}$. Let V be any KH-module. Choosing a K-basis of V, we obtain a corresponding matrix representation $\rho\colon KH \to M_n(K)$ (where $n = \dim_K V$). The point about valuation rings is that if $\mathcal{O} \subseteq K$ is any such ring with $A \subseteq \mathcal{O}$, then a basis for V can be chosen so that $\rho(h) \in M_n(\mathcal{O})$ for all $h \in H$. If this is the case, we say that V is *realized over* $\mathcal{O}$.

This is proved as follows. Take any K-basis $(v_1, \ldots, v_n)$ of V and let $\mathcal{B}$ be an A-basis for H. Let $\tilde{V}$ be the $\mathcal{O}$-submodule of V spanned by the (finite) set $\{v_i b \mid 1 \leq i \leq n, b \in \mathcal{B}\}$. Then $\tilde{V}$ is also invariant under the operation of $\mathcal{O}H$, and hence a finitely generated $\mathcal{O}H$-module. Since it is contained in a K-vector space, it is also torsion-free. So (V2) implies that $\tilde{V}$ is an $\mathcal{O}H$-lattice such that $\tilde{V} \otimes_{\mathcal{O}} K \cong V$. Thus, any $\mathcal{O}$-basis of $\tilde{V}$ is also a K-basis of V with the required property.

Note that the above argument only requires that $\mathcal{O}$ is a subring of K such that K is the field of fractions of $\mathcal{O}$ and $\mathcal{O}$ satisfies (V2). These conditions also hold, for example, for any principal ideal domain having K as field of fractions.

The following two results are first applications of the above construction.

7.3.8. Proposition. *Let V be a KH-module and A^* be the integral closure of A in K. Then we have $p_K([V])(h) \in A^*[X]$ for all $h \in H$. Thus, the map p_K of Definition 7.3.1 is in fact a map $p_K\colon R_0^+(KH) \to \mathrm{Maps}(H, A^*[X])$.*

Note that this implies, in particular, that $\chi_V(h) \in A^*$ for all $h \in H$.

Proof. Fix an element $h \in H$. Let $\mathcal{O} \subseteq K$ be any valuation ring with $A \subseteq \mathcal{O}$. By (7.3.7), there exists a basis of V such that the action of h on V is given by a matrix with coefficients in $\mathcal{O}$ with respect to that basis. This certainly implies that $p_K([V])(h) \in \mathcal{O}[X]$. Since this holds for all valuation rings $\mathcal{O}$ as above, we conclude that $p_K([V])(h) \in A^*[X]$, using (V3) in (7.3.5). □

7.3.9. Proposition (Integrality of the Schur elements). *Assume that we have a symmetrizing trace τ on H. Let V be a split simple KH-module and c_V its Schur element, with respect to the induced trace τ^K on KH. Then c_V lies in the integral closure of A in K.*

Proof. Let $\mathcal{O} \subseteq K$ be a valuation ring with $A \subseteq \mathcal{O}$. By (7.3.7), we can assume that V affords a representation $\rho\colon KH \to M_n(K)$ with $\rho(h) \in M_n(\mathcal{O})$ for all $h \in H$. Let $\mathcal{B}$ be an A-basis of H and consider the Schur relations in Corollary 7.2.2. Since $\mathcal{B}, \mathcal{B}^\vee \subseteq H$, all terms in the sum defining c_V lie in $\mathcal{O}$, and so $c_V \in \mathcal{O}$. Since this holds for all valuation rings $\mathcal{O}$ as above, we conclude that $c_V \in A^*$, again using (V3) in (7.3.5). □

7.3.10. Example. Let $A = \mathbb{Z}$, let K be an algebraically closed field of characteristic 0, and $H = A[G]$ the group algebra of a finite group G over A, with symmetrizing trace τ as in Example 7.1.5. Then KH is a split semisimple algebra. In Example 7.2.5 we have seen that $c_V = |G|/\dim_K V$ for every simple KH-module V. Now Proposition 7.3.9 shows that c_V is also an algebraic integer. Hence c_V is an integer. We have thus shown that $\dim_K V$ divides $|G|$, for all simple KH-modules V. The statements of Proposition 7.3.8 are trivial in this case: since each $g \in G$ has finite order, the eigenvalues of $\rho(g)$ are certainly algebraic integers (even roots of unity), for any representation $\rho\colon G \to GL_n(K)$.

Finally, we come to the question of defining a character table for H. Recall from Remark 7.3.3 that $\mathrm{Irr}(KH)$ (for KH split) is a linearly independent subset of $\mathrm{Hom}_K(KH, K)$. Hence, if we let $\mathrm{Irr}(KH) = \{\chi_1, \ldots, \chi_n\}$, then there exist elements $h_1, \ldots, h_n \in H$ such that the matrix of character values $(\chi_i(h_j))$ has a non-zero determinant. Now there is certainly a lot of freedom in the choice of the h_i. We impose an additional integrality condition as follows.

7.3.11. Character tables. Assume that A is integrally closed in K and that $H/[H,H]$ is a free A-module. Let $h_1, \ldots, h_n \in H$ be such that the elements $\{h_j + [H,H] \mid 1 \leq j \leq n\}$ form an A-basis of $H/[H,H]$. Then the array

$$X(H) := (\chi(h_j))_{\chi \in \mathrm{Irr}(KH),\, 1 \leq j \leq n}$$

is called a *character table* of H. Note that the entries of $X(H)$ lie in A, by Proposition 7.3.7. For $1 \leq j \leq n$, let $f_j \in \mathrm{Hom}_A(H, A)$ be the trace function defined by the condition that $f_j(h_i) = \delta_{ij}$ for all i. Then we have

$$\chi(h) = \sum_{j=1}^{n} \chi(h_j)\, f_j(h) \qquad \text{for all } \chi \in \mathrm{Irr}(KH) \text{ and all } h \in H. \tag{a}$$

(Indeed, it is enough to check this for $h = h_i$, in which case it is clear.) Assume, moreover, that KH is split semisimple. Then we have that

$$X(H) \text{ is a square matrix with a non-zero determinant.} \tag{b}$$

Indeed, we certainly have $K[H,H] = [KH, KH]$ and so $\{h_j + [KH,KH] \mid 1 \leq j \leq n\}$ is a K-basis of $KH/[KH,KH]$. It follows that the extensions of the functions f_j to KH form a basis of the space of trace functions on KH. But, by Exercise 7.4, the same holds for $\mathrm{Irr}(KH)$. Hence we have $|\mathrm{Irr}(KH)| = n$ and $X(H)$ is invertible.

If $A \neq K$, it is a non-trivial question whether H admits a character table at all. For example, this is the case for group algebras of finite groups (cf. Example 7.1.8). We obtain a well-defined character table by evaluating the irreducible characters on a set of representatives for the conjugacy classes of G; see [Isaacs 1976, p. 17]. In this case, the maps $f_j : A[G] \to A$ are given by the indicator functions on the conjugacy classes of G. We will see in Chapter 8 that character tables also exist for Iwahori–Hecke algebras associated with finite Coxeter groups.

7.4 SPECIALIZATIONS AND TITS'S DEFORMATION THEOREM

The basic assumptions of the previous section remain in force, i.e., H is an algebra over an integral domain A, finitely generated and free over A, and K is a field containing A. We assume that A is integrally closed in K. Throughout we will fix a ring homomorphism $\theta\colon A \to L$ into a field L such that L is the field of fractions of $\theta(A)$. We usually call such a ring homomorphism a *specialization* of A.

Given any homomorphism $A \to B$ of commutative rings, we consider B as an A-module via this map and write $BH := H \otimes_A B$. In particular, we apply this to the inclusion $A \subseteq K$ and the map $\theta\colon A \to L$ and obtain algebras KH and LH, respectively. What we actually want is that θ induces a map from $R_0^+(KH)$ to $R_0^+(LH)$. This will work under additional assumptions on θ. As in Section 7.3, we will make use of valuation rings. The precise setting is as follows.

7.4.1. Let $\mathcal{O} \subseteq K$ be a valuation ring with $A \subseteq \mathcal{O}$ and $J(\mathcal{O}) \cap A = \ker(\theta)$. (Note that $\ker(\theta)$ is a prime ideal since the image of θ is contained in a field.) Let k be the residue field of $\mathcal{O}$. Then the restriction of the canonical map $\pi\colon \mathcal{O} \to k$ to A has kernel $J(\mathcal{O}) \cap A = \ker(\theta)$. Since L is the field of fractions of $\theta(A)$, we may therefore regard L as a subfield of k. Thus, we have a commutative diagram

$$\begin{array}{ccccc} A & \subseteq & \mathcal{O} & \subseteq & K \\ {\scriptstyle\theta}\downarrow & & \downarrow{\scriptstyle\pi} & & \\ L & \subseteq & k & & \end{array}$$

Note that, in general, k will be strictly bigger than L (see, for example, Exercise 7.8). In order to have control over what happens between the Grothendieck groups of LH and of kH, we make the following assumption:

$$\text{(a)}\quad LH \text{ is split} \qquad \text{or} \qquad \text{(b)}\quad L = k \text{ and } L \text{ is perfect.} \qquad (*)$$

One consequence of this assumption is that then the map $d_L^k\colon R_0(LH) \to R_0(kH)$ of Lemma 7.3.4 is an isomorphism which preserves the classes of simple modules. Thus, we can and will identify $R_0(LH)$ and $R_0(kH)$.

Furthermore, $\theta\colon A \to L$ induces a map $H \to LH$, $h \mapsto h \otimes 1$. Analogously to Definition 7.3.1, we obtain a semigroup homomorphism

$$p_L : R_0^+(LH) \to \mathrm{Maps}(H, L[X]),$$

by associating with the class of an LH-module V the map which assigns to $h \in H$ the characteristic polynomial of $\rho_V(h \otimes 1)$, where ρ_V is the representation afforded by V. Now, by Remark 7.3.3, the above assumption (∗) implies that p_L is injective, i.e., the Brauer–Nesbitt Lemma 7.3.2 holds for LH.

7.4.2. Modular reduction. Let V be any KH-module and $\mathcal{O} \subseteq K$ be a valuation ring as in (7.4.1). By (7.3.7), there exists a K-basis of V such that the corresponding matrix representation $\rho : KH \to M_n(K)$ (where $n = \dim_K V$) has the property that $\rho(h) \in M_n(\mathcal{O})$ for all $h \in H$. Then that basis generates an $\mathcal{O}H$-lattice $\tilde{V}$ such that $\tilde{V} \otimes_{\mathcal{O}} K \cong V$. The k-vector space $\tilde{V} \otimes_{\mathcal{O}} k$ is a kH-module via $(v \otimes 1)(h \otimes 1) = vh \otimes 1$ ($v \in V$, $h \in H$), which we call the *modular reduction* of $\tilde{V}$. The matrix representation $\rho^k : kH \to M_n(k)$ afforded by $k\tilde{V}$ is given by

$$\rho(h \otimes 1) = (\pi(a_{ij})) \qquad \text{where } h \in H \text{ and } \rho(h) = (a_{ij}).$$

To simplify notation, we shall write

$$K\tilde{V} := \tilde{V} \otimes_{\mathcal{O}} K \quad \text{and} \quad k\tilde{V} := \tilde{V} \otimes_{\mathcal{O}} k.$$

Note that if $\tilde{V}'$ is another $\mathcal{O}H$-lattice such that $\tilde{V}' \otimes_{\mathcal{O}} K \cong V$, then $\tilde{V}$ and $\tilde{V}'$ need not be isomorphic. The same also holds for the kH-modules $\tilde{V} \otimes_{\mathcal{O}} k$ and $\tilde{V}' \otimes_{\mathcal{O}} k$. See Exercise 7.7 for a special case where we do have uniqueness.

Now we have the following basic result:

7.4.3. Theorem (Existence of decomposition maps). *Recall that we are given a ring homomorphism $\theta : A \to L$ into a field L such that L is the field of fractions of $\theta(A)$ and A is integrally closed in K. Assume also that we have chosen a valuation ring $\mathcal{O}$ as in (7.4.1) and that (∗) is satisfied. Then the following hold.*

(a) *The modular reduction in (7.4.2) induces an additive map $d_\theta : R_0^+(KH) \to R_0^+(LH)$ such that $d_\theta([K\tilde{V}]) = [k\tilde{V}]$, where $\tilde{V}$ is an $\mathcal{O}H$-lattice and $[k\tilde{V}]$ is regarded as an element of $R_0^+(LH)$ via the identification in (7.4.1).*

(b) *By Proposition 7.3.8, the map p_K has image contained in $\mathrm{Maps}(H, A[X])$. Then we have a commutative diagram*

$$\begin{array}{ccc} R_0^+(KH) & \xrightarrow{\;p_K\;} & \mathrm{Maps}(H, A[X]) \\ {\scriptstyle d_\theta}\downarrow & & \downarrow{\scriptstyle t_\theta} \\ R_0^+(LH) & \xrightarrow{\;p_L\;} & \mathrm{Maps}(H, L[X]) \end{array}$$

where $t_\theta : \mathrm{Maps}(H, A[X]) \to \mathrm{Maps}(H, L[X])$ is the map induced by θ.

(c) *The map d_θ is uniquely determined by the commutativity of the diagram in (b); in particular, d_θ only depends on θ but not on the choice of $\mathcal{O}$.*

The map d_θ will be called the *decomposition map* associated with the specialization $\theta\colon A \to L$. The matrix of that map with respect to the bases of $R_0(KH)$ and $R_0(LH)$ consisting of the classes of the simple modules is called the *decomposition matrix* associated with θ.

Proof. We begin by establishing the following relation:

$$p_L([k\tilde{V}]) = t_\theta \circ p_K([K\tilde{V}]) \qquad \text{for all } \mathcal{O}H\text{-lattices } \tilde{V} \tag{1}$$

(where we identify $R_0(LH)$ and $R_0(kH)$ as in (7.4.1)). To prove this, let $\tilde{V}$ be an $\mathcal{O}H$-lattice and $h \in H$. Choose an $\mathcal{O}$-basis of $\tilde{V}$ and let $(m_{ij}) \in M_n(\mathcal{O})$ be the matrix describing the action of h on $\tilde{V}$. By (7.4.2), the action of $h \otimes 1 \in kH$ on $k\tilde{V}$ is given by the matrix $(\pi(m_{ij}))$. Then, by definition, $p_L([k\tilde{V}])(h)$ is the characteristic polynomial of $(\pi(m_{ij}))$. On the other hand, $(t_\theta \circ p_K)([K\tilde{V}])(h)$ is computed by first taking the characteristic polynomial of (m_{ij}) and then applying θ (which is the restriction of π to A) to its coefficients. Since the two operations of applying π to a matrix with coefficients in $\mathcal{O}$ and taking the characteristic polynomial of a matrix commute with each other, the two polynomials that we just computed are equal. Thus, (1) is established.

Now let us prove (a). For this purpose, we must show that if $\tilde{V}$ and $\tilde{V}'$ are $\mathcal{O}H$-lattices such that $K\tilde{V}$ and $K\tilde{V}'$ have the same composition factors (counting multiplicities) then the classes of $k\tilde{V}$ and $k\tilde{V}'$ in $R_0(LH)$ are the same. Now, if $K\tilde{V}$ and $K\tilde{V}'$ have the same composition factors then $\rho_{K\tilde{V}}(h)$ and $\rho_{K\tilde{V}'}(h)$ are conjugate endomorphisms for every $h \in H$. So we certainly have $p_K([K\tilde{V}])(h) = p_K([K\tilde{V}'])(h)$ for all $h \in H$. Hence, by (1), we also have

$$p_L([k\tilde{V}])(h) = p_L([k\tilde{V}'])(h) \qquad \text{for all } h \in H.$$

We have already remarked in (7.4.1) that the Brauer–Nesbitt lemma holds for LH. So we conclude that $[k\tilde{V}] = [k\tilde{V}']$, as desired.

Having established the existence of d_θ, we have $[k\tilde{V}] = d_\theta([K\tilde{V}])$ for any $\mathcal{O}H$-lattice $\tilde{V}$. Hence (1) yields the commutativity of the diagram in (b).

Finally, by the Brauer–Nesbitt lemma, the map p_L is injective. Hence there exists at most one map $R_0^+(KH) \to R_0^+(LH)$ which makes the diagram in (b) commutative. This proves (c). □

7.4.4. Remark. Recall from Proposition 7.3.8 that if V is a KH-module then its character χ_V restricts to a trace function $\dot{\chi}_V\colon H \to A$. Now any A-linear map $\lambda\colon H \to A$ induces an L-linear map

$$\lambda^L\colon LH \to L, \qquad h \otimes 1 \mapsto \theta(\lambda(h)) \qquad (h \in H).$$

It is clear that if λ is a trace function then so is λ^L. Applying this to $\dot{\chi}_V$ shows that $\dot{\chi}_V^L$ is a trace function on LH. Since character values occur as coefficients in characteristic polynomials, Theorem 7.4.3 immediately implies that $\dot{\chi}_V^L$ is the character of $d_\theta([V])$. Moreover, for any simple KH-module V, we have

$$\dot{\chi}_V^L = \sum_{V'} d_{VV'} \chi_{V'},$$

where the sum is over all simple LH-modules V' (up to isomorphism) and $D = (d_{VV'})$ is the decomposition matrix associated with θ.

7.4.5. Example. Let G be a finite group and p be a rational prime number. There is a standard setting in which the assumptions of Theorem 7.4.3 are satisfied. Let $K \supseteq \mathbb{Q}$ be a finite algebraic extension, A be the ring of algebraic integers in K, and $I \subset A$ be a prime ideal containing p. Then A is a Dedekind domain and the localization of A in I is a discrete valuation ring, whose residue field L is a finite field of characteristic p (see [Curtis and Reiner 1962, §19]). In particular, L is a perfect field and so (b) in condition (7.4.1)(∗) holds. This is the set-up considered in most books on representation theory of finite groups, e.g., [Curtis and Reiner 1981] and [Feit 1982]. In [Goldschmidt 1980], a slightly different set-up is considered where K is an algebraic closure of $\mathbb{Q}$, and A is any valuation ring in K whose maximal ideal contains p; in this case, it is easily checked that the residue field of A is also algebraically closed and so (a) in (7.4.1)(∗) is satisfied. In Chapter 8, we will consider similar situations for Iwahori–Hecke algebras associated with finite Coxeter groups.

The advantage of the above general setting for decomposition maps can be seen in the following applications. Recall that $\theta\colon A \to L$ is a ring homomorphism into a field L which is the field of fractions of $\theta(A)$, and that A is integrally closed in K. If KH and LH are split, then the decomposition map d_θ of Theorem 7.4.3 exists. The following result gives a criterion for d_θ to be trivial.

7.4.6. Theorem (Tits's deformation theorem). *Assume that* KH *and* LH *are split and that* LH *is semisimple. Then the algebra* KH *is also semisimple and the decomposition map* d_θ *is an isomorphism which preserves isomorphism classes of simple modules. In particular, the map* $\mathrm{Irr}(KH) \to \mathrm{Irr}(LH)$, $\chi \mapsto \dot{\chi}^L$, *is a bijection.*

Proof. For each simple KH-module V we have an equation,

$$d_\theta([V]) = \sum_{V'} d_{V,V'}[V'] \qquad \text{with } d_{V,V'} \in \mathbb{N}_0,$$

where V' runs over the simple LH-modules (up to isomorphism). In particular, this means that $\dim_K V = \sum_{V'} d_{V,V'} \dim_L V'$ and we obtain

$$(\dim_K V)^2 = \Big(\sum_{V'} d_{V,V'} \dim_L V'\Big)^2 \geq \sum_{V'} d_{V,V'}^2 (\dim_L V')^2.$$

We now sum these inequalities over all simple KH-modules V (up to isomorphism). Since all these modules are split simple, Wedderburn's theorem shows that the left-hand side of the result will be equal to $\dim_K(KH/\operatorname{rad}(KH))$. The right-hand side equals

$$\sum_V \sum_{V'} d_{V,V'}^2 (\dim_L V')^2 = \sum_{V'} \Big(\sum_V d_{V,V'}^2\Big)(\dim_L V')^2.$$

Now we certainly have $d_\theta([KH]) = [LH]$. Hence for each V' there exists some V such that $d_{V,V'} \neq 0$, and so

$$\sum_V d_{V,V'}^2 \geq 1 \qquad \text{for all simple } LH\text{-modules } V'.$$

Using this bound, we can conclude that

$$\dim_K(KH/\operatorname{rad}(KH)) = \sum_V (\dim_K V)^2 \geq \sum_{V,V'} d_{V,V'}^2 (\dim_L V')^2 \geq \sum_{V'} (\dim_L V')^2.$$

Now we use the assumption that LH is semisimple. This implies that the right-hand side is just $\dim_L LH$, again by Wedderburn's theorem. Since $\dim_K KH = \dim_L LH$, all the inequalities above must be equalities and we have $\operatorname{rad}(KH) = 0$. Thus, KH is semisimple and, moreover, we have

$$\sum_V d_{V,V'}^2 = 1 \qquad \text{for all simple } LH\text{-modules } V'.$$

This implies that each column of the matrix $(d_{V,V'})$ has precisely one non-zero entry (which is 1). But each row certainly has some non-zero entry as well (since $d_\theta([V]) \neq 0$). Hence the matrix $(d_{V,V'})$ is a permutation matrix, which is what we wanted to prove. □

Finally, if H is symmetric, we can check as follows whether the assumption of Tits's deformation theorem is satisfied.

7.4.7. Theorem (Semisimplicity criterion). *Assume that* KH *and* LH *are split, and that* H *is symmetric with symmetrizing form* τ. *For any simple* KH*-module* V *let* $c_V \in A$ *be the Schur element with respect to the induced symmetrizing form* τ^K *on* KH. *Then* LH *is semisimple if and only if* $\theta(c_V) \neq 0$ *for all* V.

Proof. Assume first that $\theta(c_V) \neq 0$ for all simple KH-modules. Since the Schur elements themselves are non-zero, Theorem 7.2.6 shows that the algebra KH is semisimple and that we can write $\tau^K = \sum_V c_V^{-1} \chi_V$, where V runs over the simple KH-modules (up to isomorphism). As in Remark 7.4.4, the characters χ_V restrict to trace functions $\dot{\chi}_V \colon H \to A$ and the induced maps $\dot{\chi}_V^L$ are characters of LH. Thus, applying θ yields an equation of trace functions on LH,

$$\tau^L = \sum_V \theta(c_V)^{-1} \dot{\chi}_V^L.$$

By Exercise 7.6, LH must be semisimple.

Conversely, assume that $\mathsf{L}\mathsf{H}$ is semisimple. Let $\mathcal{O}$ be a valuation ring as in (7.4.1), with residue field $\mathsf{k} \supseteq \mathsf{L}$ and canonical map $\pi\colon \mathcal{O} \to \mathsf{k}$. Since LH is split and semisimple, the same holds for kH (check this!). We can now apply Tits's deformation theorem to the specialization $\pi\colon \mathcal{O} \to \mathsf{k}$. Consequently, KH is also semisimple and d_π is a bijection between isomorphism classes of simple modules for KH and kH. Let V be a simple KH-module. As in (7.3.7), there exists a K-basis of V such that the corresponding matrix representation $\rho\colon \mathsf{KH} \to M_n(\mathsf{K})$ has the property that $\rho(h) \in M_n(\mathcal{O})$ for all $h \in \mathsf{H}$. Consider the Schur relations with respect to this representation ρ. We can apply π to all terms in the sum in Corollary 7.2.2, and what we obtain are the Schur relations for the reduced matrix representation of kH. This is afforded by a kH-module V' such that $d_\pi([V]) = [V']$, and so we have $\pi(c_V) = c_{V'}$. By Theorem 7.2.6 this constant is non-zero and so $\theta(c_V) \neq 0$. □

7.5 PROPERTIES OF THE DECOMPOSITION MAP

The aim of this section is to establish some basic properties of decomposition maps, in a "standard setting" which combines the various assumptions on H that we encountered above. Thus, we are given an algebra H over an integral domain A, finitely generated and free over A. Let K be the field of fractions of A. Then we assume that the following conditions are satisfied:

- The ring A is a discrete valuation ring in K; let $\nu\colon \mathsf{K} \to \mathbb{Z} \cup \{\infty\}$ be the corresponding valuation, as in Example 7.3.6.
- Let $\theta\colon \mathsf{A} \to \mathsf{L}$ be the canonical map onto the residue field L of A. Then KH and LH are split, and KH is semisimple.
- The algebra H is symmetric, with symmetrizing form τ; see Definition 7.1.1.

In particular, the assumptions of Theorem 7.4.3 are satisfied and we have a well-defined decomposition map $d_\theta\colon R_0^+(\mathsf{KH}) \to R_0^+(\mathsf{LH})$. Moreover, we will make use of two standard results (Brauer reciprocity and the relation of blocks of the decomposition matrix with central characters) which we will not prove here; see [Feit 1982, §I.17]. Recall the notation $\mathrm{Irr}(\mathsf{KH})$ and $\mathrm{Irr}(\mathsf{LH})$ for the sets of irreducible characters of KH and LH, respectively. The decomposition matrix associated with the map d_θ is the $\mathrm{Irr}(\mathsf{KH}) \times \mathrm{Irr}(\mathsf{LH})$ matrix $D = (d_{\chi\varphi})$ with non-negative integer entries such that

$$d_\theta([V_\chi]) = \sum_{\varphi \in \mathrm{Irr}(\mathsf{LH})} d_{\chi\varphi}[V'_\varphi] \qquad \text{for } \chi \in \mathrm{Irr}(\mathsf{KH}),$$

where V_χ is a simple KH-module with character χ and V'_φ is a simple LH-module with character φ. We sometimes call the characters of KH "ordinary" and those of LH "modular". We say that $\varphi \in \mathrm{Irr}(\mathsf{LH})$ is a *modular constituent* of $\chi \in \mathrm{Irr}(\mathsf{KH})$ if $d_{\chi\varphi} \neq 0$. We use the conventions of (7.4.2) and Remark 7.4.4. By the construction of d_θ in Theorem 7.4.3, the number $d_{\chi\varphi}$ is the multiplicity of V'_φ as a composition factor in $\mathsf{L}\tilde{V}$ where $\tilde{V}$ is an H-lattice with $\mathsf{K}\tilde{V} \cong_{\mathsf{KH}} V_\chi$.

Since KH is split and semisimple, there are Schur elements $c_\chi \in K$ associated with the simple KH-modules V_χ. By Proposition 7.3.9 and Corollary 7.2.6, we have $0 \neq c_\chi \in A$ for all $\chi \in \mathrm{Irr}(KH)$.

7.5.1. Example. Let G be a finite group, K be a finite extension field of $\mathbb{Q}$ containing all the $|G|$th roots of unity, and p be a prime number. Let A be the localization of the ring of integers of K in a prime ideal containing p. Then all of the above assumptions are satisfied for the algebra $H = A[G]$.

Indeed, the ring-theoretic conditions on A have already been mentioned in Example 7.4.5. The algebras KH and LH are split, thanks to results due to Brauer; see [Curtis and Reiner 1981, §17A]. Since K has characteristic 0, the algebra KH is semisimple by Maschke's theorem (see Example 7.1.13). Finally, a symmetrizing trace for H is described in Example 7.1.5.

We will see later that we have a similar standard set-up for Iwahori–Hecke algebras associated with finite Coxeter groups.

A priori, we know nothing about the properties of the matrix D except that its entries are non-negative integers and that each row and each column has at least one non-zero entry. The aim of this section is to prove the most basic facts about D.

The rows of D describe the decomposition of $d_\theta([V_\chi])$ in the standard basis of $R_0(LH)$. An interpretation of the columns is given by the following result, which is part of Brauer's classical theory of modular representations.

7.5.2. Theorem (Brauer reciprocity). *For each $\varphi \in \mathrm{Irr}(LH)$, there exists some primitive idempotent $e_\varphi \in H$ such that*

$$[e_\varphi KH] = \sum_{\chi \in \mathrm{Irr}(KH)} d_{\chi\varphi}[V_\chi] \in R_0^+(KH).$$

Proof. Let $\hat{A}$ be the completion of A and $\hat{K} \supseteq K$ be its field of fractions. Since KH is split semisimple, we can identify $R_0(KH)$ and $R_0(\hat{K}H)$; see Lemma 7.3.4. Now, Brauer's original result (see [Feit 1982, (I.17.8)]) shows that there exists an idempotent $\hat{e}_\varphi \in \hat{A}H$ such that $[\hat{e}_\varphi \hat{K}H]$ (regarded as an element of $R_0^+(KH)$) has the required expression. But, by [Curtis and Reiner 1987, Ex. 6.16], there also exists an idempotent $e_\varphi \in H$ which is conjugate to $\hat{e}_\varphi$ in $\hat{A}H$. □

Our next aim is to prove a congruence result about the columns of D. For this purpose, recall from Proposition 7.2.8 that every $\chi \in \mathrm{Irr}(KH)$ determines a central character $\omega_\chi \colon Z(KH) \to K$.

7.5.3. Theorem (Geck–Rouquier). *For each $\varphi \in \mathrm{Irr}(LH)$, we have*

$$\sum_{\chi \in \mathrm{Irr}(KH)} \frac{d_{\chi\varphi}}{c_\chi}\omega_\chi(z) \in A \qquad \text{for all } z \in Z(KH) \cap H.$$

In particular, $\sum_\chi d_{\chi\varphi}/c_\chi \in A$ for all $\varphi \in \mathrm{Irr}(LH)$.

Proof. Let $e_\varphi \in H$ be an idempotent as in Theorem 7.5.2. We then have a K-linear map $\lambda_\varphi \colon Z(KH) \to K$, $z \mapsto \tau^K(ze_\varphi)$. We claim that

$$\lambda_\varphi = \sum_{\chi \in \mathrm{Irr}(KH)} \frac{d_{\chi\varphi}}{c_\chi} \omega_\chi.$$

By Proposition 7.2.8(b), the elements $\{\chi^* \mid \chi \in \mathrm{Irr}(KH)\}$ form a basis of $Z(KH)$. It is therefore sufficient to show that the values of the left- and right-hand sides of the above formula are the same on χ^*, for any $\chi \in \mathrm{Irr}(KH)$.

Taking $z = \chi^*$ and using Proposition 7.2.8(a), the right-hand side evaluates to $1_K d_{\chi\varphi}$. Now consider the left-hand side. We compute that

$$\begin{aligned}\lambda_\varphi(\chi^*) &= \tau^K(\chi^* e_\varphi) = \chi(e_\varphi) = 1_K \dim_K(V_\chi e_\varphi)\\ &= 1_K \dim_K \mathrm{Hom}_{KH}(e_\varphi KH, V_\chi) = 1_K d_{\chi\varphi},\end{aligned}$$

where the last two equalities hold by Exercise 7.3 and Brauer reciprocity. Hence the above claim is established.

Finally, it remains to observe that since $e_\varphi \in H$, the function λ_φ takes values in A on elements in H. □

7.5.4. Example. What does the above result mean in the case where $H = A[G]$ is a group algebra as in Example 7.5.1? We have seen in Example 7.1.8 that the centre of H is free as an A-module, with a basis given by the class sums $\hat{C}$, where C is a conjugacy class of G. Let $\chi \in \mathrm{Irr}(KG)$. Since character values are constant on conjugacy classes, we have

$$\omega_\chi(\hat{C}) = \frac{|C|\chi(g)}{\chi(1)} \qquad \text{where } g \in C.$$

By Example 7.1.13, the Schur elements are given by $c_\chi = |G|/\chi(1)$. Hence Theorem 7.5.3 implies that the expression

$$\sum_\chi \frac{d_{\chi\varphi}}{c_\chi}\omega_\chi(\hat{C}) = \sum_\chi \frac{d_{\chi\varphi}\chi(1)}{|G|}\frac{|C|\chi(g)}{\chi(1)} = \frac{1}{|C_G(g)|}\sum_\chi d_{\chi\varphi}\chi(g)$$

lies in A for all $g \in G$. For a different proof see, for example, [Goldschmidt 1980, (6.9)] or [Feit 1982, Cor. IV.2.5].

In the case of group algebras, there is another important property of the decomposition matrix: it has full rank and, even stronger, all its elementary divisors are equal to 1. The known proofs of this result use, in some form, Brauer's characterization of characters (see, for example, [Curtis and Reiner 1981, (18.12)–(18.14)]), or the Green correspondence (see [Alperin 1986]), which both heavily use the fact that the algebra under consideration is a group algebra. In general, we can still show that D has full rank, under an assumption on the centre of H.

First note that, for any subring $B \subseteq K$ with $A \subseteq B$, we have

$$Z(BH) = Z(KH) \cap BH.$$

This follows from the fact that, in order to verify that some element of an algebra lies in the centre, it is sufficient to check that this element commutes with all elements in a basis of that algebra.

7.5.5. Definition. The canonical map $H \to LH$ induces a ring homomorphism $Z(H) \otimes_A L \to Z(LH)$. We say that the centre of H is *reduction stable*, if this ring homomorphism is an isomorphism.

Example 7.5.4 shows that the centre of a group algebra is reduction stable. Later, in Remark 8.2.5, we will see that this is also the case for Iwahori–Hecke algebras. One important property of algebras with a reduction stable centre is given by the following result.

7.5.6. Theorem (Geck–Rouquier). *Assume that the centre of* H *is reduction stable. Then the columns of the decomposition matrix* D *are linearly independent over* $\mathbb{Q}$. *In particular, we have* $|\mathrm{Irr}(LH)| \leq |\mathrm{Irr}(KH)|$. *Moreover, if* $|\mathrm{Irr}(LH)| = |\mathrm{Irr}(KH)|$ *and* D *is invertible over* A, *then* LH *must be semisimple.*

Proof. Assume we have a linear relation among the columns of D, i.e., there exist integers n_φ such that

$$\sum_{\varphi \in \mathrm{Irr}(LH)} n_\varphi d_{\chi\varphi} = 0 \qquad \text{for all } \chi \in \mathrm{Irr}(KH).$$

We must show that $n_\varphi = 0$ for all φ. For this purpose, let $e_\varphi \in H$ be an idempotent as in Theorem 7.5.2, and define a K-linear map $\lambda_\varphi \colon Z(KH) \to K$ as in the proof of Theorem 7.5.3. We claim that the map

$$\lambda := \sum_\varphi n_\varphi \lambda_\varphi \colon Z(KH) \to K$$

is identically 0. Indeed, by Proposition 7.2.8(a), the elements $\{\chi^* \mid \chi \in \mathrm{Irr}(KH)\}$ form a basis of $Z(KH)$. As in the proof of Theorem 7.5.3, we compute that $\lambda(\chi^*) = \sum_\varphi n_\varphi \lambda_\varphi(\chi^*) = \sum_\varphi n_\varphi d_{\chi\varphi} 1_K = 0$. Hence $\lambda(z) = 0$ for all $z \in Z(KH)$.

Since all e_φ lie in H, we can restrict λ to an A-linear map $Z(H) \to A$, and then tensor with L. Since the centre of H is reduction stable, we obtain a map on $Z(LH)$ which is still identically 0. Explicitly, this map on $Z(LH)$ is given by

$$0 = \lambda^L(z) = \sum_\varphi n_\varphi \tau^L(z\varepsilon_\varphi) \qquad \text{for all } z \in Z(LH),$$

where ε_φ is the image of e_φ in LH. For any $\varphi \in \mathrm{Irr}(LH)$ let φ^* be the corresponding element in $Z(LH)$, defined with respect to the symmetrizing trace τ^L on LH as in Lemma 7.1.7. We compute that

$$0 = \lambda^L(\varphi^*) = \sum_{\varphi'} n_{\varphi'} \tau^L(\varphi^* \varepsilon_{\varphi'}) = \sum_{\varphi'} n_{\varphi'} \varphi(\varepsilon_{\varphi'}) = n_\varphi 1_L,$$

where the last equality holds by Exercise 7.4(a). Thus, we have shown that $n_\varphi 1_L = 0$ for all $\varphi \in \mathrm{Irr}(LH)$. If L has characteristic 0, we are done. If L has characteristic $p > 0$, we can divide all coefficients n_φ by p, and repeat the whole argument. So it follows that each n_φ is divisible by arbitrarily high powers of p. Hence we can again conclude that $n_\varphi = 0$ for all $\varphi \in \mathrm{Irr}(LH)$. Thus, the columns of D are linearly independent, as desired. In particular, we have $|\mathrm{Irr}(LH)| \leq |\mathrm{Irr}(KH)|$. Now assume that equality holds and that D is invertible over A. Then Theorem 7.5.3 implies that $1/c_\chi \in A$ for all $\chi \in \mathrm{Irr}(KH)$. Consequently, every Schur element of KH remains non-zero in L and so LH is semisimple by Theorem 7.4.7. □

An example with a decomposition matrix where the columns are not linearly independent will be given in Exercise 8.9. We will now explain how the above result can be used to compute the number of isomorphism classes of simple LH-modules in the case where H admits a character table; see (7.3.11).

7.5.7. Corollary. *Let* $\mathrm{Irr}(KH) = \{\chi_1, \ldots, \chi_n\}$ *and assume that* $(\chi_i(h_j))$ *is a character table for* H *with respect to* $h_1, \ldots, h_n \in H$. *Furthermore, assume that the columns of the decomposition matrix* D *are linearly independent over* L. *(By Theorem 7.5.6 this condition holds, for example, if the centre of* H *is reduction stable and* L *has characteristic* 0.*) Then we have:*

$$|\mathrm{Irr}(LH)| = \textit{rank of the specialized matrix } (\theta(\chi_i(h_j))).$$

Proof. Let $\mathrm{Irr}(LH) = \{\varphi_1, \ldots, \varphi_m\}$ and consider the matrix of all character values $\varphi_j(b \otimes 1)$, where $1 \leq j \leq m$ and b runs over the elements in an A-basis of H. Using Remark 7.3.3, we have

$$|\mathrm{Irr}(LH)| = \text{rank of } (\varphi_j(b \otimes 1)).$$

Now, by Remark 7.4.4, we have equations $\dot{\chi}_i^L = \sum_j d_{ij}\varphi_j$, where d_{ij} are the decomposition numbers. Since the matrix (d_{ij}) has full rank over L by assumption, we conclude that each φ_j can be expressed as a linear combination of the characters $\dot{\chi}_i^L$. Thus, if we denote the matrix of all character values $\chi_i(b)$ by Ψ, we have

$$|\mathrm{Irr}(LH)| = \text{rank of } \Psi^\theta,$$

where Ψ^θ is the matrix obtained by applying θ to all coefficients of Ψ. It remains to use the fact that we also have equations $\chi_i(b) = \sum_j f_j(b)\chi_i(h_j)$ with $f_j(b) \in A$, i.e., Ψ can be expressed as a product of the character table of H and a matrix with coefficients in A. Consequently, Ψ^θ and the specialized character table have the same rank. □

We will apply the above results in Section 11.5 to determine the number of simple modules for specialized Iwahori–Hecke algebras of exceptional type.

Finally, we turn to the block distribution of characters. For this purpose, we introduce the following notions.

7.5.8. Definition. (a) The *Brauer graph* associated with H has vertices labelled by the irreducible characters of KH, and an edge joining $\chi, \chi' \in \mathrm{Irr}(KH)$ if $\chi \neq \chi'$ and if there exists some $\varphi \in \mathrm{Irr}(LH)$ such that $d_{\chi\varphi} \neq 0 \neq d_{\chi'\varphi}$. A connected component of the Brauer graph is called a *block*. The decomposition matrix of a block B is the submatrix of D with rows and columns labelled by the irreducible characters of KH and LH, respectively, which belong to B.

(b) Let $\chi \in \mathrm{Irr}(KH)$. Recall that $0 \neq c_\chi \in A$. Let $\delta_\chi := \nu(c_\chi) \geq 0$, where $\nu\colon K \to \mathbb{Z} \cup \{\infty\}$ is the given valuation. Then δ_χ is called the *defect* of χ. If B is a block, then $\delta_B := \max\{\delta_\chi \mid \chi \text{ belongs to } B\}$ is called the defect of B.

7.5.9. Example. Consider the case where $H = A[G]$ is the group algebra of a finite group and the residue field of A has characteristic $p > 0$. By Example 7.1.13, we have $c_\chi = |G|/\chi(1)$ for all $\chi \in \mathrm{Irr}(KH)$. Then the p-part of $|G|/\chi(1)$ is usually called the p-defect of χ; see [Goldschmidt 1980]. Note that if p is ramified in A (i.e., we have $\nu(p) > 1$), then the p-defect of χ is not necessarily the same as the defect in the sense of Definition 7.5.8.

The defect of a block gives a measure of how complicated the representation theory of that block is. The simplest case is that of a block of defect 0; see Theorem 7.5.11 below.

We have the following standard result relating blocks with central characters. Recall that every $\chi \in \mathrm{Irr}(KH)$ determines a central character $\omega_\chi\colon Z(KH) \to K$. Since A is integrally closed in K, we have $\omega_\chi(z) \in A$ for all $z \in Z(H)$; see Proposition 7.3.8. Now we can state:

7.5.10. Lemma (Blocks and central characters). *Let $\chi, \chi' \in \mathrm{Irr}(KH)$. Then χ and χ' belong to the same block if and only if*

$$\theta(\omega_\chi(z)) = \theta(\omega_{\chi'}(z)) \qquad \textit{for all } z \in Z(H).$$

Proof. First assume that χ, χ' belong to the same block. It is sufficient to consider the case where χ, χ' are directly linked on the Brauer graph, i.e., there exists some $\varphi \in \mathrm{Irr}(LH)$ such that $d_{\chi\varphi} \neq 0 \neq d_{\chi'\varphi}$. Let $\tilde{V}_\chi$ be an H-lattice such that $K\tilde{V}_\chi$ affords χ. Let $z \in Z(H)$. Then z acts by the scalar $\omega_\chi(z)$ on $\tilde{V}_\chi$. Hence, $z \otimes 1$ acts by the scalar $\theta(\omega_\chi(z))$ on every modular constituent of $k\tilde{V}_\chi$. Similarly, $z \otimes 1$ acts by the scalar $\theta(\omega_{\chi'}(z))$ on every modular constituent of $k\tilde{V}_{\chi'}$ where $\tilde{V}_{\chi'}$ is an H-lattice such that $K\tilde{V}_{\chi'}$ affords χ'. Since $K\tilde{V}_\chi$ and $K\tilde{V}_{\chi'}$ have a modular constituent in common, we have $\theta(\omega_\chi(z)) = \theta(\omega_{\chi'}(z))$, as desired.

To prove the converse, we use the fact that the blocks of H correspond to central primitive idempotents in H; see [Feit 1982, (I.17.9)]. (Note that this is first proved over the completion of A and then one can reduce to A by an argument as in the proof of Theorem 7.5.2.) So, assuming that χ, χ' do not belong to the same block, we know that there exists a central idempotent $e \in H$ such that $\omega_\chi(e) = 1$ and $\omega_{\chi'}(e) = 0$. Consequently, we also have $\theta(\omega_\chi(e)) \neq \theta(\omega_{\chi'}(e))$. □

7.5.11. Theorem (Blocks of defect 0). *Let χ be an irreducible character of* KH *such that $\theta(c_\chi) \neq 0$. Then χ is an isolated vertex in the Brauer graph, and the corresponding decomposition matrix is just* (1).

Proof. Consider the element $\chi^* \in Z(KH)$. Since χ restricts to a trace function $H \to A$, we have in fact $\chi^* \in Z(H)$. By Proposition 7.2.8, we have $\dot{\omega}_\chi(\chi^*) = c_\chi$ and $\dot{\omega}_{\chi'}(\chi^*) = 0$ for any $\chi' \in \mathrm{Irr}(KH)$, $\chi' \neq \chi$. Now assume, if possible, that there exists some $\chi' \in \mathrm{Irr}(KH)$ which is linked to χ in the Brauer graph. Then Lemma 7.5.10 would imply that $0 \neq \theta(c_\chi) = \theta(\omega_\chi(\chi^*)) = \theta(\omega_{\chi'}(\chi^*)) = 0$, which is absurd. Hence χ lies in a block by itself.

It remains to show that $d_\theta([V_\chi])$ is the class of a simple module in $R_0^+(LH)$. For this purpose, we use a technique which recalls the proof of Theorem 7.4.7. By (7.4.2), there exists a basis of V_χ such that each element of H is represented by a matrix with coefficients in A with respect to that basis. Since H is symmetric, we can take an A-basis $\mathcal{B}$ of H and consider the Schur relations in Corollary 7.2.2. All terms in these relations lie in A, hence we can apply the canonical map $\theta\colon A \to L$, and what we obtain are the Schur relations with respect to the module $L\tilde{V}_\chi$ where $\tilde{V}_\chi \subseteq V_\chi$ is the H-lattice spanned by the above basis of V_χ. Since $\theta(c_\chi) \neq 0$, Lemma 7.2.3 implies that $L\tilde{V}_\chi$ is simple. □

7.5.12. Corollary. *Assume that the columns of* D *are linearly independent over* L. *Let* B *be any block of non-zero defect. Then we have*

$$|\{\psi \in \mathrm{Irr}(LH) \mid \psi \text{ belongs to } B\}| < |\{\chi \in \mathrm{Irr}(KH) \mid \chi \text{ belongs to } B\}.$$

In particular, we have $|\mathrm{Irr}(LH)| \leq |\mathrm{Irr}(KH)| - |\{B \text{ block} \mid \delta_B > 0\}|$.

Proof. Since the columns of D are linearly independent over A, we have $|\{\psi \in \mathrm{Irr}(LH) \mid \psi \text{ belongs to } B\}| \leq |\{\chi \in \mathrm{Irr}(KH) \mid \chi \text{ belongs to } B\}$. If we had equality, the same kind of argument as in the proof of Theorem 7.5.6 would show that $1/c_\chi \in A$ for all $\chi \in \mathrm{Irr}(KH)$ which belong to B. This would mean that $\delta_B = 0$, a contradiction. □

We close this section with the remark that it is also possible to obtain a number of precise results on decomposition numbers of blocks of defect 1. For example, it can be shown that all decomposition numbers in such a block are 0 or 1 and that its Brauer graph is a tree, the so-called *Brauer tree*. The proofs of these results are straightforward adaptations of Brauer's theory of blocks of defect 1 for finite groups (as presented in [Goldschmidt 1980, §11]) to the present context. In the case of Iwahori–Hecke algebras, such an adaptation is described in [Geck 1992].

7.6 BIBLIOGRAPHICAL REMARKS AND EXERCISES

In Section 7.1 we have followed the exposition in [Curtis and Reiner 1962, Section 62]. For the general theory of symmetric algebras, see also the appendix in [Broué et al. 1999]. Schur relations for symmetric algebras have certainly been known for quite some time; see the remarks in [Curtis and Reiner 1962, p. 600]. The proof of Theorem 7.2.1 is taken from [Curtis and Reiner 1962, p. 218] (where the argument is given in the case of group algebras). The connection to simple projectivity of modules appears in [Fleischmann 1990] and [Yamane 1989]; but note that we have reduced the necessary assumptions on the ground ring to a minimum. Schur elements were used as a basic tool in the study of modular representations of Iwahori–Hecke algebras in [Geck 1992].

The general setting for decomposition maps in Sections 7.3 and 7.4 is developed in [Geck and Rouquier 1996]; note that the diagram in Proposition 2.11 of that article is not quite correct (the range of p_k should be as in Theorem 7.4.3). For another, but slightly different, general setting see [Du et al. 1998]. The proof of Tits's deformation theorem follows the strategy outlined in the remarks in [Curtis and Reiner 1987, §68A, p. 643]. A semisimplicity criterion for Iwahori–Hecke algebras with equal parameters similar to that in Theorems 7.2.6 and 7.4.7 appears in [Gyoja and Uno 1989]. See [Gyoja 1995] and [Geck 1995] for the more general version above.

The exposition in Section 7.5 also follows [Geck and Rouquier 1996], except that here we have avoided the use of Henselian rings. As far as we know, Theorems 7.5.3 and 7.5.6 (although probably known) appear explicitly for the first time in [Geck and Rouquier 1996]. Theorem 7.5.3 was suggested by J. Müller, in an attempt to extend the notion of Brauer characters from finite groups to arbitrary symmetric algebras. A version of that result for Iwahori–Hecke algebras was first proved in [Geck 1993a], by a reduction to an analogous statement for finite Chevalley groups. In that article, Theorem 7.5.3 was used as a tool for computing decomposition numbers.

Exercise 7.1. Let A be any commutative ring with 1 and $n \geq 1$. Show that every trace function on $M_n(A)$ is a scalar multiple of the usual matrix trace.

Exercise 7.2. It is well-known that a unique factorization domain is integrally closed (i.e., integrally closed in its field of fractions). Using Gauss's lemma this implies, for example, that the polynomial ring over a field is integrally closed. More generally, show that if A is any integral domain and X is an indeterminate, then the integral closure of $A[X]$ in its field of fractions is $A^*[X]$, where A^* is the integral closure of A in its field of fractions. For hints, see [Atiyah and Macdonald 1969, Chap. 5, Exercise 9].

Exercise 7.3. Let H be an algebra over a commutative ring A with 1, let V be a right H-module, and $e \in H$ be an idempotent. Show that the map

$$\mathrm{Hom}_H(eH, V) \to Ve, \quad \varphi \mapsto \varphi(e),$$

is an isomorphism of A-modules.

Exercise 7.4. Let H be a finite-dimensional algebra over a field K and assume that H is split.

(a) Show that for each simple H-module V there exists an idempotent $e_V \in H$ (i.e., we have $e_V^2 = e_V \neq 0$) such that $\dim_K(Ve_V) = 1$ and $V'e_V = 0$ if V' is a simple H-module which is not isomorphic to V.

(b) Show that if H is also semisimple, then the irreducible characters of H form a basis for the space of trace functions on H.

[*Hint.* (a) Assume first that H is semisimple and let $H = \bigoplus_V H(V)$ be the Wedderburn decomposition of H. For each simple H-module V, take $e_V \in H(V)$ as the element which maps to the matrix with $(1,1)$th entry 1 and entries 0 otherwise under an isomorphism $H(V) \cong M_{n_V}(K)$. Now assume that H is not semisimple. But then $H/\operatorname{rad}(H)$ is semisimple, and the simple H-modules can also be regarded as modules for $H/\operatorname{rad}(H)$, since $\operatorname{rad}(H)$ acts as 0 on each simple module. Given V, we let $\varepsilon_V \in H/\operatorname{rad}(H)$ be an idempotent as before. Now use the "theorem on lifting idempotents" (see [Curtis and Reiner 1987, §6A]) to find an idempotent $e_V \in H$ as required.

(b) By Example 7.1.3 and Lemma 7.1.7, the dimension of $H/[H, H]$ equals the dimension of the centre of H, and this is also equal to the number of isomorphism classes of simple H-modules (by Wedderburn's theorem). Now use (a) to prove that the irreducible characters of H are linearly independent.]

Exercise 7.5. Let H_1, H_2 be finite-dimensional algebras over a field K. For $i = 1, 2$, let τ_i be a symmetrizing trace on H_i. The tensor product $H := H_1 \otimes_K H_2$ is again a finite-dimensional algebra over K. Show that the following hold.

(a) The map $\tau\colon H \to K$, $h_1 \otimes h_2 \mapsto \tau_1(h_1)\tau_2(h_2)$, is a symmetrizing trace.

(b) Assume that H_1, H_2 are split. Then H is also split. Moreover, the simple modules of H are of the form $V_1 \otimes_K V_2$, where V_i is a simple H_i-module for $i = 1, 2$. The Schur element of $V_1 \otimes_K V_2$ (with respect to τ) equals $c_{V_1} c_{V_2}$.

Exercise 7.6. Let H be a finite-dimensional symmetric algebra over a field K. Assume that the symmetrizing trace is a linear combination of characters of H. Show that H is separable, i.e., $K'H$ is semisimple for every extension field $K' \supseteq K$.

[*Hint.* Let τ be the symmetrizing trace on H. By scalar extension, it gives rise to a symmetrizing trace τ' on $K'H$, which is also a linear combination of characters of $K'H$. Now let $x \in \operatorname{rad}(K'H)$. It is enough to show that $\tau'(xy) = 0$ for all $y \in K'H$. Now use the fact that $\operatorname{rad}(K'H)$ is an ideal and that every element of $\operatorname{rad}(K'H)$ acts as a nilpotent element in each representation of $K'H$.]

Exercise 7.7. Let K be a field and $\mathcal{O} \subseteq K$ be a valuation ring with residue field k. Let H be an $\mathcal{O}$-algebra, finitely generated and free over $\mathcal{O}$. Let V be an H-lattice such that the KH-module $V \otimes_{\mathcal{O}} K$ and the kH-module $V \otimes_{\mathcal{O}} k$ are both simple. Show that if V' is an H-lattice with $V \otimes_{\mathcal{O}} K \cong_{KH} V' \otimes_{\mathcal{O}} K$, then $V \cong_{\mathcal{O}H} V'$.

[*Hint.* Using the "change of rings theorem" in [Curtis and Reiner 1987, (2.38)], there exists a non-zero H-module homomorphism $\varphi\colon V \to V'$. We can assume that the induced map $\varphi^k\colon V \otimes_{\mathcal{O}} k \to V' \otimes_{\mathcal{O}} k$ is also non-zero. Since $V \otimes_{\mathcal{O}} k$ is simple, conclude that φ^k must be an isomorphism.]

Exercise 7.8. This exercise is about the existence of certain discrete valuation rings (cf. [Matsumura 1986, Exercise 14.4]). It requires some more sophisticated results from commutative algebra.

Consider the polynomial ring $A = R[X_1, \ldots, X_n]$, where $X_1, \ldots, X_n$ are indeterminates and R is the ring of integers in some finite extension field of $\mathbb{Q}$. Let K be the field of fractions of A and $\theta\colon A \to L$ be a ring homomorphism into a field L such that L is the field of fractions of $\theta(A)$. We want to show that there exists a discrete valuation ring $\mathcal{O} \subseteq K$ such that $A \subseteq \mathcal{O}$ and $J(\mathcal{O}) \cap A = \ker(\theta)$. Proceed in the following steps:

(a) First note that the localization of A in any prime ideal is a so-called regular local ring, i.e., the minimal number of generators for its maximal ideal is given by the dimension of that ring (cf. [Matsumura 1986, §14]).

(b) Now let $A' \subseteq K$ be the localization of A in the prime ideal $\ker(\theta)$, and let $a_0, \ldots, a_d \in A'$ be generators for its maximal ideal $J(A')$, where $d + 1 = \dim A'$. Let $Y_1, \ldots, Y_d$ be indeterminates. Show that the kernel of the surjective ring homomorphism

$$\beta\colon A'[Y_1, \ldots, Y_d] \to B := A'\left[\frac{a_1}{a_0}, \ldots, \frac{a_d}{a_0}\right], \qquad Y_i \mapsto \frac{a_i}{a_0},$$

is the ideal generated by $a_0Y_1 - a_1, \ldots, a_0Y_d - a_d$.

(c) Let $I \subseteq B$ be the principal ideal generated by a_0. Show that I is a prime ideal and that $B/I \cong L[Y_1, \ldots, Y_d]$. Let $\mathcal{O} \subseteq K$ be the localization of B in I. Show that $\mathcal{O}$ is a discrete valuation ring with $A \subseteq \mathcal{O}$, $J(\mathcal{O}) \cap A = \ker(\theta)$, and with residue field $L(Y_1, \ldots, Y_d)$.

(d) For $0 \neq a \in A'$ set $\nu(a) = m$ if $a \in J(A')^m$ but $a \notin J(A')^{m+1}$. Show that ν defines the discrete valuation of K associated with $\mathcal{O}$.

[*Hint.* (a) This follows from [Matsumura 1986, Theorem 19.5] and induction on n. The ring R itself is a Dedekind domain (see [Curtis and Reiner 1962, §18 and §19]), and so the localization of R in any prime ideal is a regular local ring of dimension 1 (see [Matsumura 1986, Theorem 11.2]).

(b) This is based on the fact that the elements $\{a_0, \ldots, a_d\}$ form a so-called regular sequence in A' (see [Matsumura 1986, Theorem 14.4]). For more details see [Kunz 1979, V.5.10], where the desired assertion is explicitly proved.

(c) Using (b), it follows that $\beta^{-1}(I)$ is the ideal $J(A')[Y_1, \ldots, Y_d]$. Hence $B/I \cong L[Y_1, \ldots, Y_d]$, and so I is a prime ideal. So $L(Y_1, \ldots, Y_d)$ is the residue field of $\mathcal{O}$. Now note that B is a Noetherian ring, since it is finitely generated over the Noetherian ring A'. By Krull's principal ideal theorem (see [Matsumura 1986, Theorem 13.5]), the ideal I has height 1, and so the localization $\mathcal{O}$ is a Noetherian local ring of dimension 1 whose maximal ideal is principal. By [Matsumura 1986, Theorem 11.2], this means that $\mathcal{O}$ is a discrete valuation ring.]

Exercise 7.9. Consider the ring $\mathbb{Z}[X]$, where X is an indeterminate. Show that an ideal $\{0\} \neq \mathfrak{p} \subseteq \mathbb{Z}[X]$ is a prime ideal if and only if one of the following holds:

(a) $\mathfrak{p}$ is a principal ideal, generated by a prime number $\ell \in \mathbb{Z}$ or an irreducible polynomial $f \in \mathbb{Z}[X]$.

(b) $\mathfrak{p}$ is generated by two elements ℓ and f, where $\ell \in \mathbb{Z}$ is a prime number and $f \in \mathbb{Z}[X]$ is a monic polynomial whose reduction modulo ℓ is an irreducible polynomial in $\mathbb{F}_\ell[X]$.

[*Hint.* First check that ideals as in (a) or (b) are prime ideals. Now assume that $\mathfrak{p}$ is a prime ideal. Consider the intersection $\mathfrak{p} \cap \mathbb{Z}$. If this is $\{0\}$, consider the ideal generated by $\mathfrak{p}$ in $\mathbb{Q}[X]$. If that intersection is generated by a prime number $\ell \in \mathbb{Z}$, consider the natural map $\mathbb{Z}[X] \to \mathbb{F}_\ell[X]$ given by reduction modulo ℓ.]

Exercise 7.10. The purpose of this exercise is to establish a factorization of decomposition maps, following [Geck and Rouquier 1996] and [Geck 1998c]. Let F be an algebraic number field, R be the ring of algebraic integers in F and $A = R[v, v^{-1}]$, where v is an indeterminate. Let H be an A-algebra, finitely generated and free over A. Furthermore, let $\mathfrak{p} \subseteq R$ be a non-zero prime ideal and $\zeta \in R$ be a unit in R. Let K be the field of fractions of A and $\pi\colon R \to k$ be the canonical map, where k is the field of fractions of $R/\mathfrak{p}$. Assume that FH and kH are split. Then, by Theorem 7.4.3, we have three well-defined decomposition maps:

$$\begin{aligned} d_\zeta\colon R_0^+(KH) \to R_0^+(FH), &\quad \text{induced by the specialization } \theta_\zeta\colon A \to R,\ v \mapsto \zeta, \\ d_\mathfrak{p}\colon R_0^+(FH) \to R_0^+(kH), &\quad \text{induced by the canonical map } \pi\colon R \to k, \\ d_{\zeta,\mathfrak{p}}\colon R_0^+(KH) \to R_0^+(kH), &\quad \text{induced by the map } \pi \circ \theta_\zeta\colon A \to k. \end{aligned}$$

Show that we have $d_{\zeta,\mathfrak{p}} = d_\mathfrak{p} \circ d_\zeta$. Thus, if D_ζ, $D_\mathfrak{p}$ and $D_{\zeta,\mathfrak{p}}$ are the decomposition matrices of d_ζ, $d_\mathfrak{p}$ and $d_{\zeta,\mathfrak{p}}$, respectively, then we have $D_{\zeta,\mathfrak{p}} = D_\zeta D_\mathfrak{p}$.

[*Hint.* First note that, in each of the three cases, the hypothesis of Theorem 7.4.3 is satisfied, since FH and kH are assumed to be split and A, R are integrally closed. To prove the factorization of the decomposition maps, check that we have a commutative diagram:

$$\begin{array}{ccccc} R_0^+(KH) & \xrightarrow{\ d_\zeta\ } & R_0^+(FH) & \xrightarrow{\ d_\mathfrak{p}\ } & R_0^+(kH) \\ \downarrow{\scriptstyle p_K} & & \downarrow{\scriptstyle p_F} & & \downarrow{\scriptstyle p_k} \\ \text{Maps}(H, A[X]) & \xrightarrow{\ t_{\theta_\zeta}\ } & \text{Maps}(H, R[X]) & \xrightarrow{\ t_\pi\ } & \text{Maps}(H, k[X]) \end{array}$$

Now, we certainly have $t_\pi \circ t_{\theta_\zeta} = t_{\pi\circ\theta_\zeta}\colon \text{Maps}(H, A[X]) \to \text{Maps}(H, k[X])$. Then use the characterization of decomposition maps in Theorem 7.4.3(c) to show that the composition of the two maps of the top row is $d_{\zeta,\mathfrak{p}}$.]

8

Iwahori–Hecke algebras

In this book, Iwahori–Hecke algebras made their first appearance in Section 4.4, where we defined them as quotients of the group algebra of a braid group. We will now begin to develop the structure theory of these algebras, in the general framework of the theory of symmetric algebras (see Chapter 7).

The first crucial step is the observation that any Iwahori–Hecke algebra associated with a fixed Coxeter system (W, S) can be obtained from one "generic" algebra by the process of "specialization". In Section 8.1, we prove the basic structural properties of "generic" Iwahori–Hecke algebras associated with finite Coxeter groups. Using Tits's deformation theorem (see Section 7.4), we show that there is a bijection between their irreducible characters and those of the underlying Coxeter group.

In Section 8.2, we study the space of trace functions on Iwahori–Hecke algebras. We do this by transferring the results on the conjugacy graph in Chapter 3 to the Iwahori–Hecke algebra context. For example, this leads to a description of the centre of an Iwahori–Hecke algebra in terms of a "generic" analogue of the basis of the centre of a group algebra consisting of the class sums. As a consequence of the main results of this section, we can now define *character tables* of Iwahori–Hecke algebras and establish some of their basic properties. In fact, the search for a definition of such a table was the motivation for much of the previous work on elements with special properties in the conjugacy classes of finite Coxeter groups in Chapter 3.

We will see examples of such tables for algebras of type $I_2(m)$ in Section 8.3: they are square matrices whose entries are polynomials in (square roots of) the parameters of the algebra, and when we specialize these parameters to 1, we obtain the usual character table of the underlying finite Coxeter group.

In Section 8.4, we give a brief account of Iwahori's theorem, which shows that certain Iwahori–Hecke algebras also arise naturally in the representation theory of finite groups. In general terms, the endomorphism algebra of the permutation module of a finite group on the cosets of a subgroup is called a *Hecke algebra*. (For some historical remarks, see [Curtis and Reiner 1981, p. 281].) Such an algebra has a natural basis indexed by the double cosets of that subgroup. It was Iwahori who discovered that if we consider a Chevalley group and its action on the cosets of a Borel subgroup, then the corresponding Hecke algebra is an Iwahori–Hecke algebra as defined above. This result provides a motivation and a guideline for many aspects of the theory of "abstract" Iwahori–Hecke algebras.

8.1 GENERIC IWAHORI-HECKE ALGEBRAS

Let (W, S) be a Coxeter system and $H = H_A(W, S, \{a_s, b_s \mid s \in S\})$ be an Iwahori–Hecke algebra as in Definition 4.4.1. Recall that H has a basis $\{T_w \mid w \in W\}$ such that we have the following rules: for $s \in S$ and $w \in W$ we have

$$T_sT_w = \begin{cases} T_{sw} & \text{if } l(sw) > l(w), \\ a_sT_{sw} + b_sT_w & \text{if } l(sw) < l(w). \end{cases}$$

An analogous relation also holds for the product T_wT_s.

In order to be able to apply the full power of the results in Chapter 7, we will now show that H is a symmetric algebra if all the parameters a_s are invertible.

8.1.1. Proposition. *Define a linear map* $\tau\colon H \to A$ *by* $\tau(T_1) = 1$ *and* $\tau(T_w) = 0$ *for* $1 \neq w \in W$. *Then we have*

$$\tau(T_wT_{w'}) = \begin{cases} a_w & \text{if } w^{-1} = w', \\ 0 & \text{if } w^{-1} \neq w', \end{cases}$$

where $a_w \in A$ *is defined by* $a_w := a_{s_1} \cdots a_{s_n}$ *whenever* $w = s_1 \cdots s_n$ $(s_i \in S)$ *is a reduced expression. We have* $a_w = a_{w^{-1}}$ *for all* $w \in W$. *If, moreover, all* a_s $(s \in S)$ *are units in* A, *then* τ *is a symmetrizing trace and the basis dual to* $\{T_w \mid w \in W\}$ *is given by* $T_w^\vee = a_w^{-1}T_{w^{-1}}$ *for* $w \in W$.

Proof. First note that the map $S \to A$, $s \mapsto a_s$ satisfies the requirements of Matsumoto's theorem, and hence the map $w \mapsto a_w$ does indeed exist. Moreover, a reduced expression for w^{-1} is just given by reversing a reduced expression for w, whence the identity $a_{w^{-1}} = a_w$.

Next we prove the formula for $\tau(T_wT_{w'})$. We proceed by induction on $l(w)$. If $w = 1$, there is nothing to prove. Now let $l(w) > 0$ and choose $s \in S$ such that $l(ws) < l(w)$. Then $T_w = T_{(ws)s} = T_{ws}T_s$ and hence $T_wT_{w'} = T_{ws}T_sT_{w'}$. Now we distinguish two cases:

Case 1. $l(sw') > l(w')$. Then $T_sT_{w'} = T_{sw'}$ and so $\tau(T_wT_{w'}) = \tau(T_{ws}T_{sw'})$. Since $l(sw') > l(w')$ and $l(ws) < l(w)$, we must have $w^{-1} \neq w'$ and, hence, also $(ws)^{-1} \neq sw'$. By induction, we have $\tau(T_{ws}T_{sw'}) = 0$, and we are done.

Case 2. $l(sw') < l(w')$. Then $T_sT_{w'} = a_sT_{sw'} + b_sT_{w'}$ and hence

$$\tau(T_wT_{w'}) = a_s\tau(T_{ws}T_{sw'}) + b_s\tau(T_{ws}T_{w'}).$$

Since $l(s(ws)^{-1}) = l(w^{-1}) = l(w) > l(ws) = l((ws)^{-1})$, we have $(ws)^{-1} \neq w'$ and, hence, by induction, $\tau(T_{ws}T_{w'}) = 0$. Now consider the term $\tau(T_{ws}T_{sw'})$. If $w^{-1} \neq w'$ then we also have $(ws)^{-1} \neq sw'$ and so, by induction, $\tau(T_{ws}T_{sw'}) = 0$, as required. On the other hand, if $w^{-1} = w'$ then, by induction, $\tau(T_{ws}T_{sw'}) = a_{ws}$. Since $l((ws)s) = l(w) > l(ws)$, we have $a_{ws}a_s = a_w$, which yields that $\tau(T_wT_{w'}) = a_{ws}a_s = a_w$, as desired.

Now we can also check that τ is a trace function: since $a_w = a_{w^{-1}}$ and $w^{-1} = w'$ if and only if $(w')^{-1} = w$, we have $\tau(T_wT_{w'}) = \tau(T_{w'}T_w)$, as desired.

Finally, assume that each a_s is a unit in A. Then a_w is a unit in A for any $w \in W$. For each $w \in W$, we set $T'_w := a_w^{-1}T_{w^{-1}}$. The elements $\{T'_w\}$ also form an A-basis of H, and we have

$$\tau(T'_wT_{w'}) = a_{w'}^{-1}\tau(T_{w^{-1}}T_{w'}) = \begin{cases} 1 & \text{if } w' = w, \\ 0 & \text{if } w' \neq w. \end{cases}$$

This means that the bilinear form $H \times H \to A$, $(h, h') \mapsto \tau(hh')$, is non-degenerate over A and that $T_w^\vee = T'_w$ for all $w \in W$. □

The trace function τ may be regarded as an obvious generalization of the symmetrizing trace for group algebras in Example 7.1.5. Another interpretation for τ will be given in Corollary 8.4.7 (in the context of groups with a BN-pair).

Next we note that the construction of Iwahori–Hecke algebras is functorial in the following sense.

8.1.2. Specializations. Let A' also be a commutative ring with 1 and $\{a'_s, b'_s \mid s \in S\}$ be a collection of elements in A' such that $a'_s = a'_t$ and $b'_s = b'_t$ whenever $s, t \in S$ are conjugate in W. Then we have a corresponding Iwahori–Hecke algebra $H' = H_{A'}(W, S, \{a'_s, b'_s \mid s \in S\})$ whose standard basis we denote by $\{T'_w \mid w \in W\}$. Now assume that we have a ring homomorphism $\theta\colon A \to A'$ such that $\theta(s_s) = a'_s$ and $\theta(b_s) = b'_s$ for all $s \in S$. Then we may naturally regard A' as an A-module via θ. Using Theorem 4.4.6, we have a canonical isomorphism

$$H \otimes_A A' \to H', \qquad T_w \otimes 1 \mapsto T'_w.$$

In this situation, we call H' a *specialization* of H via $\theta\colon A \to A'$. Thus, if a_s and b_s are indeterminates in a polynomial ring, then any Iwahori–Hecke algebra associated with (W, S) is obtained from H via specialization.

8.1.3. Remark. The map $w \mapsto a_w$ will occur in a number of formulas, e.g., in formulas for character values; see Section 9.4. Note that if we are in a situation as in Example 4.4.2, where $H = A[W]$ is the group algebra of W, then $w \mapsto a_w = 1$ is just the trivial representation of W. In general, $w \mapsto a_w$ does not necessarily define a representation of H since the quadratic relations $T_s^2 = a_sT_1 + b_sT_s$ (for $s \in S$) need not be satisfied. However, this is the case precisely when $a_s^2 = a_s + b_sa_s$ for all $s \in S$. If A is an integral domain, this holds if and only if $a_s = 0$ or $b_s = a_s - 1$ ($s \in S$). In this case, we call

$$\text{ind}\colon H \to A, \qquad w \mapsto a_w,$$

the *index representation*; we also have the *sign representation*

$$\varepsilon\colon H \to A, \qquad T_s \mapsto -1 \quad (s \in S).$$

From now on, we will only consider the case where $b_s = a_s - 1$ for all $s \in S$ and where $a_s \in A$ is invertible; we will simply write $H = H_A(W, S, \{a_s \mid s \in S\})$. Note that this restriction is not essential: setting $T'_s := \lambda_s T_s$ where $\lambda_s \in A$, we have

$$(T'_s)^2 = \lambda_s^2 T_s^2 = \lambda_s a_s 1_H + \lambda_s^2 (a_s - 1) T_s = \lambda_s a_s 1_H + \lambda_s (a_s - 1) T'_s.$$

Thus, for suitable λ_s, we obtain general quadratic relations as in Definition 4.4.1.

We are now ready to define generic algebras associated with (W, S).

8.1.4. Definition. Let u_s ($s \in S$) be indeterminates over $\mathbb{C}$ such that $u_s = u_t$ if $s, t \in S$ are conjugate in W. Let $\mathbf{A} = \mathbb{Z}[u_s^{\pm 1} \mid s \in S]$ and assume that there exists a ring homomorphism $\mathbf{A} \to \mathbb{Z}$ which sends u_s to 1 for all $s \in S$. Then

$$\mathbf{H} := H_{\mathbf{A}}(W, S, \{u_s \mid s \in S\})$$

is called a *generic Iwahori–Hecke algebra* associated with (W, S). Note that we do *not* require that the u_s are independent variables; for example, they could all be equal, or powers of one fixed indeterminate. (This explains why we say "a generic algebra" and not "the generic algebra".) If u_s, u_t are independent variables whenever $s, t \in S$ are not conjugate in W, then we call $\mathbf{H}$ the *generic multi-parameter Iwahori–Hecke algebra* (see also Exercise 8.1).

By Theorem 4.4.6, $\mathbf{H}$ is free over $\mathbf{A}$ with a basis $\{T_w \mid w \in W\}$ such that

$$T_s T_w = \begin{cases} T_{sw} & \text{if } l(sw) > l(w), \\ u_s T_{sw} + (u_s - 1) T_w & \text{if } l(sw) < l(w). \end{cases}$$

By Remark 4.4.4, each T_s is invertible in $\mathbf{H}$ with inverse given by

$$T_s^{-1} = u_s^{-1} T_s + (u_s^{-1} - 1) T_1.$$

Moreover, by Proposition 8.1.1, the algebra $\mathbf{H}$ is symmetric, with symmetrizing trace $\tau\colon \mathbf{H} \to \mathbf{A}$ given by $\tau(T_1) = 1$ and $\tau(T_w) = 0$ for $1 \neq w \in W$. As before, we have the index representation $\mathrm{ind}\colon \mathbf{H} \to \mathbf{A}$, $T_s \mapsto u_s$ ($s \in S$), and the sign representation $\varepsilon\colon \mathbf{H} \to \mathbf{A}$, $T_s \mapsto -1$ ($s \in S$).

8.1.5. Remark. Let $\theta\colon \mathbf{A} \to A$ be any ring homomorphism into a commutative ring with 1, and set $a_s = \theta(u_s)$ for $s \in S$. Then we have a canonical isomorphism

$$\mathbf{H} \otimes_{\mathbf{A}} A \to H = H_A(W, S, \{a_s \mid s \in S\}), \quad T_w \otimes 1 \mapsto T_w.$$

Thus, as desired, any Iwahori–Hecke algebra associated with (W, S) can be obtained from a suitable generic Iwahori–Hecke algebra by specialization.

The condition that there exists a specialization $\theta\colon \mathbf{A} \to \mathbb{Z}$ with $\theta(u_s) = 1$ for all $s \in S$ holds, for example, if all the u_s are powers of one fixed indeterminate or if u_s, u_t are independent variables whenever $s, t \in S$ are not conjugate in W. In these cases, the specialized algebra is canonically isomorphic to the group algebra $\mathbb{Z}[W]$, which justifies speaking of $\mathbf{H}$ as a *deformation* of $\mathbb{Z}[W]$.

Recall that, by Maschke's theorem, the group algebra of a finite group over a field of characteristic 0 is semisimple; see Example 7.1.13. Thus, the above remarks show that $\mathbf{H}$ admits a semisimple specialization. We now describe a setting in which we can take advantage of the results developed in Section 7.4.

8.1.6. A setting for specialization arguments. Assume that W is finite. Let $A = \mathbb{C}$, fix an element $0 \neq q \in \mathbb{C}$ and assume that we have a specialization

$$\theta\colon \mathbf{A} \to \mathbb{C}, \quad u_s \mapsto q \quad (s \in S).$$

To simplify notation, we just write $\mathbf{H}_q := \mathbf{H} \otimes_\mathbf{A} \mathbb{C}$ to denote the specialized algebra. For example, if $q = 1$, we have $\mathbf{H}_1 = \mathbb{C}[W]$ and the specialized algebra is split semisimple.

Let $\mathbf{A}_\mathbb{C} := \mathbb{C}[u_s^{\pm 1} \mid s \in S] \supseteq \mathbf{A}$ and extend θ canonically to a surjective map $\theta_\mathbb{C}\colon \mathbf{A}_\mathbb{C} \to \mathbb{C}$. Let $\mathbf{K}$ be a finite Galois extension of $\mathbb{C}(u_s \mid s \in S)$ and $\mathbf{A}_\mathbb{C}^*$ be the integral closure of $\mathbf{A}_\mathbb{C}$ in $\mathbf{K}$. We claim that $\theta_\mathbb{C}$ has an extension to a ring homomorphism $\theta_\mathbb{C}^*\colon \mathbf{A}_\mathbb{C}^* \to \mathbb{C}$. Indeed, let $\mathcal{O} \subseteq \mathbf{K}$ be a valuation ring with maximal ideal $J(\mathcal{O})$ such that $\mathbf{A}_\mathbb{C} \subseteq \mathcal{O}$ and $J(\mathcal{O}) \cap \mathbf{A}_\mathbb{C} = \ker(\theta_\mathbb{C})$. Let k be the residue field of $\mathcal{O}$ and $\pi\colon \mathcal{O} \to k$ be the canonical map; we may regard k as an extension field of $\mathbb{C} = \theta(\mathbf{A}_\mathbb{C})$ so that $\theta_\mathbb{C}$ is the restriction of π to $\mathbf{A}_\mathbb{C}$. By property (V3) in (7.3.5), we have $\mathbf{A}_\mathbb{C}^* \subseteq \mathcal{O}$. Furthermore, every $x \in \mathbf{A}_\mathbb{C}^*$ is integral over $\mathbf{A}_\mathbb{C}$, which implies that $\pi(x)$ is algebraic over $\mathbb{C}$. Hence we have $\pi(\mathbf{A}_\mathbb{C}^*) \subseteq \mathbb{C}$, and the restriction of π to $\mathbf{A}_\mathbb{C}^*$ is a map as required.

Since the specialized algebra $\mathbf{H}_q$ is split, the assumptions of Theorem 7.4.3 are satisfied and so we have a well-defined decomposition map

$$d_{\theta_\mathbb{C}^*}\colon R_0^+(\mathbf{KH}) \to R_0^+(\mathbf{H}_q)$$

between the Grothendieck groups of $\mathbf{KH} := \mathbf{H} \otimes_\mathbf{A} \mathbf{K}$ and $\mathbf{H}_q$. On the level of characters, this works as follows. Let χ be a character of $\mathbf{KH}$, afforded by the module V say. By Proposition 7.3.8, we have $\chi(T_w) \in \mathbf{A}_\mathbb{C}^*$ for all $w \in W$ and, by Remark 7.4.4, the induced map

$$\chi_q\colon \mathbf{H}_q \to \mathbb{C}, \quad T_w \mapsto \theta_\mathbb{C}^*(\chi(T_w)),$$

is the character of $d_{\theta_\mathbb{C}^*}([V])$. Note that the assignment $\chi \mapsto \chi_q$ does not only depend on q but also on the choice of the map $\theta_\mathbb{C}^*\colon \mathbf{A}_\mathbb{C}^* \to \mathbb{C}$ as above!

8.1.7. Theorem. *Let W be a finite Coxeter group with generating set S and let $\mathbf{H}$ be a corresponding generic Iwahori–Hecke algebra. Then there exists a finite Galois extension $\mathbf{K} \supseteq \mathbb{C}(u_s \mid s \in S)$ such that the $\mathbf{K}$-algebra $\mathbf{KH}$ is split semisimple and isomorphic to the group algebra $\mathbf{K}[W]$. Moreover, if $0 \neq q \in \mathbb{C}$ is such that $\mathbf{H}_q$ is semisimple, then we have a bijection*

$$\mathrm{Irr}(\mathbf{KH}) \to \mathrm{Irr}(\mathbf{H}_q), \quad \chi \mapsto \chi_q.$$

This applies, in particular, to $q = 1$, in which case we have $\mathbf{H}_q = \mathbb{C}[W]$.

Proof. It is a general fact that every finite-dimensional associative algebra over a field becomes split after a finite field extension; see [Curtis and Reiner 1981, Prop. 7.13]. A more constructive approach for separable algebras is given in [Curtis and Reiner 1981, Prop. 7.25]; note that generic Iwahori–Hecke algebras are separable by Exercise 8.6. In any case, using the fact that we are in characteristic 0, we see that there exists a finite Galois extension $\mathbf{K} \supseteq \mathbb{C}(u_s \mid s \in S)$ such that $\mathbf{K}H$ is split. Now we place ourselves in the setting of (8.1.6) and assume that H_q is semisimple. Then all the assumptions of Tits's Theorem 7.4.6 are satisfied. Hence, $\mathbf{K}H$ is not only split but also semisimple, and the decomposition map $d_{\theta_c^*}\colon R_0^+(\mathbf{K}H) \to R_0^+(H_q)$ is an isomorphism which preserves the classes of simple modules. On the level of characters, this gives a bijection $\chi \mapsto \chi_q$ as desired.

If $q = 1$, then $H_q = \mathbb{C}[W]$ is split semisimple. The above bijection shows that, in the Wedderburn decompositions of $\mathbf{K}H$ and $\mathbb{C}[W]$, we have the same number of direct summands with the same dimensions. Since this also holds for the Wedderburn decompositions of $\mathbb{C}[W]$ and $\mathbf{K}[W]$, we see that $\mathbf{K}H \cong \mathbf{K}[W]$. □

8.1.8. Schur elements and generic degrees. We keep the above assumptions. The fact that H is symmetric implies that we have a Schur element $c_\chi \in \mathbf{K}$ associated with every $\chi \in \mathrm{Irr}(\mathbf{K}H)$. Taking into account the description of the dual basis elements in Proposition 8.1.1, the orthogonality relations in Corollary 7.2.4 now read:

$$\sum_{w \in W} \mathrm{ind}(T_w)^{-1} \chi(T_w) \chi'(T_{w^{-1}}) = \begin{cases} \chi(T_1)\, c_\chi & \text{if } \chi = \chi', \\ 0 & \text{otherwise.} \end{cases}$$

By Proposition 7.3.9, we have $c_\chi \in \mathbf{A}^*$, where $\mathbf{A}^*$ denotes the integral closure of $\mathbf{A}$ in $\mathbf{K}$. Moreover, since $\mathbf{K}H$ is semisimple, we have $c_\chi \neq 0$ for all irreducible characters χ (see Theorem 7.2.6).

The Schur element corresponding to the index representation is called the *Poincaré polynomial* of W and will be denoted by P_W. By definition, we have

$$P_W = c_{\mathrm{ind}} = \sum_{w \in W} \mathrm{ind}(T_w)\, \mathrm{ind}(T_w^\vee) = \sum_{w \in W} \mathrm{ind}(T_w) \in \mathbb{N}_0[u_s \mid s \in S].$$

If we specialize $u_s \mapsto 1$ then $\mathrm{ind}(T_w)$ specializes to 1 and so P_W specializes to $|W|$. Thus, the Poincaré polynomial P_W may be regarded as the "generic" analogue of $|W|$. Since the Schur elements are non-zero, we can define

$$D_\chi := \frac{P_W}{c_\chi} \in \mathbf{K} \qquad \text{for all } \chi \in \mathrm{Irr}(\mathbf{K}H).$$

The elements D_χ are called the *generic degrees* of W. This notation will be explained by the results in Section 8.4: if W is the Weyl group of an untwisted finite *Chevalley group* G over $\mathbb{F}_q$, the element D_χ specializes to the dimension of a simple $\mathbb{C}[G]$-module, under the specialization $\theta\colon \mathbf{A} \to \mathbb{C}$, $u_s \mapsto q$ $(s \in S)$.

Recall from Chapter 1 that we have realized W as a subgroup of $GL(V)$ in such a way that the generators in S are reflections. If (W, S) is irreducible and W is finite, the representation $W \subset GL(V)$ is absolutely irreducible (see Exercise 1.12), and hence it affords an irreducible character of W. By Theorem 8.1.7, this character is also the specialization of an irreducible character of $\mathbf{K}H$. We will now directly construct a representation of $\mathbf{K}H$ with this character.

8.1.9. Lifting the reflection representation. Let V be a $\mathbf{K}$-vector space with a basis $\{e_s \mid s \in S\}$. For each $s, t \in S$, $s \neq t$, we choose an m_{st}th root of unity $\zeta_{st} \in \mathbf{K}$ (where, as usual, m_{st} is the order of st) such that $\zeta_{st} \neq \pm 1$ if $m_{st} > 2$. We also assume that $\mathbf{K}$ is chosen so that it contains a square root $\sqrt{u_s u_t}$ for $s, t \in S$. Then let $c_{st} \in \mathbf{K}$ $(s, t \in S)$ be such that the following conditions hold:

$$\begin{aligned} c_{ss} &= u_s + 1 && \text{for all } s \in S, \\ c_{st} = c_{ts} &= 0 && \text{if } s \neq t \text{ and } m_{st} = 2, \\ c_{st}c_{ts} &= u_s + \sqrt{u_s u_t}(\zeta_{st} + \zeta_{st}^{-1}) + u_t && \text{if } s \neq t \text{ and } m_{st} > 2. \end{aligned}$$

(These conditions can be seen as "generic" analogues of the conditions on the product of two entries of a Cartan matrix in Definition 1.1.1.) For each $s \in S$ we define a linear map

$$\rho_s \colon V \to V, \qquad e_t \mapsto u_s e_t - c_{st} e_s \qquad (t \in S).$$

Our aim is to show that the assignment $T_s \mapsto \rho_s$ defines a representation of $\mathbf{K}H$.

We begin by looking at the rank 2 case. Let $s, t \in S$, $s \neq t$, and consider the parabolic subgroup $W_{st} \subseteq W$, which is a dihedral group of order $2m_{st}$. We assume that $m_{st} > 2$. Denote the corresponding parabolic subalgebra by H_{st}. The defining equation shows that the subspace $\langle e_s, e_t \rangle$ is invariant under ρ_s and ρ_t. On this subspace, the action of ρ_s and ρ_t is given by the matrices

$$M_s = \begin{bmatrix} -1 & 0 \\ -c_{st} & u_s \end{bmatrix} \quad \text{and} \quad M_t := \begin{bmatrix} u_t & -c_{ts} \\ 0 & -1 \end{bmatrix}, \quad \text{respectively.}$$

In order to show that the assignment $T_s \mapsto M_s$, $T_t \mapsto M_t$ defines a representation of $\mathbf{K}H_{st}$, it will be convenient to perform a base change so that $M_s M_t$ is a diagonal matrix. We compute that

$$\mathrm{Tr}(M_s M_t) = \sqrt{u_s u_t}(\zeta_{st} + \zeta_{st}^{-1}) \quad \text{and} \quad \det(M_s M_t) = u_s u_t.$$

This shows that both $M_s M_t$ and $M_t M_s$ have eigenvalues $\sqrt{u_s u_t}\zeta_{st}^{\pm 1}$. Note that these eigenvalues are distinct since $m_{st} > 2$, and hence $\zeta_{st} \neq \pm 1$. So there exists some $P \in GL_2(\mathbf{K})$ such that $P(M_s M_t)P^{-1}$ is a diagonal matrix with diagonal entries $\sqrt{u_s u_t}\zeta_{st}^{\pm 1}$. We let

$$M_s' := PM_sP^{-1} = \begin{bmatrix} \alpha & \beta \\ \gamma & \delta \end{bmatrix} \quad \text{and} \quad M_t' := PM_tP^{-1} \qquad \text{with } \alpha, \beta, \gamma, \delta \in \mathbf{K}.$$

Since $\det(M'_s) = \det(M_s) = -u_s$, we can invert M'_s and obtain

$$M'_t = (M'_s)^{-1}(M'_sM'_t) = -\frac{1}{u_s}\begin{bmatrix} \delta & -\beta \\ -\gamma & \alpha \end{bmatrix}(M'_sM'_t)$$
$$= \frac{1}{u_s}\begin{bmatrix} -\sqrt{u_su_t}\zeta_{st}\delta & \sqrt{u_su_t}\zeta_{st}^{-1}\beta \\ \sqrt{u_su_t}\zeta_{st}\gamma & -\sqrt{u_su_t}\zeta_{st}^{-1}\alpha \end{bmatrix}.$$

We know that $\mathrm{Tr}(M'_s) = \mathrm{Tr}(M_s) = u_s - 1$ and $\mathrm{Tr}(M'_t) = \mathrm{Tr}(M_t) = u_t - 1$. This yields two equations for the coefficients α, δ. Solving these equations, we obtain

$$\alpha = \frac{\zeta_{st}^2(u_s - 1) + \zeta_{st}\sqrt{u_su_t}(1 - u_t^{-1})}{\zeta_{st}^2 - 1}, \tag{1}$$

$$\delta = -\frac{(u_s - 1) + \zeta_{st}\sqrt{u_su_t}(1 - u_t^{-1})}{\zeta_{st}^2 - 1}. \tag{2}$$

8.1.10. Lemma. *Recall that* $m_{st} > 2$ *and* $c_{st}c_{ts} = u_s + \sqrt{u_su_t}(\zeta_{st} + \zeta_{st}^{-1}) + u_t$ *where* $\zeta_{st} \neq \pm 1$ *is an* m_{st}*th root of unity. Then the assignment*

$$T_s \mapsto M_s = \begin{bmatrix} -1 & 0 \\ -c_{st} & u_s \end{bmatrix} \quad \textit{and} \quad T_t \mapsto M_t = \begin{bmatrix} u_t & -c_{ts} \\ 0 & -1 \end{bmatrix},$$

defines a representation of $\mathbf{K}\mathbf{H}_{st}$.

Proof. We must check that the matrices M_s, M_t satisfy the defining relations for $\mathbf{H}$. The quadratic relations are clear since M_s, M_t have the correct eigenvalues. It remains to check the relation $\mathrm{Prod}(M_s, M_t; m_{st}) = \mathrm{Prod}(M_t, M_s; m_{st})$. It is sufficient to do this for M'_s and M'_t. We shall write $\zeta = \zeta_{st}$ and $m = m_{st}$.

Assume first that m is even. Then

$$\mathrm{Prod}(M'_s, M'_t; m) = (M'_sM'_t)^{m/2} \quad \text{and} \quad \mathrm{Prod}(M'_t, M'_s; m) = (M'_tM'_s)^{m/2}.$$

Now the matrix $M'_sM'_t$ is diagonal with two distinct eigenvalues $\sqrt{u_su_t}\zeta^{\pm 1}$. Since ζ is an mth root of unity, we automatically have $\zeta^{m/2} = \zeta^{-m/2}$, and so $(M'_sM'_t)^{m/2}$ is the diagonal matrix with two equal eigenvalues $\sqrt{u_su_t}^{\,m/2}\zeta^{m/2}$. Since $M'_tM'_s$ has the same eigenvalues as $M'_sM'_t$, we conclude that $(M'_tM'_s)^{m/2}$ is conjugate to a diagonal matrix with two equal eigenvalues $\sqrt{u_su_t}^{\,m/2}\zeta^{m/2}$, and hence that $(M'_sM'_t)^{m/2} = (M'_tM'_s)^{m/2}$ as desired.

Now let m be odd. Then we have $u_s = u_t =: u$, and the above formulas for α, δ simplify to $\alpha = \zeta(u-1)/(\zeta - 1)$ and $\delta = -(u-1)/(\zeta - 1)$. This yields

$$\mathrm{Prod}(M'_s, M'_t; m) = (M'_sM'_t)^{(m-1)/2}M'_s = \begin{bmatrix} u\zeta^{(m-1)/2}\alpha & u\zeta^{(m-1)/2}\beta \\ u\zeta^{-(m-1)/2}\gamma & u\zeta^{-(m-1)/2}\delta \end{bmatrix},$$

$$\mathrm{Prod}(M'_t, M'_s; m) = M'_t(M'_sM'_t)^{(m-1)/2} = \begin{bmatrix} -u\zeta^{(m+1)/2}\delta & u\zeta^{-(m+1)/2}\beta \\ u\zeta^{(m+1)/2}\gamma & -u\zeta^{-(m+1)/2}\alpha \end{bmatrix}.$$

Using the relations $\zeta^{(m-1)/2} = \zeta^{-(m+1)/2}$ and $\alpha = -\zeta\delta$, we see that these matrices are in fact equal, as desired. □

8.1.11. Theorem (Kilmoyer). *Let $(c_{st})_{s,t\in S}$ be as in (8.1.9). Then there is a unique representation $\rho\colon \mathsf{KH} \to \mathsf{End}_\mathsf{K}(V)$ such that $\rho(T_s) = \rho_s$ for all $s \in S$. This representation is absolutely irreducible if (W, S) is irreducible.*

We call ρ the *reflection representation* of $\mathbf{KH}$ with respect to $(c_{st})_{s,t\in S}$.

Proof. We must show that the maps ρ_s $(s \in S)$ satisfy the defining relations in Definition 4.4.1. First, let $s \in S$ and consider the quadratic relation $T_s^2 = u_s T_1 + (u_s - 1)T_s$. Let $V' \subseteq V$ be the subspace generated by $\{e_t - c_{st}e_s \mid t \in S, t \neq s\}$. Then we certainly have $V = \langle e_s\rangle \oplus V'$. The map ρ_s acts on $\langle e_s\rangle$ and on V' by scalar multiplication with -1 and u_s, respectively. Indeed, we have $e_s.\rho_s = u_s e_s - (u_s+1)e_s = -e_s$ and $(e_t - c_{st}e_s).\rho_s = (u_s e_t - c_{st}e_s) + c_{st}e_s = u_s e_t$. It follows that ρ_s is diagonalizable with eigenvalues -1 and u_s, and so $\rho_s^2 = u_s\,\mathrm{id}_V + (u_s - 1)\rho_s$, as desired.

Next consider the braid relation for $s, t \in S$, $s \neq t$. As in the proof of Lemma 1.1.6, we consider for each $r \in S - \{s, t\}$ the equations $c_{ss}x_s^r + c_{st}x_t^r = -c_{sr}$ and $c_{ts}x_s^r + c_{tt}x_t^r = -c_{tr}$ with unknowns x_s^r, x_t^r. The matrix of coefficients of this system of linear equations has determinant $c_{ss}c_{tt} = (u_s+1)(u_t+1) \neq 0$ if $m_{st} = 2$ and $c_{ss}c_{tt} - c_{st}c_{ts} = (\sqrt{u_s u_t} - \zeta_{st})(\sqrt{u_s u_t} - \zeta_{st}^{-1}) \neq 0$ if $m_{st} > 2$. Thus, the above system of equations has a unique solution for x_s^r, x_t^r. Then, if we let $V' \subseteq V$ be the subspace generated by $e_r' := e_r + x_s^r e_s + x_t^r e_t$ (for $r \in S, r \neq s, t$), we have a direct sum decomposition $V = \langle e_s, e_t\rangle \oplus V'$, and the choice of the x_s^r implies that it is stable under the action of ρ_s, ρ_t. More precisely, ρ_s, ρ_t act as multiplication by u_s, u_t, respectively, on V'. Hence it is sufficient to show that the restrictions of ρ_s, ρ_t to $\langle e_s, e_t\rangle$ satisfy the braid relations. If $m_{st} > 2$, this holds by Lemma 8.1.10; while if $m_{st} = 2$, this is clear, since then ρ_s and ρ_t act by diagonal matrices on that subspace.

The fact that ρ is absolutely irreducible if (W, S) is irreducible follows by an argument completely analogous to that in Exercise 1.12. □

8.1.12. Remark. Assume that W is finite.

(a) Consider the collection of elements $(c_{st})_{s,t\in S}$. We can certainly choose the roots of unity ζ_{st} so that $\zeta_{st} + \zeta_{st}^{-1} = 2\cos(2\pi/m_{st}) \in \mathbb{C}$ for all $s, t \in S$. Moreover, we can choose each c_{st} to be integral over $\mathbf{A}$: for example, we take any total ordering on S and set $c_{st} = 1$, $c_{ts} = u_s + 2\sqrt{u_s u_t}\cos(2\pi/m_{st}) + u_t$ whenever $m_{st} > 2$ and s comes before t in the chosen ordering.

The advantage of this choice is that now we can consider a specialization $\theta_\mathbb{C}^*\colon \mathbf{A}_\mathbb{C}^* \to \mathbb{C}$ such that $u_s \mapsto 1$ for all $s \in S$, as in (8.1.6). Then we see that the numbers $(\hat\theta(c_{st}))_{s,t\in S}$ are just the entries of a Cartan matrix defining (W, S). We conclude that the reflection representation of $\mathbf{KH}$ in Theorem 8.1.11 specializes to the standard representation $W \subset \mathsf{GL}(V)$ defined by the Cartan matrix $(\hat\theta(c_{st}))_{s,t\in S}$, and that we have

$$d_{\theta_\mathbb{C}^*}([V]) = [V], \qquad \text{with } d_{\theta_\mathbb{C}^*} \text{ as in Theorem 8.1.7.}$$

(b) For $0 \leq d \leq |S|$, let $\bigwedge^d V$ be the dth *exterior power* of $\mathbf{V}$. Then we also have a representation

$$\rho^{(d)}\colon \mathbf{KH} \to \mathrm{End}_K\Big(\bigwedge^d V\Big), \qquad T_s \mapsto u_s^{1-d}\rho(T_s)^{\wedge d} \quad (s \in S),$$

where $\rho(T_w)^{\wedge d}$ denotes the endomorphism induced by $\rho(T_w)$; see Exercise 8.8. Considering a specialization as in (a), we see that $\rho^{(d)}$ specializes to the exterior power of the reflection representation of W as in Section 5.1. If X is an indeterminate, we have, for all $w \in W$,

$$\det(X\,\mathrm{id}_V - \rho(T_w)) = \sum_{d=0}^{|S|} (-1)^d\, \mathrm{ind}(T_w)^{d-1}\, \mathrm{Tr}(\rho^{(d)}(T_w))\, X^{|S|-d}.$$

This follows from the general fact that the traces of the exterior powers of any linear map occur (modulo a sign) as coefficients in the characteristic polynomial of that map (see, for example, Remark 5.1.6).

8.1.13. Example. Assume that $W = \mathfrak{S}_n$ and that S consists of the standard generators $s_i = (i, i+1)$ for $1 \leq i \leq n-1$. Then all generators are conjugate and we write $u := u_{s_i}$ for all i. In this case, we can choose

$$c_{s_{i+1},s_i} = -u \quad \text{and} \quad c_{s_i,s_{i+1}} = -1 \qquad \text{for } 1 \leq i \leq n-2,$$

since $m_{s_i,s_{i+1}} = 3$ for all i. Then the action of T_{s_i} on $\mathbf{V}$ is given as follows:

$$e_{s_j}T_{s_i} = \begin{cases} -e_{s_i} & \text{if } i = j,\\ ue_{s_{i-1}} + e_{s_i} & \text{if } j = i-1,\\ ue_{s_i} + ue_{s_{i+1}} & \text{if } j = i+1,\\ ue_{s_j} & \text{otherwise.} \end{cases}$$

We have seen in Remark 8.1.12 that ρ specializes to the standard reflection representation of the underlying Coxeter group $\mathfrak{S}_n$. The corresponding partitions of n for this representation and its exterior powers are given by Proposition 5.4.12.

Let $\chi^{(d)}$ be the character afforded by $\rho^{(d)}$ for $0 \leq d \leq n-1$. Using the formula in Remark 8.1.12(b), we can compute the values of $\chi^{(d)}$ on T_{w_c} where $w_c = s_1 \cdots s_{n-1}$ is a *Coxeter element*. The matrix of $\rho(T_{w_c})$ is given by

$$M_{w_c} = \begin{bmatrix} -u^{n-2} & -u^{n-3} & & \dots & -u & -1 \\ u^{n-1} & 0 & & \dots & 0 & 0 \\ 0 & u^{n-1} & 0 & \dots & 0 & 0 \\ \vdots & & \ddots & & \vdots & \vdots \\ 0 & \dots & 0 & u^{n-1} & 0 & 0 \\ 0 & \dots & & 0 & u^{n-1} & 0 \end{bmatrix}$$

with characteristic polynomial $\sum_{d=0}^{n-1} u^{(n-2)d}X^{n-1-d}$. Thus, we obtain

$$\chi^{(d)}(T_{w_c}) = (-1)^d u^{n-d-1} \qquad \text{for } 0 \leq d \leq n-1,$$

which generalizes Exercise 5.8. We will give a different proof for the above formula in Example 9.2.9(b).

8.1.14. A generalization of the Alexander polynomial. Let $(c_{st})_{s,t\in S}$ be a collection of elements as above and consider the corresponding representation $\rho\colon \mathbf{K}H \to \mathrm{End}_K(V)$. Let $\gamma\colon \mathbf{K}H \to \mathbf{K}H$ be the automorphism of Exercise 8.2. Then $\rho^* := \rho \circ \gamma$ is a new representation of $\mathbf{K}H$ on V. It also gives rise to a representation of $\mathbf{K}H$ on $\bigwedge^d V$, in which T_s acts via $(-1)^{d-1}\rho^*(T_w)^{\wedge d}$. If we denote by $\psi^{(d)}$ the character of that representation, we have

$$\psi^{(d)}(T_w) = (-1)^{(d-1)l(w)}\,\mathrm{Tr}(\rho^*(T_w)^{\wedge d}) \qquad \text{for all } w \in W.$$

Using the relation between characteristic polynomials and exterior powers already mentioned in Remark 8.1.12(b), we have, for all $w \in W$,

$$\det(X\,\mathrm{id}_V - \rho^*(T_w)) = \sum_{d=0}^{|S|} (-1)^d(-1)^{(d-1)l(w)}\psi^{(d)}(T_w)X^{|S|-d}. \tag{a}$$

Now specialize X to $(-1)^{l(w)}$ and multiply both sides of the above equation by $(-1)^{l(w)}$. Then we obtain a map $\tilde{\psi}\colon \mathbf{K}H \to K$ such that

$$\tilde{\psi}(T_w) = (-1)^{l(w)}\det\Big(1 - (-1)^{l(w)}\rho^*(T_w)\Big) = \sum_{d=0}^{|S|} (-1)^d\psi^{(d)}(T_w). \tag{b}$$

Since the right-hand side of (b) is a linear combination of characters, we conclude that $\tilde{\psi}$ is a trace function on $\mathbf{K}H$.

Consider again the case where $W = \mathfrak{S}_n$, as in Example 8.1.13. By Exercise 8.3, there exists some $P \in \mathrm{GL}(V)$ such that

$$P\rho^*(T_w)P^{-1} = (-1)^{l(w)}\beta_n'(w) \qquad \text{for all } w \in \mathfrak{S}_n,$$

where β_n' is the reduced *Burau representation*; see (4.5.5). So (b) yields that

$$\text{the map } T_w \mapsto (-1)^{l(w)}\det(1 - \beta_n'(w)) \text{ is a trace function on } \mathbf{K}H. \tag{c}$$

This observation is used in (4.5.5) to show that the Alexander polynomial of an oriented link can be expressed in terms of the Burau representation. In this sense, (b) may be regarded as a "generalized Alexander polynomial".

8.2 CLASS POLYNOMIALS

Let W be a finite Coxeter group with generating set S. Let $\mathbf{H}$ be a generic Iwahori–Hecke algebra associated with (W, S) as in Definition 8.1.4. The aim of this section is to establish the basic results about the quotient space of $\mathbf{H}$ by the commutators, and hence about the space of trace functions on $\mathbf{H}$. Our main result is Theorem 8.2.3, which can be seen as a "generic" analogue of the corresponding results for a group algebra.

The basic problem that we have to solve is: how can we decide if two elements in $\mathbf{H}$ become equal in $\mathbf{H}/[\mathbf{H},\mathbf{H}]$? One simple criterion is to check if two elements $h, h' \in \mathbf{H}$ are conjugate by a unit in $\mathbf{H}$. Indeed, in this case, we can write $h' = xhx^{-1}$ with a unit $x \in \mathbf{H}$, and hence

$$h - h' = h - xhx^{-1} = x^{-1}xh - xhx^{-1} = [x^{-1}, xh] \in [\mathbf{H},\mathbf{H}].$$

This argument is of course valid in any associative algebra, but we can take advantage of the facts that each T_w is invertible in $\mathbf{H}$, that $\mathbf{H}$ is a "deformation" of the group algebra of W, and that we know conjugation inside W very well by the results of Chapter 3. As a first result in this direction we have:

8.2.1. Lemma. *Assume that $w, w' \in W$ are conjugate by cyclic shift or, more generally, are strongly conjugate, in the sense of Definition 3.2.4. Then T_w and $T_{w'}$ are conjugate in $\mathbf{H}$ and, hence, we have $T_w \equiv T_{w'} \bmod [\mathbf{H},\mathbf{H}]$.*

Proof. The fact that w, w' are strongly conjugate implies that $\mathbf{w}, \mathbf{w}'$ are conjugate in the braid group associated with (W, S); see Lemma 4.3.3. Since the map $\mathbf{w} \mapsto T_w$ $(w \in W)$ defines an algebra homomorphism from the group algebra of that braid group to $\mathbf{H}$, we conclude that T_w and $T_{w'}$ are also conjugate in $\mathbf{H}$. □

8.2.2. The class polynomials. We now describe an algorithm which reduces a basis element T_w $(w \in W)$ modulo $[\mathbf{H},\mathbf{H}]$. Let $\mathrm{Cl}(W)$ be the set of conjugacy classes of W. For each $C \in \mathrm{Cl}(W)$ we choose once and for all a representative $w_C \in C_{\min}$ (i.e., w_C has minimal length in C). Then we distinguish two cases.

Case 1. The given element $w \in W$ lies in $C_{\min}$ for some $C \in \mathrm{Cl}(W)$. Then

$$T_w \equiv T_{w_C} \mod [\mathbf{H},\mathbf{H}]. \tag{1}$$

This is clear by Lemma 8.2.1 and the fact that two elements of $C_{\min}$ are strongly conjugate; see Theorem 3.2.9(b).

Case 2. The element $w \in W$ does not have minimal length in its conjugacy class. Then Theorem 3.2.9(a) shows that there exist some $s \in S$ and $w' \in W$ such that w, w' are conjugate by cyclic shift and $l(sw's) < l(w')$. We claim that

$$T_w \equiv u_s T_{sw's} + (u_s - 1) T_{w's} \mod [\mathbf{H},\mathbf{H}]. \tag{2}$$

(Note that both $sw's$ and $w's$ have length strictly less than w.) To prove this, we first use Lemma 8.2.1 which shows that $T_w \equiv T_{w'} \bmod [\mathbf{H}, \mathbf{H}]$. Furthermore, we have $l(w's) < l(w')$ and $l(sw's) < l(sw')$ and so

$$\left(u_s T_{sw's} + (u_s - 1)T_{w's}\right)T_s = u_s T_{sw'} + (u_s - 1)T_{w'} = T_s T_{w'}.$$

Using induction on $l(w)$, Theorem 3.2.9 and applying repeatedly the relations (1) and (2), we see that for each $w \in W$ there exist constants $f_{w,C} \in \mathbf{A}$ such that

$$T_w \equiv \sum_{C \in \mathrm{Cl}(W)} f_{w,C} T_{w_C} \mod [\mathbf{H}, \mathbf{H}], \tag{3}$$

and $f_{w,C} = 0$ unless $l(w_C) \leq l(w)$. The $f_{w,C}$ are called *class polynomials*; they can be computed recursively using the above rules. From this recursive description we actually see that

$$f_{w,C} \in \mathbb{Z}[u_s \mid s \in S] \qquad \text{for all } w \in W,\ C \in \mathrm{Cl}(W).$$

At this stage, it is conceivable that the class polynomials might not only depend on w and C but also on the sequences of generators which are used to conjugate arbitrary elements in W to those of minimal length in their classes. The following fundamental result shows that this cannot happen:

8.2.3. Theorem. *Let* $\mathbf{H}$ *be a generic Iwahori–Hecke algebra associated with* (W, S) *over the ring* $\mathbf{A} = \mathbb{Z}[u_s^{\pm 1} \mid s \in S]$. *For each* $C \in \mathrm{Cl}(W)$ *fix* $w_C \in C_{\min}$.

(a) *For each* $w \in W$ *the class polynomials* $f_{w,C} \in \mathbf{A}$ *are uniquely determined by the equations*

$$T_w \equiv \sum_{C \in \mathrm{Cl}(W)} f_{w,C} T_{w_C} \mod [\mathbf{H}, \mathbf{H}].$$

(b) *For each* $C \in \mathrm{Cl}(W)$, *the map* $f_C : \mathbf{H} \to \mathbf{A}$, $T_w \mapsto f_{w,C}$ $(w \in W)$ *is a trace function, and we have*

$$f_C(T_w) = \delta_{C,C'} \qquad \text{for } w \in C'_{\min}.$$

Moreover, $\{f_C \mid C \in \mathrm{Cl}(W)\}$ *is a basis of the space of trace functions on* $\mathbf{H}$.

(c) *The quotient space* $\mathbf{H}/[\mathbf{H}, \mathbf{H}]$ *is a free* $\mathbf{A}$*-module. A basis is given by the images (under the natural map* $\mathbf{H} \to \mathbf{H}/[\mathbf{H}, \mathbf{H}]$*) of the elements* T_{w_C}, *for* $C \in \mathrm{Cl}(W)$. *This basis does not depend on the choice of the elements* w_C.

Proof. By equation (3) in (8.2.2), the images of the elements $\{T_{w_C} \mid C \in \mathrm{Cl}(W)\}$ generate the $\mathbf{A}$-module $\mathbf{H}/[\mathbf{H}, \mathbf{H}]$ and, by equation (1), these generating elements do not depend on the choice of the elements w_C. To complete the proof of (a) and (c), it is therefore enough to show that if we have a linear relation

$$\sum_C a_C T_{w_C} \equiv 0 \mod [\mathbf{H}, \mathbf{H}], \qquad \text{where } a_C \in \mathbf{A},$$

then $a_C = 0$ for all $C \in \mathrm{Cl}(W)$. For this purpose, consider the algebra **KH**. We may naturally regard **H** as an **A**-submodule of that algebra. Since every commutator in **H** is a commutator in **KH**, the images of the elements T_{w_C} also generate the quotient of **KH** by the commutators. Moreover, the above relation can also be regarded as a relation

$$\sum_C a_C T_{w_C} \equiv 0 \mod [\mathbf{KH}, \mathbf{KH}].$$

Since we are now working with an algebra over a field, it is sufficient to show that the quotient of **KH** by the commutators has dimension equal to $|\mathrm{Cl}(W)|$.

This can be seen as follows. By Theorem 8.1.7, the algebra **KH** is split semisimple and it is isomorphic to $\mathbf{K}[W]$. Hence the quotient spaces of these two algebras modulo their commutators have the same dimension. But in the case of a group algebra, that quotient space has dimension equal to the number of conjugacy classes (see Example 7.1.8). So we are done.

Let us now prove (b). Let χ be an irreducible character of **KH**. Being a trace function, χ is zero on all commutators. Hence, for each $w \in W$, we have

$$\chi(T_w) = \sum_{C \in \mathrm{Cl}(W)} f_{w,C}\, \chi(T_{w_C}) = \sum_{C \in \mathrm{Cl}(W)} f_C(T_w)\, \chi(T_{w_C}).$$

By Exercise 7.4(b), the irreducible characters of **KH** form a basis of the space of trace functions, and we have already seen before that this space has dimension $|\mathrm{Cl}(W)|$. Hence the matrix of all character values $(\chi(T_w))_{\chi,w}$, where $w \in W$ and $\chi \in \mathrm{Irr}(\mathbf{KH})$, has full rank. This and the above relations imply that the square matrix $(\chi(T_{w_C}))$ also has full rank. So it can be inverted and we can write

$$f_C(T_w) = \sum_{\chi \in \mathrm{Irr}(\mathbf{KH})} g(C,\chi)\, \chi(T_w) \qquad \text{for all } w \in W, \text{ where } g(C,\chi) \in \mathbf{K}.$$

In particular, this shows that each f_C is a trace function. The equation $f_C(T_w) = \delta_{C,C'}$ for $w \in C'_{\min}$ is then clear by the construction of the class polynomials. It also shows that $\{f_C\}$ forms a basis of $\mathrm{Hom}_{\mathbf{A}}(\mathbf{H}/[\mathbf{H},\mathbf{H}], \mathbf{A})$ dual to the basis of $\mathbf{H}/[\mathbf{H},\mathbf{H}]$ given by the images of the elements $\{T_{w_C}\}$. Thus, (b) is proved. □

Using the duality in (7.1.6), we now obtain a description of the centre of **H**.

8.2.4. Corollary (Geck–Rouquier). *The centre of* **H** *is a free* **A**-*module of rank equal to* $|\mathrm{Cl}(W)|$. *A basis is given by the elements*

$$z_C := \sum_{w \in W} f_{w,C} T_w^{\vee} = \sum_{w \in W} \mathrm{ind}(T_w)^{-1} f_{w,C} T_{w^{-1}}, \qquad \textit{for } C \in \mathrm{Cl}(W).$$

Proof. By Theorem 8.2.3, the trace functions $\{f_C \mid C \in \mathrm{Cl}(W)\}$ form a basis of the space of trace functions on $\mathbf{H}$. Hence the elements

$$f_C^* = \sum_{w \in W} f_C(T_w)T_w^\vee = \sum_{w \in W} f_{w,C}T_w^\vee = z_C$$

as in Lemma 7.1.7 form a basis for the centre of $\mathbf{H}$. □

The above basis for the centre of $\mathbf{H}$ can be regarded as a deformation of the usual basis of the centre of the group algebra of W as in Example 7.1.8.

8.2.5. Remark. Consider a ring homomorphism $\theta\colon \mathbf{A} \to A$ into a commutative ring A (with 1) and the corresponding algebra $H = H_A(W, S, \{a_s \mid s \in S\})$ where $\theta(u_s) = a_s$ for $s \in S$. Then note the following facts:

(a) For each $w \in W$, the relation

$$T_w \equiv \sum_C f_{w,C}T_{w_C} \quad \text{mod } [\mathbf{H}, \mathbf{H}] \qquad \text{(taken in } \mathbf{H})$$

specializes to the relation

$$T_w \equiv \sum_C \theta(f_{w,C})T_{w_C} \quad \text{mod } [H, H] \qquad \text{(taken in } H).$$

Hence the elements $\{T_{w_C} \mid C \in \mathrm{Cl}(W)\}$ generate the A-module $H/[H, H]$.

(b) For each $C \in \mathrm{Cl}(W)$, the trace function $f_C\colon \mathbf{H} \to \mathbf{A}$ in Theorem 8.2.3(b) specializes to a trace function

$$f_C^\theta\colon H \to A, \quad T_w \mapsto \theta(f_C(T_w)).$$

Since $f_C^\theta(T_w) = \delta_{C,C'}$ for $w \in C'_{\min}$, the elements $\{T_{w_C} \mid C \in \mathrm{Cl}(W)\}$ are linearly independent modulo $[H, H]$, and hence they form an A-basis of $H/[H, H]$. Moreover, the functions $\{f_C^\theta \mid C \in \mathrm{Cl}(W)\}$ form the dual basis of $\mathrm{Hom}_A(H/[H, H], A)$. As in the proof of Corollary 8.2.4, this implies that the elements

$$z_C^\theta = \sum_{w \in W} \theta(f_{w,C})T_w^\vee, \qquad \text{for } C \in \mathrm{Cl}(W),$$

form an A-basis of the centre of H. Thus, the statements in Theorem 8.2.3 and Corollary 8.2.4 also hold for H. In particular, this shows that the centre of $\mathbf{H}$ is reduction stable, in the sense of Definition 7.5.5.

8.2.6. Corollary. *In the set-up of Remark 8.2.5, every trace function* $\varphi\colon H \to A$ *is uniquely determined by its values on the elements in the set* $\{T_{w_C} \mid C \in \mathrm{Cl}(W)\}$; *we have*

$$\varphi(T_w) = \sum_{C \in \mathrm{Cl}(W)} \theta(f_{w,C})\,\varphi(T_{w_C}) \quad \textit{for all } w \in W, \tag{a}$$

$$\varphi(T_w) = \varphi(T_{w'}) \quad \textit{for } w, w' \in C_{\min}, \tag{b}$$

$$\varphi(T_w) = \varphi(T_{w^{-1}}) \quad \textit{for all } w \in W. \tag{c}$$

Conversely, given a set of elements $\{a_C \mid C \in \mathrm{Cl}(W)\}$ *in* A, *there exists a unique trace function* φ *on* H *such that* $\varphi(T_{w_C}) = a_C$ *for* $C \in \mathrm{Cl}(W)$. *Note that, in particular, the above assertions hold for trace functions on the generic algebra* $\mathbf{K}H$.

Proof. By Remark 8.2.5, we have $\varphi = \sum_C \varphi(T_{w_C}) f_C^\theta$ for any trace function φ on H. Thus, (a) and (b) follow. In particular, φ is uniquely determined by its values on the elements in the set $\{T_{w_C}\}$. Furthermore, we see that it is enough to prove (c) for the trace functions f_C^θ. Since they are obtained via specialization, it is sufficient to prove (c) for trace functions on $\mathbf{K}H$. Since $\mathrm{Irr}(\mathbf{K}H)$ forms a basis of the space of trace functions on $\mathbf{K}H$, it is enough to prove (c) for $\chi \in \mathrm{Irr}(\mathbf{K}H)$. Let $\delta\colon \mathbf{K}H \to M_n(\mathbf{K})$ be a matrix representation affording χ, where $n = \chi(T_1)$. Then, using the anti-automorphism ι of Exercise 4.8, we obtain a new representation

$$\delta'\colon \mathbf{K}H \to M_n(\mathbf{K}), \quad T_w \mapsto \delta(T_{w^{-1}})^{\mathrm{tr}}.$$

Let χ' be the character of δ'. We have $\chi'(T_w) = \mathrm{Tr}(\delta'(T_w)) = \mathrm{Tr}(\delta(T_{w^{-1}})) = \chi(T_{w^{-1}})$ for all $w \in W$. Using a specialization $u_s \mapsto 1$ ($s \in S$) as in Theorem 8.1.7, we conclude that $\chi_1'(w) = \chi_1(w^{-1})$ for all $w \in W$. But, by Corollary 3.2.14, each $w \in W$ is conjugate to its inverse. Hence we conclude that $\chi_1 = \chi_1'$ and so $\chi' = \chi$, as desired. □

Since the functions $f_C\colon T_w \mapsto f_{w,C}$ are trace functions by Theorem 8.2.3, we obtain the following summary of the main properties of the class polynomials.

8.2.7. Proposition. *The class polynomials satisfy the following conditions:*

(a) *If* $w, w' \in W$ *are strongly conjugate then* $f_{w,C} = f_{w',C}$ *for all* $C \in \mathrm{Cl}(W)$.

(b) *Let* $C \in \mathrm{Cl}(W)$. *If* $w \in W$ *and* $s \in S$ *are such that* $l(sws) < l(w)$ *then*

$$f_{sw,C} = f_{ws,C} \quad \textit{and} \quad f_{w,C} = u_s f_{sws,C} + (u_s - 1) f_{ws,C}.$$

(c) *We have* $f_{w,C} = f_{w^{-1},C}$ *for all* $w \in W$ *and* $C \in \mathrm{Cl}(W)$.

Proof. (a) and (c) hold by Corollary 8.2.6, and (b) follows by applying f_C to equation (2) in (8.2.2). □

Algorithm M (Class polynomials). Given a Coxeter system (W, S) such that W is a finite group, a set of indeterminates $\{u_s \mid s \in S\}$ such that $u_s = u_t$ whenever $s, t \in S$ are conjugate in W, and an element $w \in W$, the class polynomials $f_{w,C}$ ($C \in \mathrm{Cl}(W)$) are constructed by induction on $l(w)$.

M1. [Initialize.] Set $h \leftarrow T_w$, $Y \leftarrow \{w\}$.

M2. [Next h.] Write $h = \sum_{y \in W} a_y T_y$ with $a_y \in A$ and choose any $y \in W$ with $a_y \neq 0$. Apply Algorithm G (p. 80) to compute the cyclic shift orbit $\mathrm{Cyc}(y)$. If there exists some $z \in \mathrm{Cyc}(y)$ and some $s \in S$ such that $l(szs) < l(z)$, set

$$Y \leftarrow (Y - \{y\}) \cup \{z, zs\} \quad \text{and} \quad h \leftarrow h + a_y\Big(u_s T_z + (u_s - 1)T_{sz} - T_y\Big).$$

Otherwise, set $Y \leftarrow Y - \{y\}$ and $h \leftarrow h + a_y(T_{w_{C(y)}} - T_y)$, where $C(y)$ is the class containing y, identified using Algorithm I (p. 86).

M3. [Loop:] Repeat step M2 until $Y = \varnothing$. Then $h = \sum_{C \in \mathrm{Cl}(W)} f_{w,C} T_{w_C}$.

8.2.8. Example. Let (W, S) be of type B_4, with $S = \{t, s_1, s_2, s_3\}$ as in Proposition 1.4.3. Let u, v be indeterminates and consider the two-parameter generic Iwahori–Hecke algebra $\mathbf{H}$ associated with (W, S), i.e., we have

$$T_t^2 = uT_1 + (u-1)T_t \quad \text{and} \quad T_{s_i}^2 = vT_1 + (v-1)T_{s_i} \quad \text{for } i = 1, 2, 3.$$

Let us compute the class polynomials corresponding to $w = s_1 t s_1 s_2 s_3 s_2 s_1 t \in W$. In a first step, we find that $w \xrightarrow{s_3} y := s_1 t s_1 s_2 s_1 t$. So we have

$$T_w \equiv vT_y + (v-1)T_{ws_3} \bmod [\mathbf{H}, \mathbf{H}].$$

Now we must go on with y and $ws_3 = s_1 t s_1 s_3 s_2 s_1 t$. We have

$$y \xrightarrow{t} y' := t s_1 t s_1 s_2 s_1 \xrightarrow{s_1} x := s_1 t s_1 t s_1 s_2 = t s_1 t s_2,$$
$$ws_3 \xrightarrow{t} t s_1 t s_1 s_3 s_2 s_1 \xrightarrow{s_1} t s_1 t s_3 s_2 = s_3 x,$$

which yields expressions for T_y, T_{ws_3} modulo $[\mathbf{H}, \mathbf{H}]$. Finally, we have $x \xrightarrow{t} s_1 s_2$ and $s_3 x \xrightarrow{t} s_3 s_1 s_2$. Now we have arrived at elements of minimal length in their classes. Inserting the above expressions recursively into that for T_w, we obtain

$$\begin{aligned} T_w \equiv\, & uv^2 T_{s_1 s_2} + (u-1)v^2 T_{t s_1 s_2} + uv(v-1) T_{s_1 s_2 s_3} \\ & + v(u-1)(v-1) T_{t s_1 s_2 s_3} + v(v-1) T_{t t_1 s_2} + (v-1)^2 T_{t t_1 s_1 s_2 s_3} \bmod [\mathbf{H}, \mathbf{H}], \end{aligned}$$

where $t_1 = s_1 t s_1$. Setting $u, v \mapsto 1$, we recover the fact that w is conjugate to $s_1 s_2$.

Finally, in analogy to the case of finite groups, we can now define:

8.2.9. Definition. As above, fix $w_C \in C_{\min}$ for each $C \in \mathrm{Cl}(W)$. The matrix

$$X(\mathbf{H}) := \Big(\chi(T_{w_C})\Big)_{\substack{\chi \in \mathrm{Irr}(K\mathbf{H}) \\ C \in \mathrm{Cl}(W)}}$$

is called the *character table of the Iwahori–Hecke algebra* $\mathbf{H}$. By Corollary 8.2.6, this definition does not depend on the choice of the elements w_C; the equations

$$\chi(T_w) = \sum_{C \in \mathrm{Cl}(W)} f_{w,C}\, \chi(T_{w_C}) \qquad (\chi \in \mathrm{Irr}(KH),\ w \in W)$$

show that $X(H)$ is a character table for H in the sense of (7.3.11). Note that $X(H)$ is a square matrix whose entries lie in the ring $A_C^* \subset K$; see (8.1.6). Using a specialization $\theta_C^*\colon A_C \to \mathbb{C}$, $u_s \mapsto 1$ $(s \in S)$, as in Theorem 8.1.7, we see that

$$\Big(\theta_C^*(\chi(T_{w_C}))\Big)_{\chi,C}$$

is the character table of the finite group W. In particular, we see once more that $X(H)$ has a non-zero determinant, since the same is true for character tables of finite groups (see [Isaacs 1976]).

8.2.10. Remark. We may have $\chi(T_w) \neq \chi(T_{w'})$ even if $w, w' \in W$ are conjugate. Take, for example, $\chi = \mathrm{ind}$ and $w = s$, $w' = tst$ in the case where $(W, \{s,t\})$ is of type $I_2(m)$ for some $m \geq 3$. Here is a more interesting example: consider the two elements $w = s_1s_3s_2s_1s_4s_3$ and $w' = s_1s_2s_1s_3s_2s_1$ in $W = \mathfrak{S}_5$ from Exercise 4.5. Using Algorithm M (p. 265), we find (where $u = u_{s_i}$ for all i):

$$T_w \equiv u^2T_{s_1s_3} + u(u-1)(T_{s_1s_2s_4} + T_{s_1s_2s_3}) + (u-1)^2T_{s_1s_2s_3s_4} \mod [H,H],$$

$$T_{w'} \equiv u^2T_{s_1s_3} + u(u-1)^2T_{s_1s_2} + (u^2+1)(u-1)T_{s_1s_2s_3} \mod [H,H].$$

Hence, w, w' have the same length and are conjugate in W but $T_w, T_{w'}$ are not conjugate in H. It is an open question if the condition that $\chi(T_w) = \chi(T_{w'})$ for all $\chi \in \mathrm{Irr}(KH)$ implies that T_w and $T_{w'}$ are conjugate in H.

In the remaining chapters of this book, we will compute the character tables for all types of H. By Exercise 8.5, this problem can be reduced to the case where (W, S) is irreducible. The dihedral types will be treated in the next section.

8.3 IWAHORI–HECKE ALGEBRAS OF DIHEDRAL TYPE

Throughout this section, let $W = \langle s, t\rangle$ be a finite dihedral group where s, t are involutions and the product st has order $m \geq 3$. Let $A = \mathbb{Z}[u_s^{\pm 1}, u_t^{\pm 1}]$ where $u_s = u_t$ if m is odd, and let H be a corresponding generic Iwahori–Hecke algebra over A. Thus, H has a presentation with

generators: T_s, T_t,
relations: $T_s^2 = u_sT_1 + (u_s-1)T_s$, $\quad T_t^2 = u_tT_1 + (u_t-1)T_t$,
$\mathrm{Prod}(T_s, T_t; m) = \mathrm{Prod}(T_t, T_s; m)$.

Let K be a finite Galois extension of $\mathbb{C}(u_s, u_t)$ as in Theorem 8.1.7; assume also that K contains a square root $\sqrt{u_su_t}$. Thus, the algebra KH is split semisimple and we have a bijection between the equivalence classes of irreducible representations of KH and of W. This already implies that all irreducible representations of KH have dimension 1 or 2.

We have already seen in Lemma 8.1.10 how to construct two-dimensional representations of KH. We can now prove that these are all of them:

8.3.1. Theorem (Kilmoyer–Solomon). *Let H be a generic Iwahori–Hecke algebra associated with the dihedral group $W = \langle s, t\rangle$ of order $2m$ as above, where $m \geq 3$. Let $\zeta \in K$ be a primitive mth root of unity and, for each integer j, let $c_j, c'_j \in K$ be such that*

$$c_j c'_j = u_s + \sqrt{u_s u_t}(\zeta^j + \zeta^{-j}) + u_t.$$

Then, up to equivalence, the irreducible representations of KH are as follows:

(a) *Let m be even. There are four one-dimensional representations given by*

$$\begin{aligned} \mathrm{ind}\colon\ & T_s \mapsto u_s,\ T_t \mapsto u_t, & \varepsilon_1\colon\ & T_s \mapsto u_s,\ T_t \mapsto -1,\\ \varepsilon\colon\ & T_s \mapsto -1,\ T_t \mapsto -1, & \varepsilon_2\colon\ & T_s \mapsto -1,\ T_t \mapsto u_t, \end{aligned}$$

and $(m-2)/2$ irreducible representations of dimension 2 given by

$$\rho_j\colon T_s \mapsto \begin{bmatrix} -1 & 0 \\ c_j & u_s \end{bmatrix}, \quad T_t \mapsto \begin{bmatrix} u_t & c'_j \\ 0 & -1 \end{bmatrix} \qquad \text{for } 1 \leq j \leq (m-2)/2.$$

(b) *Let m be odd. There are two one-dimensional representations given by*

$$\mathrm{ind}\colon T_s \mapsto u_s,\ T_t \mapsto u_t, \qquad \varepsilon\colon T_s \mapsto -1,\ T_t \mapsto -1,$$

and $(m-1)/2$ irreducible representations of dimension 2 given by

$$\rho_j\colon T_s \mapsto \begin{bmatrix} -1 & 0 \\ c_j & u_s \end{bmatrix}, \quad T_t \mapsto \begin{bmatrix} u_t & c'_j \\ 0 & -1 \end{bmatrix} \qquad \text{for } 1 \leq j \leq (m-1)/2.$$

Proof. We have already remarked above that all irreducible representations of KH have dimension 1 or 2. Now the quadratic relations for T_s, T_t imply that if $\delta\colon KH \to K$ is a one-dimensional representation, then $\delta(T_s) \in \{-1, u_s\}$ and $\delta(T_t) \in \{-1, u_t\}$. Moreover, if m is odd the braid relation $\mathrm{Prod}(T_s, T_t; m_{st}) = \mathrm{Prod}(T_t, T_s; m_{st})$ immediately shows that $\delta(T_s) = \delta(T_t)$. This yields the assertion about one-dimensional representations.

In order to prove the assertions about the two-dimensional representations, we will use Lemma 8.1.10: let j be any integer such that $j \not\equiv -j \bmod m$. Then ζ^j is an mth root of unity with $\zeta^j \neq \pm 1$, and so we obtain a representation $\rho_j\colon T_s \mapsto S_j, T_t \mapsto T_j$, where

$$S_j = \begin{bmatrix} -1 & 0 \\ c_j & u_s \end{bmatrix} \quad \text{and} \quad T_j = \begin{bmatrix} u_t & c'_j \\ 0 & -1 \end{bmatrix}.$$

To prove that it is irreducible, we just need to check that the only matrices commuting with S_j and T_j are the scalar multiples of the identity matrix (since KH is semisimple). But, since $c_j, c'_j \neq 0$, a matrix commuting with S_j must be lower triangular and a matrix commuting with T_j must be upper triangular.

Hence a matrix commuting with both S_j and T_j is diagonal, and since S_j, T_j have two different eigenvalues, that matrix must be a scalar matrix.

In order to see when two such representations can possibly be equivalent we consider the trace of ρ_j on T_sT_t. We have seen in the discussion preceding Lemma 8.1.10 that $\rho_j(T_sT_t) = S_jT_j$ is conjugate to a diagonal matrix with diagonal entries $\sqrt{u_su_t}\zeta^{\pm j}$. So we obtain $\mathrm{Tr}(\rho_j(T_sT_t)) = \sqrt{u_su_t}(\zeta^j + \zeta^{-j})$. Now let j' be another integer with $j' \not\equiv -j' \bmod m$. Then

$$\zeta^j + \zeta^{-j} = \zeta^{j'} + \zeta^{-j'} \quad \Leftrightarrow \quad (\zeta^j - \zeta^{j'})(\zeta^j - \zeta^{-j'}) = 0 \quad \Leftrightarrow \quad j \equiv \pm j' \bmod m.$$

Thus, if m is even, then $\{\rho_j \mid 1 \leq j \leq (m-2)/2\}$ is a set of pairwise non-equivalent irreducible representations. In order to check that we have found all irreducible representations up to equivalence, we use Wedderburn's theorem, which implies that the sum of the squares of the degrees of the irreducible representations of a split semisimple algebra equals the dimension of that algebra. Summing the squares of the degrees that we have found so far, we obtain $4(m-2)/2 + 4 = 2m = |W|$, which implies that we have found all irreducible representations up to equivalence. The argument for m odd is completely analogous. □

8.3.2. Corollary. *Let $K_0 \subseteq K$ be a subfield containing A such that $\zeta + \zeta^{-1} \in K_0$ and $\sqrt{u_su_t} \in K_0$. Then the algebra K_0H is split semisimple.*

Proof. We use $c_j = 1$ and $c_j' = u_s + \sqrt{u_su_t}(\zeta^j + \zeta^{-j}) + u_t$ in the construction of irreducible representations in Theorem 8.3.1. Then observe that if $\zeta + \zeta^{-1} \in K_0$ then also $\zeta^j + \zeta^{-j} \in K_0$ for any integer j, and so $c_j, c_j' \in K_0$. It follows that a complete set of irreducible representations of KH can be realized over K_0, whence the assertion. □

Next we determine the values of the irreducible characters of KH. This is clear for the one-dimensional representations, so it remains to consider the two-dimensional representations. Let j be an integer such that $j \not\equiv -j \bmod m$ and $\chi_j \in \mathrm{Irr}(KH)$ be the character of ρ_j. We will only determine the values $\chi_j(T_w)$ for w running over a set of representatives of minimal length in the conjugacy classes of W. Recall from Example 3.2.8 that such a set is given by

$$\{1,\ s,\ t,\ (st)^k \mid 1 \leq k \leq m/2\} \qquad \text{if } m \text{ is even},$$
$$\{1,\ s,\ (st)^k \mid 1 \leq k \leq (m-1)/2\} \qquad \text{if } m \text{ is odd}.$$

8.3.3. Lemma. *Assume that $j \not\equiv -j \bmod m$ and let $\chi_j \in \mathrm{Irr}(KH)$ be the character of ρ_j. Then*

$$\chi_j(T_s) = u_s - 1, \quad \chi_j(T_t) = u_t - 1, \quad \chi_j(T_w) = \sqrt{u_su_t}^{\,k}(\zeta^{jk} + \zeta^{-jk})$$

where $w = (st)^k$ with $1 \leq k \leq m/2$.

Proof. We use the notation in the proof of Theorem 8.3.1. We have $\chi_j(T_s) = \mathrm{Tr}(S_j) = u_s - 1$ and $\chi_j(T_t) = \mathrm{Tr}(T_j) = u_t - 1$. Since $w = (st)^k$ is a reduced expression, we have $T_w = (T_sT_t)^k$ and hence $\rho_j(T_w) \doteq \rho_j((T_sT_t)^k) = (S_jT_j)^k$. Now S_jT_j is conjugate to a diagonal matrix with diagonal entries $\sqrt{u_su_t}\zeta^{\pm j}$. Hence $(S_jT_j)^k$ is conjugate to a diagonal matrix with diagonal entries $\sqrt{u_su_t}^k\zeta^{\pm kj}$. This yields the above formula. □

We consider the four smallest examples where $m = 3, 4, 5, 6$. For $m = 3, 5$ we let $u_s = u_t =: u$. The corresponding character tables are given in Table 8.1.

Table 8.1 Character tables for $H(A_2)$, $H(B_2)$, $H(G_2)$ and $H(I_2(5))$

$A_2 = I_2(3)$	T_1	T_s	T_{st}
ind	1	u	u^2
ε	1	-1	1
χ_1	2	$u-1$	$-u$

$I_2(5)$	T_1	T_s	T_{st}	T_{stst}
ind	1	u	u^2	u^4
ε	1	-1	1	1
χ_1	2	$u-1$	αu	$-(\alpha+1)u^2$
χ_2	2	$u-1$	$-(\alpha+1)u$	αu^2

Here, $\alpha := (-1+\sqrt{5})/2$ equals $\zeta + \zeta^4$, where ζ is a primitive fifth root of unity.

$B_2 = I_2(4)$	T_1	T_s	T_t	T_{st}	T_{w_0}
ind	1	u_s	u_t	u_su_t	$(u_su_t)^2$
ε_1	1	u_s	-1	$-u_s$	u_s^2
ε_2	1	-1	u_t	$-u_t$	u_t^2
ε	1	-1	-1	1	1
χ_1	2	u_s-1	u_t-1	0	$-2u_su_t$

$G_2 = I_2(6)$	T_1	T_s	T_t	T_{st}	T_{stst}	T_{w_0}
ind	1	u_s	u_t	u_su_t	$(u_su_t)^2$	$(u_su_t)^3$
ε_1	1	u_s	-1	$-u_s$	u_s^2	$-u_s^3$
ε_2	1	-1	u_t	$-u_t$	u_t^2	$-u_t^3$
ε	1	-1	-1	1	1	1
χ_1	2	u_s-1	u_t-1	$\sqrt{u_su_t}$	$-u_su_t$	$-2\sqrt{u_su_t}^3$
χ_2	2	u_s-1	u_t-1	$-\sqrt{u_su_t}$	$-u_su_t$	$2\sqrt{u_su_t}^3$

8.3.4. Theorem (Kilmoyer–Solomon). *The Schur elements associated with the one-dimensional representations are given by*

$$P_W = c_{\mathrm{ind}} = (u_s+1)(u_t+1)\frac{\sqrt{u_su_t}^m - 1}{u_su_t - 1} = (u_su_t)^{m/2}c_\varepsilon,$$

and, if m is even, by

$$c_{\varepsilon_1} = u_t^{-m/2}(u_s+1)(u_t+1)\frac{u_s^{m/2} - u_t^{m/2}}{u_s - u_t} = (u_t^{-1}u_s)^{m/2}c_{\varepsilon_2}.$$

If $u_s = u_t =: u$, *we have* $P_W = (u+1)(u^m - 1)/(u-1) = u^m c_\varepsilon$ *and* $c_{\varepsilon_1} = c_{\varepsilon_2} = m(u + u^{-1} + 2)/2$. *The Schur element corresponding to the two-dimensional character* χ_j *is given by*

$$c_{\chi_j} = m\frac{(u_s u_t - (\zeta^j + \zeta^{-j})\sqrt{u_s u_t} + 1)(u_s + (\zeta^j + \zeta^{-j})\sqrt{u_s u_t} + u_t)}{u_s u_t (2 - (\zeta^{2j} + \zeta^{-2j}))}.$$

If $u_s = u_t =: u$, *this expression simplifies to*

$$c_{\chi_j} = m\frac{u^2 - (\zeta^j + \zeta^{-j})u + 1}{u(2 - (\zeta^j + \zeta^{-j}))}.$$

Proof. We distinguish the cases where m is even or odd. First assume that m is even. The elements of W together with the values of ind are as follows:

w	$\mathrm{ind}(T_w)$	Index range
1	1	
$w_0 = (st)^{m/2}$	$(u_s u_t)^{m/2}$	
$(st)^k$, $(ts)^k$	$(u_s u_t)^k$	$1 \leq k \leq (m-2)/2$
$(st)^k s$	$u_s (u_s u_t)^k$	$0 \leq k \leq (m-2)/2$
$(ts)^k t$	$u_t (u_s u_t)^k$	$0 \leq k \leq (m-2)/2$

So, by definition, we have

$$P_W = 1 + (u_s u_t)^{m/2} + 2\Big(\sum_{k=1}^{(m-2)/2} (u_s u_t)^k\Big) + (u_s + u_t)\Big(\sum_{k=0}^{(m-2)/2} (u_s u_t)^k\Big).$$

Using the formula for geometric sums, the two summations evaluate to

$$u_s u_t \frac{(u_s u_t)^{(m-2)/2} - 1}{u_s u_t - 1} \quad \text{and} \quad \frac{(u_s u_t)^{m/2} - 1}{u_s u_t - 1},$$

respectively. Inserting this into the above expression yields the desired formula. The argument for the computation of c_{ε_1} is completely analogous. In fact, we obtain terms similar to those above, simply replacing u_t by u_t^{-1}. For ε_2, we have to replace u_s by u_s^{-1} and, finally, for ε, we have to replace u_s by u_s^{-1} and u_t by u_t^{-1}.

Now consider an irreducible character χ_j of degree 2. We have to compute the terms $p(w) := \mathrm{ind}(T_w)^{-1}\chi_j(T_w)\chi_j(T_{w^{-1}})$, and then sum over all $w \in W$. In order to compute character values, we use representing matrices S_j, T_j as in the proof of Theorem 8.3.1. By the remarks preceding Lemma 8.1.10, there exists an invertible matrix P such that if we set $S_j' = PS_jP^{-1}$ and $T_j' = PT_jP^{-1}$, then

$$S_j' = \begin{bmatrix} \alpha & \beta \\ \gamma & \delta \end{bmatrix} \quad \text{and} \quad S_j'T_j' := \begin{bmatrix} \sqrt{u_s u_t}\zeta^j & 0 \\ 0 & \sqrt{u_s u_t}\zeta^{-j} \end{bmatrix},$$

where $\alpha, \beta, \gamma, \delta \in \mathbf{K}$. We certainly have $p(1) = 4$. We also know that T_{w_0} acts by the scalar $\pm\sqrt{u_s u_t}^{m/2}$ (see the proof of Lemma 8.3.3), and so $p(w_0) = 4$.

Now let $w = (st)^k$ with $1 \leq k \leq (m-2)/2$. Then $\chi_j(T_w) = \mathrm{Tr}((S_j'T_j')^k)$ and this also equals $\chi_j(T_{w^{-1}})$, since $S_j'T_j'$ and $T_j'S_j'$ have the same eigenvalues. Using Lemma 8.3.3 and the above table for the values of ind, we find that

$$p(w) = \frac{1}{(u_s u_t)^k}\left(\sqrt{u_s u_t}^{\,k}(\zeta^{jk} + \zeta^{-jk})\right)^2 = \zeta^{2jk} + 2 + \zeta^{-2jk}.$$

In order to evaluate the sum over all elements w of this form, we note that $j \not\equiv -j \bmod m$ and hence

$$\sum_{k=1}^{(m-2)/2} \zeta^{2jk} = \sum_{k=1}^{(m-2)/2} \zeta^{-2jk} = -1.$$

So the contribution of these elements to c_{χ_j} is $m-4$. A similar result then also holds for the elements $w = (ts)^k$ with $1 \leq k \leq (m-2)/2$.

Next, let $w = (st)^k s$ with $1 \leq k \leq (m-2)/2$. Then $w = w^{-1}$ and so $\chi_j(T_w) = \mathrm{Tr}((S_j'T_j')^k S_j') = \chi_j(T_{w^{-1}})$, where

$$(S_j'T_j')^k S_j' = \begin{bmatrix} \sqrt{u_s u_t}\zeta^{jk} & 0 \\ 0 & \sqrt{u_s u_t}\zeta^{-jk} \end{bmatrix} \begin{bmatrix} \alpha & \beta \\ \gamma & \delta \end{bmatrix}.$$

This yields $\chi_j(T_w) = \sqrt{u_s u_t}^{\,k}(\zeta^{jk}\alpha + \zeta^{-jk}\delta)$ and so

$$p(w) = \frac{1}{(u_s u_t)^k u_s}\left(\sqrt{u_s u_t}^{\,k}(\zeta^{jk}\alpha + \zeta^{-jk}\delta)\right)^2 = \frac{1}{u_s}\left(\zeta^{2jk}\alpha^2 + 2\alpha\delta + \zeta^{-jk}\delta^2\right).$$

The sum over all elements of this form can be evaluated as above, and yields the contribution $m\alpha\delta/u_s$ to c_{χ_j}.

Finally, let $w = t(st)^k$ with $1 \leq k \leq (m-2)/2$. Then, again, $w^{-1} = w$, and so $\chi_j(T_{w^{-1}}) = \chi_j(T_w) = \mathrm{Tr}(T_j'(S_j'T_j')^k)$, where

$$T_j'(S_j'T_j')^k = \frac{1}{u_s}\begin{bmatrix} -\sqrt{u_s u_t}\zeta^{j}\delta & \sqrt{u_s u_t}\zeta^{-j}\beta \\ \sqrt{u_s u_t}\zeta^{j}\gamma & \sqrt{u_s u_t}\zeta^{-j}\alpha \end{bmatrix} \begin{bmatrix} \sqrt{u_s u_t}\zeta^{jk} & 0 \\ 0 & \sqrt{u_s u_t}\zeta^{-jk} \end{bmatrix}.$$

The corresponding term $p(w)$ and the sum over all $w \in W$ can be evaluated as before, and yields the contribution $m\alpha\delta/u_s$ to c_{χ_j}.

Collecting together all of the above contributions, we find that

$$c_{\chi_j} = \frac{1}{2}\left(4 + 4 + 2(m-4) + 2m\alpha\delta/u_s\right) = m(1 + \alpha\delta/u_s).$$

It remains to use the formulas for α and δ obtained in the remarks preceding Lemma 8.1.10. The expression $1 + \alpha\delta/u_s$ can then be transformed as claimed. This completes the proof for the case where m is even.

If m is odd, the elements of W together with the values of ind are:

w	$\mathrm{ind}(T_w)$	Index range
1	1	
$w_0 = (st)^{(m-1)/2}s$	u^m	
$(st)^k$, $(ts)^k$	u^{2k}	$1 \leq k \leq (m-1)/2$
$(st)^k s$, $(ts)^k t$	u^{2k+1}	$0 \leq k \leq (m-3)/2$

The computations for the two one-dimensional representations are completely analogous to those above (but slightly simpler). Now let χ_j be an irreducible character of degree 2. We compute the various contributions to c_{χ_j} from elements w of the form $(st)^k$, $(ts)^k$, $(st)^k s$, $t(st)^k$, respectively, much as before. Using the relation $\alpha = -\zeta^j \delta$, this yields that

$$c_{\chi_j} = m(1 - \zeta^j \delta^2/u),$$

which can be transformed as claimed using the formula $\delta = -(u-1)/(\zeta^j - 1)$. $\square$

8.3.5. Remark. Assume that we are in the equal parameter case: $u_s = u_t =: u$. Then we have

$$P_W = c_{\mathrm{ind}} = (u+1)\frac{u^m - 1}{u - 1} = u^m c_\varepsilon.$$

Moreover, it can be verified using the formulas in Theorem 8.3.4 that the Schur elements satisfy the following properties:

$$c_\chi \in \mathbb{C}[u, u^{-1}] \quad \text{and} \quad D_\chi = P_W/c_\chi \in \mathbb{C}[u] \quad \text{for all } \chi \in \mathrm{Irr}(KH).$$

Consider, for example, a character χ_j. The numerator in the formula for c_{χ_j} factorizes as $u^2 - (\zeta^j + \zeta^{-j})u + 1 = (u - \zeta^j)(u - \zeta^{-j})$. On the other hand, we have $P_W = (u+1)\prod_{i=1}^{m-1}(u - \zeta^i)$. Since $j \not\equiv -j \bmod m$, this implies that

$$D_{\chi_j} = \frac{u}{m}(u+1)(2 - (\zeta^j + \zeta^{-j}))\prod_i (u - \zeta^i),$$

where the product is over all $i \in \{1, \ldots, m-1\} - \{j, m-j\}$. Thus,

$$D_{\chi_j} = \frac{2 - (\zeta^j + \zeta^{-j})}{m} u + \text{combination of higher powers of } u.$$

For the one-dimensional characters, we obtain

$$D_{\mathrm{ind}} = 1, \qquad D_\varepsilon = u^m,$$

$$D_{\varepsilon_1} = D_{\varepsilon_2} = \frac{2u(u^m - 1)}{m(u^2 - 1)} = \frac{2}{m} u + \text{combination of higher powers of } u.$$

8.4 IWAHORI'S THEOREM

It is the purpose of this section to give a brief account of the theory of Hecke algebras associated with groups with a BN-pair. This serves both as a motivation and guideline for the "abstract" theory of Iwahori–Hecke algebras as developed in this and the following chapters. For example, the results on generic degrees and Schur elements are strongly influenced by the relations with character degrees in Proposition 8.4.4. We begin by providing some basic results about permutation modules and their endomorphism rings.

8.4.1. The Schur basis of an endomorphism ring. From now until Corollary 8.4.3, G may be any group and $B \subseteq G$ be a subgroup such that

$$|BxB/B| < \infty \qquad \text{for all } x \in G, \tag{F}$$

i.e., we require that each double coset BxB contains only finitely many cosets of B. We consider the G-set $G/B = \{Bx \mid x \in G\}$ and the corresponding permutation module $R[G/B]$, where R is any commutative ring with 1. Thus, $R[G/B]$ is a free R-module with a basis given by the cosets of B, and the action of G is extended linearly. We wish to describe the structure of the endomorphism ring

$$\mathrm{End}_G(R[G/B]) := \{\varphi \in \mathrm{End}_R(R[G/B]) \mid \varphi(v)g = \varphi(vg) \text{ for all } g \in G\},$$

which is called the *Hecke algebra* of G with respect to B. (Note that our groups act from the right but endomorphisms of $R[G/B]$ are written on the left.)

Let us introduce the following notation. For each $\varphi \in \mathrm{End}_R(R[G/B])$ we define $\dot{\varphi}\colon G/B \times G/B \to R$ by

$$\varphi(Bx') = \sum_{Bx \in G/B} \dot{\varphi}(Bx, Bx')Bx \qquad \text{for all } x' \in G.$$

Consider the action of G on $G/B \times G/B$ given by $g\colon (Bx_1, Bx_2) \mapsto (Bx_1g, Bx_2g)$. Then we have a bijection between the G-orbits on $G/B \times G/B$ and the double cosets of B in G: with a G-orbit $X \subseteq G/B \times G/B$ we associate the double coset BxB, where $x \in G$ is such that $(Bx, B) \in X$. (Note that we had to make a choice for defining this bijection; we could also have required that $(B, Bx) \in X$.) Let $D(G, B)$ be a set of double coset representatives such that $1 \in D(G, B)$. For each $x \in D(G, B)$, we define $\varepsilon_x \in \mathrm{End}_R(R[G/B])$ by the condition that

$$\dot{\varepsilon}_x(Bx_1, Bx_2) = \begin{cases} 1 & \text{if } x_1x_2^{-1} \in BxB, \\ 0 & \text{otherwise.} \end{cases}$$

(That is, $\dot{\varepsilon}_x$ is the indicator function of the G-orbit on $G/B \times G/B$ corresponding to BxB.) Note that this is well defined thanks to assumption (F): for a given coset Bx_2, there are only finitely many cosets Bx_1 for which $x_1x_2^{-1} \in BxB$.

8.4.2. Proposition (Schur). *The endomorphism ring* $\mathrm{End}_G(R[G/B])$ *is* R-*free with basis* $\mathcal{B} := \{\varepsilon_x \mid x \in D(G,B)\}$. *For any* $x, y \in D(G,B)$, *we have* $\varepsilon_x\varepsilon_y = \sum_z a_{x,y,z}\varepsilon_z$ *where*

$$a_{x,y,z} = |ByB \cap Bx^{-1}Bz/B| \qquad \textit{for } x,y,z \in D(G,B).$$

Note that $a_{x,y,z} < \infty$ and that there are only finitely many z with $a_{x,y,z} \neq 0$, thanks to assumption (F). We call $\mathcal{B}$ the *Schur basis* of $\mathrm{End}_G(R[G/B])$.

Proof. Let $\varphi \in \mathrm{End}_R(R[G/B])$. Then we have

$$\begin{aligned}
\varphi \in \mathrm{End}_G(R[G/B]) &\iff \varphi(Bx'g) = \varphi(Bx')g \quad \text{for all } x', g \in G \\
&\iff \sum_{Bx \in G/B} \dot{\varphi}(Bx, Bx'g)Bx = \sum_{Bx \in G/B} \dot{\varphi}(Bx, Bx')Bxg \\
&\iff \sum_{Bx \in G/B} \dot{\varphi}(Bxg, Bx'g)Bxg = \sum_{Bx \in G/B} \dot{\varphi}(Bx, Bx')Bxg \\
&\iff \dot{\varphi}(Bxg, Bx'g) = \dot{\varphi}(Bx, Bx') \quad \text{for all } x, x', g \in G.
\end{aligned}$$

Thus, we have $\varphi \in \mathrm{End}_G(R[G/B])$ if and only if $\dot{\varphi}$ is constant on the orbits of G on $G/B \times G/B$. This shows, first of all, that $\mathcal{B} \subseteq \mathrm{End}_G(R[G/B])$. The elements of $\mathcal{B}$ are clearly linearly independent. Now let $\varphi \in \mathrm{End}_G(R[G/B])$. We have

$$\varphi(B) = \sum_{Bx \in G/B} \dot{\varphi}(Bx, B)Bx.$$

Since this is a finite sum, there are only finitely many cosets Bx such that $\dot{\varphi}(Bx, B) \neq 0$. In particular, there are only finitely many $x \in D(G,B)$ such that $\dot{\varphi}(Bx, B) \neq 0$; let $x_1, \ldots, x_n$ be these finitely many elements of $D(G,B)$ and set $a_i := \dot{\varphi}(Bx_i, B)$. Since $\dot{\varphi}$ is constant on the orbits of G on $G/B \times G/B$, we conclude that $\varphi = \sum_{i=1}^{n} a_i\varepsilon_{x_i}$, and so $\mathcal{B}$ generates $\mathrm{End}_R(R[G/B])$.

Next we prove the multiplication rule. Fix $x, y \in D(G,B)$. It is enough to check that $\varepsilon_x\varepsilon_y(B) = \sum_z a_{x,y,z}\varepsilon_z(B)$ (since the functions on both sides are G-invariant). Now the expression $\varepsilon_x\varepsilon_y(B) = \varepsilon_x(\varepsilon_y(B))$ equals

$$\varepsilon_x\Big(\sum_{Bg \in G/B} \dot{\varepsilon}_y(Bg, B)Bg\Big) = \sum_{Bx_1 \in G/B}\Big(\sum_{Bg \in G/B} \dot{\varepsilon}_x(Bx_1, Bg)\dot{\varepsilon}_y(Bg, B)\Big)Bx_1.$$

Thus, the coefficient of Bx_1 is the number of cosets Bg such that (Bx_1, Bg) lies in the orbit of (Bx, B) and (Bg, B) lies in the orbit of (By, B). But this is equal to the number of cosets Bg such that $Bx_1g^{-1}B = BxB$ (i.e., $g \in Bx^{-1}Bx_1$) and $BgB = ByB$ (i.e., $g \in ByB$). Hence the coefficient of Bx_1 is $|ByB \cap Bx^{-1}Bx_1/B|$, which is the same as $a_{x,y,z}$ if $z \in D(G,B)$ is such that $BzB = Bx_1B$. (This also shows that there are only finitely many z with $a_{x,y,z} \neq 0$, since there are only finitely many cosets Bx_1 which occur with non-zero coefficient in the above expression.) We now consider the sum $\sum_z a_{x,y,z}\varepsilon_z$ and evaluate it on the coset B; the result is

$$\sum_{Bx_1 \in G/B} \Big(\sum_{z \in D(G,B)} a_{x,y,z}\dot{\varepsilon}_z(Bx_1, B) \Big) Bx_1.$$

Thus, we see that the coefficient of Bx_1 is $a_{x,y,z}$, where $z \in D(G,B)$ is such that (Bz, B) lies in the same orbit as (Bx_1, B), i.e., we have $BzB = Bx_1B$. So we obtain the same result as before, as desired. □

8.4.3. Corollary. *Assume that* G/B *is finite and denote by* Tr *the usual trace of the elements of* $\mathrm{End}_G(R[G/B])$ *acting on* $R[G/B]$. *Then we have*

$$\mathrm{Tr}(\varepsilon_x \varepsilon_y) = \begin{cases} |G/B|\,|BxB/B| & \text{if } ByB = Bx^{-1}B, \\ 0 & \text{otherwise.} \end{cases}$$

In particular, $\mathrm{Tr}(\varepsilon_1) = |G/B|$ *and* $\mathrm{Tr}(\varepsilon_x) = 0$ *for* $1 \neq x \in D(G,B)$.

Proof. First, let us compute the trace of ε_x on $R[G/B]$. Recall that ε_x is defined as the indicator function of the G-orbit of the pair (Bx, B) in $G/B \times G/B$. Thus, $\mathrm{Tr}(\varepsilon_x) = 0$ unless the pair (B, B) lies in that orbit. Assume that this is the case, i.e., there exists some $g \in G$ such that $Bg = B$ and $Bxg = B$. But then we have $g \in B$ and hence also $x \in B$, which is only possible for $x = 1$. Thus, we have $\mathrm{Tr}(\varepsilon_x) = 0$ if $x \neq 1$. On the other hand, the G-orbit of (B, B) contains all pairs (Bg, Bg) with $g \in G$, and so $\mathrm{Tr}(\varepsilon_1) = |G/B|$. These formulas imply that $\mathrm{Tr}(\varepsilon_x\varepsilon_y) = \sum_z a_{x,y,z}\,\mathrm{Tr}(\varepsilon_z) = a_{x,y,1}|G/B|$, and we are done, using the multiplication formula in Proposition 8.4.2. □

Finally, let $R = \mathbb{C}$ and assume that G is finite. Then every $\mathbb{C}[G]$-module is split semisimple and, hence, $H := \mathrm{End}_G(\mathbb{C}[G/B])$ is a split semisimple algebra. The space $\mathbb{C}[G/B]$ is naturally a left H-module. For each simple (left) H-module E we denote by $e_E \in H$ the corresponding centrally primitive idempotent. Then $e_E\mathbb{C}[G/B]$ is a right $\mathbb{C}[G]$-module. By Corollary 8.4.3, Tr is a symmetrizing trace on H, i.e., Tr defines a non-generate bilinear form on $\mathbb{C}[G/B]$. So Proposition 7.2.7 shows that e_E can be written as

$$e_E = \frac{1}{c_E} \sum_{x \in D(G,B)} \chi_E(\varepsilon_x)\, \varepsilon_x^\vee,$$

where χ_E is the character of E and c_E is its Schur element, i.e., we have

$$(\dim_{\mathbb{C}} E)c_E = \sum_{x \in D(G,B)} \chi_E(\varepsilon_x)\chi_E(\varepsilon_x^\vee) \qquad \text{and} \qquad c_E \neq 0.$$

8.4.4. Proposition. *In the above set-up, the right* $\mathbb{C}[G]$*-module* $e_E\mathbb{C}[G/B]$ *is isomorphic to a direct sum of* $\dim_{\mathbb{C}} E$ *copies of a simple* $\mathbb{C}[G]$*-module* V_E*. We have a direct sum decomposition of* $\mathbb{C}[G]$*-modules:*

$$\mathbb{C}[G/B] \cong \bigoplus_E (\dim_{\mathbb{C}} E)\, V_E, \qquad \dim_{\mathbb{C}} V_E = |G/B|\, c_E^{-1},$$

where E *runs over the simple* H*-modules (up to isomorphism).*

Proof. For each E we have $e_E = \sum_{i=1}^{n_E} e_{ii}^E$ where $n_E = \dim_{\mathbb{C}} E$ and $e_{ii}^E \in H$ is a primitive idempotent as in Proposition 7.2.7. Since all of our modules are semisimple, this gives rise to a decomposition $e_E\mathbb{C}[G/B] = \bigoplus_i e_{ii}^E\mathbb{C}[G/B]$, where each term is a simple $\mathbb{C}[G]$-module and all terms are isomorphic to each other (since the idempotents e_{ii}^E are conjugate in H). Thus, we have a decomposition $\mathbb{C}[G/B] \cong \bigoplus_E n_E V_E$ as desired, where $n_E V_E \cong e_E\mathbb{C}[G]$. To prove the assertion about $\dim_{\mathbb{C}} V_E$, we consider $\mathbb{C}[G/B]$ as a (left) H-module and decompose $\mathbb{C}[G/B] = \bigoplus_E m_E E$, where E runs over the simple H-modules (up to isomorphism). In order to compute the multiplicities m_E, we take characters and use Corollary 8.4.3. Thus, for any $x \in D(G,B)$ we have

$$\sum_E m_E \chi_E(\varepsilon_x) = \begin{cases} |G/B| & \text{if } x = 1, \\ 0 & \text{if } x \neq 1. \end{cases}$$

We multiply these equations by $\chi_{E'}(\varepsilon_x^\vee)$ and sum over all $x \in D(G,B)$. The result is

$$\sum_E m_E \sum_{x\in D(G,B)} \chi_E(\varepsilon_x)\chi_{E'}(\varepsilon_x^\vee) = |G/B|\,\chi_{E'}(\varepsilon_1).$$

Using the orthogonality relations in Corollary 7.2.4, this can be rewritten as

$$m_{E'} c_{E'} \dim_{\mathbb{C}} E' = \sum_E m_E \delta_{EE'} c_{E'} \dim_{\mathbb{C}} E' = |G/B| \dim_{\mathbb{C}} E',$$

and so $m_E = |G/B| c_E^{-1}$ for all simple $H_{\mathbb{C}}(G,B)$-modules E. Finally, note that since $n_E V_E = e_E\mathbb{C}[G/B]$ (where $n_E = \dim_{\mathbb{C}} E$), we can apply Exercise 7.3 and conclude that

$$n_E \dim_{\mathbb{C}} V_E = \dim_{\mathbb{C}} \mathrm{Hom}_H(He_E, \mathbb{C}[G/B]) = \sum_{E'} m_{E'} \dim_{\mathbb{C}} \mathrm{Hom}_H(He_E, E').$$

Using the fact that $He_E \cong n_E E$, the right-hand side of the above expression evaluates to $m_E n_E$. Thus, the multiplicity m_E also equals $\dim_{\mathbb{C}} V_E$. □

8.4.5. Groups with a BN-pair. We now consider the special situation where G is a (twisted or untwisted) *Chevalley group*. Any such group is a group with a BN-pair; we will not repeat here all the axioms and their consequences. For our purposes, the necessary group-theoretical properties are summarized in the following "axioms":

(1) There exist subgroups $B, N \subseteq G$ such that $B \cap N$ is normal in N and the quotient $W := N/B \cap N$ is generated by a set S of involutions.
(2) We have a double coset decomposition $G = \coprod_{w \in W} B\dot{w}B$, where $\dot{w}$ denotes a representative of $w \in W$ in N.
(3) The pair (W, S) is a Coxeter system. Denoting by $l\colon W \to \mathbb{N}_0$ the corresponding length function, we have, for any $w \in W$ and $s \in S$,

$$\dot{s}B\dot{w} \subseteq \begin{cases} B\dot{s}\dot{w}B & \text{if } l(sw) > l(w), \\ B\dot{s}\dot{w}B \cup B\dot{w}B & \text{if } l(sw) < l(w). \end{cases}$$

The group W is called the *Weyl group* of G, and the double coset decomposition in (2) is called the *Bruhat decomposition* of G. Note that we do not assume that G is a finite group. All that is needed is the finiteness condition (F) with respect to B (see (8.4.1)).

All this is best illustrated in the example where $G = GL_n(k)$ is the general linear group over a field k. In that case, we take B to be the subgroup of all upper triangular matrices in G and N to be the subgroup of all monomial matrices in G. Then $B \cap N$ consists precisely of all the diagonal matrices in G and the factor group $W = B/B \cap N$ can be naturally identified with the symmetric group $\mathfrak{S}_n$.

A convenient reference for the construction and the properties of Chevalley groups and their twisted versions is [Carter 1972b]. Statements (1) and (2) can be found in [Carter 1972b, §8.2]. The statements in (3) follow formally from the list of axioms which say that the pair (B, N) is a *Tits system* in G; see [Bourbaki 1968, Chap. IV, §2, No. 4]. Note that we may unambiguously write Bw without specifying a representative of $w \in W$ in N, since any two such representatives only differ by an element in $N \cap B$.

We consider the natural permutation action of G on the right cosets G/B, and let $R[G/B]$ be the corresponding $R[G]$-permutation module. By the Bruhat decomposition, we know that the elements $\{\dot{w} \mid w \in W\}$ form a set of representatives for the double cosets of B in G. We shall assume that

$$|B\dot{w}B/B| < \infty \qquad \text{for all } w \in W.$$

So assumption (F) is satisfied and we can apply the above results. For any $w \in W$, we denote by T_w the corresponding Schur basis element of $\mathrm{End}_G(R[G/B])$.

8.4.6. Theorem (Iwahori, Matsumoto). *The Schur basis* $\{T_w \mid w \in W\}$ *of the Hecke algebra* $\mathrm{End}_G(R[G/B])$ *satisfies the following multiplication rules:*

$$T_sT_w = \begin{cases} T_{sw} & \text{if } l(sw) > l(w), \\ q_sT_{sw} + (q_s - 1)T_w & \text{if } l(sw) < l(w), \end{cases}$$

where $q_s := |B\dot{s}B/B|$ *for all* $s \in S$.

Proof. First assume that $l(sw) > l(w)$. According to the formula in Proposition 8.4.2, we must compute the number of cosets of B which are contained in

$B\dot{w}B \cap B\dot{s}B\dot{v}$, for any $v \in W$ (note that $B\dot{s} = B\dot{s}^{-1}$). By axiom (3), we have $B\dot{s}B\dot{v} \cap B\dot{w}B \subseteq (B\dot{s}\dot{v}B \cup B\dot{v}B) \cap B\dot{w}B$. So, by the Bruhat decomposition in axiom (2), this set is empty unless $w = v$ or $w = sv$. Assume first that $w = v$. Since $l(sw) > l(w)$, we have $B\dot{s}B\dot{w} \subseteq B\dot{s}\dot{w}B$ by axiom (3) and so, again by the Bruhat decomposition, the intersection of $B\dot{s}B\dot{w}$ and $B\dot{w}B$ is empty. It remains to consider the case where $w = sv$. Then

$$\begin{aligned} B\dot{s}B\dot{s}\dot{w} \cap B\dot{w}B &= (B\dot{s}B\dot{s})\dot{w} \cap B\dot{w}B \\ &\subseteq (B \cup B\dot{s}B)\dot{w} \cap B\dot{w}B && \text{by axiom (3)} \\ &\subseteq (B\dot{w} \cup B\dot{s}B\dot{w}) \cap B\dot{w}B \\ &\subseteq (B\dot{w} \cup B\dot{s}\dot{w}B) \cap B\dot{w}B && \text{by axiom (3), } l(sw) > l(w) \\ &\subseteq B\dot{w} \cap B\dot{w}B = B\dot{w} && \text{by the Bruhat decomposition.} \end{aligned}$$

Conversely, the right-hand side is certainly contained in the left-hand side, and so the desired cardinality is $|B\dot{w} : B| = 1$. It follows that $T_sT_w = T_{sw}$.

Next consider the product T_s^2 for $s \in S$. We must compute the cardinality of the set $B\dot{s}B \cap B\dot{s}B\dot{w}$, for any $w \in W$. By axiom (3), we have $B\dot{s}B\dot{w} \cap B\dot{s}B \subseteq (B\dot{s}\dot{w}B \cup B\dot{w}B) \cap B\dot{s}B$. So, by the Bruhat decomposition, this set is empty unless $w = 1$ or $w = s$. This already shows that $T_s^2 = a_{s,s,1}T_1 + a_{s,s,s}T_s$. We have

$$a_{s,s,1} = |B\dot{s}B/B| = q_s.$$

Clearly, this also shows that the set $B\dot{s}B\dot{s}$ contains q_s cosets of B, and B is one of these cosets. But we also have $B\dot{s}B\dot{s} \subseteq B \cup B\dot{s}B$ by axiom (3) and so

$$a_{s,s,s} = |B\dot{s}B\dot{s} \cap B\dot{s}B/B| = q_s - 1.$$

Now the second multiplication rule can be proved as follows. Let $w \in W$ and $s \in S$ such that $l(sw) < l(w)$. Let $v := sw$. Then $l(sv) > l(v)$, and so $T_sT_v = T_{sv} = T_w$ by the first rule. We conclude that

$$T_sT_w = T_s^2T_v = (q_sT_1 + (q_s - 1)T_s)T_v = q_sT_{sw} + (q_s - 1)T_w,$$

as desired. □

As an immediate consequence, we obtain the following results:

8.4.7. Corollary. *Let* $\mathbf{H} = \mathbf{H}_\mathbf{A}(W, S, \{u_s \mid s \in S\})$ *be the multi-parameter generic Iwahori–Hecke algebras associated with* (W, S)*; see Definition 8.1.4. Assume that each* q_s *is invertible in* R*. Then we have*

$$\mathrm{End}_G(R[G/B]) \cong_R \mathbf{H} \otimes_\mathbf{A} R, \quad \text{via the specialization } \theta\colon \mathbf{A} \to R, u_s \mapsto q_s.$$

Assume, moreover, that G/B *is finite. Then, comparing the formulas in Proposition 8.1.1 and Corollary 8.4.3, we have* $\mathrm{Tr} = |G/B|\tau$.

Let us consider the special case where G is an untwisted Chevalley group over the finite field $\mathbb{F}_q$, where q is a prime power. Then we have

$$|B\dot{w}B/B| = q^{l(w)} \quad (w \in W) \quad \text{and} \quad [G/B] = \sum_{w \in W} q^{l(w)}.$$

In particular, $q_s = q$ for $s \in S$. This follows from a sharp form of the Bruhat decomposition; for more details see [Carter 1972b, §8.6]. (We merely mention that in general, if we consider a twisted Chevalley group over $\mathbb{F}_q$, then $q_s = q^{c_s}$ for some integer $c_s \geq 1$; see [Carter 1972b, §14.1].) Furthermore, working over $R = \mathbb{C}$, the decomposition in Proposition 8.4.4 now reads

$$\mathbb{C}[G/B] = \bigoplus_{\chi \in \mathrm{Irr}(KH)} \chi(T_1) V_\chi \quad \text{with} \quad \dim_{\mathbb{C}} V_\chi = \frac{[G/B]}{\theta_{\mathbb{C}}^*(c_\chi)} = \frac{\theta_{\mathbb{C}}^*(P_W)}{\theta_{\mathbb{C}}^*(c_\chi)}.$$

In [Benson and Curtis 1972] this is used to show that c_χ already divides P_W in $A_{\mathbb{C}}^*$ and, hence, the quotient $D_\chi = P_W/c_\chi$ lies in $A_{\mathbb{C}}^*$ and specializes to the dimension of a simple $\mathbb{C}[G]$-module. We will come back to these questions in Corollary 9.3.6.

8.4.8. Concluding remarks and generalizations. Assume first that G is finite. Instead of the permutation module on the cosets of a Borel subgroup, one considers more generally the so-called Harish-Chandra induction of cuspidal representations from Levi subgroups. If our representations are over a field of characteristic 0 then, by results due to Howlett and Lehrer, the endomorphism algebra of such an induced representation is a deformation of the twisted group algebra of a certain finite group which is an extension of a finite Coxeter group. See [Carter 1985, Chap. 10] and also [Lusztig 1984, Chap. 8] for more detailed results in the case where G arises from a connected reductive group with a connected centre. The fact that the cocycle twisting the group algebra is actually trivial is proved in [Lusztig 1984, Theorem 8.6] in the connected centre case, and extended to the general case in [Geck 1993b].

The state of knowledge in the situation where we consider representations over a field of characteristic $\ell > 0$ is less complete. See [Curtis and Reiner 1987, §72B] for an exposition of a basic part of the theory in the case where ℓ divides q (the case of defining characteristic). Surveys on recent developments in the case where ℓ does not divide q (the case of non-defining characteristic) can be found in [Dipper et al. 1999] and [Geck 1998b]. In both cases, one of the important open problems is the determination of the dimensions of the simple modules. For a solution of that problem in the case of "large" defining characteristic, see [Andersen et al. 1994]. See also the survey articles on Lusztig's conjecture in [Carter and Geck 1998] and [Carter and Saxl 1998].

In another direction, there is a similar theory for representations (over $\mathbb{C}$) of p-adic groups, where the corresponding Hecke algebras are associated with affine Weyl groups. The most far-reaching results in this area have been obtained in [Lusztig 1995]. For modular representations of p-adic groups, see [Vigneras 1996].

8.5 BIBLIOGRAPHICAL REMARKS AND EXERCISES

The origin of the theory of Hecke algebras lies in [Iwahori 1964], which contains the results in Section 8.4. Actually, Iwahori only considered untwisted finite Chevalley groups. The general case of arbitrary groups with a BN-pair is treated in [Matsumoto 1964]. Our proof of Theorem 8.4.6, using the Schur basis of the endomorphism ring of a permutation module, is taken from [Hiss 1989, §5.3]. It has the advantage that no assumptions on the ground ring are needed. For general results on permutation modules and their endomorphism algebras, see [Higman 1977]. The description of a Hecke algebra as in Section 8.4 in terms of generators and relations also appears in [Iwahori 1964, §4].

More general types of algebras, namely, the cyclotomic algebras associated with complex reflection groups, have been introduced in [Ariki and Koike 1992] and [Broué and Malle 1993]. Many results that are well known for Iwahori–Hecke algebras corresponding to finite Coxeter groups are also believed to hold for the cyclotomic algebras; see the conjectures in [Broué and Malle 1993]. A general theory of these algebras and related structures is beginning to take shape; see, for example, [Broué and Michel 1996], [Broué et al. 1998], [Malle and Mathas 1998], [Dipper et al. 1998]. See also [Malle 1996] for the explicit computation of generic degrees in dimension 2.

The application of Tits's deformation theorem in the proof of Theorem 8.1.7 is taken from [Bourbaki 1968, Chap. IV, §2, Exercise. 27]. The construction of the reflection representation of KH and the results on its exterior powers are in [Curtis et al. 1971, §9]; see also [Curtis and Reiner 1987, §67E]. The computation in Example 8.1.13 is taken from [Digne and Michel 1985, V.3.12]. The construction of the irreducible representations in type $I_2(m)$ and the computation of the corresponding Schur elements are due to [Kilmoyer and Solomon 1973]; these results can also be found in [Curtis and Reiner 1987, §67C].

The class polynomials of Section 8.2 appeared in [Geck and Pfeiffer 1993]. In that paper, we also introduced the general notion of the character table of an Iwahori–Hecke algebra. For type A_n, this had been done much earlier in [Starkey 1975] and later again in [Ram 1991]. The full version of the basic Theorem 8.2.3 only came with the work in [Geck and Rouquier 1996] on the centre of Iwahori–Hecke algebras; see also [Geck 1998d, Theorem 4.2]. The stability under reduction in Remark 8.2.5 was first proved in [Dipper and James 1987] for type A. An extension of Theorem 8.2.3 to Iwahori–Hecke algebras twisted by an automorphism of the underlying Coxeter groups is established in [Geck et al. 2000].

The automorphism j in Exercise 8.1 is due to [Green 1970, Theorem 8.1]. This paper also contains a number of results on the generic degrees of KH in the case where H arises from a finite Chevalley group as in Section 8.4.

Exercise 8.1. Let (W, S) be a Coxeter system. Show that the multi-parameter generic Iwahori–Hecke algebra H associated with (W, S) as in Definition 8.1.4 has a presentation as an associative $\mathbb{Z}$-algebra with 1 and

generators: T_s, $s \in S$, $u_s^{\pm 1}$, $s \in S$,
relations: $u_s = u_t$ if $s, t \in S$ are conjugate in W,
$u_s u_t = u_t u_s, u_s u_s^{-1} = u_s^{-1} u_s = 1$, and $u_s T_t = T_t u_s$ for $s, t \in S$,
$T_s^2 = u_s 1_H + (u_s - 1)T_s$ for all $s \in S$,
$\mathrm{Prod}(T_s, T_t; m_{st}) = \mathrm{Prod}(T_t, T_s; m_{st})$ for $s \neq t$, $m_{st} < \infty$.

Show that there exists a unique ring involution $j: \mathbf{H} \to \mathbf{H}$ (i.e., a ring automorphism of order 2) such that

$$j\Big(\sum_{w \in W} a_w T_w\Big) = \sum_{w \in W} \bar{a}_w (-1)^{l(w)} \mathrm{ind}(T_w)^{-1} T_w,$$

where $a \mapsto \bar{a}$ is the ring automorphism of $\mathbf{A}$ which takes u_s to u_s^{-1} ($s \in S$).
[*Hint*. Let $\tilde{H}$ be the algebra defined by the above presentation. Since the defining relations are satisfied in $\mathbf{H}$, we have a canonical surjective map $\tilde{H} \to \mathbf{H}$. On the other hand, show as in the proof of Lemma 4.4.3 that $\tilde{H}$ is generated by elements of the form aT_w, where $a \in \mathbf{A}$ is a monomial in u_s ($s \in S$) and $w \in W$. Then use the fact that these elements are linearly independent (over $\mathbb{Z}$) in $\mathbf{H}$. To prove the assertion about j, check the above defining relations.]

Exercise 8.2. Let $H = H_A(W, S, \{a_s \mid s \in S\})$ be an Iwahori–Hecke algebra such that $a_s \in A$ is invertible for all $s \in S$. Show that there exists a unique A-algebra automorphism $\gamma: H \to H$ such that $\gamma(T_s) = -a_s T_s^{-1}$ for all $s \in S$. Show that $\gamma(T_w) = (-1)^{l(w)} \mathrm{ind}(T_w) T_{w^{-1}}^{-1}$ for all $w \in W$.

In [Iwahori 1964, §5], this construction is attributed to O. Goldman.

[*Hint*. Check the defining relations in Definition 4.4.1.]

Exercise 8.3. The aim of this exercise is to prove the relation between the reduced Burau representation and the reflection representation ρ of the Iwahori–Hecke algebra $\mathbf{H}$ of $\mathfrak{S}_n$, which is stated in (8.1.14). By Exercise 4.11, we may consider the non-reduced and the reduced Burau representation as representations of $\mathbf{H}$, via $T_w \mapsto (-1)^{l(w)}\beta_n(w)$ and $T_w \mapsto (-1)^{l(w)}\beta_n'(w)$, respectively (where we set $t = u$). Now proceed as follows.

(a) Show that the Burau representation has two irreducible components: ε and the reduced Burau representation.
(b) Specialize $u \mapsto 1$ and show that the specialization of the Burau representation is the permutation representation of $\mathfrak{S}_n$ on the cosets of $\mathfrak{S}_{n-1}$, tensored with the sign representation ε.
(c) Use the construction of the irreducible characters in Theorem 5.4.7, the identification of the reflection character of $\mathfrak{S}_n$ in Proposition 5.4.12, and Theorem 8.1.7 to show that the character of the Burau representation must be equal to the sum of that of $\rho^* = \rho \circ \gamma$ and ε (with γ as in Exercise 8.2).

Now (a), (c) imply that the Burau representation and ρ^* have the same character. Hence the two representations are equivalent. See also [Jones 1987, (5.7)].

Exercise 8.4. Let (W, S) be a Coxeter system and $H = H_A(W, S, \{a_s, b_s \mid s \in S\})$ be a corresponding Iwahori–Hecke algebra as in Definition 4.4.1. Assume that there exists a partition $S = S_1 \cup S_2$ with $S_1 \cap S_2 = \varnothing$ such that all elements of S_1 commute with all elements of S_2. Let $W_i \subseteq W$ be the parabolic subgroup generated by S_i, so that we have $W = W_1 \times W_2$. Show that the canonical map

$$H_A(W_1, S_1, \{a_s, b_s \mid s \in S_1\}) \otimes_A H_A(W_2, S_2, \{a_s, b_s \mid s \in S_2\}) \to H$$

given by multiplication is an isomorphism of A-algebras.

Exercise 8.5. Assume that W is finite and that we have a partition $S = S_1 \cup S_2$ with $S_1 \cap S_2 = \varnothing$ such that every element of S_1 commutes with every element of S_2. Let $W_i \subseteq W$ be the parabolic subgroup generated by S_i, so that we have $W = W_1 \times W_2$. Let K be a field and H be an Iwahori–Hecke algebra over K associated with (W, S). Let $H_i \subseteq H$ be the parabolic subalgebra corresponding to W_i. By Exercise 8.4, we have $H = H_1 \otimes_K H_2$. Assume that H_1 and H_2 are split and semisimple.

(a) Show that H is also split semisimple, and that every irreducible representation of H is equivalent to a representation of the form $\delta_1 \otimes \delta_2$, where δ_i is an irreducible representation of H_i.

(b) Any $w \in W$ can be written uniquely as $w = w_1 w_2$ with $w_i \in W$. Show that if $\chi_i \in \mathrm{Irr}(H_i)$ then the map $H \to K$ given by $T_w \mapsto \chi_1(T_{w_1})\chi_2(T_{w_2})$ is an irreducible character of H, and all irreducible characters of H arise in this way. Show that the corresponding Schur element is $c_{\chi_1} c_{\chi_2}$.

(c) Assume that H_1, H_2, H are generic Iwahori–Hecke algebras. Show that, for suitable orderings of the rows and columns, the character table of H is the Kronecker product of the character tables of H_1 and H_2.

Exercise 8.6. Let $\mathbf{H}$ be a generic Iwahori–Hecke algebra associated with a Coxeter system (W, S) and assume that W is finite. Let $\mathbf{K}_0 = \mathbb{Q}(u_s \mid s \in S)$.

(a) Consider $\mathbf{K}_0\mathbf{H}$ as a right module over itself and let τ be the character of that representation. Show that the bilinear form $\beta\colon \mathbf{K}_0\mathbf{H} \times \mathbf{K}_0\mathbf{H} \to \mathbf{K}_0$ given by $\beta(h, h') := \tau(hh')$ for $h, h' \in \mathbf{K}_0\mathbf{H}$ is non-degenerate.

(b) Show that $\mathbf{K}_0\mathbf{H}$ is a separable algebra, i.e., if $\mathbf{K}$ is any field containing $\mathbf{K}_0$, then $\mathbf{KH}$ is semisimple.

[*Hint.* (a) Since the structure constants of $\mathbf{K}_0\mathbf{H}$ with respect to the basis $\{T_w \mid w \in W\}$ lie in $\mathbf{A}$, the map τ restricts to a trace function $\mathbf{H} \to \mathbf{A}$. Now consider a specialization $\theta\colon \mathbf{A} \to \mathbb{Q}$ such that $\theta(u_s) = 1$ for all $s \in S$, and check that the specialized trace function τ_θ is just the character of the regular representation of W. Now use the fact that the corresponding bilinear form on $\mathbb{Q}[W]$ is non-degenerate by Example 7.1.5.

(b) Let $h \in \mathrm{rad}(\mathbf{KH})$ and $h' \in \mathbf{H}$. Since $\mathrm{rad}(\mathbf{KH})$ is a nilpotent ideal, hh' is a nilpotent element. Hence it has trace 0 in every representation of $\mathbf{KH}$. It follows that $\tau^K(hh') = 0$, where τ^K is the canonical extension of τ. Since τ defines a non-degenerate bilinear form by (a), the same holds for τ^K. So $h = 0$, as desired.]

Exercise 8.7. Let (W, S) be a Coxeter system and assume that W is finite. Let $H = H_A(W, S, \{a_s, b_s \mid s \in S\})$ be a corresponding Iwahori–Hecke algebra as in Definition 4.4.1. (We make no further assumptions on a_s, b_s.) Consider the A-linear map $\pi\colon H \to A$ defined by $\pi(T_{w_0}) = 1$ and $\pi(T_w) = 0$ for $w \neq w_0$, where $w_0 \in W$ is the longest element. Show that $\pi(Hh) \neq 0$ and $\pi(hH) \neq 0$ for any $0 \neq h \in H$. This means that H is always a Frobenius algebra (see [Curtis and Reiner 1962, §61]), even if the a_s are not invertible in A.

[*Hint.* Let $h \neq 0$ and write $h = \sum_w h_w T_w$ with $h_w \in A$. Choose $x \in W$ of maximum possible length such that $h_x \neq 0$. Let $y := x^{-1}w_0$, $z := w_0x^{-1}$, and check that $\pi(hT_y) = \pi(T_z h) = h_x \neq 0$, using the relation $T_xT_y = T_{w_0} = T_zT_x$.]

Exercise 8.8. Let $\mathbf{H}$ be a generic Iwahori–Hecke algebra associated with a Coxeter system (W, S) where W is finite. Let $\rho\colon \mathbf{KH} \to \mathrm{End}_\mathbf{K}(\mathbf{V})$ be the reflection representation of $\mathbf{KH}$ with respect to $(\mathbf{c}_{st})_{s,t\in S}$, chosen as in Remark 8.1.12. Following [Curtis et al. 1971, §9], we define an action of $\mathbf{KH}$ on $\bigwedge^d \mathbf{V}$ $(0 \leq d \leq |S|)$ as follows. For each $s \in S$, let $\rho_s^{(d)} \in \mathrm{End}(\bigwedge^d \mathbf{V})$ be given by the formula

$$(v_1 \wedge \cdots \wedge v_d)\rho_s^{(d)} = u_s^{1-d}\Big(v_1\rho(T_s)\Big) \wedge \cdots \wedge \Big(v_d\rho(T_s)\Big) \quad (v_i \in \mathbf{V}),$$

i.e., $\rho_s^{(d)}$ is u_s^{1-d} times the map induced by $\rho(T_s)$ on $\bigwedge^d \mathbf{V}$.

(a) Show that $\rho_s^{(d)}$ is diagonalizable with eigenvalues -1 and u_s, where the multiplicities are given by $\binom{n-1}{d-1}$ and $\binom{n-1}{d}$, respectively.

(b) Show that the assignment $T_s \mapsto \rho_s^{(d)}$ $(s \in S)$ defines a representation of $\mathbf{KH}$ which specializes to the representation of W on $\bigwedge^d V$ in Section 5.1.

(c) Show that, for $0 \leq d \leq |S|$, the representations $\rho^{\wedge d}\colon \mathbf{KH} \to \mathrm{End}(\bigwedge^d \mathbf{V})$ are irreducible and pairwise inequivalent.

[*Hint.* (a) Since $\rho(T_s)$ satisfies the quadratic relations, it is diagonalizable with eigenvalues -1 and u_s. Hence, taking a basis of $\mathbf{V}$ consisting of eigenvectors of $\rho(T_s)$, the corresponding basis of $\bigwedge^d \mathbf{V}$ consists of eigenvectors of $\rho_s^{(d)}$. To count the multiplicities, argue as in Lemma 5.1.2. (b) The fact that the quadratic relations are satisfied follows from (a). The braid relations automatically hold since they hold for the maps $\rho(T_s)$ $(s \in S)$. It follows that the assignment $T_s \mapsto \rho_s^{(d)}$ $(s \in S)$ defines a representation. The last assertion is proved by applying a specialization $\theta_K^*\colon \mathbf{A}_K^* \to K$ as in (8.1.6).]

Exercise 8.9. Let (W, S) be a Coxeter system and assume that W is finite. Let $H = H_A(W, S, \{a_s \mid s \in S\})$ be a corresponding Iwahori–Hecke algebra, where we have $a_s = 0$ for all $s \in S$. This algebra is called the *0-Hecke algebra*; the generators T_s $(s \in S)$ now satisfy the quadratic relation $T_s^2 = -T_s$. This algebra has been studied in detail in [Norton 1979] and [Carter 1986].

(a) Show that $\{(-1)^{l(w)}T_w \mid w \in W\}$ is a multiplicatively closed basis for H.
(b) Let $W = \{w_1, \ldots, w_n\}$, where the labelling is chosen so that $l(w_i) \geq l(w_{i+1})$ for all i. For $1 \leq i \leq n$, let H_i be the A-submodule of H generated by all T_{w_j} with $j \leq i$. Show that $H_1 \subset H_2 \subset \cdots \subset H_n = H$ is a chain of two-sided ideals in H, where H_{i+1}/H_i is A-free of rank 1.
(c) For any subset $J \subseteq S$, define a map $\lambda_J \colon H \to A$ by $\lambda_J(T_s) = 0$ if $s \in J$ and $\lambda_J(T_s) = -1$ otherwise. Show that λ_J is a one-dimensional representation of H, and that any one-dimensional representation of H is of this form.
(d) Let A be a field. Use (a) to show that any irreducible representation of H is one-dimensional. Then use (b) to show that H has precisely $2^{|S|}$ irreducible representations. Deduce from this that H is not semisimple if $|S| > 1$.
(e) Let $\mathbf{H}'$ be the generic one-parameter Iwahori–Hecke algebra associated with (W, S) over the ring $\mathbb{C}[u]$ (where u is an indeterminate). Consider the specialization $\theta \colon \mathbb{C}[u] \to \mathbb{C}$, $u \mapsto 0$ $(s \in S)$, so that the specialized algebra H'_0 is canonically isomorphic to the 0-Hecke algebra over $\mathbb{C}$. Consider a setting similar to (8.1.6), so that we have a decomposition map between the Grothendieck groups of $\mathbf{KH}'$ and of H'_0. Show that in type A_2 with generators $\{s_1, s_2\}$, the corresponding decomposition matrix is given by

	$\lambda_{\{s_1,s_2\}}$	$\lambda_{\{s_1\}}$	$\lambda_{\{s_2\}}$	$\lambda_\varnothing$
ind	1	.	.	.
ε	.	.	.	1
χ_1	.	1	1	.

Here, the characters of $\mathbf{KH}'$ are labelled as in Table 8.1 (p. 270).

[*Hint.* To prove (e), note that $\mathbf{H}'$ is not symmetric, but this is not required to have a setting as in (8.1.6). Use the formulas in Section 8.3 to compute all the character values $\chi(T_w)$ for $\mathbf{KH}'$, and then specialize to 0.]

9

Characters of Iwahori–Hecke algebras

In the previous chapter we have seen that a generic Iwahori–Hecke algebra $\mathbf{H}$ associated with a finite Coxeter group W is isomorphic to the group algebra of W and that, moreover, we have a bijection between the irreducible characters of $\mathbf{H}$ and of W via a specialization process. We will now carry this relation further by studying in more detail the irreducible characters of $\mathbf{H}$.

Similar to the induction of representations from a subgroup of a finite group, one can also define a process of induction of representations from parabolic subalgebras of Iwahori–Hecke algebras. This will be discussed in detail in Section 9.1. The basic result is that this process of induction is compatible with specializations. This will allow us to transfer the results about induced characters for finite Coxeter groups (e.g., the results on characters of parabolic type in Chapter 6) to the case of Iwahori–Hecke algebras.

All the results which allow us to determine the character values of $\mathbf{H}$ in terms of the character values of W rely in some way on a basic observation due to Springer which determines these values for the basis element T_{w_0} where $w_0 \in W$ is the longest element. This result and its consequences are the subject of Section 9.2. In type A, we obtain a full description of the irreducible characters of $\mathbf{H}$ in terms of those of W. This result, due to Starkey in 1975, is in fact one of the oldest results on character tables of Iwahori–Hecke algebras.

Historically, much of the early work on Iwahori–Hecke algebras was centred around the problem of establishing "rationality" properties of their representations and the corresponding Schur elements (see the bibliographical comments at the end of this chapter). We have already seen that all irreducible characters of the underlying finite Coxeter group can be realized over the rational numbers—except for the non-crystallographic types $I_2(m)$, H_3 and H_4. We settle the analogous questions for Iwahori–Hecke algebras in Section 9.3, by an approach which is based on the results on "good" representatives in the conjugacy classes of finite Coxeter groups from Chapter 4. Finally, in Section 9.4, we prove a number of general results on character values of $\mathbf{H}$ and the corresponding Schur elements. In particular, we come back to the problem of attaching "a-invariants" to the irreducible characters which was discussed in an axiomatic way in Section 6.5. Here, we give an alternative definition in terms of the Schur elements and the leading coefficients of the irreducible characters of $\mathbf{H}$.

9.1 INDUCTION FROM PARABOLIC SUBALGEBRAS

In this section, we study the operation of induction of representations from parabolic subalgebras. We establish the basic functorial properties, like compatibility with specializations, Frobenius reciprocity and the Mackey formula. This will allow us to transfer the results of Chapter 6 on the characters of finite Coxeter groups to the case of Iwahori–Hecke algebras.

Let (W, S) be a Coxeter system and A be a commutative ring with 1. Let $H = H_A(W, S, \{a_s \mid s \in S\})$ be an Iwahori–Hecke algebra as in Definition 4.4.1 (see also Remark 8.1.3), where the generators T_s satisfy the quadratic relations

$$T_s^2 = a_s T_1 + (a_s - 1)T_s, \qquad s \in S.$$

Let $J \subseteq S$ and $W_J \subseteq W$ be the corresponding parabolic subgroup. Recall from (4.4.7) that we have a corresponding subalgebra

$$H_J = \langle T_w \mid w \in W_J \rangle_A \subseteq H$$

and a direct sum decomposition of (left) H_J-modules $H = \bigoplus_{x \in X_J} H_J T_x$, where X_J is the set of distinguished right coset representatives of W_J in W.

9.1.1. Definition. Let $J \subseteq S$ and $H_J \subseteq H$ be the corresponding parabolic subalgebra as above. For any H_J-module V, the *induced module* is defined by

$$\operatorname{Ind}_J^S(V) := V \otimes_{H_J} H.$$

Note that $\operatorname{Ind}_J^S(V)$ is naturally an H-module via $(v \otimes h')h = v \otimes h'h$ for $v \in V$, $h, h' \in H$. Moreover, if V is free as an A-module, then so is $\operatorname{Ind}_J^S(V)$. Thus, the process of induction sends H_J-lattices to H-lattices.

Let us be a bit more explicit about the operation of H on an induced module $\operatorname{Ind}_J^S(V)$. Since H is free as a left H_J-module, with basis given by $\{T_x \mid x \in X_J\}$, every element of $\operatorname{Ind}_J^S(V)$ can be written uniquely in the form $\sum_x v_x \otimes T_x$, where $v_x \in V$ for all $x \in X_J$. In order to describe the action of T_s $(s \in S)$ we use Deodhar's Lemma 2.1.2. For any $x \in X_J$, we have either $xs \in X_J$ or $xs = tx$ with $t \in J$. Hence we obtain, for any $v \in V$ and $x \in X_J$,

$$(v \otimes T_x).T_s = \begin{cases} v.T_t \otimes T_x & \text{if } xs = tx \text{ with } t \in J, \\ v \otimes T_{xs} & \text{if } xs \in X_J \text{ and } l(xs) > l(x), \\ a_s(v \otimes T_{xs}) + (a_s - 1)(v \otimes T_x) & \text{if } xs \in X_J \text{ and } l(xs) < l(x). \end{cases}$$

Now assume that V is A-free and let $\{v_i\}_{i \in B}$ be an A-basis of V. For any $t \in J$ and $i \in B$, we have an equation $v_i.T_t = \sum_{j \in B} a_{ij}^t v_j$, with $a_{ij}^t \in A$. Then the vectors $\{v_i \otimes T_x \mid i \in B, x \in X_J\}$ form an A-basis of $\operatorname{Ind}_J^S(V)$, and the action of T_s $(s \in S)$ is given by the equations

$$(v_i \otimes T_x).T_s = \begin{cases} \sum_{j \in B} a_{ij}^t (v_j \otimes T_x) & \text{if } xs = tx \text{ with } t \in J, \\ v_i \otimes T_{xs} & \text{if } xs \in X_J \text{ and } l(xs) > l(x), \\ a_s(v_i \otimes T_{xs}) + (a_s - 1)(v_i \otimes T_x) & \text{if } xs \in X_J \text{ and } l(xs) < l(x). \end{cases}$$

9.1.2. Example. Let (W, S) be of type A_3 with standard generators s_1, s_2, s_3 such that $s_1 s_2$ and $s_2 s_3$ have order 3. Let $q \in A$ and assume that the parameters of H are $a_{s_i} = q$ for $i = 1, 2, 3$. Let $V = A$ be the H-module defined by the representation $\mathrm{ind}: T_{s_i} \mapsto q$. Now let $J = \{s_1, s_2\}$. Then $X_J = \{1, s_3, s_3s_2, s_3s_2s_1\}$. The following table contains the result of multiplying $T_x T_{s_i}$ where $x \in X_J$:

x	s_1	s_2	s_3
1	T_{s_1}	T_{s_2}	T_{s_3}
s_3	$T_{s_1}T_{s_3}$	$T_{s_3s_2}$	$qT_1 + (q-1)T_{s_3}$
s_3s_2	$T_{s_3s_2s_1}$	$qT_{s_3} + (q-1)T_{s_3s_2}$	$T_{s_2}T_{s_3s_2}$
$s_3s_2s_1$	$qT_{s_3s_2} + (q-1)T_{s_3s_2s_1}$	$T_{s_1}T_{s_3s_2s_1}$	$T_{s_2}T_{s_3s_2s_1}$

So, taking $\{1 \otimes T_1, 1 \otimes T_{s_3}, 1 \otimes T_{s_3s_2}, 1 \otimes T_{s_3s_2s_1}\}$ as A-basis for $\mathrm{Ind}_J^S(V)$, the generators $T_{s_1}, T_{s_2}, T_{s_3}$ respectively act through the matrices

$$\begin{bmatrix} q & 0 & 0 & 0 \\ 0 & q & 0 & 0 \\ 0 & 0 & 0 & 1 \\ 0 & 0 & q & q-1 \end{bmatrix}, \quad \begin{bmatrix} q & 0 & 0 & 0 \\ 0 & 0 & 1 & 0 \\ 0 & q & q-1 & 0 \\ 0 & 0 & 0 & q \end{bmatrix}, \quad \begin{bmatrix} 0 & 1 & 0 & 0 \\ q & q-1 & 0 & 0 \\ 0 & 0 & q & 0 \\ 0 & 0 & 0 & q \end{bmatrix}.$$

We invite the reader to check directly that these matrices define a representation of H by verifying the defining relations in Definition 4.4.1.

The following result shows that the induction extends to a general operation for trace functions. For this purpose, assume that W is finite and that all parameters a_s are units in A. The decomposition $H = \bigoplus_{x \in X_J} H_J T_x$ implies that

$$T_x T_w = \sum_{y \in X_J} h^w(x, y) T_y \qquad \text{with unique } h^w(x, y) \in H_J,$$

for any $w \in W$ and $x \in X_J$. For any A-linear map $\lambda: H_J \to A$, we define

$$\mathrm{Ind}_J^S(\lambda): H \to A, \qquad T_w \mapsto \sum_{x \in X_J} \lambda(h^w(x, x)).$$

With this notation, we can now state:

9.1.3. Proposition. *Assume that λ is a trace function on H_J. Then $\mathrm{Ind}_J^S(\lambda)$ is a trace function on H. Moreover, if λ is the character of an H_J-lattice V then $\mathrm{Ind}_J^S(\lambda)$ is the character of the H-lattice $\mathrm{Ind}_J^S(V)$.*

Proof. The second statement simply follows from the definition of $\mathrm{Ind}_J^S(V)$. In order to show the first statement, we must show that $\mathrm{Ind}_J^S(\lambda)$ has value 0 on all commutators $[T_w, T_{w'}]$ $(w, w' \in W)$. For any $w, w' \in W$, we write $T_wT_{w'} = \sum_{z\in W} a^z(w,w')T_z$ with $a^z(w,w') \in A$. Then we have

$$\mathrm{Ind}_J^S(\lambda)(T_wT_{w'}) = \lambda\Big(\sum_{x\in X_J}\sum_{z\in W} a^z(w,w')h^z(x,x)\Big)$$

and a similar expression for $\mathrm{Ind}_J^S(\lambda)(T_{w'}T_w)$. Thus, it is enough to show that

$$h(w,w',z) := \sum_{x\in X_J} (a^z(w,w') - a^z(w',w))h^z(x,x) \in [H_J, H_J] \qquad (1)$$

for all $w, w', z \in W$. For this purpose, let $\mathbf{H}$ be a generic Iwahori–Hecke algebra associated with (W, S) and $\theta\colon \mathbf{A} \to A$ be a specialization such that $H = \mathbf{H} \otimes_{\mathbf{A}} A$, as in Remark 8.1.5. Then the element $h(w,w',z)$ in (1) is obtained via specialization from an analogous expression in $\mathbf{H}$ and it is enough to prove that that expression in $\mathbf{H}$ lies in $[\mathbf{H}_J, \mathbf{H}_J]$. Thus, we are reduced to proving (1) for $H = \mathbf{H}$. Now, let $\mathbf{K}$ be as in Theorem 8.1.7. Then $\mathbf{KH}_J$ is split semisimple and so every trace function on $\mathbf{KH}_J$ is a linear combination of irreducible characters. We have already remarked at the beginning of the proof that the desired assertion holds for characters. Thus, we conclude that every trace function on $\mathbf{KH}_J$ has value 0 on the element $h(w,w',z)$ in (1). In particular, this holds for the trace functions $f_{C'}\colon \mathbf{H}_J \to \mathbf{A}$ in Theorem 8.2.3 (where C' runs over the conjugacy classes of W_J). These trace functions form a basis for the space of all trace functions on $\mathbf{H}_J$ and $\mathbf{H}_J/[\mathbf{H}_J, \mathbf{H}_J]$ is free as an $\mathbf{A}$-module. Hence we have $[\mathbf{H}_J, \mathbf{H}_J] = \bigcap_{C'} \ker(f_{C'})$ which implies that (1) holds, without any further computation. □

We have the following compatibility with the symmetrizing trace on H.

9.1.4. Lemma. *Assume that all a_s $(s \in S)$ are invertible. For any $s \in S$ and $x \in X_J$, we can write uniquely $T_xT_s = \sum_{y\in X_J} h^s(x,y)T_y$ where $h^s(x,y) \in H_J$. Then we have*

$$T_sT_y^\vee = \sum_{x\in X_J} T_x^\vee h^s(x,y) \qquad \textit{for all } y \in X_J,$$

where $\{T_w^\vee \mid w \in W\}$ is defined as in Proposition 8.1.1.

Proof. We distinguish cases according to Deodhar's Lemma 2.1.2. If $xs \in X_J$ and $l(xs) > l(x)$, then $T_xT_s = T_{xs}$; if $xs \in X_J$ and $l(xs) < l(x)$, then $T_xT_s = a_sT_{xs} + (a_s - 1)T_x$; if $xs = tx$ with $t \in J$, then $l(xs) > l(x)$ and so $T_xT_s = T_tT_x$. Thus, we have

$$h^s(x,y) = \begin{cases} 1 & \text{if } y = xs \in X_J \text{ and } l(xs) > l(x),\\ a_s & \text{if } y = xs \in X_J \text{ and } l(xs) < l(x),\\ a_s - 1 & \text{if } y = x \text{ and } l(xs) < l(x),\\ T_t & \text{if } y = x \text{ and } xs = tx \text{ with } t \in J,\\ 0 & \text{otherwise.}\end{cases}$$

On the other hand, we have $T_y^\vee = \mathrm{ind}(T_y)^{-1}T_{y^{-1}}$, and $l(ys) > l(y)$ if and only if $l(sy^{-1}) > l(y^{-1})$. Writing $T_sT_y^\vee = \sum_{x\in X_J} T_x^\vee \tilde{h}^s(x,y)$ with $\tilde{h}^s(x,y) \in H_J$, we find the same values as above: if $ys \in X_J$ and $l(ys) > l(y)$, then $T_sT_y^\vee = \mathrm{ind}(T_y)^{-1}T_{sy^{-1}} = a_sT_{ys}^\vee$; if $ys \in X_J$ and $l(ys) < l(y)$, then

$$T_sT_y^\vee = \mathrm{ind}(T_y)^{-1}\Big(a_sT_{sy^{-1}} + (a_s-1)T_{y^{-1}}\Big) = T_{ys}^\vee + (a_s-1)T_y^\vee;$$

finally, if $ys = ty$ with $t \in J$, then $l(sy^{-1}) > l(y)$ and so $T_sT_y^\vee = T_y^\vee T_t$. Thus, we see that $h^s(x,y) = \tilde{h}^s(x,y)$ for all x, y. □

As in the case of group algebras, the induction of representations enjoys a number of functorial properties.

9.1.5. Functorial properties. It follows immediately from the additivity and the associativity of tensor products that the operation of induction is additive and transitive: if $J \subseteq S$ is any subset then

$$\mathrm{Ind}_J^S(V_1 \oplus V_2) \cong \mathrm{Ind}_J^S(V_1) \oplus \mathrm{Ind}_J^S(V_2) \qquad \text{(a)}$$

for all H_J-lattices V_1, V_2; furthermore, if $J \subseteq K \subseteq S$ are subsets then

$$\mathrm{Ind}_J^S(V) = \mathrm{Ind}_K^S\Big(\mathrm{Ind}_J^K(V)\Big) \qquad \text{for all } H_J\text{-lattices } V. \qquad \text{(b)}$$

Now consider the compatibility with specializations. Let B be a commutative ring with 1 and $\theta\colon A \to B$ be a ring homomorphism. We may consider B as an A-module via θ. Then, as in (8.1.2), we have a canonical isomorphism

$$BH := H \otimes_A B \to H_B(W, S, \{b_s \mid s \in S\}), \qquad T_w \otimes 1 \mapsto T_w,$$

where $b_s = \theta(a_s)$ for all $s \in S$. Now let $J \subseteq S$ be any subset and consider the parabolic subalgebras $H_J \subseteq H$ and $BH_J \subseteq BH$. Let us assume we are given an H_J-lattice V. Then tensoring with B yields a BH_J-lattice $V \otimes_A B$, which we can induce to a BH-lattice. On the other hand, we can first induce V to H and then tensor with B. Then we have an isomorphism of BH-modules:

$$\mathrm{Ind}_J^S(V) \otimes_A B \to \mathrm{Ind}_J^S(V \otimes_A B), \qquad (v \otimes T_x) \otimes b \mapsto (v \otimes b) \otimes T_x, \qquad \text{(c)}$$

where $v \in V$, $x \in X_J$ and $b \in B$. We express this by saying that *specialization commutes with induction*. To prove this, just note that the map $(v \otimes T_x) \otimes b \mapsto (v \otimes b) \otimes T_x$ is certainly well defined, and an isomorphism of B-modules. The fact that it commutes with the action of T_s ($s \in S$) is obvious from the formulas following Definition 9.1.1.

Furthermore, we also have an adjointness property between induction and restriction. This is based on the following completely general result.

9.1.6. Adjointness between $\otimes$ and Hom. Let B, C be k-algebras over a commutative ring k. Let N be a right B-module, L be a right C-module, and M be a (C, B)-bimodule (where C acts from the left and B acts from the right).

Then $L \otimes_C M$ is naturally a right B-module, where $(l \otimes m)b = l \otimes (mb)$ for $b \in B$, $l \in L$, $m \in M$. Furthermore, $\mathrm{Hom}_B(M, N)$ is naturally a right C-module, where $(\varphi c)(m) = \varphi(cm)$ for $c \in C$, $m \in M$, $\varphi \in \mathrm{Hom}_B(M, N)$.

Then we have an isomorphism of k-modules

$$\mathrm{Hom}_C(L, \mathrm{Hom}_B(M, N)) \cong_k \mathrm{Hom}_B(L \otimes_C M, N)$$

given as follows. With $f\colon L \to \mathrm{Hom}_B(M, N)$ we associate the map $L \otimes_C M \to N$, $l \otimes m \mapsto f(l)(m)$. The inverse of that map is given as follows. Let $g\colon L \otimes_C M \to N$; then we define $\tilde{g}\colon L \to \mathrm{Hom}_B(M, N)$ by $(\tilde{g}(l))(m) := g(l \otimes m)$ $(l \in L, m \in M)$.

The proof is straightforward; see [Curtis and Reiner 1981, (2.19)].

9.1.7. Proposition (Frobenius reciprocity/Nakayama relations). *Let $J \subseteq S$ be any subset, V be an H_J-module and V' an H-module. We write $\mathrm{Res}_J^S(V')$ for the H_J-module obtained by restricting the action of H to that of H_J. Then we have an isomorphism of A-modules:*

$$\mathrm{Hom}_{H_J}(V, \mathrm{Res}_J^S(V')) \cong_A \mathrm{Hom}_H(\mathrm{Ind}_J^S(V), V'). \tag{a}$$

Moreover, if each a_s $(s \in S)$ is invertible in A, then we also have an isomorphism

$$\mathrm{Hom}_{H_J}(\mathrm{Res}_J^S(V'), V) \cong_A \mathrm{Hom}_H(V', \mathrm{Ind}_J^S(V)). \tag{b}$$

Proof. (a) Consider H as an (H_J, H)-module via left multiplication by H_J and right multiplication by H. Then (9.1.6) yields

$$\mathrm{Hom}_{H_J}(V, \mathrm{Hom}_H(H, V')) \cong_A \mathrm{Hom}_H(V \otimes_{H_J} H, V').$$

Now it remains to observe that, by definition, we have $V \otimes_{H_J} H = \mathrm{Ind}_J^S(V)$ and, on the other hand, the map $\mathrm{Hom}_H(H, V') \to V'$, $\varphi \mapsto \varphi(1_H)$, is an isomorphism of H_J-modules (see Exercise 7.3).

(b) Consider H as an (H, H_J)-bimodule via left multiplication by H and right multiplication by H_J. Now (9.1.6) yields

$$\mathrm{Hom}_H(V', \mathrm{Hom}_{H_J}(H, V)) \cong_A \mathrm{Hom}_{H_J}(V' \otimes_H H, V).$$

It remains to identify the terms involved in this isomorphism. First, the map $\mathrm{Res}_J^S(V') \to V' \otimes_H H$, $v' \mapsto v' \otimes 1$, certainly is an isomorphism of H_J-modules. On the other hand, since each a_s is invertible in A, the algebra H is symmetric, with a symmetrizing form as in Proposition 8.1.1. We claim that we have an isomorphism of H-modules:

$$\Phi\colon \mathrm{Hom}_{H_J}(H, V) \to \mathrm{Ind}_J^S(V), \quad \varphi \mapsto \sum_{x \in X_J} \varphi(T_x^\vee) \otimes T_x,$$

where X_J is the set of distinguished right coset representatives of W_J in W. To prove this, we proceed as follows. Using the fact that X_J^{-1} is the set of distinguished *left* coset representatives of W_J in W (see Remark 2.1.6) and taking into account the description of the dual basis elements in Proposition 8.1.1, we also have a direct sum decomposition of right H_J-modules:

$$H = \bigoplus_{x \in X_J} T_x^\vee H_J.$$

Using this and the isomorphism $\mathrm{Hom}_{H_J}(H_J, V) \cong V$, we conclude that, given any family of elements $\{v_x \mid x \in X_J\} \subseteq V$, there exists a unique $\varphi \in \mathrm{Hom}_{H_J}(H, V)$ such that $\varphi(T_x^\vee) = v_x$ for all $x \in X_J$. Since every element of $\mathrm{Ind}_J^S(V)$ can be written uniquely in the form $\sum_{x \in X_J} v_x \otimes T_x$ (with $v_x \in V$), we now see that Φ is an isomorphism of A-modules. It remains to show that Φ is also compatible with the action of H. For this purpose, it is enough to consider the action of the generators T_s ($s \in S$). Using Lemma 9.1.4, we have, for any $\varphi \in \mathrm{Hom}_{H_J}(H, V)$,

$$\begin{aligned}
\Phi(\varphi).T_s &= \sum_{x \in X_J} \varphi(T_x^\vee) \otimes (T_x T_s) = \sum_{x \in X_J} \varphi(T_x^\vee) \otimes \Big(\sum_{y \in X_J} h^s(x,y) T_y \Big) \\
&= \sum_{x,y \in X_J} \varphi(T_x^\vee) h^s(x,y) \otimes T_y = \sum_{x,y \in X_J} \varphi(T_x^\vee h^s(x,y)) \otimes T_y \\
&= \sum_{y \in X_J} \varphi\Big(\sum_{x \in X_J} T_x^\vee h^s(x,y) \Big) \otimes T_y = \sum_{y \in X_J} \varphi(T_s T_y^\vee) \otimes T_y = \Phi(\varphi.T_s),
\end{aligned}$$

i.e., Φ commutes with the action of H, as claimed. □

The Mackey decomposition of X_J in Section 2.1 implies the following:

9.1.8. Proposition (Mackey formula). *Let $J, K \subseteq S$ be subsets and V be an H_J-module. Let X_{JK} be the set of distinguished double coset representatives for the subgroups $W_J, W_K \subseteq W$ (see Proposition 2.1.7). Then, for each $d \in X_{JK}$, the subalgebra $H_{J^d \cap K}$ acts on $V \otimes T_d \subseteq \mathrm{Ind}_J^S(V)$ by right multiplication:*

$$(v \otimes T_d) T_w = v T_{dwd^{-1}} \otimes T_d \qquad \textit{where } w \in W_{J^d \cap K} \textit{ and } v \in V.$$

Moreover, we have an isomorphism of H_K-modules

$$\mathrm{Res}_K^S \circ \mathrm{Ind}_J^S(V) \cong_{H_K} \bigoplus_{d \in X_{JK}} \mathrm{Ind}_{J^d \cap K}^K (V \otimes T_d).$$

Proof. By definition, we have $X_{JK} = X_J \cap X_K^{-1}$. Let us first check that $V \otimes T_d$ indeed is invariant under right multiplication with elements from $H_{J^d \cap K}$, and that the action is given as above. Let $w \in W_{J^d \cap K}$. Since $d \in X_K^{-1}$ is a distinguished left coset representative for W_K in W (see Remark 2.1.6), we have $l(d) + l(w) =$

$l(dw)$ which implies that $T_dT_w = T_{dw}$. On the other hand, $dwd^{-1} \in W_J$ and so $l(dwd^{-1}) + l(d) = l((dwd^{-1})d) = l(dw)$, since $d \in X_J$. It follows that $T_{dwd^{-1}}T_d = T_{dw}$ and hence $T_dT_w = T_{dwd^{-1}}T_d$, as desired.

Now recall from Lemma 2.1.9 that we have the following Mackey decomposition: $X_J = \coprod_{d \in X_{JK}} d \cdot X^K_{J^d \cap K}$, where the explicit dot means that the lengths of elements are added. So we can write each $x \in X_J$ uniquely in the form $x = dy$ where $d \in X_{JK}$ and $y \in X^K_{J^d \cap K}$ such that $l(x) = l(d) + l(y)$. This implies a direct sum decomposition of A-modules:

$$\mathrm{Ind}_J^S(V) = \bigoplus_{x \in X_J} V \otimes T_x = \bigoplus_{d \in X_{JK}} V_d \qquad \text{where} \qquad V_d := \bigoplus_{y \in X^K_{J^d \cap K}} (V \otimes T_d)T_y.$$

We have checked above that $V \otimes T_d$ is invariant under right multiplication with $H_{J^d \cap K}$. Moreover, V_d is certainly free as an A-module and the map

$$\mathrm{Ind}^K_{J^d \cap K}(V \otimes T_d) \to V_d, \qquad (v \otimes T_d) \otimes T_y \mapsto v \otimes T_dT_y,$$

is an isomorphism which is compatible with the action of H_K. □

9.1.9. Induction and restriction for generic Iwahori–Hecke algebras. Let H be a generic Iwahori–Hecke algebra as in Section 8.1. We fix a non-zero element $q \in \mathbb{C}$ and place ourselves in a setting as in (8.1.6). Then we have a map $\chi \mapsto \chi_q$ which takes a character χ of $\mathbf{K}H$ to a character χ_q of H_q, which is compatible with integral linear combinations of characters.

Now let $J \subseteq S$ and consider the induction and restriction of characters between $\mathbf{K}H_J$ and $\mathbf{K}H$. The above compatibility and (9.1.5)(c) certainly imply that

$$\mathrm{Res}_J^S(\chi)_q = \mathrm{Res}_J^S(\chi_q) \qquad \text{for all characters } \chi \text{ of } \mathbf{K}H, \tag{a}$$

$$\mathrm{Ind}_J^S(\psi)_q = \mathrm{Ind}_J^S(\psi_q) \qquad \text{for all characters } \psi \text{ of } \mathbf{K}H_J. \tag{b}$$

Now assume that $q = 1$, so that we have a bijection $\mathrm{Irr}(\mathbf{K}H) \leftrightarrow \mathrm{Irr}(W)$, $\chi \leftrightarrow \chi_1$, as in Theorem 8.1.7. For any $\psi \in \mathrm{Irr}(\mathbf{K}H_J)$ and $\chi \in \mathrm{Irr}(\mathbf{K}H)$, denote by $m(\chi, \psi)$ the multiplicity of ψ in the restriction of χ to $\mathbf{K}H_J$. Thus, we have

$$\mathrm{Res}_J^S(\chi) = \sum_{\psi \in \mathrm{Irr}(\mathbf{K}H_J)} m(\chi, \psi)\, \psi. \tag{c}$$

The above compatibility results immediately imply that we have the same relations after specialization:

$$\mathrm{Res}_J^S(\chi_1) = \sum_{\psi \in \mathrm{Irr}(\mathbf{K}H_J)} m(\chi, \psi)\, \psi_1. \tag{c'}$$

Finally, by Frobenius reciprocity, we also have

$$m(\chi, \psi) = \text{ multiplicity of } \chi \text{ in } \mathrm{Ind}_J^S(\psi). \tag{d}$$

Thus, the induce/restrict matrices for $\mathbf{K}H$ are the same as those for W.

9.1.10. Corollary. *In the above set-up, assume that* $\chi \in \mathrm{Irr}(\mathbf{K}\mathbf{H})$ *is such that the specialized character* $\chi_1 \in \mathrm{Irr}(W)$ *is of parabolic type with respect to* $J \subseteq S$*; see Definition 6.3.1. Then* χ *occurs with multiplicity* 1 *in the induced character* $\mathrm{Ind}_J^S(\mathrm{ind}_J)$*, where* ind_J *is the index character of* $\mathbf{K}\mathbf{H}_J$.

Proof. Since $(\mathrm{ind}_J)_1$ is the trivial character of W_J, this is clear by (9.1.9). □

9.2 CHARACTER VALUES AND THE LONGEST ELEMENT

Let $\mathbf{H}$ be a generic Iwahori–Hecke algebra associated with a finite Coxeter group W, as in Definition 8.1.4. We let $\mathbf{K}$ be a sufficiently large finite Galois extension of $\mathbb{C}(u_s \mid s \in S)$, so that $\mathbf{K}\mathbf{H}$ is a split semisimple algebra and we have a bijection between $\mathrm{Irr}(\mathbf{K}\mathbf{H})$ and $\mathrm{Irr}(W)$, as in Theorem 8.1.7. In general, we cannot expect that a character value at T_w is completely determined by the specialized character value at w. But it turns out that this is the case for special choices of w. We place ourselves in the set-up of (8.1.6) where $\theta\colon \mathbf{A} \to \mathbb{C}$ is a specialization such that $\theta(u_s) = 1$ for all $s \in S$. Recall that $\mathbf{A}_\mathbb{C} = \mathbb{C}[u_s^{\pm 1} \mid s \in S] \subseteq \mathbf{K}$ and that we denote by $\mathbf{A}_\mathbb{C}^*$ the integral closure of $\mathbf{A}_\mathbb{C}$ in $\mathbf{K}$. We fix a corresponding extension $\theta_\mathbb{C}^*\colon \mathbf{A}_\mathbb{C}^* \to \mathbb{C}$, which induces a bijection

$$\mathrm{Irr}(\mathbf{K}\mathbf{H}) \to \mathrm{Irr}(W), \qquad \chi \mapsto \chi_1,$$

where $\chi_1(w) = \theta_\mathbb{C}^*(\chi(T_w))$ for $w \in W$; see Theorem 8.1.7.

9.2.1. Lemma. *For* $\chi \in \mathrm{Irr}(\mathbf{K}\mathbf{H})$ *and* $s \in S$ *we have* $\chi(T_s) = h_s u_s - k_s$ *where*

$$h_s = \frac{1}{2}(\chi_1(1) + \chi_1(s)) \quad \textit{and} \quad k_s = \frac{1}{2}(\chi_1(1) - \chi_1(s)).$$

Proof. The quadratic relation $T_s^2 = u_s T_1 + (u_s - 1)T_s$ also holds in a representation affording χ. Hence the only possible eigenvalues of T_s in that representation are u_s (with multiplicity h_s, say) and -1 (with multiplicity k_s, say). Then $\chi(T_s) = h_s u_s - k_s$ and $\chi(T_1) = h_s + k_s$. Specialization via $\theta_\mathbb{C}^*$ also yields

$$\chi_1(s) = \theta_\mathbb{C}^*(\chi(T_s)) = \theta_\mathbb{C}^*(h_s u_s - k_s) = \theta(h_s u_s - k_s) = h_s - k_s.$$

Solving these equations yields the above expressions for h_s and k_s. □

We say that $s, t \in S$ are *equivalent* if s, t are conjugate in W. Let $S' \subseteq S$ be a set of representatives for the corresponding equivalence classes of S. Then $\mathbf{A} = \mathbb{Z}[u_s^{\pm 1} \mid s \in S']$. Now let $w_0 \in W$ be the longest element and take a reduced expression $w_0 = s_1 \cdots s_N$ with $s_i \in S$. For $s \in S'$ let N_s be the number of $i \in \{1, \ldots, N\}$ such that s, s_i are conjugate in W. By Exercise 1.15, the numbers N_s are independent of the choice of a reduced expression for w_0.

9.2.2. Theorem (Springer). *Let* $\chi \in \mathrm{Irr}(\mathbf{K}\mathbf{H})$*. The element* $T_{w_0}^2$ *is central in* $\mathbf{H}$ *and it acts on a simple module affording* χ *by the scalar*

$$z_\chi := \prod_{s \in S'} u_s^{f_s}, \quad \textit{where} \quad f_s := N_s\left(1 + \frac{\chi_1(s)}{\chi_1(1)}\right) \in \mathbb{Z}.$$

(See also Proposition 9.4.3 for an alternative description of z_χ.)

Proof. By Definition 4.4.1, the map $w \mapsto T_w$ ($w \in W$) is an algebra homomorphism from the monoid algebra of the braid monoid $B^+ = B^+(W, S)$ to $\mathbf{H}$. Then $T_{w_0}^2$ is the image of Δ^2, where $\Delta := w_0$ is the fundamental element in B^+. We conclude that $T_{w_0}^2$ is central in $\mathbf{H}$, since Δ^2 is central in B^+ by Lemma 4.1.9.

Now let $\chi \in \mathrm{Irr}(K\mathbf{H})$ and $\delta\colon K\mathbf{H} \to M_m(K)$ be a representation affording χ. Since $T_{w_0}^2$ is central, $\delta(T_{w_0}^2)$ is a scalar multiple of the identity matrix (by Schur's lemma). Let $z_\chi \in K$ denote this scalar, so that we have $\det(\delta(T_{w_0}^2)) = z_\chi^m$. Taking a reduced expression $w_0 = s_1 \cdots s_N$ ($s_i \in S$), we can write

$$z_\chi^m = \det(\delta(T_{w_0}^2)) = \det(\delta(T_{s_1}))^2 \cdots \det(\delta(T_{s_N}))^2.$$

Thus, in order to compute z_χ, we are reduced to computing the determinants of $\delta(T_s)$ for $s \in S$ and evaluating the product over $s_1, \ldots, s_N$.

Now, if $s, t \in S$ are conjugate in W then the corresponding basis elements T_s and T_t are also conjugate in $\mathbf{H}$. Indeed, by Exercise 1.10, it is sufficient to consider the case where $m_{st} \geq 3$ is odd. In this case, the braid relation reads

$$T_s \,\mathrm{Prod}(T_t, T_s; m_{st} - 1) = \mathrm{Prod}(T_t, T_s; m_{st} - 1) T_t,$$

and so $\mathrm{Prod}(T_t, T_s; m_{st} - 1)$ conjugates T_s to T_t. We now obtain

$$z_\chi^m = \det(\delta(T_{s_1}))^2 \cdots \det(\delta(T_{s_N}))^2 = \prod_{s \in S'} (\det(\delta(T_s))^{2N_s}.$$

By definition of h_s, k_s, the matrix $\delta(T_s)$ has precisely h_s eigenvalues u_s and k_s eigenvalues -1, and so $\det(\delta(T_s)) = (-1)^{k_s} u_s^{h_s}$. This yields

$$z_\chi^m = \prod_{s \in S'} \Big((-1)^{k_s} u_s^{h_s}\Big)^{2N_s} = \prod_{s \in S'} u_s^{2N_s h_s} = \prod_{s \in S'} u_s^{m f_s},$$

where the last equality follows from the formula for h_s in Lemma 9.2.1 and the defining formula for f_s. Now N_s is the number of $i \in \{1, \ldots, N\}$ such that s_i is conjugate to s. By Exercise 1.15, N_s in fact equals the size of the conjugacy class of s in W. By a standard result on irreducible characters of finite groups (see [Isaacs 1976, Theorem 3.7]), the expression $N_s \chi_1(s)/m$ is an algebraic integer. Since it is also a rational number, we have $N_s \chi_1(s)/m \in \mathbb{Z}$ and, hence,

$$f_s = N_s\Big(1 + \chi_1(s)/m\Big) = N_s + N_s \chi_1(s)/m \quad \in \mathbb{Z}.$$

Now we can write

$$z_\chi^m = \prod_{s \in S'} u_s^{m f_s} = \Big(\prod_{s \in S'} u_s^{f_s}\Big)^m, \quad \text{and so} \quad z_\chi = \zeta \prod_{s \in S'} u_s^{f_s},$$

where $\zeta \in \mathbb{C}$ is an mth root of unity. It remains to show that $\zeta = 1$. For this purpose, we assume (as we may) that δ has the property that $\delta(T_w) \in M_m(\mathcal{O})$

for all $w \in W$, where $\mathcal{O} \subseteq \mathbf{K}$ is a valuation ring as in (8.1.6). Under the canonical map from $\mathcal{O}$ onto its residue field, the element $T_{w_0}^2$ specializes to the element $w_0^2 = 1 \in W$, and hence the corresponding matrix $\delta(T_{w_0}^2)$ specializes to the identity matrix. Hence we have $1 = \theta_C^*(z_\chi) = \theta_C^*(\zeta) = \zeta$, as desired. □

9.2.3. Example. Assume that all parameters are equal; we write $u := u_s$ for $s \in S$. Then the constant z_χ in Theorem 9.2.2 is given by

$$z_\chi = u^{f_\chi} \qquad \text{where } f_\chi = \sum_{s \in S'} N_s\left(1 + \frac{\chi_1(s)}{\chi_1(1)}\right).$$

Given the character table of W, the constants f_χ are easily computed.

(a) Assume that (W, S) is irreducible and of exceptional type $I_2(m)$ $(m \geq 5)$, H_3, H_4, F_4, E_6, E_7 or E_8. Using the tables in Appendix C, we find the following complete list of cases where f_χ is odd:

Type	Characters	Respective values for f_χ
H_3	$4_r'$, 4_r	15, 15
H_4	16_{rr}, 16_r, $16_r'$, $16_{rr}'$	75, 75, 45, 45
E_7	$512_a'$, 512_a	63, 63
E_8	4096_z, 4096_x, $4096_x'$, $4096_z'$	135, 135, 105, 105

Note that, in these examples, w_0 is central (see Exercise 1.13). The above list of characters is precisely the list of *exceptional characters* in [Gyoja 1984, Def. 2.2]. (Also compare with the list in (6.3.6)(b).) We will see in Example 9.2.9(a) below that then the character value $\chi(T_{w_0})$ involves a square root of u. By the results of Section 9.3, these will turn out to be the only cases where this can happen.

(b) Assume that (W, S) is irreducible. For $0 \leq d \leq |S|$, let $\chi^{(d)} \in \mathrm{Irr}(\mathbf{K}H)$ be the character afforded by the dth exterior power of the reflection representation, as constructed in Remark 8.1.12(b). By Exercise 8.8, we have $\chi^{(d)}(1) = \binom{|S|}{d}$ and $\chi^{(d)}(s) = \binom{|S|-1}{d} - \binom{|S|-1}{d-1}$ for all $s \in S$. Hence we obtain in this case

$$f_{\chi^{(d)}} = 2l(w_0)\frac{|S| - d}{|S|} = h(|S| - d) \qquad \text{for } 0 \leq d \leq |S|,$$

where h denotes the Coxeter number of W and where we use Theorem 1.5.8.

In order to obtain character formulas on basis elements T_w for more general choices of w, we will use the results on "good" representatives in the classes of W from Chapter 4. On the level of $\mathbf{H}$, these results are formulated as follows:

9.2.4. Remark. Let C be a conjugacy class and $C_{\min}$ the set of elements of minimal length in C. Assume that the elements in C have order d, say. Consider the associated *braid monoid* $B^+(W, S)$ and recall that we have a map $W \to B^+(W, S)$, $w \mapsto \mathbf{w}$. By Theorem 4.3.5, there exists some $w \in C_{\min}$ such that

$$\boldsymbol{w}^{d} = \boldsymbol{w}_{J_1}^{d_1} \cdots \boldsymbol{w}_{J_r}^{d_r} \qquad (\text{in } B^+(W,S)) \tag{a}$$

where $J_1 \supset \cdots \supset J_r$ is a decreasing chain of subsets of S, and the d_i are positive even integers. Here, we denote by w_J the longest element in the parabolic subgroup $W_J \subseteq W$. Since, by Definition 4.4.1, the assignment $\boldsymbol{w} \mapsto T_w$ ($w \in W$) induces an $\mathbf{A}$-algebra homomorphism $\mathbf{A}[B^+(W,S)] \to \mathbf{H}$, we also obtain

$$T_w^d = T_{w_{J_1}}^{d_1} \cdots T_{w_{J_r}}^{d_r} \qquad (\text{in } \mathbf{H}). \tag{b}$$

Now note that for each subset $J \subseteq S$, the element $T_{w_J}^2$ is central in the parabolic subalgebra $\mathbf{H}_J$ (see Theorem 9.2.2). Hence, as in Remark 4.3.2(b), we see that all terms in the above product commute with each other.

9.2.5. Theorem (Geck–Michel). *Let $\delta\colon \mathbf{K}\mathbf{H} \to M_m(\mathbf{K})$ be any representation of $\mathbf{K}\mathbf{H}$, and $w \in C_{\min}$ for some conjugacy class C of W. Let d be the order of w. Then $\delta(T_w)^d$ is diagonalizable and the eigenvalues are monomials in the indeterminates $\{u_s\}_{s\in S'}$. Consequently, for $\mathbf{K}$ large enough, the eigenvalues of $\delta(T_w)$ are dth roots of unity times fractional powers of these monomials in $\mathbf{K}$.*

Proof. First we consider the special case where $w = w_J$ for some subset $J \subseteq S$. We show that $\delta(T_{w_J}^2)$ is diagonalizable and the eigenvalues are monomials in $\{u_s \mid s \in J \cap S'\}$. To see this, consider the restriction of δ to the corresponding parabolic subalgebra $\mathbf{K}\mathbf{H}_J$. This restriction splits up into a direct sum of irreducible constituents, and Theorem 9.2.2 implies that $T_{w_J}^2$ is represented by a scalar matrix in each such irreducible constituent. This implies, in particular, that $\delta(T_{w_J}^2)$ is diagonalizable with eigenvalues of the desired form.

Now consider any class C and $w \in C_{\min}$. By case 1 in (8.2.2), all basis elements T_x ($x \in C_{\min}$) are conjugate by units in $\mathbf{K}\mathbf{H}$. In particular, the matrices $\delta(T_x)$ ($x \in C_{\min}$) have the same eigenvalues (counting multiplicities). Moreover, either all or none of them are diagonalizable. We can therefore assume from the outset that our given element $w \in C_{\min}$ is chosen as in Remark 9.2.4, so that

$$T_w^d = T_{w_{J_1}}^{d_1} \cdots T_{w_{J_r}}^{d_r}, \qquad \text{where } J_1 \supset \cdots \supset J_r \text{ and the } d_i \text{ are even.}$$

Now all terms in the product commute with each other. It follows that if we apply our representation δ to the above relation, then the representing matrices on these terms also commute with each other. We have seen before that a matrix $\delta(T_{w_J}^2)$ is always diagonalizable. Hence so are all the matrices $\delta(T_{w_{J_i}}^{d_i})$, since the d_i are even. Moreover, the eigenvalues are monomials in the parameters of $\mathbf{K}\mathbf{H}_{J_i}$, i.e., in $\{u_s \mid s \in J_i \cap S'\}$. It remains to use the fact that the product of a collection of pairwise commuting and individually diagonalizable matrices is again diagonalizable, and the eigenvalues are the products of the respective eigenvalues of the individual matrices. □

The above proof shows in fact how to obtain the exact shape of the absolute values of the eigenvalues. This will be made more explicit in Chapter 11 where we systematically use these methods to obtain information about the character values of Iwahori–Hecke algebras of exceptional type. The above result already has the following theoretical consequence.

9.2.6. Corollary. *Let $e \geq 1$ be the exponent of W, i.e., e is minimal with the property that $w^e = 1$ for all $w \in W$. Assume that K is large enough so that, for each $s \in S'$, there exists some $v_s \in K$ with $v_s^e = u_s$. Then we have*

$$\chi(T_w) \in \mathbb{Z}[\zeta, v_s \mid s \in S'] \qquad \textit{for all } \chi \in \mathrm{Irr}(KH) \textit{ and } w \in W,$$

where $\zeta \in \mathbb{C}$ is an eth root of unity.

Proof. Let $\chi \in \mathrm{Irr}(KH)$. First assume that $w \in W$ has minimal length in some conjugacy class of W. Let $d \geq 1$ be the order of w. By Theorem 9.2.5, any eigenvalue of T_w in a representation affording χ is a dth root of unity times a monomial in v_s ($s \in S'$). Hence we have $\chi(T_w) \in \mathbb{Z}[\zeta, v_s \mid s \in S']$, as desired.

Now consider any $w \in W$. Using the class polynomials in Corollary 8.2.6, we have $\chi(T_w) = \sum_C f_{w,C}\chi(T_{w_C})$, where $w_C \in C_{\min}$. Since $f_{w,C} \in \mathbb{Z}[u_s \mid s \in S'] \subseteq \mathbb{Z}[\zeta, v_s \mid s \in S']$, this implies again the desired result about $\chi(T_w)$. □

9.2.7. Normalization of $\theta_{\mathbb{C}}^*$. Let $e \geq 1$ and $\{v_s \mid s \in S'\}$ be as in Corollary 9.2.6. Then the specialization $\theta_{\mathbb{C}}^*\colon A_{\mathbb{C}}^* \to \mathbb{C}$ in (8.1.6) can be chosen such that

$$\theta_{\mathbb{C}}^*(v_s) = 1 \qquad \text{for all } s \in S'.$$

This can be seen as follows. Recall that we started with a specialization $\theta\colon A \to \mathbb{C}$ such that $\theta(u_s) = 1$ for all $s \in S'$, and we then constructed $\theta_{\mathbb{C}}^*$ using a suitable valuation ring in K. Now let $A_{\mathbb{C}}^e := \mathbb{C}[v_s \mid s \in S'] \subseteq K$. Every v_s is integral over A; hence we certainly have $A_{\mathbb{C}}^e \subseteq A_{\mathbb{C}}^*$. Since $A_{\mathbb{C}}^e$ is still a polynomial ring, we can extend θ to a specialization $\theta_{\mathbb{C}}^e\colon A_{\mathbb{C}}^e \to \mathbb{C}$ such that $\theta_{\mathbb{C}}^e(v_s) = 1$ for all $s \in S'$. Then we choose a valuation ring $\mathcal{O} \subseteq K$ such that $A_{\mathbb{C}}^e \subseteq \mathcal{O}$ and $J(\mathcal{O}) \cap A_{\mathbb{C}}^e = \ker(\theta_{\mathbb{C}}^e)$. Arguing as in (8.1.6), we see that the restriction to $A_{\mathbb{C}}^*$ of the canonical map from $\mathcal{O}$ to its residue field is a specialization as required.

Fixing $\theta_{\mathbb{C}}^*$ in this way, we now obtain explicit formulas for certain character values, using the bijection $\mathrm{Irr}(KH) \leftrightarrow \mathrm{Irr}(W)$, $\chi \leftrightarrow \chi_1$, of Theorem 8.1.7.

9.2.8. Proposition (Broué–Michel). *Assume that $w \in W$ has order $d \geq 1$ and satisfies $T_w^d = T_{w_0}^{2r}$ for some $r \geq 1$. Let $f_s \in \mathbb{Z}$ ($s \in S'$) be as in Theorem 9.2.2. Then we have*

$$\chi(T_w) = \chi_1(w) \prod_{s \in S'} v_s^{f_s r e/d} \qquad \textit{for all } \chi \in \mathrm{Irr}(KH).$$

In particular, we have $\chi(T_w) = 0$ if $\chi_1(w) = 0$.

Proof. Let $\lambda_1, \ldots, \lambda_m \in A_C^*$ be the eigenvalues of T_w in a representation affording χ. Then $\lambda_1^d, \ldots, \lambda_m^d$ are the eigenvalues of $T_{w_0}^{2r}$. Now recall from Theorem 9.2.2 that $T_{w_0}^2$ acts by the scalar $\prod_{s \in S'} v_s^{ef_s}$. Hence we have

$$\lambda_i^d = \prod_{s \in S'} v_s^{f_s re} = \alpha^d \quad \text{for all } i, \quad \text{where } \alpha := \prod_{s \in S'} v_s^{f_s re/d}.$$

It follows that there exist dth roots of unity $\zeta_i \in \mathbb{C}$ such that $\lambda_i = \zeta_i \alpha$ for all i. Thus, we have $\chi(T_w) = (\sum_i \zeta_i)\alpha$. But, specialization via θ_C^* yields

$$\chi_1(w) = \theta_C^*(\chi(T_w)) = \sum_i \theta_C^*(\lambda_i) = \sum_i \zeta_i \theta_C^*(\alpha) = \sum_i \zeta_i,$$

whence the assertion. Note that the last equality holds by (9.2.7). □

9.2.9. Example. (a) Let $w_0 \in W$ be the longest element. Then w_0 has order 2 and the assumptions of Proposition 9.2.8 are trivially satisfied. So we have

$$\chi(T_{w_0}) = \chi_1(w_0) \prod_{s \in S'} \sqrt{u_s}^{f_s} \qquad \text{for all } \chi \in \mathrm{Irr}(\mathbf{K}H),$$

where $\sqrt{u_s} \in \mathbf{K}$ are square roots of u_s with $\theta_C^*(\sqrt{u_s}) = 1$, for all $s \in S'$.

(b) Let $w_c \in W$ be a *Coxeter element*, i.e., w_c is the product of the generators in S in any given order. By Proposition 4.3.4, we have $\mathbf{w}_c^h = \mathbf{w}_0^2$ in the braid monoid, where $h \geq 1$ is the order of w_c. Hence we also have $T_{w_c}^h = T_{w_0}^2$ in H, and Proposition 9.2.8 can be applied. In particular, it follows that

$$\chi(T_{w_c}) = 0 \quad \text{for any } \chi \in \mathrm{Irr}(\mathbf{K}H) \text{ with } \chi_1(w_c) = 0.$$

Assume now that (W, S) is irreducible and that all parameters of $\mathbf{K}H$ are equal to u. Let $\chi^{(d)}$ be the character of the dth exterior power of the reflection representation of $\mathbf{K}H$. Then we find, using Example 9.2.3:

$$\chi^{(d)}(T_{w_c}) = \chi^{(d)}(w_c) u^{|S|-d} \qquad \text{for } 0 \leq d \leq |S|.$$

In type A, the above formula was proved in a different way in Example 8.1.13.

We are now in a position to determine the character tables of the generic Iwahori–Hecke algebras of type H_3 and A_{n-1}, in the sense of Definition 8.2.9.

9.2.10. The character table for type H_3. Let (W, S) be of type H_3, where $S = \{s_1, s_2, s_3\}$ is such that $s_1 s_2$ has order 5 and $s_2 s_3$ has order 3. All s_i are conjugate and we write $u = u_{s_i}$ for $i = 1, 2, 3$. By Appendix B, we have the following representatives of minimal length in the classes of W (where we simply write 1 for s_1, 12 for $s_1 s_2$ etc.):

$w_C \in C_{min}$	1_W	1	12	13	23	123	1212	12123	121232123	w_0
order d	1	2	5	2	3	10	5	6	10	2
$\mathbf{w}_C^d$						$\mathbf{w}_0^2$		$\mathbf{w}_0^2$	$\mathbf{w}_0^6$	$\mathbf{w}_0^2$

Thus, each w_C either lies in a proper parabolic subgroup (and the character values on T_{w_C} are determined using restriction to a suitable parabolic subalgebra), or Proposition 9.2.8 applies. Hence we can easily determine the whole character table of $\mathbf{H}$; see Exercise 9.7 for details. The result is printed in Table 9.1, where the irreducible characters are denoted as in Table C.1. (Table 9.1 can also be computed using the matrix representations constructed in (11.1.8).)

Table 9.1 The character table of the generic Iwahori–Hecke algebra of type H_3

χ	1	12	13	23	123	1212	12123	121232123	w_0
$1_r'$	-1	1	1	1	-1	1	-1	-1	-1
1_r	u	u^2	u^2	u^2	u^3	u^4	u^5	u^9	u^{15}
$5_r'$	$2u-3$	$-u+1$	u^2-2u+2	$-2u+1$	.	$-u^2+1$	u^2	.	$-5u^6$
5_r	$3u-2$	u^2-u	$2u^2-2u+1$	u^2-2u	.	u^4-u^2	$-u^3$	.	$5u^9$
3_s	$u-2$	$\alpha u+1$	$-2u+1$	$-u+1$	$-\alpha u$	$\overline{\alpha}u^2+1$	.	$-\overline{\alpha}u^3$	$3u^5$
$\overline{3}_s$	$u-2$	$\overline{\alpha}u+1$	$-2u+1$	$-u+1$	$-\overline{\alpha}u$	αu^2+1	.	$-\alpha u^3$	$3u^5$
$3_s'$	$2u-1$	$u^2+\alpha u$	u^2-2u	u^2-u	αu^2	$u^4+\overline{\alpha}u^2$	.	$\overline{\alpha}u^6$	$-3u^{10}$
$\overline{3}_s'$	$2u-1$	$u^2+\overline{\alpha}u$	u^2-2u	u^2-u	$\overline{\alpha}u^2$	$u^4+\alpha u^2$	.	αu^6	$-3u^{10}$
$4_r'$	$2u-2$	$-u$	u^2-2u+1	u^2-u+1	$\sqrt{u}^3$	$-u^2$	$-\sqrt{u}^5$	$\sqrt{u}^9$	$-4\sqrt{u}^{15}$
4_r	$2u-2$	$-u$	u^2-2u+1	u^2-u+1	$-\sqrt{u}^3$	$-u^2$	$\sqrt{u}^5$	$-\sqrt{u}^9$	$4\sqrt{u}^{15}$

where $\alpha=(-1+\sqrt{5})/2$, $\overline{\alpha}=(-1-\sqrt{5})/2$ and the missing entries are 0.

Now let $W=\mathfrak{S}_n$ with generators $s_i=(i,i+1)$ for $1\leq i\leq n-1$. All s_i are conjugate and we write $u=u_{s_i}$ for all i. The following rule shows how the character table of $\mathbf{H}$ can be obtained from that of $\mathfrak{S}_n$. For any partition $\lambda\vdash n$, we have a corresponding Young subgroup $\mathfrak{S}_\lambda\subseteq\mathfrak{S}_n$; see (5.4.3). Let $\rho_\lambda\colon\mathfrak{S}_\lambda\to\mathsf{GL}(V_\lambda)$ be the standard reflection representation of $\mathfrak{S}_\lambda$. For any $\nu\vdash n$, let C_ν be the conjugacy class containing the elements of cycle type ν, and let $w_\nu\in C_\nu$ be a representative of minimal length in that class. By Example 3.1.16, we can assume that w_ν is a Coxeter element in the Young subgroup $\mathfrak{S}_\nu$. With this notation, we can now state:

9.2.11. Theorem (Starkey's rule). *For any $\chi\in\mathsf{Irr}(K\mathbf{H})$ and $\lambda\vdash n$, we have*

$$\chi(T_{w_\lambda})=\sum_{\nu\vdash n}\chi_1(w_\nu)\,t_\lambda^\nu\quad\textit{where}\quad t_\lambda^\nu:=\frac{|C_\nu\cap\mathfrak{S}_\lambda|}{|\mathfrak{S}_\lambda|}\det(u\,\mathrm{id}_{V_\lambda}-\rho_\lambda(w_\nu)).$$

Proof. We first consider the case where $\lambda=[n]$, i.e., $w_c:=w_{[n]}$ is a Coxeter element in $\mathfrak{S}_n$ itself and $\rho:=\rho_{[n]}\colon\mathfrak{S}_n\to\mathsf{GL}(V)$ is the reflection representation of $\mathfrak{S}_n$. In this case, the above identity reads:

$$\chi(T_{w_c})=\frac{1}{|\mathfrak{S}_n|}\sum_{w\in\mathfrak{S}_n}\chi_1(w)\det(u\,\mathrm{id}_V-\rho(w)),\qquad\text{for all }\chi\in\mathsf{Irr}(K\mathbf{H}).\qquad(*)$$

To prove that $(*)$ holds, we evaluate both sides independently and compare the results. For this purpose, we will use the action of $\mathfrak{S}_n$ on the dth exterior power of V (for $0 \leq d \leq n-1$). Denote by $\chi^{(d)}$ the corresponding irreducible character of $\mathbf{K}\mathbf{H}$, as constructed in Remark 8.1.12(b). First consider the right-hand side of $(*)$. By Remark 5.1.6, we have

$$\det(u\,\mathrm{id}_V - \rho(w)) = \sum_{d=0}^{n-1} (-1)^d \chi_1^{(d)}(w)\, u^{n-1-d}.$$

Inserting this into the right-hand side of $(*)$ and using the orthogonality relations for the irreducible characters, we obtain that the right-hand side of $(*)$ equals

$$\begin{aligned}
&\frac{1}{|\mathfrak{S}_n|} \sum_{w \in \mathfrak{S}_n} \chi_1(w) \Big(\sum_{d=0}^{n-1} (-1)^d \chi_1^{(d)}(w)\, u^{n-1-d} \Big) \\
&= \sum_{d=0}^{n-1} (-1)^d u^{n-1-d} \Big(\frac{1}{|\mathfrak{S}_n|} \sum_{w \in \mathfrak{S}_n} \chi_1(w) \chi_1^{(d)}(w) \Big) \\
&= \begin{cases} (-1)^d u^{n-1-d} & \text{if } \chi = \chi^{(d)} \text{ for some } 0 \leq d \leq n-1, \\ 0 & \text{otherwise.} \end{cases}
\end{aligned}$$

which also equals the left-hand side, by Example 9.2.9(b) and Exercise 5.8.

Now consider the general case, where λ is any partition of n. We claim that $(*)$ extends to direct products of symmetric groups and, hence, to $\mathfrak{S}_\lambda$, i.e., if we denote by $\mathbf{H}_\lambda \subseteq \mathbf{H}$ the parabolic subalgebra corresponding to $\mathfrak{S}_\lambda$, we have

$$\psi(T_{w_\lambda}) = \frac{1}{|\mathfrak{S}_\lambda|} \sum_{w \in \mathfrak{S}_\lambda} \psi_1(w) \det(u\,\mathrm{id}_{V_\lambda} - \rho_\lambda(w)) \quad \text{for all } \psi \in \mathrm{Irr}(\mathbf{K}\mathbf{H}_\lambda). \quad (**)$$

Indeed, if λ has parts $\lambda_1, \ldots, \lambda_r > 0$, then $\mathfrak{S}_\lambda \cong \mathfrak{S}_{\lambda_1} \times \cdots \times \mathfrak{S}_{\lambda_r}$ and $w_\lambda \in \mathfrak{S}_\lambda$ can be written as a product of Coxeter elements of $\mathfrak{S}_{\lambda_1}, \ldots, \mathfrak{S}_{\lambda_r}$. Furthermore, ρ_λ is the direct sum of the reflection representations of $\mathfrak{S}_{\lambda_1}, \ldots, \mathfrak{S}_{\lambda_r}$. Consequently, if we consider the right-hand side of $(*)$ for each factor $\mathfrak{S}_{\lambda_i}$ and multiply all these expressions together, we obtain the right-hand side of $(**)$. A similar result also holds for the left-hand side, using Exercise 8.5. Thus, $(**)$ is proved.

Now we can argue as follows. For $\chi \in \mathrm{Irr}(\mathbf{K}\mathbf{H})$ and $\psi \in \mathrm{Irr}(\mathbf{K}\mathbf{H}_\lambda)$ denote by $m(\chi, \psi)$ the multiplicity of ψ in the restriction of χ to $\mathbf{K}\mathbf{H}_\lambda$. Using the compatibility with specializations in (9.1.9), we now obtain

$$\begin{aligned}
\chi(T_{w_\lambda}) &= \sum_{\psi \in \mathrm{Irr}(\mathbf{K}\mathbf{H}_\lambda)} m(\chi, \psi) \Big(\frac{1}{|\mathfrak{S}_\lambda|} \sum_{w \in \mathfrak{S}_\lambda} \psi_1(w) \det(u\,\mathrm{id}_{V_\lambda} - \rho_\lambda(w)) \Big) \\
&= \frac{1}{|\mathfrak{S}_\lambda|} \sum_{w \in \mathfrak{S}_\lambda} \Big(\sum_{\psi \in \mathrm{Irr}(\mathbf{K}\mathbf{H}_\lambda)} m(\chi, \psi)\, \psi_1(w) \Big) \det(u\,\mathrm{id}_{V_\lambda} - \rho_\lambda(w)),
\end{aligned}$$

where the sum in parentheses equals $\chi_1(w)$. Finally, note that if $w, w' \in \mathfrak{S}_\lambda$ are conjugate in $\mathfrak{S}_n$ then $\rho_\lambda(w)$ and $\rho_\lambda(w')$ are conjugate linear transformations. Hence $\det(u\,\mathrm{id}_{V_\lambda} - \rho_\lambda(w))$ only depends on the conjugacy class of w in $\mathfrak{S}_n$. □

The point about the above result is that the polynomials $t_\chi^\vee$ can be computed in a purely combinatorial way; see Exercise 9.6. Hence, we have an efficient way to determine the character table of the Iwahori–Hecke algebra of $\mathfrak{S}_n$, once we know the table for $\mathfrak{S}_n$ itself. We will come back to this in Chapter 10.

9.3 SPLITTING FIELDS FOR IWAHORI–HECKE ALGEBRAS

We keep the set-up of the previous section. We now turn to the problem of characterizing subfields of $\mathbf{K}$ over which all irreducible characters of $\mathbf{KH}$ can be realized. Recall that if $\mathbf{K}_0 \subseteq \mathbf{K}$ is a subfield and χ is a character of $\mathbf{KH}$, we say that χ can be *realized over* $\mathbf{K}_0$ if there exists a $\mathbf{K}_0\mathbf{H}$-module M such that $M \otimes_{\mathbf{K}_0} \mathbf{K}$ affords χ; in particular, all character values $\chi(T_w)$ lie in $\mathbf{K}_0$. By the general results in [Curtis and Reiner 1981, (7.15)] (for example), a subfield $\mathbf{K}_0 \subseteq \mathbf{K}$ is a *splitting field* for $\mathbf{H}$ if all $\chi \in \mathrm{Irr}(\mathbf{KH})$ can be realized over $\mathbf{K}_0$.

The results about type $I_2(m)$ in Corollary 8.3.2 and the computations in Example 9.2.3 already indicate that, in general, such a field will have to contain a splitting field for W and square roots of the parameters u_s $(s \in S)$. The aim of this section is to show that this is indeed sufficient. Our main tools are:

- the results on characters of parabolic type of W (see Section 6.3), together with the "commutation of induction and specialization" in (9.1.9);
- Corollary 9.2.6, which shows that each character value $\chi(T_w)$ lies in an extension of $\mathbb{C}(u_s \mid s \in S)$ which is generated by roots of u_s $(s \in S)$.

We assume throughout that $\mathbf{K}$ and the specialization $\theta_{\mathbb{C}}^*\colon \mathbf{A}_{\mathbb{C}}^* \to \mathbb{C}$ are chosen as in (9.2.7). In particular, $\mathbf{K}$ contains roots of the parameters u_s of sufficiently large degree. By Theorem 8.1.7, we have a bijection $\mathrm{Irr}(\mathbf{KH}) \leftrightarrow \mathrm{Irr}(W)$, $\chi \leftrightarrow \chi_1$.

A priori, if we have a subfield of $\mathbf{K}$ which contains all character values $\chi(T_w)$, then that subfield is not necessarily a splitting field for $\mathbf{H}$. (A general discussion of these matters, involving the Schur index, can be found in Chapter 9 of [Curtis and Reiner 1987].) But, in our case, we have the following result.

9.3.1. Proposition. *Let $\chi \in \mathrm{Irr}(\mathbf{KH})$ and $\mathbf{K}' \subseteq \mathbf{K}$ be a subfield with $u_s \in \mathbf{K}'$ $(s \in S)$ and such that the following conditions hold:*

(a) *$\mathbf{K}'$ contains a subfield $K' \subseteq \mathbb{Q}(\cos(2\pi/m_{st}) \mid s,t \in S)$ such that the specialized character $\chi_1 \in \mathrm{Irr}(W)$ can be realized over K'.*

(b) *We have $\chi(T_w) \in \mathbf{K}'$ for all $w \in W$.*

Then χ can be realized over $\mathbf{K}'$.

Proof. By Exercise 8.5, it is sufficient to assume that (W,S) is irreducible. Now assume that this is the case. We distinguish two cases.

Case 1. The specialized character χ_1 is of parabolic type. Then, by Corollary 9.1.10, χ occurs with multiplicity 1 in $\mathsf{Ind}_J^S(\mathsf{ind}_J)$, for some $J \subseteq S$. Since ind_J is certainly realized over $\mathbf{K}'$, the same holds by construction for $\mathsf{Ind}_J^S(\mathsf{ind}_J)$. The assertion now follows from Exercise 9.2.

Case 2. The specialized character χ_1 is not of parabolic type. Then, by (6.3.6)(a), (W, S) is of type H_4 and $\chi_1 = 48_{rr}$. In this case, we can argue as follows. First note that, by condition (a), we must have $\sqrt{5} \in \mathbf{K}'$. For, by Theorem 6.3.8, $\mathbb{Q}(\sqrt{5}) = \mathbb{Q}(\cos(2\pi/m_{st}) \mid s,t \in S)$ is a splitting field for W but 48_{rr} cannot be realized over $\mathbb{Q}$. Now we proceed as follows. Let $W_J \subset W$ be the parabolic subgroup of type H_3. By a computation using the ordinary character tables of W and W_J, we find that the restriction of 48_{rr} to W_J decomposes as $3_s + \overline{3}_s + 3_s' + \overline{3}_s' + 2 \cdot (4_r + 4_r' + 5_r + 5_r')$. Thus, by Frobenius reciprocity, we see that χ_1 occurs with multiplicity 1 in the character obtained by inducing the reflection representation of W_J to W. By (9.1.9), we have a similar statement for the induction of characters from $\mathbf{K}H_J$ to $\mathbf{K}H$. Hence, again by Exercise 9.2, it is enough to show that the reflection representation of $\mathbf{K}H_J$ can be realized over $\mathbf{K}'$. For this purpose, we choose the constants (c_{st}) needed in the construction of that representation as in Remark 8.1.12(a). All parameters are equal, say $u = u_s$, and hence $\sqrt{u_s u_t} = u$ already lies $\mathbf{K}'$. Furthermore, the elements $2\cos(2\pi/m_{st})$ certainly lie in $\mathbf{K}'$ for $m_{st} \in \{2, 3\}$. But, as we have seen above, this also holds for $m_{st} = 5$, in which case $2\cos(2\pi/5) = (\sqrt{5} - 1)/2$. □

The next two results show that the degree of the field in which all character values $\chi(T_w)$ lie is bounded in terms of information about the characters of W.

9.3.2. Lemma. *Let $\chi \in \mathsf{Irr}(\mathbf{K}H)$ and $K \subseteq \mathbb{C}$ be a finite Galois extension of $\mathbb{Q}$ such that $\chi_1(w) \in K$ for all $w \in W$. Then $\chi(T_w) \in K[v_s \mid s \in S']$ for all $w \in W$, with $\{v_s \mid s \in S'\}$ as in Corollary* 9.2.6.

Proof. By Corollary 9.2.6, we know that $\chi(T_w) \in \overline{\mathbb{Q}}[v_s \mid s \in S']$, where $\overline{\mathbb{Q}}$ denotes the field of all algebraic numbers in $\mathbb{C}$. We must show that the coefficients of $\chi(T_w)$ (as a polynomial in the v_s) already lie in the subfield K of $\overline{\mathbb{Q}}$. This will be done by a Galois-theoretic argument.

We begin with any field automorphism φ_1 of $\overline{\mathbb{Q}}$ which fixes K elementwise. Then we can extend φ_1 to an automorphism φ of $\overline{\mathbb{Q}}(v_s \mid s \in S')$ by setting $\varphi(v_s) = v_s$ for all $s \in S'$. We have the following commutation rule:

$$\varphi_1(\theta^*_{\overline{\mathbb{Q}}}(f)) = \theta^*_{\overline{\mathbb{Q}}}(\varphi(f)) \qquad \text{for all } f \in \overline{\mathbb{Q}}[v_s \mid s \in S'].$$

Now consider the bijection between $\mathsf{Irr}(\mathbf{K}H)$ and $\mathsf{Irr}(W)$ given by $\chi \mapsto \chi_1$, where $\chi_1(w) = \theta^*_{\overline{\mathbb{Q}}}(\chi(T_w))$ for $w \in W$. We apply this to χ and the character χ^φ defined by the formula $\chi^\varphi(T_w) = \varphi(\chi(T_w))$ for $w \in W$. Note that, by Proposition 9.3.1, χ can be realized over $\overline{\mathbb{Q}}(v_s \mid s \in S')$ and so Exercise 9.1 shows that $\chi^\varphi \in \mathsf{Irr}(\mathbf{K}H)$. We can apply the above commutation rule and obtain for any $w \in W$:

$$(\chi^\varphi)_1(w) = \theta^*_{\overline{\mathbb{Q}}}(\varphi(\chi(T_w))) = \varphi_1(\theta^*_{\overline{\mathbb{Q}}}(\chi(T_w))) = \varphi_1(\chi_1(w)) = \chi_1(w),$$

where the last equality holds by our assumption on K. Since $\chi \mapsto \chi_1$ is a bijection, we have $\chi^\varphi = \chi$ and so $\varphi(\chi(T_w)) = \chi(T_w)$ for all w. By the definition of φ, this implies that φ_1 fixes all coefficients of $\chi(T_w)$. Since this holds for all φ_1 as above and $K \supseteq \mathbb{Q}$ is a Galois extension, we conclude that $\chi(T_w) \in K[v_s \mid s \in S']$. □

9.3.3. Lemma. *Let $K' \subseteq K$ be a subfield such that $u_s \in K'$ for all $s \in S$. Let $\chi \in \mathrm{Irr}(KH)$ and set $K'(\chi) := K'(\chi(T_w) \mid w \in W) \subseteq K$. Furthermore, for each $J \subseteq S$, we assume given a subset $\mathcal{S}_J \subseteq \mathrm{Irr}(KH_J)$ such that each $\psi \in \mathcal{S}_J$ can be realized over K'. Then the degree $[K'(\chi) : K']$ is less than or equal to the cardinality of the set*

$$\{\chi' \in \mathrm{Irr}(KH) \mid m(\chi,\psi) = m(\chi',\psi) \text{ for all } J \subseteq S \text{ and all } \psi \in \mathcal{S}_J\}$$

where $m(\chi,\psi)$ is defined as in (9.1.9).

Proof. Let $K_1 \subseteq K$ be a subfield such that K_1 is a finite Galois extension of K' containing all v_s $(s \in S')$ and a primitive $|W|$th root of unity. Then K_1W is split by Theorem 6.3.8. Furthermore, by Proposition 9.3.1 and Corollary 9.2.6, K_1H is still split and we have $K'(\chi) \subseteq K_1$. So we can work entirely inside K_1. Set $m := [K'(\chi) : K']$. Since $K_1 \supseteq K'$ is a finite Galois extension, there exist m different embeddings $\varphi_1, \ldots, \varphi_m \colon K'(\chi) \to K_1$ which fix K' elementwise. Each φ_i can be extended to an automorphism $\tilde{\varphi}_i \colon K_1 \to K_1$.

For $1 \leq i \leq m$, we define $\chi^i \colon K_1H \to K_1$ by $\chi^i(T_w) = \tilde{\varphi}_i(\chi(T_w))$ for $w \in W$. Then Exercise 9.1 shows that χ^i is again an irreducible character of K_1H. Moreover, $\chi^i = \chi^j$ implies that $\varphi_i(\chi(T_w)) = \varphi_j(\chi(T_w))$ for all $w \in W$, and hence $i = j$. Thus, the characters $\chi^1, \ldots, \chi^m$ are pairwise different.

Now let $J \subseteq S$, $\psi \in \mathcal{S}_J$ and consider the induced character $\mathrm{Ind}_J^S(\psi)$. Since ψ_J is afforded by a module ove K', the same holds (by the construction of induced modules) for $\mathrm{Ind}_J^S(\psi)$. So, each $\tilde{\varphi}_i$ fixes that character and, hence, permutes its irreducible constituents. Consequently, all χ^i have the same multiplicity in $\mathrm{Ind}_J^S(\psi)$. Thus, m cannot be bigger than the cardinality of the above set. □

9.3.4. Example. (a) Consider the following special case of Lemma 9.3.3. Let

$$K' = \mathbb{Q}(u_s \mid s \in S) \qquad \text{and} \qquad \mathcal{S}_J = \{\mathrm{ind}_J\} \quad \text{for all } J \subseteq S,$$

where ind_J denotes the index character of KH_J. Then ind_J can certainly be realized over K'. As in Section 6.3, we set, for any $\chi \in \mathrm{Irr}(KH)$ and $J \subseteq S$:

$$n(J,\chi) = \text{multiplicity of } \chi \text{ in } \mathrm{Ind}_J^S(\mathrm{ind}_J) = \text{multiplicity of } \chi_1 \text{ in } 1_{W_J}^W,$$

where the second equality holds by (9.1.9). Now Lemma 9.3.3 implies that if $\chi \in \mathrm{Irr}(KH)$ is uniquely determined by the multiplicities $n(J,\chi)$ for all $J \subseteq S$, then χ is rational in the sense that all character values $\chi(T_w)$ lie in $\mathbb{Q}(u_s \mid$

$s \in S$). By Theorem 6.3.3 and (6.3.6), this applies to all irreducible characters for (W, S) of type A_n, B_n, D_n, F_4, E_6, E_7 and E_8, which are not listed in the table in Example 9.2.3(a). Hence, all these characters are realized over $\mathbf{K}'$; see Proposition 9.3.1. For the remaining types or characters, we can always conclude that $[\mathbf{K}'(\chi) : \mathbf{K}'] \leq 2$, except for the characters of degree 2 in type $I_2(m)$ and the four characters of degree 24 in type H_4; see (6.3.6) and the last column in the tables in Appendix C.

(b) Assume that (W, S) is of type H_4. Then one can check that the four irreducible characters of degree 24 have the same multiplicities in all the induced characters $\mathrm{Ind}_J^S(\mathrm{ind}_J)$ $(J \subseteq S)$. To deal with these characters, we argue as follows.

All generators $s \in S$ are conjugate and we write $u := u_s$ $(s \in S)$. Let $I \subseteq S$ be the unique subset such that W_I is the dihedral group of order 10. Now we set $\mathbf{K}' := K(u)$ (where $K \subseteq \mathbb{C}$ is any field containing the fifth roots of unity), $\mathcal{S}_I = \mathrm{Irr}(\mathbf{K}\mathbf{H}_I)$ and $\mathcal{S}_J = \{\mathrm{ind}_J\}$ for $J \neq I$. By Corollary 8.3.2, the assumptions of Lemma 9.3.3 are satisfied. Using a computation with the ordinary character tables of W and W_I and the compatibility results in (9.1.9), we obtain the following multiplicities $m(\chi, \psi)$ where $\chi \in \mathrm{Irr}(\mathbf{K}\mathbf{H})$ has degree 24 and $\psi \in \mathrm{Irr}(\mathbf{K}\mathbf{H}_I)$:

$m(\chi,\psi)$	ind_I	ε_I	ψ_1	ψ_2
24_s, 24_t	2	2	4	6
$\overline{24}_s$, $\overline{24}_t$	2	2	6	4

where ψ_1, ψ_2 are the two characters of degree 2 of $\mathbf{K}\mathbf{H}_I$. Hence, Lemma 9.3.3 implies that $[\mathbf{K}'(\chi) : \mathbf{K}] \leq 2$ for any $\chi \in \mathrm{Irr}(\mathbf{K}\mathbf{H})$ of degree 24. The exact fields in which the character values in type H_4 lie will be determined in Section 11.2.

The above results already show that if K is a splitting field for W then we can find a splitting field for $\mathbf{H}$ of degree at most 2 over $K(u_s \mid s \in S)$. But what is that extension of degree 2? The answer is given by the following result.

9.3.5. Theorem (Benson–Curtis, Lusztig, Alvis–Lusztig, Digne–Michel). *Let W be a finite Coxeter group with generating set S, and let $\mathbf{H}$ be a corresponding generic Iwahori–Hecke algebra. Let*

$$\mathbf{K}_0 := K(\sqrt{u_s} \mid s \in S) \subseteq \mathbf{K}, \quad \textit{where } K \subseteq \mathbb{C} \textit{ is a splitting field for } W.$$

(For the determination of K, see Theorem 6.3.8.) Then every irreducible character of $\mathbf{K}\mathbf{H}$ can be realized over $\mathbf{K}_0$, and so the algebra $\mathbf{K}_0\mathbf{H}$ is split. Moreover, for every $\chi \in \mathrm{Irr}(\mathbf{K}\mathbf{H})$, we have

$$\mathrm{ind}(T_{w_0})c_\chi \in R[\sqrt{u_s} \mid s \in S] \quad \textit{and} \quad \chi(T_w) \in R[\sqrt{u_s} \mid s \in S] \quad (w \in W),$$

where c_χ denotes the Schur element and R is the ring of algebraic integers in K.

Proof. First note that we may assume without loss of generality that (W, S) is irreducible. For, if this is not the case, then $\mathbf{K}\mathbf{H}$ can be written as the tensor product of the Iwahori–Hecke algebras of the irreducible factors of (W, S); see Exercise 8.4. If all factors in this tensor product decomposition are split over $\mathbf{K}_0$, then so is the algebra $\mathbf{H}$ itself; see Exercise 8.5.

So, from now on, (W, S) is irreducible. Let $\{v_s \mid s \in S'\}$ be as in Corollary 9.2.6. Then we can assume that $\sqrt{u_s} = v_s^{e/2}$ for $s \in S'$, where e is the exponent of W. We now proceed in three steps.

Step 1. We consider the special case where K contains all eth roots of unity and show that all character values $\chi(T_w)$ lie in $K(\sqrt{u_s} \mid s \in S')$. For type $I_2(m)$, this is clear by Lemma 8.3.3. The types A_n, B_n, D_n, F_4 and E_6 have already been considered in Example 9.3.4(a). It remains to consider the types E_7, E_8, H_3 and H_4. In these cases, all parameters u_s are equal, and we write $v := v_s$, $u := u_s$ for all $s \in S'$. Thus, all the values $\chi(T_w)$ lie in $K(v)$, where $v^e = u$. Since K contains all eth roots of unity, the extension $K(u) \subseteq K(v)$ is a Galois extension with a cyclic Galois group of order e. It follows that for each divisor d of e, there is exactly one intermediate field of degree d over $K(u)$, namely, $K(v^{e/d})$. Hence, in order to show that all values $\chi(T_w)$ lie in $K(\sqrt{u})$, it is enough to show that these values lie in an extension of degree at most 2 over $K(u)$. But this has been shown in Example 9.3.4.

Step 2. Now let $K \subseteq \mathbb{C}$ be any splitting field for W. Then K contains all character values $\chi_1(w)$ ($\chi \in \operatorname{Irr}(\mathbf{KH})$, $w \in W$) and, by Theorem 6.3.8, we can in fact assume without loss of generality that K is the field which is generated over $\mathbb{Q}$ by those character values. Then $K \supseteq \mathbb{Q}$ is a finite Galois extension. We claim that $\mathbf{K}_0 = K(\sqrt{u_s} \mid s \in S')$ is a splitting field for $\mathbf{H}$. Indeed, by step 1, all $\chi(T_w)$ lie in the field $K_1(\sqrt{u_s} \mid s \in S')$, where K_1 is generated over K by a primitive eth root of unity. Combining this with Lemma 9.3.2, we deduce that

$$\chi(T_w) \in K_1(\sqrt{u_s} \mid s \in S') \cap K[v_s \mid s \in S'] = K[\sqrt{u_s} \mid s \in S']$$

for all $w \in W$ and all $\chi \in \operatorname{Irr}(\mathbf{KH})$. Now the claim follows from Proposition 9.3.1.

Step 3. Finally, consider the ring of integers $R \subseteq K$. Then $R[\sqrt{u_s} \mid s \in S']$ is integrally closed in $\mathbf{K}_0$ by Exercise 7.2. Hence all character values $\chi(T_w)$ lie in $R[\sqrt{u_s} \mid s \in S']$ by the general integrality result in Proposition 7.3.8. As far as c_χ is concerned, we can use a similar argument and conclude that $c_\chi \in R[\sqrt{u_s}^{\pm 1} \mid s \in S']$. (We need the inverses of the parameters so that the dual basis elements $T_w^\vee$ exist.) It remains to consider the defining formula for c_χ in (8.1.8) and to note that $\operatorname{ind}(T_w)$ divides $\operatorname{ind}(T_{w_0})$ in $\mathbf{A}$, since we have $l(w_0) = l(w) + l(w^{-1}w_0)$ and so $\operatorname{ind}(T_{w_0}) = \operatorname{ind}(T_w)\operatorname{ind}(T_{w^{-1}w_0})$ for all $w \in W$. □

We now discuss a number of applications and refinements of the above result.

First, we consider the special case where all parameters of $\mathbf{H}$ are equal; write $u = u_s$ ($s \in S$). Furthermore, we assume that W is *crystallographic*, i.e., given a prime power q, there exists a finite untwisted *Chevalley group* G over the field with q elements with Weyl group W; see Section 8.4. Let $B \subseteq G$ be a Borel subgroup. Then, by Corollary 8.4.7 (and using the notation of (8.1.6)), we have

$$\mathrm{End}_G(\mathbb{C}[G/B]) \cong_{\mathbb{C}} \mathbf{H}_q, \quad \text{via the specialization } \theta\colon \mathbf{A} \to \mathbb{C}, u \mapsto q.$$

This interpretation allows us to prove the following result.

9.3.6. Corollary (Benson–Curtis). *In the above setting, we have*

$$D_\chi = \frac{P_W}{c_\chi} = \frac{1}{r_\chi} f_\chi \in \mathbb{Q}[u] \qquad \text{for all } \chi \in \mathrm{Irr}(\mathbf{K}\mathbf{H}),$$

where $f_\chi \in \mathbb{Z}[u]$ is a monic polynomial and r_χ is a positive integer.

Proof. We set $v = \sqrt{u}$ and work in the polynomial ring $\mathbb{Q}[v]$. Since W is crystallographic, $\mathbb{Q}$ is a splitting field for W (see Theorem 6.3.8). Hence, by Theorem 9.3.5, we have $v^{2N} c_\chi \in \mathbb{Z}[v]$, where $N = l(w_0)$. The crucial ingredient for the following arguments is that, by Corollary 8.4.7, we have a decomposition

$$\mathbb{C}[G/B] = \sum_{\chi \in \mathrm{Irr}(\mathbf{K}\mathbf{H})} \chi(T_1)\, V_\chi \qquad \text{with} \qquad \dim_{\mathbb{C}} V_\chi = \frac{\theta^*_{\mathbb{C}}(P_W)}{\theta^*_{\mathbb{C}}(c_\chi)}. \tag{$*$}$$

Now we proceed in three steps.

Step 1. We claim that $v^{2N} c_\chi$ divides $v^{2N} P_W$ in $\mathbb{Q}[v]$. This is seen as follows. By the division algorithm for polynomials, there exist $f, g \in \mathbb{Q}[v]$ such that

$$v^{2N} P_W = f v^{2N} c_\chi + g \quad \text{where } g = 0 \text{ or } \deg(g) < \deg(v^{2N} c_\chi).$$

Furthermore, let $m \neq 0$ be an integer such that $mf, mg \in \mathbb{Z}[v]$. Now we assume that $q = p^2$ for some prime number p and fix $\theta^*_{\mathbb{C}}$ such that $\theta^*_{\mathbb{C}}(v) = p$. Applying $\theta^*_{\mathbb{C}}$ to the above identity yields:

$$m \dim_{\mathbb{C}} V_\chi \overset{(*)}{=} m \frac{P_W(p)}{c_\chi(p)} = m \frac{p^{2N} P_W(p)}{p^{2N} c_\chi(p)} = m f(p) + \frac{m g(p)}{p^{2N} c_\chi(p)}.$$

Now, the left-hand side and $mf(p)$ are integers. Hence so is $mg(p)/(p^{2N} c_\chi(p))$. But, if we had $g \neq 0$, then $mg/(v^{2N} c_\chi)$ would be a quotient of two polynomials where the denominator has a degree strictly bigger than that of the numerator. Hence, for a large value of $q = p^2$, the quotient $mg(p)/(p^{2N} c_\chi(p)) \in \mathbb{Q}$ would be non-zero but arbitrarily small; in particular, it would not be an integer, giving a contradiction. Hence, $g = 0$ and we have shown that $D_\chi \in \mathbb{Q}[v]$.

Step 2. We claim that $D_\chi \in \mathbb{Q}[u]$. Now, by step 1, we have $D_\chi \in \mathbb{Q}[v]$. So we can write $D_\chi = f + gv$ with $f, g \in \mathbb{Q}[u]$. We assume that q is a prime number and fix $\theta^*_{\mathbb{C}}$ such that $\theta^*_{\mathbb{C}}(v) = \sqrt{q} \in \mathbb{R} - \mathbb{Q}$. Applying $\theta^*_{\mathbb{C}}$ to the above identity yields $D_\chi(\sqrt{q}) = f(q) + g(q)\sqrt{q}$, where the left-hand side is an integer by $(*)$

and $f(q), g(q) \in \mathbb{Q}$. If we had $g \neq 0$, we would choose q large enough such that $g(q) \neq 0$ and then also conclude that $\sqrt{q} \in \mathbb{Q}$, giving a contradiction.

Step 3. Finally, let r_χ be the least positive integer such that $f_\chi := r_\chi D_\chi \in \mathbb{Z}[v]$. We have $c_\chi = (r_\chi P_W)/f_\chi \in \mathbb{Z}[v]$ and P_W is monic. Hence, since $\mathbb{Z}[v]$ is a factorial ring, we conclude that f_χ divides P_W and so f_χ is monic. □

9.3.7. Remark. (a) The only types which are not covered by the above result are $I_2(m)$ ($m \geq 5$, $m \neq 6$), H_3 and H_4. In these cases, it is still true that $D_\chi \in \mathbb{R}[u]$ for all $\chi \in \mathrm{Irr}(\mathbf{K}H)$. But the only way to prove this seems to be by explicitly computing the generic degrees. For type $I_2(m)$, we have already done this in Remark 8.3.5; for types H_3 and H_4, see the tables in Appendix E.

(b) If we consider a generic Iwahori–Hecke algebra with unequal parameters, then it is no longer true that the generic degrees are polynomials in these parameters (or their square roots). The smallest example is given by type $B_2 = I_2(4)$. Using Theorem 8.3.4, we obtain for the character χ_1:

$$D_{\chi_1} = \frac{P_W}{c_{\chi_1}} = u_s u_t \frac{(u_s+1)(u_t+1)}{u_s+u_t}.$$

9.3.8. Integrality of irreducible representations. Let R be a ring as in Theorem 9.3.5. Then it can be shown that every $\chi \in \mathrm{Irr}(\mathbf{K}H)$ is afforded by a representation in which each T_w ($w \in W$) is represented by a matrix with coefficients in $R[\sqrt{u_s}^{\pm 1} \mid s \in S]$. But the proof requires some highly sophisticated and deep methods. (In this book, this result will only be used in the proof of Corollary 9.3.9.)

Indeed, using Exercise 8.5, it suffices to consider the case where (W, S) is irreducible. If all parameters u_s are equal then every irreducible character of $\mathbf{K}H$ is afforded by a so-called *W-graph* (see Section 11.1). This is shown by a general argument in [Gyoja 1984] for finite Weyl groups, using the results of [Lusztig 1981a]. The types H_3 and H_4 have been settled by Alvis and Alvis–Lusztig; see (11.1.8) and Section 11.2. For type $I_2(m)$, we have realized the irreducible characters in a form as required in Theorem 8.3.1. It remains to consider the types B_n and F_4 with unequal parameters. Representations in the desired form for type B_n have been constructed in [Dipper et al. 1995, Theorems 4.20, 4.22]. For type F_4, we will construct W-graphs for the irreducible characters in Section 11.3.

The above results can be used to settle the question of semisimplicity and splitting fields for specialized Iwahori–Hecke algebras.

9.3.9. Corollary. *Let* R *be as in Theorem 9.3.5 and* $\theta_0\colon R[\sqrt{u_s}^{\pm 1} \mid s \in S] \to L$ *be a ring homomorphism into a field* L. *Then the specialized algebra* LH *is split semisimple if and only if* $\theta_0(c_\chi) \neq 0$ *for all* $\chi \in \mathrm{Irr}(\mathbf{K}H)$.

Proof. We assume first that LH is split semisimple. Then we can apply Theorem 7.4.7 and so $\theta_0(c_\chi) \neq 0$ for all $\chi \in \mathrm{Irr}(\mathbf{K}H)$. Conversely, assume that the latter condition is satisfied. Now the problem is that we do not yet know if LH is split. But, using (9.3.8), we can proceed as follows. Let $\mathcal{O}$ be the subring of K_0 (defined as in Theorem 9.3.5) generated by $R[\sqrt{u_s}^{\pm 1}]$ and all elements $1/c_\chi$ for $\chi \in \mathrm{Irr}(KH)$. Then θ_0 can be naturally extended to a map from $\mathcal{O}$ to L. Now every irreducible character of KH can be realized over $\mathcal{O}$. So Proposition 7.2.7 shows that we have a basis $\{e_{ij}^\chi\}$ of KH which is adapted to the Wedderburn decomposition of KH and such that each e_{ij}^χ already lies in $\mathcal{O}H$. But the latter property means that we already have an isomorphism of $\mathcal{O}$-algebras $\mathcal{O}H \to \bigoplus_\chi M_{d_\chi}(\mathcal{O})$, where $\chi \in \mathrm{Irr}(KH)$ and d_χ denotes the degree of χ. Applying the canonical extension of θ_0 to $\mathcal{O}$ yields that we also have an isomorphism $LH \to \bigoplus_\chi M_{d_\chi}(L)$. Hence, LH is split semisimple, as desired. □

9.4 CHARACTER VALUES AND SCHUR ELEMENTS

Keeping the set-up of the previous section, we now derive a number of general properties of the character values and the Schur elements of KH. These properties will play a basic role in the determination of the character tables for the exceptional types in Chapter 11. At the end of this section, we will also define "a-invariants" of the irreducible characters of KH, thus taking up the issues discussed earlier in Section 6.5.

We begin by studying the effect of the A-algebra automorphism $\gamma\colon H \to H$ of Exercise 8.2, which is given by $\gamma(T_s) = -u_s T_s^{-1}$ for $s \in S$. We have a canonical extension to an automorphism $\gamma^K\colon KH \to KH$. If follows that if $\delta\colon KH \to M_m(K)$ is any representation then the composition of δ with γ^K also is a representation. We call δ^* the *dual representation* of δ. If χ is the character of δ, then the character of δ^* will be denoted by χ^* and called the *dual character*.

9.4.1. Proposition. *Let $\chi \in \mathrm{Irr}(KH)$ and χ^* be the corresponding dual character. (We certainly have $\chi^* \in \mathrm{Irr}(KH)$.) Then the following hold:*

(a) *Let $a \mapsto \bar{a}$ be the unique $\mathbb{C}$-algebra automorphism of $\mathbb{C}[\sqrt{u_s} \mid s \in S]$ which takes $\sqrt{u_s}$ to $\sqrt{u_s}^{-1}$ for any $s \in S$. Then*

$$\chi^*(T_w) = (-1)^{l(w)}\,\mathrm{ind}(T_w)\overline{\chi(T_w)} \qquad \text{for all } w \in W.$$

(b) *Consider the bijection $\mathrm{Irr}(KH) \leftrightarrow \mathrm{Irr}(W)$, $\chi \leftrightarrow \chi_1$. Then $(\chi^*)_1 = \chi_1 \otimes \varepsilon$, where ε is the sign character of W.*

Proof. Recall that K is a finite Galois extension of $\mathbb{C}(\sqrt{u_s} \mid s \in S)$. By Theorem 9.3.5, we may in fact assume that we have equality. Let $\mathcal{O} \subseteq K$ be a valuation ring as in (8.1.6), and $\delta\colon KH \to M_m(K)$ be a representation affording χ such that $\delta(T_w) \in M_m(\mathcal{O})$ for all $w \in W$. Let $\delta^k\colon k[W] \to M_m(k)$ be the corresponding specialized representation, where k is the residue field of $\mathcal{O}$. (Note that k is some extension of $\mathbb{C}$.) Now, for any $s \in S$, the matrix

$$\delta^*(T_s) = \delta(\gamma(T_s)) = -u_s\delta(T_s^{-1}) = -\delta(T_s) + (u_s - 1)\delta(T_1)$$

specializes to the matrix $-\delta^k(s)$. But the representation of W given by $s \mapsto -\delta^k(s)$ (for $s \in S$) is nothing but the tensor product of δ^k with ε, hence it affords $\chi_1 \otimes \varepsilon$. This implies (b).

In order to prove (a), we use the ring automorphism j for the generic multiparameter Iwahori–Hecke algebra introduced in Exercise 8.1. By specialization, it also yields a ring automorphism j of our algebra $\mathbf{H}$ (which may have equal parameters). Then j can be extended canonically to a $\mathbb{C}$-linear ring automorphism of $\mathbf{KH}$, so that we have

$$j\Big(\sum_{w\in W} a_w T_w\Big) = \sum_{w\in W} (-1)^{l(w)}\, \mathrm{ind}(T_w)^{-1} \bar{a}_w T_w \qquad (a_w \in \mathbf{K}).$$

Since j is semilinear with respect to $a \mapsto \bar{a}$, it follows that the map $\delta^j : \mathbf{KH} \to M_m(\mathbf{K})$ defined by the formula $\delta^j(T_w) = (\bar{a}_{ij})$, where $\delta(j(T_w)) = (a_{ij})$, is a representation of $\mathbf{KH}$. Let χ^j denote its character; we have

$$\chi^j(T_w) = \overline{\chi(j(T_w))} = (-1)^{l(w)}\overline{\mathrm{ind}(T_w)^{-1}\chi(T_w)} = (-1)^{l(w)}\, \mathrm{ind}(T_w)\overline{\chi(T_w)}$$

for all $w \in W$. If we specialize $\sqrt{u_s} \mapsto 1$, the expression on the right-hand side specializes to $(-1)^{l(w)}\chi_1(w)$ which is nothing but the value of $\chi_1 \otimes \varepsilon$ at $w \in W$. Hence χ^* and χ^j specialize to the same character of W, which implies (a). □

9.4.2. Corollary. *Let $\chi \in \mathrm{Irr}(\mathbf{KH})$. Then, for any $w \in W$, $\chi(T_w)$ is a polynomial in $\mathbb{C}[\sqrt{u_s} \mid s \in S]$ whose total degree is less than or equal to that of* $\mathrm{ind}(T_w)$.

Proof. By Proposition 9.4.1(a), the expression $(-1)^{l(w)}\,\mathrm{ind}(T_w)\overline{\chi(T_w)}$ equals the character value $\chi^*(T_w)$. Hence that expression also lies in $\mathbb{C}[\sqrt{u_s} \mid s \in S]$ (see Theorem 9.3.5). So the total degree of $\chi(T_w)$ cannot exceed that of $\mathrm{ind}(T_w)$. □

9.4.3. Proposition. *Let $\chi \in \mathrm{Irr}(\mathbf{KH})$ and $z_\chi \in \mathbf{K}$ be as in Theorem 9.2.2, i.e., we have $\chi(T_{w_0}^2) = z_\chi \chi(T_1)$. Then we have*

$$c_{\chi^*} = \bar{c}_\chi = \mathrm{ind}(T_{w_0})\, z_\chi^{-1}\, c_\chi,$$

where $a \mapsto \bar{a}$ is the map which takes $\sqrt{u_s}$ to $\sqrt{u_s}^{-1}$ for any $s \in S$.

Proof. The first equality is proved as follows. By Corollary 8.2.6(c), we have $\chi(T_w) = \chi(T_{w^{-1}})$. So, the defining formula in (8.1.8) reads

$$\chi(T_1)c_\chi = \sum_{w\in W} \mathrm{ind}(T_w)^{-1}\chi(T_w)^2.$$

Now we use the formula in Proposition 9.4.1(a) and compute that

$$\begin{aligned}\chi^*(T_1) &= \sum_{w\in W} \mathrm{ind}(T_w)^{-1}\chi^*(T_w)^2 = \sum_{w\in W} \mathrm{ind}(T_w)\overline{\chi(T_w)}^2 \\ &= \overline{\sum_{w\in W} \mathrm{ind}(T_w)^{-1}\chi(T_w)^2} = \chi(T_1)\bar{c}_\chi.\end{aligned}$$

Since $\chi(T_1) = \chi^*(T_1)$, we conclude that $c_{\chi^*} = \bar{c}_\chi$, as desired.

To prove the second equality, we consider a representation $\delta\colon \mathbf{KH} \to M_m(\mathbf{K})$ affording χ. Using a matrix version of the defining formula for the Schur elements in Theorem 7.2.1, we obtain the following identity:

$$\sum_{w \in W} \operatorname{ind}(T_w)^{-1}\delta(T_w)\delta(T_{w^{-1}}) = c_\chi \operatorname{Tr}(I_m)I_m = mc_\chi I_m,$$

where I_m denotes the $m \times m$ identity matrix. Now consider the longest element $w_0 \in W$. By Lemma 1.5.3, we have $l(w_0) = l(w_0w^{-1}) + l(w)$ and, similarly, $l(w_0) = l(w^{-1}) + l(ww_0)$ for all $w \in W$, which implies that

$$T_{w_0} = T_{w_0w^{-1}}T_w \qquad \text{and} \qquad T_{w_0} = T_{w^{-1}}T_{ww_0}. \tag{$*$}$$

Inserting this into the above equation yields that

$$\begin{aligned} mc_\chi I_m &= \sum_{w \in W} \operatorname{ind}(T_w)^{-1}\delta(T_{w_0w^{-1}}^{-1})\delta(T_{w_0}^2)\delta(T_{ww_0}^{-1}) \\ &= z_\chi \sum_{w \in W} \operatorname{ind}(T_w)^{-1}\delta(T_{w_0w^{-1}}^{-1})\delta(T_{ww_0}^{-1}), \end{aligned}$$

where we used the fact that $T_{w_0}^2$ acts by the scalar z_χ in δ. Using the defining equation for the dual representation δ^*, we see that

$$\delta(T_x^{-1}) = (-1)^{l(x)} \operatorname{ind}(T_x)^{-1}\delta^*(T_{x^{-1}}) \qquad \text{for all } x \in W.$$

Applying this to the terms in the above sum, we obtain

$$mc_\chi I_m = z_\chi \sum_{w \in W} \operatorname{ind}(T_w)^{-1} \operatorname{ind}(T_{w_0w^{-1}})^{-1}\delta^*(T_{ww_0}) \operatorname{ind}(T_{ww_0})^{-1}\delta^*(T_{w_0w^{-1}}).$$

Now, the relations ($*$) also imply that $\operatorname{ind}(T_{w_0}) = \operatorname{ind}(T_{w_0w^{-1}})\operatorname{ind}(T_w)$. Hence the right-hand side of the above equation can be rewritten as follows:

$$z_\chi \operatorname{ind}(T_{w_0})^{-1} \sum_{w \in W} \operatorname{ind}(T_{ww_0})^{-1}\delta^*(T_{ww_0})\delta^*(T_{w_0w^{-1}}).$$

If we make the change of variable $w \to xw_0$, we see that the above sum is just the defining equation for the Schur element of χ^*. Hence we conclude that $mc_\chi I_m = z_\chi \operatorname{ind}(T_{w_0})^{-1} mc_{\chi^*} I_m$, which yields the desired formula. □

The following result provides a tool for checking that a given list of polynomials coincides with the Schur elements.

9.4.4. Lemma. *The Schur elements* c_χ *are uniquely determined by the following system of equations:*

$$\sum_{\chi \in \operatorname{Irr}(\mathbf{KH})} c_\chi^{-1}\chi(T_w) = \begin{cases} 1 & \text{if } w = 1 \\ 0 & \text{if } w \neq 1. \end{cases}$$

Here, it suffices to take one element w *from each conjugacy class of* W.

Proof. Consider the identity $\tau = \sum_\chi c_\chi^{-1}\chi$ (see Theorem 7.2.6). Since $\tau(T_1) = 1$ and $\tau(T_w) = 0$ for $w \neq 1$, the Schur elements satisfy the above equations. Now let $\{w\}$ be a set of representatives of the conjugacy classes of W. Then the corresponding matrix of values $(\chi(T_w))_{\chi,w}$ is a square matrix and it specializes to the ordinary character table of W. Hence it is an invertible matrix. It follows that the above system of equations has a unique solution. □

Let us apply the above results to the case where (W, S) is of type A_{n-1}, i.e., we have $W = \mathfrak{S}_n$ with respect to the generators $s_i = (i, i+1)$ for $1 \leq i \leq n-1$. Then all parameters for $\mathbf{H}$ are equal and we write $u = u_{s_i}$ for all i. Using Starkey's rule in Theorem 9.2.11, we will deduce a formula for the Schur elements for $\mathbf{H}$ in terms of the *Molien series* for the characters of $\mathfrak{S}_n$ (see Chapter 5). To fix notation, for $\chi \in \mathrm{Irr}(K\mathbf{H})$ we denote by $P_\chi \in \mathbb{Q}(u)$ the Molien series of the specialized character $\chi_1 \in \mathrm{Irr}(\mathfrak{S}_n)$ with respect to the standard reflection representation $\rho\colon \mathfrak{S}_n \to \mathrm{GL}(V)$. By Proposition 5.2.4, we have

$$P_\chi = \frac{1}{|\mathfrak{S}_n|} \sum_{w \in \mathfrak{S}_n} \frac{\chi_1(w^{-1})}{\det(\mathrm{id}_V - u \cdot \rho(w))}.$$

Recalling the definition of P_W and D_χ from (8.1.8), we can now state:

9.4.5. Proposition (Generic degrees in type A). *For $\chi \in \mathrm{Irr}(K\mathbf{H})$, we have*

$$D_\chi = (1-u)^{n-1} P_W P_\chi \quad \textit{and} \quad c_\chi^{-1} = (1-u)^{n-1} P_\chi.$$

In particular, a_χ (as in Definition 9.4.8) equals b_{χ_1} (as in Definition 5.4.4), and we have $D_\chi = u^{a_\chi} +$ combination of higher powers of u (see also Corollary 9.3.6). Moreover, the generic degrees equal the fake degrees in type A (cf. (5.3.3)).

(In Theorem 10.5.2, we will find an explicit combinatorial formula for D_χ.)

Proof. Let the notation be as in Theorem 9.2.11. By Lemma 9.4.4, it is enough to prove the following relation for any partition $\lambda \vdash n$:

$$\sum_{\chi \in \mathrm{Irr}(K\mathbf{H})} P_\chi \chi(T_{w_\lambda}) = \begin{cases} (1-u)^{1-n} & \text{if } \lambda = [1^n], \\ 0 & \text{otherwise.} \end{cases}$$

But, using Starkey's rule and Molien's formula, the left-hand side evaluates to

$$\begin{aligned} &\sum_{\chi \in \mathrm{Irr}(K\mathbf{H})} P_\chi \Big(\frac{1}{|\mathfrak{S}_\lambda|} \sum_{w \in \mathfrak{S}_\lambda} \chi_1(w) \det(u\, \mathrm{id}_{V_\lambda} - \rho_\lambda(w)) \Big) \\ &= \frac{1}{|\mathfrak{S}_n||\mathfrak{S}_\lambda|} \sum_{w \in \mathfrak{S}_\lambda} \sum_{y \in \mathfrak{S}_n} \frac{\det(u\, \mathrm{id}_{V_\lambda} - \rho_\lambda(w))}{\det(\mathrm{id}_V - u \cdot \rho(y^{-1}))} \Big(\sum_{\chi \in \mathrm{Irr}(K\mathbf{H})} \chi_1(w)\chi_1(y^{-1}) \Big) \\ &= \frac{1}{|\mathfrak{S}_\lambda|} \sum_{w \in \mathfrak{S}_\lambda} \frac{\det(u\, \mathrm{id}_{V_\lambda} - \rho_\lambda(w))}{\det(\mathrm{id}_V - u \cdot \rho(w^{-1}))}, \end{aligned}$$

where the last equality holds by using the second orthogonality relations for the irreducible characters of $\mathfrak{S}_n$. Now, since $\det(\rho(w)) = (-1)^{l(w)}$, we can write

$\det(\mathrm{id}_V - u\cdot\rho(w^{-1})) = (-1)^{l(w)+n-1}\det(u\,\mathrm{id}_V - \rho(w))$; furthermore, Exercise 9.6 yields $\det(u\,\mathrm{id}_V - \rho(w)) = (u-1)^{r(\lambda)-1}\det(u\,\mathrm{id}_{V_\lambda} - \rho_\lambda(w))$, where $r(\lambda)$ is the number of parts of λ. Inserting this into the above expressions, we obtain

$$\sum_{\chi\in\mathrm{Irr}(\mathbf{KH})} P_\chi \chi(T_{w_\lambda}) = \frac{1}{|\mathfrak{S}_\lambda|}\Big(\sum_{w\in\mathfrak{S}_\lambda}(-1)^{l(w)}\Big)(-1)^{n-1}(u-1)^{1-r(\lambda)}.$$

But the above sum is 0 for $\lambda \neq [1^n]$ and it evaluates to $(1-u)^{1-n}$ for $\lambda = [1^n]$, as desired. The remaining assertions are then clear by the above formula. □

Finally, we consider the induction of characters from parabolic subalgebras. Recall from (9.1.9) that this operation commutes with specialization.

9.4.6. Lemma. *Let $J \subseteq S$ and $\mathbf{H}_J \subseteq \mathbf{H}$ the corresponding parabolic subalgebra. For any $\psi \in \mathrm{Irr}(\mathbf{KH}_J)$ and $\chi \in \mathrm{Irr}(\mathbf{KH})$ let $m(\chi,\psi)$ be the multiplicity of χ in $\mathrm{Ind}_J^S(\psi)$. Then we have*

$$c_\psi^{-1} = \sum_{\chi\in\mathrm{Irr}(\mathbf{KH})} m(\chi,\psi)c_\chi^{-1} \qquad \textit{for all } \psi\in\mathrm{Irr}(\mathbf{KH}_J). \tag{a}$$

Furthermore, if V is a simple $\mathbf{KH}$-module affording $\chi \in \mathrm{Irr}(\mathbf{KH})$, we have

$$c_\chi\Big(\sum_{\psi\in\mathrm{Irr}(\mathbf{KH}_J)} m(\chi,\psi)c_\psi^{-1}\Big) = c_\chi^J, \tag{b}$$

where $c_\chi^J \in \mathbf{K}$ is the scalar by which $I_J^S(\mathrm{id}_V)$ acts on V; see Exercise 9.3.

Proof. (a) By Frobenius reciprocity, we have $\chi(T_w) = \sum_\psi m(\chi,\psi)\psi(T_w)$ for any $w \in W_J$. Hence, by Lemma 9.4.4 (applied to $\mathbf{KH}$), we have, for any $w \in W_J$,

$$\sum_\psi\Big(\sum_\chi m(\chi,\psi)c_\chi^{-1}\Big)\psi(T_w) = \sum_\chi c_\chi^{-1}\chi(T_w) = \begin{cases} 1 & \text{if } w=1,\\ 0 & \text{if } w\neq 1.\end{cases}$$

So the assertion follows by again using Lemma 9.4.4 (applied to $\mathbf{KH}_J$).

(b) Consider the representation $\rho\colon \mathbf{KH} \to \mathrm{End}_\mathbf{K}(V)$ affording χ. Since $\mathbf{KH}_J$ is semisimple, we can assume that $\mathrm{Res}_J^S(V) = V_1 \oplus \cdots \oplus V_n$, where V_i are simple $\mathbf{KH}_J$-modules; for each i, let $\psi_i \in \mathrm{Irr}(\mathbf{KH}_J)$ be the character afforded by V_i. We choose $\varphi_i \in \mathrm{End}_\mathbf{K}(V_i)$ such that $c_{\psi_i}\mathrm{Tr}(\varphi_i) = 1$ and set $\varphi := \bigoplus_i \varphi_i \in \mathrm{End}_\mathbf{K}(V)$. (Note that φ_i certainly exists since all Schur elements of $\mathbf{KH}_J$ are non-zero.) Let $I_J(\varphi)$ be the operator in Definition 7.1.9, defined with respect to $\mathbf{KH}_J$. Since that operator certainly is compatible with direct sums, Theorem 7.2.1 yields

$$I_J(\varphi) = \bigoplus_{i=1}^n I_J(\varphi_i) = \bigoplus_{i=1}^n c_{\psi_i}\mathrm{Tr}(\varphi_i)\mathrm{id}_{V_i} = \bigoplus_{i=1}^n \mathrm{id}_{V_i} = \mathrm{id}_V.$$

On the other hand, the operator $I(\varphi)$ (with respect to $\mathbf{KH}_J$) can be evaluated as follows. For any $w \in W$, we write $w = w'x$ with $w' \in W_J$ and $x \in X_J$. Then we have $l(w) = l(w') + l(x)$ and so $T_w = T_{w'}T_x$ and $T_w^\vee = \mathrm{ind}(T_w)^{-1}T_{w^{-1}} = T_x^\vee T_{w'}^\vee$. Hence we obtain:

$$\begin{aligned} I(\varphi)(v) &= \sum_{w \in W} \varphi(vT_w^\vee)T_w = \sum_{x \in X_J} \Big(\sum_{w' \in W_J} \varphi(vT_x^\vee T_{w'}^\vee)T'_w \Big) T_x \\ &= \sum_{x \in X_J} I_J(\varphi)(vT_x^\vee)T_x = \sum_{x \in X_J} (vT_x^\vee)T_x = I_J^S(\mathrm{id}_V)(v) = c_\chi^J v \end{aligned}$$

for all $v \in V$. Thus, Theorem 7.2.1 and the definition of φ yield the identity $c_\chi^J \,\mathrm{id}_V = I(\varphi) = c_V \,\mathrm{Tr}(\varphi)\,\mathrm{id}_V = c_V(\sum_i 1/c_{\psi_i})\,\mathrm{id}_V$. Collecting equal terms in the sum, we obtain the desired formula. □

We close this section by defining "a-invariants" of the irreducible characters of **KH**, thus coming back to a subject that we encountered in Section 6.5. Now these a-invariants will be defined in terms of the Schur elements of **KH**. For this purpose, we introduce the following notation. Let

$$\Gamma = \Big\{ \prod_{s \in S} \sqrt{u_s}^{\,m_s} \mid m_s \in \mathbb{Z} \Big\},$$

considered as an abelian group under multiplication. We assume that Γ is an ordered group, i.e., we assume that we are given a multiplicatively closed subset $\Gamma_+ \subseteq \Gamma - \{1\}$ such that we have $\Gamma = \Gamma_+ \amalg \{1\} \amalg \Gamma_-$, where $\Gamma_- = \{g^{-1} \mid g \in \Gamma_+\}$. We also assume that $\sqrt{u_s} \in \Gamma_+$ for all $s \in S$. Of course, there may be several ways to define Γ_+, but we fix one choice throughout what follows.

Now, given any Laurent polynomial $0 \neq f \in \mathbb{R}[\sqrt{u_s}^{\pm 1} \mid s \in S]$, we can write uniquely $f = r_0 g_0 + \cdots + r_n g_n$ where $0 \neq r_i \in \mathbb{R}$ and $g_i \in \Gamma$, such that $g_i g_0^{-1} \in \Gamma_+$ for all $i > 0$. (Note that any finite subset of Γ has a unique maximum and a unique minimum.) For short, we will sometimes just write

$$f = r_0 g_0 + \text{higher terms}.$$

Furthermore, for $w \in W$ we define $v_w \in \Gamma_+$ by the condition that $v_w^2 = \mathrm{ind}(T_w)$.

For example, if there exist some $v \in \mathbf{K}$ and integers $e_s \geq 1$ such that $\sqrt{u_s} = v^{e_s}$ for all $s \in S$, then we have $\Gamma \subseteq \{v^m \mid m \in \mathbb{Z}\}$ and $\Gamma_+ \subseteq \{v^m \mid m > 0\}$.

9.4.7. Lemma (Lusztig). *Let* $\chi \in \mathrm{Irr}(\mathbf{KH})$. *Then the set*

$$\{\alpha \in \Gamma \mid \alpha v_w^{-1}\chi(T_w) \in \mathbb{R}[\sqrt{u_s} \mid s \in S] \text{ for all } w \in W\}$$

has a unique minimal element, which we denote by α_χ. *We have*

$$\alpha_\chi \in \Gamma_+ \cup \{1\} \quad \text{and} \quad c_\chi = r_\chi \alpha_\chi^{-2} + \text{higher terms, where } r_\chi \in \mathbb{R},\ r_\chi > 0.$$

Proof. Let $w \in W$. Then, for suitable $m_w \in \Gamma$ and $0 \neq r_w \in \mathbb{R}$, we can write

$$v_w^{-1}\chi(T_w) = r_w m_w^{-1} + \text{higher terms}.$$

Since $\Gamma_+^{-1} = \Gamma_-$, we have $\alpha_\chi = \max\{m_w\}$. (Note again that any finite subset of Γ has a unique maximum and a unique minimum.) Now consider the Schur element c_χ. As in the proof of Proposition 9.4.3, we have the following formula:

$$\chi(T_1)c_\chi = \sum_{w\in W} \operatorname{ind}(T_w)^{-1}\chi(T_w)^2 = \sum_{w\in W}(v_w^{-1}\chi(T_w))^2.$$

We insert the above expression for $v_w^{-1}\chi(T_w)$ and obtain

$$\begin{aligned}\chi(T_1)c_\chi &= \sum_{w\in W}(r_w m_w^{-1} + \text{higher terms})^2 = \sum_{w\in W}\left(r_w^2 m_w^{-2} + \text{higher terms}\right)\\ &= \Big(\sum_{\substack{w\in W\\ m_w=\alpha_\chi}} r_w^2\Big)\alpha_\chi^{-2} + \text{higher terms}.\end{aligned}$$

Since r_w are non-zero real numbers, there are no cancellations in the term $\sum_w r_w^2$ of the above formula. So we have $c_\chi = r_\chi\alpha_\chi^{-2} + \text{higher terms}$, where $r_\chi > 0$.

Finally, we show that $\alpha_\chi \in \Gamma_+ \cup \{1\}$. Set $\alpha := \min\{\alpha_\chi \mid \chi \in \operatorname{Irr}(KH)\}$. We want to show that $\alpha = 1$. For this purpose, we consider the equation $\tau^K = \sum_\chi c_\chi^{-1}\chi$; see Theorem 7.2.6. Evaluating at T_1 and multiplying by α^{-2} yields

$$\begin{aligned}\alpha^{-2} &= \sum_{\chi\in\operatorname{Irr}(KH)} \alpha^{-2}c_\chi^{-1}\chi(1) = \sum_{\chi\in\operatorname{Irr}(KH)}\left(\frac{\chi(1)}{r_\chi}\alpha^{-2}\alpha_\chi^2 + \text{higher terms}\right)\\ &= \Big(\sum_{\substack{\chi\in\operatorname{Irr}(KH)\\ \alpha_\chi=\alpha}} \frac{\chi(1)}{r_\chi}\Big) + \text{combination of elements in } \Gamma_+.\end{aligned}$$

Now there are no cancellations in the term $\sum_\chi(\chi(1)/r_\chi)$, since $r_\chi > 0$ for all χ. Hence the right-hand side of the above formula contains a non-trivial constant term. Hence so does the left-hand side, which forces $\alpha = 1$. □

9.4.8. Definition (Lusztig). For $\chi \in \operatorname{Irr}(KH)$, let α_χ be as in Lemma 9.4.7. Then the constant term of $\alpha_\chi v_w^{-1}\chi(T_w) \in \mathbb{R}[\sqrt{u_s} \mid s \in S]$ will be denoted by $c_{w,\chi}$ and called the *leading coefficient* of $\chi(T_w)$. Moreover, writing

$$\alpha_\chi = \prod_{s\in S'} \sqrt{u_s}^{\,a_\chi^s} \qquad \text{with } a_\chi^s \in \mathbb{Z}, \text{ and } S' \subseteq S \text{ as in Theorem 9.2.2},$$

the tuple $a_\chi := (a_\chi^s)_{s\in S'}$ is called the *generalized a-invariant* of χ. If all parameters u_s are equal, we write $v = \sqrt{u_s}$ $(s \in S)$. Then we have $\alpha_\chi = v^{a_\chi}$ for some $a_\chi \in \mathbb{N}_0$, which is called the *a-invariant* of χ.

9.4.9. Example. *Generalized a-invariants for type* $I_2(m)$. Let $W = \langle s, t\rangle$ be a dihedral group where st has order $m \geq 3$. The case of equal parameters has been already considered in Remark 8.3.5. Now assume that m is even and that u_s, u_t are independent indeterminates. We fix an ordering of Γ where

$$\Gamma_+ = \{\sqrt{u_s}^k \sqrt{u_t}^l \mid k, l \in \mathbb{Z}, k > 0\} \cup \{\sqrt{u_t}^l \mid l > 0\}.$$

Then, using the notation of Section 8.3 and Theorem 8.3.4, we find that

$$\begin{array}{rclcrcl}
\alpha_{\mathrm{ind}} & = & 1 & \qquad & r_{\mathrm{ind}} & = & 1 \\
\alpha_{\varepsilon} & = & (v_s v_t)^{m/2} & & r_{\varepsilon} & = & 1 \\
\alpha_{\varepsilon_1} & = & v_t & & r_{\varepsilon_1} & = & 1 \\
\alpha_{\varepsilon_2} & = & (v_s v_t^{-1})^{m/2} v_t & & r_{\varepsilon_2} & = & 1 \\
\alpha_{\chi_j} & = & v_s & & r_{\chi_j} & = & m/(2 - \zeta^{2j} - \zeta^{-2j})
\end{array}$$

Here are some non-zero leading coefficients $c_{w,\chi}$:

$$\begin{array}{rclcrcl}
\mathrm{ind}(T_1) & = & 1 & \qquad & c_{1,\mathrm{ind}} & = & 1 \\
\varepsilon(T_{w_0}) & = & 1 & & c_{w_0,\varepsilon} & = & 1 \\
\varepsilon_1(T_t) & = & -1 & & c_{t,\varepsilon_1} & = & -1 \\
\varepsilon_2(T_{tw_0}) & = & (-1)^{m/2} v_t^{m-2} & & c_{tw_0,\varepsilon_2} & = & (-1)^{m/2} \\
\chi_j(T_s) & = & u_s - 1 & & c_{s,\chi_j} & = & -1
\end{array}$$

It takes a little more effort to find all elements w for which the corresponding leading coefficients are non-zero. For example, using the computations in the proof of Theorem 8.3.4, it can be checked that $c_{w,\chi_j} \neq 0$ for all $w \in W$ such that $l(w)$ is odd, except for t and sw_0. Then the orthogonality relations in Exercise 9.8 show that these are all the elements w for which we have $c_{w,\chi_j} \neq 0$.

9.4.10. Remark. Consider the a-invariants $a_\chi \in \mathbb{N}_0$ which are given in terms of the equal parameter case. After the explicit computation of generic degrees and Schur elements, it can be checked that the numbers a_χ are precisely Lusztig's a-invariants that we introduced in Definition 6.5.7; see also Remark 6.5.14. For type $I_2(m)$, this follows from Remark 8.3.5; for type A_{n-1}, we have just seen this in Proposition 9.4.5 above. For the remaining classical and exceptional types, this is seen from the explicit results in Chapter 10 and the tables on a-invariants and generic degrees in Appendices C and E.

Consequently, the numbers a_χ satisfy all the properties in Theorem 6.5.13. Note, however, that some of these properties can be proved by general arguments using the definition via Lemma 9.4.7. For example, the properties in (6.5.1)(a) and Definition 6.5.5(b) immediately follow from Lemma 9.4.6 and Proposition 9.4.3, respectively. The verification of Definition 6.5.5(a) still requires a case by case argument.

If, moreover, W is *crystallographic*, then the leading coefficients have the following remarkable interpretation (see [**Lusztig 1987b**, §3]). There exists an algebra J over $\mathbb{Z}$ (whose definition involves the Kazhdan–Lusztig basis of $\mathbf{H}$), which is free with a natural basis $\{t_w \mid w \in W\}$ such that the following hold:

(i) *There exists a homomorphism of* A*-algebras* $\phi\colon \mathbf{H} \to \mathrm{J} \otimes_{\mathbb{Z}} \mathsf{A}$ *such that the induced map* $\phi_{\mathsf{K}}\colon \mathsf{K}\mathbf{H} \to \mathrm{J} \otimes_{\mathbb{Z}} \mathsf{K}$ *is an isomorphism.*

(ii) *Let* μ *be an irreducible character of* $\mathrm{J} \otimes_{\mathbb{Z}} \mathsf{K}$*. Then* $\chi_\mu := \mu \circ \phi_{\mathsf{K}}$ *is an irreducible character of* $\mathsf{K}\mathbf{H}$ *and we have* $\mu(t_w) = \pm c_{w,\chi_\mu}$ *for all* $w \in W$.

It may be conjectured that an analogous statement also holds in the case of unequal parameters, but it does not seem to be possible to carry over the proofs in [Lusztig 1987b]: these proofs require deep positivity properties of the Kazhdan–Lusztig basis of $\mathbf{H}$ (see [Springer 1982]), which definitely do not hold in the unequal parameter case; see the example in [Lusztig 1983, p. 106].

9.5 BIBLIOGRAPHICAL REMARKS AND EXERCISES

The Mackey formula and the Frobenius reciprocity laws for Iwahori–Hecke algebras in the general setting of Section 9.1 are due to [Dipper and James 1986, §2]. Proposition 9.1.3 does not seem to have appeared explicitly in the literature before. An explicit formula for the case of a Coxeter element and the induced index representation can be found in [Pfeiffer 1994b, (1.3)]. Exercise 9.3 (for type A) can be found in [Dipper and Du 1993, §1].

Springer's Theorem 9.2.2 is contained in [Curtis 1975]. There, it was used to show that the characters of degree 512 in type E_7 and of degree 4096 in type E_8 have values which involve a square root of the parameter of the Iwahori–Hecke algebra, as in Example 9.2.3. The fact that this result can also be used to determine further character values as in Proposition 9.2.8 is implicitly contained in the computation in [Lusztig 1984, p. 320]. It has also been used and generalized to cyclotomic algebras in [Broué and Malle 1993]; for a full discussion see [Broué and Michel 1996, §4 E]. Theorem 9.2.6 appears in [Geck and Michel 1997]. Starkey's rule was obtained in [Starkey 1975], a Ph. D. thesis under the direction of J. A. Green. Our proof here follows [Geck 1999]. For a generalization to Ariki–Koike algebras, see [Shoji 1999, (7.10)].

It may be helpful to recall some of the history of the results about splitting fields of Iwahori–Hecke algebras. The origin of these results lies in the paper [Benson and Curtis 1972], where crystallographic Coxeter groups were considered (but the exceptional characters in type E_7 and E_8 were excluded from the discussion) and where the parameters of the corresponding Iwahori–Hecke algebra were powers of one fixed indeterminate. Springer's argument in [Curtis 1975] then showed that these exceptional characters indeed involve square roots of the parameters. In [Lusztig 1981a], an explicit isomorphism is constructed between the group algebra of a finite Weyl group W and the corresponding one-parameter generic Iwahori–Hecke algebra, where the ground field contains a square root of that parameter. The proof uses the theory of cells and some deep properties of Kazhdan–Lusztig polynomials. The non-crystallographic types H_3 and H_4 have been settled by explicitly constructing all irreducible representations in terms of so-called W-graphs in [Lusztig 1981a] and [Alvis and Lusztig 1982]. Finally, the case of two different, independent parameters in types B_n and F_4 has been considered in [Digne and Michel 1985, Theorem II.3.1] as far as the character values are concerned; using an easy argument about Schur indices, the corresponding result about splitting fields is mentioned explicitly in [Geck 1994].

Our arguments in this chapter are different in one respect, namely that we can use the results on "good" elements to obtain the first (and crucial) restriction on splitting fields in Corollary 9.2.6. This slightly simplifies some of the remaining arguments (which otherwise are essentially those of [Benson and Curtis 1972] and [Digne and Michel 1985]); we thus also have an independent proof for types H_3 and H_4. A general semisimplicity criterion as in Corollary 9.3.9 appeared in [Gyoja 1995, Theorem 2.9] and [Geck 1995, §2.5]; for earlier work on this problem, see [Green 1970], [Fleischmann 1990], [Gyoja and Uno 1989]. The proof in [Benson and Curtis 1972] of Corollary 9.3.6 uses Hilbert's irreducibility theorem; we were able to avoid that here since we first determined splitting fields for $\mathbf{H}$.

Splitting fields for cyclotomic Hecke algebras associated with complex reflection groups have been determined in [Malle 2000].

The properties of character values and generic degrees in Section 9.4 are a summary of known results, which can also be found (in some cases only for one-parameter algebras) in [Curtis and Reiner 1987, §68C] and [Lusztig 1981b, §1]. The automorphism j and the argument used in the proof of Proposition 9.4.1 are due to [Green 1970, §8]. The argument in the proof of Proposition 9.4.3 appears in [Lusztig 1984, §5.12]. The formula in Lemma 9.4.6(b) (for type B_n) can be found in [Hoefsmit 1974, §3.2]. The (generalized) a-invariants were introduced in [Lusztig 1979a] for one-parameter generic Iwahori–Hecke algebras; the general definitions are proposed in [Lusztig 1983, §7]. The leading coefficients are studied in detail in [Lusztig 1987b]. These notions play a basic role in the representation theory of reductive groups over finite fields; see [Lusztig 1984].

Exercise 9.1. Let H be a finite-dimensional associative algebra over a field K_0. Let $K \supseteq K_0$ be an extension field and $KH = H \otimes_{K_0} K$. Let $\sigma\colon K \to K$ be a field automorphism which fixes K_0 elementwise. Let $\delta\colon KH \to M_d(K)$ be a representation of KH, and define $\delta^\sigma\colon KH \to M_d(K)$ by $\delta^\sigma(h) = (\sigma(a_{ij}))$ where $h \in H$ and $\delta(h) = (a_{ij})$. Show that δ^σ is again a representation of KH, and that δ^σ is irreducible if and only if δ is. Furthermore, if χ is the character of δ then the map $\chi^\sigma\colon KH \to K$, $h \mapsto \sigma(\chi(h))$, is the character of δ^σ.

Exercise 9.2. Let H be a finite-dimensional symmetric algebra over a field K_0. Let K be an extension field of K_0 and assume that KH is split semisimple. Let $\mathcal{B}$ be a fixed basis of H. We say that a KH-module V is realizable over K_0 if there exists an H-module V_0 such that $V \cong V_0 \otimes_{K_0} K$. Show that a simple KH-module V is realizable over K_0 if

(a) we have $\chi_V(b) \in K_0$ for all $b \in \mathcal{B}$ (where χ_V is the character of V), and
(b) there exists an H-module V' such that V has multiplicity 1 in $V' \otimes_{K_0} K$.

[*Hint.* Let $e_V \in KH$ be the central primitive idempotent corresponding to V. By Proposition 7.2.7 and condition (a), we have in fact $e_V \in H$. Let V' be as in (b), and check that $V'e_V$ is an H-module such that $V \cong (V'e_V) \otimes_{K_0} K$.]

Exercise 9.3. The purpose of this exercise is to prove a relative version of Lemma 7.1.9 for Iwahori–Hecke algebras. Let $H = H_A(W, S, \{a_s \mid s \in S\})$ be an Iwahori–Hecke algebra as in Section 9.1, where W is finite. Assume that all a_s $(s \in S)$ are invertible in A. Let V, V' be (right) H-modules. For any $J \subseteq S$ and $\varphi \in \mathrm{Hom}_{H_J}(V, V')$, we define $I_J^S(\varphi) \in \mathrm{Hom}_A(V, V')$ by

$$I_J^S(\varphi)(v) := \sum_{x \in X_J} \varphi(vT_x^\vee)T_x \qquad (v \in V).$$

Show that we have $I_J^S(\varphi) \in \mathrm{Hom}_H(V, V')$.

[*Hint.* It is sufficient to show that $I_J^S(\varphi)$ commutes with the action of T_s for any $s \in S$. We have $I_J^S(\varphi)(vT_s) = \sum_{x \in X_J} \varphi(vT_sT_x^\vee)T_x$. Now use Lemma 9.1.4.]

Exercise 9.4. Let $H = H_A(W, S, \{a_s \mid s \in S\})$ be an Iwahori–Hecke algebra as in Section 9.1, where W is finite. Assume that all a_s $(s \in S)$ are invertible in A. For any subset $J \subseteq S$, let A_J be the free H_J-module of rank 1 defined by the restriction of the index representation from H to H_J. Let $J, K \subseteq S$. Use the Mackey formula and Frobenius reciprocity to show that

$$\mathrm{Hom}_H(\mathrm{Ind}_J^S(A_J), \mathrm{Ind}_K^S(A_K))$$

is an A-module which is finitely generated and free of rank $|X_{JK}|$ over A.

Exercise 9.5. Let $\mathbf{H}$ be a generic Iwahori–Hecke algebra associated with a Coxeter system (W, S), where W is finite. Let $\{z_C \mid C \in \mathrm{Cl}(W)\}$ be the central elements as in Corollary 8.2.4. Show that, for any $\chi, \chi' \in \mathrm{Irr}(K\mathbf{H})$, we have

$$\sum_{C \in \mathrm{Cl}(W)} \chi'(T_{w_C})\omega_\chi(z_C) = \begin{cases} c_\chi & \text{if } \chi = \chi', \\ 0 & \text{if } \chi \neq \chi', \end{cases}$$

where ω_χ is the central character associated with χ (see Proposition 7.2.8).

[*Hint.* Consider the orthogonality relations for the irreducible characters of $K\mathbf{H}$ and write $\chi'(T_w) = \sum_C f_{w,C}\chi'(T_{w_C})$. Note that $\chi(z_C) = \chi(1)\omega_\chi(z_C)$.]

Exercise 9.6. Consider the polynomials t_λ^ν occurring in Theorem 9.2.11. Show that if $C_\nu \cap \mathfrak{S}_\lambda \neq \varnothing$ and ν has non-zero parts $\nu_1, \nu_2, \ldots$, then

$$\det(u\,\mathrm{id}_{V_\lambda} - \rho_\lambda(w_\nu)) = (u-1)^{-r(\lambda)} \prod_{i \geq 1} (u^{\nu_i} - 1),$$

where $r(\lambda)$ is the number of non-zero parts of λ. Let $\lambda^1, \ldots, \lambda^m$ be all partitions of n and set $t_{ij} = t_{\lambda_j}^{\lambda^i}$. Show that the matrix (t_{ij}) has an upper triangular shape with non-zero entries along the diagonal, if we order the partitions such that $a^*(\lambda^1) \leq a^*(\lambda^2) \leq \cdots \leq a^*(\lambda^m)$. Compute examples for $n = 2, 3, 4$. If you want to check your computation, see [Carter 1986, §4].

[*Hint.* Use the results on characteristic polynomials in (3.4.3). To see the triangular shape of (t_{ij}), argue as in step 1 of the proof of Theorem 5.4.5.]

Exercise 9.7. Let (W, S) be of type H_3. The purpose of this exercise is to provide the details for the computation of the character table in Table 9.1 (p. 300). Use the following information about the induce/restrict matrices for the parabolic subgroups of type $A_1 \times A_1$, A_2 and $I_2(5)$:

H_3		$I_2(5)$				A_2			$A_1 \times A_1$			
χ	f_χ	ind	ε	χ_1	χ_2	(1^3)	$(2,1)$	(3)	$\varepsilon \otimes \varepsilon$	$\varepsilon \otimes$ ind	ind $\otimes \varepsilon$	ind $\otimes$ ind
$\chi_{(1,15)}$	0	.	1	.	.	1	.	.	1	.	.	.
$\chi_{(1,0)}$	30	1	.	.	.	.	.	1	.	.	.	1
$\chi_{(5,5)}$	12	.	1	1	1	1	2	.	2	1	1	1
$\chi_{(5,2)}$	18	1	.	1	1	.	2	1	1	1	1	2
$\chi_{(3,6)}$	10	.	1	1	.	1	1	.	1	1	1	.
$\chi_{(3,8)}$	10	.	1	.	1	1	1	.	1	1	1	.
$\chi_{(3,1)}$	20	1	.	1	.	.	1	1	.	1	1	1
$\chi_{(3,3)}$	20	1	.	.	1	.	1	1	.	1	1	1
$\chi_{(4,3)}$	15	.	.	1	1	1	1	1	1	1	1	1
$\chi_{(4,4)}$	15	.	.	1	1	1	1	1	1	1	1	1
	f_ψ	10	0	5	5	0	3	6	0	2	2	4

(The second column contains the constants f_χ as in Example 9.2.3; the last row contains these constants for the irreducible characters of the respective parabolic subalgebra.) Using this information, proceed as follows.

(a) Assume that w_C lies in one of the above parabolic subgroups. The character values of that subalgebra are contained in Example 8.1. Then we have $\chi(T_w) = \sum_\psi m(\chi, \psi)\psi(T_w)$, where $m(\chi, \psi)$ runs over the entries in the appropriate column of the above table.

(b) If w_C lies in a cuspidal class, we need the character values of W, the constants f_χ from the second column of the above table, and the formula in Proposition 9.2.8. In these cases, the values $\chi(T_{w_C})$ are just monomials.

Also determine the corresponding Schur elements by solving the corresponding system of equations as in Lemma 9.4.4. The results are printed in Table E.2.

Exercise 9.8. Consider the leading coefficients $c_{w,\chi}$ as in Definition 9.4.8. Show that, for $\chi, \chi' \in \mathrm{Irr}(\mathbf{K}H)$, we have the following orthogonality relations.

$$\sum_{w \in W} c_{w,\chi} c_{w,\chi'} = \begin{cases} r_\chi \chi(1) & \text{if } \chi = \chi', \\ 0 & \text{otherwise.} \end{cases}$$

[*Hint.* Consider the orthogonality relations for χ, χ' (see Corollary 7.2.4). Multiply both sides by $\alpha_\chi \alpha_{\chi'}$ and determine the constant terms.]

10
Character values in classical types

We now turn to the programme of determining the character tables of all the irreducible types: the exceptional types to come in the next chapter, the classical types in this. The main result for each series of irreducible Coxeter groups is a formula which determines the value of the irreducible character χ on the basis element T_w of the Iwahori–Hecke algebra, where w is an element of minimal length in its conjugacy class in the group. The prototype formula is the classical Murnaghan–Nakayama formula for the character values of the symmetric group, which expresses a particular value recursively, in terms of character values of smaller symmetric groups and the combinatorics of partitions and Young diagrams, labelling both the irreducible characters and the conjugacy classes of the group in question.

In that spirit we derive, for each of the three series, a formula which takes as its arguments the partitions or pairs of partitions which label a given irreducible character (see Chapter 5) and a given conjugacy class (see Chapter 3). The formula then describes in terms of these data a polynomial in the parameters of the Iwahori–Hecke algebra which gives the corresponding entry in the character table. These results are due to Halverson–Ram, Pfeiffer and Ram.

These formulas are derived for type A in Section 10.2, for type B in Section 10.3, and for type D in Section 10.4. By specializing the parameters of the Iwahori–Hecke algebra to 1 we also obtain formulas for the character values of the Coxeter group. This includes as case A the aforementioned Murnaghan–Nakayama rule.

The argument for determining the character values relies on Hoefsmit's explicit construction of matrix representations affording the irreducible characters in type B. That construction will be briefly discussed in Section 10.1, leaving some of the proofs as exercises.

In Section 10.5, we discuss the determination of the generic degrees for Iwahori–Hecke algebras of classical type A, B and D. We present Steinberg's proof for type A and show, following Hoefsmit and Benson–Gay, how the generic degrees for type D are derived from those of type B. We will also state Hoefsmit's formulas for type B and give sketches for three different proofs.

In the final section, we explain without proof the main ingredients of the Lascoux–Leclerc–Thibon conjecture, which is concerned with the problem of computing the decomposition numbers of Iwahori–Hecke algebras of type A associated with specializations of the parameter to roots of unity in characteristic 0.

10.1 HOEFSMIT'S MATRICES

Throughout the chapter we will work inside the generic Iwahori–Hecke algebra $\mathbf{H}(B_n)$ of type B_n with parameters u and v associated with the Coxeter group W_n. This has as a parabolic subalgebra the generic algebra $\mathbf{H}(\mathfrak{S}_n)$ of type A_{n-1} with parameter u associated with the Coxeter group $\mathfrak{S}_n$. In Section 10.4 we consider the specialization $\mathbf{H}(B_n)'$ obtained from the algebra $\mathbf{H}(B_n)$ via $v \mapsto 1$. This specialized algebra contains as a subalgebra the generic algebra $\mathbf{H}(D_n)$ of type D_n with parameter u associated with the Coxeter group W_n'.

10.1.1. The generic algebra $\mathbf{H}(B_n)$. Consider the Coxeter group W_n of type B_n, generated by $S = \{t, s_1, \ldots, s_{n-1}\}$ as in Section 1.4, with Coxeter diagram

B_n $(n \geq 2)$ 

where i stands for the generator s_i, $i = 1, \ldots, n-1$, and 0 for t. Let $\mathbf{H}$ be the generic multi-parameter Iwahori–Hecke algebra associated with (W_n, S) with parameters $u = u_{s_1} = \cdots = u_{s_{n-1}}$ and $v = u_t$; see Definition 8.1.4. By Proposition 9.3.1 and Example 9.3.4(a), every field containing the parameters is a splitting field for a generic Iwahori–Hecke algebra of classical type. So let $\mathbf{K} = \mathbb{C}(u, v)$ and define $\mathbf{H}(B_n) := \mathbf{K}\mathbf{H}$. Then $\mathbf{H}(B_n)$ is a split semisimple algebra.

As a $\mathbf{K}$-algebra, $\mathbf{H}(B_n)$ is generated by T_t and $S_i = T_{s_i}$ for $i = 1, \ldots, n-1$, subject to the relations

$$\begin{aligned} T_t^2 &= (v-1)T_t + v, && \\ S_i^2 &= (u-1)S_i + u, && i = 1, \ldots, n-1, \\ T_tS_1T_tS_1 &= S_1T_tS_1T_t, && \\ T_tS_i &= S_iT_t, && i \geq 2, \\ S_iS_{i+1}S_i &= S_{i+1}S_iS_{i+1}, && i = 1, \ldots, n-2, \\ S_iS_j &= S_jS_i, && |i-j| > 1. \end{aligned}$$

We also introduce, for $i = 0, \ldots, n-1$, elements

$$T_i = S_i \cdots S_1T_tS_1 \cdots S_i \in \mathbf{H}(B_n)$$

so that $T_i = T_{t_i}$ where $t_i \in W_n$ is as in (1.4.1).

10.1.2. Irreducible characters of $\mathbf{H}(B_n)$. In Theorem 5.5.6 we saw that the irreducible characters of a Coxeter group of type B_n are parametrized by the pairs of partitions $\lambda = (\lambda^0, \lambda^1)$ such that $|\lambda^0| + |\lambda^1| = n$. Since $\mathbf{H}(B_n)$ is split semisimple, the requirements for the specialization setting in (8.1.6) are satisfied. So Theorem 8.1.7 yields a bijection $\mathrm{Irr}(\mathbf{H}(B_n)) \xrightarrow{\sim} \mathrm{Irr}(W_n)$, $\chi \mapsto \chi_{u=v=1}$.

10.1.3. Standard tableaux and hooks. Recall from (5.4.2) that we visualize a partition λ by a Young diagram, consisting of λ_i boxes in row i. In the same way, we visualize and identify a pair of partitions $\lambda = (\lambda^0, \lambda^1)$ by a pair of Young diagrams.

We also need generalizations of the notions involved in the Littlewood–Richardson rule (6.1.6). If $\mu = (\mu^0, \mu^1)$ is another pair of partitions, we write $\mu \subseteq \lambda$ if $\mu^0 \subseteq \lambda^0$ and $\mu^1 \subseteq \lambda^1$. Further, if $\mu \subseteq \lambda$ then $\lambda - \mu$ is the pair of skew diagrams $\lambda^0 - \mu^0$ and $\lambda^1 - \mu^1$.

We call $\lambda - \mu$ a *border strip* (or simply a *strip*) if it does not contain any 2×2 blocks. We denote by cc^λ_μ the number of connected components of $\lambda - \mu$. If additionally $\lambda - \mu$ is connected, we call it a *rim hook* or simply a *hook*. In that case, we denote by l^λ_μ the *leg length* of the hook $\lambda - \mu$, which is by definition the number of rows covered by $\lambda - \mu$ in the diagram λ, minus cc^λ_μ.

A *tableau* of shape $\lambda - \mu$ is a pair $\mathbb{T} = (\mathbb{T}^0, \mathbb{T}^1)$ of tableaux where $\mathbb{T}^0$ is of shape $\lambda^0 - \mu^0$ and $\mathbb{T}^1$ is of shape $\lambda^1 - \mu^1$. The *weight* of such a tableau $\mathbb{T}$ is again the sequence $(a_1, a_2, \ldots)$ where a_i is the number of entries i in $\mathbb{T}$. The *word* $w(\mathbb{T})$ is the sequence obtained by reading the entries of $\mathbb{T}$ row by row from right to left, first in $\mathbb{T}^0$ then in $\mathbb{T}^1$.

For a pair of partitions $\lambda = (\lambda^0, \lambda^1)$ of total size n we say that $\mathbb{T}$ is a *standard tableau* of shape λ if $\mathbb{T}$ is a tableau of shape λ and weight $(1, 1, \ldots)$, i.e., where each $i = 1, \ldots, n$ occurs exactly once as an entry. We denote the set of all standard tableaux of shape λ by $\mathsf{ST}(\lambda)$. If, further, μ is a pair of partitions of $n - k$ for some $k < n$ such that $\mu \subseteq \lambda$ then a standard tableaux of shape $\lambda - \mu$ is a tableau of shape $\lambda - \mu$ which contains each of $n - k + 1, \ldots, n$ exactly once as an entry. We denote by $\mathsf{ST}(\lambda - \mu)$ the set of all standard tableaux of shape $\lambda - \mu$.

10.1.4. Matrices. Let $\lambda = (\lambda^0, \lambda^1)$ be a pair of partitions of total size n and denote by $\mathsf{ST}(\lambda)$ the set of all standard tableaux of shape λ. For $\mathbb{T} = (\mathbb{T}^0, \mathbb{T}^1) \in \mathsf{ST}(\lambda)$ and $m = 1, \ldots, n$ we define the content of $\mathbb{T}$ at m as

$$\mathsf{ct}(\mathbb{T} : m) := \begin{cases} u^{j-i}v & \text{if } \mathbb{T}^0 \text{ contains the entry } m \\ -u^{j-i} & \text{if } \mathbb{T}^1 \text{ contains the entry } m \end{cases} \tag{1}$$

in row i and column j. Note that, the natural action of $\mathfrak{S}_n$ on the set $\{1, \ldots, n\}$ induces an action of $\mathfrak{S}_n$ on the set of fillings of the Young diagram of shape λ with the numbers $\{1, \ldots, n\}$. A standard tableau under this action in general is not mapped to a standard tableau. But we always have

$$\mathsf{ct}(\mathbb{T}.w : m.w) = \mathsf{ct}(\mathbb{T} : m) \quad \text{for any } w \in \mathfrak{S}_n.$$

For $i = 1, \ldots, n-1$, define

$$m_i(\mathbb{T}) = (u - 1) \Big/ \left(1 - \frac{\mathsf{ct}(\mathbb{T} : i)}{\mathsf{ct}(\mathbb{T} : i+1)}\right). \tag{2}$$

Note that $\mathrm{ct}(\mathbb{T}:1) = -1$ or ν. Moreover, if i and $i+1$ are in the same component $\mathbb{T}^k$ of $\mathbb{T}$ the value of $m_i(\mathbb{T})$ does not depend on the component. It then just depends on the relative positions of i and $i+1$ in $\mathbb{T}^k$.

Table 10.1 Representing matrices for $\mathbf{H}(B_2)$

λ	$\mathsf{ST}(\lambda)$	T_t	S_1
$(11.\varnothing)$	$\{(\begin{smallmatrix}1\\2\end{smallmatrix},\varnothing)\}$	ν	-1
(1.1)	$\{(1,2),(2,1)\}$	$\begin{bmatrix}\nu & 0\\ 0 & -1\end{bmatrix}$	$\frac{1}{\nu+1}\begin{bmatrix}u-1 & u+\nu\\ u\nu+1 & u\nu-\nu\end{bmatrix}$
$(\varnothing.11)$	$\{(\varnothing,\begin{smallmatrix}1\\2\end{smallmatrix})\}$	-1	-1
$(2,\varnothing)$	$\{(12,\varnothing)\}$	ν	u
$(\varnothing,2)$	$\{(\varnothing,12)\}$	-1	u

10.1.5. Theorem (Hoefsmit). *Let $\lambda = (\lambda^0,\lambda^1)$ be a pair of partitions of total size n and let V^λ be the $\mathbf{K}$-vector space with basis $\mathsf{ST}(\lambda)$. Then, for each generator of $\mathbf{H}(B_n)$, we define an action on V^λ as follows. For $\mathbb{T} \in \mathsf{ST}(\lambda)$ set $m_i'(\mathbb{T}) = m_i(\mathbb{T}) + 1$ and*

$$\mathbb{T}.T = \mathrm{ct}(\mathbb{T}:1)\mathbb{T} \qquad \text{and} \qquad \mathbb{T}.S_i = m_i(\mathbb{T})\mathbb{T} + m_i'(\mathbb{T})\mathbb{T}',$$

for $i = 1,\ldots,n-1$, where $\mathbb{T}' = \mathbb{T}s_i$ if $\mathbb{T}s_i \in \mathsf{ST}(\lambda)$ and $\mathbb{T}' = 0$ otherwise. Then the following hold:

(i) *V^λ is an $\mathbf{H}(B_n)$-module via this action.*

(ii) *The $\mathbf{H}(B_n)$-modules V^λ, where λ runs through the pairs of partitions of total size n, form up to isomorphism a complete set of pairwise non-isomorphic irreducible $\mathbf{H}(B_n)$-modules.*

(iii) *The character of $\mathbf{H}(B_n)$ on $V = V^\lambda$ is χ_λ.*

Proof. The proofs of (i), (ii), (iii) proceed by induction on n, based on the fact that we have the following direct sum decomposition of V^λ. By construction we have, for $n \geq 2$,

$$V^\lambda = \bigoplus_{d\in I(\lambda^0)} V^{((\lambda^0)^{(d)},\lambda^1)} \oplus \bigoplus_{d\in I(\lambda^1)} V^{(\lambda^0,(\lambda^1)^{(d)})} \tag{1}$$

as $\mathbf{H}(B_{n-1})$-module, where for a partition μ with parts $\mu_1 \geq \cdots \geq \mu_r > 0$, as in (6.1.8), $I(\mu)$ is the set of all $d \in \{1,\ldots,r\}$ such that $d = r$ or $d < r$ and $\mu_d > \mu_{d+1}$ and, for $d \in I(\mu)$, $\mu^{(d)}$ is the partition of $n-1$ which is obtained from μ by decreasing μ_d by 1.

(i) To prove that the above action of the generators on V^λ indeed defines a representation is a matter of checking all the defining relations for $\mathbf{H}(B_n)$.

For the details of such a calculation, see [Hoefsmit 1974, Theorem 2.2.7] or [Ariki and Koike 1992, Theorem 3.7].

Now we prove (ii) and (iii) simultaneously, by induction on n. The case $n = 2$ is verified explicitly by the data in Table 10.1. The labelling of the characters in this table is easily matched with the labelling provided in Section 5.5.

By equation (1) the representing matrices are compatible with the branching rule (6.1.8) and (6.1.9).

Induction on n shows that this implies the fact that the modules V^λ are absolutely irreducible and pairwise inequivalent. The characters χ_λ are then identified using Exercise 6.2. For more details, see again [Hoefsmit 1974, Theorem 2.2.14] and [Ariki and Koike 1992, Theorem 3.10]. □

The central property of these actions is the fact that the elements T_i act diagonally on each V^λ.

10.1.6. Theorem (Hoefsmit). *Let* $\mathbb{T} \in \mathsf{ST}(\lambda)$ *for some pair of partitions* λ *of total size* n, *and let* $i \in \{0, \ldots, n-1\}$. *Then* $\mathbb{T}.T_i = u^i\,\mathrm{ct}(\mathbb{T} : i{+}1)\mathbb{T}$.

Proof. For $i = 0$ this is the definition of the action of $T_0 = T_t$. For $i > 0$ we have $T_i = S_i T_{i-1} S_i$ and thus

$$\mathbb{T}.T_i = (m_i(\mathbb{T})\mathbb{T} + m_i'(\mathbb{T})\mathbb{T}').T_{i-1}S_i$$

where $\mathbb{T}' = \mathbb{T}.s_i$ if $\mathbb{T}s_i \in \mathsf{ST}(\lambda)$ and $\mathbb{T}' = 0$ otherwise.

By induction on i, we have

$$\mathbb{T}.T_{i-1} = u^{i-1}\,\mathrm{ct}(\mathbb{T} : i)\mathbb{T} \quad \text{and} \quad \mathbb{T}'.T_{i-1} = u^{i-1}\,\mathrm{ct}(\mathbb{T}' : i)\mathbb{T}'$$

if $\mathbb{T}' \neq 0$.

If $\mathbb{T}s_i \notin \mathsf{ST}(\lambda)$ then the entries i and $i+1$ are contained either in the same row or in the same column of either $\mathbb{T}^0$ or $\mathbb{T}^1$. If i and $i{+}1$ lie in the same row then $\mathrm{ct}(\mathbb{T} : i{+}1) = u\,\mathrm{ct}(\mathbb{T} : i)$ and $m_i(\mathbb{T}) = (u-1)/(1-u^{-1}) = u$. If i and $i{+}1$ lie in the same column then $\mathrm{ct}(\mathbb{T} : i{+}1) = u^{-1}\,\mathrm{ct}(\mathbb{T} : i)$ and $m_i(\mathbb{T}) = -1$. In any case, we have

$$\mathbb{T}.T_i = \mathbb{T}.S_i T_{i-1} S_i = m_i(\mathbb{T})^2 u^{i-1}\,\mathrm{ct}(\mathbb{T} : i)\mathbb{T} = u^i\,\mathrm{ct}(\mathbb{T} : i{+}1)\mathbb{T},$$

as desired.

Otherwise, if $\mathbb{T}' = \mathbb{T}s_i \in \mathsf{ST}(\lambda)$, we have $\mathrm{ct}(\mathbb{T}' : i) = \mathrm{ct}(\mathbb{T} : i{+}1)$. Moreover, $\mathbb{T}.T_i = \mathbb{T}S_i T_{i-1} S_i = (m_i(\mathbb{T})\mathbb{T} + m_i'(\mathbb{T})\mathbb{T}')T_{i-1}S_i$ where, by induction on i, $\mathbb{T}T_{i-1} = u^{i-1}\,\mathrm{ct}(\mathbb{T} : i)\mathbb{T}$ and $\mathbb{T}'T_{i-1} = u^{i-1}\,\mathrm{ct}(\mathbb{T}' : i)\mathbb{T}'$. Therefore,

$$\begin{aligned}
\mathbb{T}.T_i &= \mathbb{T}.S_i T_{i-1} S_i \\
&= u^{i-1}\left(m_i(\mathbb{T})\,\mathrm{ct}(\mathbb{T} : i)\mathbb{T} + m_i'(\mathbb{T})\,\mathrm{ct}(\mathbb{T}' : i)\mathbb{T}'\right) S_i \\
&= u^{i-1}\left(m_i(\mathbb{T})^2\,\mathrm{ct}(\mathbb{T} : i) + m_i'(\mathbb{T})m_i'(\mathbb{T}')\,\mathrm{ct}(\mathbb{T}' : i)\right)\mathbb{T} \\
&\quad + u^{i-1} m_i'(\mathbb{T})\left(m_i(\mathbb{T})\,\mathrm{ct}(\mathbb{T} : i) + m_i(\mathbb{T}')\,\mathrm{ct}(\mathbb{T}' : i)\right)\mathbb{T}'
\end{aligned}$$

and the result follows in this case from $m_i(\mathbb{T})\,\mathrm{ct}(\mathbb{T}:i) + m_i(\mathbb{T}')\,\mathrm{ct}(\mathbb{T}':i) = 0$ and $m_i(\mathbb{T})^2\,\mathrm{ct}(\mathbb{T}:i) = \big(u - m_i'(\mathbb{T})m_i'(\mathbb{T}')\big)\,\mathrm{ct}(\mathbb{T}':i)$; see Exercise 10.1. □

We record the following property of the representing matrices for future reference.

10.1.7. Lemma. *Let λ be a pair of partitions of total size n and let w be any product of distinct generators in $\{s_1, \ldots, s_{n-1}\}$. Then, for $\mathbb{T} \in \mathsf{ST}(\lambda)$,*

$$\mathbb{T}.T_w = \sum_{w' \leq w} \nu_{w'}\mathbb{T}.w' \quad \textit{for certain coefficients } \nu_{w'} \in \mathbf{K},$$

where the summation is over all subexpressions w' of w such that $\mathbb{T}.w' \in \mathsf{ST}(\lambda)$. Moreover, if $w = a_1 \ldots a_l$ for certain $a_j = s_{i_j}$ and $w' = b_1 \ldots b_l$ where either $b_j = a_j$ or $b_j = 1$, $j = 1, \ldots, l$, and if $\mathbb{T}.w' \in \mathsf{ST}(\lambda)$ then the coefficient $\nu_{w'}$ of $\mathbb{T}.w'$ is a product

$$\nu_{w'} = c_1 \cdots c_l, \quad \textit{where} \quad c_j = \begin{cases} m'_{i_j}(\mathbb{T}.b_1 \cdots b_{j-1}) & \textit{if } b_j = a_j, \\ m_{i_j}(\mathbb{T}.b_1 \cdots b_{j-1}) & \textit{if } b_j = 1. \end{cases}$$

Proof. By the definition of the action of $\mathbf{H}(B_n)$ on V^λ in (10.1.4) the image of a tableau $\mathbb{T}$ under S_i is a $\mathbf{K}$-linear combination of $\mathbb{T}$ and $\mathbb{T}.s_i$. The claim thus follows by induction on the length of w. □

We terminate the discussion of the representing matrices with the observation that certain subalgebras of $\mathbf{H}(B_n)$ allow a decomposition of the module V^λ.

10.1.8. Proposition. *Let $\lambda = (\lambda^0, \lambda^1)$ be a pair of partitions of total size n and let $k \leq n$. Further, define a map*

$$\Upsilon\colon \mathsf{ST}(\lambda) \to \coprod_{\mu \subseteq \lambda} \mathsf{ST}(\mu) \times \mathsf{ST}(\lambda - \mu),$$

where the union ranges over all pairs $\mu = (\mu^0, \mu^1)$ of partitions of total size $n-k$, as follows. For each $\mathbb{T} \in \mathsf{ST}(\lambda)$ let $\Upsilon(\mathbb{T}) = (\mathbb{T}_1, \mathbb{T}_2)$ be the decomposition of $\mathbb{T}$ into a tableau $\mathbb{T}_1$ of shape μ, containing the entries $1, \ldots, n-k$ in their respective positions in $\mathbb{T}$, and a tableau $\mathbb{T}_2$ of shape $\lambda - \mu$, containing the entries $n-k+1, \ldots, n$ in their respective positions in $\mathbb{T}$. Then the following hold:

(i) *The map Υ is a bijection.*
(ii) *If $V^{\lambda-\mu}$ is the $\mathbf{K}$-vector space with basis $\mathsf{ST}(\lambda-\mu)$ then, as $\mathbf{K}$-vector space,*

$$V^\lambda \cong \bigoplus_\mu V^\mu \otimes V^{\lambda-\mu},$$

with an explicit isomorphism given by extending Υ linearly.

(iii) $V^{\lambda-\mu}$ *is a module for the subalgebra* $H_k^* := \langle T_{n-k}, S_{n-k+1}, \ldots, S_{n-1}\rangle$ *of* $H(B_n)$ *via*

$$\mathbb{T}.T_{n-k} = u^{n-k}\mathrm{ct}(\mathbb{T} : n-k+1) \quad \textit{and} \quad \mathbb{T}.S_i = m_i(\mathbb{T})\mathbb{T} + m_i'(\mathbb{T})\mathbb{T}'$$

for $i = n-k+1, \ldots, n-1$, *where* $\mathbb{T}' = \mathbb{T}.s_i$ *if* $\mathbb{T}.s_i \in \mathrm{ST}(\lambda - \mu)$ *and* $\mathbb{T}' = 0$ *otherwise.*

(iv) *The map* Υ *induces an* $H(B_{n-k}) \otimes H_k^*$*-module isomorphism from* V^λ *to* $\bigoplus_\mu V^\mu \otimes V^{\lambda-\mu}$.

(v) *We denote the character of* H_k^* *on* $V^{\lambda-\mu}$ *by* $\chi_{\lambda-\mu}$ *and suppose that* $h \in H(B_n)$ *has the form* $h = h_1 h_2$ *for some* $h_1 \in H(B_{n-k})$ *and some* $h_2 \in H_k^*$. *Then*

$$\chi_\lambda(h_1 h_2) = \sum_{\mu \subseteq \lambda} \chi_\mu(h_1)\chi_{\lambda-\mu}(h_2)$$

where the sum is over all pairs of partitions μ *of total size* $n-k$ *with* $\mu \subseteq \lambda$.

Proof. In every standard tableau $\mathbb{T}$ of shape λ, the boxes containing the entries $1, \ldots, n-k$ form a tableau of shape μ for some pair of partitions μ of total size $n-k$ such that $\mu \subseteq \lambda$ and the entries $n-k+1, \ldots, n-1$ in $\mathbb{T}$ form a tableau of shape $\lambda - \mu$. Conversely, a standard tableau of shape μ and a standard tableau of shape $\lambda - \mu$ are uniquely combined into a standard tableau of shape λ. This shows (i).

The remaining points are straightforward consequences of this observation and the definition of the action in (10.1.4). □

10.2 CHARACTER VALUES OF TYPE A

Keeping the set-up of the previous section, we now turn from representing matrices to character values. As a first result, we will derive a formula for elements of a particular shape in $H(B_n)$ which allows us then to completely describe the character values of the algebra $H(\mathfrak{S}_n)$ of type A_{n-1}.

10.2.1. The generic algebra $H(\mathfrak{S}_n)$. The algebra $H(B_n)$ contains the generic algebra $H(\mathfrak{S}_n) = \langle S_i \mid i = 1, \ldots, n-1\rangle$ of type A_{n-1} as a parabolic subalgebra. The relations between the generators S_i are listed among those for $H(B_n)$ in (10.1.1).

The irreducible characters of $\mathfrak{S}_n$ are by Theorem 5.4.5 parametrized by the partitions of n. By Theorem 8.1.7 we also have a bijection $\mathrm{Irr}(H(\mathfrak{S}_n)) \xrightarrow{\sim} \mathrm{Irr}(\mathfrak{S}_n)$, $\chi \mapsto \chi_{u=1}$.

Moreover we see from the construction in (5.5.4), for each partition λ of n, that the characters $\chi_{(\lambda,\varnothing)}$ and $\chi_{(\varnothing,\lambda)}$ in $\mathrm{Irr}(W_n)$ restrict to the character $\chi_\lambda \in \mathrm{Irr}(\mathfrak{S}_n)$. Since specialization is compatible with restriction (9.1.9), the same is true for the corresponding characters of the algebras $H(B_n)$ and $H(\mathfrak{S}_n)$.

Let $k \leq n$ be fixed and $h \in H(B_{n-k})$. We will assume that the character values on h are known and proceed to determine the character values on elements of the form $h \cdot S_{n-k+1} \cdots S_{n-1}$.

By Proposition 10.1.8, we have

$$\chi_\lambda(h \cdot S_{n-k+1} \cdots S_{n-1}) = \sum_\mu \chi_\mu(h)\, \chi_{\lambda-\mu}(S_{n-k+1} \cdots S_{n-1}).$$

On the other hand, since $S_{n-k+1} \cdots S_{n-1}$ is an element of the subalgebra generated by $\{S_{n-k+1}, \ldots, S_{n-1}\}$ of $H(B_n)$ which is isomorphic to $H(\mathfrak{S}_k)$, the character value $\chi_{\lambda-\mu}(S_{n-k+1} \cdots S_{n-1})$ is a linear combination of the character values of $S_1 \cdots S_{k-1}$ in $\mathrm{Irr}(H(\mathfrak{S}_k))$.

10.2.2. Lemma. *Let λ be a pair of partitions of total size n and let μ be a pair of partitions of total size $n-k$ such that $\mu \subseteq \lambda$. Suppose that*

$$\mathrm{Res}^{H_k^*}_{H(\mathfrak{S}_k)}(\chi_{\lambda-\mu}) = \sum_{\delta \vdash k} c^\lambda_{\mu,\delta} \chi_\delta,$$

for certain $c^\lambda_{\mu,\delta} \in \mathbb{N}_0$. Then, if $\delta = [k-r, 1^r]$ for some $r \leq k$ the coefficient $c^\lambda_{\mu,\delta}$ is the number of tableaux $\mathbb{T}$ of shape $\lambda - \mu$ and weight δ such that $w(\mathbb{T})$ is a lattice permutation; see the Littlewood–Richardson rule (6.1.6).

Proof. We have, by Lemma 6.1.3, on the level of groups,

$$\mathrm{Res}^{W_n}_{W_{n-k} \times W_k}(\chi_\lambda) = \sum_{\mu,\nu} c^{\lambda^0}_{\mu^0,\nu^0} c^{\lambda^1}_{\mu^1,\nu^1} (\chi_\mu \boxtimes \chi_\nu),$$

where the sum is over all pairs of partitions μ of total size $n-k$ and all pairs of partitions ν of total size k and where the $c^{\lambda^i}_{\mu^i,\nu^i}$ are Littlewood–Richardson coefficients. Moreover, by Lemma 6.1.4,

$$\mathrm{Res}^{W_k}_{\mathfrak{S}_k}(\chi_\nu) = \sum_{\delta \vdash k} c^\delta_{\nu^0,\nu^1} \chi_\delta$$

with Littlewood–Richardson coefficients $c^\delta_{\nu^0,\nu^1}$. Therefore,

$$c^\lambda_{\mu,\delta} = \sum_{\nu=(\nu^0,\nu^1)} c^{\lambda^0}_{\mu^0,\nu^0} c^{\lambda^1}_{\mu^1,\nu^1} c^\delta_{\nu^0,\nu^1},$$

summed over all pairs of partitions ν of total size k. Now since δ is a hook, and $c^\delta_{\nu^0,\nu^1} > 0$ only if $\nu^i \subseteq \delta$, $i = 0, 1$, both ν^0 and ν^1 are hooks as well. Using Exercise 10.4, we get that $\sum_{\nu=(\nu^0,\nu^1)} c^{\lambda^0}_{\mu^0,\nu^0} c^{\lambda^1}_{\mu^1,\nu^1} c^\delta_{\nu^0,\nu^1}$ is the number of tableaux $\mathbb{T}$ of shape $\lambda - \mu$ and weight δ such that $w(\mathbb{T})$ is a lattice permutation. ☐

10.2.3. Lemma. *Let* $\lambda = (\lambda^0, \lambda^1)$ *be a pair of partitions of total size* n *and let* $\mu = (\mu^0, \mu^1)$ *be a pair of partitions of total size* $n-k$ *such that* $\mu \subseteq \lambda$. *Then*

$$\chi_{\lambda-\mu}(S_{n-k+1}\cdots S_{n-1}) = \begin{cases} (u-1)^{cc^\lambda_\mu - 1}(-1)^{l^\lambda_\mu} u^{k-l^\lambda_\mu - cc^\lambda_\mu}, & \text{if } \lambda-\mu \text{ is a strip,} \\ 0 & \text{otherwise.} \end{cases}$$

Proof. We have $\chi_{\lambda-\mu} = \sum_\delta c^\lambda_{\mu\delta}\chi_\delta$ where the sum is over all partitions δ of k, for certain coefficients $c^\lambda_{\mu\delta}$.

By Exercise 5.8 and Example 9.2.9(b) we have $\chi_\delta(S_1\cdots S_{k-1}) = 0$ unless δ has the form $[k-r, 1^r]$ for some $r \le k$, in which case $\chi_\delta(S_1\cdots S_{k-1}) = (-1)^r u^{k-r}$. Therefore

$$\chi_{\lambda-\mu}(S_{n-k+1}\cdots S_{n-1}) = \sum_{r=0}^{k-1} c^\lambda_{\mu,[k-r,1^r]}(-1)^r u^{k-r-1}.$$

By Lemma 10.2.2 the multiplicity $c^\lambda_{\mu\delta}$ in the case $\delta = [k-r, 1^r]$ for some $r \le k$ equals the number of tableaux $\mathbb{T}$ of shape $\lambda - \mu$ and weight δ such that $w(\mathbb{T})$ is a lattice permutation. For $\delta = [k-r, 1^r]$ this means placing $k-r$ symbols 1 together with the symbols $2, \ldots, r+1$ in a diagram of shape $\lambda - \mu$ in such a way that the entries are increasing from left to right along each row and strictly increasing down each column. This cannot be done in such a way that $w(\mathbb{T})$ is a lattice permutation if $\lambda - \mu$ contains a 2×2 block. Therefore $c^\lambda_{\mu,[k-r,1^r]} = 0$ unless $\lambda - \mu$ is a strip.

Now suppose that $\lambda - \mu$ is a strip. We then write

$$G(\lambda,\mu) := \sum_{r=0}^{k-1} c^\lambda_{\mu,[k-r,1^r]}(-1)^r u^{k-r-1} = \sum_{\mathbf{T}} (-1)^{r(\mathbf{T})} u^{k-r(\mathbf{T})-1}$$

where the sum is over all tableaux $\mathbb{T}$ of shape $\lambda - \mu$ such that $w(\mathbb{T})$ is a lattice permutation and where $r(\mathbb{T}) = r$ for a tableau $\mathbb{T}$ with weight $\delta = [k-r, 1^r]$.

The connected components of the strip $\lambda - \mu$ are hooks, the first of which has the following shape.

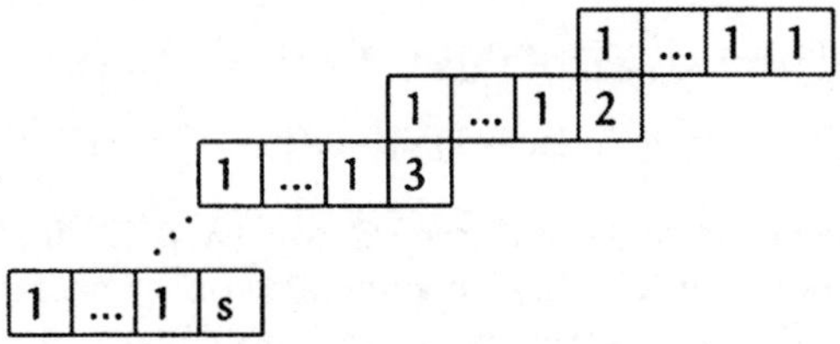

If $\lambda - \mu$ is connected there is only one tableau $\mathbb{T}$ of this shape such that $w(\mathbb{T})$ is a lattice permutation with $r(\mathbb{T}) = l^\lambda_\mu = s - 1$. Thus

$$G(\lambda,\mu) = (-1)^{l^\lambda_\mu} u^{k-l^\lambda_\mu - 1},$$

which proves the claim in this case.

If $\lambda - \mu$ has more than one connected component then the remaining components have one of the following shapes.

(1)

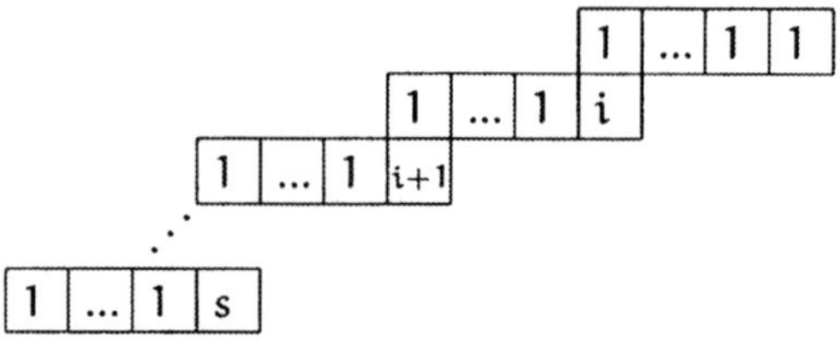

(2)

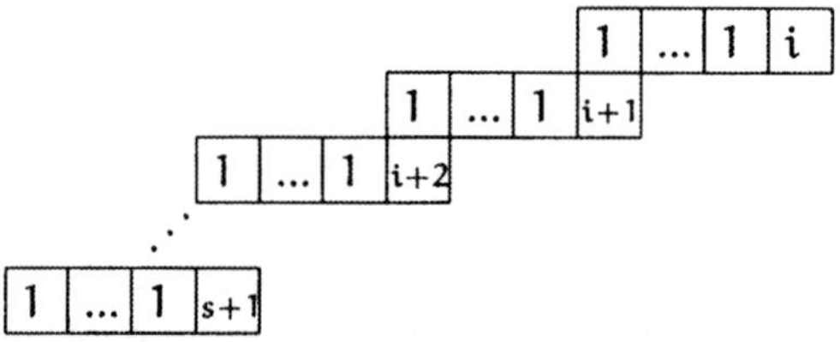

Now consider, for a suitable pair of partitions μ', the diagram $\lambda - \mu'$ consisting of all but the last connected components of $\lambda - \mu$. By induction on the number of connected components cc^{λ}_{μ} we have

$$G(\lambda, \mu') = \sum_{\mathbb{T}'} (-1)^{r(\mathbb{T}')} u^{k - r(\mathbb{T}') - 1} = (u-1)^{cc^{\lambda}_{\mu'} - 1} (-1)^{l^{\lambda}_{\mu'}} u^{k' - cc^{\lambda}_{\mu'} - l^{\lambda}_{\mu'}},$$

where $k' = |\lambda - \mu'|$ and the sum is over all tableaux $\mathbb{T}'$ of shape $\lambda - \mu'$ such that $w(\mathbb{T}')$ is a lattice permutation. By construction, $\lambda - \mu'$ has one connected component less than $\lambda - \mu$, thus $cc^{\lambda}_{\mu'} = cc^{\lambda}_{\mu} - 1$.

Corresponding to each tableau $\mathbb{T}'$ such that $w(\mathbb{T}')$ is a lattice permutation in the sum $G(\lambda, \mu')$ there are exactly two tableaux $\mathbb{T}_1$ and $\mathbb{T}_2$ of shape $\lambda - \mu$. Here both $\mathbb{T}_1$ and $\mathbb{T}_2$ consist of $\mathbb{T}'$ and another component, which for $\mathbb{T}_1$ has the form (1) and for $\mathbb{T}_2$ has the form (2). Thus $r(\mathbb{T}_2) = r(\mathbb{T}_1) + 1$. Their contribution to the sum $G(\lambda, \mu)$ is

$$\begin{aligned} (-1)^{r(\mathbb{T}_1)} u^{k - r(\mathbb{T}_1) - 1} + (-1)^{r(\mathbb{T}_2)} u^{k - r(\mathbb{T}_2) - 1} &= (u-1)(-1)^{r(\mathbb{T}_1)} u^{k - r(\mathbb{T}_1) - 2} \\ = \ (-1)^{r(\mathbb{T}')} u^{k' - r(\mathbb{T}') - 1} \cdot (u-1) \cdot (-1)^{r(\mathbb{T}_1) - r(\mathbb{T}')} u^{k - k' - r(\mathbb{T}_1) + r(\mathbb{T}') - 1}, \end{aligned}$$

which is a multiple of the contribution of $\mathbb{T}'$ to $G(\lambda, \mu')$. Finally, $r(\mathbb{T}_1) - r(\mathbb{T}') = l^{\lambda}_{\mu} - l^{\lambda}_{\mu'}$ is independent of $\mathbb{T}'$, whence summation over all tableaux $\mathbb{T}'$ of shape $\lambda - \mu'$ such that $w(\mathbb{T}')$ is a lattice permutation yields

$$\begin{aligned} G(\lambda, \mu) &= G(\lambda, \mu')(u-1)(-1)^{l^{\lambda}_{\mu} - l^{\lambda}_{\mu'}} u^{k - k' - l^{\lambda}_{\mu} + l^{\lambda}_{\mu'} - 1} \\ &= (u-1)^{c^{\lambda}_{\mu} - 1} (-1)^{l^{\lambda}_{\mu}} u^{k - l^{\lambda}_{\mu} - c^{\lambda}_{\mu}}, \end{aligned}$$

proving the claim in this case as well. □

We can now formulate a recursion formula for the character values of $\mathbf{H}$. By convention, $\chi_{(\emptyset,\emptyset)}$ is the trivial character of the trivial group.

10.2.4. Proposition. *Let $\lambda = (\lambda^0, \lambda^1)$ be a pair of partitions of total size n and let $w \in W_n$ be of the form $w = w' s_{n-k+1} \cdots s_{n-1}$ for some $w' \in W_{n-k}$. Then*

$$\chi_\lambda(T_w) = \sum_{\mu \subseteq \lambda} (u-1)^{cc^\lambda_\mu - 1}(-1)^{l^\lambda_\mu} u^{k - l^\lambda_\mu - cc^\lambda_\mu} \chi_\mu(T_{w'}),$$

where the sum is taken over all pairs of partitions μ of total size $n-k$ such that $\lambda - \mu$ is a strip.

Proof. Since $T_{w'} \in \mathbf{H}(B_{n-k})$ and $S_{n-k+1} \cdots S_{n-1} \in \mathbf{H}^*_k$, we have by Proposition 10.1.8(v),

$$\chi_\lambda(T_w) = \sum_{\mu \subseteq \lambda} \chi_\mu(T_{w'}) \chi_{\lambda-\mu}(S_{n-k+1} \cdots S_{n-1}),$$

summed over all pairs of partitions μ of total size $n-k$ with $\mu \subseteq \lambda$, where $\chi_{\lambda-\mu}(S_{n-k+1} \cdots S_{n-1})$ is determined by Lemma 10.2.3 as 0 unless $\lambda - \mu$ is a strip. □

As a consequence of this proposition and the fact that $\{w \in \mathfrak{S}_n \mid w = w' s_{n-k+1} \cdots s_{n-1}$ for some $w' \in \mathfrak{S}_{n-k}\}$ contains a set of representatives of the conjugacy classes of $\mathfrak{S}_n$ (see Example 3.1.16) we obtain a recursion formula that completely determines the character table of $\mathbf{H}(\mathfrak{S}_n)$.

10.2.5. Theorem. *Let λ be a partition of n and let $w \in \mathfrak{S}_n$ be of the form $w = w' s_{n-k+1} \cdots s_{n-1}$ for some $w' \in \mathfrak{S}_{n-k}$. Then*

$$\chi_\lambda(T_w) = \sum_{\mu \subseteq \lambda} (u-1)^{cc^\lambda_\mu - 1}(-1)^{l^\lambda_\mu} u^{k - l^\lambda_\mu - cc^\lambda_\mu} \chi_\mu(T_{w'}),$$

where the sum is taken over all partitions μ of $n-k$ such that $\lambda - \mu$ is a strip.

Proof. This follows immediately from Proposition 10.2.4 since, for every partition λ of n, we have $\chi_\lambda = \mathrm{Res}^{\mathbf{H}(B_n)}_{\mathbf{H}(\mathfrak{S}_n)}(\chi_{(\lambda,\emptyset)})$, as noted in (10.2.1). □

10.2.6. Example. Table 10.2 shows the character table of the Iwahori–Hecke algebra $\mathbf{H}(\mathfrak{S}_4)$ of type A_3. The characters are labelled by the partitions of 4. For each class a standard representative is listed.

As a further corollary we also obtain the classical Murnaghan–Nakayama formula for the character values of the symmetric group $\mathfrak{S}_n$.

10.2.7. Theorem. *Let λ be a partition of n and let $w \in \mathfrak{S}_n$ be of the form $w = w' s_{n-k+1} \cdots s_{n-1}$ for some $w' \in \mathfrak{S}_{n-k}$. Then*

$$\chi_\lambda(w) = \sum_{\mu \subseteq \lambda} (-1)^{l^\lambda_\mu} \chi_\mu(w'),$$

where the sum is taken over all partitions μ of $n-k$ such that $\lambda - \mu$ is a hook.

Table 10.2 The character table of the generic algebra $H(\mathfrak{S}_4)$

	$[1^4]$	$[2,1^2]$	$[2^2]$	$[3,1]$	$[4]$
χ	1	s_1	s_1s_3	s_1s_2	$s_1s_2s_3$
$[1^4]$	1	-1	1	1	-1
$[2,1^2]$	3	$u-2$	$-2u+1$	$-u+1$	u
$[2^2]$	2	$u-1$	u^2+1	$-u$	0
$[3,1]$	3	$2u-1$	u^2-2u	u^2-u	$-u^2$
$[4]$	1	u	u^2	u^2	u^3

Proof. Set $u = 1$ in Theorem 10.2.5 and use the fact that then the summand for $\mu \subseteq \lambda$ is zero if $\lambda - \mu$ has more than one connected component. □

10.3 CHARACTER VALUES OF TYPE B

As the next step in our programme to determine the character tables of the classical types we prove another formula for elements of a specific shape in $H(B_n)$. This will complete the description of the character values of $H(B_n)$. We start with the character table of the Coxeter group W_n.

10.3.1. Theorem. *Let $\lambda = (\lambda^0, \lambda^1)$ be a pair of partitions such that $|\lambda^0| + |\lambda^1| = n$ and let k be such that $1 \leq k \leq n$. Further let $w' \in W_{n-k}$ and let $w = w'b^{\epsilon}_{n-k,k}$ where the block $b^{\epsilon}_{n-k,k}$ of length k and starting at $n-k$ is as defined in (3.4.2). Then the value of χ_λ on w is given by*

$$\chi_\lambda(w) = \sum_{\mu \subseteq \lambda} \epsilon^{f^\lambda_\mu} (-1)^{l^\lambda_\mu} \chi_\mu(w'),$$

where the sum is taken over all pairs of partitions $\mu = (\mu^0, \mu^1)$ of total size $n-k$ and such that $\lambda - \mu$ is a hook, and where $f^\lambda_\mu = i$ if the hook is contained in the component λ^i of λ.

Proof. Let $n_i = |\lambda_i|$, $i = 0, 1$ and let X be the set of distinguished right coset representatives of $\mathfrak{S}_{n_0} \times \mathfrak{S}_{n_1}$ in $\mathfrak{S}_n$. Then X is a set of coset representatives of $W_{n_0} \times W_{n_1}$ in W_n. Using the set-up of (5.5.4), we have

$$\chi_\lambda(w) = \sum_x (\chi_{\lambda^0} \boxtimes (\varepsilon' \otimes \chi_{\lambda^1}))\,({}^x w)$$

where the sum ranges over all coset representatives $x \in X$ with ${}^x w \in W_{n_0} \times W_{n_1}$.

Let $x \in X$ be such that ${}^x w \in W_{n_0} \times W_{n_1}$ and write ${}^x w = (w_0(x), w_1(x))$. Moreover, let $b = b^{\epsilon}_{n-k,k}$. Since x is order-preserving on the sets $\{1, \ldots, n_0\}$ and $\{n_0+1, \ldots, n_0+n_1\}$ (see Exercise 10.5) we have either $\{n-k+1, \ldots, n\}.x^{-1} = \{n_0-k+1, \ldots, n_0\}$ and ${}^x b = b^{\epsilon}_{n_0-k,k}$ or $\{n-k+1, \ldots, n\}.x^{-1} = \{n-k+1, \ldots, n\}$ and ${}^x b = b$

Accordingly, let $X_0 = \{x \in X \mid {}^x b \neq b\}$ and $X_1 = \{x \in X \mid {}^x b = b\}$. Then X_1 is the set of distinguished right coset representatives x of $\mathfrak{S}_{n_0} \times \mathfrak{S}_{n_1-k}$ in $\mathfrak{S}_{n-k}$ such that ${}^x w' \in W_{n_0} \times W_{n_1-k}$ and X_0 is in bijection to a set of distinguished right coset representatives x of $\mathfrak{S}_{n_0} \times \mathfrak{S}_{n_1-k}$ in $\mathfrak{S}_{n-k}$. Moreover,

$$\chi_\lambda(w) = \sum_{i=0,1} \sum_{x \in X_i} \left(\chi_{\lambda^0}(w_0(x)) \cdot (\varepsilon' \otimes \chi_{\lambda^1})(w_1(x))\right).$$

If ${}^x b \neq b$ then $w_0(x) = w_0'(x){}^x b$ for some $w_0'(x) \in W_{n_0-k}$ and, by Theorem 10.2.7,

$$\chi_{\lambda^0}(\overline{w_0'(x){}^x b}) = \sum_{\mu^0} (-1)^{l^{\lambda^0}_{\mu^0}} \chi_{\mu^0}(\overline{w_0'(x)}),$$

where $\overline{w}$ denotes the image in $\mathfrak{S}_n$ of $w \in W_n$ under the natural projection to $\mathfrak{S}_n$ and where the sum is over all partitions μ^0 of $|\lambda^0| - k$ such that $\mu^0 \subseteq \lambda^0$ and $\lambda^0 - \mu^0$ is a hook.

Similarly, if ${}^x b = b$ then $w_1(x) = w_1'(x) b$ for some $w_1'(x) \in W_{n_1-k}$ and

$$\begin{aligned}(\varepsilon' \otimes \chi_{\lambda^1})(w_1'(x)b) &= \varepsilon'(b)\, \varepsilon'(w_1'(x))\, \chi_{\lambda^1}(\overline{w_1'(x)b}) \\ &= \epsilon\varepsilon'(w_1'(x)) \sum_{\mu^1} (-1)^{l^{\lambda^1}_{\mu^1}} \chi_{\mu^1}(\overline{w_1'(x)}),\end{aligned}$$

where the sum is over all partitions μ^1 of $|\lambda^1| - k$ such that $\mu^1 \subseteq \lambda^1$ and $\lambda^1 - \mu^1$ is a hook.

In total we get

$$\begin{aligned}\chi_\lambda(w) &= \sum_{x \in X_0} \Big(\sum_{\mu^0} (-1)^{l^{\lambda^0}_{\mu^0}} \chi_{\mu^0}(\overline{w_0'(x)}) \cdot (\varepsilon' \otimes \chi_{\lambda^1})(w_1(x)) \Big) \\ &\quad + \sum_{x \in X_1} \chi_{\lambda^0}(w_0(x)) \cdot \Big(\epsilon\varepsilon'(w_1'(x)) \sum_{\mu^1} (-1)^{l^{\lambda^1}_{\mu^1}} \chi_{\mu^1}(\overline{w_1'(x)}) \Big) \\ &= \sum_{\mu = (\mu^0, \mu^1)} \epsilon^{f^\lambda_\mu} (-1)^{l^\lambda_\mu} \chi_\mu(w'),\end{aligned}$$

since in the case $f^\lambda_\mu = 0$ and $x \in X_0$ we have ${}^x w' = w_0'(x) w_1(x)$ and

$$\chi_\mu(w') = \sum_{x \in X_0} \chi_{\mu^0}(w_0'(x)) \cdot (\varepsilon' \otimes \chi_{\lambda^1})(w_1(x)),$$

and in the case $f^\lambda_\mu = 1$ and $x \in X_1$ we have ${}^x w' = w_0(x) w_1'(x)$ and

$$\chi_\mu(w') = \sum_{x \in X_1} \chi_{\lambda^0}(w_0(x)) \cdot (\varepsilon' \otimes \chi_{\mu^1})(w_1'(x)).$$

□

10.3.2. Hooks and the content of a partition. Given a pair of partitions $\lambda = (\lambda^0, \lambda^1)$ of total size n, we define the *content* $\mathrm{ct}(\lambda)$ as

$$\mathrm{ct}(\lambda) = \mathrm{ct}(\mathbb{T}:1)\cdots\mathrm{ct}(\mathbb{T}:n)$$

for any standard tableau $\mathbb{T} \in \mathrm{ST}(\lambda)$. By the definition of $\mathrm{ct}(\mathbb{T}:m)$ in (10.1.4) this is indeed independent of the choice of $\mathbb{T}$.

Further, for some $k < n$, let μ be a pair of partitions of total size $n-k$ such that $\mu \subseteq \lambda$ and consider the action of the element $T_{n-k}\cdots T_{n-1} \in H_k^*$ on $V^{\lambda-\mu}$. By Theorem 10.1.6, we have

$$\mathbb{T}.T_{n-k}\cdots T_{n-1} = \prod_{i=n-k}^{n-1}\left(u^i\,\mathrm{ct}(\mathbb{T}:i{+}1)\right)\mathbb{T} = \frac{\mathrm{ct}(\lambda)}{\mathrm{ct}(\mu)}u^{kn-k(k+1)/2}\mathbb{T}$$

for every $\mathbb{T} \in \mathrm{ST}(\lambda-\mu)$. We then define $e_\mu^\lambda \in \mathbb{Q}$ such that ke_μ^λ is the exponent of u in that scalar, i.e.,

$$\frac{\mathrm{ct}(\lambda)}{\mathrm{ct}(\mu)}u^{kn-k(k+1)/2} = v^{|\lambda^0-\mu^0|}(-1)^{|\lambda^1-\mu^1|}u^{ke_\mu^\lambda}$$

We then have

$$\mathbb{T}.T_{n-k}\cdots T_{n-1} = u^{ke_\mu^\lambda}v^{|\lambda^0-\mu^0|}(-1)^{|\lambda^1-\mu^1|}\mathbb{T}$$

for any standard tableau $\mathbb{T}$ of shape $\lambda-\mu$.

Finally, if $\lambda-\mu$ is a hook and $\mathbb{T} \in \mathrm{ST}(\lambda-\mu)$ then $\{\mathrm{ct}(\mathbb{T}:i) \mid i = n-k,\ldots,n\} = \{(-1)^{f_\mu^\lambda}v^{1-f_\mu^\lambda}u^{d-i} \mid i = 0,\ldots,k-1\}$ for some integer d. It follows that in this case $e_\mu^\lambda = n-k+d \in \mathbb{N}_0$.

Again, fix $k \leq n$ and consider the subalgebras $H(B_{n-k})$ and H_k^* of $H(B_n)$ generated by $T_t, S_1, \ldots, S_{n-k-1}$ and $T_{n-k}, S_{n-k+1}, \ldots, S_{n-1}$, respectively.

10.3.3. Proposition. *Let $\lambda = (\lambda^0, \lambda^1)$ be a pair of partitions of total size n, let $k < n$ and let $w' \in W_{n-k}$. Then*

$$\chi_\lambda(T_{w'}T_{n-k}S_{n-k+1}\cdots S_{n-1}) = \sum_\mu v^{1-f_\mu^\lambda}(-1)^{f_\mu^\lambda}(-1)^{l_\mu^\lambda}u^{e_\mu^\lambda}\chi_\mu(T_{w'})$$

where the sum is taken over all pairs of partitions μ of total size $n-k$ such that $\lambda-\mu$ is a hook and where e_μ^λ is defined as in (10.3.2).

Proof. We first note two properties of the character values of the group. Let λ be a pair of partitions of total size n and let $w_1 \in \langle t, s_1, \ldots, s_{n-k-1}\rangle$ and $w_2 \in \langle t_{n-k}, s_{n-k+1}, \ldots, s_{n-1}\rangle$. Then, by specializing Proposition 10.1.8(v), we have

$$\chi_\lambda(w_1w_2) = \sum_\mu \chi_\mu(w_1)\chi_{\lambda-\mu}(w_2),$$

where the sum is taken over all pairs of partitions μ of total size $n-k$ with $\mu \subseteq \lambda$.

Further, if μ is such a pair of partitions of total size $n-k$ with $\mu \subseteq \lambda$ then, by Theorem 10.3.1,

$$\chi_{\lambda-\mu}(t^i_{n-k}s_{n-k+1}\cdots s_{n-1}) = \begin{cases} (-1)^{f^\lambda_\mu}(-1)^{l^\lambda_\mu} & \text{if } \lambda-\mu \text{ is a hook,} \\ 0 & \text{otherwise.} \end{cases} \qquad (*)$$

It follows from Lemma 4.3.8(ii) that $(T_{n-k}S_{n-k+1}\cdots S_{n-1})^k = T_{n-k}\cdots T_{n-1}$. Next we show that

$$\chi_{\lambda-\mu}(T_{n-k}S_{n-k+1}\cdots S_{n-1}) = \begin{cases} (-1)^{f^\lambda_\mu}v^{1-f^\lambda_\mu}(-1)^{l^\lambda_\mu}u^{e^\lambda_\mu} & \text{if } \lambda-\mu \text{ is a hook,} \\ 0 & \text{otherwise.} \end{cases}$$

Set

$$\rho = ((-1)^{|\lambda^1-\mu^1|}v^{|\lambda^0-\mu^0|})^{1/k}u^{e^\lambda_\mu}$$

where the exponent of (-1) is the number of boxes in component 1, and the exponent of v is the number of boxes in component 0.

Then ρ^k is the only eigenvalue of $T_{n-k}\cdots T_{n-1} = (T_{n-k}S_{n-k+1}\cdots S_{n-1})^k$ on $V^{\lambda-\mu}$ which acts by (10.3.2) as the scalar ρ^k on $V^{\lambda-\mu}$.

Therefore the eigenvalues of $T_{n-k}S_{n-k+1}\cdots S_{n-1}$ are of the form $\zeta^j\rho$ for a kth root of unity ζ and certain $j \in \{0,\ldots,k-1\}$. Hence

$$\chi_{\lambda-\mu}(T_{n-k}S_{n-k+1}\cdots S_{n-1}) = Z\rho$$

for a certain sum Z of kth roots of unity.

Now ρ specializes to some $2k$th root of unity η, and by equation $(*)$ above, we have for the character value of the group

$$\eta Z = \chi_{\lambda-\mu}(t_{n-k}s_{n-k+1}\cdots s_{n-1}) = \begin{cases} \pm 1 & \text{if } \lambda-\mu \text{ is a hook,} \\ 0 & \text{otherwise.} \end{cases}$$

Thus we must have $Z = 0$ if $\lambda-\mu$ is not a hook and we conclude that in this case $\chi_{\lambda-\mu}(T_{n-k}S_{n-k+1}\cdots S_{n-1}) = 0$ as well.

Otherwise, if $\lambda-\mu$ is a hook then all boxes of the diagram $\lambda-\mu$ lie in the same component of the diagram λ and we have either $\rho = u^{e^\lambda_\mu}v$ or $\rho = -u^{e^\lambda_\mu}$ which specialize to 1 and -1. It then follows from $(*)$ that $Z = (-1)^{l^\lambda_\mu}$. □

We thus obtain a complete description of the character table of $H(B_n)$. The case $n = 3$ is shown in Table 10.3 on p. 336.

10.3.4. Theorem. *Let* $\lambda = (\lambda^0, \lambda^1)$ *be a pair of partitions of total size* n *and let* (α, β) *be a pair of compositions such that* $|\alpha| + |\beta| = n$. *Let*

$$w = w_{(\alpha,\beta)} = b^{\epsilon_1}_{m_1,e_1}\cdots b^{\epsilon_r}_{m_r,e_r} \in W_n$$

be the element in signed block form as defined in equation (2) *in* (3.4.2).

Table 10.3 The character table of the generic algebra $H(B_3)$

	$(1^3.\varnothing)$	$(1^2.1)$	(1.1^2)	$(\varnothing.1^3)$	$(21.\varnothing)$	(1.2)	(2.1)	$(\varnothing.12)$	$(3.\varnothing)$	$(\varnothing.3)$
χ	1	t_0	t_0t_1	$t_0t_1t_2$	s_1	t_0s_1	t_0s_2	$t_0t_1s_2$	s_1s_2	$t_0s_1s_2$
$(1^3.\varnothing)$	1	v	v^2	v^3	-1	$-v$	$-v$	$-v^2$	1	v
$(1^2.1)$	3	$2v-1$	v^2-2uv	$-3u^2v^2$	$u-2$	$-v$	$uv-v+1$	uv	$1-u$	0
(1.1^2)	3	$v-2$	$-2uv+1$	$3u^2v$	$u-2$	1	$-v-u+1$	uv	$1-u$	0
$(\varnothing.1^3)$	1	-1	1	-1	-1	1	1	-1	1	-1
$(21.\varnothing)$	2	$2v$	$u^2v^2+v^2$	$2u^3v^3$	$u-1$	$uv-v$	$uv-v$	0	$-u$	$-uv$
(1.2)	3	$v-2$	$-2uv+u^2$	$3u^4v$	$2u-1$	$-u$	$uv-u+1$	$-u^2v$	u^2-u	0
(2.1)	3	$2v-1$	u^2v^2-2uv	$-3u^4v^2$	$2u-1$	uv	$uv-v-u$	$-u^2v$	u^2-u	0
$(\varnothing.21)$	2	-2	u^2+1	$-2u^3$	$u-1$	$1-u$	$1-u$	0	$-u$	u
$(3.\varnothing)$	1	v	u^2v^2	u^6v^3	u	uv	uv	u^3v^2	u^2	u^2v
$(\varnothing.3)$	1	-1	u^2	$-u^6$	u	$-u$	$-u$	u^3	u^2	$-u^2$

Further let $k = e_r$ and let $w' = b^{\epsilon_1}_{m_1,e_1} \cdots b^{\epsilon_{r-1}}_{m_{r-1},e_{r-1}}$. Then the value of χ_λ on T_w is given by

$$\chi_\lambda(T_w) = \sum_{\mu} (u-1)^{cc^\lambda_\mu - 1}(-1)^{l^\lambda_\mu} u^{k - l^\lambda_\mu - cc^\lambda_\mu} \chi_\mu(T_{w'}),$$

where the sum is taken over all double partitions μ of $n-k$, such that $\lambda - \mu$ is a strip if $\alpha \neq \varnothing$, and

$$\chi_\lambda(T_w) = \sum_{\mu} v^{1-f^\lambda_\mu}(-1)^{f^\lambda_\mu}(-1)^{l^\lambda_\mu} u^{e^\lambda_\mu} \chi_\mu(T_{w'})$$

where the sum is taken over all double partitions μ of $n-k$, such that $\lambda - \mu$ is a hook in the case $\alpha = \varnothing$ and where e^λ_μ is defined as in (10.3.2).

Proof. This follows from Propositions 10.2.4 and 10.3.3. □

10.4 CHARACTER VALUES OF TYPE D

In this section we study the representations of the generic Iwahori–Hecke algebra $H(D_n)$ corresponding to the Coxeter group of type D_n and derive a character formula that determines the character table of $H(D_n)$.

10.4.1. The generic algebra $H(D_n)$. Let $n \geq 4$ and denote by $H(B_n)'$ the Iwahori–Hecke algebra of type B_n with parameters u and 1. Then $H(B_n)'$ is obtained from the generic algebra $H(B_n)$ via the specialization $v \mapsto 1$ and, hence, satisfies the relations in (10.1.1), where the first relation now has the form

$$T_t^2 = 1.$$

Define $U = T_t S_1 T_t = T_{ts_1t} \in H(B_n)'$. Then $U^2 = T_t S_1^2 T_t = (u-1)U + u$ and we have the following relations:

$$\begin{aligned} U^2 &= (u-1)U + u \\ S_i^2 &= (u-1)S_i + u && i = 1, \ldots, n-1 \\ US_i &= S_i U && i \neq 2 \\ US_2U &= S_2US_2 \\ S_iS_{i+1}S_i &= S_{i+1}S_iS_{i+1}, && i = 1, \ldots, n-2, \\ S_iS_j &= S_jS_i, && |i-j| > 1. \end{aligned}$$

This shows, first of all, that the subspace $H(D_n) := \langle U, S_1, \ldots, S_{n-1} \rangle \subseteq H(B_n)'$ is a subalgebra whose generators satisfy the defining relations for the generic algebra of type D_n. Since we also have $H(B_n)' = H(D_n) \oplus T_t H(D_n)$, a dimension argument shows that $H(D_n)$ is in fact the generic algebra of type D_n.

10.4.2. Irreducible characters of $H(D_n)$. In (5.6.1) we saw that the irreducible characters of a Coxeter group of type D_n are easily described in terms of the restriction of characters from type B_n and using some basic results from Clifford theory. We have a similar description on the level of Iwahori–Hecke algebras. This is based on the following facts.

First, consider the algebra $H(B_n)'$. If we specialize $u \mapsto 1$, the corresponding specialized algebra is the group algebra of W_n. Applying Tits's Theorem 7.4.6 and arguing as in the proof of Theorem 8.1.7, we conclude that $H(B_n)'$ is semisimple and that we have a bijection $\mathrm{Irr}(H(B_n)') \xrightarrow{\sim} \mathrm{Irr}(W_n)$, $\chi \mapsto \chi_{u=1}$.

Now consider the subalgebra $H(D_n) \subset H(B_n)'$. Again, by Theorem 8.1.7, we have a bijection $\mathrm{Irr}(H(D_n)) \xrightarrow{\sim} \mathrm{Irr}(W_n')$, $\chi \mapsto \chi_{u=1}$. On the other hand, we can restrict characters from $H(B_n')$ to $H(D_n)$ and from W_n to W_n'. These operations commute, i.e., we have

$$\left(\mathrm{Res}_{H(D_n)}^{H(B_n)'}(\chi)\right)_{u=1} = \mathrm{Res}_{W_n'}^{W_n}(\chi_{u=1}) \qquad \text{for all } \chi \in \mathrm{Irr}(H(B_n')).$$

(This follows from two facts: (a) the bijection in Theorem 8.1.7 is defined in terms of a decomposition map; and (b) the decomposition map is compatible with restriction to subalgebras.) Thus the restriction of characters from $H(B_n)'$ to $H(D_n)$ follows the same pattern as that on the level of the groups $W_n' \subset W_n$.

Recall from Section 5.5 that the irreducible characters of W_n are labelled by the pairs of partitions of total size n. Let (λ, μ) be a pair of partitions with $|\lambda| + |\mu| = n$ and let $\chi_{(\lambda,\mu)}$ be the corresponding character in $\mathrm{Irr}(H(B_n)')$. Denote by $\chi'_{(\lambda,\mu)}$ its restriction to $H(D_n)$. Transferring the results from (5.6.1), we have two cases:

(a) If $\lambda \neq \mu$ then we have $\chi'_{(\lambda,\mu)} = \chi'_{(\mu,\lambda)} \in \mathrm{Irr}(H(D_n))$.

(b) If $\lambda = \mu$ then $\chi'_{(\lambda,\lambda)} = \chi_{(\lambda,+)} + \chi_{(\lambda,-)}$ where $\chi_{(\lambda,+)}$ and $\chi_{(\lambda,-)}$ are two distinct characters in $\mathrm{Irr}(H(D_n))$.

Thus, as for the group, $\mathrm{Irr}(H(D_n))$ consists of the characters $\chi'_{(\lambda,\mu)} = \chi'_{(\mu,\lambda)}$ for pairs of partitions (λ, μ) with $\lambda \neq \mu$ and, in the case of even n, also of the characters $\chi_{(\lambda,\pm)}$ for the partitions λ of $n/2$.

The aim of this section is to determine the character table of $H(D_n)$. In the light of Definition 8.2.9, this requires the values of the characters in $\mathrm{Irr}(H(D_n))$ on a set of elements T_w where w is a representative of minimal length of a conjugacy class of W'_n.

Such a set of representatives is given in Proposition 3.4.12 as

$$\{w_{(\alpha,\beta)} \mid (\alpha, \beta) \in \mathcal{P}_2\} \amalg \{w_{(\alpha,\varnothing)} \mid \alpha \in \mathcal{P}_1\} \amalg \{w^t_{(\alpha,\varnothing)} \mid \alpha \in \mathcal{P}_1\}$$

where $\mathcal{P}_1$ is the set of all decreasing compositions $\alpha = (\alpha_1, \ldots, \alpha_l)$ of n with α_i even for all i, and $\mathcal{P}_2$ is the set of pairs (α, β) consisting of a decreasing composition α and an increasing composition β such that $|\alpha| + |\beta| = n$, that β has an even number of parts and that $\alpha \notin \mathcal{P}_1$.

10.4.3. Proposition. *Let (λ, μ) be a pair of partitions with $\lambda \neq \mu$ and such that $|\lambda| + |\mu| = n$ and consider the corresponding irreducible character $\chi'_{(\lambda,\mu)} \in \mathrm{Irr}(H(D_n))$. Furthermore, let (α, β) be a pair of compositions where β has an even number of parts, and consider the corresponding element $w = w_{(\alpha,\beta)} \in W'_n$, as defined in (3.4.2). Then we have*

$$\chi'_{(\lambda,\mu)}(T_w) = \chi'_{(\lambda,\mu)}(T_{twt}) = \chi_{(\lambda,\mu)}(T_w)_{v=1}.$$

Proof. The value of $\chi_{(\lambda,\mu)}$ on $T_{w_{(\alpha,\beta)}}$ is determined by specializing the formula in Theorem 10.3.4 under $v \mapsto 1$. Moreover, we have $T_{tw_{(\alpha,\beta)}t} = T_t^{-1} T_{w_{(\alpha,\beta)}} T_t$, since $T_t^2 = 1$ in $H(B'_n)$, whence $\chi_{(\lambda,\mu)}(T_{tw_{(\alpha,\varnothing)}t}) = \chi_{(\lambda,\mu)}(T_{w_{(\alpha,\varnothing)}})$. □

This completely determines the part of the character table of $H(D_n)$ corresponding to the pairs of partitions $\lambda \neq \mu$. We are left with the task of determining the character values of $\chi_{(\lambda,\pm)}$ on the set of representatives of minimal length. For this, we will work with the specializations under $v \mapsto 1$ of the matrix representations in (10.1.4).

10.4.4. Matrices. Let (λ, μ) be a pair of partitions such that $|\lambda| + |\mu| = n$. Denote by $\mathrm{ST}(\lambda, \mu)$ the set of pairs of standard tableaux $\mathbb{T} = (\mathbb{T}^0, \mathbb{T}^1)$ of shape (λ, μ). By (10.1.4)(1), we have for $\mathbb{T} \in \mathrm{ST}(\lambda, \mu)$ and $1 \leq m \leq n$ in the current setting with $v = 1$ that

$$\mathrm{ct}(\mathbb{T} : m) = (-1)^l u^{j-i}$$

if the box containing m in $(\mathbb{T} = \mathbb{T}^0, \mathbb{T}^1)$ is in position (i, j) of $\mathbb{T}^l$. Note that $\mathrm{ct}(\mathbb{T} : 1) = (-1)^l$ for each $\mathbb{T} \in \mathrm{ST}(\lambda, \mu)$ since 1 always occupies position $(1, 1)$ in the component $\mathbb{T}^l$ containing it.

Let $V^{(\lambda,\mu)}$ be the K-vector space and $H(B_n)'$-module with basis $ST(\lambda,\mu)$. By (10.1.4), the operation of $H(D_n)$ on $V^{(\lambda,\mu)}$ is given by

$$\mathbb{T}.U = m_1(\mathbb{T})\mathbb{T} - m_1'(\mathbb{T})\mathbb{T}',$$

where $\mathbb{T}' = \mathbb{T}s_1$ if $\mathbb{T}s_1 \in ST(\lambda,\mu)$ and $\mathbb{T}' = 0$ otherwise, and

$$\mathbb{T}.S_i = m_i(\mathbb{T})\mathbb{T} + m_i'(\mathbb{T})\mathbb{T}',$$

for $i = 1, \ldots, n-1$, where $\mathbb{T}' = \mathbb{T}s_i$ if $\mathbb{T}s_i \in ST(\lambda,\mu)$ and $\mathbb{T}' = 0$ otherwise.

10.4.5. Proposition. *For a double partition* (λ,μ) *of* n *consider the linear map* $\sigma\colon V^{(\lambda,\mu)} \to V^{(\mu,\lambda)}$ *defined by*

$$(\mathbb{T}^0, \mathbb{T}^1).\sigma = (\mathbb{T}^1, \mathbb{T}^0)$$

for every pair of tableaux $(\mathbb{T}^0, \mathbb{T}^1) \in ST(\lambda,\mu)$. *Then* σ *is an* $H(D_n)$*-module isomorphism.*

Proof. We need to show that $(v.h).\sigma = (v.\sigma).h$ for all $v \in V^{(\lambda,\mu)}$ and all $h \in H(D_n)$. It is sufficient to show this for the basis $ST(\lambda,\mu)$ of $V^{(\lambda,\mu)}$ and for a generating set of $H(D_n)$. So let $\mathbb{T} \in ST(\lambda,\mu)$ and note that

$$ct(\mathbb{T}.\sigma : m) = -ct(\mathbb{T} : m)$$

for all $m \in \{1, \ldots, n\}$. It follows that

$$\frac{ct(\mathbb{T}.\sigma : i)}{ct(\mathbb{T}.\sigma : i+1)} = \frac{ct(\mathbb{T} : i)}{ct(\mathbb{T} : i+1)}$$

and hence, $m_i(\mathbb{T}.\sigma) = m_i(\mathbb{T})$ for all $i = 1, \ldots, n-1$. Furthermore, we have that $(\mathbb{T}.s_i).\sigma = (\mathbb{T}.\sigma).s_i$ for all $i = 1, \ldots, n-1$ and that $\mathbb{T}.s_1 \notin ST(\lambda,\mu)$ if and only if $(\mathbb{T}.\sigma).s_i \notin ST(\mu,\lambda)$. This finally yields $(\mathbb{T}.\sigma).U = (\mathbb{T}.U).\sigma$ and $(\mathbb{T}.\sigma).S_i = (\mathbb{T}.S_i).\sigma$ for all $i = 1, \ldots, n-1$. □

10.4.6. From now until the end of the section we assume that n is even and that λ is a partition of $n/2$. Then σ acts as an endomorphism on $V = V^{(\lambda,\lambda)}$. Since $\sigma^2 = 1$ the possible eigenvalues of σ are ± 1. Let

$$V^{(\lambda,+)} = \ker(\sigma - \mathrm{id}_V) \quad \text{and} \quad V^{(\lambda,-)} = \ker(\sigma + \mathrm{id}_V)$$

be the subspaces of $V^{(\lambda,\lambda)}$ where σ acts as 1 and as -1, respectively. Then, since σ commutes with the action of $H(D_n)$ on V by Proposition 10.4.5 and since σ is neither id_V nor $-\mathrm{id}_V$, both $V^{(\lambda,+)}$ and $V^{(\lambda,-)}$ are non-zero $H(D_n)$-modules. Moreover,

$$V^{(\lambda,\lambda)} = V^{(\lambda,+)} \oplus V^{(\lambda,-)}. \tag{1}$$

We define $\chi_{(\lambda,+)}$ to be the character of $V^{(\lambda,+)}$ and $\chi_{(\lambda,-)}$ to be that of $V^{(\lambda,-)}$. (We will see in (10.4.10) below that this choice of sign is consistent with that in Proposition 5.6.3.)

In order to find the character values of $\chi_{(\lambda,\pm)}$ we will determine their difference. We have by (1), for all $h \in H(D_n)$,

$$\chi_{(\lambda,\lambda)}(h) = \chi_{(\lambda,+)}(h) + \chi_{(\lambda,-)}(h) \quad \text{and} \quad \chi_{(\lambda,\lambda)}(h\sigma) = \chi_{(\lambda,+)}(h) - \chi_{(\lambda,-)}(h),$$

since σ acts as 1 on $V^{(\lambda,+)}$ and as -1 on $V^{(\lambda,-)}$. (Here $\chi_{(\lambda,\lambda)}(h\sigma)$ is the trace of the product of the endomorphisms h and σ on $V^{(\lambda,\lambda)}$.) It follows that

$$\chi_{(\lambda,+)}(h) = \frac{\chi_{(\lambda,\lambda)}(h) + \chi_{(\lambda,\lambda)}(h\sigma)}{2} \quad \text{and} \quad \chi_{(\lambda,-)}(h) = \frac{\chi_{(\lambda,\lambda)}(h) - \chi_{(\lambda,\lambda)}(h\sigma)}{2},$$

for all $h \in H(D_n)$. Now note that we have the following relation between the character values of $\chi_{(\lambda,\pm)}$:

$$\chi_{(\lambda,+)}(T_{twt}) = \chi_{(\lambda,-)}(T_w) \qquad \text{for any } w \in W_n'. \tag{2}$$

In particular, we have $\chi_{(\lambda,\pm)}(T_w) = \frac{1}{2}\chi_{(\lambda,\lambda)}(T_w)$ if $w, twt \in W_n'$ are conjugate in W_n' and have minimal length in their class. This is seen as follows. The map $T_w \mapsto T_{twt} = T_tT_wT_t$ defines an algebra automorphism of $H(D_n)$ (note that $T_t^2 = 1$). Hence the assignment ψ: $T_w \mapsto \chi_{(\lambda,+)}(T_{twt})$ defines an irreducible character of $H(D_n)$. If we set $u = 1$ and use the results in (5.6.1), we see that both ψ and $\chi_{(\lambda,-)}$ specialize to the same character of W_n'. Hence (2) holds. Furthermore, note that if $w, twt \in W_n'$ are conjugate in W_n' and have minimal length in their class then T_w and T_{twt} are also conjugate in $H(D_n)$.

So it remains to evaluate $\chi_{(\lambda,\pm)}$ on those representatives $w_{(\alpha,\beta)}$ which are not conjugate to $tw_{(\alpha,\beta)}t$ in W_n'. By Proposition 3.4.12, these are elements of the form $w_{(\alpha,\varnothing)}$ or $tw_{(\alpha,\varnothing)}t$ where α is a composition of n having only even parts. This means

$$w_{(\alpha,\varnothing)} = s_1b_2s_3b_4\cdots b_{n-2}s_{n-1}, \tag{3}$$

where $b_i = 1$ or s_i for all even indices $i = 2, 4, \ldots, n-2$.

10.4.7. Theorem. *Let λ be a partition of $n/2$, let $\alpha = (\alpha_1, \ldots, \alpha_m)$ be a composition of n with all parts α_i even, and let $w = w_{(\alpha,\varnothing)}$ as defined in (3.4.2). Moreover, let $\alpha' = (\alpha_1', \ldots, \alpha_m')$ be the composition of $n/2$ obtained from α by setting $\alpha_i' = \alpha_i/2$. Then*

$$\chi_{(\lambda,+)}(T_w) - \chi_{(\lambda,-)}(T_w) = (u+1)^m\chi_\lambda(T_v')$$

where $v = w_{(\alpha',\varnothing)} \in \mathfrak{S}_{n/2}$ and T_v' is the corresponding element in the algebra $H(\mathfrak{S}_{n/2})$ with parameter u^2.

Proof. As the trace of $T_w\sigma$ on $V = V^{(\lambda,\lambda)}$, the value of $\chi_{(\lambda,\lambda)}(T_w\sigma)$ is the sum

$$\chi_{(\lambda,\lambda)}(T_w\sigma) = \mathrm{Tr}_V(T_w\sigma) = \sum_{\mathbb{T}\in ST(\lambda,\lambda)} a^{T_w}_{\mathbb{T},\mathbb{T}.\sigma}, \tag{1}$$

where $a^{T_w}_{\mathbb{T},\mathbb{T}'}$ denotes the coefficient of $\mathbb{T}'$ in $\mathbb{T}.T_w = \sum_{\mathbb{T}'} a^{T_w}_{\mathbb{T},\mathbb{T}'}\mathbb{T}'$.

Step 1. Let $R = \{s_1, s_3, \dots, s_{n-1}\} \subseteq S$ and, as usual, denote by $w_R = s_1 s_3 \cdots s_{n-1}$ the longest element of the parabolic subgroup W_R of W_n. We show that $a^{T_w}_{\mathbb{T},\mathbb{T}.\sigma} = 0$ unless $\mathbb{T}$ satisfies $\mathbb{T}.\sigma = \mathbb{T}.w_R$.

By Lemma 10.1.7, $\mathbb{T}.T_w$ is a linear combination of those tableaux $\mathbb{T}.w'$ where w' is a subexpression of w and $\mathbb{T}.w' \in ST(\lambda,\lambda)$. Now σ permutes, for every row i and every column j, the two entries in $\mathbb{T}^0$ and $\mathbb{T}^1$ in this position, whence the permutation induced by σ on $\{1,\dots,n\}$ has cycle structure $[2^{n/2}]$. But the only subexpression of $w = w_{(\alpha,\varnothing)}$ (which for every choice of α is a subexpression of the Coxeter element $s_1 s_2 \cdots s_{n-1}$ of $\mathfrak{S}_n$) with cycle structure $[2^{n/2}]$ is $w_R = s_1 s_3 \cdots s_{n-1} = (1,2)(3,4)\cdots(n-1,n)$. Therefore, for every row i and every column j, the entries contained in $\mathbb{T}^0$ and $\mathbb{T}^1$ in this position are $\{2k-1, 2k\}$.

Step 2. Denote by $ST'(\lambda,\lambda) = \{\mathbb{T} \in ST(\lambda,\lambda) \mid \mathbb{T}.\sigma = \mathbb{T}.w_R\}$ the set of those tableaux that contribute to the trace $\mathrm{Tr}_V(T_w\sigma)$. Further, let $y_1 = 1$ and for $k > 1$ let $y_k = (s_k s_{k+1} \cdots s_{2k-2}) y_{k-1}$. Then $y_{n/2}$ maps $\{1, 2, \dots, \frac{n}{2}\}$ to $\{1, 3, \dots, n-1\}$ (in the given order) and $\{\frac{n}{2}+1, \frac{n}{2}+2, \dots, n\}$ to $\{2, 4, \dots, n\}$ (in that order). We show that the map

$$\Psi\colon ST(\lambda) \times 2^R \to ST'(\lambda,\lambda),\ (\mathbb{T}, L) \mapsto (\mathbb{T}.bw_L, \mathbb{T}.bw_{R-L}),$$

which maps a pair consisting of a standard tableau $\mathbb{T}$ of shape λ and a subset L of R to a pair of tableaux in $ST'(\lambda,\lambda)$, is a bijection.

First, we note that $\Psi(\mathbb{T}, L) = (\mathbb{T}.bw_L, \mathbb{T}.bw_{R-L}) \in ST'(\lambda,\lambda)$ since we have $(\mathbb{T}.bw_L).w_R = \mathbb{T}.bw_{R-L}$ for all $L \subseteq R$ and all $\mathbb{T} \in ST(\lambda)$.

Next suppose that $\Psi(\mathbb{T}, L) = \Psi(\mathbb{T}', L')$ for some $\mathbb{T}, \mathbb{T}' \in ST(\lambda)$ and some $L, L' \subseteq R$. Then from $\mathbb{T}'.b = (\mathbb{T}.b).w_L w_{L'}$ it follows that $w_L = w_{L'}$ since both $\mathbb{T}.b$ and $\mathbb{T}'.b$ contain the numbers $\{1, 3, \dots, n-1\}$ and w_L and $w_{L'}$ are products of factors $(2k-1, 2k)$ for certain $k \in \{1, 2, \dots, \frac{n}{2}\}$. But then also $\mathbb{T} = \mathbb{T}'$.

Finally, let $(\mathbb{T}^0, \mathbb{T}^1) \in ST'(\lambda,\lambda)$ and construct a pre-image $(\mathbb{T}, L)$ as follows. For each $k \in \{1, 2, \dots, \frac{n}{2}\}$ let $\mathbb{T} \in ST(\lambda)$ contain the entry k in the position of the entries $2k-1$ and $2k$ in $\mathbb{T}^0$ and $\mathbb{T}^1$ and let L be such that it contains s_{2k-1} if and only if $2k$ is contained in $\mathbb{T}^0$. Then $\Psi(\mathbb{T}, L) = (\mathbb{T}^0, \mathbb{T}^1)$. Therefore, Ψ is both injective and surjective.

Since $\Psi(\mathbb{T}, L).\sigma = \Psi(\mathbb{T}, R - L)$, we conclude that

$$\chi_{(\lambda,\lambda)}(T_w\sigma) = \sum_{\mathbb{T}\in ST'(\lambda,\lambda)} a^{T_w}_{\mathbb{T},\mathbb{T}.\sigma} = \sum_{\mathbb{T}\in ST(\lambda)} \sum_{L\subseteq R} a^{T_w}_{\Psi(\mathbb{T},L),\Psi(\mathbb{T},R-L)}. \tag{2}$$

Step 3. Fix $\mathbb{T} = \Psi(\mathbb{T}', L) \in \mathsf{ST}'(\lambda, \lambda)$ and set $\mathbb{T}_j = \mathbb{T}.s_1 s_3 \cdots s_k$ where k is the largest odd index $\le j$. Then, by Proposition 10.1.7, we have

$$a^{T_w}_{\mathbb{T},\mathbb{T}.\sigma} = M_0(\mathbb{T}) \cdot M_1(\mathbb{T}) \cdots M_{n-1}(\mathbb{T})$$

where we set $M_0(\mathbb{T}) = 1$ and where $M_j(\mathbb{T})$ is the coefficient of $\mathbb{T}_j$ in $\mathbb{T}_{j-1}.T_{b_j}$. For $j = 2, 4, \ldots, n-2$ set $\delta_j = 1$ if both j and $j+2$ lie in the same component $\mathbb{T}^1$ or $\mathbb{T}^0$ of $\mathbb{T}$ and $\delta_j = -1$ otherwise. We evaluate $M_j(\mathbb{T})$ as follows:

(i) If j is odd then $M_j(\mathbb{T}) = \frac{1}{2}(u+1)$.
(ii) If j is even and $b_j = 1$ then $M_j(\mathbb{T}) = 1$.
(iii) If j is even and $b_j = s_j$ then

$$M_j(\mathbb{T}) = (u-1) \Bigg/ \left(1 - \delta_j \frac{\mathrm{ct}(\mathbb{T}' : \frac{j}{2})}{\mathrm{ct}(\mathbb{T}' : \frac{j}{2} + 1)}\right).$$

For (i) note that if j is odd then $M_j(\mathbb{T})$ is the coefficient of $\mathbb{T}_j$ in $\mathbb{T}_{j-1}.S_j$. Thus by (10.4.4), $M_j(\mathbb{T}) = m_j'(\mathbb{T}_{j-1}) = m_j(\mathbb{T}_{j-1})+1$. Moreover, $\mathbb{T}_{j-1} = \mathbb{T}s_1 s_3 \cdots s_{j-2}$, whence the positions of j and $j+1$ in $\mathbb{T}_{j-1}$ are the same as in $\mathbb{T}$. Furthermore, $j+1 = j.w_R$. Hence

$$\frac{\mathrm{ct}(\mathbb{T}_{j-1} : j)}{\mathrm{ct}(\mathbb{T}_{j-1} : j+1)} = \frac{\mathrm{ct}(\mathbb{T} : j)}{\mathrm{ct}(\mathbb{T} : j.w_R)} = \frac{\mathrm{ct}(\mathbb{T} : j)}{-\mathrm{ct}(\mathbb{T} : j)} = -1.$$

Therefore, $M_j(\mathbb{T}) = m_j'(\mathbb{T}_{j-1}) = \frac{1}{2}(u-1) + 1 = \frac{1}{2}(u+1)$.

For (ii) and (iii) we have that $\mathbb{T}_j = \mathbb{T}_{j-1}$ and $M_j(\mathbb{T})$ is the coefficient of $\mathbb{T}_{j-1}$ in $\mathbb{T}_{j-1}.b_j$. If $b_j = 1$ then clearly $M_j(\mathbb{T}) = 1$. So suppose $b_j = s_j$. Then, by (10.4.4), $M_j(\mathbb{T}) = m_j(\mathbb{T}_{j-1})$. Moreover, $\mathbb{T}_{j-1} = \mathbb{T}s_1 s_3 \cdots s_{j-1}$, whence the position of $j+1$ is the same in $\mathbb{T}$ and $\mathbb{T}_{j-1}$ while $j-1 = j.s_{j-1}$ has in $\mathbb{T}$ the position of j in $\mathbb{T}_{j-1}$. Furthermore, $j.s_{j-1} = j.w_R$ and $j+1 = (j+2).w_R$. Thus

$$\frac{\mathrm{ct}(\mathbb{T}_{j-1} : j)}{\mathrm{ct}(\mathbb{T}_{j-1} : j+1)} = \frac{\mathrm{ct}(\mathbb{T} : j.s_{j-1})}{\mathrm{ct}(\mathbb{T} : j+1)} = \frac{\mathrm{ct}(\mathbb{T} : j.w_R)}{\mathrm{ct}(\mathbb{T} : (j+2).w_R)} = \frac{\mathrm{ct}(\mathbb{T} : j)}{\mathrm{ct}(\mathbb{T} : j+2)}.$$

Now $\mathbb{T} = \Psi(\mathbb{T}', L)$. This implies $\mathrm{ct}(\mathbb{T} : j) = \epsilon_j\, \mathrm{ct}(\mathbb{T}' : \frac{j}{2})$, where $\epsilon_j = 1$ or -1 depending on whether j is in $\mathbb{T}^0$ or in $\mathbb{T}^1$ (which is equivalent to $s_{j-1} \notin L$ or $s_{j-1} \in L$), and similarly $\mathrm{ct}(\mathbb{T} : j+2) = \epsilon_{j+2}\, \mathrm{ct}(\mathbb{T}' : \frac{j}{2} + 1)$. Therefore, if $b_j = s_j$ then

$$M_j(\mathbb{T}) = m_j(\mathbb{T}_j) = (u-1) \Bigg/ \left(1 - \epsilon_j \epsilon_{j+2} \frac{\mathrm{ct}(\mathbb{T}' : \frac{j}{2})}{\mathrm{ct}(\mathbb{T}' : \frac{j}{2} + 1)}\right).$$

Consequently, we have

$$a^{T_w}_{\mathbb{T},\mathbb{T}} = \frac{1}{2}(u+1) \cdot M_2(\mathbb{T}) \cdot \frac{1}{2}(u+1) \cdot M_4(\mathbb{T}) \cdots M_{n-2}(\mathbb{T}) \cdot \frac{1}{2}(u+1) \qquad (3)$$

where, for j even, $M_j(\mathbb{T}) = 1$ or $M_j(\mathbb{T}) = m_j(\mathbb{T}_j)$ as above depending on whether $b_j = 1$ or $b_j = s_j$.

Step 4. Suppose that j is even. We evaluate the sum $M_j(\mathbb{T}) + M_j(\mathbb{T}.s_{j-1})$ as follows:

If $b_j = 1$ then we have $M_j(\mathbb{T}) + M_j(\mathbb{T}.s_{j-1}) = 2$. If $b_j = s_j$ then

$$M_j(\mathbb{T}) + M_j(\mathbb{T}.s_{j-1}) = \frac{(u-1)}{1-Q} + \frac{(u-1)}{1+Q} = \frac{2(u-1)}{1-Q^2}$$

where

$$Q = \delta_j \frac{\mathrm{ct}(\mathbb{T}' : \frac{j}{2})}{\mathrm{ct}(\mathbb{T}' : \frac{j}{2}+1)}.$$

Thus $\frac{1}{2}(u+1)(M_j(\mathbb{T}) + M_j(\mathbb{T}.s_{j-1})) = u+1$ if $b_j = 1$ and

$$\frac{u+1}{2}\left(M_j(\mathbb{T}) + M_j(\mathbb{T}.s_{j-1})\right) = (u^2-1) \Bigg/ \left(1 - \frac{\mathrm{ct}(\mathbb{T}' : \frac{j}{2})^2}{\mathrm{ct}(\mathbb{T}' : \frac{j}{2}+1)^2}\right),$$

for $b_j = s_j$, which is the coefficient of $\mathbb{T}' \in \mathsf{ST}(\lambda)$ in $\mathbb{T}'.S'_{j/2}$ and where $S'_{j/2}$ is the generator corresponding to $s_{j/2}$ of the algebra $H(\mathfrak{S}_{n/2})$ with parameter u^2.

Step 5. Finally, let $\nu = w_{(\alpha',\varnothing)} \in \mathfrak{S}_{n/2}$ and let T'_ν be the corresponding element in the algebra $H(\mathfrak{S}_{n/2})$ with parameter u^2. Let $\mathbb{T}' \in \mathsf{ST}(\lambda)$. Then, since $\Psi(\mathbb{T}', L).s_{2k-1} = \Psi(\mathbb{T}', L - \{s_{2k-1}\} \cup \{s_{2k-1}\} - L)$, and $M_{2k}(\Psi(\mathbb{T}', L))$ depends only on whether $s_{2k-1} \in L$ or not, we can evaluate the sum $\sum_{L \subseteq R} a^{T_w}_{\Psi(\mathbb{T}',L),\Psi(\mathbb{T}',R-L)}$ as follows:

$$\begin{aligned}
\sum_{L\subseteq R} a^{T_w}_{\Psi(\mathbb{T}',L),\Psi(\mathbb{T}',R-L)} &= \sum_{L\subseteq R} \prod_{k=0}^{n/2-1} \frac{u+1}{2} M_{2k}(\Psi(\mathbb{T}', L)) \\
&= \prod_{k=0}^{n/2-1} \frac{u+1}{2} \left(M_{2k}(\Psi(\mathbb{T}', \varnothing)) + M_{2k}(\Psi(\mathbb{T}', \{s_{2k-1}\}))\right) \\
&= (u+1)^m a'^{T'_\nu}_{\mathbb{T}',\mathbb{T}'},
\end{aligned}$$

where $a'^{T'_\nu}_{\mathbb{T}',\mathbb{T}'}$ is the coefficient of $\mathbb{T}' \in \mathsf{ST}(\lambda)$ in $\mathbb{T}'.T'_\nu$.

Summing over all tableaux $\mathbb{T}' \in \mathsf{ST}(\lambda)$, we then get

$$\begin{aligned}
\chi_{(\lambda,\lambda)}(T_w\sigma) &= \sum_{\mathbb{T}'\in\mathsf{ST}(\lambda)} \Big(\sum_{L\subseteq R} a^{T_w}_{\Psi(\mathbb{T}',L),\Psi(\mathbb{T}',R-L)}\Big) \\
&= (u+1)^m \sum_{\mathbb{T}'\in\mathsf{ST}(\lambda)} a'^{T'_\nu}_{\mathbb{T}',\mathbb{T}'} = (u+1)^m \chi_\lambda(T'_\nu),
\end{aligned}$$

as desired. □

10.4.8. By Proposition 10.4.3, (10.4.6)(2), and Theorem 10.4.7 the character table of $\mathbf{H}(D_n)$ is completely determined. Which parts of the character table are constructed by which formula is illustrated as follows:

	$w = w_{(\alpha,\beta)}$, $(\alpha,\beta) \in \mathcal{P}_2$	$w = w_{(\alpha,\varnothing)}$, $\alpha \in \mathcal{P}_1$	$tw_{(\alpha,\varnothing)}t$, $\alpha \in \mathcal{P}_1$
$\chi'_{(\lambda,\mu)}$, $\lambda \neq \mu$	$\chi_{(\lambda,\mu)}(T_w)$		
$\chi_{(\lambda,+)}$, $\lambda \vdash n/2$	$\frac{1}{2}\chi_{(\lambda,\lambda)}(T_w)$	$+$	$-$
		$\frac{1}{2}\left(\chi_{(\lambda,\lambda)}(T_w) \pm (u+1)^{l(\alpha)}\chi_\lambda(T'_v)\right)$	
$\chi_{(\lambda,-)}$, $\lambda \vdash n/2$		$-$	$+$

Here v is an element of cycle structure α' in $\mathfrak{S}_{n/2}$ if $\alpha = 2\alpha'$ and T'_v is the corresponding basis element of the generic algebra $\mathbf{H}(\mathfrak{S}_{n/2})$ with parameter u^2, as in Theorem 10.4.7. The case $n = 4$ is shown in Table 10.4 on p. 346.

As a corollary we obtain the character table of the Coxeter group W'_n.

10.4.9. Theorem. *Let $n \geq 4$. Then the character table of the Coxeter group W'_n of type D_n is completely determined as follows. If $w \in W'_n$ and (λ, μ) is a pair of partitions of total size n such that $\lambda \neq \mu$ then*

$$\chi'_{(\lambda,\mu)}(w) = \chi_{(\lambda,\mu)}(w).$$

In the case of even n, if λ is a partition of $n/2$ then

$$\chi_{(\lambda,+)}(w) + \chi_{(\lambda,-)}(w) = \chi_{(\lambda,\lambda)}(w).$$

Moreover, if w is conjugate to twt in W'_n then $\chi_{(\lambda,+)}(w) = \chi_{(\lambda,-)}(w)$. Otherwise, if w and twt are not conjugate in W'_n then w is an element of cycle structure $(2\alpha, \varnothing)$ and we have

$$\chi_{(\lambda,+)}(w) - \chi_{(\lambda,-)}(w) = 2^{l(\alpha)}\chi_\lambda(v)$$

where $v \in \mathfrak{S}_{n/2}$ is an element of cycle structure α.

Proof. Specializing $u = 1$, this follows from Proposition 10.4.3, (10.4.6)(2) and Theorem 10.4.7. □

10.4.10. Identifying the splitting characters. Assume that n is even and let $\lambda \vdash n/2$. Then we have two irreducible characters $\chi_{(\lambda,\pm)}$ in type D_n. In this section, these two characters were defined using the (± 1)-eigenspace decomposition of the module $V^{(\lambda,\lambda)}$ while, in Proposition 5.6.3, they were characterized in terms of the j-induction from suitable parabolic subgroups. In both cases, $\chi_{(\lambda,\pm)}$ occur as constituents in the restriction of the character $\chi_{(\lambda,\lambda)}$ from type B_n to D_n. However, this does not yet show that the signs agree. We now prove that this is indeed the case. First note that it is sufficient to work with the Coxeter groups W'_n and W_n. (The corresponding result on the level of Iwahori–Hecke algebras then follows by specialization.) Assume that the sign for labelling $\chi_{(\lambda,\pm)}$ has been chosen such that the character formula in Theorem 10.4.9 holds. On the other hand, by Proposition 6.1.5, we have

$$\left(\chi_{(\lambda,+)} - \chi_{(\lambda,-)}, 1_{\mathfrak{S}_{2\lambda}}^{W'_n}\right)_{W'_n} = \pm 1$$

and, in order to fix the sign as desired, it suffices to show that the above scalar product is non-negative. Now, by Frobenius reciprocity and Theorem 10.4.9, that scalar product equals

$$\left(\chi_{(\lambda,+)} - \chi_{(\lambda,-)}, 1_{\mathfrak{S}_{2\lambda}}^{\mathfrak{S}_n}\right)_{\mathfrak{S}_n} = \sum_{\gamma \vdash n/2} \frac{1}{|C_{\mathfrak{S}_n}(w_{2\gamma})|} 2^{l(\gamma)} \chi_\lambda(w_\gamma)\, 1_{\mathfrak{S}_{2\lambda}}^{\mathfrak{S}_n}(w_{2\gamma}),$$

where $w_\gamma \in \mathfrak{S}_{n/2}$ is an element of cycle type γ. Thus, we are reduced to working in $\mathfrak{S}_n$. Now, a straightforward computation shows that

$$|C_{\mathfrak{S}_n}(w_{2\gamma})| = 2^{l(\gamma)} |C_{\mathfrak{S}_{n/2}}(w_\gamma)|.$$

Using also the identity for permutation characters in Exercise 5.11 (with $f = 2$), the above scalar product formula can be rewritten in the following form:

$$\left(\chi_{(\lambda,+)} - \chi_{(\lambda,-)}, 1_{\mathfrak{S}_{2\lambda}}^{W'_n}\right)_{W'_n} = \sum_{\gamma \vdash n/2} \frac{1}{|C_{\mathfrak{S}_{n/2}}(w_\gamma)|} \chi_\lambda(w_\gamma)\, 1_{\mathfrak{S}_\lambda}^{\mathfrak{S}_{n/2}}(w_\gamma).$$

Now the result on the right-hand side is 1 by Theorem 5.4.7.

10.5 ON THE GENERIC DEGREES OF THE CLASSICAL TYPES

In this section, we discuss the determination of the Schur elements and the generic degrees for Iwahori–Hecke algebras of classical type A, B and D; recall the basic definitions from Section 7.2 and (8.1.8). We begin with the Poincaré polynomials for these types.

10.5.1. Poincaré polynomials. The Poincaré polynomials for the classical types are as follows:

$$P_{A_{n-1}}(u) = \prod_{i=1}^{n} (1 + u + u^2 + \cdots + u^{i-1}),$$

$$P_{B_n}(u, v) = P_{A_{n-1}}(u) \prod_{i=1}^{n} (1 + u^{i-1} v),$$

$$P_{D_n}(u) = P_{A_{n-1}}(u) \prod_{i=1}^{n-1} (1 + u^i).$$

These formulas are proved by induction on n as follows. Consider, for example, type B_n. By definition, we have $P_{B_n}(u, v) = \sum_{w \in W_n} \mathrm{ind}(T_w)$. For $n = 1$, we have

Table 10.4 The character table of the generic algebra $\mathbf{H}(D_4)$

	$(1^4.\varnothing)$	$(1^2.1^2)$	$(\varnothing,1^4)$	$(21^2.\varnothing)$	(1.21)	(2.1^2)	$(2^2.+)$	$(2^2.-)$	$(\varnothing.2^2)$	$(31.\varnothing)$	$(\varnothing.31)$	$(4.+)$	$(4.-)$
χ	1	u_1	$u_1u_2u_3$	s_1	u_1s_2	u_1s_3	s_1s_3	us_3	$s_1u_2s_3$	s_1s_2	$u_1s_2s_3$	$s_1s_2s_3$	us_2s_3
$(1^2.+)$	3	$-2u+1$	$3u^4$	$u-2$	u	$-u^2+u-1$	u^2+2	$-2u+1$	$-u^2$	$-u+1$	0	-1	u
$(1^2.-)$	3	$-2u+1$	$3u^4$	$u-2$	u	$-u^2+u-1$	$-2u+1$	u^2+2	$-u^2$	$-u+1$	0	u	-1
(1.1^3)	4	$-2u+2$	$-4u^3$	$u-3$	$u-1$	$3u-1$	$-2u+2$	$-2u+2$	0	$-u+2$	$-u$	$u-1$	$u-1$
$(\varnothing.1^4)$	1	1	1	-1	-1	-1	1	1	1	1	1	-1	-1
$(1^2.2)$	6	u^2-4u+1	$6u^6$	$3u-3$	$-u^2+u$	$-3u^2+3u$	u^2-4u+1	u^2-4u+1	$2u^3$	u^2-2u+1	0	$-u^2+u$	$-u^2+u$
(1.21)	8	$2(u-1)^2$	$-8u^6$	$4u-4$	$-u^2+u$	$(u-1)^3$	$2(u-1)^2$	$2(u-1)^2$	0	u^2-3u+1	u^2	$-u^2+u$	$-u^2+u$
$(\varnothing.21^2)$	3	u^2+2	$3u^4$	$u-2$	-1	$-u^2+u-1$	$-2u+1$	$-2u+1$	$-u^2$	$-u+1$	0	u	u
$(2.+)$	3	u^2-2u	$3u^8$	$2u-1$	$-u^2$	u^3-u^2+u	$2u^2+1$	u^2-2u	$-u^4$	u^2-u	0	u^3	$-u^2$
$(2.-)$	3	u^2-2u	$3u^8$	$2u-1$	$-u^2$	u^3-u^2+u	u^2-2u	$2u^2+1$	$-u^4$	u^2-u	0	$-u^2$	u^3
$(\varnothing.2^2)$	2	u^2+1	$2u^6$	$u-1$	0	u^3-1	u^2+1	u^2+1	$2u^3$	$-u$	$-u^2$	0	0
(1.3)	4	$2u^2-2u$	$-4u^9$	$3u-1$	u^3-u^2	u^3-3u^2	$2u^2-2u$	$2u^2-2u$	0	$2u^2-u$	$-u^3$	u^3-u^2	u^3-u^2
$(\varnothing.31)$	3	$2u^2+1$	$3u^8$	$2u-1$	u^3	u^3-u^2+u	u^2-2u	u^2-2u	$-u^4$	u^2-u	0	$-u^2$	$-u^2$
$(\varnothing.4)$	1	u^2	u^{12}	u	u^3	u^3	u^2	u^2	u^6	u^2	u^4	u^3	u^3

The representatives of minimal length in the classes of W_4' are taken from Example 3.4.13. The difference characters of the splitting characters have the following values on splitting classes (cf. Theorem 10.4.9):

	$(2^2.+)$	$(2^2.-)$	$(4.+)$	$(4.-)$
$\chi_{(1^2.+)}-\chi_{(1^2.-)}$	$(u+1)^2$	$-(u+1)^2$	$-(u+1)$	$u+1$
$\chi_{(2.+)}-\chi_{(2.-)}$	$(u+1)^2$	$-(u+1)^2$	$u^2(u+1)$	$-u^2(u+1)$

$W_1 = \{1, t\}$ and so $P_{B_1}(u, v) = 1 + v$, as desired. Now let $n > 1$ and assume that the above formula has already been proved for B_{n-1}. We consider the maximal parabolic subgroup $W_{n-1} \subset W_n$. Then we have $W_n = W_{n-1} \cdot X_n$, where X_n is the set of distinguished coset representatives of W_{n-1} in W_n. For any $w \in W_{n-1}$ and $x \in X_n$, we have $T_{wx} = T_w T_x$. This yields the factorization

$$P_{B_n}(u, v) = \sum_{w \in W_{n-1}} \sum_{x \in X_n} \mathrm{ind}(T_w)\,\mathrm{ind}(T_x) = P_{B_{n-1}}(u, v)\Big(\sum_{x \in X_n} \mathrm{ind}(T_x)\Big).$$

Using the explicit description of X_n from Example 2.2.5, we find that

$$\sum_{x \in X_n} \mathrm{ind}(T_x) = (1 + u^{n-1}v)(1 + u + u^2 + \cdots + u^{n-1}).$$

By induction, this gives the desired expression. The proofs for type A_{n-1} and D_n are analogous, using Examples 2.2.4 and 2.2.6 instead of Example 2.2.5.

Let us now turn to the generic degrees. It will be convenient to introduce the following notation. For any integer $m \geq 1$, we set

$$[m] := 1 + u + \cdots + u^{m-1} = \frac{u^m - 1}{u - 1} \quad \text{and} \quad [0] := 0.$$

Moreover, we set $[m]! := [1][2] \cdots [m]$ and $[0]! = 1$. Finally, let

$$\Delta(Z, u) := \prod_{\substack{k, l \in Z \\ k > l}} (u^k - u^l) \qquad \text{for any finite subset } Z \subset \mathbb{N}_0.$$

Now we are ready to state the following result on the generic degrees in type A.

10.5.2. Theorem (Steinberg). *Let λ be a partition of n. We associate with λ a β-set X, as in (6.4.1). Let $|X| = b$. Then the generic degree of the irreducible character of $H(\mathfrak{S}_n)$ labelled by λ is given by*

$$D_\lambda = \frac{(u-1)^n\,[n]!\,\Delta(X, u)}{u^{b(b-1)(b-2)/6}\Big(\prod_{k \in X} (u-1)^k [k]!\Big)}.$$

Note that, by Proposition 9.4.5, this also yields a formula for the fake degrees.

Proof. The idea of the argument is as follows. First, we show that the generic degrees D_λ are determined by a system of linear equations involving the *Kostka numbers* (see step 1). Using the formula in Exercise 6.5(b), we can invert this system of linear equations and obtain an expression for D_λ as the determinant of a matrix with polynomial coefficients (see step 2). Then it remains to transform this determinant into the desired expression (see step 3).

Step 1. Let μ be any composition of n and λ be a partition of n. We denote by $\kappa_{\lambda\mu}$ the multiplicity with which the irreducible character of $\mathfrak{S}_n$ labelled by λ occurs in the permutation character $\mathrm{Ind}_{\mathfrak{S}_\mu}^{\mathfrak{S}_n}(1_{\mathfrak{S}_\mu})$. (In the remarks following Theorem 5.4.7, $\kappa_{\lambda\mu}$ is called a Kostka number.) Now the trivial character of $\mathfrak{S}_\mu$ corresponds to the index character, denoted by ind_μ, of the parabolic subalgebra $H(\mathfrak{S}_\mu) \subseteq H(\mathfrak{S}_n)$. By (8.1.8), the Schur element of ind_μ is the Poincaré polynomial of $H(\mathfrak{S}_\mu)$, which we denote by P_μ. Now the compatibility of induction and specialization (9.1.9) and Lemma 9.4.6 show that we have the following identity:

$$P_\mu^{-1} = \sum_{\lambda \vdash n} \kappa_{\lambda\mu} c_\lambda^{-1},$$

where c_λ denotes the Schur element of the irreducible character of $H(\mathfrak{S}_n)$ labelled by λ; we have $c_\lambda = P_{A_{n-1}}(u)/D_\lambda = [n]!/D_\lambda$. Since $\mathfrak{S}_\mu \cong \mathfrak{S}_{\mu_1} \times \cdots \times \mathfrak{S}_{\mu_r}$, we also have a corresponding factorization of P_μ (see Exercise 8.5). Using the formula for the Poincaré polynomial in type A from (10.5.1), we obtain

$$\sum_{\lambda \vdash n} \kappa_{\lambda\mu} D_\lambda = \frac{[n]!}{[\mu_1]!\,[\mu_2]! \cdots [\mu_r]!} \qquad \text{where} \qquad \mu = (\mu_1, \mu_2, \ldots, \mu_r).$$

Step 2. For the following discussion, we assume that λ has n parts, $\lambda_1 \geq \lambda_2 \geq \ldots \geq \lambda_n$ say. (If necessary, we add to λ some parts equal to 0.) Using the notation and the formula in Exercise 6.5(b), we have

$$D_\lambda = \sum_{\nu \vdash n} \Big(\sum_{w \in \mathfrak{S}_n} \varepsilon(w)\, \kappa_{\nu\lambda_w} \Big) D_\nu = \sum_{w \in \mathfrak{S}_n} \varepsilon(w) \Big(\sum_{\nu \vdash n} \kappa_{\nu\lambda_w} D_\nu \Big).$$

Inserting the formula obtained in step 1, this can be further evaluated to

$$\begin{aligned} D_\lambda &= \sum_{w \in \mathfrak{S}_n} \varepsilon(w) \frac{[n]!}{[\lambda_1 - 1 + 1.w]! \cdots [\lambda_n - n + n.w]!} \\ &= [n]!\, \det(m_{ij}) \qquad \text{with} \quad m_{ij} = \frac{1}{[\lambda_i - i + j]!} \quad (1 \leq i, j \leq n), \end{aligned}$$

where we use the convention that $[k]! = 0$ if $k < 0$. Now we set $\alpha_i = \lambda_i + n - i$ for all i, so that we have $X = \{\alpha_1 > \alpha_2 > \ldots > \alpha_n\}$. Upon multiplying the first row of (m_{ij}) by $[\alpha_1]!$, the second by $[\alpha_2]!$ and so on, we find that $([\alpha_1]! \cdots [\alpha_n]!) \det(m_{ij}) = \det(m'_{ij})$ where

$$m'_{ij} = \underbrace{[\alpha_i][\alpha_i - 1] \cdots [\alpha_i - n + j + 1]}_{n - j \text{ factors}} \qquad \text{for } 1 \leq i, j \leq n.$$

(Recall the convention that $[k] = 0$ for $k < 0$.)

Step 3. Now we obtain a closed formula for the determinant in step 2. We set $x_i := [\alpha_i]$ and $y_i := [i-1]$ for $1 \leq i \leq n$. Then we have, for any $1 \leq k \leq n$:

$$x_i - y_k = \frac{1}{u-1}(u^{\alpha_i} - u^{k-1}) = \frac{u^{k-1}}{u-1}(u^{\alpha_i - k+1} - 1) = u^{k-1}[\alpha_i - k + 1].$$

This yields the following expression for the coefficients m'_{ij} introduced in step 2:

$$m'_{ij} = u^{-\binom{n-j}{2}}(x_i - y_1)(x_i - y_2)\cdots(x_i - y_{n-j}).$$

Since $\binom{n-1}{2} + \binom{n-2}{2} + \cdots + 1 = n(n-1)(n-2)/6$, we obtain, using Exercise 10.7,

$$\det(m'_{ij}) = u^{-n(n-1)(n-2)/6} \prod_{1\leq i<j\leq n} \Big([\alpha_i] - [\alpha_j]\Big).$$

Combining the above expression with step 2, we find the following closed formula:

$$D_\lambda = \frac{[n]! \displaystyle\prod_{1\leq i<j\leq n} ([\alpha_i] - [\alpha_j])}{u^{n(n-1)(n-2)/6} \displaystyle\prod_{i=1}^{n} [\alpha_i]!}.$$

Now we have $[\alpha_i] - [\alpha_j] = (u^{\alpha_i} - u^{\alpha_j})/(u-1)$ and so

$$\prod_{1\leq i<j\leq n} ([\alpha_i] - [\alpha_j]) = (u-1)^{-n(n-1)/2}\Delta(X, u).$$

Also note that

$$\prod_{k\in X}(u-1)^k[k]! = (u-1)^{\sum_i \alpha_i}\prod_i [\alpha_i]!$$

where $\sum_i \alpha_i = n(n-1)/2 + \sum_i \lambda_i = n(n+1)/2$. So we obtain exactly the desired expression, where we have used a β-set X with $b = n$. Finally, it is easily checked that the right-hand side of the formula for D_λ is invariant under the shift operation for β-sets. □

We now state the following result on the generic degrees for type B_n.

10.5.3. Theorem (Hoefsmit). *Let (λ, μ) be a pair of partitions with $|\lambda|+|\mu| = n$. As in (6.4.2), we associate with (λ, μ) a symbol $\Lambda = \binom{X}{Y}$ of rank n and defect 1. Let $|X| = b+1$, $|Y| = b$. Then the generic degree of the irreducible character of $H(B_n)$ labelled by (λ, μ) is given by*

$$D_\Lambda = \frac{u^{b(b+1)/2}(u-1)^n P_{B_n}(u,v)\,\Delta(X,u)\,\Delta(Y,u) \displaystyle\prod_{(k,l)\in X\times Y} (u^{k-1}v + u^l)}{u^{m(b)}v^{b(b-1)/2}\Big(\displaystyle\prod_{k\in X}(u-1)^k P_{B_k}(u,v)\Big)\Big(\displaystyle\prod_{l\in Y}(u-1)^l P_{B_l}(u,v)\Big)(u+v)^b}$$

where we set $m(b) := \binom{2b-1}{2} + \binom{2b-3}{2} + \cdots = b(b-1)(4b+1)/6$.

We do not give a proof here but only add the following comments.

10.5.4. Remarks on the proof of Theorem 10.5.3.

(1) The original proof in [Hoefsmit 1974] is purely combinatorial. It proceeds by an induction on n, which is based on the branching rule for type B_n (see (6.1.9)) and the factorization in Lemma 9.4.6(b). In this set-up, one has to compute explicitly the "relative" Schur elements appearing in Exercise 9.3, with respect to the parabolic subgroup $W_{n-1} \subset W_n$. Hoefsmit achieves this in a long series of steps, by evaluating explicitly the matrix representations for the irreducible characters of $H(B_n)$ (see Section 10.1) on the basis elements corresponding to the distinguished coset representatives of W_{n-1} in W_n.

(2) A completely different proof is given in [Lusztig 1977, (9.6)]. This proof is based on the representation theory of the finite unitary groups $G = U_N(q)$; it involves the Deligne–Lusztig theory of characters of finite groups of Lie type and, thus, ultimately rests on deep results from algebraic geometry (see [Carter 1985]). The basic argument is the following. On the one hand, Lusztig classifies the so-called "unipotent characters" of G. He shows that they are naturally parametrized by the partitions of N and that their degrees are obtained from the generic degrees of type A_{N-1} by setting u to $-q$ in the formulas in Theorem 10.5.2. On the other hand, the unipotent characters can also be described in terms of so-called "Harish-Chandra series" and then the degrees are determined by the formula in [Lusztig 1984, Cor. 8.7]. Now, given n and an integer $s \geq 0$, Lusztig shows that one can choose N such that G has a Harish-Chandra series of unipotent characters in bijection with the irreducible characters of $H(B_n)$. Moreover, the degrees of the unipotent characters in that series are determined via the generic degrees for type B_n where we set $v = q^{2s+1}$ and $u = q^2$. The identification of the two parametrizations of the unipotent characters of G is described explicitly in the appendix of [Fong and Srinivasan 1990]. Thus, one obtains explicit formulas for the generic degrees in type B_n, where the parameters are specialized to $v = q^{2s+1}$ and $u = q^2$. However, for fixed n, we can let run q over all prime powers and s over all non-negative integers. Thus, the above specializations are sufficient to determine the polynomial expressions in Theorem 10.5.3.

(3) A recent new proof can be found in [Iancu 2000]. This is based on the observation that the symmetrizing trace τ on $H(B_n)$ is a special example of a *Markov trace* in the sense of [Geck and Lambropoulou 1997, §4]. (This observation is inspired by [Ram and Remmel 1997], where type A is considered.) These Markov traces depend on two parameters and they are linear combinations of the irreducible characters of $\mathrm{Irr}(H(B_n))$, where the coefficients are called "weights". These weights (for special choices of the parameters) have been determined in [Orellana 1999, Theorem 5.5], generalizing and extending an earlier argument for Markov traces in type A (see [Wenzl 1988, §3]). Thus, once the weights are determined for *all* choices of the parameters, one also obtains explicit formulas for the c_χ using the identity $\tau = \sum_\chi (1/c_\chi)\chi$.

(4) The formulas in Theorem 10.5.3 have (as yet conjectural) generalizations to the cyclotomic algebras associated with the wreath products $C_r \wr \mathfrak{S}_n$. The conjectural formulas for the generic degrees were found by [Malle 1995]. It is likely that the strategy outlined in (3) might also be used to prove these conjectures.

We shall now take Theorem 10.5.3 for granted and discuss some consequences.

10.5.5. The case of equal parameters. Assume that $u = v$. Then we have seen in Corollary 9.3.6 that the generic degrees are actually polynomials in u (and not just fractions of polynomials). Using Theorem 10.5.3, we find that

$$D_\Lambda = 2^{|X \cap Y| - |Y|} u^{a(\Lambda)} + \text{combination of higher powers of } u,$$

with $a(\Lambda)$ as in Definition 6.4.3. Indeed, let us write $X = \{\alpha_1 > \alpha_2 > \ldots > \alpha_{b+1}\}$ and $Y = \{\beta_1 > \beta_2 > \ldots > \beta_b\}$. Then the maximal power of u dividing the numerator in the formula for D_Λ has exponent

$$b(b+1)/2 + \sum_{i=1}^{b+1} (i-1)\alpha_i + \sum_{j=1}^{b} (j-1)\beta_j + \sum_{i=1}^{b+1} \sum_{j=1}^{b} \min\{\alpha_i, \beta_j\}.$$

On the other hand, the maximal power of u dividing the denominator in that formula has exponent $m(b) + b(b-1)/2 + b$. Hence, using Exercise 6.7, we see that the difference of these two expressions is $a(\Lambda)$, as desired. The coefficient of $u^{a(\Lambda)}$ is determined by a similar discussion. Thus, we see that the a-invariants for type B_n, as defined in Section 6.5, are given in terms of the generic degrees, as stated in Remark 9.4.10.

Finally, the generic degrees for type D_n are derived from those of type B_n, using the embedding of the generic algebra $H(D_n)$ into an Iwahori–Hecke algebra of type B_n. More precisely, as in Section 10.4, let $H(B_n)'$ be the Iwahori–Hecke algebra of type B_n, where we set $v = 1$. Thus, we have the quadratic relations

$$T_t^2 = T_1, \qquad T_t T_w = T_{tw} \quad \text{and} \quad T_w T_t = T_{wt} \quad \text{for all } w \in W_n.$$

Then the generic algebra $H(D_n)$ can be identified with the subalgebra of $H(B_n)'$ which is generated by $T_{ts_1t}, T_{s_1}, \ldots, T_{s_{n-1}}$. Recall from (10.4.2) that the irreducible characters of $H(B_n)'$ are parametrized by pairs of partitions (λ, μ) with $|\lambda| + |\mu| = n$. Then the irreducible characters of $H(D_n)$ are determined as follows.

- If $\lambda \neq \mu$, then the restriction of the irreducible character of $H(B_n)'$ labelled by (λ, μ) is an irreducible character of $H(D_n)$, which we also label by (λ, μ). Moreover, the characters of $H(B_n)'$ labelled by (λ, μ) and (μ, λ) have the same restriction to $H(D_n)$.
- If $\lambda = \mu$, then the restriction of the irreducible character of $H(B_n)'$ labelled by (λ, λ) splits into the sum of two irreducible characters of $H(D_n)$ which are labelled as $(\lambda, \pm)$.

10.5.6. Proposition (Hoefsmit, Benson–Gay). *Let (λ, μ) be a pair of partitions with $|\lambda|+|\mu| = n$. Let $D'_{(\lambda,\mu)}$ be the generic degree of the irreducible character of $H(B_n)'$ which is labelled by (λ, μ). (A formula for $D'_{(\lambda,\mu)}$ is obtained by setting $v = 1$ in Hoefsmit's formula in Theorem 10.5.3.) Then the following hold:*

(a) *If $\lambda \neq \mu$ then $D'_{(\lambda,\mu)}$ is the generic degree of the irreducible character of $H(D_n)$ labelled by (λ, μ).*

(b) *If $\lambda = \mu$, then the two irreducible characters of $H(D_n)$ labelled as $(\lambda, \pm)$ have the same generic degree, and this is equal to $\frac{1}{2}D'_{(\lambda,\lambda)}$.*

We remark that this result establishes a relation between the generic degrees for type B_n and D_n but it is does not require the explicit knowledge of the generic degrees for type B_n.

Proof. First note that the usual symmetrizing trace τ on $H(D_n)$ is just the restriction of the usual symmetrizing trace on $H(B_n)'$. Now let $w \in W'_n$ and consider the identity:

$$\tau(T_w) = \sum_{(\lambda,\mu)} c^{-1}_{(\lambda,\mu)} \chi'_{(\lambda,\mu)}(T_w),$$

where (λ, μ) runs over all pairs of partitions with $|\lambda|+|\mu| = n$, $\chi_{(\lambda,\mu)}$ denotes the corresponding irreducible character of $H(B_n)'$, and $c_{(\lambda,\mu)}$ is the Schur element of $\chi_{(\lambda,\mu)}$. We split the above sum into two pieces, namely, the sum of all terms with $\lambda = \mu$ and the sum of all terms with $\lambda \neq \mu$. Then, since $w \in W'_n$, we can use the relations in (10.4.2) to evaluate the terms in those sums as follows:

$$\tau(T_w) = \sum_{\lambda=\mu} c^{-1}_{(\lambda,\lambda)} \Big(\chi_{(\lambda,+)}(T_w) + \chi_{(\lambda,-)}(T_w)\Big)$$
$$+ 2 \sum_{\lambda\neq\mu} c^{-1}_{(\lambda,\mu)} \chi_{(\lambda,\mu)}(T_w),$$

where the second sum runs over the *unordered* pairs of partitions λ, μ with $\lambda \neq \mu$. Now the uniqueness statement in Lemma 9.4.4 implies the following:

(a) If $\lambda \neq \mu$, then $c_{(\lambda,\mu)}/2$ is the Schur element of the irreducible character $\chi'_{(\lambda,\mu)}$ of $H(D_n)$.

(b) If $\lambda = \mu$, then the Schur elements of the irreducible characters $\chi_{(\lambda,+)}$ and $\chi_{(\lambda,-)}$ of $H(D_n)$ are the same and equal to $c_{(\lambda,\lambda)}$.

To complete the proof, it remains to observe the identity $P_{B_n}(u, 1) = 2P_{D_n}(u)$ and to use the definition of the generic degrees in (8.1.8). □

10.5.7. The a-invariants for type D_n. Using Proposition 10.5.6 and the formula for type B_n in Theorem 10.5.3, we obtain an explicit formula for the generic degrees in type D_n as follows. Let (λ, μ) be a pair of partitions with $|\lambda|+|\mu| = n$. As in (6.4.2), we associate with (λ, μ) a symbol $\Lambda = \binom{X}{Y}$ of rank n and defect 0.

Let $|X| = |Y| = b$. Then the generic degree of the unique irreducible character of $\mathbf{H}(D_n)$ labelled by (λ, μ) (for $\lambda \neq \mu$) or of the two irreducible characters labelled by $(\lambda, \pm)$ (for $\lambda = \mu$) is given by

$$D'_\Lambda = \frac{(u-1)^n P_{D_n}(u)\,\Delta(X,u)\,\Delta(Y,u) \displaystyle\prod_{(k,l)\in X\times Y} (u^k + u^l)}{2^c u^{m'(b)} \Big(\displaystyle\prod_{k\in X}\prod_{h=1}^{k} (u^{2h}-1)\Big)\Big(\displaystyle\prod_{l\in Y}\prod_{h=1}^{l} (u^{2h}-1)\Big)}$$

where $m'(b) := \binom{2b-2}{2} + \binom{2b-4}{2} + \cdots = b(b-1)(4b-5)/6$ and $c = b-1$ (for $\lambda \neq \mu$) or $c = b$ (for $\lambda = \mu$). (To obtain the above formula, we have to set $v = 1$ in the formula for type B_n and take into account that now we have associated with (λ, μ) a symbol of defect 0.) Arguing as in (10.5.5), we see that

$$D'_\Lambda = 2^{|X\cap Y|-c} u^{a(\Lambda)} + \text{combination of higher powers of } u,$$

with $a(\Lambda)$ defined as in (6.4.3). Thus, the a-invariants for type D_n, as defined in Section 6.5, are also given in terms of the generic degrees.

10.5.8. Example. Let us consider the character afforded by the standard reflection representation of a generic Iwahori–Hecke algebra of type A_{n-1}, B_n or D_n. The labelling of the character of that representation is determined in Propositions 5.4.12, 5.5.7 and Remark 5.6.6, respectively. Then we obtain the following formulas for the corresponding generic degrees:

Type	Label	Generic degree
A_{n-1}	$[n-1, 1]$	$u(1+u+u^2+\cdots+u^{n-2})$
B_n	$([n-1],[1])$	$\dfrac{v(1+u+u^2+\cdots+u^{n-1})(1+u^{n-2}v)}{1+u^{-1}v}$
D_n	$([n-1],[1])$	$\dfrac{u(1+u+u^2+\cdots+u^{n-1})(1+u^{n-2})}{1+u}$

10.6 THE LASCOUX–LECLERC–THIBON CONJECTURE

In this final section, we consider the generic Iwahori–Hecke algebra $\mathbf{H}(\mathfrak{S}_n)$ associated with the symmetric group $\mathfrak{S}_n$. We consider a specialization θ_e which sends the parameter u to a primitive eth root of unity (for some $e \geq 1$), and let D_e be the corresponding decomposition matrix (see Theorem 7.4.3).

According to the *Lascoux–Leclerc–Thibon conjecture*, the problem of computing D_e can be translated to that of computing the canonical basis (in the sense of Kashiwara and Lusztig) of a certain highest weight module for an affine Kac–Moody algebra. This was subsequently proved by [Ariki 1996] and yields a purely combinatorial algorithm for computing D_e. The purpose of this section is to describe that algorithm (without proof).

10.6.1. The decomposition matrix D_e. Recall from Section 5.4 that the irreducible characters of $\mathfrak{S}_n$ have a natural parametrization by the partitions of n. Thus, via the bijection in Theorem 8.1.7, we obtain a natural parametrization

$$\mathrm{Irr}(H(\mathfrak{S}_n)) = \{\chi_\lambda \mid \lambda \vdash n\}. \tag{a}$$

Now consider the specialized algebra $\mathbb{C}H(\mathfrak{S}_n)$ under the map θ_e. The simple modules of $\mathbb{C}H(\mathfrak{S}_n)$ are parametrized as follows.

A partition $\mu = [1^{m_1}, 2^{m_2}, \dots]$ (in exponential notation) is called *e-regular* if we have $m_i < e$ for all i. By [Dipper and James 1986, Theorem 7.6], the simple $\mathbb{C}H(\mathfrak{S}_n)$-modules are parametrized by the e-regular partitions of n. Thus, writing $\mathrm{Irr}(\mathbb{C}H(\mathfrak{S}_n)) = \{\psi_\mu \mid \mu \vdash n \text{ e-regular}\}$, the coefficients of $D_e = (d_{\lambda\mu})$ are given by the following equations (see Remark 7.4.4):

$$\dot{\chi}_\lambda^{\theta_e} = \sum_{\substack{\mu \vdash n \\ \text{e-regular}}} d_{\lambda\mu}\, \psi_\mu \qquad \text{for all } \lambda \vdash n, \tag{b}$$

where $\dot{\chi}_\lambda^{\theta_e}(T_w) = \theta_e(\chi_\lambda(T_w))$ for all $w \in \mathfrak{S}_n$. Moreover, Dipper and James show that the labelling of the ψ_μ can be chosen in such a way that we have

$$d_{\lambda\mu} = \begin{cases} 1 & \text{if } \lambda = \mu, \\ 0 & \text{if } \mu \not\trianglelefteq \lambda. \end{cases} \tag{c}$$

In fact, given the labelling in (a), the conditions in (c) uniquely determine the labelling of the irreducible characters of $\mathbb{C}H(\mathfrak{S}_n)$ by e-regular partitions of n.

A fully self-contained introduction to modular characters and decomposition numbers of $H(\mathfrak{S}_n)$ can be found in [Mathas 1999]. The Lascoux–Leclerc–Thibon conjecture is concerned with the problem of computing the coefficients $d_{\lambda\mu}$. To state it, we need some preparations.

10.6.2. The Fock space $\mathcal{F}$. For any $n \geq 0$, let $\mathcal{F}_n$ be the $\mathbb{Q}(u)$-vector space with a basis consisting of all the partitions of n; by convention, $\mathcal{F}_0$ has as basis the empty partition, $\varnothing$. The *Fock space* is the direct sum of all these spaces:

$$\mathcal{F} := \bigoplus_{n \geq 0} \mathcal{F}_n.$$

Given an integer $e \geq 1$, we wish to define certain linear operators $f_r \colon \mathcal{F} \to \mathcal{F}$ for $0 \leq r \leq e-1$. For this purpose, we need the following definitions. Let $\lambda \vdash n$ and consider the corresponding Young diagram (see (5.4.2)).

The *e-residue* of a box which occupies the jth position in the ith row, say, is the unique $r \in \{0, \dots, e-1\}$ such that $r \equiv j - i \pmod e$.

Next, if we can remove the rightmost box in a row and the result is still a Young diagram (for a partition of $n-1$) we call that box a *removable box*. (For example, the rightmost box in the last row is always removable.)

Finally, if we can add a new box to some row and the result is again a Young diagram (for a partition of $n+1$), we call that row an *indent row*. (For example, we can always add a box to the first row; note that we can to add a new row consisting of one box to the bottom of the diagram.) The e-residue of such a row is the e-residue of the new box that we can add.

Now we are able to define the desired maps f_r. Let $\lambda \vdash n$ and $0 \leq r \leq e-1$. Suppose that the ith row of the Young diagram of λ is an indent row whose e-residue is r. Then we let $N(\lambda, r, i)$ be the number of indent rows strictly above i whose e-residue is r, minus the number of removable boxes in rows above i whose e-residue is r. We define

$$\lambda.f_r = \sum_i u^{N(\lambda,r,i)} \lambda_{[i]},$$

where the sum is over all indent rows i of λ with e-residue r, and $\lambda_{[i]}$ denotes the partition of $n+1$ obtained by adding a new box to the ith row. Thus, by definition, $\mathcal{F}_n$ is mapped into $\mathcal{F}_{n+1}$ under f_r, for all n, r.

10.6.3. The subspace $\mathcal{K} \subseteq \mathcal{F}$. For $k \geq 1$, we let $[k]_u := (u^k - u^{-k})/(u - u^{-1})$ and define new operators $f_r^{(k)} \colon \mathcal{F} \to \mathcal{F}$ by the inductive rule:

$$f_r^{(1)} := f_r \qquad \text{and} \qquad f_r^{(k)} = f_r^{(k-1)}/[k]_u \quad (k \geq 2).$$

Let $\mathcal{K}_n \subseteq \mathcal{F}_n$ be the subspace generated by all elements of the form $\varnothing.f_\iota^\kappa := \varnothing.f_{i_1}^{(k_1)} \cdots f_{i_n}^{(k_n)}$, where $\iota = (i_1, \ldots, i_n)$ and $\kappa = (k_1, \ldots, k_n)$ are any sequences of integers with $0 \leq i_1, \ldots, i_n \leq e-1$ and $k_1, \ldots, k_n \geq 1$. Now we have the following statement, which is proved by a purely combinatorial argument. For any e-regular partition $\nu \vdash n$, there exist sequences ι, κ as above such that

$$\varnothing.f_\iota^\kappa = \nu + \mathbb{Z}[u^{\pm 1}]\text{-linear combination of } \lambda \vdash n, \text{ where } \lambda \trianglelefteq \nu,\ \lambda \neq \nu. \qquad \text{(a)}$$

Here, $\trianglelefteq$ denotes the usual dominance order. Note that $\varnothing.f_\iota^\kappa$ depends on the choice of ι, κ. However, if we just make some choice for each e-regular $\nu \vdash n$ and denote the corresponding element in $\mathcal{K}_n$ by K_ν, we obtain

$$\{K_\nu \mid \nu \vdash n \text{ e-regular}\} \text{ is a basis of } \mathcal{K}_n. \qquad \text{(b)}$$

For more details, see the proof in [Lascoux et al. 1996, §6], which also yields an algorithm for finding suitable sequences ι, κ for each ν.

Now, using an interpretation of $\mathcal{F}$ as a module for an affine Kac–Moody algebra and applying the theory of *canonical bases* (due to Lusztig and Kashiwara; see [Lusztig 1993]), we have the following result.

10.6.4. Theorem (The canonical basis of $\mathcal{K} = \bigoplus_{n \geq 0} \mathcal{K}_n$). *For any $n \geq 0$ and any e-regular partition $\mu \vdash n$ there exists a unique element $G_\mu \in \mathcal{K}_n$ satisfying the following two conditions:*

(a) *If we express* G_μ *with respect to the "standard basis"* $\{\lambda\}$ *of* $\mathcal{F}_n$, *we have* $G_\mu = \sum_\lambda d_{\lambda\mu}(u)\,\lambda$, *where* $d_{\lambda\mu}(u) \in \mathbb{Z}[u]$ *are such that* $d_{\mu\mu} = 1$ *and* $d_{\lambda\mu}(0) = 0$ *for* $\lambda \neq \mu$. *Moreover, we have* $d_{\lambda\mu} = 0$ *unless* $\lambda \trianglelefteq \mu$.

(b) *If* $\{K_\nu\}$ *is any basis of* $\mathcal{K}_n$ *as in (10.6.3)(b), we have* $G_\mu = \sum_\nu h_{\nu\mu}(u)\,K_\nu$ *where* $h_{\nu\mu}(u) \in \mathbb{Z}[u, u^{-1}]$ *are such that* $h_{\nu\mu}(u^{-1}) = h_{\nu\mu}(u)$.

The basis $\{G_\mu \mid \mu$ *e-regular*$\}$ *is called the canonical basis of* $\mathcal{K}$.

The proof in [Lascoux et al. 1996, §6] is constructive: it shows that the coefficients $h_{\nu\mu}(u)$ can be computed by a "triangular algorithm".

10.6.5. Example. Let $e = 2$ and $n = 5$. Among the seven partitions of 5, the following are 2-regular: [5], [41] and [32]. The first step consists of constructing elements K_ν corresponding to these three partitions. Without even knowing the algorithm described by Lascoux, Leclerc and Thibon for solving this problem, we can just try several sequences ι, κ and compute the corresponding elements $\varnothing.f_\iota^\kappa$. We find:

	$K_{[5]}$	$K_{[41]}$	$K_{[32]}$
κ:	11111	1211	122
ι:	01010	0101	010
[5]	1	.	.
[41]	.	1	.
[32]	1	.	1
[311]	$2u$	.	u
[221]	u^2	.	u^2
[2111]	.	u	.
[11111]	u^2	.	.

Now we have to construct the elements G_μ. By Theorem 10.6.4(a), G_μ is a linear combination of μ and partitions λ with $\lambda \trianglelefteq \mu$ (where the coefficient of λ is a multiple of u). On the other hand, by Theorem 10.6.4(b), G_μ is also a linear combination of the K_ν where the coefficients are invariant under $u \mapsto u^{-1}$. The fact that the above table has a triangular shape implies that we can proceed as follows. We begin with the last column, corresponding to $\mu = [32]$. Then the above two conditions imply that we have $G_{[32]} = K_{[32]}$. Now we consider the next column (from the right), corresponding to $\mu = [41]$. We conclude again that $G_{[41]} = K_{[41]}$. Finally consider the first column, corresponding to $\mu = [5]$. Now the coefficient of $\lambda = [32]$ is 1, so the condition in Theorem 10.6.4(a) is not satisfied. But, if we subtract the third column from the first column, all conditions are satisfied. Thus, we have $G_{[5]} = K_{[5]} - K_{[32]}$.

Of course, for bigger values of n, more work has to be done but the principle is the same. First, we construct elements K_ν such that we have a triangular matrix as above. The last column of that matrix must correspond to a canonical basis element. Then we consider the next to last column, subtract a suitable multiple of the last column and so on. All this is implemented in Mathas's GAP package SPECHT, and in Veigneau's package ACE; see [Veigneau 1996].

Let us finally come back to D_e, the decomposition matrix as in (10.6.1). The following result, which is Conjecture 6.9(ii) in [Lascoux et al. 1996], shows that D_e is determined by G_μ:

10.6.6. Theorem (Ariki). *For any e-regular partition* $\mu \vdash n$, *let us write*

$$G_\mu = \sum_{\lambda \vdash n} d_{\lambda\mu}(u)\,\lambda,$$

as in Theorem 10.6.4, i.e., $d_{\lambda\mu}(u)$ *are the coefficients expressing the canonical basis of* $\mathcal{K}$ *in terms of the standard basis of the Fock space* $\mathcal{F}$. *Then we have* $d_{\lambda\mu} = d_{\lambda\mu}(1)$, *where* $d_{\lambda\mu}$ *are the coefficients of the decomposition matrix* D_e.

This is Theorem 4.4 in [Ariki 1996]; it is an extremely powerful result. The proof involves an enormous number of highly sophisticated methods and results concerning representations of affine Hecke algebras, canonical bases, and geometric interpretations using intersection cohomology.

We also remark that Ariki proves in fact a more general result, which also applies to Iwahori–Hecke algebras of type B_n and even to the cyclotomic Hecke algebras associated with the wreath products $C_r \wr \mathfrak{S}_n$ for any $r \geq 1$.

10.7 BIBLIOGRAPHICAL REMARKS AND EXERCISES

The seminormal representations in (10.1.4) are due to [Hoefsmit 1974]. For type A, they have also been constructed, for example, in [Dipper and James 1987] and [Wenzl 1988]. In [Ariki and Koike 1992] and [Ariki 1995] this construction is generalized to more general types of generic algebras, namely, the cyclotomic algebras associated with the wreath products $C_r \wr \mathfrak{S}_n$ for any $r \geq 2$. (Type B_n is the special case where $r = 2$.) These algebras are defined in [Ariki and Koike 1992] and [Broué and Malle 1993]. For a generalization of seminormal representations to all types of Iwahori–Hecke algebras, see [Ram 1997]. Moreover, integral versions of the irreducible representations for types A and B are constructed in [Dipper and James 1986] and [Dipper et al. 1995]; see also the remarks in (9.3.8) on this question of "integrality".

The classical Murnaghan–Nakayama formula for the characters of the symmetric group has been proved by Murnaghan and later, using the language of hooks, by Nakayama; see [Macdonald 1995, I, §7, Ex. 5] and the reference given there. For a completely different approach see, for example, [Greene 1992]. The formula for the Iwahori–Hecke algebra of type A appears in [Ram 1991]. Here we follow [Pfeiffer 1994b] and [Pfeiffer 1996] for the proof. This derivation of the formula from the Littlewood–Richardson rule parallels the proof of the Murnaghan–Nakayama formula in [Kerber 1991].

Character formulas for the Coxeter groups of type B and D are contained in [Stembridge 1989], where the formula for type B is a special case of such a formula for wreath products of cyclic groups and symmetric groups. The formulas for the Iwahori–Hecke algebras of type B have been developed independently in [Pfeiffer 1995] and in [Halverson and Ram 1996]; the latter article also contains a proof of the formula for type D. (That formula for type D appeared in [Pfeiffer 1995], too, but the proof was not complete. The gap is now filled using Proposition 10.4.5, which is taken from [Halverson and Ram 1996].) In [Halverson and Ram 1998] these results are generalized to the cyclotomic algebras mentioned above. See also [Shoji 1999].

The character formulas for the Iwahori–Hecke algebras of classical types have been implemented in CHEVIE. This is based on the implementation of the character formulas for the Coxeter groups of classical type in GAP which is described in detail in [Pfeiffer 1994a].

The generic degrees for type A have been determined in [Steinberg 1951], as the degrees of the irreducible characters of $G = GL_n(q)$ which occur in the permutation representation on the cosets of a Borel subgroup $B \subseteq G$ (cf. Section 8.4). Our proof is essentially that of Steinberg; the only difference is that we avoid the use of the group G in step 1 of the proof of Theorem 10.5.2. A completely different approach is developed in [Murphy 1992], using idempotents affording the irreducible representations. Yet another proof is given in [Ram and Remmel 1997], using Ocneanu's trace function that we met in Section 4.5. Part (b) of Proposition 10.5.6 is due to [Benson and Gay 1977], while (a) is contained in [Hoefsmit 1974]. For the proof of Proposition 10.5.6, we use the argument given in [Malle 1995, 5.13]. The identification of the maximal power of u dividing the generic degrees can be found in [Lusztig 1979a]. Example 10.5.8 appeared in [Hoefsmit 1974, §3.4]. The formulas for the generic degrees in classical types can also be found in [Carter 1985, §13.5]. "Hook formulas" in the case of equal parameters are derived in [Olsson 1986].

For further reading on the Lascoux–Leclerc–Thibon conjecture and an overview of Ariki's proof, see the survey article [Geck 1998c] and the references therein.

Exercise 10.1. Using the notation of Section 10.1, show that

(i) $m_i(\mathbb{T})\,\mathrm{ct}(\mathbb{T}:i) + m_i(\mathbb{T}')\,\mathrm{ct}(\mathbb{T}':i) = 0$;

(ii) $m_i(\mathbb{T})^2\,\mathrm{ct}(\mathbb{T}:i) = \big(u - m_i'(\mathbb{T})m_i'(\mathbb{T}')\big)\,\mathrm{ct}(\mathbb{T}':i)$;

(iii) $m_i(\mathbb{T}) + m_i(\mathbb{T}') = u - 1$.

Exercise 10.2. Let λ be a pair of partitions of total size n, let V^λ be the vector space with basis consisting of all standard tableaux of shape λ and define the action of the generators $T_t, S_1, \ldots, S_{n-1}$ of $H(B_n)$ on V^λ as in Theorem 10.1.5.

(i) Show that the representing matrices satisfy all the relations of $H(B_n)$ in (10.1.1).

(ii) Show that the representations of $H(B_n)$ on the spaces V^λ, where λ runs through the pairs of partitions of total size n, are pairwise inequivalent and that they form up to isomorphism a complete set of irreducible representations of $\mathbf{H}(B_n)$.

[*Hint.* Use induction on n and the branching rule in equation (3) of (10.1.4).]

Exercise 10.3. Let $\mathbf{A} = \mathbb{Z}[u^{\pm 1}, v^{\pm 1}]$ and $\theta \colon \mathbf{A} \to L$ be a specialization into a field L such that L is the field of fractions of the image of θ. Let $e \geq 1$ be minimal such that $1+\theta(u)+\theta(u)^2+\cdots+\theta(u)^{e-1} = 0$. (If no such e exists, we set $e = \infty$.) Show that the following hold:

(a) We have $\theta(c_\chi) \neq 0$ for all irreducible characters χ of $\mathbf{H}(B_n)$ if and only if $e > n$ and $\theta(v+u^i) \neq 0$ for $-(n-1) \leq i \leq n-1$. Here, $c_\chi = P_{B_n}(u,v)/D_\chi \in \mathbf{A}$ is the Schur element corresponding to χ.

(b) If the conditions in (a) are satisfied, then Hoefsmit's matrices as defined in (10.1.4) can be specialized under θ and yield a complete set of pairwise non-equivalent irreducible representations for the specialized algebra $L\mathbf{H}(B_n)$.

(c) $L\mathbf{H}(B_n)$ is split semisimple if and only if the conditions in (a) are satisfied.

Check that the conditions in (a) are satisfied, for example, if $u = v = 1$ and the characteristic of L does not divide the order of W_n. The statement in (c) is contained in [Dipper and James 1992, Theorem 5.5] (but the proof is different).

[*Hint.* First note that $c_\chi \in \mathbf{A}$, since $\mathbb{Q}(u,v)$ is a splitting field for $\mathbf{H}(B_n)$. (a) is checked using the formula in Theorem 10.5.3. For (b) and (c), argue as in the proof of Corollary 9.3.9, but using Hoefsmit's matrices instead of (9.3.8).]

Exercise 10.4. For partitions α, β, γ, denote by $\mathfrak{C}^{\alpha}_{\beta,\gamma}$ the set of tableaux $\mathbb{T}$ of shape $\alpha - \beta$ and weight γ such that $w(\mathbb{T})$ is a lattice permutation. Suppose that $\delta = [k-r, 1^r]$ is a hook. Further, let λ, μ and ν be pairs of partitions such that $\mu \subseteq \lambda$. Construct a bijection between

$$\coprod_{\nu=(\nu^0,\nu^1)} \left\{ (\mathbb{T}_0, \mathbb{T}_1, \mathbb{T}_2) \mid \mathbb{T}_0 \in \mathfrak{C}^{\lambda^0}_{\mu^0,\nu^0},\ \mathbb{T}_1 \in \mathfrak{C}^{\lambda^1}_{\mu^1,\nu^1},\ \mathbb{T}_2 \in \mathfrak{C}^{\delta}_{\nu^0,\nu^1} \right\}$$

and the set of all tableaux $(\mathbb{T}'_0, \mathbb{T}'_1)$ of shape $\lambda - \mu$ and weight δ.

[*Hint.* Observe that, since δ is a hook for every choice of $\nu_0 \subseteq \delta$, there are at most two possible choices for ν_1 such that $\mathfrak{C}^{\delta}_{\nu^0,\nu^1} \neq \varnothing$ and in any such case $|\mathfrak{C}^{\delta}_{\nu^0,\nu^1}| = 1$.]

Exercise 10.5. Consider the Coxeter group $W = \mathfrak{S}_n$ generated by the set $S = \{s_1, \ldots, s_{n-1}\}$ where $s_i = (i, i+1)$. Further, let $k \leq n$, let $J = S - \{s_k\}$ and let $X = X_J$ be the set of distinguished right coset representatives of W_J in W.

(i) Show that, for $w \in W$, we have $w \in X$ if and only if x is an order-preserving map from $\{1, \ldots, k\}$ to $\{1, \ldots, k\}.x$ and from $\{k+1, \ldots, n\}$ to $\{k+1, \ldots, n\}.x$.

(ii) Show that the map $x \mapsto \{1, \ldots, k\}.x$ is a bijection between the set X of coset representatives and the set of subsets of size k of $\{1, \ldots, n\}$.

(iii) Set $x_0 = (s_k \cdots s_1)(s_{k+1} \cdots s_2) \cdots (s_{n-1} \cdots s_{n-k})$. Show that, as permutation, $x_0 = (1, n-k+1)(2, n-k+2) \cdots (k, n)$ and that x_0 is the longest coset representative in X.

(iv) Let λ be a partition of k and let μ be a partition of $n-k$. Show that the map $(\mathbb{T}^0, \mathbb{T}^1, x) \mapsto (\mathbb{T}^0.x, \mathbb{T}^1.x_0x)$ establishes a bijection between $\mathsf{ST}(\lambda) \times \mathsf{ST}(\mu) \times X$ and $\mathsf{ST}(\lambda, \mu)$.

[*Hint.* Clearly, $W_J \cong \mathfrak{S}_k \times \mathfrak{S}_{n-k}$. For $x \in X$ and $i < n$, suppose that $i.x > (i+1).x$ for some $i \neq k$ and show that $l(s_i x) < l(x)$.]

Exercise 10.6. Let χ_{reg} be the character of the regular representation of $\mathbf{H}(\mathfrak{S}_n)$. Show that its values are given by the following formula:

$$\chi_{\mathrm{reg}}(T_{w_\alpha}) = [\mathfrak{S}_n : \mathfrak{S}_\alpha](u-1)^{n-r} \qquad \text{for all } \alpha = [\alpha_1, \ldots, \alpha_r] \vdash n.$$

The above formula is due to [Ram and Remmel 1997].

[*Hint.* We have $\chi_{\mathrm{reg}} = \sum_\chi \chi(1)\chi$ where the sum is over all $\chi \in \mathrm{Irr}(\mathbf{H}(\mathfrak{S}_n))$. Assume first that $\alpha = [n]$. In this case, evaluate that sum explicitly, using Example 9.2.9(b) and Exercise 5.8. If $\alpha \neq [n]$, then w_α is a Coxeter element in the proper parabolic subgroup $\mathfrak{S}_\alpha \subset \mathfrak{S}_n$. In this case, use the Mackey formula to show that the restriction of χ_{reg} to $\mathbf{H}(\mathfrak{S}_\alpha)$ is $[\mathfrak{S}_n : \mathfrak{S}_\alpha]$ times the character of the regular representation of $\mathbf{H}(\mathfrak{S}_\alpha)$.]

Exercise 10.7. The result of this exercise is used in the proof of Theorem 10.5.2. We consider elements $x_1, \ldots, x_n$ and $y_1, \ldots, y_n$ in a commutative ring and set

$$A_{ij} := \underbrace{(x_i - y_1)(x_i - y_2) \cdots (x_i - y_{n-j})}_{n-j \text{ factors}} \qquad \text{for } 1 \leq i, j \leq n.$$

Show that we have $\det(A_{ij}) = \prod_{1 \leq k < l \leq n}(x_k - x_l)$.

[*Hint.* Reduce to the known formula for the Vandermonde determinant.]

11

Computing character values and generic degrees

In this chapter, we describe methods for computing Schur elements and character tables of generic Iwahori–Hecke algebras of any type. These methods are partly presented in the form of algorithms which have been implemented in CHEVIE. As an application, we obtain the Schur elements and character tables for the exceptional types H_3, H_4, F_4, E_6, E_7 and E_8. Together with the results in Section 8.3 (type $I_2(m)$) and Chapter 10 (classical types), this completes the determination of Schur elements and character tables for all types.

In Section 11.1, we begin with the problem of actually constructing representations affording the irreducible characters of an Iwahori–Hecke algebra. It is a remarkable fact that, in many cases, the irreducible representations have been found to be afforded by Kazhdan–Lusztig W-graphs. The most general result in this direction is due to [Gyoja 1984]: every irreducible representation of a one-parameter generic Iwahori–Hecke algebra associated with a *crystallographic* finite Coxeter group is afforded by a W-graph. (The proof requires some deep results on Kazhdan–Lusztig cells from [Lusztig 1981a].) We will describe W-graphs for the types H_3 and H_4 (which are not covered by the above result); these results are due to Alvis and Lusztig. In Section 11.3, we also describe W-graphs for type F_4 (with unequal parameters); these results are new.

In Section 11.4, we describe methods for computing the generic degrees and the Schur elements of a generic Iwahori–Hecke algebra. We present a systematic algorithmic approach which yields complete results for all exceptional types in a straightforward manner. Finally, in Section 11.5, we present methods for computing character values of generic Iwahori–Hecke algebras. This uses in an essential way the results on good elements in Section 9.2. The tables are electronically available through the CHEVIE package of GAP. Since they may become quite large (e.g., for type E_8), we do not print them here.

As a main application, we can determine explicitly the distribution of the irreducible characters into blocks and the number of irreducible characters for non-semisimple specialized Iwahori–Hecke algebras of exceptional type.

Throughout this chapter, (W, S) is a Coxeter system with W finite, and $\mathbf{H}$ is a corresponding generic Iwahori–Hecke algebra over the ring $\mathbf{A} = \mathbb{Z}[u_s^{\pm 1} \mid s \in S]$. Note that, as in Definition 8.1.4 and Remark 8.1.5, we only require that there is a specialization $\mathbf{A} \to \mathbb{Z}$ such that $u_s \mapsto 1$ for all $s \in S$, but we do not require that the parameters u_s are actually independent for non-conjugate generators. Furthermore, let $K \subseteq \mathbb{C}$ be a sufficiently large subfield and set $\mathbf{K} := K(\sqrt{u}_s \mid s \in S)$. Then, by Theorem 9.3.5 and Theorem 8.1.7, the algebra $\mathbf{KH}$ is split semisimple and isomorphic to the group algebra $\mathbf{K}[W]$.

11.1 W-GRAPHS AND REPRESENTING MATRICES

In this section, we are concerned with the problem of constructing representations affording the various irreducible characters of $\mathbf{KH}$. Note that, for a general associative algebra, one cannot expect any "canonical" solution to this problem. In Theorem 8.1.11, we found an explicit construction of the reflection representation of $\mathbf{KH}$, in which the action of the generators T_s ($s \in S$) is given by formulas of a particular simple form. The following definition may be regarded as a formal generalization of that construction.

11.1.1. Definition (Kazhdan–Lusztig). A *W-graph* for $\mathbf{H}$ consists of the following data:

(a) a set $\mathfrak{C}$ together with a map I which assigns to each $x \in \mathfrak{C}$ a set $I(x) \subseteq S$;
(b) a collection of elements $\{\mu^s_{x,y}\}$ in $\mathbf{K}$, where $x, y \in \mathfrak{C}$ and $s \in S$ are such that $s \in I(x)$ and $s \notin I(y)$.

These data are subject to the following requirements. First, the elements in (b) should be invariant under the automorphism of $\mathbf{K}$ which is given by $\sqrt{u}_s \mapsto \sqrt{u}_s^{\,-1}$ for all $s \in S$. Next, let V be a $\mathbf{K}$-vector space with a basis $\{e_y \mid y \in \mathfrak{C}\}$ and, for each $s \in S$, define a linear map $\rho_s : V \to V$ by

$$\begin{aligned} e_y &\mapsto u_s e_y + \sum_{\substack{x \in \mathfrak{C} \\ s \in I(x)}} \sqrt{u}_s \mu^s_{x,y} e_x && \text{if } s \notin I(y), \\ e_y &\mapsto -e_y && \text{if } s \in I(y). \end{aligned}$$

Then we require that the assignment $T_s \mapsto \rho_s$ defines a representation of $\mathbf{KH}$. We say that the above data define an *integral W-graph* if all elements in (b) are Laurent polynomials in $\sqrt{u}_s$ ($s \in S$), whose coefficients are algebraic integers.

11.1.2. Remark. (a) A more accurate motivation for the above definition is provided by a certain standard W-graph with underlying set $\mathfrak{C} = W$, whose construction is due to [Kazhdan and Lusztig 1979] in the one-parameter case and [Lusztig 1983] in the multi-parameter case; see Theorem 11.1.10 below.

(b) The above definition also makes sense for infinite Coxeter groups, if we require that, for any $s \in S$ and $y \in \mathfrak{C}$ there are only finitely many $x \in \mathfrak{C}$ such that $s \in I(x)$, $s \notin I(y)$ and $\mu^s_{x,y} \neq 0$. For example, this is the case for the Kazhdan–Lusztig W-graph in Theorem 11.1.10.

(c) The original definition of W-graphs in [Kazhdan and Lusztig 1979] is only concerned with the case of equal parameters. In that case, one can also require that the μ-constants lie in K (i.e., they are constant polynomials). In general, it does not seem to be possible to preserve this condition; see, for example, the irreducible representations in type $I_2(m)$ with unequal parameters in Theorem 8.3.1.

Here are some examples of W-graphs which exist for all types of (W, S).

11.1.3. Example. The representation ind: $T_s \mapsto u_s$ $(s \in S)$ is afforded by the W-graph which consists of a singleton set $\mathfrak{C} = \{x\}$ with $I(x) = \varnothing$; we need not specify any μ-value. Similarly, the representation $\varepsilon\colon T_s \mapsto -1$ $(s \in S)$ is afforded by the W-graph which consists of a singleton set $\mathfrak{C} = \{x\}$ with $I(x) = S$.

Now consider the reflection representation ρ of $\mathbf{K}\mathbf{H}$, as defined in (8.1.9). Assume that K is large enough such that, for any $s, t \in S$, $s \neq t$, we have $2\cos(2\pi/m_{st}) \in K$, where m_{st} is the order of st. We now define a W-graph for $\mathbf{H}$ as follows. Let $\mathfrak{C} = S$. For $s \in S$, let $I(s) \subseteq S$ be the singleton set $\{s\}$. The μ-function is given as follows. We choose a total order $\leq$ on S. Let $s, t \in S$, $s < t$. If s, t commute with each other, the corresponding μ-values are all 0. Otherwise, we define $\mu^s_{s,t}, \mu^t_{t,s} \in \mathbf{K}$ by

$$\mu^s_{s,t} = 1 \qquad \text{and} \qquad \mu^t_{t,s} = \sqrt{u}_s\sqrt{u}_t^{\,-1} + 2\cos(2\pi/m_{st}) + \sqrt{u}_t\sqrt{u}_s^{\,-1}.$$

The action of the corresponding linear maps ρ_s $(s \in S)$ on V can be expressed in the single formula $e_t \mapsto (u_s + 1)e_s + \sqrt{u}_s\mu^s_{s,t}e_t$. Thus, if we set $c_{ts} = -\sqrt{u}_s\mu^s_{s,t}$ and $c_{st} = -\sqrt{u}_t\mu^t_{t,s}$, we have

$$c_{st}c_{ts} = u_s + 2\cos(2\pi/m_{st})\sqrt{u_s u_t} + u_t \qquad (\text{if } m_{st} > 2),$$

and we see that the linear maps ρ_s act by the same formula as in the definition of the reflection representation. Hence the above data indeed define a W-graph.

11.1.4. Remark. Assume that we are given data $(\mathfrak{C}, I, \{\mu^s_{x,y}\})$ satisfying conditions (a) and (b) in Definition 11.1.1. Then it is a purely combinatorial matter to decide if these data define a W-graph and, if this is the case, whether or not the corresponding representation of $\mathbf{K}\mathbf{H}$ is irreducible. Indeed, the linear maps ρ_s given by the formulas in Definition 11.1.1 extend to a representation of $\mathbf{K}\mathbf{H}$ if and only if they satisfy the defining relations:

$$\rho_s^2 = u_s\,\mathrm{id}_V + (u_s - 1)\rho_s \qquad \text{for all } s \in S,$$

$$\underbrace{\rho_s \circ \rho_t \circ \cdots}_{m_{st}\text{ factors}} = \underbrace{\rho_t \circ \rho_s \circ \cdots}_{m_{st}\text{ factors}} \qquad \text{for all } s, t \in S,\ s \neq t,$$

where, as usual, m_{st} denotes the order of $st \in S$. Assume now that this is the case, and let $\rho\colon \mathbf{K}\mathbf{H} \to \mathrm{End}_K(V)$ be the corresponding representation, with character χ. Then, for any $w \in W$, we can determine $\rho(T_w)$ and $\chi(T_w)$, simply by taking a reduced expression $w = s_1 \cdots s_n$ $(s_i \in S)$, forming the product of the corresponding maps ρ_{s_i}, and setting $\chi(T_w) = \mathrm{Tr}(\rho(T_w))$.

Finally, specializing $\sqrt{u}_s \mapsto 1$ $(s \in S)$, we obtain a character χ_1 of W; see (8.1.6). Computing $\chi(T_w)$ as w varies over a complete set of representatives (of minimal length) in the conjugacy classes of W, we can also check if ρ is irreducible or not: by Tits's deformation theorem, it is enough to check the analogous property for χ_1, which can be done using the orthogonality relations.

11.1.5. Decomposing W-graphs. Assume that the data $(\mathfrak{C}, I, \{\mu^s_{x,y}\})$ define a W-graph for $\mathbf{H}$. We define a pre-order $\prec_W$ on $\mathfrak{C}$ as follows. For $x, y \in \mathfrak{C}$ $(x \neq y)$ we write $x \prec_W y$ if there exists a sequence $x = x_0, \ldots, x_m = y$ with $x_i \in \mathfrak{C}$ for all i and a sequence $s_1, \ldots, s_n$ with $s_i \in S$ for all i, such that

$$s_i \in I(x_{i-1}), \quad s_i \notin I(x_i) \quad \text{and} \quad \mu^{s_i}_{x_{i-1},x_i} \neq 0 \qquad \text{for } 1 \leq i \leq m.$$

The pre-order $\prec_W$ gives rise to an equivalence relation $\sim_W$ on $\mathfrak{C}$: we have $x \sim_W y$ if and only if either $x = y$ or $x \prec_W y$ and $y \prec_W x$. The equivalence classes of $\mathfrak{C}$ under $\sim_W$ are called the *cells* of $\mathfrak{C}$. Let $\mathfrak{C}' \subseteq \mathfrak{C}$ be such a cell. Then, by restricting I and $\{m^s_{x,y}\}$ to $\mathfrak{C}'$, we again obtain a W-graph, with underlying set $\mathfrak{C}'$.

This is seen as follows. Write $\mathfrak{C} = \mathfrak{C}_1 \amalg \ldots \amalg \mathfrak{C}_n$, where $\mathfrak{C}_i$ are the equivalence classes under $\sim_W$ and the labelling is chosen to be compatible with $\prec_W$, i.e., for $x \in \mathfrak{C}_i$ and $y \in \mathfrak{C}_j$, we have $x \prec_W y \Rightarrow i < j$. Then, for any $s \in S$, the matrix of ρ_s with respect to the basis $\{e_y \mid y \in \mathfrak{C}\}$ has a block lower triangular shape, with blocks corresponding to the equivalences classes $\mathfrak{C}_i$. Hence, the projection onto the diagonal block corresponding to $\mathfrak{C}_i$ determines a representation on the subspace $\langle e_y \mid y \in \mathfrak{C}_i\rangle \subseteq V$ in which the action of T_s $(s \in S)$ is given by the required formulas.

11.1.6. Algebraically conjugate W-graphs. Let $\chi \in \mathrm{Irr}(\mathbf{KH})$ and assume that χ is afforded by a W-graph consisting of data $(\mathfrak{C}, I, \{\mu^s_{x,y}\})$ as in Definition 11.1.1. Now let $\sigma\colon K \to K$ be a field automorphism. Then σ canonically extends to an automorphism of $\mathbf{K}$ such that $\sigma(\sqrt{u_s}) = \sqrt{u_s}$ for all $s \in S$. Via σ, we obtain a character $\chi^\sigma \in \mathrm{Irr}(\mathbf{KH})$ which is given by the formula (see Exercise 9.1):

$$\chi^\sigma(T_w) = \sigma(\chi(T_w)) \qquad \text{for all } w \in W.$$

Now consider the data $(\mathfrak{C}, I, \{\sigma(\mu^s_{x,y})\})$. We claim that they define a W-graph which affords χ^σ. Indeed, let V be a $\mathbf{K}$-vector space with basis $\mathfrak{C}$ as in Definition 11.1.1 and $\rho\colon \mathbf{KH} \to \mathrm{End}_{\mathbf{K}}(V)$ be the corresponding representation. Then ρ^σ is a representation in which T_s acts by similar formulas as in Definition 11.1.1, but where each $\mu^s_{x,y}$ has been replaced by $\sigma(\mu^s_{x,y})$.

11.1.7. The dual W-graph. Let $\chi \in \mathrm{Irr}(\mathbf{KH})$. Then we have a corresponding dual character $\chi^* \in \mathrm{Irr}(\mathbf{KH})$ as in Proposition 9.4.1. (Under the specialization $\sqrt{u_s} \mapsto 1$, the operation $\chi \mapsto \chi^*$ corresponds to tensoring with the sign character.) Assume now that χ is afforded by a W-graph consisting of data $(\mathfrak{C}, I, \{\mu^s_{x,y}\})$ as in Definition 11.1.1. Let $\rho\colon \mathbf{KH} \to \mathrm{End}_{\mathbf{K}}(V)$ be the corresponding representation of $\mathbf{KH}$. We wish to describe a W-graph affording the dual representation ρ^*, with character χ^*. Recall that ρ^* is defined via composition with the algebra automorphism $\gamma\colon \mathbf{K}H \to \mathbf{K}H$, $T_s \mapsto -u_s T_s^{-1}$ $(s \in S)$. Since $T_s^{-1} = u_s^{-1}T_s + (u_s^{-1} - 1)T_1$, the action of $\gamma(T_s)$ on V is given as follows:

$$e_y \mapsto -u_s^{-1}e_y - \sum_{\substack{x\in\mathfrak{C}\\ s\in I(x)}} \sqrt{u_s}^{-1}\mu^s_{x,y}e_x \quad \text{if } s \notin I(y),$$

$$e_y \mapsto e_y \qquad\qquad \text{if } s \in I(y).$$

Now consider the dual space $V^* = \mathrm{Hom}_{\mathbf{K}}(V, \mathbf{K})$ and let $(\ ,\ )$ be the natural pairing $V^* \times V \to \mathbf{K}$. Then the above right action of $\mathbf{K}H$ on V (obtained via composing the original action with γ) induces a natural left action on V^*, which we denote by $f \mapsto h{*}f$, for $f \in V^*$ and $h \in \mathbf{K}H$. Let $\{f_y \mid y \in \mathfrak{C}\}$ be the dual basis of V^*. By evaluating both sides of the identity $(f_y, e_x.\gamma(T_s)) = (T_s{*}f_y, e_x)$ for any $x, y \in \mathfrak{C}$, we see that this left action of T_s on V^* is given by

$$f_y \mapsto u_s f_y - \sum_{\substack{x\in\mathfrak{C}\\ s\notin I(x)}} \sqrt{u_s}\mu^s_{y,x}f_x \quad \text{if } s \in I(y),$$

$$f_y \mapsto -f_y \qquad\qquad \text{if } s \notin I(y).$$

Now we have an anti-involution on $\mathbf{K}H$ given by $T_w \mapsto T_{w^{-1}}$ $(w \in W)$; see Exercise 4.8. By composition with that anti-involution, we can turn the above left action of $\mathbf{K}H$ on V^* into a right action. Thus, if we let

$$\hat{I}(x) := S - I(x) \quad \text{and} \quad \hat{\mu}^s_{x,y} = -\mu^s_{y,x},$$

then the above right action of $\mathbf{K}H$ on V^* arises from the W-graph $(\mathfrak{C}, \hat{I}, \{\hat{\mu}^s_{x,y}\})$, which is called the *dual W-graph*. Let $\hat{\rho}\colon \mathbf{K}H \to \mathrm{End}_{\mathbf{K}}(V^*)$ be the corresponding representation; then we have

$$\rho^*(T_w) = \hat{\rho}(T_{w^{-1}})^{tr} \qquad \text{for all } w \in W.$$

Consequently, ρ^* and $\hat{\rho}$ have the same character. (Recall from Corollary 8.2.6(c) that T_w and $T_{w^{-1}}$ have the same trace in any representation of $\mathbf{K}H$.) Thus, the dual W-graph affords the dual character χ^*.

11.1.8. W-graphs for type H_3. Let (W, S) be of type H_3, where the labelling of the elements in S is given by the following Dynkin diagram:

H_3 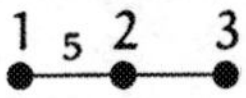

All generators are conjugate; we write $u = u_{s_i}$ for $i = 1, 2, 3$. Let

$$K = \mathbb{Q}(\lambda) \qquad \text{with} \qquad \lambda = \frac{1}{2}(1 + \sqrt{5}).$$

There are 10 irreducible characters; they are denoted as in Table C.1. W-graphs for 1_r, $3'_s$, $4'_r$ and 5_r are shown in Figure 11.1.

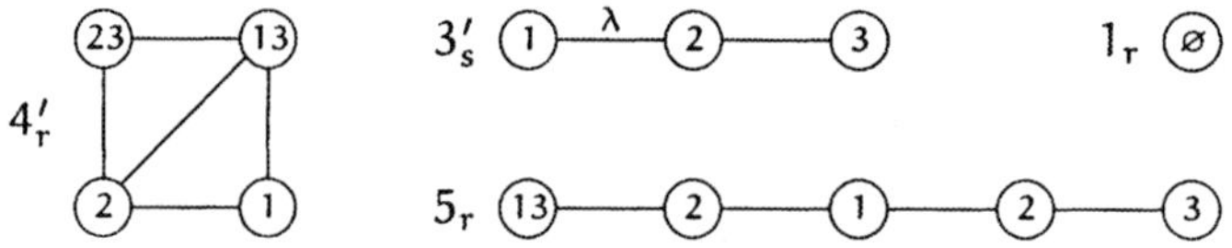

Fig. 11.1. W-graphs for type H_3 (Lusztig).

The remaining characters and corresponding W-graphs are obtained by algebraic conjugation or duality; see (11.1.6) and (11.1.7).

To read the graphs in Figure 11.1, we use the following conventions. For each vertex x, the numbers inscribed in that vertex correspond to the generators which lie in the set $I(x)$. The μ-function is given as follows. Let $x \neq y$ be two vertices. If x, y are connected by an unlabelled segment (or by a segment labelled λ) then $\mu^s_{x,y} = 1$ (or $\mu^s_{x,y} = \lambda$) for all possible s; if x, y are not connected, then $\mu^s_{x,y} = 0$ for all possible s.

In order to check that the graphs in Figure 11.1 are indeed W-graphs, one has to write down the corresponding matrices for all T_{s_i} and verify the defining relations; see Remark 11.1.4. For example, the matrices for $4'_r$ are given by

$$T_{s_1} \mapsto \begin{bmatrix} u & \sqrt{u} & 0 & 0 \\ 0 & -1 & 0 & 0 \\ 0 & \sqrt{u} & u & \sqrt{u} \\ 0 & 0 & 0 & -1 \end{bmatrix}, T_{s_2} \mapsto \begin{bmatrix} -1 & 0 & 0 & 0 \\ \sqrt{u} & u & \sqrt{u} & 0 \\ 0 & 0 & -1 & 0 \\ 0 & 0 & \sqrt{u} & u \end{bmatrix}, T_{s_3} \mapsto \begin{bmatrix} -1 & 0 & 0 & 0 \\ 0 & -1 & 0 & 0 \\ \sqrt{u} & \sqrt{u} & u & 0 \\ 0 & \sqrt{u} & 0 & u \end{bmatrix}$$

Furthermore, we check that the representations afforded by the above graphs are irreducible by the procedure described in Remark 11.1.4. In doing so, we also obtain the character table of $\mathbf{K}\mathbf{H}$. Once that table is known, we can verify using Lemma 9.4.4 that the polynomials in Table E.2 are the generic degrees for $\mathbf{H}$.

The character table was computed by a different method in (9.2.10). From Table 9.1 (p. 300), we can also determine the exact field extension of $\mathbb{Q}(u)$ in which all character values $\chi(T_w)$ lie. We denote that field by $K_0(\chi)$. Note that, in order to determine $K_0(\chi)$, it is sufficient to look at the values of χ which appear in the character table. (This follows by using the class polynomials and the equations $\chi(T_w) = \sum_C f_{w,C}\chi(T_{w_C})$.) The results are as follows:

χ	$K_0(\chi)$
$1_r, 1'_r, 5_r, 5'_r$	$\mathbb{Q}(u)$
$3_s, \overline{3}_s, 3'_s, \overline{3}'_s$	$\mathbb{Q}(\sqrt{5}, u)$
$4'_r, 4'_r$	$\mathbb{Q}(\sqrt{u})$

Since all irreducible characters have Schur index 1 (see Proposition 9.3.1), each χ can even be realized over $K_0(\chi)$.

We close this section with the construction of the Kazhdan–Lusztig W-graph. For the remainder of this section, (W, S) may be any Coxeter system. Let Γ and

$\Gamma_+ \subset \Gamma$ be as in the remarks preceding Lemma 9.4.7. Furthermore, for $w \in W$ we define $\nu_w \in \Gamma_+$ by the condition that $\nu_w^2 = \mathsf{ind}(T_w)$. Let $\leq$ be the Bruhat–Chevalley order on W; see Exercise 1.7.

11.1.9. Definition (M-polynomials). Let $w \in W$ and suppose we are given a collection of elements $h_{y,z} \in \mathbb{Z}[\Gamma]$ for all $y, z \in W$ with $y \leq z \leq w$. Then, for any $y \in W$ and $t \in S$ such that $ty < y < w < tw$, we define the corresponding *M-polynomial* $M_{y,w}^t \in \mathbb{Z}[\Gamma]$ recursively by the conditions

$$\sum_{\substack{y \leqslant z < w \\ tz < z}} h_{y,z} M_{z,w}^t - \nu_t h_{y,w} \in \mathbb{Z}[\Gamma_-], \tag{M1}$$

$$\overline{M}_{y,w}^t = M_{y,w}^t, \tag{M2}$$

where $\mathbb{Z}[\Gamma_-] = \{\sum_{g \in \Gamma} a_g g \in \mathbb{Z}[\Gamma] \mid a_g \neq 0 \Rightarrow g \in \Gamma_-\}$ and $a \mapsto \bar{a}$ is the involution of $\mathbb{Z}[\Gamma]$ which takes g to g^{-1} for any $g \in \Gamma$. Note that, indeed, $M_{y,w}^t$ is uniquely determined: if y is given and $M_{z,w}^t$ has been already determined for all $z \in W$ with $y < z < w$ and $tz < z$, then (M1) uniquely determines the coefficient of any $g \in \Gamma - \Gamma_-$ in $M_{y,w}^t$; the remaining coefficients are then determined by (M2). If there exists no $z \in W$ such that $y < z < tw$ and $tz < z$, then (M1) yields the initial condition $M_{y,w}^t - \nu_t h_{y,w} \in \mathbb{Z}[\Gamma_-]$.

To state the following basic result, we write $\tilde{T}_w = \nu_w^{-1} T_w$ for $w \in W$ and extend the involution $a \mapsto \bar{a}$ of $\mathbb{Z}[\Gamma]$ to a map $\mathbf{H}_\Gamma \to \mathbf{H}_\Gamma$, $h \mapsto \bar{h}$, by the formula

$$\overline{\sum_{w \in W} a_w \tilde{T}_w} = \sum_{w \in W} \bar{a}_w \tilde{T}_{w^{-1}}^{-1} \qquad (a_w \in \mathbb{Z}[\Gamma]).$$

Thus, if j, γ are the automorphisms introduced in Exercises 8.1 and 8.2, we have $\bar{h} = j(\gamma(h)) = \gamma(j(h))$ for all $h \in \mathbf{H}_\Gamma$. So $h \mapsto \bar{h}$ is in fact a ring involution.

11.1.10. Theorem (Kazhdan–Lusztig, Lusztig).

(a) *For each $w \in W$, there exists a unique $C'_w \in \mathbf{KH}$ such that*

$$\overline{C}'_w = C'_w \qquad \textit{and} \qquad C'_w \in \tilde{T}_w + \sum_{y \in W} \mathbb{Z}[\Gamma_-]\, \tilde{T}_y. \tag{$*$}$$

Writing $C'_w = \sum_y P^*_{y,w} \tilde{T}_y$ with $P^*_{y,w} \in \mathbb{Z}[\Gamma]$, we have $P^*_{w,w} = 1$ and $P^*_{y,w} = 0$ unless $y \leq w$. The elements $\{C'_w \mid w \in W\}$ form a basis of $\mathbf{KH}$.

(b) *Let $M_{x,y}^s$ be the M-polynomials with respect to $P^*_{x,y}$ and consider the following data:*

$$\begin{aligned} \mathfrak{C} &:= W, \\ I(x) &:= \{s \in S \mid l(sx) < l(x)\} \quad (x \in W), \\ \mu_{x,y}^s &:= \begin{cases} 1 & \textit{if } x = sy > y, \\ -(-1)^{l(x)+l(y)} M_{x,y}^s & \textit{if } sx < x < y < sy, \\ 0 & \textit{otherwise} \end{cases} \end{aligned}$$

(where $x, y \in W$, $s \in S$). Then $(\mathfrak{C}, I, \{\mu^s_{x,y}\})$ defines a W-graph in the sense of Definition 11.1.1.

The basis $\{C'_w\}$ is called the *Kazhdan–Lusztig basis* of **K**H. The above W-graph is called the *Kazhdan–Lusztig W-graph*, and its cells (in the sense of (11.1.5)) are called the *Kazhdan–Lusztig cells* of W.

Proof. We proceed in three steps.

Step 1. First we show that if we can prove the existence of elements C'_w having all of the properties in (a), then C'_w is uniquely determined by (∗). This is seen as follows. Since $C'_y \in \tilde{T}_y + \sum_{x<y} \mathbb{Z}[\Gamma_-]\tilde{T}_x$ for all $y \in W$, a simple induction shows that we also have $\tilde{T}_y \in C'_y + \sum_{x<y} \mathbb{Z}[\Gamma_-]C'_x$ for all $y \in W$. In particular, this proves that the elements $\{C'_w\}$ form a basis of **K**H.

Now fix $w \in W$ and suppose we had another element $C''_w \in \mathbf{K}H$ satisfying (∗). Then we have

$$C'_w - C''_w = \sum_{y \in W} c_y \tilde{T}_y = \sum_{y \in W} d_y C'_y \qquad \text{with all } c_y, d_y \in \mathbb{Z}[\Gamma_-].$$

Applying the map $h \mapsto \overline{h}$ to this equation yields $C'_w - C''_w = \sum_{y \in W} \overline{d}_y C'_y$. (Note that $\overline{C}'_w = C'_w$ and $\overline{C}''_w = C''_w$.) Since $\{C'_y\}$ are linearly independent, we conclude that $d_y = \overline{d}_y$ and so $d_y \in \mathbb{Z}[\Gamma_-] \cap \mathbb{Z}[\Gamma_+] = \{0\}$ for all $y \in W$. Hence we have $C'_w = C''_w$, as desired.

Step 2. We prove the existence of C'_w by induction on $l(w)$. At the same time we show that we have the following rules, where $t \in S$ is such that $tw < w$:

$$(\tilde{T}_t + v_t^{-1}\tilde{T}_1)\, C'_{tw} = C'_w + \sum_{\substack{z<tw \\ tz<z}} M^t_{z,tw}\, C'_z, \tag{C1$'$}$$

$$\tilde{T}_t\, C'_w = v_t C'_w. \tag{C2$'$}$$

To start the induction, we set $C'_1 := \tilde{T}'_1$ and $C'_s := \tilde{T}_s + v_s^{-1}\tilde{T}_1$ for $s \in S$. Using the quadratic relation $\tilde{T}_s^2 = \tilde{T}_1 + (v_s - v_s^{-1})\tilde{T}_s$, it follows that $\overline{C}'_s = C'_s$; moreover, C'_s is a linear combination of $\tilde{T}_1$ and $\tilde{T}_s$ and we have $\tilde{T}_s C'_s = v_s C'_s$.

Now let $w \neq 1$ and assume that C'_y has already been constructed for all $y \in W$ with $l(y) < l(w)$, satisfying (C1$'$) and (C2$'$). Moreover, we assume that $C'_y = \sum_{x \in W} P^*_{x,y}\tilde{T}_x$ where $P^*_{y,y} = 1$ and $P^*_{x,y} = 0$ unless $x \leq y$. Let $M^s_{x,y}$ be the corresponding M-polynomials. Let $s \in S$ be such that $sw < w$ and set

$$C'_w := C'_s C'_{sw} - \sum_{\substack{z<sw \\ sz<z}} M^s_{z,sw}\, C'_z.$$

We check that (∗) is satisfied. First note that since all terms on the right-hand side are invariant under $h \mapsto \overline{h}$, so is C'_w. Now write all the above expressions as linear combinations of $\{\tilde{T}_y \mid y \in W\}$. Then the coefficient of $\tilde{T}_y$ in C'_w is given by

$$f_y := P^*_{sy,sw} + v_s^c P^*_{y,sw} - \sum_{\substack{y \leqslant z < sw \\ sz < z}} P^*_{y,z} M^s_{z,sw},$$

where $c = 1$ if $sy < y$ and $c = -1$ if $sy > y$. If $sy < y$, we have $f_y \in P^*_{sy,sw} + \mathbb{Z}[\Gamma_-]$ by (M1). Thus, by induction, f_y has the required form if $y = w$ or if $sy < y$. Now assume that $sy > y$. Then $s(sy) < sy$ and so we already know that $f_{sy} \in \mathbb{Z}[\Gamma_-]$. On the other hand, consider the identity (C2′) for z and s, where $y \leq z < sw$, $sz < z$. Comparing coefficients in $\tilde{T}_s C'_z = v_s C'_z$, we find that $P^*_{y,z} = v_s^{-1} P^*_{sy,z}$. Hence we have

$$f_y = P^*_{sy,sw} + v_s^{-1} P^*_{y,sw} - \sum_{\substack{y \leqslant z < sw \\ sz < z}} v_s^{-1} P^*_{sy,z} M^s_{z,sw}.$$

But this is the same as $v_s^{-1} f_{sy}$ and we already know that the latter expression lies in $\mathbb{Z}[\Gamma_-]$. Thus, we have shown that C'_w satisfies (∗). Furthermore, assume that $f_y \neq 0$. Then one of the terms $P^*_{sy,sw}$, $P^*_{y,sw}$ or $P^*_{sy,z}$ ($z < sw$) must be non-zero. But this implies that $y \leq w$, by induction. (Indeed, the only case which is not trivial is the case where $sy < sw$ and $sy < y$. In that case, sy is a subexpression of sw. So, since $l(sy) < l(y)$ and $l(sw) < l(w)$, we also have $y = s(sy) < s(sw) = w$.)

It remains to prove the multiplication rules (C1′) and (C2′). Take any $t \in S$ with $tw < w$. Since C'_w is uniquely determined by (∗), the defining equation for C'_w remains valid when we replace s by t. Thus, (C1′) holds. On the other hand, to prove (C2′), we multiply (C1′) from the left with $\tilde{T}_t - v_t\tilde{T}_1$. By the quadratic relation for $\tilde{T}_t$, the left-hand side of the resulting identity is 0. On the other hand, by induction, we have $(\tilde{T}_t - v_t\tilde{T}_1)C'_z = 0$ for all $z \in W$ with $tz < z < tw$. So we also have $(\tilde{T}_t - v_t\tilde{T}_1)C'_w = 0$, as desired.

Step 3. As we have already remarked in step 1, the elements $\{C'_w \mid w \in W\}$ form a basis of **K**H. The multiplication rules in step 2 do not yet have quite the correct form. To transform them, we consider the ring automorphism j introduced in Exercise 8.1. With respect to the basis $\{\tilde{T}_w \mid w \in W\}$, this is given by the formula:

$$j\Big(\sum_{w \in W} a_w \tilde{T}_w\Big) = \sum_{w \in W} \overline{a}_w \varepsilon_w \tilde{T}_w, \qquad \text{where } \varepsilon_w = (-1)^{l(w)}.$$

Now set $C_w := \varepsilon_w j(C'_w)$ for $w \in W$. Then we have $C_w = \sum_{y \leqslant w} \varepsilon_w \varepsilon_y \overline{P}^*_{y,w} \tilde{T}_y$. In particular, this shows that $\{C_w \mid w \in W\}$ is a basis of **K**H. Now, applying j to (C1′) (with sw replaced by w) and (C2′), we obtain

$$(\tilde{T}_s - v_s\tilde{T}_1)C_w = C_{sw} - \sum_{\substack{z<w \\ sz<z}} \varepsilon_z\varepsilon_w M^s_{z,w} C_z \qquad \text{if } sw > w, \tag{C1}$$

$$\tilde{T}_s C_w = -v_s^{-1} C_w \qquad \text{if } sw < w. \tag{C2}$$

Note that $C_s = \tilde{T}_s - v_s\tilde{T}_1$. Finally, in order to obtain a "right-handed" version of the above formulas, we apply the anti-involution ι of Exercise 4.8. This yields

$$\overline{\iota(C_w)} = \iota(C_w) \quad \text{and} \quad \iota(C_w) = \sum_{y \leq w} \varepsilon_w \varepsilon_y \overline{P}^*_{y,w} \tilde{T}_{y^{-1}}.$$

The elements $\{\iota(C_w) \mid w \in W\}$ also form a basis of $\mathbf{K}H$. With respect to that basis, the linear map of $\mathbf{K}H$ given by right multiplication with T_s ($s \in S$) is given by formulas as required in Definition 11.1.1. □

11.1.11. Kazhdan–Lusztig polynomials. The expression for f_y in the proof of Theorem 11.1.10 actually yields a recursion formula for the polynomials $P^*_{y,w}$. First, we have $P^*_{w,w} = 1$ for all $w \in W$ and $P^*_{y,w} = 0$ unless $y \leq w$. Now let $y, w \in W$ and $s \in S$ be such that $y < w$ and $sw < w$. Then we have

$$P^*_{y,w} = v_s P^*_{y,sw} + P^*_{sy,sw} - \sum_{\substack{y \leqslant z < sw \\ sz < z}} P^*_{y,z} M^s_{z,sw} \qquad \text{if } sy < y, \tag{P1}$$

$$P^*_{y,w} = v_s^{-1} P^*_{sy,w} \qquad \text{if } sy > y, \tag{P2}$$

where $M^s_{z,sw}$ are the M-polynomials with respect to $P^*_{z,sw}$ If we set $P_{y,w} := v_w v_y^{-1} P^*_{y,w}$ then $P_{y,w}$ is a polynomial in the variables u_s ($s \in S$) with constant term 1; see Exercise 11.1. These polynomials are the famous *Kazhdan–Lusztig polynomials*, which were introduced in [Kazhdan and Lusztig 1979] in the case of equal parameters and in [Lusztig 1983] in general. For further properties of the $P_{y,w}$ and $M^t_{y,w}$ in the equal parameter case, see Exercise 11.2.

Some examples are given in Exercises 11.4 for type $I_2(m)$. In the case of equal parameters, the CHEVIE command `LeftCells` computes the Kazhdan–Lusztig cells of a finite Coxeter group (of rank up to 5, say).

11.1.12. Example (Lusztig). Let $s, t \in S$, $s \neq t$. Assume that s, t do not commute with each other, and that $v_s > v_t$. (This situation occurs, for example, in type $I_2(m)$ with $m \geq 4$ even.) Applying (P1) and (P2) repeatedly, we obtain

$$P^*_{1,sts} = v_s^{-2} v_t^{-1} - v_s^{-2} v_t.$$

In particular, negative coefficients may occur! (This phenomenon cannot happen in the case of equal parameters; see the bibliographical remarks.) Note that it was crucial to assume that $v_s > v_t$. If $v_s = v_t$, we obtain $P^*_{1,sts} = v^{-3}$.

11.2 W-GRAPHS FOR THE IRREDUCIBLE CHARACTERS OF TYPE H_4

Let (W, S) be of type H_4, where the labelling of the elements in S is given by the following diagram:

H_4 

All generators are conjugate; we write $u = u_{s_i}$ for $i = 1, 2, 3, 4$. Let

$$K = \mathbb{Q}(\lambda) \qquad \text{with} \qquad \lambda = \frac{1}{2}(1+\sqrt{5}),$$

There are 34 irreducible characters; they are denoted as in Table C.2. In Figure 11.2*, we print explicitly W-graphs for the characters 1_r, 4_t, 6_s, 8_r, 8_{rr}, 9_s, 10_r, 16_r, 16_{rr}, 16_t, 18_r, 25_r and 36_{rr}.

The characters $\overline{24}_t$, $\overline{24}_s$, $\overline{30}_s$, 40_r and 48_{rr} are given by W-graphs which can be described using a single general pattern, as follows. Consider the last graph in Figure 11.2; we denote it by Γ. For each vertex x of Γ, we denote by I_x the subset of S indicated in that table. We shall assume that for each vertex x we are given a totally ordered set $X_x = \{x_1, \ldots, x_n\}$, with n depending on x. Let $\tilde{\Gamma} := \coprod_{x \in \Gamma} X_x$. For each $x_i \in X_x$, we define $I(x_i)$ to be I_x. We define the μ-function on $\tilde{\Gamma}$ as follows. For $x_i \neq x_j$ inside the same set X_x, we set $\mu^s_{x_i, x_j} = 0$. Now let $x \neq y$ be two vertices of Γ and let $X_x = \{x_1, \ldots, x_n\}$, $X_y = \{y_1, \ldots, y_m\}$. If x, y are connected in Γ by an unlabelled segment, then we require that $n = m$ and we set $\mu^s_{x_i, y_j} = \delta_{ij}$ for all i, j. If x, y are connected in Γ by a segment labelled by a letter $A, B, \ldots$ other than Λ, then that letter represents a matrix (μ_{ij}) with $\mu^s_{y_i, x_j} = \mu_{ij}$ for all i, j (assuming $I_x \subseteq I_y$ and $s \in I_y$). If x, y are connected in Γ by a segment labelled by Λ, then Λ represents a matrix (λ_{ij}) with $\mu^{s_1}_{x_i, y_j} = \mu^{s_2}_{y_j, x_i} = \lambda_{ij}$ for all i, j (assuming $s_1 \in I_x$, $s_2 \in I_y$). If x, y are not connected then we set $\mu^s_{x_i, y_j} = \mu^t_{y_j, x_i} = 0$ for all i, j.

The remaining characters and corresponding W-graphs are obtained by algebraic conjugation or duality; see (11.1.6) and (11.1.7). Following the procedure in Remark 11.1.4, we check that the representations afforded by the above graphs are irreducible and determine the character table of $\mathbf{H}$. Once that table is known, we can verify using Lemma 9.4.4 that the polynomials in Table E.3 are the generic degrees for $\mathbf{H}$ (as defined in (8.1.8)). Finally, we can now also determine the fields $K_0(\chi)$ (defined as in (11.1.8)):

χ	$K_0(\chi)$
1_r, $1_r'$, 8_r, 8_{rr}, 10_r, 18_r, 25_r, $25_r'$, 36_{rr}, $36_{rr}'$, 40_r, 48_{rr}	$\mathbb{Q}(u)$
4_t, $4_t'$, $\overline{4}_t$, $\overline{4}_t'$, 6_s, $\overline{6}_s$, 9_s, $9_s'$, $\overline{9}_s$, $\overline{9}_s'$, 16_t, $\overline{16}_t$, 24_s, $\overline{24}_s$, 24_t, $\overline{24}_t$, 30_s, $\overline{30}_s$	$\mathbb{Q}(\sqrt{5}, u)$
16_r, $16_r'$, 16_{rr}, $16_{rr}'$	$\mathbb{Q}(\sqrt{u})$

By Proposition 9.3.1, every $\chi \neq 48_{rr}$ can be realized over $K_0(\chi)$ and 48_{rr} can be realized over $\mathbb{Q}(\sqrt{5}, u)$. However, note that 48_{rr} cannot be realized over $\mathbb{Q}(u)$. Indeed, if this were the case then 48_{rr} could also be realized over $\mathbb{Q}[u]$ (since $\mathbb{Q}[u]$ is a principal ideal domain; see the remarks in (7.3.7)) and so the specialized character of W could be realized over $\mathbb{Q}$, contradicting Theorem 6.3.8(b).

*The figures are reproduced (with some minor changes) from [Alvis and Lusztig 1982] and [Alvis and Lusztig 1994] with the kind permission of W. de Gruyter and the authors.

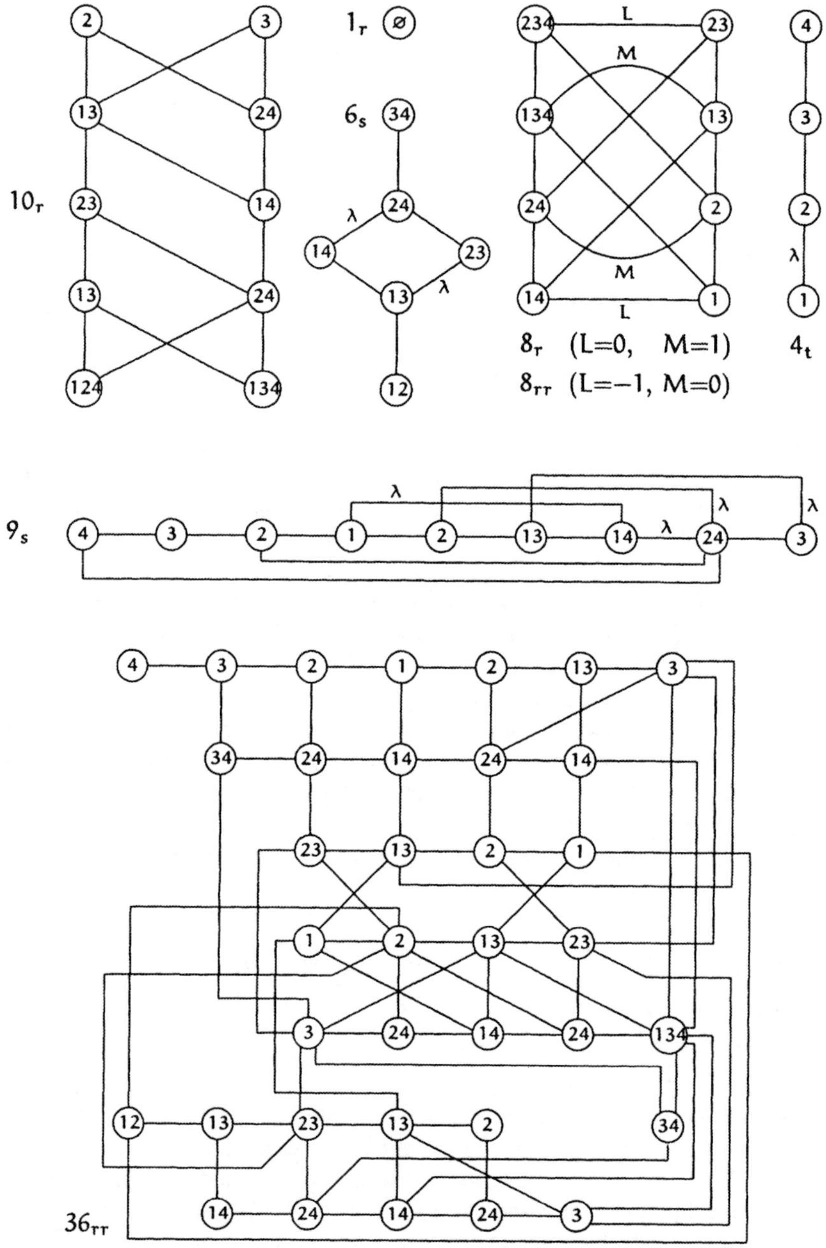

Fig. 11.2. W-graphs for type H_4 (Alvis and Lusztig).

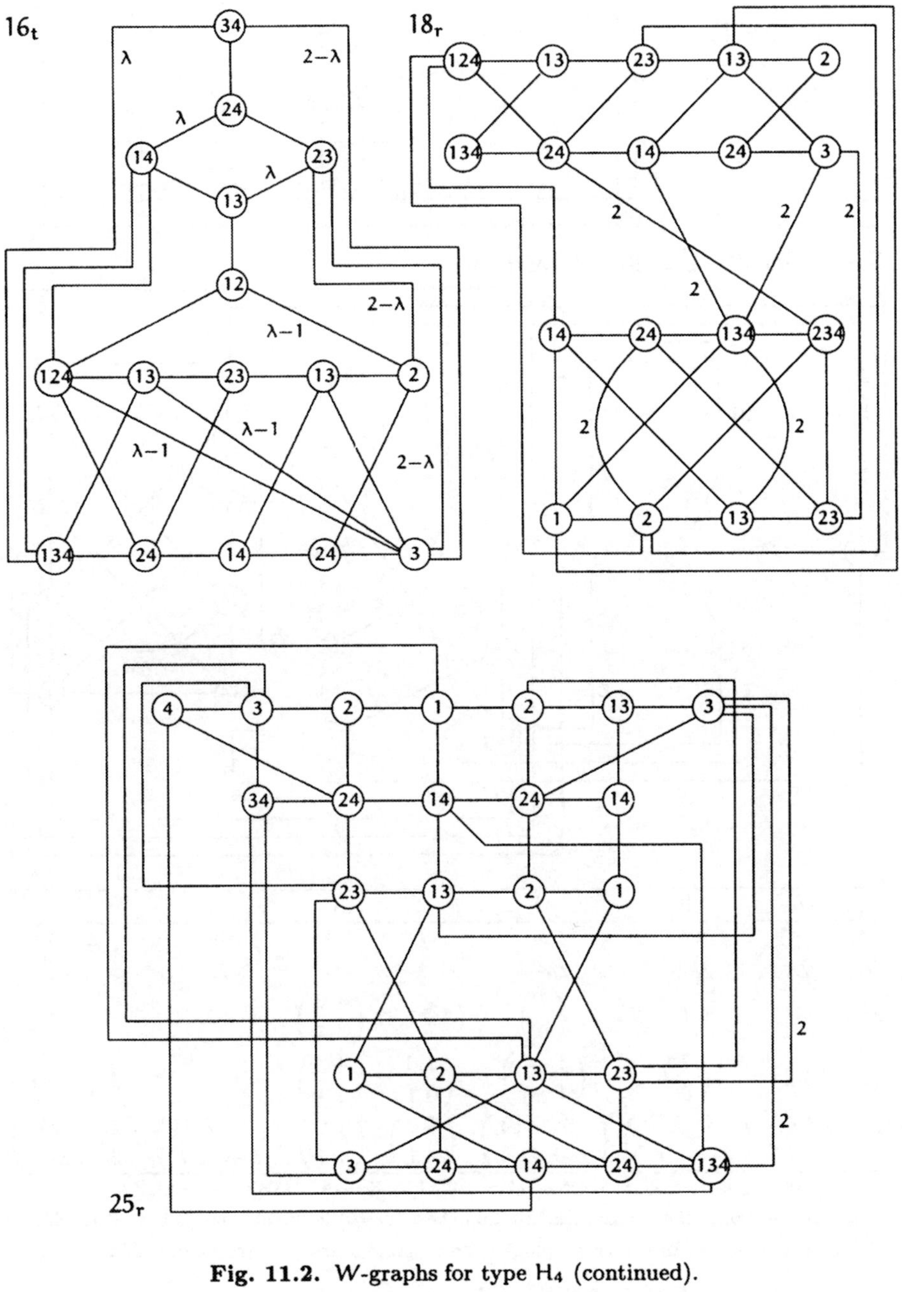

Fig. 11.2. *W*-graphs for type H_4 (continued).

16_{rr} $(\delta = 1)$

16_{r} $(\delta = -1)$

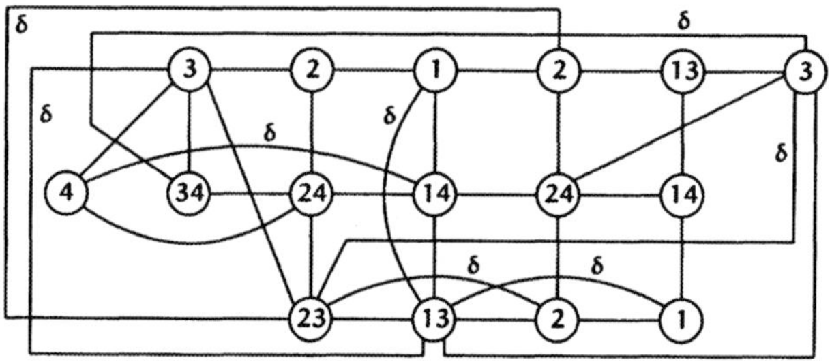

Pattern for $\overline{24}_t$, $\overline{24}_s$, $\overline{30}_s$, 40_r and 48_{rr}:

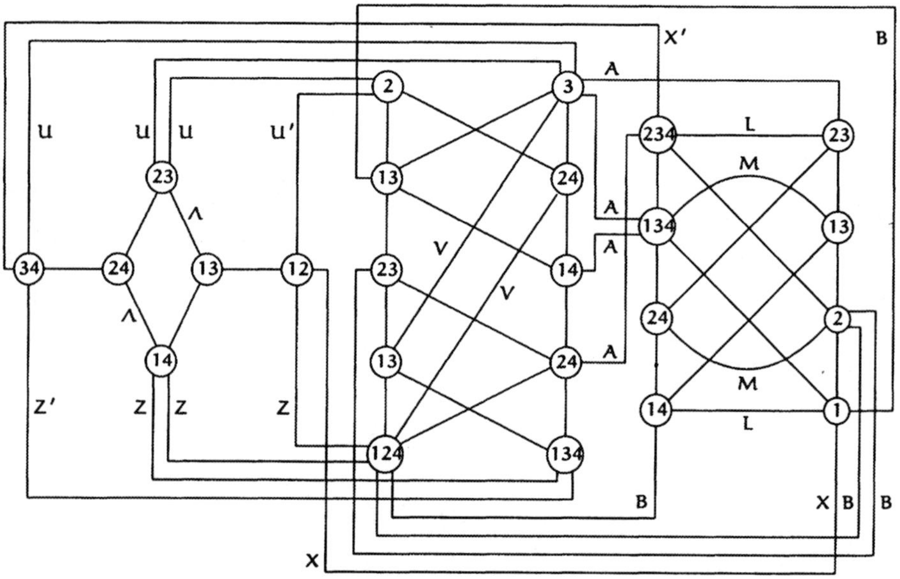

χ	A	B	U	Z	Λ	L	V
$\overline{24}_t$	$\lambda+1$	$\lambda+2$	λ	λ	λ	$\lambda+1$	$2\lambda+1$
$\overline{24}_s$	2λ	$2\lambda+1$	$2\lambda-1$	λ	λ	2λ	$3\lambda+1$
$\overline{30}_s$	1	1	$\begin{pmatrix}1\\0\end{pmatrix}$	$(1\ \ 2+\lambda)$	$\begin{pmatrix}1&0\\1&1\end{pmatrix}$	$-\lambda$	1
40_r	$(3\ \ 2)$	$\begin{pmatrix}2\\0\end{pmatrix}$	$\begin{pmatrix}1&4\\1&-1\end{pmatrix}$	$\begin{pmatrix}1&0\\0&1\end{pmatrix}$	$\begin{pmatrix}1&0\\1&1\end{pmatrix}$	2	$\begin{pmatrix}2&3\\1&-1\end{pmatrix}$
48_{rr}	$\begin{pmatrix}1&0\\0&1\end{pmatrix}$	$\begin{pmatrix}3&\lambda+1\\2-\lambda&1\end{pmatrix}$	$\begin{pmatrix}0&\lambda+1\\2-\lambda&\lambda-1\end{pmatrix}$	$\begin{pmatrix}1&0\\0&1\end{pmatrix}$	$\begin{pmatrix}1&0\\1&1\end{pmatrix}$	$\begin{pmatrix}1&\lambda+1\\2-\lambda&-1\end{pmatrix}$	$\begin{pmatrix}2-\lambda&2\lambda\\2-\lambda&\lambda-1\end{pmatrix}$

$M = L+1,\quad Z' = Z\Lambda^{tr},\quad U' = \Lambda^{tr}U,\quad X = \Lambda^{-1}UB,\quad X' = AZ\Lambda^{-1}$

Thus, each vertex in the pattern for 48_{rr} and the 6 (or 16) leftmost vertices in that for $\overline{30}_s$ (or 40_r) give two vertices in the *W*-graph. All the remainder give one vertex in the *W*-graph.

Fig. 11.2. *W*-graphs for type H_4 (continued).

11.3 W-GRAPHS FOR THE IRREDUCIBLE CHARACTERS OF TYPE F_4

Let (W, S) be of type F_4, where the labelling of the elements in S is given by the following Dynkin diagram:

F_4 

We consider the generic Iwahori–Hecke algebra $\mathbf{H}$ over $\mathbf{A} = \mathbb{Z}[u^{\pm 1}, v^{\pm 1}]$ such that we have the following quadratic relations:

$$T_{s_i}^2 = uT_1 + (u-1)T_{s_i} \text{ for } i = 1,2 \text{ and } T_{s_i}^2 = vT_1 + (v-1)T_{s_i} \text{ for } i = 3,4.$$

There are 25 irreducible characters; they are denoted as in Table C.3.

11.3.1. Kazhdan–Lusztig cells in type F_4. Let $\Gamma = \{\sqrt{u}^m \sqrt{v}^n \mid m, n \in \mathbb{Z}\}$ and assume that $\sqrt{u} < \sqrt{v}$. Using an implementation of the recursion formulas in (11.1.11) in CHEVIE, we have computed the polynomials $P_{y,w}$ ($y \leq w$) and the cells of W (where Exercise 11.3 provides a substantial reduction). Each of these cells gives rise to a representation of $\mathbf{KH}$ and so, by specializing $u, v \mapsto 1$, also a representation of W. The results of these computations can be summarized as follows. Let $\mathfrak{C}$ be a Kazhdan–Lusztig cell in W and $d := |\mathfrak{C}|$. Then there are four cases.

(a) We have $d \in \{1, 2, 4, 8, 9\}$ and the corresponding cell representation affords an irreducible character. (There are d cells giving rise to the same character.) All irreducible characters except $4_1, 6_1, 6_2, 12, 16$ arise in this way.
(b) We have $d = 20$ and the corresponding cell representation affords the character $16 + 4_1$. (There are four cells of this kind.)
(c) We have $d = 34$ and the corresponding cell representation affords the character $16 + 12 + 6_1$. (There are six cells of this kind.)
(d) We have $d = 34$ and the corresponding cell representation affords the character $16 + 12 + 6_2$. (There are six cells of this kind.)

Using a MAPLE implementation of the "meat-axe" (see [Parker 1984]), we were able to decompose the cell representations of $\mathbf{KH}$ into irreducibles. This step was the most delicate one: there is no a priori guarantee that the irreducible constituents that we computed in that way have any particularly nice form. But as a result, we have obtained W-graphs for all irreducible characters of $\mathbf{KH}$.

Figure 11.3 shows W-graphs for the characters 1_r, 4_t, 6_s, 8_r, 8_{rr}, 9_s, 10_r, 16_r, 16_{rr}, 16_t, 18_r, 25_r and 36_{rr}. (W-graphs for the remaining characters are obtained by duality, as described in (11.1.7).) We use the following conventions. The nodes of the graph correspond to the elements of $\mathfrak{C}$. For each $x \in \mathfrak{C}$, the set $I(x)$ is described inside the corresponding node, where we write 1, 2, 3, 4 for s_1, s_2, s_3, s_4, respectively. The μ-function is determined as follows. Let $x, y \in \mathfrak{C}$, $x \neq y$, and $s \in S$ be such that $s \in I(x)$, $s \notin I(y)$. First of all, we have $\mu_{x,y}^s = 0$ if x and y are not connected in the graph. Now assume that there is an edge between x and y. Then the value of $\mu_{x,y}^s$ is as follows.

(1) We have $\mu^s_{x,y} = 0$ if one of the following conditions is satisfied:
   (1a) The edge between x and y has an arrow pointing towards y.
   (1b) $I(x)$ consists of three elements, $I(y)$ consists of one element, and $s = s_1$.
   (1c) We have $s = s_2$, $I(x)$ contains s_2, s_4 and $I(y)$ contains s_3 but not s_4.

(2) If (1a)–(1c) do not hold, then $\mu^s_{x,y} = 1$, with the following exceptions:
   (2a) If there is an integer labelling the edge between x and y, then $\mu^s_{x,y}$ equals that integer.
   (2b) If $s \in \{s_3, s_4\}$ and if there is a label on the edge between x and y which is not an integer, then $\mu^s_{x,y}$ is equal to that label, where we define:

$$a := \sqrt{u}\sqrt{v}^{-1} + \sqrt{v}\sqrt{u}^{-1}, \qquad b := \sqrt{u}^2\sqrt{v}^{-1} + \sqrt{v}\sqrt{u}^{-2},$$

$$f := \sqrt{u} + \sqrt{u}^{-1}, \qquad g := \sqrt{u}^2 + \sqrt{u}^{-2}, \qquad h := \sqrt{u}^3 + \sqrt{u}^{-3}.$$

For example, when we consider the graph corresponding to $\chi''_{(9,6)}$, we let $\mathfrak{C} = \{1, \ldots, 9\}$; then the non-zero μ-values are as follows:

$y \in \mathfrak{C}$	$I(y)$	$\mu^s_{x,y} \neq 0$
1	$\{3,4\}$	$\mu^{s_1}_{2,1} = 1$
2	$\{1,3,4\}$	$\mu^{s_2}_{3,2} = 1$
3	$\{2,4\}$	$\mu^{s_1}_{2,3} = \mu^{s_3}_{4,3} = 1, \quad \mu^{s_3}_{1,3} = b, \quad \mu^{s_3}_{2,3} = a$
4	$\{3\}$	$\mu^{s_4}_{3,4} = \mu^{s_4}_{5,4} = \mu^{s_2}_{6,4} = 1, \quad \mu^{s_4}_{1,4} = h, \quad \mu^{s_4}_{2,4} = g$
5	$\{3,4\}$	$\mu^{s_1}_{2,5} = \mu^{s_2}_{7,5} = 1$
6	$\{2,3\}$	$\mu^{s_4}_{7,6} = \mu^{s_1}_{8,6} = 1$
7	$\{2,4\}$	$\mu^{s_3}_{6,7} = \mu^{s_1}_{9,7} = 1, \quad \mu^{s_3}_{5,7} = a$
8	$\{1,3\}$	$\mu^{s_2}_{6,8} = \mu^{s_4}_{9,8} = 1$
9	$\{1,4\}$	$\mu^{s_2}_{7,9} = \mu^{s_3}_{8,9} = 1, \quad \mu^{s_3}_{2,9} = a$

Following the procedure in Remark 11.1.4, we check that the representations afforded by the above graphs are irreducible and determine the character table of $\mathbf{H}$. Once that table is known, we can also verify using Lemma 9.4.4 that the polynomials in Table 11.1 are the Schur elements for $\mathbf{H}$.

11.3.2. Remark. Consider the cells of W in (11.3.1). We say that two cells are equivalent if they can be linked by a chain of cells such that the representations corresponding to any two consecutive cells in that chain have an irreducible character in common. For each equivalence class of cells, there is a unique irreducible character which occurs with multiplicity 1 in the characters afforded by the various cells in that class. This can be seen as an analogue of a *special character*, as defined in [Lusztig 1979a] in the case of equal parameters. From Table 11.1 we see that the Schur elements of all irreducible characters occurring in the character of a given cell have the same *generalized a-invariant*, in the sense of Definition 9.4.8. Such a property is known to hold in the one-parameter case (see [Lusztig 1987b, (3.4)]) and has been suggested to hold in general in [Lusztig 1983, §7].

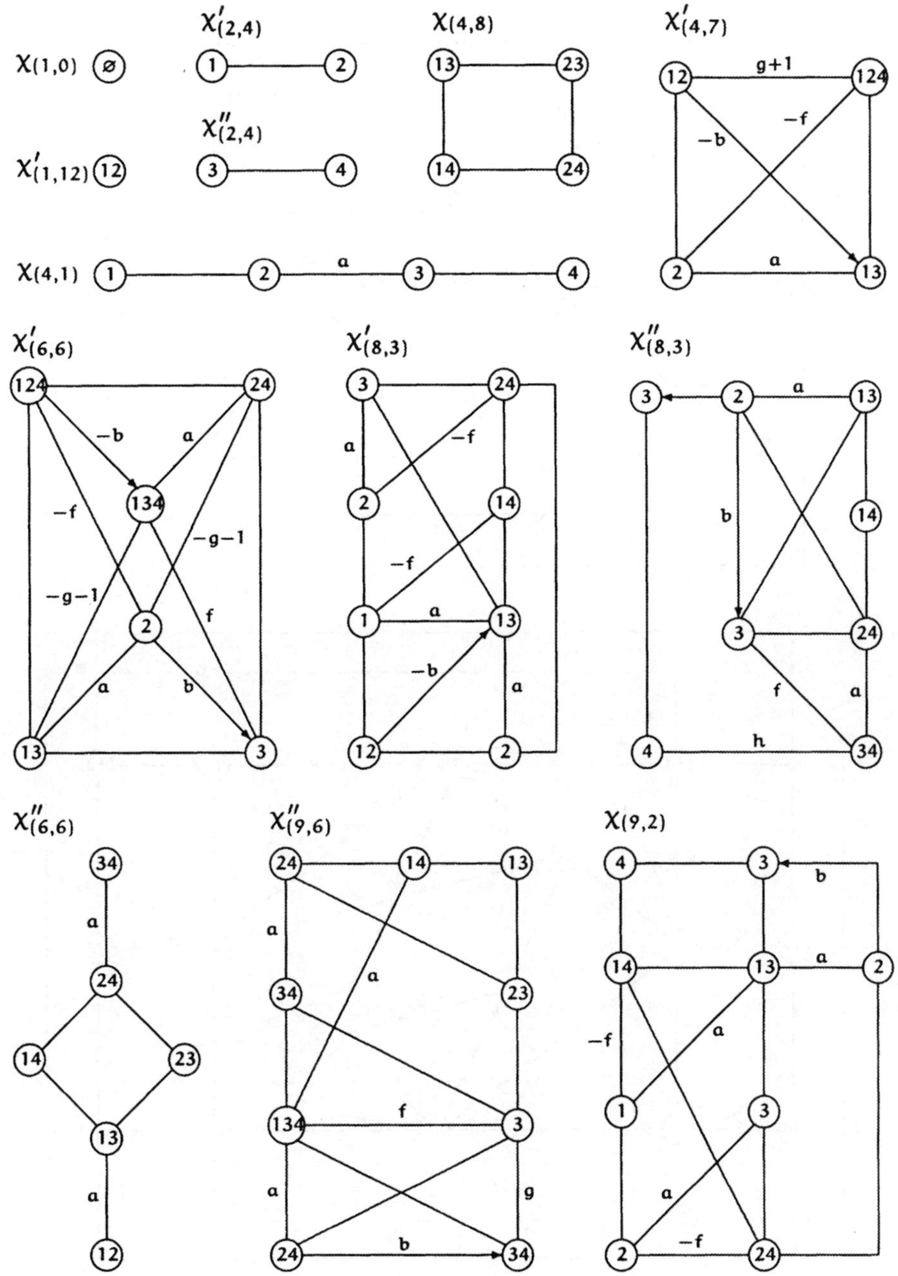

Fig. 11.3. *W*-graphs for type F_4.

$\chi_{(12,4)}$

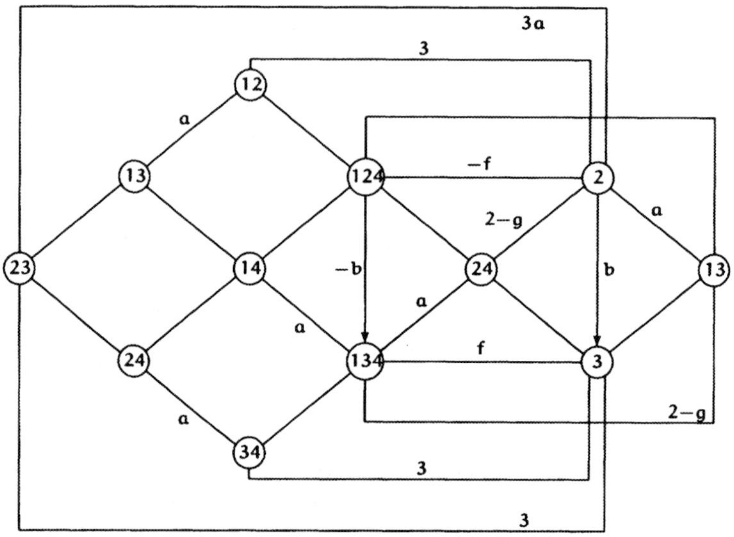

$\chi_{(16,5)}$

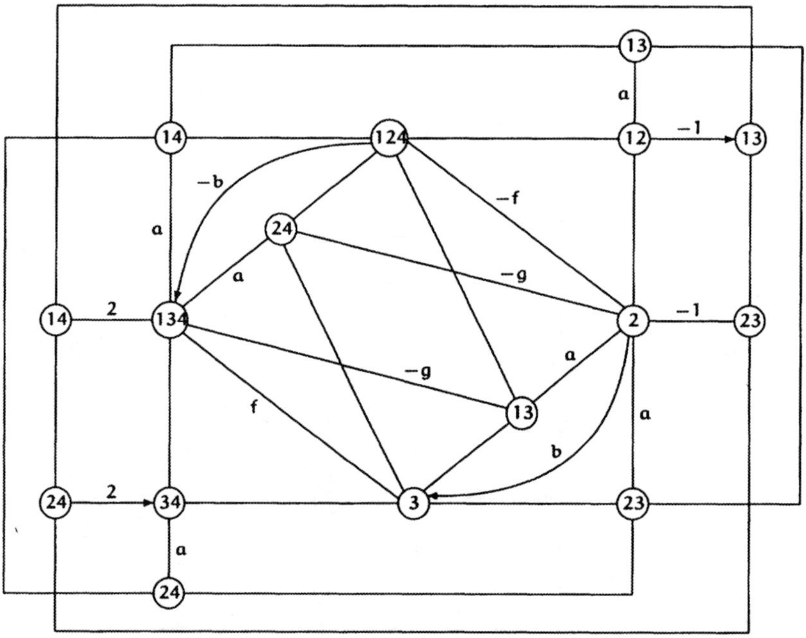

Fig. 11.3. *W*-graphs for type F_4 (continued).

Table 11.1 The Schur elements for the two-parameter generic algebra of type F_4

χ		Schur element c_χ
$\chi_{(1,0)}$	1_1	$(u+1)(v+1)(u^2+u+1)(v^2+v+1)(uv^2+1)(u^2v^2+1)(u^2v+1)(u^2v^2-uv+1)(1+uv)^2$
$\chi''_{(1,12)}$	1_2	$v^{-12}(u+1)(v+1)(u^2+u+1)(v^2+v+1)(u+v^2)(u^2+v^2)(u^2-uv+v^2)(u^2+v)(u+v)^2$
$\chi'_{(1,12)}$	1_3	$u^{-12}(u+1)(v+1)(u^2+u+1)(v^2+v+1)(u+v^2)(u^2+v^2)(u^2-uv+v^2)(u^2+v)(u+v)^2$
$\chi_{(1,24)}$	1_4	$(uv)^{-12}(u+1)(v+1)(u^2+u+1)(v^2+v+1)(uv^2+1)(u^2v^2+1)(u^2v+1)(u^2v^2-uv+1)(1+uv)^2$
$\chi''_{(2,4)}$	2_1	$v^{-3}(u+1)^2(u^2+u+1)(u^2-u+1)(v^2+v+1)(1+uv)(u+v)(u^2v+1)(u^2+v)$
$\chi'_{(2,16)}$	2_2	$u^{-12}v^{-3}(u+1)^2(u^2+u+1)(u^2-u+1)(v^2+v+1)(1+uv)(u+v)(u^2v+1)(u^2+v)$
$\chi'_{(2,4)}$	2_3	$u^{-3}(u^2+u+1)(v+1)^2(v^2+v+1)(v^2-v+1)(u+v^2)(uv^2+1)(1+uv)(u+v)$
$\chi''_{(2,16)}$	2_4	$u^{-3}v^{-12}(u^2+u+1)(v+1)^2(v^2+v+1)(v^2-v+1)(u+v^2)(uv^2+1)(1+uv)(u+v)$
$\chi_{(4,8)}$	4_1	$2(uv)^{-3}(u^2+u+1)(v^2+v+1)(u+v)^2(1+uv)^2$
$\chi_{(9,2)}$	9_1	$(uv)^{-2}(u+1)(v+1)(u+v^2)(u^2v^2+1)(u^2+v)(1+uv)^2$
$\chi''_{(9,6)}$	9_2	$u^{-2}v^{-6}(u+1)(v+1)(uv^2+1)(u^2v+1)(u^2+v^2)(u+v)^2$
$\chi'_{(9,6)}$	9_3	$u^{-6}v^{-2}(u+1)(v+1)(uv^2+1)(u^2v+1)(u^2+v^2)(u+v)^2$
$\chi_{(9,10)}$	9_4	$(uv)^{-6}(u+1)(v+1)(u+v^2)(u^2v^2+1)(u^2+v)(1+uv)^2$
$\chi'_{(6,6)}$	6_1	$3(uv)^{-3}(u+1)^2(v+1)^2(u^2-uv+v^2)(1+uv)^2$
$\chi''_{(6,6)}$	6_2	$3(uv)^{-3}(u+1)^2(v+1)^2(u^2v^2-uv+1)(u+v)^2$
$\chi_{(12,4)}$	12	$6(uv)^{-3}(u^2-u+1)(v^2-v+1)(u+v)^2(1+uv)^2$
$\chi_{(4,1)}$	4_2	$(uv)^{-1}(u+1)(v+1)(u^2+u+1)(v^2+v+1)(u+v)(u^2v^2+1-uv)(1+uv)^2$
$\chi''_{(4,7)}$	4_3	$u^{-1}v^{-7}(u+1)(v+1)(u^2+u+1)(v^2+v+1)(1+uv)(u^2-uv+v^2)(u+v)^2$
$\chi'_{(4,7)}$	4_4	$u^{-7}v^{-1}(u+1)(v+1)(u^2+u+1)(v^2+v+1)(1+uv)(u^2-uv+v^2)(u+v)^2$
$\chi_{(4,13)}$	4_5	$(uv)^{-7}(u+1)(v+1)(u^2+u+1)(v^2+v+1)(u+v)(u^2v^2-uv+1)(1+uv)^2$
$\chi''_{(8,3)}$	8_1	$u^{-1}v^{-3}(u+1)^2(u^2+u+1)(u^2-u+1)(v^2+v+1)(u+v^2)(uv^2+1)$
$\chi'_{(8,9)}$	8_2	$u^{-7}v^{-3}(u+1)^2(u^2+u+1)(u^2-u+1)(v^2+v+1)(u+v^2)(uv^2+1)$
$\chi'_{(8,3)}$	8_3	$u^{-3}v^{-1}(u^2+u+1)(v+1)^2(v^2+v+1)(v^2-v+1)(u^2+v)(u^2v+1)$
$\chi''_{(8,9)}$	8_4	$u^{-3}v^{-7}(u^2+u+1)(v+1)^2(v^2+v+1)(v^2-v+1)(u^2+v)(u^2v+1)$
$\chi_{(16,5)}$	16	$2(uv)^{-3}(u^2+u+1)(v^2+v+1)(u^2v^2+1)(u^2+v^2)$

11.4 ON THE COMPUTATION OF THE GENERIC DEGREES

Let (W, S) be a finite Coxeter system and $\mathbf{H}$ be a corresponding generic Iwahori–Hecke algebra over $\mathbf{A} = \mathbb{Z}[u_s^{\pm 1} \mid s \in S]$. Let $\mathbf{K}$ be the field of fractions of $\mathbb{R}[\sqrt{u_s}^{\pm 1} \mid s \in S]$. Then, by Theorem 9.3.5, the algebra $\mathbf{KH}$ is split semisimple. Recall from (8.1.8) that the generic degrees are defined by

$$D_\chi = \frac{P_W}{c_\chi} \qquad \text{for all } \chi \in \mathrm{Irr}(\mathbf{KH}),$$

where P_W is the Poincaré polynomial and c_χ is the Schur element:

$$P_W = \sum_{w \in W} \mathrm{ind}(T_w) \qquad \text{and} \qquad \chi(T_1)c_\chi = \sum_{w \in W} \mathrm{ind}(T_w)^{-1}\chi(T_w)^2.$$

Thus, D_χ and c_χ can be computed in principle, once the character values $\chi(T_w)$ are known. However, we have not yet obtained that information and, in fact, we wish to use the generic degrees in the computation of the character values in the following section. Therefore, we are looking for an algorithm which requires only knowledge about the irreducible characters of the underlying group W. We begin by stating precisely what we shall require.

11.4.1. We assume that the following information about W is given:

(a) the character table of W;
(b) good representatives of minimal length in the conjugacy classes of W;
(c) induce/restrict matrices for all maximal parabolic subgroups of W.

The information in (a) is obtained using the algorithm described in (5.3.5). For (b), see the results in Section 4.3 and Appendix B. For any $J \subseteq S$, the corresponding induce/restrict matrix (as required in (c)) contains the multiplicities with which the irreducible characters of W_J occur in the restrictions of the irreducible characters of W to W_J. In order to compute them, we need to know the character tables of W_J, W and the fusion map from conjugacy classes of W_J to conjugacy classes of W. The latter map is determined using Algorithm I (p. 86).

The idea of the algorithm is to find the generic degrees as the unique set of solutions of a system of linear equations.

11.4.2. A system of linear relations. First consider the information contained in the induce/restrict matrices in (11.4.1)(c). By an inductive hypothesis, we can assume that the generic degrees for all proper parabolic subalgebras of $\mathbf{H}$ are known. Note that, if $S = \{s\}$, then W only has the trivial and the sign character, whose generic degrees are 1 and u_s, respectively; moreover, if (W, S) is of dihedral type, the generic degrees are known by Theorem 8.3.4.

Now let $J \subseteq S$, $J \neq S$, and consider the corresponding parabolic subalgebra of $\mathbf{H}_J$. Furthermore, for $\chi \in \mathrm{Irr}(\mathbf{KH})$ and $\psi \in \mathrm{Irr}(\mathbf{KH}_J)$, denote by $m_J(\chi, \psi)$ the multiplicity of ψ in the restriction of χ to $\mathbf{KH}_J$. By the compatibility results in

(9.1.9), these coefficients are the same as in the corresponding induce/matrix in (11.4.1)(c). Since $c_\chi = P_W/D_\chi$ and $c_\psi = P_{W_J}/D_\psi$, we obtain by Lemma 9.4.6 the following relations:

$$\sum_{\chi \in \mathrm{Irr}(KH)} m_J(\chi_1, \psi_1)\, D_\chi = \frac{P_W}{P_{W_J}} D_\psi \qquad \text{for all } \psi \in \mathrm{Irr}(KH_J). \tag{a}$$

(Of course, it is enough to consider only maximal parabolic subalgebras.) In order to obtain further relations, we use the scalars z_χ in Theorem 9.2.2, which themselves are computed using the values of the irreducible characters of W on the elements in S. The duality operation in Proposition 9.4.3 yields:

$$z_\chi\, D_\chi - \mathrm{ind}(T_{w_0})\, D_{\chi^*} = 0 \qquad \text{for all } \chi \in \mathrm{Irr}(KH), \tag{b}$$

where $\chi^* \in \mathrm{Irr}(KH)$ is determined by $(\chi^*)_1 = \varepsilon \otimes \chi_1$; see Proposition 9.4.1.

Finally, the information in (11.4.1)(b) gives rise to the following set of relations. Let $w \in W$ be a representative of minimal length in some conjugacy class of W. We say that w is *1-good* if there exists some $r \geq 1$ such that $T_w^d = T_{w_0}^{2r}$, where $d \geq 1$ denotes the order of w. Then we know that $\chi(T_w)$ is a monomial times $\chi_1(w)$. More precisely, if $\chi_1(w) \neq 0$, then—by Proposition 9.2.8 and Theorem 9.3.5—the scalar z_χ must have a dth root in K, which we may then take to be a monomial in the variables $\sqrt{u_s}$ $(s \in S)$ and denote by $z_\chi^{1/d}$. In this case, we have $\chi(T_w) = \chi_1(w) z_\chi^{1/d}$. If $\chi_1(w) = 0$, then Proposition 9.2.8 shows that $\chi(T_w) = 0$. Combining these relations with Lemma 9.4.4, we obtain

$$\sum_{\substack{\chi \in \mathrm{Irr}(KH) \\ \chi_1(w) \neq 0}} \chi_1(w)\, z_\chi^{1/d}\, D_\chi = 0 \qquad \text{for every 1-good } w \in W \text{ (of order d)}. \tag{c}$$

Now the relations in (a), (b) and (c) constitute a system of linear equations of which the generic degrees D_χ are a solution. A priori, there seems to be no reason why that system of linear equations should have full rank. However, this is the case for (W, S) of type H_3, H_4, F_4, E_6, E_7 and E_8. Thus, by verifying that the polynomials in the tables in Appendix E satisfy these relations, we have a proof that those polynomials are the generic degrees.

11.4.3. Example. *The generic degrees in type* $I_2(5)$. In this case, the character table and the generic degrees of H are known by the results in Section 8.3. As an illustration, let us write down explicitly the linear relations obtained by (11.4.2). We have four irreducible characters denoted by ind, ε, χ_1, χ_2. The relations are:

$$\begin{array}{lclclclcl}
D_{\mathrm{ind}} & & & + & D_{\chi_1} & + & D_{\chi_2} & = & P_W/P_{W_{\{s\}}} \\
 & & D_\varepsilon & + & D_{\chi_1} & + & D_{\chi_2} & = & uP_W/P_{W_{\{s\}}} \\
u^{20}D_{\mathrm{ind}} & - & u^{10}D_\varepsilon & & & & & = & 0 \\
u^2D_{\mathrm{ind}} & + & D_\varepsilon & + & \alpha u D_{\chi_1} & - & (\alpha+1)uD_{\chi_2} & = & 0 \\
u^4D_{\mathrm{ind}} & + & D_\varepsilon & - & (\alpha+1)u^2D_{\chi_1} & + & \alpha u^2 D_{\chi_2} & = & 0
\end{array}$$

where $\alpha = (-1+\sqrt{5})/2$. Here, the first two relations arise from the induce/restrict matrix with respect to one of two maximal parabolic subgroups of type A_1. (The first relation corresponds to the index character of that subgroup, the second to the sign character; see the formula in (6.3.5).) The third relation arises from the fact that $\mathrm{ind}^* = \varepsilon$; note that we have $\chi_i^* = \chi_i$ for $i = 1, 2$, so we do not obtain any new relations from these characters. The last two relations arise from (11.4.2)(c) applied to the 1-good elements st and $stst$ in W. One checks that the above system has rank 5.

11.4.4. Generic degrees for the exceptional types. Let (W, S) be of type H_3, H_4, F_4, E_6, E_7 or E_8 and $\mathbf{H}$ be the corresponding one-parameter generic Iwahori–Hecke algebra, where we set $v := \sqrt{u_s}$ for all $s \in S$. We set up a system of linear equations as described in (11.4.2), in three steps: first, by taking all relations as in (11.4.2)(a), then all relations as in (11.4.2)(b) and, finally, all relations as in (11.4.2)(c). The following table shows how the rank of the resulting system increases at each step and reaches its desired value after the third step.

Type	$\lvert \mathrm{Irr}(W) \rvert$	Rank after step (1)	(2)	(3)
H_3	10	6	10	10
H_4	34	14	22	34
F_4	25	16	21	25
E_6	25	20	24	25
E_7	60	48	60	60
E_8	112	82	106	112

A simple verification shows that the polynomials in the tables in Appendix E satisfy these relations. Thus, they must be the generic degrees for these types. (If one actually wants to compute these polynomials, one can proceed by specializing v to sufficiently many integers and solving the specialized systems of equations; thus, a polynomial solution is obtained by interpolation.)

11.4.5. Corollary. *Let W be a finite Coxeter group with generating set S and assume that all parameters of $\mathbf{H}$ are even powers of one indeterminate v. Then, for any $\chi \in \mathrm{Irr}(K\mathbf{H})$, there exists an algebraic integer r_χ and an integer $a_\chi \geq 0$ such that $v^{a_\chi} c_\chi / r_\chi$ is a monic polynomial all of whose roots are roots of unity. Moreover, if all parameters of $\mathbf{H}$ are equal, then both the Schur element c_χ and the generic degree D_χ are polynomials in v^2.*

Proof. By Exercise 8.5, we may assume without loss of generality that (W, S) is irreducible. Then the assertion can be checked case by case from the results in Section 8.3 (for type $I_2(m)$), Chapter 10 (for the classical types) and the tables in Appendix E (for the remaining exceptional types). Note that the statement on c_χ and D_χ in the equal parameter case has been proved by a general argument for the crystallographic types in Corollary 9.3.6. □

11.5 ON THE COMPUTATION OF CHARACTER VALUES

In this section, we assume for technical simplicity that all parameters of $\mathbf{H}$ are (positive) even powers of one fixed indeterminate v, that is,

$$u_s = v^{2c_s} \qquad \text{for all } s \in S, \text{ where } c_s \in \mathbb{N} \ (c_s \neq 0).$$

(Note that this is no serious restriction, since the multi-parameter case can be interpolated by letting the c_s vary.) To simplify notation, we tacitly assume that the scalars in the ground ring over which $\mathbf{H}$ is defined are extended from $\mathbb{Z}$ to $\mathbb{C}$; thus, we consider $\mathbf{H}$ as an algebra over $A_{\mathbb{C}} = \mathbb{C}[v, v^{-1}]$. Let us fix an element $w \in W$ which has minimal length in its conjugacy class. We shall now describe algorithms which allow the computation of $\chi(T_w)$ for all $\chi \in \mathrm{Irr}(K\mathbf{H})$. We assume that the information specified in (11.4.1) is given for W and, also, for all parabolic subgroups of W.

First, if w is contained in a proper parabolic subgroup $W_J \subset W$, then we can assume, by an inductive hypothesis, that all character values for $K\mathbf{H}_J$ are already know. Hence, we have

$$\chi(T_w) = \sum_{\psi \in \mathrm{Irr}(K\mathbf{H}_J)} m(\chi, \psi)\, \psi(T_w) \qquad \text{with } m(\chi, \psi) \text{ as in (9.1.9).}$$

Now consider the case where w lies in a cuspidal conjugacy class, i.e., no conjugate of w lies in a proper parabolic subgroup of W. By Corollary 9.4.2, each $\chi^i(T_w)$ is a polynomial in v of total degree at most l, where $l \geq 1$ is such that $\mathrm{ind}(T_w) = v^l$. So we can write

$$\chi(T_w) = \sum_{j=1}^{l} a_{\chi,j}\, v^j \qquad \text{where } a_{\chi,j} \in \mathbb{R} \text{ for all } j. \tag{$*$}$$

The idea is to set up a system of linear equations of which the collection of numbers $(a_{\chi,j})$ is the unique set of solutions.

11.5.1. Rational characters. Assume that the specialized character χ_1 is of parabolic type and uniquely determined by its multiplicities in the induced characters $1_{W_J}^W$ for all $J \subseteq S$. Then, by Example 9.3.4(a), χ is rational in the sense that $\chi(T_y) \in \mathbb{C}[v^2]$ for all $y \in S$. Thus, we have

$$a_{\chi,j} = 0 \qquad \text{for all odd } j.$$

For all types of (W, S), the characters χ satisfying the above conditions are known by the results in Section 6.3.

The next (considerably more substantial) set of relations comes from the existence of good representatives in the conjugacy classes of W. By Theorem 9.2.5, the eigenvalues of T_w in a representation affording χ are fractional powers of v. The powers occurring in this way can be determined explicitly, and this will tell us that some coefficients $a_{\chi,j}$ have to be 0. This is made precise as follows.

11.5.2. Good elements. Assume now that w is a good element in the sense of Definition 4.3.1. Thus, denoting by $d \geq 1$ the order of w, we have a relation

$$T_w^d = T_{w_{J_1}}^{d_1} \cdots T_{w_{J_r}}^{d_r}, \qquad J_1 \supset \cdots \supset J_r, \qquad d_i \geq 2 \text{ even}, \tag{a}$$

where all terms commute with each other and w_J denotes the longest element in W_J, for any $J \subseteq S$. For technical reasons, we shall assume that the last subset J_r is the empty set; moreover, we can assume that $J_1 = S$ (possibly with $d_1 = 0$).

Now consider pairs $\gamma = (J, \psi)$ where $J \subseteq S$ and $\psi \in \mathrm{Irr}(\mathbf{K}H_J)$. For such a pair γ, we denote by z_γ the scalar by which $T_{w_J}^2$ acts in an irreducible representation of $\mathbf{K}H_J$ which affords ψ; we have

$$z_\gamma = v^{2f_\gamma} \qquad \text{where } f_\gamma \in \mathbb{Z} \text{ is determined by Theorem 9.2.2.} \tag{b}$$

Now let $\gamma = (J, \psi)$ and $\gamma' = (J', \psi')$ be two such pairs. We write $\gamma \xrightarrow{m} \gamma'$ if $J' \subseteq J$ and $m = m(\psi, \psi')$ is the multiplicity of ψ' in the restriction of ψ to $\mathbf{K}H_{J'}$. (Note again that, by (9.1.9), the numbers $m(\psi, \psi')$ can be computed from the induce/restrict matrices for W_J and $W_{J'}$.) We consider chains

$$\mathcal{C}\colon \gamma_1 = (S, \chi) \xrightarrow{m_1} \gamma_2 \xrightarrow{m_2} \cdots \xrightarrow{m_{r-1}} \gamma_r \quad \text{with} \quad \begin{cases} m_k \neq 0 \text{ for all } k, \\ \gamma_k = (J_k, \psi_k) \text{ for all } k. \end{cases}$$

Note that, by our above convention, we always have $\gamma_r = (\varnothing, \mathrm{ind})$. Each chain $\mathcal{C}$ determines an eigenvalue $v^{f_\mathcal{C}}$ of T_w^d, with multiplicity $m_1 \cdots m_{r-1}$ (and all eigenvalues arise in this way), where

$$f_\mathcal{C} := d_1 f_{\gamma_1} + \cdots + d_r f_{\gamma_r}. \tag{c}$$

Now, Theorem 9.2.5 implies that if $a_{\chi,j} \neq 0$ for some j, then v^{jd} is among the eigenvalues of T_w^d. So the above discussion yields that

$$a_{\chi,j} = 0 \qquad \text{if there is no chain } \mathcal{C} \text{ with } f_\mathcal{C} = jd. \tag{d}$$

11.5.3. The specialization $v \mapsto 1$. This yields a bijection between $\mathrm{Irr}(\mathbf{K}H)$ and $\mathrm{Irr}(W)$; see Theorem 8.1.7. Thus, we obtain the following linear relation:

$$\sum_{j=1}^{l} a_{\chi,j} = \chi_1(w). \tag{a}$$

There is a particular case where this completely determines $\chi(T_w)$, namely, the case where $r = 1$ in (11.5.2)(a). Then we have

$$\chi(T_w) = \chi_1(w) v^{d_1 f_\gamma / d} \qquad \text{where } \gamma = (S, \chi). \tag{b}$$

By Example 9.2.9, this can always be applied, for example, if w is the longest element or a Coxeter element in W.

We shall now consider relations which involve two or more irreducible characters of **KH**. Let us write $\mathrm{Irr}(\mathbf{KH}) = \{\chi^1, \ldots, \chi^n\}$. To simplify notation, we also write a_j^i instead of $a_{\chi_i, j}$. Thus, we have

$$\chi^i(T_w) = \sum_{j=0}^{l} a_j^i v^j.$$

11.5.4. Duality. Consider the duality operation on characters as in Proposition 9.4.1. For each i, let i^* be such that $\chi_{i^*} = \chi_i^*$. Note that the map $i \mapsto i^*$ is determined by the character table of W; see Proposition 9.4.1(b). Furthermore, the identity $\chi^*(T_w) = (-1)^{l(w)} \mathrm{ind}(T_w)\overline{\chi(T_w)}$ in Proposition 9.4.1(a) gives rise to the following relation:

$$a_j^{i^*} = (-1)^{l(w)} a_{l-j}^i \qquad \text{for all } i, j.$$

11.5.5. Generic degrees. For each i, let D_i be the generic degree of χ_i. By definition, we have that each D_i is a rational function in v. Let $P' \in \mathbb{R}[v]$ be of minimal possible degree such that $D_i' := P'D_i \in \mathbb{R}[v]$ for all i. Then, for some $m \geq 1$, we can write $D_i' = \sum_{k=0}^{m} d_k^i v^i$ where $d_k^i \in \mathbb{R}$. Now consider the equation $\sum_i D_\chi \chi(T_w) = 0$ of Lemma 9.4.4. Inserting the above expressions for $\chi_i(T_w)$ and D_i', we obtain the following set of relations:

$$\sum_{\substack{j,k \geq 0 \\ j+k=e}} \sum_{i=1}^{n} d_k^i a_j^i = 0 \qquad \text{for } e = 0, 1, 2, \ldots, m + l,$$

where we set $d_k^i = 0$ and $a_j^i = 0$ for $k > m$ and $j > l$.

11.5.6. Induction from parabolic subalgebras. Let $J \subseteq S$ and consider the corresponding parabolic subalgebra $\mathbf{KH}_J$. For any $\psi \in \mathrm{Irr}(\mathbf{KH}_J)$, we have

$$\sum_{i=1}^{n} m(\chi^i, \psi)\chi^i(T_w) = \mathrm{Ind}_J^S(\psi)(T_w) \qquad \text{with } m(\chi^i, \psi) \text{ as in (9.1.9).}$$

Thus, if we can compute the right-hand side of the above identity, we obtain linear relations for the character values $\chi^i(T_w)$. Now, by definition, we have

$$\mathrm{Ind}_J^S(\psi)(T_w) = \sum_{x \in X_J} \psi(h^w(x, x)),$$

where $h^w(x, y) \in H_J$ (for $x, y \in X_J$) is determined by $T_x T_w = \sum_{y \in X_J} h^w(x, y) T_y$. Then we write $h^w(x, x) = \sum_{w' \in W} a^w(x, w') T_{w'}$ with $a^w(x, w') \in \mathbf{A}$ and use the class polynomials with respect to W_J. This yields the following formula:

$$\psi(h^w(x,x)) = \sum_{C \in \mathrm{Cl}(W_J)} \Big(\sum_{w' \in W_J} a^w(x,w')\, f_{w',C} \Big)\, \psi(T_{w_C}),$$

where $w'_C \in C_{\min}$ for all $C \in \mathrm{Cl}(W_J)$. For practical purposes, this is not very efficient. Below, we describe an alternative method which also yields relations via induction from parabolic subalgebras.

The underlying idea of that alternative method is the following. Suppose we have a subset $J \subseteq S$ such that if we specialize v to $\zeta \in \mathbb{C}$, then the specialized parabolic subalgebra is not semisimple. (By Theorem 7.4.7, this property can be checked using the Schur elements of $\mathbf{K}\mathbf{H}_J$.) Then we have a corresponding decomposition matrix whose columns are linearly independent over $\mathbb{C}$ by Theorem 7.5.3. (Note that the centre of $\mathbf{H}_J$ is reduction stable by Remark 8.2.5.) So, by Theorem 7.5.3 and Corollary 7.5.7, the specialized character table of $\mathbf{K}\mathbf{H}_J$ no longer has full rank. Thus, the rows of that table must satisfy some non-trivial linear relation and any such relation "induces" a non-trivial relation among the specializations of the χ^i. More precisely, this is done as follows.

11.5.7. Specializations to roots of unity. Let $J \subseteq S$, $J \neq S$. By an inductive hypothesis, we can assume that the character table of $\mathbf{K}\mathbf{H}_J$ is already known. Now let $e \geq 2$ and $\zeta_{2e} \in \mathbb{C}$ be a primitive 2eth root of unity such that some Schur element of $\mathbf{K}\mathbf{H}_J$ becomes 0 under the specialization $\theta_e \colon A_\mathbb{C} \to \mathbb{C}$, $v \mapsto \zeta_{2e}$. Then, by the above discussion, the specialized character table of $\mathbf{K}\mathbf{H}_J$ no longer has full rank. So there exist complex numbers ξ_ψ which are not all 0 and such that we have

$$\sum_{\psi \in \mathrm{Irr}(\mathbf{K}\mathbf{H}_J)} \xi_\psi \theta_e(\psi(T_y)) = 0 \qquad \text{for all } y \in W_J. \tag{a}$$

Now, by restriction, each ψ gives rise to a trace function on $\mathbf{H}_J$. Then, by Proposition 9.1.3, $\sum_\psi \xi_\psi \operatorname{Ind}_J^S(\psi)$ is a trace function on $\mathbf{H}$. But Ind_J^S is compatible with specializations (see the defining equation). So we have

$$\sum_{\psi \in \mathrm{Irr}(\mathbf{K}\mathbf{H}_J)} \xi_\psi \theta_e(\operatorname{Ind}_J^S(\psi)(T_w)) = 0. \tag{b}$$

Using the notation of (11.4.2) and inserting the expression $\chi^i(T_w) = \sum_j a^i_j v^j$, we obtain the following linear relation:

$$\sum_{j=0}^{l} \sum_{i=1}^{n} \Big(\sum_{\psi \in \mathrm{Irr}(\mathbf{K}\mathbf{H}_J)} m_J(\chi^i, \psi)\, \xi_\psi\, \zeta_{2e}^j \Big)\, a^i_j = 0. \tag{c}$$

Here, we actually obtain several relations when we let (ξ_ψ) run over a basis of the null space of the specialized character table of $\mathbf{K}\mathbf{H}_J$. (Again, it is enough to consider only maximal parabolic subalgebras.)

There is yet another way of obtaining relations from specializations of v to roots of unity over $\mathbb{Q}$, which uses p-blocks of W.

11.5.8. Blocks of W. Let p be a prime and $\zeta_p \in \mathbb{C}$ be a primitive pth root of unity. We consider the specialization $\theta_p \colon A_{\mathbb{C}} \to \mathbb{C}$ defined by $\theta_p(v) = \zeta_p$. Assume that we have a subset $\mathcal{A} \subseteq \mathrm{Irr}(KH)$ satisfying the following conditions:

(a) The specialized characters $\{\chi_1 \mid \chi \in \mathcal{A}\}$ of W form a p-block.
(b) The matrix of values $(\chi_1(w))_{\chi \in \mathcal{A}, w \in W_{p'}}$ has rank $|\mathcal{A}| - 1$, where $W_{p'}$ denotes the set of elements in W whose order is prime to p.
(c) We have $\theta_p(c_\chi) = 0$ for all $\chi \in \mathcal{A}$.

(The condition in (b) is rather restrictive: the results on blocks of defect 1 in [Goldschmidt 1980, §11] imply that it is satisfied, for example, if all characters χ_1 are rational-valued and have p-defect 1. This is the only situation we will consider. In any case, note that (a) and (b) can be checked by explicit computation using the character table of W.) Now, by condition (b), there exist complex numbers n_χ ($\chi \in \mathcal{A}$), which are not all 0, such that

$$\sum_{\chi \in \mathcal{A}} n_\chi \chi_1(w) = 0 \qquad \text{for all } w \in W \text{ of order prime to } p.$$

Note that $(n_\chi)_{\chi \in \mathcal{A}}$ can be computed explicitly using the ordinary character table of W. We claim that we also have

$$\sum_{\chi \in \mathcal{A}} n_\chi \, \theta_p(\chi(T_w)) = 0 \qquad \text{for all } w \in W.$$

This is seen as follows. As in Exercise 11.5, we have a factorization $D_p = \mathbf{D}_p D'$, where D_p is the usual p-modular decomposition matrix of W, $\mathbf{D}_p$ is the decomposition matrix of $\mathbf{H}$ associated with the specialization θ_p, and D' is a matrix whose coefficients are non-negative integers. This factorization and condition (a) imply that $\mathcal{A}$ is a union of blocks of $\mathbf{H}$ with respect to θ_p. (Indeed, the equation $D_p = \mathbf{D}_q D'$ means that the columns of D_p are non-negative integral linear combinations of the columns of $\mathbf{D}_q$. Then it remains to use the definition of blocks in Definition 7.5.8.) It follows that we also have a factorization $D_{\mathcal{A},p} = \mathbf{D}_{\mathcal{A},p} D'_{\mathcal{A}}$, where the subscript $\mathcal{A}$ indicates that we only consider submatrices with rows and columns labelled by the simple modules belonging to $\mathcal{A}$. Now the condition on the rank of the matrix of values in (b) means that $D_{\mathcal{A},p}$ has $|\mathcal{A}|$ rows, $|\mathcal{A}| - 1$ columns and rank $|\mathcal{A}| - 1$; moreover, if we consider the row vector $\underline{n} = (n_\chi)_{\chi \in \mathcal{A}}$, then we have $\underline{n} D_{\mathcal{A},p} = 0$. (This follows from the interpretation of the decomposition numbers in terms of Brauer characters of finite groups in [Goldschmidt 1980, §6].) In particular, we have $\mathrm{rank}(\mathbf{D}_{\mathcal{A},p}) \geq |\mathcal{A}| - 1$. By Theorem 7.5.6 and Remark 8.2.5, the columns of $\mathbf{D}_{\mathcal{A},p}$ are linearly independent over $\mathbb{Q}$. Now, condition (c) means that the blocks (with respect to θ_p) which are contained in $\mathcal{A}$ all have positive defect. So, using Corollary 7.5.12, we see that $\mathcal{A}$ consists precisely of one block with respect to θ_p and $\mathbf{D}_{\mathcal{A},p}$ has $|\mathcal{A}| - 1$ columns. We can now conclude that D' is a square matrix with a non-zero determinant. Thus, we have $\underline{n}D_{\mathcal{A},p}D'_{\mathcal{A}} = \underline{n}D_{\mathcal{A},p} = 0$ and so $\underline{n}D_{\mathcal{A},p} = 0$. This in turn implies that $\sum_{\chi\in\mathcal{A}} n_\chi \theta_p(\chi(T_w)) = 0$ for all $w \in W$.

The following example illustrates the above methods in a case where we have already computed all character values by different methods in Chapter 10.

11.5.9. Example. Let (W, S) be of type B_3, with generators $S = \{t, s_1, s_2\}$ as in Section 1.4. We assume that the parameters of $\mathbf{H}$ are $u_t = v^{20}$, $u_{s_1} = u_{s_2} = v^2$.

There are three cuspidal classes: the class of the longest element w_0, the class containing the Coxeter elements and the class containing $w = ts_1ts_1s_2$. For the first two classes, the character values are determined via (11.5.3)(b). Now consider the third class. We have

$$w^4 = w_0^2 w_{\{t\}}^2, \qquad \text{and} \qquad \text{ind}(T_w) = v^{46}.$$

Let $J = \{t\}$. Here is some information about the ordinary characters of W and W_J:

i	1	2	3	4	5	6	7	8	9	10
χ^i	$(111.\varnothing)$	(11.1)	(1.11)	$(\varnothing.111)$	$(21.\varnothing)$	(1.2)	(2.1)	$(\varnothing.21)$	$(3.\varnothing)$	$(\varnothing.3)$
i^*	10	6	7	9	8	2	3	5	4	1
$\chi_1^i(w)$	-1	1	1	-1	0	-1	-1	0	1	1
z_χ	v^{120}	v^{88}	v^{48}	1	v^{132}	v^{56}	v^{96}	v^{12}	v^{144}	v^{24}
$m(\chi, \varepsilon_J)$	0	1	2	1	0	2	1	2	0	1
$m(\chi, 1_J)$	1	2	1	0	2	1	2	0	1	0

where ε_J and 1_J are the sign character and the trivial character of $W_J \cong \mathfrak{S}_2$. We have $z_{\text{ind}_J} = v^{40}$ and $z_{\varepsilon_J} = 1$. Hence, by the procedure described in (11.5.2), the character values $\chi^i(T_w)$ are as follows, where $a_j^i \in \mathbb{C}$ are unknown:

$$\chi^1(T_w) = a_1^1 v^{40}, \qquad \chi^4(T_w) = a_1^4, \qquad \chi^5(T_w) = a_1^5 v^{43},$$
$$\chi^8(T_w) = a_1^8 v^3, \qquad \chi^9(T_w) = a_1^9 v^{46}, \qquad \chi^{10}(T_w) = a_1^{10} v^6$$
$$\chi^2(T_w) = a_1^2 v^{22} + a_2^2 v^{32}, \qquad \chi^3(T_w) = a_1^3 v^{12} + a_2^3 v^{22},$$
$$\chi^6(T_w) = a_1^6 v^{14} + a_2^6 v^{24}, \qquad \chi^7(T_w) = a_1^7 v^{24} + a_2^7 v^{34}$$

Now, for $i = 1, 4, 5, 8, 9, 10$, the specialization $v \mapsto 1$ immediately yields that $a_1^i = \chi_1^i(w)$; see (11.5.3)(a). For the remaining characters, we have eight unknown numbers to determine. First, setting $v \mapsto 1$, we obtain the relations

$$a_1^2 + a_2^2 = 1, \qquad a_1^3 + a_2^3 = 1, \qquad a_1^6 + a_2^6 = -1, \qquad a_1^7 + a_2^7 = -1.$$

Next, by (11.5.4), the duality operation yields the relations

$$a_1^2 = -a_2^6, \qquad a_2^2 = -a_1^6, \qquad a_1^3 = -a_2^7, \qquad a_2^3 = -a_1^7.$$

Thus, up to now, we have six relations for eight unknowns. In order to obtain two more relations, we now use the induction from the parabolic subgroup of type B_2. Its character table has been already determined in Section 8.3. We have:

	T_1	T_{s_1}	$T_{ts_1ts_1}$	T_t	T_{ts_1}
$\psi_{(\varnothing.2)}$	1	v^2	v^4	-1	$-v^2$
$\psi_{(1.1)}$	2	v^2-1	$-2v^{22}$	$v^{20}-1$	0
$\psi_{(\varnothing.11)}$	1	-1	1	-1	1
$\psi_{(2.\varnothing)}$	1	v^2	v^{44}	v^{20}	v^{22}
$\psi_{(11.\varnothing)}$	1	-1	v^{40}	v^{20}	$-v^{20}$

Specializing $v \mapsto \zeta_4$, the above matrix has rank 2. This yields three relations:

$$\psi_{(\varnothing.11)} - \psi_{(11.\varnothing)} \equiv 0 \bmod \zeta_4,$$
$$\psi_{(\varnothing.2)} - \psi_{(2.\varnothing)} \equiv 0 \bmod \zeta_4,$$
$$\psi_{(11.\varnothing)} + \psi_{(\varnothing.2)} - \psi_{(1.1)} \equiv 0 \bmod \zeta_4.$$

Using the induce/restrict matrix in Example 6.4.8, this yields the following relations among the χ^i:

$$\chi^3 + \chi^4 - \chi^6 - \chi^{10} \equiv 0 \bmod \zeta_4,$$
$$\chi^1 + \chi^2 - \chi^7 - \chi^9 \equiv 0 \bmod \zeta_4,$$
$$\chi^1 - \chi^3 + \chi^5 - \chi^7 + \chi^8 + \chi^{10} \equiv 0 \bmod \zeta_4.$$

Inserting the above expressions for the character values $\chi^i(T_w)$, we obtain:

$$-a_1^2 + a_2^2 - a_1^7 + a_2^7 = 0,$$
$$a_1^3 - a_2^3 + a_1^6 - a_2^6 = 0,$$
$$-a_1^3 + a_2^3 - a_1^7 + a_2^7 = 2.$$

Now our system of equations has a unique solution: $a_2^2 = a_1^3 = a_1^6 = a_2^7 = 0$, $a_1^2 = a_2^3 = -a_2^6 = a_1^7 = 1$. Hence the character values on T_w are given by

$$\chi^2(T_w) = v^{22}, \quad \chi^3(T_w) = v^{22}, \quad \chi^6(T_w) = -v^{24}, \quad \chi^7(T_w) = -v^{24}.$$

Thus, we have recovered the values in Table 10.3 (p. 336) by a completely different method.

The following example illustrates that the information provided by (11.5.2) is rather strong: it drastically restricts the possibilities for the monomials which can occur with a non-zero coefficient in a character value.

11.5.10. Example. We assume that (W, S) is of type E_6. All $s \in S$ are conjugate and we write $u_s = v^2$ ($s \in S$). The table in Appendix B shows that there is only one cuspidal class of W for which we have $r > 1$ in (11.5.2)(a). This class has a representative w of minimal length such that

$$T_w^6 = T_{w_0}^2 T_{w_J}^4 \qquad \text{where } J = \{s_2, s_4\}.$$

Now recall from Theorem 9.2.5 that for each $\chi \in \mathrm{Irr}(K\mathbf{H})$, the eigenvalues of T_w in χ are fractional powers of v (up to roots of unity). In Table 11.2 (which appeared in [Geck and Michel 1997]) we give the list of "monomial parts" of the eigenvalues (with their multiplicities in brackets).

Table 11.2 Monomial parts of the eigenvalues for type E_6

χ	Absolute values
1_p	v^{28} (1)
$1_p'$	1 (1)
10_s	v^{12} (1), v^{14} (8), v^{16} (1)
6_p	v^{22} (2), v^{24} (4)
$6_p'$	v^4 (4), v^6 (2)
20_s	v^{12} (4), v^{14} (12), v^{16} (4)
15_p	v^{16} (1), v^{18} (8), v^{20} (6)
$15_p'$	v^8 (6), v^{10} (8), v^{12} (1)
15_q	v^{18} (10), v^{20} (5)
$15_q'$	v^8 (5), v^{10} (10)
20_p	v^{20} (10), v^{22} (10)
$20_p'$	v^6 (10), v^8 (10)
24_p	v^{14} (2), v^{16} (16), v^{18} (6)
$24_p'$	v^{10} (6), v^{12} (16), v^{14} (2)
30_p	v^{16} (1), v^{18} (18), v^{20} (11)
$30_p'$	v^8 (11), v^{10} (18), v^{12} (1)
60_s	v^{12} (8), v^{14} (44), v^{16} (8)
80_s	v^{12} (12), v^{14} (56), v^{16} (12)
90_s	v^{12} (15), v^{14} (60), v^{16} (15)
60_p	v^{14} (4), v^{16} (42), v^{18} (14)
$60_p'$	v^{10} (14), v^{12} (42), v^{14} (4)
64_p	v^{15} (4), v^{17} (40), v^{19} (20)
$64_p'$	v^9 (20), v^{11} (40), v^{13} (4)
81_p	$v^{40/3}$(9), $v^{46/3}$(54), $v^{52/3}$(18)
$81_p'$	$v^{32/3}$(18), $v^{38/3}$(54), $v^{44/3}$(9)

11.5.11. Character values for the exceptional types. Let (W, S) be of type H_3, H_4, F_4, E_6, E_7 or E_8 and $\mathbf{H}$ be the corresponding one-parameter generic Iwahori–Hecke algebra, where $v := \sqrt{u_s}$ for all $s \in S$.

For a fixed good element w of minimal length in a cuspidal class of W, we set up a system of linear equations as described above. The first step consists of applying (11.5.1) and (11.5.2). Then we can write

$$\chi^i(T_w) = \sum_{j=1}^{l(i)} a_j^i v^{e_j(i)} \qquad \text{with unknowns } a_j^i,$$

where the exponents $e_j(i)$ satisfy the following conditions:

(a) for given i, j there exists a chain $\mathcal{C}$ as in (11.5.2), whose first term is (S, χ^i) and such that $e_j(i) = f_{\mathcal{C}}/d$;

(b) each $e_j(i)$ is an integer; moreover, if χ^i satisfies the condition in (11.5.1) then all $e_j(i)$ $(1 \leq j \leq l(i))$ are even.

The total number of unknowns at this stage (which is $\sum_{i,j} e_j(i)$) is printed in the third column of Table 11.3. Then we proceed in five steps, applying in turn the relations in (11.5.3), (11.5.4), (11.5.7), (11.5.8) (for primes p which divide $|W|$ to the first power only) and (11.5.5). Table 11.3 shows how the rank of the resulting system increases. (The GAP programs implementing these algorithms are available in the contributions directory of the CHEVIE homepage.) It turns out that in each case the rank eventually equals the number of irreducible characters. (We have no a priori explanation for this.) The full character tables are available electronically through the CHEVIE package of GAP; the tables for type F_4, E_6 and E_7 are printed explicitly in [Geck 1994], [Geck 1995].

11.5.12. An application: modular characters. Let $0 \neq \zeta \in \mathbb{C}$ and consider the specialization $\theta_\zeta \colon A_\mathbb{C} \to \mathbb{C}$, $v \mapsto \zeta$. Let $\mathbf{H}_\zeta$ be the corresponding specialized algebra and denote by $\mathbf{D}_\zeta$ the decomposition matrix associated with θ_ζ; see Theorem 7.4.3. If $\mathbf{H}_\zeta$ is semisimple then, by Tits's Theorem 7.4.6, $\mathbf{D}_\zeta$ is a permutation matrix. So let us now assume that $\mathbf{H}_\zeta$ is not semisimple. By Theorem 7.4.7 this is the case precisely when there exists some $\chi \in \mathrm{Irr}(\mathbf{KH})$ such that $\theta_\zeta(c_\chi) = 0$. We shall now make the further assumption that all parameters of $\mathbf{H}$ are equal to v^2. Then, by Corollary 11.4.5, ζ must be a root of unity. We assume that ζ has order $2e$, for some $e > 1$. (Note that, for $e = 1$, we have $\mathbf{H}_\zeta = \mathbb{C}[W]$.) Now, what can we say about the decomposition matrix $\mathbf{D}_\zeta$?

By definition, $\mathbf{D}_\zeta$ has rows labelled by $\mathrm{Irr}(\mathbf{KH})$ and columns labelled by $\mathrm{Irr}(\mathbf{H}_\zeta)$. Moreover, for a suitable ordering of the rows and columns, $\mathbf{D}_\zeta$ has a block diagonal shape, corresponding to the distribution of the simple modules into blocks; see Definition 7.5.8. Once the character table of $\mathbf{KH}$ is known, we can explicitly compute $|\mathrm{Irr}(\mathbf{H}_\zeta)|$ (i.e., the number of columns of $\mathbf{D}_\zeta$) and the block distribution of $\mathrm{Irr}(\mathbf{KH})$. The result is as follows.

11.5.13. Proposition. *Consider a specialized algebra $\mathbf{H}_\zeta$ as above, where $\zeta \in \mathbb{C}$ is a primitive $2e$th root of unity and $\mathbf{H}$ is of exceptional type H_3, H_4, F_4, E_6, E_7 or E_8. Then $|\mathrm{Irr}(\mathbf{H}_\zeta)|$ is given by the following table. (There are only entries for $e > 1$ if $\mathbf{H}_\zeta$ is not semisimple.)*

e	1	2	3	4	5	6	7	8	9	10	12	14	15	18	20	24	30
H_3	10	5	8		8	9				9							
H_4	34	18	26	30	24	30				29	33		32		33		33
F_4	25	8	15	19		20		24			24						
E_6	25	8	13	19	23	20		24	24		24						
E_7	60	12	32	40	54	43	58	56	58	57	58	59		59			
E_8	112	23	52	69	96	75	108	100	106	104	102	110	110	109	111	111	111

Moreover, the block distribution of $\mathrm{Irr}(\mathbf{KH})$ (with respect to the specialization θ_ζ) is given by the tables in Appendix F.

Table 11.3 Computing character values for type H_4, F_4, E_6, E_7, E_8

Type	class	w^d	$\#a_j^i$	$v \mapsto 1$	dual	$v \mapsto \zeta_{2e}$	p-block	D_χ
H_4	no. 19	$w_0^2 w_{12}^4$	76	32	54	64	–	76
	no. 21	$w_0^2 w_{12}^8$	76	32	54	62	–	76
	no. 25	$w_0^2 w_{23}^4$	76	28	52	63	–	76
	no. 27	$w_0^2 w_{12}^2$	76	32	54	64	–	76
	no. 31	$w_0^6 w_{12}^4$	76	32	54	64	–	76
F_4	D_4	$w_0^2 w_{34}^4$	43	23	33	42	–	43
	C_3+A_1	$w_0^2 w_{12}^4$	43	23	33	42	–	43
	$A_3+\tilde{A}_1$	$w_0^2 w_{23}^2$	41	21	31	40	–	41
E_6	A_5+A_1	$w_0^2 w_{24}^4$	53	21	37	52	$52(p=5)$	53
E_7	D_4+3A_1	$w_0^2 w_{24567}^4$	184	60	122	182	$184(p=5)$	–
	$D_6(a_2)+A_1$	$w_0^2 w_{24}^4$	126	46	86	125	$126(p=5)$	–
	$2A_3+A_1$	$w_0^2 w_{257}^2$	112	58	85	112	–	–
	D_6+A_1	$w_0^2 w_{24}^8$	42	42	–	–	–	–
	A_5+A_2	$w_0^2 w_{2345}^2$	148	60	104	147	$148(p=5)$	–
	A_7	$w_0^2 w_{257}^2 w_2^4$	84	58	71	84	–	–
	$E_7(a_2)$	$w_0^2 w_{257}^2$	52	52	–	–	–	–
	$E_7(a_3)$	$w_0^6 w_{45}^4$	42	42	–	–	–	–
E_8	D_4+4A_1	$w_0^2 w_{123456}^4$	538	110	324	497	$502(p=7)$	538
	$E_7(a_4)+A_1$	$w_0^2 w_{24}^4$	276	96	186	270	$270(p=7)$	276
	$2D_4$	$w_0^2 w_{2478}^4$	436	96	266	424	$424(p=7)$	436
	$2A_3+2A_1$	$w_0^2 w_{2345}^2$	350	108	229	341	$341(p=7)$	350
	D_6+2A_1	$w_0^2 w_{3456}^8$	246	84	165	234	$234(p=7)$	246
	$E_6(a_2)+A_2$	$w_0^2 w_{2345}^2$	380	112	246	371	$371(p=7)$	380
	$A_5+A_2+A_1$	$w_0^2 w_{234578}^2 w_{78}^2$	530	112	321	517	$518(p=7)$	530
	$D_8(a_1)$	$w_0^2 w_{4578}^4$	226	92	159	220	$220(p=7)$	226
	D_8	$w_0^2 w_2^{12}$	40	40	–	–	–	–
	A_7+A_1	$w_0^2 w_{2345}^2 w_{25}^4$	300	108	204	291	$291(p=7)$	300
	A_8	$w_0^2 w_{13}^4$	94	94	–	–	–	–
	$E_8(a_4)$	$w_0^2 w_{34}^4$	94	94	–	–	–	–
	$D_5(a_1)+A_3$	$w_0^4 w_{123456}^2$	278	110	194	275	$275(p=7)$	278
	E_6+A_2	$w_0^2 w_{2345}^6$	292	92	192	283	$283(p=7)$	292
	$E_8(a_7)$	$w_0^2 w_{2345}^2$	206	104	155	202	$202(p=7)$	206
	$E_7(a_2)+A_1$	$w_0^2 w_{2345}^2 w_{24}^8$	378	104	241	368	$368(p=7)$	378
	$D_8(a_2)$	$w_0^6 w_{2456}^4 w_5^{20}$	160	84	122	155	$155(p=7)$	160

The table records the number of unknowns in the system of linear equations for the character values $\chi(T_w)$ (see (11.5.11)) and the rank of that system after using the specialization $v \mapsto 1$, the duality $\chi \mapsto \chi^*$, induction from parabolic subgroups in combination with all specializations $v \mapsto \zeta_{2e}$ (where $P_W(\zeta_{2e}) = 0$), p-blocks of W and generic degrees.

Proof. To compute the number of simple modules for $\mathbf{H}_\zeta$, we proceed as follows. The centre of $\mathbf{H}$ is reduction stable by Remark 8.2.5. Moreover, the character table of $\mathbf{H}$ is a character table in the sense of (7.3.11). So Theorem 7.5.6 implies that $|\operatorname{Irr}(\mathbf{H}_\zeta)|$ equals the rank of the specialized character table, i.e., we have

$$|\operatorname{Irr}(\mathbf{H}_\zeta)| = \text{rank of the matrix } (\theta_\zeta(\chi(T_{w_C})))_{\chi\in\operatorname{Irr}(K\mathbf{H}),C\in\operatorname{Cl}(W)}.$$

We have computed these ranks using GAP.

The block distribution of the irreducible characters of $K\mathbf{H}$ can be computed via Lemma 7.5.10, using the *central characters* ω_χ. For this purpose, we have to compute explicitly the matrix of values $\omega_\chi(z_C)$, where $\chi \in \operatorname{Irr}(K\mathbf{H})$ and z_C are the central elements of Corollary 8.2.4. Now, by Exercise 9.5, we have

$$\sum_{C\in\operatorname{Cl}(W)} \chi'(T_{w_C})\omega_\chi(z_C) = \begin{cases} c_\chi & \text{if } \chi = \chi', \\ 0 & \text{if } \chi \neq \chi'. \end{cases}$$

This means that the matrix $(\omega_\chi(z_C))$ can be obtained by inverting the character table of $\mathbf{H}$ and multiplying by the diagonal matrix with diagonal entries given by the Schur elements of $K\mathbf{H}$. Note that, from a computational point of view, this is quite a hard problem. We solved it by specializing v to a sufficiently large number of different values in various finite fields and using interpolation and the Chinese remainder theorem. Still, for type E_8, we needed around 150 MB of main memory and 1 day of CPU time to perform these computations in GAP. (The programs are available in the contributions directory of the CHEVIE homepage.) Note that, once a certain matrix of polynomials is found, it can be checked whether it equals $(\omega_\chi(z_C))$ by verifying the above orthogonality relations. □

11.6 BIBLIOGRAPHICAL REMARKS AND EXERCISES

The basic source for the results on W-graphs (in the case of equal parameters) is [Kazhdan and Lusztig 1979], which contains the construction of the basis $\{C'_w\}$ (appearing in the proof of Theorem 11.1.10), the notion of W-graphs, and a far-reaching conjecture about the relevance of these constructions to the problem of computing multiplicities in Verma modules. The constructions in general are given in [Lusztig 1983]. The fact that the Bruhat–Chevalley order is not required to characterize the basis $\{C'_w\}$ is mentioned in [Lusztig 1991, p. 158]. See also the elegant exposition in [Sörgel 1997, Section 2] (in the equal parameter case).

The polynomials $P_{y,w} = v_w v_y^{-1} P^*_{y,w}$ discussed in Section 11.1 have remarkable positivity properties in the case of equal parameters and where (W, S) is crystallographic in the sense of [Lusztig 1985, (3.1.1)], i.e., where $m_{st} \in \{2,3,4,6,\infty\}$ for all $s,t \in S$. Then a deep geometric argument shows that the coefficients of $P_{y,w}$ are non-negative integers; see [Kazhdan and Lusztig 1980]. There is even a stronger positivity property for the basis $\{C'_w\}$ defined in the proof of Theorem 11.1.10: if we express $C'_x C'_y$ as a linear combination of C'_z, then the coefficients will lie in $\mathbb{N}_0[v^{\pm 1}]$. In the case where W is finite, this result follows by interpreting the multiplication in **H** in terms of a multiplication of complexes in a derived category of constructible sheaves over the manifold G/B, where G is a connected reductive group with Weyl group W and $B \subseteq G$ is a Borel subgroup; see [Springer 1982]. (For general crystallographic groups, see the remarks in [Lusztig 1985, §3].)

These positivity results have strong implications for the structure of **H** and important consequences in the representation theory of finite Chevalley groups; see Chapters 5 and 12 in [Lusztig 1984] and also [Curtis 1988]. The partition of $\mathrm{Irr}(W)$ into families (as defined in Section 6.5) can also be interpreted in terms of Kazhdan–Lusztig cells. This result, due originally to Barbasch–Vogan (see [Lusztig 1984, (5.25)] and the references therein) has been established in [Lusztig 1986] in a more direct way. For further reading, we also recommend [Lusztig 1987b] and [Lusztig 1991].

We remark that the coefficients of $P_{y,w}$ are also known to be non-negative for non-crystallographic finite Coxeter groups (but only after explicitly computing them): for the dihedral types $I_2(m)$ this is an easy exercise (see [Humphreys 1990, Example 7.12(a)]); for types H_3 and H_4, see [Alvis 1987]. Note that the situation is different in the case of unequal parameters; see Example 11.1.12 (which is taken from [Lusztig 1983, p. 106]). The complete determination of all Kazhdan–Lusztig polynomials and M-polynomials for type $I_2(m)$ with unequal parameters (see Exercise 11.4) does not seem to have appeared in print before. (See [Xi 1994, Example 1.21] for part of it.)

The irreducible representations of a generic Iwahori–Hecke algebra **KH** are afforded by W-graphs in the following cases. For type A_n, this appears in [Kazhdan and Lusztig 1979]. The special feature of type A_n is that the "cell" representations (in the sense of Kazhdan and Lusztig) are all irreducible; see also [Garsia and McLarnan 1988, Fact 8, p. 54] and [Naruse 1989]. A similar result is established in [Lusztig 1983] for type B_n, for certain choices of the parameters. The case of one-parameter generic Iwahori–Hecke algebras associated with finite Weyl groups (of any type) is settled by [Gyoja 1984], using some deep results from [Lusztig 1981a]. A uniqueness result about W-graphs is obtained in [Gyoja 1996]. The results in (11.1.8) about W-graphs in type H_3 are contained in [Lusztig 1981a, §5]. For type H_4, see [Alvis and Lusztig 1994]. The definition of the dual W-graph appears in [Alvis and Lusztig 1982]. Our results on W-graphs for type F_4 with independent parameters are new; special choices for the parameters have were considered in [Bremke 1994]. The W-graphs for equal parameters were determined by H. Naruse (private communication). Recently, H. Naruse has determined W-graphs for the irreducible representations in type E_6 (private communication, July 1998). There has been much progress in actually computing Kazhdan–Lusztig cells (in the one-parameter case); see [Chen and Shi 1998] and the references therein.

The generic degrees for the exceptional types have been computed in a number of steps: for types E_6, E_7 by [Surowski 1978], for type E_8 by [Benson 1979], for type F_4 by [Lusztig 1979b, §3] (using some preliminary work by Surowski), for type H_3 by [Lusztig 1982, §13] and for type H_4 by [Alvis and Lusztig 1982]. The use of 1-good elements in (11.4.2) yields a systematic algorithmic approach which considerably simplifies the original computations. Tables with generic degrees can also be found in [Carter 1985, §13.5].

The character tables for the generic Iwahori–Hecke algebras of type F_4 (with unequal parameters), E_6 and E_7 were determined in [Geck 1994] and [Geck 1995], before the methods involving good elements became available. The table for E_8 was computed for the first time in [Geck and Michel 1997], using the techniques described in Section 11.5. The tables for types H_3 and H_4 are easily computed using the representations afforded by the W-graphs of Lusztig and Alvis–Lusztig.

The decomposition matrices of Iwahori–Hecke algebras of exceptional type under specializations at roots of unity over $\mathbb{Q}$ have been completely determined in [Geck and Lux 1991], [Bremke 1994] and [McDonough and Pallikaros 1999] (for type F_4), [Geck 1993a] (for type E_6), [Geck 1995], [Müller 1995] (for type E_7) and [Müller 1997] (for type H_3, H_4); partial results for type E_8 can be found in [Müller 1995]. See also [Geck 1992] and [Geck 1998a] where the Brauer trees of blocks of defect 1 are completely determined. Results on the number of simple modules (analogous to those in Proposition 11.5.13) for type A can be found in [Dipper and James 1986]; for the cyclotomic algebras associated with the complex reflection groups $C_r \rtimes \mathfrak{S}_n$ (which include type B_n), see [Dipper et al. 1995] and [Ariki and Mathas 2000]. In the latter case, the proofs are very deep and rely on Ariki's solution of the Lascoux–Leclerc–Thibon conjecture (see Section 10.6).

In a different direction, decomposition numbers for a specialization of the form $\mathbb{Z}[v, v^{-1}] \to \mathbb{F}_p[v, v^{-1}]$ (where p is a prime number) have been studied in [Fleischmann 1990], [Fleischmann 1993] and [Gyoja 1996].

Exercise 11.1. In the set-up of Theorem 11.1.10, let $y, w \in W$ be such that $y \leq w$. Show that $P_{y,w} := v_w v_y^{-1} P^*_{y,w}$ is a polynomial in the variables v_s^2 $(s \in S)$, with constant term 1. Moreover, assuming that $s \in S$ is such that $sy < y < w < sw$, show that $v_s v_w v_y^{-1} M^s_{y,w}$ is a polynomial in v_t^2 $(t \in S)$, with constant term 0.

[*Hint.* Use induction on $l(w) - l(y)$ and the recursion formulas (P1) and (P2).]

Exercise 11.2. In the set-up of Theorem 11.1.10, assume that all v_s are equal; we write $v = v_s$ $(s \in S)$. Then $\Gamma = \{v^m \mid m \in \mathbb{Z}\}$ and $\Gamma_+ = \{v^m \mid m > 0\}$. Let $y, w \in W$ and $s \in S$ be such that $sy < y < w < sw$. Show that $M^s_{y,w} \in \mathbb{Z}$ and that this is the coefficient of $v^{l(w)-l(y)-1}$ in the polynomial $P_{y,w}$; see Exercise 11.1. (This coefficient is denoted $\mu(y, w)$ in [Kazhdan and Lusztig 1979, (1.2)].) In particular, we have $M^s_{y,w} = 0$ if $l(w) - l(y)$ is even.

[*Hint.* To prove that $M^s_{y,w} \in \mathbb{Z}$, proceed by induction on $l(w) - l(y)$. If there exists no z such that $y < z < sw$ and $sz < z$, the initial condition (M1) means that $M^s_{y,w} - vP^*_{y,w} \in v^{-1}\mathbb{Z}[v^{-1}]$. But $vP^*_{y,w}$ is a linear combination of negative powers of v plus a constant term. Hence $M^s_{y,w}$ must be that constant term. The argument for the induction step is similar. To prove the second statement, note that $M^s_{y,w}$ is the coefficient of v^{-1} in $P^*_{y,w}$.]

Exercise 11.3. The following result gives a necessary condition for two elements to lie in the same Kazhdan–Lusztig cell. (This is a quite efficient tool for practical purposes.) For any $w \in W$, define $\mathcal{R}(w) := \{s \in S \mid ws < w\}$ and call this the *right descent set* of w. Show that we have the implication

$$y \prec_W w \Rightarrow \mathcal{R}(w) \subseteq \mathcal{R}(y) \qquad \text{for all } y, w \in W$$

(with $\prec_W$ defined as in (11.1.5)). In particular, we have $\mathcal{R}(y) = \mathcal{R}(w)$ if $y \sim_W w$.

[*Hint.* See [Xi 1994, Cor. 1.20] or imitate the original proof in the one-parameter case in [Kazhdan and Lusztig 1979, (2.4)].]

Exercise 11.4. *Kazhdan–Lusztig polynomials in type* $I_2(m)$. Let $W = \langle s, t\rangle$ be a dihedral group, where st has order $m \geq 3$ (or $m = \infty$). Assume first that $v_s = v_t$. Show that then we have $P_{y,w} = 1$ for all $y \leq w$ and determine the corresponding cells.

Now assume that $m \geq 4$ is even (or $m = \infty$) and that v_s, v_t are independent indeterminates over $\mathbb{Z}$ such that $v_s > v_t$. Show that the polynomials $P_{y,w}$ are given as follows. For $y \leq w$, write $v_w v_y^{-1} = v_s^{m_s} v_t^{m_t}$ with $m_s, m_t \geq 0$. Then

$$P_{y,w} = \begin{cases} \sum_{i=0}^{m_t} (-1)^i v_t^{2i} & \text{if } \mathcal{L}(w) = \mathcal{R}(w) = \{s\} \text{ and } y \leq tsw < sw, \\ 1 + v_t^2 & \text{if } \mathcal{L}(w) = \mathcal{R}(w) = \{t\} \text{ and } y \leq stw < tw, \\ 1 & \text{otherwise.} \end{cases}$$

In particular, we always have $P_{y,w} \in \mathbb{Z}[v_t^2]$. The M-polynomials are given by

$$M^s_{y,w} = \begin{cases} v_s v_t^{-1} + v_s^{-1} v_t & \text{if } l(w) = l(y) + 1,\ sy < y < w < sw, \\ 1 & \text{if } l(w) = l(y) + 3,\ sy < y < w < sw. \end{cases}$$

All other M-polynomials are 0. Here $\mathcal{L}(w) := \{r \in S \mid l(rw) < l(w)\}$ and $\mathcal{R}(w) := \{r \in S \mid l(wr) > l(w)\}$. Determine the corresponding cells.

[*Hint.* Let $y \leq w$ and proceed by a double induction on $l(w)$ and on $l(w) - l(y)$. If $w = 1$ or $y = w$, there is nothing to prove. Now assume that $y < w$. First settle explicitly the cases where $l(w) - l(y) \in \{1, 2, 3\}$. Then assume that $l(w) - l(y) \geq 4$ and distinguish a number of cases according to whether $sw > w$, $sw < w$, $tw < w$ or $tw > w$, and similarly for y. Assume, for example, that $sw < w$ and $sy < y$. Then the critical point in the evaluation of $P_{y,w}$ is the sum

$$\sum_{\substack{y \leqslant z < sw \\ sz < z}} P_{y,w}(v_w v_z^{-1} M^s_{z,sw})$$

occurring in the recursion formula in Theorem 11.1.10. But, by induction, we already know that $M^s_{z,sw} = 0$ unless $l(sw) - l(z) \in \{1, 3\}$. Moreover, for any given length less than m, there is precisely one element z such that $sz < z$. Hence the above sum has just three terms, and each of these terms can be evaluated by induction. If $tw < w$ and $ty < y$, the situation is even simpler since then we always have $\mu^t_{z,tw} = 0$. Finally, similar rules apply to the computation of the M-polynomials.]

Exercise 11.5. Let W be a finite Coxeter group with generators S and $\mathbf{H}$ be a corresponding generic Iwahori–Hecke algebra over the ring $\mathbf{A}_R = R[v, v^{-1}]$, where v is an indeterminate and R is the ring of integers in some sufficiently large algebraic number field F. In particular, we assume that F is a splitting field for W. We also assume that the parameters of $\mathbf{H}$ are of the form $u_s = v^{2c_s}$ with $c_s \geq 1$ for all $s \in S$. Let $\mathbf{K}$ be the field of fractions of $\mathbf{A}_R$. Then, by Theorem 9.3.5, the algebra $\mathbf{KH}$ is split semisimple and, by Theorem 8.1.7, we have a bijection $\mathrm{Irr}(\mathbf{KH}) \leftrightarrow \mathrm{Irr}(W)$ (induced by the specialization $v \mapsto 1$).

Now let p be a prime number and $q = p^r$ for some $r \geq 0$. Consider the specialization $\theta_q \colon \mathbf{A}_R \to R$, $v \mapsto \zeta_q$, where $\zeta_q \in R$ is a primitive qth root of unity. We assume that F is large enough such that $F\mathbf{H}$ is split. Let $\mathbf{D}_q$ be the corresponding decomposition matrix, relating the simple modules of $\mathbf{KH}$ with the simple modules of $F\mathbf{H}$. On the other hand, let D_p be the usual p-modular decomposition matrix of W, relating the simple modules of $F[W]$ with the simple modules of $k[W]$, where k is the residue field of the localization of R in a prime ideal containing p. (Note that k has characteristic p.) Show that if the labelling of the rows of $\mathbf{D}_q$ corresponds to the labelling of the rows of D_p via the above bijection $\mathrm{Irr}(\mathbf{KH}) \leftrightarrow \mathrm{Irr}(W)$, then we have a factorization

$$D_p = \mathbf{D}_q D'$$

where D' is a matrix whose entries are non-negative integers.

[*Hint.* Let $\mathfrak{p} \subset R$ be a prime ideal with $p \in \mathfrak{p}$ and $\pi \colon R \to k$ the canonical map, where k is the field of fractions of $R/\mathfrak{p}$. Choosing F large enough, we can also assume that $k\mathbf{H}$ is split. Now we are in the setting of Exercise 7.10 and so we have a factorization of decomposition maps: $d_{\pi\circ\theta_q} = d_p \circ d_{\theta_q}$. But, since k has characteristic p, the image of v under $\pi\circ\theta_q$ is 1. So the map $\pi\circ\theta_q$ is independent of q. Thus, we obtain the same result for $r = 0$ or for $r \geq 1$. But, if $q = 1$, then d_p is the usual p-modular decomposition map of W and d_{θ_1} produces the above bijection $\mathrm{Irr}(\mathbf{KH}) \leftrightarrow \mathrm{Irr}(W)$. See also Example 3.2 in [Geck 1998c].]

Appendix: Tables for the exceptional types

A COXETER CLASSES AND PARABOLIC TABLES OF MARKS

This section contains information about the Coxeter classes and the parabolic tables of marks of the Coxeter groups of exceptional types (cf. Sections 2.3 and 2.4).

Different Coxeter classes of the same type are distinguished by dashes. A direct product of Coxeter groups of type A like $A_{n_1} \times \cdots \times A_{n_r}$ is abbreviated as $A_{n_1,\ldots,n_r}$. The same names are used to label the rows and (implicitly) the columns of the tables of marks.

1	120					
A_1	60	4				
$A_{1,1}$	30	4	2			
A_2	20	4	.	2		
$I_2(5)$	12	4	.	.	2	
H_3	1	1	1	1	1	1

Fig. A.1. The parabolic table of marks for H_3.

1	14400									
A_1	7200	120								
$A_{1,1}$	3600	120	8							
A_2	2400	120	.	12						
$I_2(5)$	1440	120	.	.	20					
$A_{2,1}$	1200	80	8	6	.	2				
$I_2(5) \times A_1$	720	72	8	.	10	.	2			
A_3	600	60	4	12	.	.	.	2		
H_3	120	30	4	6	10	.	.	.	2	
H_4	1	1	1	1	1	1	1	1	1	1

Fig. A.2. The parabolic table of marks for H_4.

1	1152											
A_1'	576	48										
A_1''	576	.	48									
$A_{1,1}$	288	24	24	4								
A_2'	192	48	.	.	12							
A_2''	192	.	48	.	.	12						
B_2	144	24	24	.	.	.	8					
$A_{2,1}'$	96	24	8	4	6	.	.	2				
$A_{2,1}''$	96	8	24	4	.	6	.	.	2			
B_3'	24	12	6	2	6	.	4	.	.	2		
B_3''	24	6	12	2	.	6	4	.	.	.	2	
F_4	1	1	1	1	1	1	1	1	1	1	1	1

Fig. A.3. The parabolic table of marks for F_4.

1	51840																
A_1	25920	720															
$A_{1,1}$	12960	720	48														
A_2	8640	720	.	72													
$A_{1,1,1}$	6480	540	72	.	12												
$A_{2,1}$	4320	480	48	36	.	6											
A_3	2160	360	24	72	.	.	8										
$A_{2,1,1}$	2160	300	56	18	12	6	.	2									
$A_{2,2}$	1440	240	48	24	.	12	.	.	12								
$A_{3,1}$	1080	210	36	36	6	6	4	.	.	2							
A_4	432	120	24	36	.	6	8	.	.	.	2						
D_4	270	90	18	36	6	.	12	.	.	.	.	6					
$A_{2,2,1}$	720	140	40	12	12	8	.	4	6	.	.	.	2				
$A_{4,1}$	216	66	20	18	6	6	4	2	.	2	1	.	.	1			
A_5	72	30	12	12	2	6	4	.	6	2	2	.	.	.	2		
D_5	27	15	7	9	3	3	5	1	.	1	2	3	.	.	.	1	
E_6	1	1	1	1	1	1	1	1	1	1	1	1	1	1	1	1	1

Fig. A.4. The parabolic table of marks for E_6.

1	2903040							
A_1	1451520	23040						
$A_{1,1}$	725760	23040	768					
A_2	483840	23040	.	1440				
$A'_{1,1,1}$	362880	17280	1152	.	1152			
$A''_{1,1,1}$	362880	17280	1152	.	.	96		
$A_{2,1}$	241920	15360	768	720	.	.	48	
A_3	120960	11520	384	1440	.	.	.	96
$A_{1,1,1,1}$	181440	11520	1152	.	576	144	.	.
$A_{2,1,1}$	120960	9600	896	360	.	96	48	.
$A_{2,2}$	80640	7680	768	480	.	.	96	.
$A'_{3,1}$	60480	6720	576	720	576	.	48	48
$A''_{3,1}$	60480	6720	576	720	.	48	48	48
A_4	24192	3840	384	720	.	.	48	96
D_4	15120	2880	288	720	.	48	.	144
$A_{2,1,1,1}$	60480	5760	768	180	192	144	36	.
$A_{2,2,1}$	40320	4480	640	240	.	96	64	.
$A_{3,1,1}$	30240	3840	512	360	288	72	48	24
$A_{3,2}$	20160	2880	448	300	.	48	72	16
$A_{4,1}$	12096	2112	320	360	.	48	48	48
$D_4 \times A_1$	7560	1560	240	360	144	48	24	72
A'_5	4032	960	192	240	192	.	48	48
A''_5	4032	960	192	240	.	16	48	48
D_5	1512	480	112	180	.	24	24	60
$A_{3,2,1}$	10080	1600	320	150	96	72	46	8
$A_{4,2}$	4032	832	192	132	.	48	40	16
$A_{5,1}$	2016	512	128	120	96	24	32	24
$D_5 \times A_1$	756	252	72	90	24	24	18	30
A_6	576	192	64	60	.	16	24	16
D_6	126	60	26	30	24	8	12	14
E_6	56	32	16	20	.	8	8	12
E_7	1	1	1	1	1	1	1	1

Fig. A.5. The parabolic table of marks for E_7.

48																							
.	8																						
.	.	24																					
.	.	.	48																				
.	.	.	.	8																			
.	.	.	.	.	12																		
.	.	.	.	.	.	48																	
48	12	.	.		.	.	12																
.	16	12	.		.	.	.	4															
24	8	.	24	4	.	.	.	.	4														
.	4	24	.	8	.	.	.	.	.	4													
.	8	.	.	8	6	.	.	.	.	.	2												
24	.	.	24	8	.	24	.	.	.	.	.	8											
.	.	12	48	.	12	.	.	.	.	.	.	.	12										
.	.	12	.	8	12	.	.	.	.	.	.	.	.	4									
.	4	.	.	4	12	24	.	.	.	.	.	.	.	.	4								
24	14	12	8	4	.	.	6	4	4	2	.	.	.	.	.	2							
.	12	12	.	8	2	.	.	4	.	4	2	.	.	.	.	.	2						
8	8	6	24	4	6	.	.	2	4	.	2	.	6	.	.	.	.	2					
12	6	.	6	6	6	12	6	.	2	.	2	4	.	.	2	.	.	.	2				
.	4	12	.	8	6	.	.	.	.	4	2	.	.	4	.	.	.	.	.	2			
6	2	6	12	4	6	6	.	.	2	2	.	4	6	2	2	.	.	.	.	.	2		
.	4	2	.	4	6	8	.	2	.	.	2	.	.	2	4	.	.	.	.	.	.	2	
1	1	1	1	1	1	1	1	1	1	1	1	1	1	1	1	1	1	1	1	1	1	1	1

Fig. A.5. The parabolic table of marks for E_7 (continued).

1	696729600												
A_1	348364800	2903040											
$A_{1,1}$	174182400	2903040	46080										
A_2	116121600	2903040	.	103680									
$A_{1,1,1}$	87091200	2177280	69120	.	2304								
$A_{2,1}$	58060800	1935360	46080	51840	.	1440							
A_3	29030400	1451520	23040	103680	.	.	3840						
$A_{1,1,1,1}$	43545600	1451520	69120	.	4608	.	.	384					
$A_{2,1,1}$	29030400	1209600	53760	25920	2304	1440	.	.	96				
$A_{2,2}$	19353600	967680	46080	34560	.	2880	.	.	.	288			
$A_{3,1}$	14515200	846720	34560	51840	1152	1440	1920	.	.	.	96		
A_4	5806080	483840	23040	51840	.	1440	3840	.	.	.	.	240	
D_4	3628800	362880	17280	51840	1152	.	5760	.	.	.	.	.	1152
$A_{2,1,1,1}$	14515200	725760	46080	12960	3840	1080	.	384	144	.	.	.	.
$A_{2,2,1}$	9676800	564480	38400	17280	2304	1920	.	.	192	144	.	.	.
$A_{3,1,1}$	7257600	483840	30720	25920	2304	1440	960	192	96	.	96	.	.
$A_{3,2}$	4838400	362880	26880	21600	1152	2160	640	.	48	288	96	.	.
$A_{4,1}$	2903040	266112	19200	25920	1152	1440	1920	.	96	.	96	120	.
$D_4\times A_1$	1814400	196560	14400	25920	1440	720	2880	192	.	.	144	.	576
A_5	967680	120960	11520	17280	384	1440	1920	.	.	144	96	240	.
D_5	362880	60480	6720	12960	576	720	2400	.	48	.	48	240	576
$A_{2,2,1,1}$	4838400	322560	28160	8640	3072	1200	.	384	224	72	.	.	.
$A_{3,2,1}$	2419200	201600	19200	10800	1920	1380	320	192	168	144	64	.	.
$A_{4,1,1}$	1451520	145152	13824	12960	1536	1080	960	192	144	.	96	60	.
$A_{3,3}$	1209600	120960	13440	8640	1152	1440	320	96	96	288	96	.	.
$A_{4,2}$	967680	104832	11520	9504	1152	1200	640	.	144	144	96	40	.
$D_4\times A_2$	604800	75600	8640	9180	1056	900	960	192	36	144	144	.	192
$A_{5,1}$	483840	64512	7680	8640	768	960	960	64	96	72	96	120	.
$D_5\times A_1$	181440	31752	4320	6480	624	540	1200	96	72	.	84	120	288
A_6	138240	24192	3840	4320	384	720	640	.	48	144	96	120	.
D_6	30240	7560	1560	2160	240	360	560	48	24	72	72	120	144
E_6	13440	4032	960	1440	192	240	480	.	48	24	48	120	192
$A_{4,2,1}$	483840	56448	7424	4752	1152	732	320	192	152	72	64	20	.
$A_{4,3}$	241920	32256	4992	3024	768	660	192	96	120	144	64	10	.
$A_{6,1}$	69120	12672	2304	2160	384	420	320	64	72	72	64	60	.
$D_5\times A_2$	60480	11592	2080	2214	432	330	400	96	70	36	68	40	96
A_7	17280	4032	960	864	192	240	160	16	48	72	48	40	.
$E_6\times A_1$	6720	2072	544	720	144	140	240	32	40	12	36	60	96
D_7	2160	756	252	270	72	90	90	24	18	36	30	30	24
E_7	240	126	60	72	26	30	40	8	12	12	14	20	24
E_8	1	1	1	1	1	1	1	1	1	1	1	1	1

Fig. A.6. The parabolic table of marks for E_8.

24																											
.	24																										
.	.	16																									
.	.	.	16																								
.	.	.	.	12																							
.	.	.	.	.	48																						
.	.	.	.	.	.	24																					
.	.	.	.	.	.	.	48																				
48	24	.	.	.	.	.	.	8																			
12	24	16	8	.	.	.	.	.	4																		
24	.	16	.	12	.	.	.	.	.	4																	
.	.	16	32	.	.	.	.	.	.	.	8																
.	24	.	16	12	.	.	.	.	.	.	.	4															
12	.	.	24	.	48	.	.	.	.	.	.	.	12														
.	12	16	.	12	.	12	.	.	.	.	.	.	.	4													
12	.	8	.	12	24	.	24	.	.	.	.	.	.	.	4												
.	.	.	16	12	.	24	.	.	.	.	.	.	.	.	.	4											
.	.	8	8	.	24	24	24	.	.	.	.	.	.	.	.	.	8										
.	12	.	.	12	.	12	48	.	.	.	.	.	.	.	.	.	.	12									
36	24	16	8	8	.	.	.	8	4	4	.	2	.	.	.	.	.	.	2								
12	24	16	24	6	.	.	.	.	4	2	8	4	.	.	.	.	.	.	.	2							
12	12	16	8	12	.	12	.	.	4	4	.	.	.	4	.	2	.	.	.	.	2						
18	12	8	10	12	24	.	8	4	2	.	.	4	6	.	4	.	.	.	.	.	.	2					
.	12	8	16	12	.	12	.	.	.	.	4	4	.	4	.	4	.	.	.	.	.	.	2				
12	8	8	.	12	8	6	24	4	.	4	.	.	.	2	4	.	.	6	.	.	.	.	.	2			
6	.	6	12	6	12	12	6	.	.	2	4	.	6	.	2	4	4	.	.	.	.	.	.	.	2		
2	6	4	4	6	6	8	12	.	2	.	.	2	.	2	2	2	4	6	.	.	.	.	.	.	.	2	
1	1	1	1	1	1	1	1	1	1	1	1	1	1	1	1	1	1	1	1	1	1	1	1	1	1	1	1

Fig. A.6. The parabolic table of marks for E_8 (continued).

Table A.1 Coxeter classes of parabolic subgroups in H_3, H_4, F_4 and E_6

$J \subseteq H_4$	$\lvert W_J \rvert$	p_J	$\lvert X_{JJJ} \rvert$
1	1	1	14400
A_1	2	4	120
$A_{1,1}$	4	3	8
A_2	6	2	12
$I_2(5)$	10	1	20
$A_{2,1}$	12	1	2
$I_2(5) \times A_1$	20	1	2
A_3	24	1	2
H_3	120	1	2
H_4	14400	1	1

$J \subseteq F_4$	$\lvert W_J \rvert$	p_J	$\lvert X_{JJJ} \rvert$
1	1	1	1152
A_1'	2	2	48
A_1''	2	2	48
$A_{1,1}$	4	3	4
A_2'	6	1	12
A_2''	6	1	12
B_2	8	1	8
$A_{2,1}'$	12	1	2
$A_{2,1}''$	12	1	2
B_3'	48	1	2
B_3''	48	1	2
F_4	1152	1	1

$J \subseteq H_3$	$\lvert W_J \rvert$	p_J	$\lvert X_{JJJ} \rvert$
1	1	1	120
A_1	2	3	4
$A_{1,1}$	4	1	2
A_2	6	1	2
$I_2(5)$	10	1	2
H_3	120	1	1

$J \subseteq E_6$	$\lvert W_J \rvert$	p_J	$\lvert X_{JJJ} \rvert$
1	1	1	51840
A_1	2	6	720
$A_{1,1}$	4	10	48
A_2	6	5	72
$A_{1,1,1}$	8	5	12
$A_{2,1}$	12	10	6
A_3	24	5	8
$A_{2,1,1}$	24	5	2
$A_{2,2}$	36	1	12
$A_{3,1}$	48	4	2
A_4	120	4	2
D_4	192	1	6
$A_{2,2,1}$	72	1	2
$A_{4,1}$	240	2	1
A_5	720	1	2
D_5	1920	2	1
E_6	51840	1	1

For each Coxeter class of parabolic subgroups W_J, $J \subseteq S$, we list the type and the order of W_J, the number p_J of conjugates of J in S and the size of X_{JJJ}, the index of W_J in its normalizer in W.

Table A.2 Coxeter classes of parabolic subgroups in E_7 and E_8

$J \subseteq E_7$	$\lvert W_J \rvert$	p_J	$\lvert X_{JJJ} \rvert$
1	1	1	2903040
A_1	2	7	23040
$A_{1,1}$	4	15	768
A_2	6	6	1440
$A'_{1,1,1}$	8	1	1152
$A''_{1,1,1}$	8	10	96
$A_{2,1}$	12	18	48
A_3	24	6	96
$A_{1,1,1,1}$	16	2	48
$A_{2,1,1}$	24	12	8
$A_{2,2}$	36	4	24
$A'_{3,1}$	48	2	48
$A''_{3,1}$	48	9	8
A_4	120	5	12
D_4	192	1	48
$A_{2,1,1,1}$	48	1	12
$A_{2,2,1}$	72	3	4
$A_{3,1,1}$	96	3	4
$A_{3,2}$	144	3	4
$A_{4,1}$	240	5	2
$D_4 \times A_1$	384	1	8
A'_5	720	1	12
A''_5	720	2	4
D_5	1920	2	4
$A_{3,2,1}$	288	1	2
$A_{4,2}$	720	1	2
$A_{5,1}$	1440	1	2
$D_5 \times A_1$	3840	1	2
A_6	5040	1	2
D_6	23040	1	2
E_6	51840	1	2
E_7	2903040	1	1

$J \subseteq E_8$	$\lvert W_J \rvert$	p_J	$\lvert X_{JJJ} \rvert$
1	1	1	696729600
A_1	2	8	2903040
$A_{1,1}$	4	21	46080
A_2	6	7	103680
$A_{1,1,1}$	8	21	2304
$A_{2,1}$	12	28	1440
A_3	24	7	3840
$A_{1,1,1,1}$	16	7	384
$A_{2,1,1}$	24	28	96
$A_{2,2}$	36	8	288
$A_{3,1}$	48	20	96
A_4	120	6	240
D_4	192	1	1152
$A_{2,1,1,1}$	48	8	24
$A_{2,2,1}$	72	8	24
$A_{3,1,1}$	96	10	16
$A_{3,2}$	144	10	16
$A_{4,1}$	240	12	12
$D_4 \times A_1$	384	2	48
A_5	720	4	24
D_5	1920	2	48
$A_{2,2,1,1}$	144	2	8
$A_{3,2,1}$	288	4	4
$A_{4,1,1}$	480	4	4
$A_{3,3}$	576	2	8
$A_{4,2}$	720	4	4
$D_4 \times A_2$	1152	1	12
$A_{5,1}$	1440	3	4
$D_5 \times A_1$	3840	3	4
A_6	5040	3	4
D_6	23040	1	8
E_6	51840	1	12
$A_{4,2,1}$	1440	1	2
$A_{4,3}$	2880	1	2
$A_{6,1}$	10080	1	2
$D_5 \times A_2$	11520	1	2
A_7	40320	1	2
$E_6 \times A_1$	103680	1	2
D_7	322560	1	2
E_7	2903040	1	2
E_8	696729600	1	1

B GOOD REPRESENTATIVES OF CUSPIDAL CLASSES

In this section we list good representatives for the cuspidal classes of the exceptional types of Coxeter groups; cf. Section 4.3. Throughout we will use a labelling of Coxeter graphs as in Table 1.1 on p. 16 and write the element $s_1 s_2 s_3 \cdots$ as $123\cdots$ for short. We will also write $\Delta_{123\ldots}$ for Δ_J if $J = \{1, 2, 3, \ldots\}$ and $\Delta = w_0$.

Table B.1 Cuspidal classes of H_3

C	$\lvert C\rvert$	$\lvert C_{min}\rvert$	d	$l(w)$	$w \in C_{min}$	w^d
6	$2^2 \cdot 3$	4	10	3	123	Δ^2
8	$2^2 \cdot 5$	6	6	5	12123	Δ^2
9	$2^2 \cdot 3$	6	10	9	121232123	Δ^6
10	1	1	2	15	w_0	Δ^2

Table B.2 Cuspidal classes of H_4

C	$\lvert C\rvert$	$\lvert C_{min}\rvert$	d	$l(w)$	$w \in C_{min}$	w^d
11	$2^5 \cdot 3 \cdot 5$	8	30	4	1234	Δ^2
14	$2^4 \cdot 3^2 \cdot 5$	12	20	6	121234	Δ^2
15	$2^5 \cdot 3 \cdot 5$	18	15	8	12123243	Δ^2
17	$2^4 \cdot 3 \cdot 5^2$	22	12	10	1212321234	Δ^2
18	$2^3 \cdot 3$	24	10	12	121213212343	Δ^2
19	$2^5 \cdot 3^2$	34	10	14	12132121321234	$\Delta^2\Delta_{12}^4$
21	$2^4 \cdot 3^2$	12	10	16	1212132121321234	$\Delta^2\Delta_{12}^8$
22	$2^5 \cdot 3 \cdot 5$	24	15	16	1212321234321234	Δ^4
23	$2^4 \cdot 3^2 \cdot 5$	38	20	18	121213212134321234	Δ^6
24	$2^3 \cdot 5$	40	6	20	12121321213214321234	Δ^2
25	$2^4 \cdot 5^2$	36	6	22	1212132121321234321234	$\Delta^2\Delta_{23}^4$
26	$2^3 \cdot 3$	24	5	24	121213212134321213212343	Δ^2
27	$2^5 \cdot 3^2$	56	5	26	12123212132123432121321234	$\Delta^2\Delta_{12}^2$
28	$2^5 \cdot 3 \cdot 5$	40	30	28	1212132123432121321234321234	Δ^{14}
29	$2^2 \cdot 3 \cdot 5$	60	4	30	121213212132124321213214321234	Δ^2
30	$2^3 \cdot 3$	24	10	36	121213212132432121321234321213212343	Δ^6
31	$2^4 \cdot 3^2$	36	10	38	12121321213212343212132123432121321234	$\Delta^6\Delta_{12}^4$
32	$2^3 \cdot 5$	40	3	40	1212132121321243212132124321213214321234	Δ^2
33	$2^3 \cdot 3$	24	5	48	1212132121321· ·432121321234321213212343213212343	Δ^4
34	1	1	2	60	w_0	Δ^2

Table B.3 Cuspidal classes of F_4

C	$\lvert C\rvert$	$\lvert C_{min}\rvert$	d	$l(w)$	$w \in C_{min}$	w^d
F_4	$2^5 \cdot 3$	8	12	4	1234	Δ^2
B_4	$2^4 \cdot 3^2$	14	8	6	123234	Δ^2
$F_4(a_1)$	2^4	16	6	8	12132343	Δ^2
D_4	2^5	8	6	10	1232343234	$\Delta^2\Delta_{34}^4$
$C_3 + A_1$	2^5	8	6	10	1213213234	$\Delta^2\Delta_{12}^4$
$D_4(a_1)$	$2^2 \cdot 3$	12	4	12	121321343234	Δ^2
$A_3 + \tilde{A}_1$	$2^2 \cdot 3^2$	16	4	14	12132132343234	$\Delta^2\Delta_{23}^2$
$A_2 + \tilde{A}_2$	2^4	16	3	16	1213213432132343	Δ^2
$4A_1$	1	1	2	24	w_0	Δ^2

Table B.4 Cuspidal classes of E_6

C	$\lvert C\rvert$	$\lvert C_{min}\rvert$	d	$l(w)$	$w \in C_{min}$	w^d
E_6	$2^5 \cdot 3^3 \cdot 5$	32	12	6	123456	Δ^2
$E_6(a_1)$	$2^7 \cdot 3^2 \cdot 5$	80	9	8	12342546	Δ^2
$E_6(a_2)$	$2^4 \cdot 3^2 \cdot 5$	144	6	12	123142345465	Δ^2
$A_5 + A_1$	$2^5 \cdot 3^2 \cdot 5$	48	6	14	12342345423456	$\Delta^2\Delta_{24}^4$
$3A_2$	$2^4 \cdot 5$	80	3	24	123142314542314565423456	Δ^2

Table B.5 Cuspidal classes of E_7

C	$\lvert C\rvert$	$\lvert C_{min}\rvert$	d	$l(w)$	$w \in C_{min}$	w^d
E_7	$2^9 \cdot 3^2 \cdot 5 \cdot 7$	64	18	7	1234567	Δ^2
$E_7(a_1)$	$2^9 \cdot 3^4 \cdot 5$	160	14	9	123425467	Δ^2
$E_7(a_2)$	$2^7 \cdot 3^3 \cdot 5 \cdot 7$	280	12	11	12342546576	$\Delta^2\Delta_{257}^2$
$E_7(a_3)$	$2^9 \cdot 3^3 \cdot 7$	366	30	13	1234254234567	$\Delta^6\Delta_{24}^4$
$D_6 + A_1$	$2^8 \cdot 3^3 \cdot 7$	96	10	15	123423454234567	$\Delta^2\Delta_{24}^8$
A_7	$2^5 \cdot 3^4 \cdot 5 \cdot 7$	316	8	17	12342345423456576	$\Delta^2\Delta_{257}^2\Delta_2^4$
$E_7(a_4)$	$2^6 \cdot 5 \cdot 7$	800	6	21	123142314354234· ·654765	Δ^2
$D_6(a_2) + A_1$	$2^7 \cdot 3 \cdot 5 \cdot 7$	708	6	23	1231423143542314354· ·6576	$\Delta^2\Delta_{13}^4$
$A_5 + A_2$	$2^6 \cdot 3^2 \cdot 5 \cdot 7$	420	6	25	1231423145423145· ·654234567	$\Delta^2\Delta_{2345}^2$
$D_4 + 3A_1$	$2^5 \cdot 3 \cdot 7$	32	6	31	1234234542345· ·654234567654234567	$\Delta^2\Delta_{24567}^4$
$2A_3 + A_1$	$2^2 \cdot 3^3 \cdot 5 \cdot 7$	360	4	33	123142345423145· ·654234567654234567	$\Delta^2\Delta_{257}^2$
$7A_1$	1	1	2	63	w_0	Δ^2

For each class C its size $|C|$ is given, the number $|C_{min}|$ of elements of minimal length, a good representative $w \in C_{min}$, together with its length $l(w)$ and order d. The label of C is the address of the class in the CHEVIE list (for H_3 and H_4) or the label from [Carter 1972a].

Table B.6 Cuspidal classes of E_8

C	|C|	$\|C_{min}\|$	d	$l(w)$	$w \in C_{min}$	w^d
E_8	$2^{13}.3^4.5.7$	128	30	8	12345678	Δ^2
$E_8(a_1)$	$2^{11}.3^4.5^2.7$	320	24	10	1234254678	Δ^2
$E_8(a_2)$	$2^{12}.3^5.5.7$	624	20	12	123425465478	Δ^2
$E_8(a_4)$	$2^{13}.3^2.5^2.7$	732	18	14	12342542345678	$\Delta^2\Delta^4_{24}$
$E_8(a_5)$	$2^{13}.3^4.5.7$	1516	15	16	1231423454657658	Δ^2
E_7+A_1	$2^{12}.3^2.5^2.7$	192	18	16	1234234542345678	$\Delta^2\Delta^{16}_{24}$
D_8	$2^{12}.3^5.5^2$	852	14	18	123423454234565768	$\Delta^2\Delta^{12}_{2}$
$E_8(a_3)$	$2^9.3^3.5^2.7$	2696	12	20	12314234542365476548	Δ^2
$D_8(a_1)$	$2^{11}.3^3.5^2.7$	2040	12	22	1234234542345654765876	$\Delta^2\Delta^4_{4578}$
$E_8(a_7)$	$2^9.3^3.5^2.7$	2360	12	22	1231423454316542345678	$\Delta^2\Delta^2_{2345}$
$E_8(a_6)$	$2^{11}.3^4.7$	3370	10	24	123142314542345654765876	Δ^2
$E_7(a_2)+A_1$	$2^{10}.3^3.5^2.7$	1758	12	24	123142314354316542345678	$\Delta^2\Delta^2_{2345}\Delta^8_{24}$
E_6+A_2	$2^9.3^3.5^2.7$	840	12	26	12314231454231456542345678	$\Delta^2\Delta^6_{2345}$
$D_8(a_2)$	$2^{12}.3^4.5.7$	4996	30	26	12342354234654276542345678	$\Delta^6\Delta^4_{2456}\Delta^{20}_{5}$
A_8	$2^{13}.3^2.5^2.7$	2816	9	28	1231423145423145654234567687	$\Delta^2\Delta^4_{34}$
$D_8(a_3)$	$2^8.3^4.5^2.7$	7748	8	30	123142314354231465423476548765	Δ^2
D_6+2A_1	$2^{10}.3^4.7$	256	10	32	12342345423456542345676542345678	$\Delta^2\Delta^8_{2456}$
A_7+A_1	$2^7.3^5.5^2.7$	2080	8	34	1231423454231456542345676542345678	$\Delta^2\Delta^2_{2345}\Delta^4_{25}$
$E_8(a_8)$	$2^7.5.7$	4480	6	40	1231423143542314565423145676542345678765	Δ^2
$E_7(a_4)+A_1$	$2^9.5^2.7$	11592	6	42	123142314354231435426542314567654234567876	$\Delta^2\Delta^4_{34}$
$2D_4$	$2^9.3.5^2.7$	4070	6	44	12314231435423456542314567654231435465768765	$\Delta^2\Delta^4_{1367}$
$E_6(a_2)+A_2$	$2^8.3^2.5^2.7$	16374	6	44	12314231454231436542314354265431765423456787	$\Delta^2\Delta^2_{2345}$
$A_5+A_2+A_1$	$2^9.3^2.5^2.7$	3752	6	46	1231423145423145654234567654234567876542345678	$\Delta^2\Delta^2_{234578}\Delta^2_{78}$
$D_5(a_1)+A_3$	$2^8.3^3.5^2.7$	15134	12	46	1231423154231654317654231435426543176542345678	$\Delta^4\Delta^2_{123456}$
$2A_4$	$2^{11}.3^4.7$	7952	5	48	123142314542314565423145676542314567876542345678	Δ^2
$2D_4(a_1)$	$2^4.3^3.5.7$	15120	4	60	1231423143· ·54231435426542345765423143548765423143542654765876	Δ^2
D_4+4A_1	$2^6.5.7$	56	6	64	1231423143· ·542314354265423143542654317654231435426543176542345678	$\Delta^2\Delta^4_{123456}$
$2A_3+2A_1$	$2^3.3^3.5^2.7$	1260	4	66	1231423143· ·54231435426542314354265431765423143542654317876542345678	$\Delta^2\Delta^2_{2345}$
$4A_2$	$2^7.5.7$	4480	3	80	12314231435423143 5· ·654231435426765423143542654317876542314354265431765423456788765	Δ^2
$8A_1$	1	1	2	120	w_0	Δ^2

Note that in general the chain $J_1 \supset J_2 \supset \cdots \supset J_r$ of subsets of S is not uniquely determined by the conjugacy class.

The representatives in Tables B.1–B.6 have been chosen as the lexicographically smallest good element in their class.

The following tables contain the characteristic polynomials of the elements of W in the natural reflection representation of W. Recall from Section 1.1 that the reflection representation is defined in terms of a Cartan matrix corresponding to (W, S). We assume that the Cartan matrices are those determined by the diagrams in (1.3.7). Given any $w \in W$, the characteristic polynomial of w can be computed explicitly using the defining formula for the action of each $s \in S$ and a reduced expression for w. Furthermore, note that if the conjugacy class of w is not cuspidal, then the class of w will be cuspidal in W_J for some $J \subset S$. Using Exercise 3.11, we see that the characteristic polynomial of w is obtained from that with respect to W_J by multiplication with $(q-1)^{|S-J|}$.

We now list these polynomials for a complete set of representatives for the cuspidal classes of W, in the case where W is of exceptional type. For types H_3, H_4, see Table B.7; for types F_4, E_6, E_7 and E_8, see Table B.8. (The results for type $I_2(m)$ are given in (5.3.4).) The classes are labelled in the same way as in the previous tables. By inspection of the tables, we see that—except for two classes in type F_4—elements in different cuspidal classes have different characteristic polynomials.

Table B.7 Characteristic polynomials of cuspidal classes for types H_3, H_4

Type H_3	
6	$\Phi_2\Phi_{10,a}$
8	$\Phi_2\Phi_6$
9	$\Phi_2\Phi_{10,b}$
10	Φ_2^3

Type H_4			
11	$\Phi_{30,b}$	25	$\Phi_2^2\Phi_6$
14	$\Phi_{20,a}$	26	$\Phi_{5,b}^2$
15	$\Phi_{15,a}$	27	Φ_5
17	Φ_{12}	28	$\Phi_{30,a}$
18	$\Phi_{10,a}^2$	29	Φ_4^2
19	Φ_{10}	30	$\Phi_{10,b}^2$
21	$\Phi_2^2\Phi_{10,a}$	31	$\Phi_2^2\Phi_{10,b}$
22	$\Phi_{15,b}$	32	Φ_3^2
23	$\Phi_{20,b}$	33	$\Phi_{5,a}^2$
24	Φ_6^2	34	Φ_2^4

Here Φ_d denotes the dth cyclotomic polynomial in q (see Table E.1 below).

Table B.8 Characteristic polynomials of cuspidal classes for types F_4, E_6, E_7, E_8

Type F_4		
2	$4A_1$	Φ_2^4
5	D_4	$\Phi_2^2\Phi_6$
6	$D_4(a_1)$	Φ_4^2
8	C_3+A_1	$\Phi_2^2\Phi_6$
9	$A_2+\tilde{A}_2$	Φ_3^2
10	$F_4(a_1)$	Φ_6^2
11	F_4	Φ_{12}
24	$A_3+\tilde{A}_1$	$\Phi_2^2\Phi_4$
25	B_4	Φ_8

Type E_6		
4	$3A_2$	Φ_3^3
10	$E_6(a_2)$	$\Phi_3\Phi_6^2$
12	A_5+A_1	$\Phi_2^2\Phi_3\Phi_6$
14	$E_6(a_1)$	Φ_9
15	E_6	$\Phi_3\Phi_{12}$

Type E_7		
31	$7A_1$	Φ_2^7
36	D_4+3A_1	$\Phi_2^5\Phi_6$
37	$E_7(a_4)$	$\Phi_2\Phi_6^3$
38	$D_6(a_2)+A_1$	$\Phi_2^3\Phi_6^2$
39	$2A_3+A_1$	$\Phi_2^3\Phi_4^2$
44	D_6+A_1	$\Phi_2^3\Phi_{10}$
47	A_5+A_2	$\Phi_2\Phi_3^2\Phi_6$
52	$E_7(a_1)$	$\Phi_2\Phi_{14}$
54	A_7	$\Phi_2\Phi_4\Phi_8$
55	E_7	$\Phi_2\Phi_{18}$
59	$E_7(a_2)$	$\Phi_2\Phi_6\Phi_{12}$
60	$E_7(a_3)$	$\Phi_2\Phi_6\Phi_{10}$

Type E_8		
2	$8A_1$	Φ_2^8
6	$2D_4(a_1)$	Φ_4^4
9	D_4+4A_1	$\Phi_2^6\Phi_6$
10	$4A_2$	Φ_3^4
11	$E_8(a_8)$	Φ_6^4
13	$E_7(a_4)+A_1$	$\Phi_2^2\Phi_6^3$
15	$2D_4$	$\Phi_2^4\Phi_6^2$
17	$2A_3+2A_1$	$\Phi_2^4\Phi_4^2$
21	$D_8(a_3)$	Φ_8^2
24	D_6+2A_1	$\Phi_2^4\Phi_{10}$
25	$2A_4$	Φ_5^2
26	$E_8(a_6)$	Φ_{10}^2
29	$E_6(a_2)+A_2$	$\Phi_3^2\Phi_6^2$
32	$E_8(a_3)$	Φ_{12}^2
34	$A_5+A_2+A_1$	$\Phi_2^2\Phi_3^2\Phi_6$
39	$D_8(a_1)$	$\Phi_4^2\Phi_{12}$
42	D_8	$\Phi_2^2\Phi_{14}$
45	A_7+A_1	$\Phi_2^2\Phi_4\Phi_8$
47	E_7+A_1	$\Phi_2^2\Phi_{18}$
48	A_8	$\Phi_3\Phi_9$
49	$E_8(a_4)$	$\Phi_6\Phi_{18}$
52	$E_8(a_2)$	Φ_{20}
54	$D_5(a_1)+A_3$	$\Phi_2^2\Phi_4^2\Phi_6$
55	E_6+A_2	$\Phi_3^2\Phi_{12}$
56	$E_8(a_7)$	$\Phi_6^2\Phi_{12}$
58	$E_7(a_2)+A_1$	$\Phi_2^2\Phi_6\Phi_{12}$
61	$E_8(a_1)$	Φ_{24}
63	$D_8(a_2)$	$\Phi_2^2\Phi_6\Phi_{10}$
64	$E_8(a_5)$	Φ_{15}
65	E_8	Φ_{30}

Here Φ_d denotes the dth cyclotomic polynomial in q (see Table E.1 below).

C IRREDUCIBLE CHARACTERS

In this section, we list tables containing information about the complex irreducible characters of the finite Coxeter groups of exceptional type. In particular, these tables define a labelling of the irreducible characters which is used throughout the book. We also present some additional information concerning the b-invariants and the values on the generators $s \in S$. (The methods for computing this information are described in (5.3.5).) Moreover, we give the values of Lusztig's a-function and the partition of $\mathrm{Irr}(W)$ into families (see (6.5.10)) and, for each $\chi \in \mathrm{Irr}(W)$, we print the unique maximal subset J (see (6.3.6)) such that χ occurs with multiplicity 1 in the induced character $1_{W_J}^W$ (if J exists).

For type $I_2(m)$ the relevant information is given in (5.3.4), (6.3.5), (6.5.10).

The information contained in the tables of this and the following section is used in an essential way in the proof of Theorem 5.3.8 (about splitting fields for W), Theorem 6.5.13 (about families and a-invariants) and Theorem 9.3.5 (about splitting fields for the corresponding generic Iwahori–Hecke algebra).

In the tables below, we use the following conventions. The numbers in the first column correspond to the various families $\mathcal{F} \subseteq \mathrm{Irr}(W)$. Two irreducible characters $\chi, \chi' \in \mathrm{Irr}(W)$ lie in the same family if and only if the corresponding numbers in the first column are equal. For all types, we use the labelling of the generators in S as given by the Dynkin diagrams in Table 1.2. In the last column, we simply write, for example, 123 for the subset $J = \{s_1, s_2, s_3\} \subseteq S$. The notational conventions for the characters (except for types H_3 and H_4) are taken from [Lusztig 1984, Chap. 4], which in turn go back to Kondo and Frame.

Table C.1 Irreducible characters for type H_3

$\mathcal{F}$	χ	a_χ	b_χ	$\chi(s)$	$J \subseteq S$
1	$\chi_1 = 1_r$	0	0	1	123
2	$\chi_2 = 3_s'$	1	1	1	23
2	$\chi_3 = \overline{3}_s'$	1	3	1	23
3	$\chi_4 = 5_r$	2	2	1	23
4	$\chi_5 = 4_r'$	3	3	0	23
4	$\chi_6 = 4_r$	3	4	0	23
5	$\chi_7 = 5_r'$	5	5	-1	13
6	$\chi_8 = 3_s$	6	6	-1	1
6	$\chi_9 = \overline{3}_s$	6	8	-1	1
7	$\chi_{10} = 1_r'$	15	15	-1	$\varnothing$

There are 10 irreducible characters. Two characters whose labels differ by a prime are obtained one from another by tensoring with the sign character (which is $1_r'$); moreover, the longest element w_0 (which is central) acts as -1 on the character to which the prime is attached. All character values lie in the field $\mathbb{Q}(\sqrt{5})$, and a bar denotes algebraic conjugation in that field. A character is rational if and only if its label has a subscript r.

Table C.2 Irreducible characters for type H_4

$\mathcal{F}$	χ	a_χ	b_χ	$\chi(s)$	$J \subseteq S$
1	$\chi_1 = 1_r$	0	0	1	1234
2	$\chi_3 = 4_t$	1	1	2	234
2	$\chi_5 = \overline{4}_t$	1	7	2	234
3	$\chi_{11} = 9_s$	2	2	3	234
3	$\chi_{13} = \overline{9}_s$	2	6	3	234
4	$\chi_{18} = 16_{rr}$	3	3	4	123
4	$\chi_{20} = 16_r$	3	6	4	123
5	$\chi_{27} = 25_r$	4	4	5	123
6	$\chi_{31} = 36_{rr}$	5	5	6	123
7	$\chi_{26} = \overline{24}_s$	6	6	0	234
7	$\chi_{24} = \overline{24}_t$	6	7	0	234
7	$\chi_{33} = 40_r$	6	8	0	234
7	$\chi_{34} = 48_{rr}$	6	9	0	—
7	$\chi_{22} = 18_r$	6	10	0	234
7	$\chi_{29} = 30_s$	6	10	0	234
7	$\chi_{30} = \overline{30}_s$	6	10	0	234
7	$\chi_{23} = 24_t$	6	11	0	234
7	$\chi_{16} = 16_t$	6	11	0	134
7	$\chi_7 = 6_s$	6	12	0	23
7	$\chi_9 = 8_r$	6	12	0	234
7	$\chi_{15} = 10_r$	6	12	0	134
7	$\chi_{25} = 24_s$	6	12	0	234
7	$\chi_{17} = \overline{16}_t$	6	13	0	134
7	$\chi_{10} = 8_{rr}$	6	13	0	234
7	$\chi_8 = \overline{6}_s$	6	20	0	23
8	$\chi_{32} = 36'_{rr}$	15	15	−6	134
9	$\chi_{28} = 25'_r$	16	16	−5	134
10	$\chi_{21} = 16'_r$	18	18	−4	23
10	$\chi_{19} = 16'_{rr}$	18	21	−4	23
11	$\chi_{12} = 9'_s$	22	22	−3	13
11	$\chi_{14} = \overline{9}'_s$	22	26	−3	13
12	$\chi_4 = 4'_t$	31	31	−2	1
12	$\chi_6 = \overline{4}'_t$	31	37	−2	1
13	$\chi_2 = 1'_r$	60	60	−1	∅

There are 34 irreducible characters. The numbering of the characters refers to that in [Alvis and Lusztig 1982] (which is the same as that in CHEVIE). Two characters whose labels differ only by a prime are obtained one from another by tensoring with the sign character (which is $1'_r$). All character values lie in the field $\mathbb{Q}(\sqrt{5})$, and a bar denotes algebraic conjugation in that field. A character is rational if and only if its label contains a subscript r or rr. The longest element w_0 (which is central) acts as $+1$ on a character whose label contains a subscript r or s. If the label contains rr or t, then w_0 acts as -1.

The character 48_{rr} is exceptional in two ways: firstly, it is rational-valued but it cannot be realized over $\mathbb{Q}$ (see Exercise 6.10); secondly, it is not of parabolic type (see Definition 6.3.1). The multiplicities of 48_{rr} in $1_{W_J}^W$ are given as follows, where J runs over a set of representatives for the Coxeter classes of subsets of S:

J	∅	1	12	13	23	123	124	134	234	1234
$(48_{rr}, 1_{W_J}^W)_W$	48	24	4	12	8	0	2	4	2	0

Table C.3 Irreducible characters for type F_4

$\mathcal{F}$	χ		a_χ	s_1	s_3	$J \subseteq S$
1	1_1	$\chi_{(1,0)}$	0	1	1	1234
2	4_2	$\chi_{(4,1)}$	1	2	2	234
2	2_1	$\chi''_{(2,4)}$	1	2	0	124
2	2_3	$\chi'_{(2,4)}$	1	0	2	234
3	9_1	$\chi_{(9,2)}$	2	3	3	234
4	8_1	$\chi''_{(8,3)}$	3	4	0	134
5	8_3	$\chi'_{(8,3)}$	3	0	4	234
6	12	$\chi_{(12,4)}$	4	0	0	134
6	16	$\chi_{(16,5)}$	4	0	0	134
6	9_2	$\chi''_{(9,6)}$	4	3	−3	124
6	6_2	$\chi''_{(6,6)}$	4	0	0	34
6	9_3	$\chi'_{(9,6)}$	4	−3	3	134
6	6_1	$\chi'_{(6,6)}$	4	0	0	134
6	4_4	$\chi'_{(4,7)}$	4	−2	2	134
6	4_3	$\chi''_{(4,7)}$	4	2	−2	124
6	4_1	$\chi_{(4,8)}$	4	0	0	23
6	1_3	$\chi'_{(1,12)}$	4	−1	1	34
6	1_2	$\chi''_{(1,12)}$	4	1	−1	12
7	8_2	$\chi'_{(8,9)}$	9	−4	0	34
8	8_4	$\chi''_{(8,9)}$	9	0	−4	13
9	9_4	$\chi_{(9,10)}$	10	−3	−3	13
10	4_5	$\chi_{(4,13)}$	13	−2	−2	3
10	2_2	$\chi'_{(2,16)}$	13	−2	0	3
10	2_4	$\chi''_{(2,16)}$	13	0	−2	1
11	1_4	$\chi_{(1,24)}$	24	−1	−1	∅

There are 25 irreducible characters. The subscript in the third column is the pair consisting of the degree and the b-invariant of a character. The columns labelled by s_1, s_3 contain the character values on those elements.

Table C.4 Irreducible characters for type E_6

$\mathcal{F}$	χ	a_χ	b_χ	$\chi(s)$	$J \subseteq S$
1	1_p	0	0	1	123456
2	6_p	1	1	4	13456
3	20_p	2	2	10	13456
4	30_p	3	3	10	13456
4	15_q	3	4	5	13456
4	15_p	3	5	5	2345
5	64_p	4	4	16	12346
6	60_p	5	5	10	12346
7	81_p	6	6	9	12356
8	24_p	6	6	4	12356
9	80_s	7	7	0	12356
9	90_s	7	8	0	1356
9	60_s	7	8	0	12356
9	10_s	7	9	0	12356
9	20_s	7	10	0	134
10	$81'_p$	10	10	−9	1235
11	$60'_p$	11	11	−10	1235
12	$24'_p$	12	12	−4	1356
13	$64'_p$	13	13	−16	123
14	$30'_p$	15	15	−10	125
14	$15'_q$	15	16	−5	125
14	$15'_p$	15	17	−5	13
15	$20'_p$	20	20	−10	12
16	$6'_p$	25	25	−4	1
17	$1'_p$	36	36	−1	∅

There are 25 irreducible characters. A prime attached to a character means that character tensored with the sign character (which is $1'_p$).

Table C.5 Irreducible characters for type E_7

$\mathcal{F}$	χ	a_χ	b_χ	$\chi(s)$	$J \subseteq S$
1	1_a	0	0	1	1234567
2	$7'_a$	1	1	5	234567
3	27_a	2	2	15	234567
3	$21'_b$	3	3	11	134567
5	$56'_a$	3	3	24	234567
5	35_b	3	4	15	234567
5	21_a	3	6	9	24567
6	120_a	4	4	40	134567
6	$105'_a$	4	5	35	124567
6	$15'_a$	4	7	5	134567
7	$189'_b$	5	5	51	134567
8	168_a	6	6	40	124567
9	105_b	6	6	25	134567
10	210_a	6	6	50	124567
11	$189'_c$	7	7	39	123467
12	$315'_a$	7	7	45	124567
12	280_b	7	8	40	124567
12	$70'_a$	7	9	10	124567
12	$280'_a$	7	9	40	123567
12	$35'_a$	7	13	5	2457
13	405_a	8	8	45	123467
13	$216'_a$	8	9	24	123467
13	189_a	8	10	21	23457
14	$378'_a$	9	9	30	123467
15	210_b	10	10	10	123467
16	420_a	10	10	20	123567
16	$336'_a$	10	11	16	13467
16	84_a	10	12	4	123567
17	$512'_a$	11	11	0	123567
17	512_a	11	12	0	123567
18	105_c	12	12	5	24567
19	$210'_b$	13	13	−10	123567
20	$420'_a$	13	13	−20	13467
20	336_a	13	14	−16	12457
20	$84'_a$	13	15	−4	13467
21	378_a	14	14	−30	13467
22	$105'_c$	15	15	−5	12357
23	$405'_a$	15	15	−45	12457
23	216_a	15	16	−24	12457
23	$189'_a$	15	17	−21	2457
24	315_a	16	16	−45	12356
24	$280'_b$	16	17	−40	12356
24	70_a	16	18	−10	12357
24	280_a	16	18	−40	12357
24	35_a	16	22	−5	257
25	189_c	20	20	−39	2457
26	$168'_a$	21	21	−40	1356
27	$210'_a$	21	21	−50	1235
28	$105'_b$	21	21	−25	12357
29	189_b	22	22	−51	1235
30	$120'_a$	25	25	−40	1257
30	105_a	25	26	−35	123
30	15_a	25	28	−5	1257
31	56_a	30	30	−24	125
31	$35'_b$	30	31	−15	125
31	$21'_a$	30	33	−9	13
32	21_b	36	36	−11	257
33	$27'_a$	37	37	−15	12
34	7_a	46	46	−5	1
35	$1'_a$	63	63	−1	∅

There are 60 irreducible characters. A prime attached to a character means that the longest element w_0 acts as -1. Two characters which have the same label except for a prime are obtained one from another by tensoring with the sign character (which is $1'_a$).

Table C.6 Irreducible characters for type E_8

$\mathcal{F}$	χ	a_χ	b_χ	$\chi(s)$	$J \subseteq S$
1	1_x	0	0	1	12345678
2	8_z	1	1	6	2345678
3	35_x	2	2	21	2345678
4	112_z	3	3	56	2345678
4	84_x	3	4	42	2345678
4	28_x	3	8	14	234578
5	210_x	4	4	84	2345678
5	160_z	4	7	64	1245678
5	50_x	4	8	20	2345678
6	560_z	5	5	196	2345678
7	567_x	6	6	189	1345678
8	700_x	6	6	210	2345678
8	400_z	6	7	120	2345678
8	300_x	6	8	90	1245678
9	1400_z	7	7	350	1345678
9	1344_x	7	8	336	1345678
9	448_z	7	9	112	1345678
9	1008_z	7	9	252	1245678
9	56_z	7	19	14	23457
10	1400_x	8	8	280	1345678
10	1575_x	8	10	315	1245678
10	1050_x	8	10	210	1345678
10	175_x	8	12	35	1345678
10	350_x	8	14	70	234578
11	3240_z	9	9	594	1345678
12	2268_x	10	10	378	1245678
12	972_x	10	12	162	1245678
12	1296_z	10	13	216	1235678
13	2240_x	10	10	336	1345678
13	1400_{zz}	10	11	210	1345678
13	840_z	10	13	126	1234678
14	4096_z	11	11	512	1245678
14	4096_x	11	12	512	1245678

$\mathcal{F}$	χ	a_χ	b_χ	$\chi(s)$	$J \subseteq S$
15	525_x	12	12	105	1235678
16	4200_x	12	12	420	1245678
16	3360_z	12	13	336	1245678
16	840_x	12	14	84	1234578
17	4536_z	13	13	378	1245678
18	2800_z	13	13	280	1234678
18	700_{xx}	13	16	70	1235678
18	2100_x	13	16	210	1235678
19	2835_x	14	14	189	1245678
20	6075_x	14	14	405	134567
21	4200_z	15	15	210	134567
22	5600_z	15	15	280	1234678
22	3200_x	15	16	160	1234678
22	2400_z	15	17	120	234578
23	4480_y	16	16	0	1234678
23	7168_w	16	17	0	1234678
23	3150_y	16	18	0	1234678
23	4536_y	16	18	0	1235678
23	4200_y	16	18	0	1234678
23	5670_y	16	18	0	1235678
23	1344_w	16	19	0	1235678
23	2016_w	16	19	0	1234678
23	5600_w	16	19	0	1235678
23	420_y	16	20	0	1235678
23	1134_y	16	20	0	234578
23	1400_y	16	20	0	234578
23	2688_y	16	20	0	1235678
23	1680_y	16	22	0	123567
23	168_y	16	24	0	234578
23	448_w	16	25	0	23457
23	70_y	16	32	0	2345
24	2100_y	20	20	0	134678
25	$4200'_z$	21	21	−210	1235678

Table C.6 Irreducible characters for type E_8 (continued)

$\mathcal{F}$	χ	a_χ	b_χ	$\chi(s)$	$J \subseteq S$	$\mathcal{F}$	χ	a_χ	b_χ	$\chi(s)$	$J \subseteq S$
26	$5600_z'$	21	21	−280	123467	36	$1050_x'$	32	34	−210	123568
26	$3200_x'$	21	22	−160	123468	36	$175_x'$	32	36	−35	123568
26	$2400_z'$	21	23	−120	123567	36	$350_x'$	32	38	−70	1356
27	$6075_x'$	22	22	−405	134678	37	$525_x'$	36	36	−105	12357
28	$2835_x'$	22	22	−189	1235678	38	$1400_z'$	37	37	−350	12356
29	$4536_z'$	23	23	−378	123468	38	$1344_x'$	37	38	−336	12356
30	$4200_x'$	24	24	−420	134678	38	$1008_z'$	37	39	−252	12357
30	$3360_z'$	24	25	−336	123468	38	$448_z'$	37	39	−112	12357
30	$840_x'$	24	26	−84	134678	38	$56_z'$	37	49	−14	134
31	$2800_z'$	25	25	−280	123567	39	$700_x'$	42	42	−210	12357
31	$700_{xx}'$	25	28	−70	123567	39	$400_z'$	42	43	−120	12357
31	$2100_x'$	25	28	−210	123568	39	$300_x'$	42	44	−90	1356
32	$4096_x'$	26	26	−512	123567	40	$567_x'$	46	46	−189	1235
32	$4096_z'$	26	27	−512	123567	41	$560_z'$	47	47	−196	1235
33	$2240_x'$	28	28	−336	123567	42	$210_x'$	52	52	−84	1257
33	$1400_{zz}'$	28	29	−210	123567	42	$160_z'$	52	55	−64	123
33	$840_z'$	28	31	−126	13467	42	$50_x'$	52	56	−20	1257
34	$2268_x'$	30	30	−378	123568	43	$112_z'$	63	63	−56	125
34	$972_x'$	30	32	−162	123568	43	$84_x'$	63	64	−42	125
34	$1296_z'$	30	33	−216	12457	43	$28_x'$	63	68	−14	13
35	$3240_z'$	31	31	−594	13467	44	$35_x'$	74	74	−21	12
36	$1400_x'$	32	32	−280	123568	45	$8_z'$	91	91	−6	1
36	$1575_x'$	32	34	−315	12457	46	$1_x'$	120	120	−1	∅

There are 112 irreducible characters. Two characters whose label differ only by a prime are obtained one from another by tensoring with the sign character (which is $1_x'$). The longest element w_0 acts as $+1$ on a character whose label contains a subscript x or y. If the label contains z or w, then w_0 acts as -1. Characters whose label contains y or w are unchanged under tensoring with the sign character.

D TRUNCATED INDUCTION

The following tables contain information about the J-*induction* (with respect to Lusztig's a-function; see Section 6.5) of characters from maximal parabolic subgroups of finite Coxeter groups of exceptional type. We specify a maximal parabolic subgroup by giving its type. For every component of type A, we only print the J-induction of the sign character. (By (6.5.8), this is enough to describe the J-induction in type A.) For type $I_2(m)$, the relevant information is contained in (6.5.10) and (6.3.5). Together with the results on the classical types in Section 6.1, this yields a complete description of the J-induction.

Table D.1 J-induction from maximal parabolic subgroups in type H_3

$I_2(5) \subset H_3$:	$J(1) = 1_r$,	$J(\chi_1) = 3'_s$,	$J(\chi_2) = \overline{3}'_s$,	$J(\varepsilon) = 5'_r$

$A_2 \subset H_3$	$J(\varepsilon) = 4'_r + 4_r$
$A_1 \times A_1 \subset H_3$	$J(\varepsilon) = 5_r$

Table D.2 J-induction from maximal parabolic subgroups in type H_4

$H_3 \subset H_4$
$J(1_r) = 1_r, \quad J(3'_s) = 4_t, \quad J(\overline{3}'_s) = \overline{4}_t, \quad J(5_r) = 9_s + \overline{9}_s,$
$J(4'_r) = 16_{rr}, \quad J(4_r) = 16_r, \quad J(5'_r) = 36_{rr}, \quad J(1'_r) = 36'_{rr},$
$J(3_s) = \overline{24}_s + \overline{24}_t + 40_r + 48_{rr} + 30_s + \overline{30}_s + 16_t + 6_s,$
$J(\overline{3}'_s) = 40_r + 48_{rr} + 30_s + \overline{30}_s + 24_t + 24_s + \overline{16}_t + \overline{6}_s$

$I_2(5) \times A_1 \subset H_4$
$J(1, \varepsilon) = 4_t + \overline{4}_t, \quad J(\chi_1, 11) = 9_s, \quad J(\chi_2, \varepsilon) = \overline{9}_s,$
$J(\varepsilon, \varepsilon) = \overline{24}_s + \overline{24}_t + 2 \cdot 40_r + 2 \cdot 48_{rr} + 18_r + 30_s + \overline{30}_s + 24_t + 16_t + 10_r + 24_s + \overline{16}_t$

$A_3 \subset H_4$	$J(\varepsilon) = \overline{24}_s + \overline{24}_t + 40_r + 2 \cdot 48_{rr} + 18_r + 30_s + \overline{30}_s + 24_t + 8_r + 24_s + 8_{rr}$
$A_2 \times A_1 \subset H_4$	$J(\varepsilon) = 25_r$

Table D.3 J-induction from maximal parabolic subgroups in type F_4

$C_3 \subset F_4$
$J(3.\varnothing) = 1_1$, $J(2.1) = 4_2$, $J(21.\varnothing) = 2_1$, $J(\varnothing.3) = 2_3$, $J(1.2) = 9_1$, $J(11.1) = 8_1$, $J(1.11) = 12 + 16 + 9_2 + 6_2$, $J(\varnothing.21) = 16 + 9_3 + 4_1$, $J(111.\varnothing) = 9_2 + 4_3 + 1_2$, $J(\varnothing.111) = 8_4$

$B_3 \subset F_4$
$J(3.\varnothing) = 1_1$, $J(2.1) = 4_2$, $J(21.\varnothing) = 2_3$, $J(\varnothing.3) = 2_1$, $J(1.2) = 9_1$, $J(11.1) = 8_3$, $J(1.11) = 12 + 16 + 6_2 + 9_3$, $J(\varnothing.21) = 16 + 9_2 + 4_1$, $J(111.\varnothing) = 9_3 + 4_4 + 1_3$, $J(\varnothing.111) = 8_2$

$\tilde{A}_2 \times A_1 \subset F_4$	$J(\varepsilon) = 12 + 16 + 9_2 + 6_1 + 4_3$
$A_2 \times \tilde{A}_1 \subset F_4$	$J(\varepsilon) = 12 + 16 + 9_3 + 6_1 + 4_4$

The parabolic subgroups are uniquely determined by giving their type if we also fix a choice of long and short roots. We use the convention of Table 1.2 (p. 20). Thus, a subgroup of type C_3 (or of type B_3) is generated by $\{s_2, s_3, s_4\}$ (or by $\{s_1, s_2, s_3\}$). Moreover, we use the convention that the subgroup of type $\tilde{A}_2 \times A_1$ (or $A_2 \times \tilde{A}_1$) is generated by $\{s_1, s_3, s_4\}$ (or $\{s_1, s_2, s_4\}$).

Table D.4 J-induction from maximal parabolic subgroups in type E_6

$D_5 \subset E_6$
$J(\varnothing.5) = 1_p$, $J(1.4) = 6_p$, $J(2.3) = 20_p$, $J(\varnothing.41) = 20_p$, $J(1.31) = 30_p$, $J(11.3) = 15_p$, $J(\varnothing.32) = 15_q$, $J(2.21) = 64_p$, $J(1.22) = 60_p$, $J(11.21) = 81_p$, $J(\varnothing.311) = 81_p$, $J(1.211) = 80_s + 90_s$, $J(111.2) = 90_s + 20_s$, $J(\varnothing.221) = 60_s$, $J(11.111) = 81'_p$, $J(\varnothing.2111) = 24'_p$, $J(1.1111) = 64'_p$, $J(\varnothing.11111) = 20'_p$

$A_5 \subset E_6$	$J(\varepsilon) = 30'_p + 15'_q$
$A_4 \times A_1 \subset E_6$	$J(\varepsilon) = 60'_p$
$A_2 \times A_2 \times A_1 \subset E_6$	$J(\varepsilon) = 80_s + 60_s + 10_s$

Table D.5 J-induction from maximal parabolic subgroups in type E_7

$E_6 \subset E_7$
$J(1_p) = 1_a$, $J(6_p) = 7'_a$, $J(20_p) = 27_a$, $J(30_p) = 56'_a$, $J(15_q) = 35_b$,
$J(15_p) = 21_a$, $J(64_p) = 120_a + 105'_a$, $J(60_p) = 189'_b$, $J(81_p) = 210_a$,
$J(24_p) = 168_a$, $J(80_s) = 315'_a$, $J(90_s) = 280'_a$, $J(60_s) = 280_b$,
$J(10_s) = 70'_a$, $J(20_s) = 35'_a$, $J(81'_p) = 420_a + 336'_a$, $J(60'_p) = 512'_a + 512_a$,
$J(24'_p) = 105_c$, $J(64'_p) = 420'_a + 336_a$, $J(30'_p) = 405'_a$, $J(15'_q) = 216_a$,
$J(15'_p) = 189'_a$, $J(20'_p) = 189_c$, $J(6'_p) = 120'_a + 105_a$, $J(1'_p) = 21_b$

$D_6 \subset E_7$
$J(\varnothing.6) = 1_a$, $J(1.5) = 7'_a$, $J(2.4) = 27_a$, $J(\varnothing.51) = 27_a$, $J(3.+) = 56'_a + 35_b$,
$J(3.-) = 21'_b$, $J(1.41) = 56'_a$, $J(11.4) = 21_a$, $J(\varnothing.42) = 35_b$,
$J(2.31) = 120_a$, $J(21.3) = 105'_a$, $J(\varnothing.33) = 15'_a$, $J(1.32) = 189'_b$,
$J(11.31) = 210_a$, $J(2.22) = 168_a$, $J(\varnothing.411) = 210_a$, $J(21.-) = 189'_c$,
$J(21.+) = 315'_a + 280_b + 280'_a$, $J(1.311) = 315'_a + 280'_a$, $J(\varnothing.321) = 280_b$,
$J(111.3) = 280'_a + 35'_a$, $J(11.22) = 405_a + 216'_a$, $J(2.211) = 405_a + 189_a$,
$J(1.221) = 378'_a$, $J(11.211) = 420_a$, $J(111.21) = 336'_a$, $J(\varnothing.222) = 84_a$,
$J(\varnothing.3111) = 105_c$, $J(1.2111) = 420'_a$, $J(\varnothing.2211) = 84'_a$, $J(1111.2) = 336_a$,
$J(111.+) = 405'_a + 216_a$, $J(111.-) = 105'_c$, $J(11.1111) = 315_a + 280'_b + 280_a$,
$J(\varnothing.21111) = 189_c$, $J(1.11111) = 210'_a$, $J(\varnothing.111111) = 56_a + 35'_b$

$D_5 \times A_1 \subset E_7$
$J(\varnothing.5, \varepsilon) = 7'_a$, $J(1.4, \varepsilon) = 27_a$, $J(2.3, \varepsilon) = 56'_a + 35_b$, $J(\varnothing.41, \varepsilon) = 21'_b$,
$J(1.31, \varepsilon) = 120_a$, $J(11.3, \varepsilon) = 105'_a$, $J(\varnothing.32, \varepsilon) = 15'_a$, $J(2.21, \varepsilon) = 189'_b$,
$J(1.22, \varepsilon) = 105_b$, $J(11.21, \varepsilon) = 315'_a + 280_b + 280'_a$, $J(\varnothing.311, \varepsilon) = 189'_c$,
$J(1.211, \varepsilon) = 405_a$, $J(111.2, \varepsilon) = 189_a$, $J(\varnothing.221, \varepsilon) = 216'_a$,
$J(11.111, \varepsilon) = 512'_a + 512_a$, $J(\varnothing.2111, \varepsilon) = 420'_a + 84'_a$, $J(1.1111, \varepsilon) = 378_a$,
$J(\varnothing.11111, \varepsilon) = 168'_a$

$A_6 \subset E_7$	$J(\varepsilon) = 105'_b$
$A_5 \times A_1 \subset E_7$	$J(\varepsilon) = 315_a + 280'_b + 70_a$
$A_3 \times A_2 \times A_1 \subset E_7$	$J(\varepsilon) = 210_b$
$A_4 \times A_2 \subset E_7$	$J(\varepsilon) = 210'_b$

Table D.6 J-induction from maximal parabolic subgroups in type E_8

$E_7 \subset E_8$

$J(1_a) = 1_x, \quad J(7'_a) = 8_z, \quad J(27_a) = 35_x, \quad J(56'_a) = 112_z,$
$J(21'_b) = 112_z + 84_x, \quad J(35_b) = 84_x, \quad J(21_a) = 28_x, \quad J(120_a) = 210_x,$
$J(105'_a) = 160_z, \quad J(15'_a) = 50_x, \quad J(189'_b) = 560_z, \quad J(168_a) = 700_x + 300_x,$
$J(105_b) = 700_x + 400_z, \quad J(210_a) = 567_x, \quad J(189'_c) = 1400_z + 1344_x + 1008_z,$
$J(315'_a) = 1400_z, \quad J(280_b) = 1344_x, \quad J(70'_a) = 448_z, \quad J(280'_a) = 1008_z,$
$J(35'_a) = 56_z, \quad J(405_a) = 1400_x + 1575_x, \quad J(216'_a) = 1050_x,$
$J(189_a) = 1575_x + 350_x, \quad J(378'_a) = 3240_z, \quad J(210_b) = 2240_x + 1400_{zz},$
$J(420_a) = 2268_x, \quad J(336'_a) = 1296_z, \quad J(84_a) = 972_x, \quad J(512'_a) = 4096_z,$
$J(512_a) = 4096_x, \quad J(105_c) = 525_x, \quad J(420'_a) = 2800_z, \quad J(210'_b) = 4536_z,$
$J(336_a) = 2100_x, \quad J(84'_a) = 700_{xx}, \quad J(378_a) = 6075_x, \quad J(405'_a) = 5600_z,$
$J(105'_c) = 5600_z + 3200_x, \quad J(216_a) = 3200_x, \quad J(189'_a) = 2400_z,$
$J(315_a) = 4480_y + 7168_w + 4536_y + 5670_y,$
$J(280'_b) = 7168_w + 4200_y + 5600_w + 2688_y, \quad J(70_a) = 3150_y + 2016_w,$
$J(280_a) = 4536_y + 5670_y + 5600_w + 1400_y + 1680_y,$
$J(35_a) = 1680_y + 448_w + 70_y, \quad J(189_c) = 2100_y, \quad J(168'_a) = 5600'_z + 3200'_x,$
$J(210'_a) = 5600'_z + 2400'_z, \quad J(105'_b) = 4200'_z, \quad J(189_b) = 6075'_x,$
$J(120'_a) = 2800'_z, \quad J(105_a) = 2100'_x, \quad J(15_a) = 700'_{xx}, \quad J(56_a) = 2268'_x,$
$J(35'_b) = 972'_x, \quad J(21'_a) = 1296'_z, \quad J(21_b) = 525'_x,$
$J(27'_a) = 1400'_z + 1344'_x + 1008'_z, \quad J(7_a) = 567'_x, \quad J(1'_a) = 112'_z + 84'_x$

$E_6 \times A_1 \subset E_8$

$J(1_p, \varepsilon) = 8_z, \quad J(6_p, \varepsilon) = 35_x, \quad J(20_p, \varepsilon) = 112_z + 84_x, \quad J(30_p, \varepsilon) = 210_x,$
$J(15_q, \varepsilon) = 50_x, \quad J(15_p, \varepsilon) = 160_z, \quad J(64_p, \varepsilon) = 560_z,$
$J(60_p, \varepsilon) = 700_x + 400_z, \quad J(81_p, \varepsilon) = 1400_z + 1344_x + 1008_z,$
$J(24_p, \varepsilon) = 1400_z + 1344_x + 448_z, \quad J(80_s, \varepsilon) = 1400_x, \quad J(90_s, \varepsilon) = 1575_x,$
$J(60_s, \varepsilon) = 1050_x, \quad J(10_s, \varepsilon) = 175_x, \quad J(20_s, \varepsilon) = 350_x,$
$J(81'_p, \varepsilon) = 4096_z + 4096_x, \quad J(60'_p, \varepsilon) = 4200_x + 3360_z,$
$J(24'_p, \varepsilon) = 2800_z + 700_{xx}, \quad J(64'_p, \varepsilon) = 6075_x,$
$J(30'_p, \varepsilon) = 4480_y + 7168_w + 3150_y + 4536_y + 5670_y + 5600_w,$
$J(15'_q, \varepsilon) = 7168_w + 4200_y + 2016_w + 2688_y,$
$J(15'_p, \varepsilon) = 5670_y + 5600_w + 1134_y + 1680_y + 448_w,$
$J(20'_p, \varepsilon) = 5600'_z + 3200'_x, \quad J(6'_p, \varepsilon) = 4096'_x + 4096'_z,$
$J(1'_p, \varepsilon) = 1400'_z + 1344'_x + 448'_z$

Table D.6 J-induction from maximal parabolic subgroups in type E_8 (cont'd)

$D_7 \subset E_8$
$J(\varnothing.7) = 1_x$, $J(1.6) = 8_z$, $J(2.5) = 35_x$, $J(\varnothing.61) = 35_x$, $J(3.4) = 112_z + 84_x$,
$J(1.51) = 112_z$, $J(11.5) = 28_x$, $J(\varnothing.52) = 84_x$, $J(2.41) = 210_x$,
$J(21.4) = 160_z$, $J(\varnothing.43) = 50_x$, $J(1.42) = 560_z$, $J(3.31) = 560_z$,
$J(2.32) = 700_x$, $J(\varnothing.511) = 567_x$, $J(11.41) = 567_x$, $J(22.3) = 300_x$,
$J(1.33) = 400_z$, $J(1.411) = 1400_z + 1008_z$, $J(21.31) = 1400_z + 1344_x + 1008_z$,
$J(\varnothing.421) = 1344_x$, $J(111.4) = 1008_z + 56_z$, $J(11.32) = 1400_x + 1575_x + 1050_x$,
$J(2.311) = 1400_x + 1575_x$, $J(211.3) = 1575_x + 350_x$, $J(\varnothing.331) = 1050_x$,
$J(21.22) = 3240_z$, $J(1.321) = 3240_z$, $J(2.221) = 2240_x + 840_z$,
$J(11.311) = 2268_x$, $J(111.31) = 1296_z$, $J(\varnothing.322) = 972_x$,
$J(21.211) = 4096_z + 4096_x$, $J(\varnothing.4111) = 525_x$, $J(11.221) = 4200_x$,
$J(1.222) = 840_x$, $J(111.22) = 3360_z$, $J(1.3111) = 2800_z$,
$J(\varnothing.3211) = 700_{xx}$, $J(1111.3) = 2100_x$, $J(2.2111) = 6075_x$,
$J(1.2211) = 4200_z$, $J(111.211) = 5600_z + 3200_x$,
$J(11.2111) = 4480_y + 7168_w + 4536_y + 4200_y + 5670_y + 5600_w + 1400_y$,
$J(1111.21) = 7168_w + 4536_y + 5670_y + 5600_w + 2688_y + 1680_y$,
$J(\varnothing.2221) = 4200_y + 1344_w + 168_y$, $J(\varnothing.31111) = 2100_y$, $J(111.1111) = 4200'_z$,
$J(1.21111) = 5600'_z$, $J(11111.2) = 2400'_z$, $J(\varnothing.22111) = 3200'_x$,
$J(11.11111) = 4200'_x + 3360'_z$, $J(\varnothing.211111) = 2268'_x + 972'_x$,
$J(1.111111) = 3240'_z$, $J(\varnothing.1111111) = 700'_x + 400'_z$

$D_5 \times A_2 \subset E_8$
$J(\varnothing.5, \varepsilon) = 112_z + 28_x$, $J(1.4, \varepsilon) = 210_x + 160_z$, $J(\varnothing.41, \varepsilon) = 560_z$,
$J(2.3, \varepsilon) = 560_z$, $J(1.31, \varepsilon) = 700_x$, $J(\varnothing.32, \varepsilon) = 400_z$, $J(11.3, \varepsilon) = 300_x$,
$J(2.21, \varepsilon) = 1400_z + 1344_x + 448_z$, $J(1.22, \varepsilon) = 1400_x + 1050_x + 175_x$,
$J(11.21, \varepsilon) = 3240_z$, $J(\varnothing.311, \varepsilon) = 3240_z$, $J(1.211, \varepsilon) = 2240_x$,
$J(\varnothing.221, \varepsilon) = 1400_{zz}$, $J(111.2, \varepsilon) = 840_z$, $J(11.111, \varepsilon) = 4536_z$,
$J(1.1111, \varepsilon) = 4480_y + 7168_w + 3150_y + 4200_y + 5670_y + 1344_w + 5600_w + 1134_y$,
$J(\varnothing.2111, \varepsilon) = 4200_z$, $J(\varnothing.11111, \varepsilon) = 4536'_z$

$A_7 \subset E_8$	$J(\varepsilon) = 2240'_x + 1400'_{zz}$
$A_6 \times A_1 \subset E_8$	$J(\varepsilon) = 2835'_x$
$A_4 \times A_3 \subset E_8$	$J(\varepsilon) = 4480_y + 7168_w + 3150_y + 4200_y + 1344_w + 2016_w + 420_y$
$A_4 \times A_2 \times A_1 \subset E_8$	$J(\varepsilon) = 2835_x$

E GENERIC DEGREES

In this section, let $\mathbf{H}$ be a generic one-parameter Iwahori–Hecke algebra of a finite Coxeter group $W = \langle S \rangle$, where all parameters u_s $(s \in S)$ are equal to an indeterminate u. Then, by Corollary 9.3.6 and Section 11.4 the *generic degrees* of $\mathbf{H}$ are polynomials in u. We now list the generic degrees explicitly for all exceptional types, as computed by the algorithm described in Section 11.4. Note that we have $a_\chi D_\chi^* = a_{\chi^*} D_\chi$ by Proposition 9.4.1. In each case, we also print the Poincaré polynomial $P_W = \sum_{w \in W} u^{l(w)}$. For type $I_2(m)$ see Theorem 8.3.4.

Table E.1 Cyclotomic polynomials

Φ_1	$u-1$	Φ_{20}	$u^8-u^6+u^4-u^2+1$
Φ_2	$u+1$	Φ_{24}	u^8-u^4+1
Φ_3	u^2+u+1	Φ_{30}	$u^8+u^7-u^5-u^4-u^3+u+1$
Φ_4	u^2+1	$\Phi_{5,a}$	$u^2-\overline{\alpha}u+1$
Φ_5	$u^4+u^3+u^2+u+1$	$\Phi_{5,b}$	$u^2-\alpha u+1$
Φ_6	u^2-u+1	$\Phi_{10,a}$	$u^2+\overline{\alpha}u+1$
Φ_7	$u^6+u^5+u^4+u^3+u^2+u+1$	$\Phi_{10,b}$	$u^2+\alpha u+1$
Φ_8	u^4+1	$\Phi_{15,a}$	$u^4+\overline{\alpha}u^3-\overline{\alpha}u^2+\overline{\alpha}u+1$
Φ_9	u^6+u^3+1	$\Phi_{15,b}$	$u^4+\alpha u^3-\alpha u^2+\alpha u+1$
Φ_{10}	$u^4-u^3+u^2-u+1$	$\Phi_{20,a}$	$u^4+\overline{\alpha}u^2+1$
Φ_{12}	u^4-u^2+1	$\Phi_{20,b}$	$u^4+\alpha u^2+1$
Φ_{14}	$u^6-u^5+u^4-u^3+u^2-u+1$	$\Phi_{30,a}$	$u^4-\overline{\alpha}u^3-\overline{\alpha}u^2-\overline{\alpha}u+1$
Φ_{15}	$u^8-u^7+u^5-u^4+u^3-u+1$	$\Phi_{30,b}$	$u^4-\alpha u^3-\alpha u^2-\alpha u+1$
Φ_{18}	u^6-u^3+1		$\alpha=(-1+\sqrt{5})/2$, $\overline{\alpha}=(-1-\sqrt{5})/2$

(The definition of $\Phi_{5,a}, \Phi_{10,a}, \ldots$ is taken from [Alvis and Lusztig 1982].)

Table E.2 Generic degrees for type H_3

χ	D_χ	deg D_χ	χ^*	a_{χ^*}
1_r	1	0	$1_r'$	15
$3_s'$	$\frac{1}{3+\alpha}u\Phi_3\Phi_6\Phi_{5,b}\Phi_{10,b}$	9	3_s	6
$\overline{3}_s'$	$\frac{1}{3+\overline{\alpha}}u\Phi_3\Phi_6\Phi_{5,a}\Phi_{10,a}$	9	$\overline{3}_s$	6
5_r	$u^2\Phi_5\Phi_{10}$	10	$5_r'$	5
$4_r'$	$\frac{1}{2}u^3\Phi_2^3\Phi_6\Phi_{10}$	12	4_r	3

The Poincaré polynomial is $P_W = \Phi_2^3\Phi_3\Phi_5\Phi_6\Phi_{10}$.

Table E.3 Generic degrees for type H_4

χ	D_χ	deg D_χ	χ^*	a_{χ^*}
1_r	1	0	$1_r'$	60
4_t	$\frac{1}{3+\alpha}u\Phi_4^2\Phi_{12}\Phi_{20}\Phi_{5,a}\Phi_{10,a}\Phi_{15,a}\Phi_{30,a}$	29	$4_t'$	31
$\overline{4}_t$	$\frac{1}{3+\overline{\alpha}}u\Phi_4^2\Phi_{12}\Phi_{20}\Phi_{5,b}\Phi_{10,b}\Phi_{15,b}\Phi_{30,b}$	29	$\overline{4}_t'$	31
9_s	$\frac{1}{3+\alpha}u^2\Phi_3^2\Phi_6^2\Phi_{12}\Phi_{15}\Phi_{30}\Phi_{5,a}\Phi_{10,a}\Phi_{20,b}$	38	$9_s'$	22
$\overline{9}_s$	$\frac{1}{3+\overline{\alpha}}u^2\Phi_3^2\Phi_6^2\Phi_{12}\Phi_{15}\Phi_{30}\Phi_{5,b}\Phi_{10,b}\Phi_{20,a}$	38	$\overline{9}_s'$	22
16_{rr}	$\frac{1}{2}u^3\Phi_2^3\Phi_4^2\Phi_6^2\Phi_{10}^2\Phi_{12}\Phi_{20}\Phi_{30}$	42	$16_{rr}'$	18
16_r	$\frac{1}{2}u^3\Phi_2^3\Phi_4^2\Phi_6^2\Phi_{10}^2\Phi_{12}\Phi_{20}\Phi_{30}$	42	$16_r'$	18
25_r	$u^4\Phi_5^2\Phi_{10}^2\Phi_{15}\Phi_{20}\Phi_{30}$	44	$25_r'$	16
36_{rr}	$u^5\Phi_3^2\Phi_4^2\Phi_6^2\Phi_{12}\Phi_{15}\Phi_{20}\Phi_{30}$	45	$36_{rr}'$	15
$\overline{24}_s$	$\frac{1}{600-960\overline{\alpha}}u^6\Phi_2^4\Phi_3^2\Phi_4^2\Phi_{12}\Phi_{15}\Phi_{20}\Phi_{30}\Phi_{5,a}^2\Phi_{10,b}^2$	54	self-dual	
$\overline{24}_t$	$\frac{1}{60-90\overline{\alpha}}u^6\Phi_2^4\Phi_3^2\Phi_6^2\Phi_{10}^2\Phi_{12}\Phi_{20}\Phi_{30}\Phi_{5,a}^2\Phi_{15,b}$	54	self-dual	
40_r	$\frac{1}{40}u^6\Phi_2^4\Phi_4^2\Phi_5^2\Phi_6^2\Phi_{12}\Phi_{15}\Phi_{20}\Phi_{30}$	54	self-dual	
48_{rr}	$\frac{1}{12}u^6\Phi_2^4\Phi_3^2\Phi_4^2\Phi_6^2\Phi_{10}^2\Phi_{15}\Phi_{20}\Phi_{30}$	54	self-dual	
18_r	$\frac{1}{10}u^6\Phi_3^2\Phi_4^2\Phi_5\Phi_{10}^2\Phi_{12}\Phi_{15}\Phi_{20}\Phi_{30}$	54	self-dual	
30_s	$\frac{1}{30-30\alpha}u^6\Phi_3^2\Phi_4^2\Phi_5^2\Phi_6^2\Phi_{12}\Phi_{20}\Phi_{30}\Phi_{10,a}^2\Phi_{15,b}$	54	self-dual	
$\overline{30}_s$	$\frac{1}{30-30\overline{\alpha}}u^6\Phi_3^2\Phi_4^2\Phi_5^2\Phi_6^2\Phi_{12}\Phi_{20}\Phi_{30}\Phi_{10,b}^2\Phi_{15,a}$	54	self-dual	
24_t	$\frac{1}{60-90\alpha}u^6\Phi_2^4\Phi_3^2\Phi_6^2\Phi_{10}^2\Phi_{12}\Phi_{20}\Phi_{30}\Phi_{5,b}^2\Phi_{15,a}$	54	self-dual	
16_t	$\frac{1}{20}u^6\Phi_2^4\Phi_4^2\Phi_6^2\Phi_{10}^2\Phi_{12}\Phi_{15}\Phi_{30}\Phi_{5,b}^2\Phi_{20,b}$	54	self-dual	
6_s	$\frac{1}{60+30\alpha}u^6\Phi_3^2\Phi_4^2\Phi_6^2\Phi_{10}^2\Phi_{12}\Phi_{15}\Phi_{20}\Phi_{5,b}^2\Phi_{30,a}$	54	self-dual	
8_r	$\frac{1}{8}u^6\Phi_2^4\Phi_4^2\Phi_6^2\Phi_{10}^2\Phi_{12}\Phi_{15}\Phi_{20}\Phi_{30}$	54	self-dual	
10_r	$\frac{1}{10}u^6\Phi_4^2\Phi_5^2\Phi_6^2\Phi_{10}\Phi_{12}\Phi_{15}\Phi_{20}\Phi_{30}$	54	self-dual	
24_s	$\frac{1}{600-960\alpha}u^6\Phi_2^4\Phi_3^2\Phi_4^2\Phi_{12}\Phi_{15}\Phi_{20}\Phi_{30}\Phi_{5,b}^2\Phi_{10,a}^2$	54	self-dual	
$\overline{16}_t$	$\frac{1}{20}u^6\Phi_2^4\Phi_4^2\Phi_6^2\Phi_{10}^2\Phi_{12}\Phi_{15}\Phi_{30}\Phi_{5,a}^2\Phi_{20,a}$	54	self-dual	
8_{rr}	$\frac{1}{10}u^6\Phi_2^4\Phi_5\Phi_6^2\Phi_{10}^2\Phi_{12}\Phi_{15}\Phi_{20}\Phi_{30}$	54	self-dual	
$\overline{6}_s$	$\frac{1}{60+30\overline{\alpha}}u^6\Phi_3^2\Phi_4^2\Phi_6^2\Phi_{10}^2\Phi_{12}\Phi_{15}\Phi_{20}\Phi_{5,a}^2\Phi_{30,b}$	54	self-dual	

The Poincaré polynomial is $P_W = \Phi_2^4\Phi_3^2\Phi_4^2\Phi_5^2\Phi_6^2\Phi_{10}^2\Phi_{12}\Phi_{15}\Phi_{20}\Phi_{30}$.

Table E.4 Generic degrees for type F_4

χ	D_χ	deg D_χ	χ^*	a_{χ^*}
1_1	1	0	1_4	24
4_2	$\frac{1}{2}u\Phi_2^2\Phi_6^2\Phi_8$	11	4_5	13
2_1	$\frac{1}{2}u\Phi_4\Phi_8\Phi_{12}$	11	2_2	13
2_3	$\frac{1}{2}u\Phi_4\Phi_8\Phi_{12}$	11	2_4	13
9_1	$u^2\Phi_3^2\Phi_6^2\Phi_{12}$	14	9_4	10
8_1	$u^3\Phi_4^2\Phi_8\Phi_{12}$	15	8_2	9
8_3	$u^3\Phi_4^2\Phi_8\Phi_{12}$	15	8_4	9
12	$\frac{1}{24}u^4\Phi_2^4\Phi_3^2\Phi_8\Phi_{12}$	20	self-dual	
16	$\frac{1}{4}u^4\Phi_2^4\Phi_4^2\Phi_6^2\Phi_{12}$	20	self-dual	
9_2	$\frac{1}{8}u^4\Phi_3^2\Phi_4^2\Phi_8\Phi_{12}$	20	9_3	4
6_2	$\frac{1}{12}u^4\Phi_3^2\Phi_4^2\Phi_6^2\Phi_8$	20	self-dual	
6_1	$\frac{1}{3}u^4\Phi_3^2\Phi_6^2\Phi_8\Phi_{12}$	20	self-dual	
4_4	$\frac{1}{4}u^4\Phi_2^2\Phi_4\Phi_6^2\Phi_8\Phi_{12}$	20	4_3	4
4_1	$\frac{1}{8}u^4\Phi_2^4\Phi_6^2\Phi_8\Phi_{12}$	20	self-dual	
1_3	$\frac{1}{8}u^4\Phi_4^2\Phi_6^2\Phi_8\Phi_{12}$	20	1_2	4

The Poincaré polynomial is $P_W = \Phi_2^4\Phi_3^2\Phi_4^2\Phi_6^2\Phi_8\Phi_{12}$.

Table E.5 Generic degrees for type E_6

χ	D_χ	deg D_χ	χ^*	a_{χ^*}
1_p	1	0	$1_p'$	36
6_p	$u\Phi_8\Phi_9$	11	$6_p'$	25
20_p	$u^2\Phi_4\Phi_5\Phi_8\Phi_{12}$	16	$20_p'$	20
30_p	$\frac{1}{2}u^3\Phi_4^2\Phi_5\Phi_9\Phi_{12}$	21	$30_p'$	15
15_q	$\frac{1}{2}u^3\Phi_5\Phi_8\Phi_9\Phi_{12}$	21	$15_q'$	15
15_p	$\frac{1}{2}u^3\Phi_5\Phi_6^2\Phi_8\Phi_9$	21	$15_p'$	15
64_p	$u^4\Phi_2^3\Phi_4^2\Phi_6^2\Phi_8\Phi_{12}$	23	$64_p'$	13
60_p	$u^5\Phi_4\Phi_5\Phi_8\Phi_9\Phi_{12}$	25	$60_p'$	11
81_p	$u^6\Phi_3^3\Phi_6^2\Phi_9\Phi_{12}$	26	$81_p'$	10
24_p	$u^6\Phi_4^2\Phi_8\Phi_9\Phi_{12}$	24	$24_p'$	12
80_s	$\frac{1}{6}u^7\Phi_2^4\Phi_5\Phi_8\Phi_9\Phi_{12}$	29	self-dual	
90_s	$\frac{1}{3}u^7\Phi_3^3\Phi_5\Phi_6^2\Phi_8\Phi_{12}$	29	self-dual	
60_s	$\frac{1}{2}u^7\Phi_4^2\Phi_5\Phi_8\Phi_9\Phi_{12}$	29	self-dual	
10_s	$\frac{1}{3}u^7\Phi_5\Phi_6^2\Phi_8\Phi_9\Phi_{12}$	29	self-dual	
20_s	$\frac{1}{6}u^7\Phi_4^2\Phi_5\Phi_6^2\Phi_8\Phi_9$	29	self-dual	

The Poincaré polynomial is $P_W = \Phi_2^4\Phi_3^3\Phi_4^2\Phi_5\Phi_6^2\Phi_8\Phi_9\Phi_{12}$.

Table E.6 Generic degrees for type E_7

χ	D_χ	$\deg D_\chi$	χ^*	a_{χ^*}
1_a	1	0	$1'_a$	63
$7'_a$	$u\Phi_7\Phi_{12}\Phi_{14}$	17	7_a	46
27_a	$u^2\Phi_3^2\Phi_6^2\Phi_9\Phi_{12}\Phi_{18}$	26	$27'_a$	37
$56'_a$	$\frac{1}{2}u^3\Phi_2^4\Phi_6^2\Phi_7\Phi_{10}\Phi_{14}\Phi_{18}$	33	56_a	30
$21'_b$	$u^3\Phi_7\Phi_9\Phi_{14}\Phi_{18}$	27	21_b	36
35_b	$\frac{1}{2}u^3\Phi_5\Phi_7\Phi_8\Phi_{12}\Phi_{14}\Phi_{18}$	33	$35'_b$	30
21_a	$\frac{1}{2}u^3\Phi_7\Phi_8\Phi_9\Phi_{10}\Phi_{12}\Phi_{14}$	33	$21'_a$	30
120_a	$\frac{1}{2}u^4\Phi_2^4\Phi_5\Phi_6^2\Phi_9\Phi_{10}\Phi_{14}\Phi_{18}$	38	$120'_a$	25
$105'_a$	$\frac{1}{2}u^4\Phi_5\Phi_7\Phi_8\Phi_9\Phi_{10}\Phi_{12}\Phi_{18}$	38	105_a	25
$15'_a$	$\frac{1}{2}u^4\Phi_5\Phi_8\Phi_9\Phi_{10}\Phi_{12}\Phi_{14}\Phi_{18}$	38	15_a	25
$189'_b$	$u^5\Phi_3^2\Phi_6^2\Phi_7\Phi_9\Phi_{12}\Phi_{14}\Phi_{18}$	41	189_b	22
168_a	$u^6\Phi_4^2\Phi_7\Phi_8\Phi_9\Phi_{12}\Phi_{14}\Phi_{18}$	42	$168'_a$	21
105_b	$u^6\Phi_5\Phi_7\Phi_9\Phi_{10}\Phi_{12}\Phi_{14}\Phi_{18}$	42	$105'_b$	21
210_a	$u^6\Phi_5\Phi_7\Phi_8\Phi_9\Phi_{10}\Phi_{14}\Phi_{18}$	42	$210'_a$	21
$189'_c$	$u^7\Phi_3^2\Phi_6^2\Phi_7\Phi_9\Phi_{12}\Phi_{14}\Phi_{18}$	43	189_c	20
$315'_a$	$\frac{1}{6}u^7\Phi_3^3\Phi_5\Phi_7\Phi_8\Phi_{10}\Phi_{12}\Phi_{14}\Phi_{18}$	47	315_a	16
280_b	$\frac{1}{2}u^7\Phi_2^4\Phi_5\Phi_6^3\Phi_7\Phi_{10}\Phi_{12}\Phi_{14}\Phi_{18}$	47	$280'_b$	16
$70'_a$	$\frac{1}{3}u^7\Phi_5\Phi_7\Phi_8\Phi_9\Phi_{10}\Phi_{12}\Phi_{14}\Phi_{18}$	47	70_a	16
$280'_a$	$\frac{1}{3}u^7\Phi_4^2\Phi_5\Phi_7\Phi_8\Phi_9\Phi_{10}\Phi_{14}\Phi_{18}$	47	280_a	16
$35'_a$	$\frac{1}{6}u^7\Phi_5\Phi_6^3\Phi_7\Phi_8\Phi_9\Phi_{10}\Phi_{12}\Phi_{14}$	47	35_a	16
405_a	$\frac{1}{2}u^8\Phi_3^3\Phi_5\Phi_6^2\Phi_8\Phi_9\Phi_{12}\Phi_{14}\Phi_{18}$	48	$405'_a$	15
$216'_a$	$\frac{1}{2}u^8\Phi_2^4\Phi_3^2\Phi_6^3\Phi_9\Phi_{10}\Phi_{12}\Phi_{14}\Phi_{18}$	48	216_a	15
189_a	$\frac{1}{2}u^8\Phi_3^2\Phi_6^3\Phi_7\Phi_8\Phi_9\Phi_{10}\Phi_{12}\Phi_{18}$	48	$189'_a$	15
$378'_a$	$u^9\Phi_3^2\Phi_6^2\Phi_7\Phi_8\Phi_9\Phi_{12}\Phi_{14}\Phi_{18}$	49	378_a	14
210_b	$u^{10}\Phi_5\Phi_7\Phi_8\Phi_9\Phi_{10}\Phi_{12}\Phi_{14}\Phi_{18}$	50	$210'_b$	13
420_a	$\frac{1}{2}u^{10}\Phi_4^2\Phi_5\Phi_7\Phi_8\Phi_9\Phi_{12}\Phi_{14}\Phi_{18}$	50	$420'_a$	13
$336'_a$	$\frac{1}{2}u^{10}\Phi_2^4\Phi_6^2\Phi_7\Phi_8\Phi_9\Phi_{10}\Phi_{14}\Phi_{18}$	50	336_a	13
84_a	$\frac{1}{2}u^{10}\Phi_4^2\Phi_7\Phi_8\Phi_9\Phi_{10}\Phi_{12}\Phi_{14}\Phi_{18}$	50	$84'_a$	13
$512'_a$	$\frac{1}{2}u^{11}\Phi_2^7\Phi_4^2\Phi_6^3\Phi_8\Phi_{10}\Phi_{12}\Phi_{14}\Phi_{18}$	52	512_a	11
105_c	$u^{12}\Phi_5\Phi_7\Phi_9\Phi_{10}\Phi_{12}\Phi_{14}\Phi_{18}$	48	$105'_c$	15

The Poincaré polynomial is $P_W = \Phi_2^7\Phi_3^3\Phi_4^2\Phi_5\Phi_6^3\Phi_7\Phi_8\Phi_9\Phi_{10}\Phi_{12}\Phi_{14}\Phi_{18}$.

Table E.7 Generic degrees for type E_8

χ	D_χ	$\deg D_\chi$	χ^*	a_{χ^*}
1_x	1	0	$1'_x$	120
8_z	$u\Phi_4^2\Phi_8\Phi_{12}\Phi_{20}\Phi_{24}$	29	$8'_z$	91
35_x	$u^2\Phi_5\Phi_7\Phi_{10}\Phi_{14}\Phi_{15}\Phi_{20}\Phi_{30}$	46	$35'_x$	74
112_z	$\frac{1}{2}u^3\Phi_2^4\Phi_6^2\Phi_7\Phi_8\Phi_{10}^2\Phi_{14}\Phi_{18}\Phi_{24}\Phi_{30}$	57	$112'_z$	63
84_x	$\frac{1}{2}u^3\Phi_4^2\Phi_7\Phi_8\Phi_9\Phi_{12}\Phi_{14}\Phi_{20}\Phi_{24}\Phi_{30}$	57	$84'_x$	63
28_x	$\frac{1}{2}u^3\Phi_4^2\Phi_7\Phi_8\Phi_{12}\Phi_{14}\Phi_{15}\Phi_{18}\Phi_{20}\Phi_{24}$	57	$28'_x$	63
210_x	$\frac{1}{2}u^4\Phi_5\Phi_7\Phi_8^2\Phi_9\Phi_{10}^2\Phi_{15}\Phi_{20}\Phi_{24}\Phi_{30}$	68	$210'_x$	52
160_z	$\frac{1}{2}u^4\Phi_2^4\Phi_4^2\Phi_5\Phi_6^2\Phi_{10}^2\Phi_{12}\Phi_{14}\Phi_{15}\Phi_{18}\Phi_{20}\Phi_{30}$	68	$160'_z$	52
50_x	$\frac{1}{2}u^4\Phi_5^2\Phi_8^2\Phi_{10}\Phi_{14}\Phi_{15}\Phi_{18}\Phi_{20}\Phi_{24}\Phi_{30}$	68	$50'_x$	52
560_z	$u^5\Phi_4^2\Phi_5\Phi_7\Phi_8^2\Phi_{10}\Phi_{12}\Phi_{14}\Phi_{15}\Phi_{20}\Phi_{24}\Phi_{30}$	73	$560'_z$	47
567_x	$u^6\Phi_3^3\Phi_6^3\Phi_7\Phi_9\Phi_{12}^2\Phi_{14}\Phi_{15}\Phi_{18}\Phi_{24}\Phi_{30}$	74	$567'_x$	46
700_x	$\frac{1}{2}u^6\Phi_4^2\Phi_5^2\Phi_7\Phi_8\Phi_{10}^2\Phi_{12}\Phi_{15}\Phi_{18}\Phi_{20}\Phi_{24}\Phi_{30}$	78	$700'_x$	42
400_z	$\frac{1}{2}u^6\Phi_2^4\Phi_5^2\Phi_6^2\Phi_8\Phi_{10}^2\Phi_{14}\Phi_{15}\Phi_{18}\Phi_{20}\Phi_{24}\Phi_{30}$	78	$400'_z$	42
300_x	$\frac{1}{2}u^6\Phi_4^2\Phi_5^2\Phi_8\Phi_9\Phi_{10}^2\Phi_{12}\Phi_{14}\Phi_{15}\Phi_{20}\Phi_{24}\Phi_{30}$	78	$300'_x$	42
1400_z	$\frac{1}{6}u^7\Phi_4^2\Phi_5^2\Phi_7\Phi_8^2\Phi_9\Phi_{12}^2\Phi_{14}\Phi_{18}\Phi_{20}\Phi_{24}\Phi_{30}$	83	$1400'_z$	37
1344_x	$\frac{1}{2}u^7\Phi_2^4\Phi_4^2\Phi_6^2\Phi_7\Phi_8\Phi_9\Phi_{10}^2\Phi_{12}\Phi_{14}\Phi_{18}\Phi_{20}\Phi_{24}\Phi_{30}$	83	$1344'_x$	37
448_z	$\frac{1}{3}u^7\Phi_4^4\Phi_7\Phi_8^2\Phi_9\Phi_{12}\Phi_{14}\Phi_{15}\Phi_{18}\Phi_{20}\Phi_{24}\Phi_{30}$	83	$448'_z$	37
1008_z	$\frac{1}{3}u^7\Phi_3^3\Phi_4^2\Phi_6^3\Phi_7\Phi_8^2\Phi_{12}^2\Phi_{14}\Phi_{15}\Phi_{20}\Phi_{24}\Phi_{30}$	83	$1008'_z$	37
56_z	$\frac{1}{6}u^7\Phi_4^2\Phi_7\Phi_8^2\Phi_9\Phi_{10}^2\Phi_{12}^2\Phi_{14}\Phi_{15}\Phi_{18}\Phi_{20}\Phi_{24}$	83	$56'_z$	37
1400_x	$\frac{1}{6}u^8\Phi_4^4\Phi_5^2\Phi_7\Phi_9\Phi_{10}^2\Phi_{14}\Phi_{15}\Phi_{18}\Phi_{20}\Phi_{24}\Phi_{30}$	88	$1400'_x$	32
1575_x	$\frac{1}{3}u^8\Phi_3^3\Phi_5^2\Phi_6^3\Phi_7\Phi_{10}^2\Phi_{12}^2\Phi_{14}\Phi_{15}\Phi_{20}\Phi_{24}\Phi_{30}$	88	$1575'_x$	32
1050_x	$\frac{1}{2}u^8\Phi_5^2\Phi_7\Phi_8^2\Phi_9\Phi_{10}^2\Phi_{14}\Phi_{15}\Phi_{18}\Phi_{20}\Phi_{24}\Phi_{30}$	88	$1050'_x$	32
175_x	$\frac{1}{3}u^8\Phi_5^2\Phi_7\Phi_9\Phi_{10}^2\Phi_{12}^2\Phi_{14}\Phi_{15}\Phi_{18}\Phi_{20}\Phi_{24}\Phi_{30}$	88	$175'_x$	32
350_x	$\frac{1}{6}u^8\Phi_5^2\Phi_7\Phi_8^2\Phi_9\Phi_{10}^2\Phi_{12}^2\Phi_{14}\Phi_{15}\Phi_{18}\Phi_{20}\Phi_{30}$	88	$350'_x$	32
3240_z	$u^9\Phi_3^3\Phi_4^2\Phi_5\Phi_6^3\Phi_8\Phi_9\Phi_{10}\Phi_{12}^2\Phi_{15}\Phi_{18}\Phi_{20}\Phi_{24}\Phi_{30}$	89	$3240'_z$	31
2268_x	$\frac{1}{2}u^{10}\Phi_3^3\Phi_4^2\Phi_6^4\Phi_7\Phi_8\Phi_9\Phi_{12}^2\Phi_{15}\Phi_{18}\Phi_{20}\Phi_{24}\Phi_{30}$	90	$2268'_x$	30
2240_x	$\frac{1}{2}u^{10}\Phi_2^4\Phi_4^2\Phi_5\Phi_6^2\Phi_7\Phi_8\Phi_{10}^2\Phi_{12}\Phi_{14}\Phi_{15}\Phi_{18}\Phi_{20}\Phi_{24}\Phi_{30}$	92	$2240'_x$	28
1400_{zz}	$\frac{1}{2}u^{10}\Phi_4^2\Phi_5^2\Phi_7\Phi_8^2\Phi_{10}\Phi_{12}^2\Phi_{14}\Phi_{15}\Phi_{18}\Phi_{20}\Phi_{24}\Phi_{30}$	92	$1400'_{zz}$	28
972_x	$\frac{1}{2}u^{10}\Phi_3^4\Phi_4^2\Phi_6^3\Phi_8\Phi_9\Phi_{12}^2\Phi_{14}\Phi_{15}\Phi_{18}\Phi_{20}\Phi_{24}\Phi_{30}$	90	$972'_x$	30
840_z	$\frac{1}{2}u^{10}\Phi_4^2\Phi_5\Phi_7\Phi_8^2\Phi_9\Phi_{10}^2\Phi_{12}^2\Phi_{14}\Phi_{15}\Phi_{20}\Phi_{24}\Phi_{30}$	92	$840'_z$	28
1296_z	$\frac{1}{2}u^{10}\Phi_2^4\Phi_3^3\Phi_6^4\Phi_8\Phi_9\Phi_{10}^2\Phi_{12}^2\Phi_{14}\Phi_{15}\Phi_{18}\Phi_{24}\Phi_{30}$	90	$1296'_z$	30
4096_z	$\frac{1}{2}u^{11}\Phi_2^7\Phi_4^4\Phi_6^4\Phi_8^2\Phi_{10}^2\Phi_{12}^2\Phi_{14}\Phi_{18}\Phi_{20}\Phi_{24}\Phi_{30}$	94	$4096'_z$	26
4096_x	$\frac{1}{2}u^{11}\Phi_2^7\Phi_4^4\Phi_6^4\Phi_8^2\Phi_{10}^2\Phi_{12}^2\Phi_{14}\Phi_{18}\Phi_{20}\Phi_{24}\Phi_{30}$	94	$4096'_x$	26

Table E.7 Generic degrees for type E_8 (continued)

χ	D_χ	deg D_χ	χ^*	a_{χ^*}
525_x	$u^{12}\Phi_5^2\Phi_7\Phi_9\Phi_{10}^2\Phi_{14}\Phi_{15}\Phi_{18}\Phi_{20}\Phi_{24}\Phi_{30}$	84	$525'_x$	36
4200_x	$\frac{1}{2}u^{12}\Phi_4^2\Phi_5^2\Phi_7\Phi_8^2\Phi_9\Phi_{10}\Phi_{12}\Phi_{14}\Phi_{15}\Phi_{18}\Phi_{20}\Phi_{24}\Phi_{30}$	96	$4200'_x$	24
3360_z	$\frac{1}{2}u^{12}\Phi_2^4\Phi_5\Phi_6^2\Phi_7\Phi_8^2\Phi_9\Phi_{10}^2\Phi_{14}\Phi_{15}\Phi_{18}\Phi_{20}\Phi_{24}\Phi_{30}$	96	$3360'_z$	24
840_x	$\frac{1}{2}u^{12}\Phi_4^2\Phi_5\Phi_7\Phi_8^2\Phi_9\Phi_{10}^2\Phi_{12}\Phi_{14}\Phi_{15}\Phi_{18}\Phi_{20}\Phi_{24}\Phi_{30}$	96	$840'_x$	24
4536_z	$u^{13}\Phi_3^3\Phi_4^2\Phi_6^3\Phi_7\Phi_8\Phi_9\Phi_{12}^2\Phi_{14}\Phi_{15}\Phi_{18}\Phi_{20}\Phi_{24}\Phi_{30}$	97	$4536'_z$	23
2800_z	$\frac{1}{2}u^{13}\Phi_2^4\Phi_5^2\Phi_6^4\Phi_7\Phi_8\Phi_{10}^2\Phi_{14}\Phi_{15}\Phi_{18}\Phi_{20}\Phi_{24}\Phi_{30}$	95	$2800'_z$	25
700_{xx}	$\frac{1}{2}u^{13}\Phi_4^2\Phi_5^2\Phi_7\Phi_8\Phi_{10}^2\Phi_{12}^2\Phi_{14}\Phi_{15}\Phi_{18}\Phi_{20}\Phi_{24}\Phi_{30}$	95	$700'_{xx}$	25
2100_x	$\frac{1}{2}u^{13}\Phi_4^2\Phi_5^2\Phi_7\Phi_8\Phi_9\Phi_{10}^2\Phi_{12}^2\Phi_{14}\Phi_{15}\Phi_{20}\Phi_{24}\Phi_{30}$	95	$2100'_x$	25
2835_x	$u^{14}\Phi_3^3\Phi_5\Phi_6^3\Phi_7\Phi_9\Phi_{10}\Phi_{12}^2\Phi_{14}\Phi_{15}\Phi_{18}\Phi_{20}\Phi_{24}\Phi_{30}$	98	$2835'_x$	22
6075_x	$u^{14}\Phi_3^4\Phi_5^2\Phi_6^4\Phi_9\Phi_{10}^2\Phi_{12}^2\Phi_{15}\Phi_{18}\Phi_{20}\Phi_{24}\Phi_{30}$	98	$6075'_x$	22
5600_z	$\frac{1}{2}u^{15}\Phi_4^4\Phi_5^2\Phi_7\Phi_8^2\Phi_{10}^2\Phi_{12}^2\Phi_{15}\Phi_{18}\Phi_{20}\Phi_{24}\Phi_{30}$	99	$5600'_z$	21
4200_z	$u^{15}\Phi_4^2\Phi_5^2\Phi_7\Phi_8\Phi_9\Phi_{10}^2\Phi_{12}\Phi_{14}\Phi_{15}\Phi_{18}\Phi_{20}\Phi_{24}\Phi_{30}$	99	$4200'_z$	21
3200_x	$\frac{1}{2}u^{15}\Phi_2^4\Phi_4^2\Phi_5^2\Phi_6^2\Phi_8^2\Phi_{10}^2\Phi_{12}\Phi_{14}\Phi_{15}\Phi_{18}\Phi_{20}\Phi_{24}\Phi_{30}$	99	$3200'_x$	21
2400_z	$\frac{1}{2}u^{15}\Phi_4^4\Phi_5^2\Phi_8^2\Phi_9\Phi_{10}^2\Phi_{12}^2\Phi_{14}\Phi_{15}\Phi_{20}\Phi_{24}\Phi_{30}$	99	$2400'_z$	21
4480_y	$\frac{1}{120}u^{16}\Phi_2^8\Phi_5^2\Phi_7\Phi_8^2\Phi_9\Phi_{12}^2\Phi_{14}\Phi_{15}\Phi_{18}\Phi_{20}\Phi_{24}\Phi_{30}$	104	self-dual	
7168_w	$\frac{1}{12}u^{16}\Phi_2^8\Phi_4^4\Phi_6^2\Phi_7\Phi_9\Phi_{10}^2\Phi_{12}\Phi_{14}\Phi_{15}\Phi_{18}\Phi_{20}\Phi_{24}\Phi_{30}$	104	self-dual	
3150_y	$\frac{1}{6}u^{16}\Phi_3^3\Phi_5^2\Phi_6^4\Phi_7\Phi_8^2\Phi_{12}^2\Phi_{14}\Phi_{15}\Phi_{18}\Phi_{20}\Phi_{24}\Phi_{30}$	104	self-dual	
4536_y	$\frac{1}{24}u^{16}\Phi_3^4\Phi_4^4\Phi_7\Phi_8^2\Phi_9\Phi_{10}^2\Phi_{14}\Phi_{15}\Phi_{18}\Phi_{20}\Phi_{24}\Phi_{30}$	104	self-dual	
4200_y	$\frac{1}{8}u^{16}\Phi_4^4\Phi_5^2\Phi_7\Phi_8^2\Phi_9\Phi_{12}^2\Phi_{14}\Phi_{15}\Phi_{18}\Phi_{20}\Phi_{24}\Phi_{30}$	104	self-dual	
5670_y	$\frac{1}{30}u^{16}\Phi_3^4\Phi_5^2\Phi_7\Phi_8^2\Phi_9\Phi_{10}^2\Phi_{12}^2\Phi_{14}\Phi_{18}\Phi_{20}\Phi_{24}\Phi_{30}$	104	self-dual	
1344_w	$\frac{1}{4}u^{16}\Phi_2^4\Phi_4^2\Phi_6^2\Phi_7\Phi_8^2\Phi_9\Phi_{10}^2\Phi_{12}\Phi_{14}\Phi_{15}\Phi_{18}\Phi_{20}\Phi_{24}\Phi_{30}$	104	self-dual	
2016_w	$\frac{1}{6}u^{16}\Phi_2^4\Phi_3^3\Phi_4^2\Phi_6^4\Phi_7\Phi_{10}^2\Phi_{12}^2\Phi_{14}\Phi_{15}\Phi_{18}\Phi_{20}\Phi_{24}\Phi_{30}$	104	self-dual	
5600_w	$\frac{1}{6}u^{16}\Phi_2^4\Phi_4^2\Phi_5^2\Phi_6^4\Phi_7\Phi_9\Phi_{10}^2\Phi_{12}^2\Phi_{14}\Phi_{18}\Phi_{20}\Phi_{24}\Phi_{30}$	104	self-dual	
420_y	$\frac{1}{5}u^{16}\Phi_5^2\Phi_7\Phi_8^2\Phi_9\Phi_{10}^2\Phi_{12}^2\Phi_{14}\Phi_{15}\Phi_{18}\Phi_{20}\Phi_{24}\Phi_{30}$	104	self-dual	
1134_y	$\frac{1}{6}u^{16}\Phi_3^4\Phi_6^3\Phi_7\Phi_8^2\Phi_9\Phi_{10}^2\Phi_{12}^2\Phi_{14}\Phi_{15}\Phi_{20}\Phi_{24}\Phi_{30}$	104	self-dual	
1400_y	$\frac{1}{24}u^{16}\Phi_4^4\Phi_5^2\Phi_6^4\Phi_7\Phi_8^2\Phi_9\Phi_{14}\Phi_{15}\Phi_{18}\Phi_{20}\Phi_{24}\Phi_{30}$	104	self-dual	
2688_y	$\frac{1}{8}u^{16}\Phi_2^8\Phi_6^4\Phi_7\Phi_8^2\Phi_9\Phi_{10}^2\Phi_{14}\Phi_{15}\Phi_{18}\Phi_{20}\Phi_{24}\Phi_{30}$	104	self-dual	
1680_y	$\frac{1}{20}u^{16}\Phi_4^4\Phi_5^2\Phi_7\Phi_8^2\Phi_9\Phi_{10}^2\Phi_{12}^2\Phi_{14}\Phi_{15}\Phi_{18}\Phi_{24}\Phi_{30}$	104	self-dual	
168_y	$\frac{1}{8}u^{16}\Phi_4^4\Phi_7\Phi_8^2\Phi_9\Phi_{10}^2\Phi_{12}^2\Phi_{14}\Phi_{15}\Phi_{18}\Phi_{20}\Phi_{24}\Phi_{30}$	104	self-dual	
448_w	$\frac{1}{12}u^{16}\Phi_2^4\Phi_4^2\Phi_6^4\Phi_7\Phi_8^2\Phi_9\Phi_{10}^2\Phi_{12}^2\Phi_{14}\Phi_{15}\Phi_{18}\Phi_{20}\Phi_{30}$	104	self-dual	
70_y	$\frac{1}{30}u^{16}\Phi_5^2\Phi_6^4\Phi_7\Phi_8^2\Phi_9\Phi_{10}^2\Phi_{12}^2\Phi_{14}\Phi_{15}\Phi_{18}\Phi_{20}\Phi_{24}$	104	self-dual	
2100_y	$u^{20}\Phi_5^2\Phi_7\Phi_8^2\Phi_9\Phi_{10}^2\Phi_{14}\Phi_{15}\Phi_{18}\Phi_{20}\Phi_{24}\Phi_{30}$	100	self-dual	

$$P_W = \Phi_2^8\Phi_3^4\Phi_4^4\Phi_5^2\Phi_6^4\Phi_7\Phi_8^2\Phi_9\Phi_{10}^2\Phi_{12}^2\Phi_{14}\Phi_{15}\Phi_{18}\Phi_{20}\Phi_{24}\Phi_{30}.$$

F BLOCKS AT ROOTS OF UNITY

Let $\mathbf{H}$ be a generic one-parameter Iwahori–Hecke algebra associated with a finite Coxeter system (W, S), where all parameters u_s $(s \in S)$ are equal to v^2, for an indeterminate v. As in (11.5.12), let $e \geq 1$ and consider a specialization $\theta_e\colon v \mapsto \zeta_{2e}$ where $\zeta_{2e} = \exp(\pi\sqrt{-1}/e) \in \mathbb{C}$ is a primitive 2eth root of unity. The corresponding blocks will be called Φ_e*-blocks* of $\mathbf{H}$. In this section, we list the distribution of the irreducible characters of $\mathbf{KH}$ into Φ_e-blocks for all exceptional types and all e such that the specialized algebra is not semisimple.

In the tables below, we use the following conventions. The characters inside a block are ordered according to increasing a-invariant (see Definition 6.5.7). Furthermore, we only print the characters which belong to blocks of positive defect. (By Theorem 7.5.11, a character of defect 0 forms a block by itself.) Finally, if $\{\chi_1, \ldots, \chi_m\}$ is a subset of $\mathrm{Irr}(\mathbf{KH})$ which forms a Φ_e-block, then the characters $\{\chi_1^*, \ldots, \chi_m^*\}$ also form a Φ_e-block, where χ_i^* are the dual characters as in Proposition 9.4.1. (This follows from the fact that the duality operation is defined using an algebra automorphism.) We will only print explicitly one of the sets $\{\chi_i\}$ or $\{\chi_i^*\}$.

F.1. Remark. Consider a subset $\{\chi_1, \ldots, \chi_m\}$ of $\mathrm{Irr}(\mathbf{KH})$ which forms a Φ_e-block of defect 1. For each i, let a_i be Lusztig's a-invariant for χ_i; see Definition 6.5.7. Then the tables below show that we have $a_i \neq a_j$ for $i \neq j$. So we can choose the labelling such that $a_1 < a_2 < \ldots < a_m$.

Now assume first that (W, S) is of crystallographic type. Then, by [Geck 1992, Theorem 9.6], all decomposition numbers for the characters χ_i are 0 or 1 and the corresponding *Brauer graph* is a straight line. Furthermore, by [Geck 1998a, Theorem 4.4], the above ordering of the characters also determines the ordering of the characters on that straight line. Thus, the Brauer graph is given by

This is proved by a general argument using Lusztig's asymptotic algebra J (see [Lusztig 1987a]) and requires the deep positivity results which are mentioned in the bibliographical remarks to Chapter 11. The analogous results for blocks of defect 1 for the non-crystallographic types $I_2(m)$, H_3 and H_4 are established in [Müller 1997], by explicit computation of the decomposition numbers. Thus, the tables below already determine the decomposition matrices for all blocks of defect 1.

F.2. Remark. By inspection of the tables, we obtain the following result. Assume (W, S) is of exceptional type H_3, H_4, F_4, E_6, E_7 or E_8. Let $e \geq 1$ and assume that $\chi, \chi' \in \mathrm{Irr}(\mathbf{KH})$ belong to the same block with respect to θ_e, i.e., we have $\theta_e(\omega_\chi(z)) = \theta_e(\omega_{\chi'}(z))$ for all $z \in Z(\mathbf{H})$. Then χ, χ' have the same defect, i.e., the maximal power of $v - \zeta_{2e}$ dividing c_χ is the same as the maximal power of $v - \zeta_{2e}$ dividing $c_{\chi'}$. A conceptual proof of this fact for the crystallographic types is given in [Geck 1992, Prop. 7.4].

Table F.1 Blocks of positive defect of $\mathsf{H}(\mathsf{H}_3)$ at roots of unity

e	d	Φ_e-blocks of characters of defect d
2	3	$\{1_r, \overline{3}'_s, 3'_s, 5_r, 5'_r, \overline{3}_s, 3_s, 1'_r\}$
3	1	$\{1_r, 5_r, 4'_r\}$ + dual
5	1	$\{1_r, 4'_r, \overline{3}_s\}$ + dual
6	1	$\{1_r, 5_r, 5'_r, 1'_r\}$
10	1	$\{1_r, 3'_s, 3_s, 1'_r\}$

Table F.2 Blocks of positive defect of $\mathsf{H}(\mathsf{H}_4)$ at roots of unity

e	d	Φ_e-blocks of characters of defect d
2	1	$\{16_r, 16'_{rr}\}$ + dual
	4	$\{1_r, \overline{4}_t, 4_t, 9_s, \overline{9}_s, 25_r, 36_{rr}, 6_s, 10_r, 30_s, 18_r, \overline{6}_s, \overline{30}_s, 36'_{rr}, 25'_r, \overline{9}'_s, 9'_s, 4'_t, \overline{4}'_t, 1'_r\}$
3	2	$\{1_r, 16_{rr}, 25_r, 40_r, 10_r, 8_r, 25'_r, 16'_{rr}, 1'_r\}$,
		$\{\overline{4}_t, 4_t, 16_r, \overline{16}_t, 16_t, 8_{rr}, 16'_r, \overline{4}'_t, 4'_t\}$
4	2	$\{1_r, \overline{9}_s, 9_s, 25_r, \overline{24}_t, 24_t, 8_{rr}, 25'_r, \overline{9}'_s, 9'_s, 1'_r\}$
5	1	$\{\overline{9}_s, 18_r, \overline{9}'_s\}$, $\{\overline{4}_t, 8_{rr}, \overline{4}'_t\}$
	2	$\{1_r, 9_s, 16_{rr}, \overline{24}_s, 8_r, \overline{6}_s, 16'_{rr}, 9'_s, 1'_r\}$,
		$\{4_t, 16_r, 36_{rr}, 48_{rr}, \overline{24}_t, \overline{16}_t, 36'_{rr}, 16'_r, 4'_t\}$
6	2	$\{1_r, \overline{4}_t, 4_t, 25_r, \overline{24}_s, 24_s, 18_r, 25'_r, \overline{4}'_t, 4'_t, 1'_r\}$
10	1	$\{4_t, 9_s, 10_r, 9'_s, 4'_t\}$
	2	$\{1_r, \overline{4}_t, \overline{9}_s, 36_{rr}, 40_r, \overline{30}_s, \overline{24}_s, 36'_{rr}, \overline{9}'_s, \overline{4}'_t, 1'_r\}$
12	1	$\{1_r, 25_r, 48_{rr}, 25'_r, 1'_r\}$
15	1	$\{1_r, 16_{rr}, 30_s, 16'_{rr}, 1'_r\}$, $\{\overline{4}_t, 16_r, \overline{24}_t, 16'_r, \overline{4}'_t\}$
20	1	$\{1_r, 9_s, 16_t, 9'_s, 1'_r\}$
30	1	$\{1_r, 4_t, 6_s, 4'_t, 1'_r\}$

Table F.3 Blocks of positive defect of $H(F_4)$ at roots of unity

e	d	Φ_e-blocks of characters of defect d
2	2	$\{4_2, 4_3, 4_4, 4_5\}$
	4	$\{1_1, 2_3, 2_1, 9_1, 8_1, 8_3, 1_3, 6_1, 1_2, 9_3, 6_2, 9_2, 8_2, 8_4, 9_4, 2_4, 2_2, 1_4\}$
3	2	$\{1_1, 2_1, 2_3, 1_2, 1_3, 4_1, 2_2, 2_4, 1_4\}$, $\{4_2, 8_1, 8_3, 4_3, 4_4, 16, 8_2, 8_4, 4_5\}$
4	1	$\{2_1, 4_3, 2_4\}$ + dual
	2	$\{1_1, 4_2, 9_1, 4_1, 6_1, 12, 9_4, 4_5, 1_4\}$
6	2	$\{1_1, 2_1, 2_3, 8_1, 8_3, 9_2, 9_3, 12, 8_2, 8_4, 2_2, 2_4, 1_4\}$
8	1	$\{1_1, 9_1, 16, 9_4, 1_4\}$
12	1	$\{1_1, 4_2, 6_2, 4_5, 1_4\}$

Table F.4 Blocks of positive defect of $H(E_6)$ at roots of unity

e	d	Φ_e-blocks of characters of defect d
2	1	$\{64_p, 64'_p\}$
	4	$\{1_p, 6_p, 20_p, 15_p, 30_p, 15_q, 60_p, 24_p, 81_p, 10_s, 60_s, 20_s, 90_s, 81'_p, 60'_p, 24'_p, 15'_q, 15'_p, 30'_p, 20'_p, 6'_p, 1'_p\}$
3	3	$\{1_p, 6_p, 20_p, 15_p, 30_p, 15_q, 64_p, 60_p, 24_p, 10_s, 60_s, 20_s, 80_s, 60'_p, 24'_p, 64'_p, 15'_q, 15'_p, 30'_p, 20'_p, 6'_p, 1'_p\}$
4	1	$\{20_p, 60_p, 60'_p, 20'_p\}$
	2	$\{1_p, 6_p, 15_q, 15_p, 81_p, 90_s, 80_s, 10_s, 81'_p, 15'_q, 15'_p, 6'_p, 1'_p\}$
5	1	$\{1_p, 24_p, 81'_p, 64'_p, 6'_p\}$ + dual
6	2	$\{1_p, 6_p, 20_p, 30_p, 15_q, 60_p, 24_p, 60_s, 80_s, 60'_p, 24'_p, 15'_q, 30'_p, 20'_p, 6'_p, 1'_p\}$
8	1	$\{1_p, 30_p, 81_p, 81'_p, 30'_p, 1'_p\}$
9	1	$\{1_p, 20_p, 64_p, 90_s, 64'_p, 20'_p, 1'_p\}$
12	1	$\{1_p, 6_p, 15_p, 20_s, 15'_p, 6'_p, 1'_p\}$

Table F.5 Blocks of positive defect of $\mathsf{H}(\mathsf{E}_7)$ at roots of unity

e	d	Φ_e-blocks of characters of defect d
2	3	$\{56'_a, 120_a, 280_b, 216'_a, 336'_a, 336_a, 216_a, 280'_b, 120'_a, 56_a\}$
	7	$\{1_a, 7'_a, 27_a, 21_a, 35_b, 21'_b, 15'_a, 105'_a, 189'_b, 168_a, 105_b, 210_a, 315'_a, 280'_a,$
		$189'_c, 70'_a, 35'_a, 189_a, 405_a, 378'_a, 210_b, 420_a, 84_a, 105_c, 84'_a, 420'_a, 210'_b,$
		$378_a, 105'_c, 405'_a, 189'_a, 315_a, 35_a, 70_a, 280_a, 189_c, 168'_a, 105'_b, 210'_a,$
		$189_b, 15_a, 105_a, 21'_a, 35'_b, 21_b, 27'_a, 7_a, 1'_a\}$
3	1	$\{189'_c, 216'_a, 27'_a\}$, $\{189'_b, 378'_a, 189'_a\}$ + duals
	3	$\{1_a, 21_a, 35_b, 120_a, 105_b, 210_a, 168_a, 280_b, 210_b, 84_a, 420_a, 512'_a, 105_c,$
		$336_a, 35_a, 70_a, 280_a, 15_a, 105_a, 56_a, 21_b, 7_a\}$ + dual
4	2	$\{1_a, 56'_a, 210_a, 105_b, 405_a, 189_a, 336'_a, 315_a, 70_a, 35_a, 189_b, 120'_a, 21_b\}$,
		$\{7'_a, 105'_a, 15'_a, 280_b, 189'_c, 378'_a, 210'_b, 216_a, 105'_c, 35'_b, 21'_a, 27'_a\}$ + duals
5	1	$\{1_a, 84_a, 216_a, 189_b, 56_a\}$, $\{7'_a, 378'_a, 512_a, 168'_a, 27'_a\}$,
		$\{21'_b, 189'_c, 336'_a, 189'_a, 21'_a\}$ + duals
6	1	$\{189'_c, 405_a, 378_a, 189_b, 27'_a\}$ + dual,
		$\{56'_a, 120_a, 336'_a, 336_a, 120'_a, 56_a\}$
	3	$\{1_a, 7'_a, 21_a, 35_b, 21'_b, 15'_a, 105'_a, 105_b, 168_a, 210_a, 70'_a,$
		$280'_a, 315'_a, 84_a, 210_b, 420_a, 105_c, 420'_a, 84'_a, 210'_b, 105'_c, 315_a,$
		$70_a, 280_a, 210'_a, 105'_b, 168'_a, 15_a, 105_a, 21'_a, 35'_b, 21_b, 7_a, 1'_a\}$
7	1	$\{1_a, 27_a, 120_a, 405_a, 512'_a, 216_a, 15_a\}$ + dual
8	1	$\{1_a, 105_b, 216'_a, 280'_b, 189_b, 21_b\}$, $\{7'_a, 120_a, 189'_c, 105'_c, 56_a, 27'_a\}$ + duals
9	1	$\{1_a, 35_b, 280_b, 512'_a, 315_a, 56_a, 7_a\}$ + dual
10	1	$\{1_a, 21'_b, 189'_c, 420_a, 405'_a, 189_b, 35'_b\}$ + dual,
		$\{7'_a, 27_a, 168_a, 378'_a, 378_a, 168'_a, 27'_a, 7_a\}$
12	1	$\{1_a, 56'_a, 210_a, 336'_a, 280_a, 120'_a, 21_b\}$ + dual
14	1	$\{1_a, 27_a, 105'_a, 189_a, 189'_a, 105_a, 27'_a, 1'_a\}$
18	1	$\{1_a, 7'_a, 21_a, 35'_a, 35_a, 21'_a, 7_a, 1'_a\}$

Table F.6 Blocks of positive defect of $\mathbf{H}(\mathsf{E}_8)$ at roots of unity

e	d	Φ_e-blocks of characters of defect d
2	1	$\{4096_z, 4096'_x\}$ + dual
	4	$\{112_z, 160_z, 400_z, 1344_x, 1296_z, 2240_x, 3360_z, 2800_z, 3200_x, 448_w, 5600_w,$
		$2016_w, 1344_w, 3200'_x, 3360'_z, 2800'_z, 2240'_x, 1296'_z, 1344'_x, 400'_z, 160'_z, 112'_z\}$
	8	$\{1_x, 8_z, 35_x, 84_x, 28_x, 210_x, 50_x, 560_z, 300_x, 700_x, 567_x, 1400_z, 56_z, 448_z,$
		$1008_z, 1400_x, 350_x, 1575_x, 175_x, 1050_x, 3240_z, 972_x, 2268_x, 840_z, 1400_{zz},$
		$4200_x, 840_x, 525_x, 4536_z, 2100_x, 700_{xx}, 6075_x, 2835_x, 2400_z, 5600_z,$
		$4200_z, 420_y, 70_y, 3150_y, 1134_y, 4200_y, 5670_y, 1680_y, 168_y, 1400_y, 4536_y,$
		$2100_y, 4200'_z, 2400'_z, 5600'_z, 2835'_x, 6075'_x, 4536'_z, 840'_x, 4200'_x, 2100'_x,$
		$700'_{xx}, 840'_z, 1400'_{zz}, 972'_x, 2268'_x, 3240'_z, 1575'_x, 1400'_x, 350'_x, 175'_x, 1050'_x,$
		$525'_x, 448'_z, 56'_z, 1008'_z, 1400'_z, 700'_x, 300'_x, 567'_x, 560'_z, 210'_x, 50'_x, 28'_x, 84'_x,$
		$35'_x, 8'_z, 1'_x\}$
3	1	$\{567_x, 2835'_x, 2268'_x\}$, $\{1296_z, 4536'_z, 3240'_z\}$ + duals,
		$\{1008_z, 2016_w, 1008'_z\}$, $\{1575_x, 3150_y, 1575'_x\}$
	4	$\{1_x, 35_x, 28_x, 84_x, 210_x, 50_x, 300_x, 700_x, 1344_x, 1050_x, 175_x, 1400_x, 350_x,$
		$2240_x, 4096_z, 4200_x, 840_x, 525_x, 2100_x, 700_{xx}, 3200_x, 420_y, 168_y, 1400_y,$
		$4200_y, 4480_y, 70_y, 1680_y, 2688_y, 2100_y, 3200'_x, 840'_x, 4200'_x, 700'_{xx},$
		$2100'_x, 4096'_z, 2240'_x, 175'_x, 1400'_x, 1050'_x, 350'_x, 525'_x, 1344'_x, 300'_x, 700'_x,$
		$50'_x, 210'_x, 28'_x, 84'_x, 35'_x, 1'_x\}$,
		$\{8_z, 112_z, 160_z, 560_z, 400_z, 1400_z, 56_z, 448_z, 840_z, 1400_{zz}, 4096_x, 3360_z,$
		$2800_z, 5600_z, 2400_z, 4200_z, 448_w, 5600_w, 7168_w, 1344_w, 2400'_z, 5600'_z,$
		$4200'_z, 3360'_z, 2800'_z, 4096'_x, 1400'_{zz}, 840'_z, 448'_z, 56'_z, 1400'_z, 400'_z,$
		$560'_z, 160'_z, 112'_z, 8'_z\}$
4	2	$\{8_z, 560_z, 1344_x, 1400_{zz}, 840_z, 4536_z, 3200_x, 4200'_z, 2240'_x, 3240'_z,$
		$1400'_z, 1008'_z, 56'_z\}$ + dual,
		$\{28_x, 160_z, 300_x, 972_x, 840_x, 700_{xx}, 1344_w, 840'_x, 700'_{xx}, 972'_x,$
		$300'_x, 160'_z, 28'_x\}$,
		$\{84_x, 700_x, 2268_x, 4200_x, 2100_x, 5600_w, 2016_w, 448_w, 4200'_x, 2100'_x,$
		$2268'_x, 700'_x, 84'_x\}$
	4	$\{1_x, 35_x, 112_z, 50_x, 210_x, 400_z, 567_x, 175_x, 1575_x, 350_x, 1050_x, 1296_z,$
		$525_x, 3360_z, 2800_z, 2835_x, 6075_x, 1134_y, 420_y, 2688_y, 4480_y, 3150_y, 70_y,$
		$5670_y, 2100_y, 2835'_x, 6075'_x, 3360'_z, 2800'_z, 1296'_z, 350'_x, 175'_x, 1575'_x,$
		$1050'_x, 525'_x, 400'_z, 567'_x, 50'_x, 210'_x, 112'_z, 35'_x, 1'_x\}$

Table F.6 Blocks of positive defect of $\mathbf{H}(\mathsf{E}_8)$ at roots of unity (continued)

5	1	$\{210_x, 2240_x, 2835_x, 840'_x, 35'_x\}$, $\{160_z, 840_z, 3360'_z, 3240'_z, 560'_z\}$ + duals
	2	$\{1_x, 84_x, 28_x, 567_x, 1344_x, 2268_x, 972_x, 4096_z, 168_y, 2688_y, 1134_y, 4536_y, 4096'_z, 972'_x, 2268'_x, 1344'_x, 567'_x, 28'_x, 84'_x, 1'_x\}$, $\{8_z, 112_z, 56_z, 448_z, 1008_z, 1296_z, 4096_x, 4536_z, 2016_w, 448_w, 1344_w, 7168_w, 4536'_z, 4096'_x, 1296'_z, 448'_z, 56'_z, 1008'_z, 112'_z, 8'_z\}$
6	1	$\{972_x, 2835_x, 4536'_z, 3240'_z, 567'_x\}$ + dual, $\{1008_z, 1575_x, 1134_y, 1575'_x, 1008'_z\}$
	2	$\{112_z, 160_z, 400_z, 1344_x, 2240_x, 3360_z, 3200_x, 1344_w, 7168_w, 3200'_x, 3360'_z, 2240'_x, 1344'_x, 400'_z, 160'_z, 112'_z\}$
	4	$\{1_x, 8_z, 35_x, 84_x, 28_x, 50_x, 210_x, 560_z, 300_x, 700_x, 448_z, 1400_z, 56_z, 1400_x, 175_x, 1050_x, 350_x, 840_z, 1400_{zz}, 840_x, 4200_x, 525_x, 2100_x, 700_{xx}, 2400_z, 4200_z, 5600_z, 4536_y, 420_y, 1680_y, 4200_y, 5670_y, 168_y, 4480_y, 2100_y, 2400'_z, 4200'_z, 5600'_z, 840'_x, 4200'_x, 700'_{xx}, 2100'_x, 840'_z, 1400'_{zz}, 1050'_x, 175'_x, 350'_x, 1400'_x, 525'_x, 1400'_z, 56'_z, 448'_z, 300'_x, 700'_x, 560'_z, 210'_x, 50'_x, 28'_x, 84'_x, 35'_x, 8'_z, 1'_x\}$
7	1	$\{1_x, 300_x, 4096_z, 6075_x, 3200'_x, 972'_x, 50'_x\}$ + dual, $\{8_z, 160_z, 1296_z, 2400_z, 4096'_x, 3240'_z, 400'_z\}$ + dual
8	1	$\{8_z, 2240_x, 4536_z, 4200'_z, 3240'_z, 1344'_x\}$ + dual, $\{28_x, 1296_z, 2100_x, 2800'_z, 2268'_x, 300'_x\}$ + dual, $\{112_z, 700_x, 972_x, 700'_{xx}, 400'_z, 84'_x\}$ + dual
	2	$\{1_x, 35_x, 160_z, 567_x, 175_x, 1400_x, 1575_x, 525_x, 2835_x, 6075_x, 7168_w, 2016_w, 5600_w, 6075'_x, 2835'_x, 175'_x, 1575'_x, 1400'_x, 525'_x, 567'_x, 160'_z, 35'_x, 1'_x\}$
9	1	$\{1_x, 50_x, 700_{xx}, 3200'_x, 4096'_z, 1575'_x, 28'_x\}$ + dual, $\{160_z, 1008_z, 2800_z, 5600'_z, 4096'_x, 560'_z, 112'_z\}$ + dual, $\{8_z, 400_z, 1400_{zz}, 2016_w, 1400'_{zz}, 400'_z, 8'_z\}$, $\{35_x, 700_x, 2240_x, 3150_y, 2240'_x, 700'_x, 35'_x\}$
10	1	$\{50_x, 1400_{zz}, 2835_x, 4200'_x, 3240'_z, 560'_z, 35'_x\}$ + dual
	2	$\{1_x, 8_z, 84_x, 28_x, 567_x, 448_z, 1008_z, 1400_z, 2268_x, 972_x, 4536_z, 1400_y, 4200_y, 4480_y, 3150_y, 4536'_z, 972'_x, 2268'_x, 448'_z, 1008'_z, 1400'_z, 567'_x, 28'_x, 84'_x, 8'_z, 1'_x\}$

Table F.6 Blocks of positive defect of $\mathbf{H}(\mathsf{E}_8)$ at roots of unity (continued)

12	1	$\{8_z, 560_z, 1344_x, 3200_x, 4200'_z, 2240'_x, 448'_z\}$ + dual,
		$\{28_x, 160_z, 300_x, 840_x, 1344_w, 840'_x, 300'_x, 160'_z, 28'_x\}$,
		$\{84_x, 700_x, 4200_x, 7168_w, 4200'_x, 700'_x, 84'_x\}$,
	2	$\{1_x, 35_x, 112_z, 210_x, 50_x, 400_z, 1050_x, 1400_x, 525_x, 3360_z, 2800_z,$
		$1400_y, 2688_y, 4536_y, 2100_y, 3360'_z, 2800'_z, 1050'_x, 1400'_x, 525'_x, 400'_z,$
		$50'_x, 210'_x, 112'_z, 35'_x, 1'_x\}$
14	1	$\{1_x, 700_x, 3240_z, 6075_x, 5600'_z, 2268'_x, 210'_x, 8'_z\}$ + dual
15	1	$\{1_x, 84_x, 1344_x, 4096_z, 5670_y, 4096'_z, 1344'_x, 84'_x, 1'_x\}$,
		$\{8_z, 112_z, 1400_z, 4096_x, 5600_w, 4096'_x, 1400'_z, 112'_z, 8'_z\}$
18	1	$\{1_x, 210_x, 1008_z, 2100_x, 2400'_z, 1575'_x, 560'_z, 84'_x\}$ + dual,
		$\{8_z, 35_x, 300_x, 840_z, 1134_y, 840'_z, 300'_x, 35'_x, 8'_z\}$
20	1	$\{1_x, 112_z, 567_x, 1296_z, 1680_y, 1296'_z, 567'_x, 112'_z, 1'_x\}$
24	1	$\{1_x, 35_x, 160_z, 350_x, 448_w, 350'_x, 160'_z, 35'_x, 1'_x\}$
30	1	$\{1_x, 8_z, 28_x, 56_z, 70_y, 56'_z, 28'_x, 8'_z, 1'_x\}$

References

The numbers in braces show the pages on which the references are cited.

[Alperin 1986] J. L. Alperin, *The Green correspondence and the Brauer lift*, J. Algebra **104** (1986), 78–79. {241}

[Alperin 1987] J. L. Alperin, *Brauer's induction theorem and the symmetric groups*, Comm. Algebra **15** (1987), 47–51. {173}

[Alvis 1987] D. Alvis, *The left cells of the Coxeter group of type* H_4, J. Algebra **107** (1987), 160–168. {394}

[Alvis and Lusztig 1982] D. Alvis and G. Lusztig, *The representations and generic degrees of the Hecke algebra of type* H_4, J. Reine Angew. Math. **336** (1982), 201–212; see also [Alvis and Lusztig 1994]. {317, 371, 394, 395, 412, 422, 435}

[Alvis and Lusztig 1994] D. Alvis and G. Lusztig, *Corrections to 'The representations and generic degrees of the Hecke algebra of type* H_4*'*, J. Reine Angew. Math. **449** (1994), 217–218; see [Alvis and Lusztig 1982]. {371, 394, 435}

[Andersen et al. 1994] H. H. Andersen, J. C. Jantzen and W. Soergel, *Representations of quantum groups at a* p*th root of unity and of semisimple groups in characteristic* p*: independence of* p, Astérisque **220** (1994). {280}

[Ariki 1995] S. Ariki, *Representation theory of a Hecke algebra of* $G(r, p, n)$, J. Algebra **177** (1995), 164–185. {357}

[Ariki 1996] S. Ariki, *On the decomposition numbers of the Hecke algebra of* $G(r, 1, n)$, J. Math. Kyoto Univ. **36** (1996), 789–808. {353, 357}

[Ariki and Koike 1992] S. Ariki and K. Koike, *A Hecke algebra of* $(\mathbb{Z}/r\mathbb{Z}) \wr S_n$ *and construction of its irreducible representations*, Adv. Math. **106** (1992), 216–243. {281, 325, 357}

[Ariki and Mathas 2000] S. Ariki and A. Mathas, *The number of simple modules of the Hecke algebras of type* $G(r, 1, n)$, Math. Z. (2000), to appear. {395}

[Artin 1947] E. Artin, *Theory of braids*, Ann. of Math. **48** (1947), 101–126. {105, 111}

[Atiyah and Macdonald 1969] M. F. Atiyah and I. G. Macdonald, *Introduction to commutative algebra*, Addison-Wesley, 1969. {246}

[Atkinson 1986] M. D. Atkinson, *A new proof of a theorem of Solomon*, Bull. London Math. Soc. **18** (1986), 351–354. {68}

[Atkinson 1992] M. D. Atkinson, *Solomon's descent algebra revisited*, Bull. London Math. Soc. **24** (1992), 545–551. {68}

[Atkinson and van Willigenburg 1997] M. D. Atkinson and S. J. van Willigenburg, *The* p*-modular descent algebra of the symmetric group*, Bull. London Math. Soc. **29** (1997), 407–414. {68}

[Benard 1971] M. Benard, *On the Schur indices of characters of the exceptional Weyl groups*, Ann. of Math. **94** (1971), 89–107. {173}

[Benson 1979] C. T. Benson, *The generic degrees of the irreducible characters of* E_8, Comm. Algebra **7** (1979), 1199–1209. {394}

[Benson and Curtis 1972] C. T. Benson and C. W. Curtis, *On the degrees and rationality of certain characters of finite Chevalley groups*, Trans. Amer. Math. Soc. **165** (1972), 251–273; see also [Curtis 1975]. {v, 177, 212, 280, 317, 318, 437}

[Benson and Gay 1977] C. T. Benson and D. A. Gay, *On dimension functions of the generic algebra of type* D_n, J. Algebra **45** (1977), 435–438. {358}

[Benson and Grove 1973] C. T. Benson and L. C. Grove, *The Schur indices of the reflection group* I_4, J. Algebra **27** (1973), 574–578. {173, 212, 215}

[Benson and Grove 1985] C. T. Benson and L. C. Grove, *Finite reflection groups (2nd edition)*, Graduate Texts in Mathematics, vol. 99, Springer-Verlag, 1985. {1, 172}

[Bergeron et al. 1992] F. Bergeron, N. Bergeron, R. B. Howlett and D. E. Taylor, *A decomposition of the descent algebra of a finite Coxeter group*, J. Algebraic Combin. **1** (1992), 23–44. {68}

[Bessis 1997] D. Bessis, *Sur le corps de définition d'un groupe de réflexions complexe*, Comm. Algebra **25** (1997), 2703–2716. {173}

[Beynon and Lusztig 1978] W. M. Beynon and G. Lusztig, *Some numerical results on the characters of exceptional Weyl groups*, Math. Proc. Cambridge Philos. Soc. **84** (1978), 417–426. {149, 173}

[Birman 1993] J. S. Birman, *New points of view in knot theory*, Bull. Amer. Math. Soc. **28** (1993), 253–287. {133}

[Bleher et al. 1997] F. Bleher, M. Geck and W. Kimmerle, *Automorphisms of integral group rings of finite Coxeter groups and Iwahori–Hecke algebras*, J. Algebra **197** (1997), 615–655. {34, 103, 173}

[Borcherds 1998] R. E. Borcherds, *Coxeter groups, Lorentzian lattices, and K3 surfaces*, Internat. Math. Res. Notices **19** (1998), 1011–1031. {56}

[Bourbaki 1968] N. Bourbaki, *Groupes et algèbres de Lie, chap. 4, 5 et 6*, Hermann, Paris, 1968. {v, 1, 17, 30, 33, 34, 36, 100, 133, 149, 172, 278, 281}

[Bremke 1994] K. Bremke, *The decomposition numbers of Hecke algebras of type* F_4 *with unequal parameters*, Manuscripta Math. **83** (1994), 331–346. {394, 395}

[Bremke and Malle 1997] K. Bremke and G. Malle, *Reduced words and a length function for* $G(e, 1, n)$, Indag. Math. **8** (1997), 453–469. {34}

[Brieskorn 1971] E. Brieskorn, *Die Fundamentalgruppe des Raumes der regulären Orbits einer endlichen komplexen Spiegelungsgruppe*, Invent. Math. **12** (1971), 57–61. {111}

[Brieskorn and Saito 1972] E. Brieskorn and K. Saito, *Artin-Gruppen und Coxeter-Gruppen*, Invent. Math. **17** (1972), 245–271. {111, 133}

[Brink and Howlett 1993] B. Brink and R. B. Howlett, *A finiteness property and an automatic structure for Coxeter groups*, Math. Ann. **296** (1993), 179–190. {67}

[Brink and Howlett 1999] B. Brink and R. B. Howlett, *Normalizers of parabolic subgroups in Coxeter groups*, Invent. Math. **136** (1999), 323–351. {56}

[Broué and Malle 1993] M. Broué and G. Malle, *Zyklotomische Heckealgebren*, Astérisque **212** (1993), 119–189. {172, 281, 317, 357}

[Broué and Michel 1996] M. Broué and J. Michel, *Sur certains éléments réguliers des groupes de Weyl et les variétés de Deligne–Lusztig associées*, in [Cabanes 1996], pp. 73–139. {100, 281, 317}

[Broué et al. 1998] M. Broué, G. Malle and R. Rouquier, *Complex reflection groups, braid groups, Hecke algebras*, J. Reine Angew. Math. **500** (1998), 127–190. {ix, 133, 172, 281}

[Broué et al. 1999] M. Broué, G. Malle and J. Michel, *Towards Spetses. I*, Transform. Groups **4** (1999), 157–218. {172, 246}

[Cabanes 1996] M. Cabanes (ed.), *Finite reductive groups, related structures and representations*, Progr. Math., vol. 141, Birkhäuser, Basel, 1996. {436, 439, 441, 442, 443}

[Carter 1972a] R. W. Carter, *Conjugacy classes in the Weyl group*, Compositio Math. **25** (1972), 1–59. {v, 87, 100, 104, 407}

[Carter 1972b] R. W. Carter, *Simple groups of Lie type*, Wiley, New York, 1972; reprinted 1989 as Wiley Classics Library Edition. {19, 30, 100, 149, 278, 280}

[Carter 1985] R. W. Carter, *Finite groups of Lie type: Conjugacy classes and complex characters*, Wiley, New York, 1985; reprinted 1993 as Wiley Classics Library Edition. {v, 149, 152, 280, 350, 358, 395}

[Carter 1986] R. W. Carter, *Representation theory of the 0-Hecke algebra*, J. Algebra **104** (1986), 89–103. {284, 320}

[Carter and Geck 1998] R. W. Carter and M. Geck (eds.), *Representations of reductive groups*, Publications of the Newton Institute, Cambridge University Press, 1998. {280}

[Carter and Saxl 1998] R. W. Carter and J. Saxl (eds.), *Algebraic groups and their representations*, NATO Sci. Ser. C Math. Phys. Sci., vol. 517, Kluwer Academic Publishers Group, Dordrecht, 1998. {280}

[Casselman 1994] W. A. Casselman, *Machine calculations in Weyl groups*, Invent. Math. **116** (1994), 95–108. {67}

[Char et al. 1991] B. W. Char, K. O. Geddes, G. H. Gonnet, M. B. Monagan and S. M. Watt, *Maple V, Language reference manual*, Springer-Verlag, 1991; see also `http://daisy.waterloo.ca`. {x}

[Chen and Shi 1998] Y. Chen and J.-Y. Shi, *Left cells in the Weyl group of type* E_7, Comm. Algebra **26** (1998), 3837–3852. {394}

[Cohen 1976] A. Cohen, *Finite complex reflection groups*, Ann. Sci. École Norm. Sup. (4) **9** (1976), 379–436; see also the erratum in *ibid.* **11** (1978), 613. {172}

[Crowell and Fox 1977] R. H. Crowell and R. H. Fox, *Introduction to knot theory*, Grad. Texts in Math., vol. 57, Springer-Verlag, 1977; reprint of the 1963 original. {127, 133, 136}

[Curtis 1975] C. W. Curtis, *Corrections and additions to 'On the degrees and rationality of certain characters of finite Chevalley groups'*, Trans. Amer. Math. Soc. **202** (1975), 405–406; see [Benson and Curtis 1972]. {317, 435}

[Curtis 1988] C. W. Curtis, *Representations of Hecke algebras*, Astérisque **168** (1988), 13–60. {394}

[Curtis and Reiner 1962] C. W. Curtis and I. Reiner, *Representation theory of finite groups and associative algebras*, Wiley, New York, 1962; reprinted 1988 as Wiley Classics Library Edition. {230, 237, 246, 248, 284}

[Curtis and Reiner 1981] C. W. Curtis and I. Reiner, *Methods of representation theory, vol. I*, Wiley, New York, 1981; reprinted 1990 as Wiley Classics Library Edition. {33, 218, 228, 230, 232, 237, 240, 241, 250, 255, 291, 302}

[Curtis and Reiner 1987] C. W. Curtis and I. Reiner, *Methods of representation theory, vol. II*, Wiley, New York, 1987; reprinted 1994 as Wiley Classics Library Edition. {v, 1, 60, 240, 246, 247, 248, 280, 281, 302, 318}

[Curtis et al. 1971] C. W. Curtis, N. Iwahori and R. W. Kilmoyer, *Hecke algebras and characters of parabolic type of finite groups with BN-pairs*, Inst. Hautes Études Sci. Publ. Math. **40** (1971), 81–116. {172, 212, 281, 284}

[Deligne 1972] P. Deligne, *Les immeubles des groupes de tresses généralisés*, Invent. Math. **17** (1972), 273–302. {133}

[Deligne 1997] P. Deligne, *Action du groupe de tresses sur une catégorie*, Invent. Math. **128** (1997), 159–175. {133}

[Deligne and Lusztig 1976] P. Deligne and G. Lusztig, *Representations of reductive groups over finite fields*, Ann. of Math. **103** (1976), 103–161. {34}

[Deodhar 1977] V. V. Deodhar, *Some characterizations of Bruhat ordering on a Coxeter group and determination of the relative Möbius function*, Invent. Math. **39** (1977), 187–198. {67}

[Deodhar 1982] V. V. Deodhar, *On the root system of a Coxeter group*, Comm. Algebra **10** (1982), 611–630. {68}

[Deodhar 1989] V. V. Deodhar, *A note on subgroups generated by reflections in Coxeter groups*, Arch. Math. (Basel) **53** (1989), 543–546. {68}

[Digne and Michel 1985] F. Digne and J. Michel, *Fonctions $\mathcal{L}$ des variétés de Deligne–Lusztig et descente de Shintani*, Mém. Soc. Math. France **20** (1985). {281, 318}

[Dipper and Du 1993] R. Dipper and J. Du, *Trivial and alternating source modules of Hecke algebras of type* A, Proc. London Math. Soc. (3) **66** (1993), 479–506. {317}

[Dipper and James 1986] R. Dipper and G. D. James, *Representations of Hecke algebras of general linear groups*, Proc. London Math. Soc. (3) **52** (1986), 20–52. {317, 354, 357, 395}

[Dipper and James 1987] R. Dipper and G. D. James, *Blocks and idempotents of Hecke algebras of general linear groups*, Proc. London Math. Soc. (3) **54** (1987), 57–82. {281, 357}

[Dipper and James 1992] R. Dipper and G. D. James, *Representations of Hecke algebras of type* B_n, J. Algebra **146** (1992), 454–481. {359}

[Dipper et al. 1995] R. Dipper, G. D. James and G. E. Murphy, *Hecke algebras of type* B_n *at roots of unity*, Proc. London Math. Soc. (3) **70** (1995), 505–528. {308, 357, 395}

[Dipper et al. 1998] R. Dipper, G. D. James and A. Mathas, *Cyclotomic* q*-Schur algebras*, Math. Z. **229** (1998), 385–416. {281}

[Dipper et al. 1999] R. Dipper, M. Geck, G. Hiss and G. Malle, *Representations of Hecke algebras and finite groups of Lie type*, Algorithmic algebra and number theory (B. Matzat, G. M. Greuel and G. Hiss, eds.), Springer-Verlag, 1999, pp. 331–378. {280}

[du Cloux 1990] F. du Cloux, *Un algorithme de forme normale pour les groupes de Coxeter*, Tech. report, Centre de Mathématiques, École Polytechnique, April 1990. {67}

[du Cloux 1999] F. du Cloux, *A transducer approach to Coxeter groups*, J. Symbolic Comput. **27** (1999), 1–14. {67}

[Du et al. 1998] J. Du, B. Parshall and L. Scott, *Stratifying endomorphism algebras associated to Hecke algebras*, Trans. Amer. Math. Soc. **203** (1998), 169–210. {246}

[Dyer 1987] M. Dyer, *Hecke algebras and reflections in Coxeter groups*, Ph. D. thesis, University of Sydney, 1987. {34, 68}

[Dyer 1990] M. Dyer, *Reflection subgroups of Coxeter systems*, J. Algebra **135** (1990), 57–73. {68}

[Epstein et al. 1992] D. B. A. Epstein, J. W. Cannon, D. F. Holt, S. V. F. Levy, M. S. Paterson and W. P. Thurston, *Word processing in groups*, Jones and Bartlett Publishers, Boston, MA, 1992. {67, 69}

[Feit 1982] W. Feit, *The representation theory of finite groups*, North-Holland, Amsterdam, 1982. {237, 239, 240, 241, 244}

[Fleischmann 1990] P. Fleischmann, *Projective simple modules of symmetric algebras and their specializations with applications to Hecke algebras*, Arch. Math. (Basel) **55** (1990), 247–258. {246, 318, 395}

[Fleischmann 1993] P. Fleischmann, *On Lusztig's isomorphism theorems for Hecke algebras and representation theory*, J. London Math. Soc. (2) **47** (1993), 450–472. {395}

[Fong and Srinivasan 1990] P. Fong and B. Srinivasan, *Brauer trees in classical groups*, J. Algebra **131** (1990), 179–225. {350}

[Frame 1951] J. S. Frame, *The classes and characters of the groups of 27 lines and 28 bitangents*, Ann. Mat. Pura Appl. (4) **32** (1951), 83–119. {172}

[Frame 1970] J. S. Frame, *The characters of the Weyl group* E_8, Computational problems in abstract algebra (J. Leech, ed.), Pergamon, Oxford, 1970, pp. 111–130. {172}

[Frobenius 1900] F. G. Frobenius, *Über die Charaktere der symmetrischen Gruppe*, Sitzungsber. Preuss. Akad. Wiss. Berlin (1900), 516–534. {172}

[Fulton 1997] W. Fulton, *Young tableaux*, London Math. Soc. Stud. Texts, vol. 35, Cambridge University Press, 1997. {172, 182, 212, 213, 214}

[Garsia and McLarnan 1988] A. M. Garsia and T. J. McLarnan, *Relations between Young's natural and the Kazhdan–Lusztig representations of* S_n, Adv. Math. **69** (1988), 32–92. {394}

[Garsia and Reutenauer 1989] A. M. Garsia and C. Reutenauer, *A decomposition of Solomon's descent algebra*, Adv. Math. **77** (1989), 189–262. {68}

[Garside 1969] F. A. Garside, *The braid group and other groups*, Quart. J. Math. Oxford **20** (1969), 235–254. {133}

[Gasharov 1998] V. Gasharov, *A short proof of the Littlewood–Richardson rule*, European J. Combin. **19** (1998), 451–453. {212}

[Geck 1992] M. Geck, *Brauer trees of Hecke algebras*, Comm. Algebra **20** (1992), no. 10, 2937–2973. {245, 246, 395, 428}

[Geck 1993a] M. Geck, *The decomposition numbers of the Hecke algebra of type* E_6, Math. Comp. **61** (1993), 889–899. {246, 395}

[Geck 1993b] M. Geck, *A note on Harish-Chandra induction*, Manuscripta Math. **80** (1993), 393–401. {280}

[Geck 1994] M. Geck, *On the character values of Iwahori–Hecke algebras of exceptional type*, Proc. London Math. Soc. (3) **68** (1994), 51–76. {318, 391, 395}

[Geck 1995] M. Geck, *Beiträge zur Darstellungstheorie von Iwahori–Hecke Algebren*, Aachener Beitr. Math., vol. 11, Verlag der Augustinus Buchhandlung, Aachen, 1995. {x, 246, 318, 391, 395}

[Geck 1998a] M. Geck, *Kazhdan–Lusztig cells and decomposition numbers*, Represent. Theory **2** (1998), 264–277 (electronic). {395, 428}

[Geck 1998b] M. Geck, *Modular Harish-Chandra series, Hecke algebras and (generalized) q-Schur algebras*; Preprint (1998), to appear in the Proceedings of the Symposium on Modular Representation Theory of Finite Groups held in Charlottesville VA, May 1998. {280}

[Geck 1998c] M. Geck, *Representations of Hecke algebras at roots of unity*, Astérisque **252** (1998), Séminare Bourbaki (1997/98), Exp. 836, pp. 33–55. {249, 358, 397}

[Geck 1998d] M. Geck, *Trace functions on Hecke algebras*, Banach Center Publ., Polish Acad. Sci. **42** (1998), 87–109. {133, 281}

[Geck 1999] M. Geck, *The character table of the Iwahori–Hecke algebra of the symmetric group: Starkey's rule*, C. R. Acad. Sci. Paris **329** (1999), 361–366. {317}

[Geck and Lambropoulou 1997] M. Geck and S. Lambropoulou, *Markov traces and knot invariants related to Iwahori–Hecke algebras of type* B, J. Reine Angew. Math. **482** (1997), 191–213. {133, 350}

[Geck and Lux 1991] M. Geck and K. Lux, *The decomposition numbers of the Hecke algebra of type* F_4, Manuscripta Math. **70** (1991), 285–306. {395}

[Geck and Malle 2000] M. Geck and G. Malle, *On the existence of a unipotent support for the irreducible characters of finite groups of Lie type*, Trans. Amer. Math. Soc. **352** (2000), 429–456. {177}

[Geck and Michel 1997] M. Geck and J. Michel, *"Good" elements in finite Coxeter groups and representations of Iwahori–Hecke algebras*, Proc. London Math. Soc. (3) **74** (1997), 275–305. {v, 133, 317, 390, 395}

[Geck and Pfeiffer 1993] M. Geck and G. Pfeiffer, *On the irreducible characters of Hecke algebras*, Adv. Math. **102** (1993), 79–94. {v, 100, 281}

[Geck and Rouquier 1996] M. Geck and R. Rouquier, *Centers and simple modules for Iwahori–Hecke algebras*, in [Cabanes 1996], pp. 251–272. {246, 249, 281}

[Geck et al. 1996] M. Geck, G. Hiss, F. Lübeck, G. Malle and G. Pfeiffer, CHEVIE — *A system for computing and processing generic character tables*, Appl. Algebra Engrg. Comm. Comput. **7** (1996), 175–210. {v, x, 101}

[Geck et al. 2000] M. Geck, S. Kim and G. Pfeiffer, *Minimal length elements in twisted conjugacy classes of finite Coxeter groups*, J. Algebra (2000), to appear. {101, 281}

[Goldschmidt 1980] D. M. Goldschmidt, *Lectures on character theory*, Publish or Perish, Berkeley, 1980. {137, 145, 154, 156, 160, 190, 216, 231, 237, 241, 244, 245, 387}

[Goldschmidt 1993] D. M. Goldschmidt, *Group characters, symmetric functions, and the Hecke algebra*, Univ. Lecture Ser., vol. 4, Amer. Math. Soc., Providence, RI, 1993. {173}

[Green 1970] J. A. Green, *The Steinberg characters of finite Chevalley groups*, Math. Z. **117** (1970), 272–288. {281, 318}

[Greene 1992] C. Greene, *A rational-function identity related to the Murnaghan–Nakayama formula for the characters of* S_n, J. Algebraic Combin. **1** (1992), 235–255. {357}

[Grove 1974] L. C. Grove, *The characters of the hecatonicosahedroidal group*, J. Reine Angew. Math. **265** (1974), 160–169. {172}

[Gyoja 1984] A. Gyoja, *On the existence of a W-graph for an irreducible representation of a Coxeter group*, J. Algebra **86** (1984), 422–438. {296, 308, 361, 394}

[Gyoja 1995] A. Gyoja, *Modular representation theory over a ring of higher dimension with applications to Hecke algebras*, J. Algebra **174** (1995), 553–572. {246, 318}

[Gyoja 1996] A. Gyoja, *Cells and modular representations of Hecke algebras*, Osaka J. Math. **33** (1996), 307–341. {394, 395}

[Gyoja and Uno 1989] A. Gyoja and K. Uno, *On the semisimplicity of Hecke algebras*, J. Math. Soc. Japan **41** (1989), no. 1, 75–79. {246, 318}

[Halverson and Ram 1996] T. Halverson and A. Ram, *Murnaghan–Nakayama rules for characters of Iwahori–Hecke algebras of classical type*, Trans. Amer. Math. Soc. **348** (1996), 3976–3995. {v, 358}

[Halverson and Ram 1998] T. Halverson and A. Ram, *Murnaghan–Nakayama rules for characters of Iwahori–Hecke algebras of the complex reflection groups* $G(r, p, n)$, Canad. J. Math.

50 (1998), 167–192. {358}

[Higman 1977] D. G. Higman, *Lectures on permutation representations*, Vorlesungen Math. Inst. Giessen, Heft 4, 1977. {281}

[Hiller 1982] H. Hiller, *Geometry of Coxeter groups*, Res. Notes Mathematics, vol. 54, Pitman, Boston, Mass., 1982. {1}

[Hiss 1989] G. Hiss, *Zerlegungszahlen endlicher Gruppen vom Lie-Typ in nicht-definierender Charakteristik*, Habilitationsschrift, RWTH Aachen, 1989. {281}

[Hoefsmit 1974] P. N. Hoefsmit, *Representations of Hecke algebras of finite groups with BN pairs of classical type*, Ph.D. thesis, University of British Columbia, Vancouver, 1974. {318, 325, 350, 357, 358}

[Howlett 1980] R. B. Howlett, *Normalizers of parabolic subgroups of reflection groups*, J. London Math. Soc. (2) **21** (1980), 62–80. {56, 68}

[Humphreys 1990] J. E. Humphreys, *Reflection groups and Coxeter groups*, Cambridge Stud. Adv. Math., vol. 29, Cambridge University Press, 1990. {1, 17, 20, 33, 394}

[Iancu 2000] L. Iancu, *Markov traces and generic degrees in type* B_n, Preprint (2000). {133, 350}

[Isaacs 1976] I. M. Isaacs, *Character theory of finite groups*, Academic Press, 1976; reprinted 1994 by Dover Publications. {137, 156, 165, 169, 186, 217, 223, 234, 267, 295}

[Iwahori 1964] N. Iwahori, *On the structure of a Hecke ring of a Chevalley group over a finite field*, J. Fac. Sci. Univ. Tokyo **10** (1964), 215–236. {34, 133, 281, 282}

[James and Kerber 1981] G. D. James and A. Kerber, *The representation theory of the symmetric group*, Encyclopedia of Mathematics, vol. 16, Addison-Wesley, 1981. {173, 212}

[Johnson 1990] D. L. Johnson, *Presentations of groups*, London Math. Soc. Stud. Texts, vol. 15, Cambridge University Press, 1990. {40}

[Jones 1987] V. F. R. Jones, *Hecke algebra representations of braid groups and link polynomials*, Ann. of Math. **126** (1987), 335–388. {131, 133, 134, 282}

[Kac 1985] V. G. Kac, *Infinite dimensional Lie algebras*, Cambridge University Press, 1985. {17, 33}

[Kazhdan and Lusztig 1979] D. A. Kazhdan and G. Lusztig, *Representations of Coxeter groups and Hecke algebras*, Invent. Math. **53** (1979), 165–184. {362, 370, 393, 394, 395, 396}

[Kazhdan and Lusztig 1980] D. A. Kazhdan and G. Lusztig, *Schubert varieties and Poincaré duality*, Proc. Sympos. Pure Math. **34** (1980), 185–203. {393}

[Kerber 1991] A. Kerber, *Algebraic combinatorics via finite group actions*, Bibliographisches Institut, Mannheim, 1991. {357}

[Kilmoyer and Solomon 1973] R. Kilmoyer and L. Solomon, *On the theorem of Feit–Higman*, J. Combin. Theory Ser. A **15** (1973), 310–322. {281}

[Kim 1998] S. Kim, *Mots de longueur maximale dans une classe de conjugaison d'un groupe symétrique, vu comme groupe de Coxeter*, C. R. Acad. Sci. Paris **327** (1998), 617–622. {101}

[Knuth 1968] D. E. Knuth, *The art of computer programming, volume 1: Fundamental algorithms*, Addison-Wesley, 1968; third edition, 1997. {x}

[Kondo 1965] T. Kondo, *The characters of the Weyl group of type* F_4, J. Fac. Sci. Univ. Tokyo **11** (1965), 145–153. {172}

[Kunz 1979] E. Kunz, *Einführung in die kommutative Algebra und algebraische Geometrie*, Vieweg Studium, vol. 46, Vieweg, Braunschweig, Wiesbaden, 1979. {248}

[Lascoux et al. 1996] A. Lascoux, B. Leclerc and J. Y. Thibon, *Hecke algebras at roots of unity and crystal bases of quantum affine algebras*, Comment. Math. Phys. **181** (1996), 205–263. {355, 356, 357}

[Ledermann 1987] W. Ledermann, *Introduction to group characters. Second edition*, Cambridge University Press, 1987. {162, 172, 176}

[Lusztig 1977] G. Lusztig, *Irreducible representations of finite classical groups*, Invent. Math. **43** (1977), 125–175. {149, 175, 212, 350}

[Lusztig 1979a] G. Lusztig, *A class of irreducible representations of a finite Weyl group*, Indag. Math. **41** (1979), 323–335. {v, 142, 149, 172, 211, 212, 214, 318, 358, 376}

[Lusztig 1979b] G. Lusztig, *Unipotent representations of a finite Chevalley group of type* E_8, Quart. J. Math. Oxford **30** (1979), 315–338. {395}

[Lusztig 1981a] G. Lusztig, *On a theorem of Benson and Curtis*, J. Algebra **71** (1981), 490–498. {308, 317, 361, 394}

[Lusztig 1981b] G. Lusztig, *Unipotent characters of the symplectic and odd orthogonal groups over a finite field*, Invent. Math. **64** (1981), 263–296. {212, 318}

[Lusztig 1982] G. Lusztig, *A class of irreducible representations of a finite Weyl group II*, Indag. Math. **44** (1982), 219–226. {v, 212, 395}

[Lusztig 1983] G. Lusztig, *Left cells in Weyl groups*, Lie Group Representations I (R. L. R. Herb and J. Rosenberg, eds.), Lecture Notes in Math., vol. 1024, Springer-Verlag, 1983, pp. 99–111. {317, 318, 362, 370, 376, 393, 394}

[Lusztig 1984] G. Lusztig, *Characters of reductive groups over a finite field*, Ann. of Math. Stud., vol. 107, Princeton University Press, 1984. {v, 153, 173, 177, 211, 212, 214, 215, 280, 317, 318, 350, 394, 411}

[Lusztig 1985] G. Lusztig, *Cells in affine Weyl groups*, Algebraic groups and related topics (Kyoto/Nagoya, 1983), Adv. Stud. Pure Math., vol. 6, North-Holland, Amsterdam, 1985, pp. 255–287. {212, 393, 394}

[Lusztig 1986] G. Lusztig, *Sur les cellules gauches des groupes de Weyl*, C. R. Acad. Sci. Paris **302** (1986), 5–8. {394}

[Lusztig 1987a] G. Lusztig, *Cells in affine Weyl groups II*, J. Algebra **109** (1987), 536–548. {428}

[Lusztig 1987b] G. Lusztig, *Leading coefficients of character values of Hecke algebras*, Proc. Sympos. Pure Math. **47** (1987), 235–262. {212, 316, 317, 318, 376, 394}

[Lusztig 1991] G. Lusztig, *Intersection cohomology methods in representation theory*, Proceedings of the International Congress of Mathematics, Kyoto, Japan, 1990 (I. Satake, ed.), Springer-Verlag, 1991, pp. 155–174. {ix, 393, 394}

[Lusztig 1993] G. Lusztig, *Introduction to quantum groups*, Progr. Math., vol. 110, Birkhäuser, Basel, 1993. {355}

[Lusztig 1995] G. Lusztig, *Classification of unipotent representations of simple* p*-adic groups*, Internat. Math. Res. Notices **11** (1995), 517–589. {280}

[Lusztig and Spaltenstein 1979] G. Lusztig and N. Spaltenstein, *Induced unipotent classes*, J. London Math. Soc. (2) **19** (1979), 41–52. {68, 172}

[Macdonald 1972] I. G. Macdonald, *Some irreducible representations of Weyl groups*, Bull. London Math. Soc. **4** (1972), 148–150. {172, 173}

[Macdonald 1995] I. G. Macdonald, *Symmetric functions and Hall polynomials*, 2nd edition, Oxford University Press, 1995. {162, 172, 173, 175, 181, 182, 196, 212, 213, 214, 357}

[Malle 1995] G. Malle, *Unipotente Grade imprimitiver komplexer Spiegelungsgruppen*, J. Algebra **177** (1995), 768–826. {212, 351, 358}

[Malle 1996] G. Malle, *Degrés relatifs des algèbres cyclotomiques associées aux groupes de réflexions complexes de dimension deux*, in [Cabanes 1996], pp. 311–332. {281}

[Malle 2000] G. Malle, *On the rationality and fake degrees of characters of cyclotomic algebras*, J. Math. Sci. Univ. Tokyo (2000), to appear. {173, 212, 318}

[Malle and Mathas 1998] G. Malle and A. Mathas, *Symmetric cyclotomic Hecke algebras*, J. Algebra **205** (1998), 275–293. {281}

[Mathas 1999] A. Mathas, *Iwahori–Hecke algebras and Schur algebras of the symmetric group*, Univ. Lecture Ser., vol. 15, Amer. Math. Soc., Providence, RI, 1999. {354}

[Matsumoto 1964] H. Matsumoto, *Générateurs et relations des groupes de Weyl généralisés*, C. R. Acad. Sci. Paris **258** (1964), 3419–3422. {34, 281}

[Matsumura 1986] H. Matsumura, *Commutative ring theory*, Cambridge Stud. Adv. Math., vol. 8, Cambridge University Press, 1986. {231, 232, 248, 249}

[McDonough and Pallikaros 1999] T. P. McDonough and C. A. Pallikaros, *On the irreducible representations of the specializations of the generic Hecke algebra of type* F_4, J. Algebra **218** (1999), 654–671. {395}

[Michel 1999] J. Michel, *A note on words in braid monoids*, J. Algebra **215** (1999), 366–377. {133}

[Moody 1993] J. A. Moody, *The faithfullness question for the Burau representation*, Proc. Amer. Math. Soc. **119** (1993), 671–679. {134}

[Müller 1995] J. Müller, *Zerlegungszahlen für generische Iwahori–Hecke Algebren von exzeptionellem Typ*, Dissertation, RWTH Aachen, 1995. {395}

[Müller 1997] J. Müller, *Decomposition numbers for generic Iwahori–Hecke algebras of noncrystallographic type*, J. Algebra **189** (1997), 125–149. {395, 428}

[Murphy 1992] G. E. Murphy, *On the representation theory of the symmetric groups and associated Hecke algebras*, J. Algebra **152** (1992), 492–513. {358}

[Naruse 1989] H. Naruse, *On an isomorphism between Specht module and left cell of* S_n, Tokyo J. Math. **12** (1989), 247–267. {394}

[Norton 1979] P. N. Norton, *0-Hecke algebras*, J. Austral. Math. Soc. **27** (1979), 337–357. {284}

[Olsson 1986] J. B. Olsson, *Remarks on symbols, hooks and degrees of unipotent characters*, J. Combin. Theory Ser. A **42** (1986), 223–238. {212, 358}

[Orellana 1999] R. C. Orellana, *Weights of Markov traces on Hecke algebras*, J. Reine Angew. Math. **508** (1999), 157–178. {133, 350}

[Orlik and Solomon 1983] P. Orlik and L. Solomon, *Coxeter arrangements*, Proc. Sympos. Pure Math., vol. 40, 1983, pp. 269–291. {68}

[Pahlings 1976] H. Pahlings, *Characterization of groups by their character tables II*, Comm. Algebra **4** (1976), 111–153. {173}

[Parker 1984] R. A. Parker, *The computer calculation of modular characters (the meat-axe)*, Computational Group Theory (M. D. Atkinson, ed.), Academic Press, 1984. {x, 375}

[Pfeiffer 1994a] G. Pfeiffer, *Character tables of Weyl groups in* GAP, Bayreuther Math. Schr. **47** (1994), 165–222. {358}

[Pfeiffer 1994b] G. Pfeiffer, *Young characters on Coxeter basis elements of Iwahori–Hecke algebras and a Murnaghan–Nakayama formula*, J. Algebra **168** (1994), 525–535. {317, 357}

[Pfeiffer 1995] G. Pfeiffer, *Charakterwerte von Iwahori–Hecke-Algebren von klassischem Typ*, Aachener Beitr. Math., vol. 14, Verlag der Augustinus Buchhandlung, Aachen, 1995. {v, x, 358}

[Pfeiffer 1996] G. Pfeiffer, *Character values of Iwahori–Hecke algebras of type* B, in [Cabanes 1996], pp. 333–360. {357}

[Pfeiffer 1997] G. Pfeiffer, *The subgroups of* M_{24}*, or how to compute the table of marks of a finite group*, Experiment. Math. **6** (1997), 247–270. {64}

[Ram 1991] A. Ram, *A Frobenius formula for the characters of the Hecke algebras*, Invent. Math. **106** (1991), 461–488. {281, 357}

[Ram 1997] A. Ram, *Seminormal representations of Weyl groups and Iwahori–Hecke algebras*, Proc. London Math. Soc. (3) **75** (1997), 99–133. {357}

[Ram and Remmel 1997] A. Ram and J. B. Remmel, *Applications of the Frobenius formulas for the characters of the symmetric group and the Hecke algebras of type* A, J. Algebraic Combin. **6** (1997), 59–87. {350, 358, 360}

[Richardson 1982] R. W. Richardson, *Conjugacy classes of involutions in Coxeter groups*, Bull. Austral. Math. Soc. **26** (1982), 1–15. {68, 101}

[Schneider 1990] G. J. A. Schneider, *Dixon's character table algorithm revisited*, J. Symbolic Comput. **9** (1990), 601–606. {151, 172}

[Schönert 1995] M. Schönert, *GAP — Groups, Algorithms, and Programming*, Lehrstuhl D für Mathematik, RWTH Aachen, Germany, 5th edition, 1995; electronically available at `http://www.math.rwth-aachen.de/LDFM/homes/GAP/WWW`. {x}

[Shephard and Todd 1954] G. C. Shephard and J. A. Todd, *Finite unitary reflection groups*, Canad. J. Math. **6** (1954), 274–304. {172}

[Shi 1997] J.-Y. Shi, *The enumeration of Coxeter elements*, J. Algebraic Combin. **6** (1997), 161–171. {101}

[Shoji 1979] T. Shoji, *On the Springer representations of the Weyl groups of classical algebraic groups*, Comm. Algebra **7** (1979), 1713–1745. {212}

[Shoji 1999] T. Shoji, *A Frobenius formula for the characters of Ariki–Koike algebras*, Preprint. {317, 358}

[Shoji 2000] T. Shoji, *Length functions for* $G(r, p, n)$, Adv. Stud. Pure Math. (2000), to appear. {34}

[Sims 1994] C. C. Sims, *Computing with finitely presented groups*, Encyclopedia of Mathematics, vol. 48, Cambridge University Press, 1994. {67, 70}

[Solomon 1976] L. Solomon, *A Mackey formula in the group ring of a Coxeter group*, J. Algebra **41** (1976), 255–268. {68, 103}

[Sörgel 1997] W. Sörgel, *Kazhdan–Lusztig-Polynome und eine Kombinatorik für Kipp-Moduln*, Represent. Theory **1** (1997), 37–68 (electronic). {393}

[Specht 1935] W. Specht, *Die irreduziblen Darstellungen der symmetrischen Gruppe*, Math. Z. **39** (1935), 696–711. {173}

[Specht 1937] W. Specht, *Darstellungstheorie der Hyperoktaedergruppe*, Math. Z. **42** (1937), 629–640. {100, 173}

[Springer 1974] T. A. Springer, *Regular elements of finite reflection groups*, Invent. Math. **25** (1974), 159–198. {172}

[Springer 1982] T. A. Springer, *Quelques applications de la cohomologie d'intersection*, Astérisque **92–93** (1982), Séminaire Bourbaki (1981/82), Exp. 589. {317, 394}

[Starkey 1975] A. J. Starkey, *Characters of the generic Hecke algebra of a system of BN-pairs*, Ph.D. thesis, University of Warwick, July 1975. {281, 317}

[Steinberg 1951] R. Steinberg, *A geometric approach to the representations of the full linear group over a Galois field*, Trans. Amer. Math. Soc. **71** (1951), 274–282. {358}

[Steinberg 1968] R. Steinberg, *Endomorphisms of linear algebraic groups*, Mem. Amer. Math. Soc., no. 80, 1968. {68, 100, 104, 172}

[Stembridge 1989] J. R. Stembridge, *On the eigenvalues of representations of reflection groups and wreath products*, Pacific J. Math. **140** (1989), 353–396. {357}

[Sturmfels 1993] B. Sturmfels, *Algorithms in invariant theory*, Texts Monogr. Symbol. Comput., Springer-Verlag, Vienna, 1993. {172}

[Surowski 1978] D. B. Surowski, *Degrees of irreducible characters of* (B, N)*-pairs of types* E_6 *and* E_7, Trans. Amer. Math. Soc. **243** (1978), 235–249. {394}

[Tits 1984] J. Tits, *Lectures on buildings and arithmetic groups*, typewritten manuscript, Yale University, 1984. {34}

[Traczyk 1998] P. Traczyk, *A new proof of Markov's braid theorem*, Banach Center Publ., Polish Acad. Sci. **42** (1998), 409–419. {128}

[Veigneau 1996] S. Veigneau, *ACE — An algebraic combinatorics environment for the computer algebra system MAPLE*, User's reference manual, Version 2.0, Institut Gaspard Monge, 1996; electronically available at `http://weyl.univ-mlv.fr/~sv/ACE`. {357}

[Vigneras 1996] M. F. Vigneras, *À propos d'une conjecture de Langlands modulaire*, in [Cabanes 1996], pp. 415–452. {280}

[Vogel 1990] P. Vogel, *Representation of links by braids: a new algorithm*, Comment. Math. Helv. **65** (1990), 104–113. {128}

[Wenzl 1988] H. Wenzl, *Hecke algebras of type* A *and subfactors*, Invent. Math. **92** (1988), 349–383. {133, 350, 357}

[Xi 1994] N. Xi, *Representations of Affine Hecke algebras*, Lecture Notes in Math., vol. 1587, Springer-Verlag, 1994. {394, 396}

[Yamane 1989] H. Yamane, *Irreducible projective modules of the Hecke algebra of a finite Coxeter group*, J. Algebra **127** (1989), 373–384. {246}

[Young 1977] A. Young, *The collected papers of Alfred Young (1873–1940). With a foreword by G. de B. Robinson and a biography by H. W. Turnbull*, Mathematical Expositions, vol. 21, University of Toronto Press, Toronto, 1977. {100, 173}

[Zelevinsky 1981] A. V. Zelevinsky, *Representations of finite classical groups. A Hopf algebra approach*, Lecture Notes in Math., vol. 869, Springer-Verlag, 1981. {212}

Index

$\mathcal{A}$-families, 203
a-invariant, 195, 315, 376
Alexander polynomial, 131
Artin–Tits braid group, 110

b-invariant, 142
β-set, 193
block, 244
border strip, 323
braid monoid, 11, 106, 296
braid relations, 106
branching rule, 182, 183
Brauer graph, 244, 428
Brieskorn–Saito normal form, 109
Bruhat decomposition, 278
Bruhat–Chevalley order, 13, 35
Burau representation, 132, 260
Burnside ring, 61

canonical bases, 355
Cartan matrix, 4
Cartan matrix of finite type, 16
cells, 364
central character, 184, 227, 393
character, 219
character table, 233, 266
Chevalley group, 19, 255, 277, 306
class polynomials, 262
commutator, 220
complex reflection, 146
composition, 56, 154
confluent rewriting system, 70
conjugacy graph, 79
conjugate partition, 155
content, 334
coset graph, 47
Coxeter class, 54
Coxeter element, 29, 74, 259, 299
Coxeter graph, 16, 188
Coxeter group, 14
Coxeter number, 29
Coxeter system, 14
crystallographic, 19, 306, 316, 361
cuspidal class, 74
cycle structure, 79
cyclic shift, 74

decomposition map, 236
defect, 244
deformation, 253
degrees, 149
Deligne normal form, 108
distinguished representatives
  of double cosets, 42
  of right cosets, 39
dominance order, 157
dual character, 309
dual representation, 309
dual symbol, 196
dual W-graph, 365
Dynkin diagram, 19

e-regular, 354
e-residue, 354
empty partition, 57
exceptional characters, 296
exchange condition, 13, 35
exterior power, 138, 259

fake degrees, 149, 347
Fock space, 354
fundamental element, 109
fundamental roots, 5

G-set, 60
generic degrees, 255, 307, 422
generic Iwahori–Hecke algebra, 253
good element, 117

Hecke algebra, 250, 274
hook, 323

indecomposable Cartan matrix, 15
index representation, 252
induced module, 287
irreducible Coxeter system, 15
Iwahori–Hecke algebra, 122

J-induction, 203, 417
j-induction, 145
Jones polynomial, 131

Kazhdan–Lusztig basis, 368
Kazhdan–Lusztig cells, 368
Kazhdan–Lusztig polynomials, 370
knots and links, 127, 131, 136

Kostka numbers, 159, 213, 347

Lascoux–Leclerc–Thibon conjecture, 353
lattice, 219
lattice permutation, 182
leading coefficient, 315
leg length, 323
length function, 5
Littlewood–Richardson rule, 181
longest coset representative, 45
longest element, 26
LR coefficients, 178

M-polynomial, 367
Mackey system, 68
mark, 61, 63
Markov relations, 128
Markov trace, 133, 350
Maschke's theorem, 223
minimal length
  in a conjugacy class, 74
modular constituent, 239
Molien series, 142, 312

negative block, 92
negative root, 9
Newton power sum, 162, 176
non-crystallographic type, 19
normal form, 69

parabolic Burnside ring, 63
parabolic closure, 101
parabolic subalgebra, 127
parabolic subgroup, 14
parabolic submonoid, 108
parabolic table of marks, 63
parabolic type, 190
partition, 56
permutation character, 76
Φ_e-blocks, 428
Poincaré polynomial, 255
positive block, 92
positive root, 9
prefix, 40, 45
projective, 221
pure braid monoid, 117

quadratic relations, 106

rank, 5
realized over, 232, 302
reduced expression, 5
reduced root system, 17
reduction stable, 242
reflection, 4, 37
reflection group, 5
reflection representation, 14, 35, 258
reflection subgroups, 68
restricted coset graph, 52
right coset representatives, 41
right descent set, 396
rim hook, 323
root system, 5
roots, 5

saturated, 204
Schur basis, 275
Schur element, 223
Schur function, 162
Schur's lemma, 223
self-opposed, 56
shift operation, 194
shifted coset graph, 71
ShortLex, 69
sign representation, 6, 252
signed block form, 93
simple roots, 5
skew diagram, 181
special character, 210
special symbol, 195
specialization, 234, 252
split simple, 218
splitting field, 302
standard tableau, 323
strip, 323
strongly conjugate, 80
subexpression, 12
suffix, 40, 46
symbol, 194
symmetric algebra, 142, 218
symmetrizing trace, 218

table of marks, 62
tableau, 182, 213, 323
terminal, 81
Tits system, 278
trace function, 218

valuation, 231

W-graph, 308, 362, 368
Wedderburn's theorem, 218
weight, 182, 323
Weyl group, 19, 278
word, 182, 323

Young diagram, 155
Young subgroup, 155

Lightning Source UK Ltd.
Milton Keynes UK
UKOW04n1039270414

230648UK00001B/27/P

9 780198 502500